The Systematics Association Special Volume Series

Neotropical Savannas and Seasonally Dry Forests

Plant Diversity, Biogeography, and Conservation

The Systematics Association Special Volume Series

Series Editor

Alan Warren

Department of Zoology, The Natural History Museum,
Cromwell Road, London SW7 5BD, UK.

The Systematics Association promotes all aspects of systematic biology by organizing conferences and workshops on key themes in systematics, publishing books and awarding modest grants in support of systematics research. Membership of the Association is open to internationally based professionals and amateurs with an interest in any branch of biology including palaeobiology. Members are entitled to attend conferences at discounted rates, to apply for grants and to receive the newsletters and mailed information; they also receive a generous discount on the purchase of all volumes produced by the Association.

The first of the Systematics Association's publications *The New Systematics* (1940) was a classic work edited by its then-president Sir Julian Huxley, that set out the problems facing general biologists in deciding which kinds of data would most effectively progress systematics. Since then, more than 70 volumes have been published, often in rapidly expanding areas of science where a modern synthesis is required.

The *modus operandi* of the Association is to encourage leading researchers to organize symposia that result in a multi-authored volume. In 1997 the Association organized the first of its international Biennial Conferences. This and subsequent Biennial Conferences, which are designed to provide for systematists of all kinds, included themed symposia that resulted in further publications. The Association also publishes volumes that are not specifically linked to meetings and encourages new publications in a broad range of systematics topics.

Anyone wishing to learn more about the Systematics Association and its publications should refer to our website at www.systass.org.

Other Systematics Association publications are listed after the index for this volume.

The Systematics Association Special Volume Series 69

Neotropical Savannas and Seasonally Dry Forests

Plant Diversity, Biogeography, and Conservation

Edited by

R. Toby Pennington

Royal Botanic Garden
Edinburgh, U.K.

Gwilym P. Lewis

Royal Botanic Gardens
Kew, U.K.

James A. Ratter

Royal Botanic Garden
Edinburgh, U.K.

CRC Press is an imprint of the
Taylor & Francis Group, an **informa** business

A TAYLOR & FRANCIS BOOK

Published in 2006 by
CRC Press
Taylor & Francis Group
6000 Broken Sound Parkway NW, Suite 300
Boca Raton, FL 33487-2742

First issued in paperback 2019

No claim to original U.S. Government works

ISBN 13: 978-0-367-45358-9 (pbk)
ISBN 13: 978-0-8493-2987-6 (hbk)

Visit the Taylor & Francis Web site at
http://www.taylorandfrancis.com

and the CRC Press Web site at
http://www.crcpress.com

Library of Congress Card Number 2005028775

Library of Congress Cataloging-in-Publication Data

Neotropical savannas and dry forests : diversity, biogeography, and conservation / editors, R. Toby
    Pennington and James A. Ratter.
        p. cm. -- (Systematics association special volumes ; no. 69)
    Includes bibliographical references and index.
    ISBN 0-8493-2987-6 (alk. paper)
    1. Forest ecology--Tropics. 2. Savanna ecology--Tropics. I. Pennington, Toby. II. Ratter, J. A. III.
Systematics Association special volume ; no. 69

QK936.N44 2006
577.30913--dc22   2005028775

Preface

This book is the result of a plant diversity symposium that formed part of a conference, *Tropical savannas and seasonally dry forests: ecology, environment and development*, held at the Royal Botanic Garden Edinburgh from 14 to 20 September 2003. The conference was attended by 150 delegates from 25 countries worldwide.

From the outset our intention was to focus on neotropical dry vegetation, where our combined expertise lies, and to address the imbalance between the study of rain forests and seasonally dry ecosystems by discussion, and then by publishing data on seasonally dry forests and savannas. Invited speakers were thus chosen with the aim of covering these ecosystems across the whole of the Neotropics. However, like many conference proceedings, and despite our best intentions, this book is not comprehensive or even in its coverage and some geographical areas have slipped through the net. We hope, however, that it will serve to stimulate more research and, more importantly, the urgently needed conservation of the often neglected seasonally dry forests and savannas of the Neotropics.

We would like to thank: the Systematics Association for their generous sponsorship of the plant diversity symposium; the Royal Society of Edinburgh and the Royal Botanic Gardens Kew Legume Donation Fund for their support of speakers attending the symposium; the Royal Botanic Garden Edinburgh for provision of the venue and a reception; John-Paul Shirreffs for help with graphic design; and Sadie Watson, the Edinburgh Centre for Tropical Forests and Maureen Warwick for administrative help before and during the meeting.

We were assisted in the production of this book by 25 reviewers whose expertise and time were greatly appreciated. We would also like to thank Sam Bridgewater, Hélène Citerne, Kim Howell, and Carol Notman for assistance in editing the manuscripts, and all the authors for their contributions.

Toby Pennington
Gwilym Lewis
James Ratter

Contributors

Luzmila Arroyo
Herbario del Oriente
Museo de Historia Natural Noel Kempff
  Mercado
Casilla 2489
Santa Cruz, Bolivia
larroyo@museonoelkempff.org

Thomas H. Atkinson
Dow AgroSciences
5005 Red Bluff Road
Austin, Texas 78702, USA
thatkinson@dow.com

Gerardo Aymard
Universidad Nacional Experimental de los
  Llanos Ezéquiel Zamora
3323 Mesa de Cavacas, Guanare
Edo. Portuguesa, Venezuela
gaymard@cantv.net

Bastian Bise
Conservatoire et Jardin botaniques de la Ville
  de Genève
Chemin de l'Impératrice 1
CP 60, CH-1292 Chambésy
Genève, Switzerland
bastian.bise@cjb.ville-ge.ch

Evan Bowen-Jones
Fauna and Flora International
Great Eastern House
Tenison Road
Cambridge CB1 2TT, UK
evan.bowen-jones@fauna-flora.org

Samuel Bridgewater
Department of Botany
Natural History Museum
Cromwell Road
London, SW7 5BD, UK
s.bridgewater@nhm.ac.uk

Judith Caballero
Herbario del Oriente
Museo de Historia Natural Noel Kempff
  Mercado
Casilla 2489
Santa Cruz, Bolivia
jcaballero@museonoelkempff.org

Sofia Caetano
Molecular Genetics and Phylogenetics
  Laboratory
Conservatoire et Jardin botaniques de la Ville
  de Genève
Chemin de l'Impératrice 1
CP 60, 1292 Chambésy
Genève, Switzerland
Sofia.Caetano@cjb.ville-ge.ch

Clément Calenge
UMR CNRS 5558
Laboratoire de Biométrie et Biologie Evolutive
Université Claude Bernard Lyon 1
43 Boulevard du 11 novembre 1918
F-69622 Villeurbanne Cedex, France
calenge@biomserv.univ-lyon1.fr

Cyrille Chatelain
Conservatoire et Jardin botaniques de la Ville
  de Genève
Chemin de l'Impératrice 1
CP 60, CH-1292 Chambésy
Genève, Switzerland
cyrille.chatelain@cjb.ville-ge.ch

Ezequial Chavez
Herbario del Oriente
Museo de Historia Natural Noel Kempff
  Mercado
Casilla 2489
Santa Cruz, Bolivia
echavez@museonoelkempff.org

Lisete Correa
Departamento de Geografía
Museo de Historia Natural Noel Kempff
  Mercado
Casilla 2489
Santa Cruz, Bolivia
lcorrea@museonoelkempff.org

Giselda Durigan
Floresta Estadual de Assis
Caixa Postal 104, 19800-000 Assis
São Paulo State, Brazil
giselda@femanet.com.br

Christopher William Fagg
Departamento de Engenharia Florestal
CP 04357, Universidade de Brasília
70 919 970, Brasília, DF, Brazil

Jeanine Maria Felfili
Departamento de Engenharia Florestal
CP 04357, Universidade de Brasília
70 919 970, Brasília, DF, Brazil
felfili@unb.br

Maria Cristina Felfili
IBAMA
Brasília, DF, Brazil

Thomas W. Gillespie
Department of Geography
University of California, Los Angeles
Los Angeles, California 90095-1524, USA
tg@geog.ucla.edu

Marco Antonio González
Grupo Autónomo para la Investigación
  Ambiental A.C.
Crespo 520-A
Centro Oaxaca
68000 Oaxaca, México
gaia@spersaoaxaca.com.mx

James E. Gordon
Center for Tropical Plant Conservation
Fairchild Tropical Botanic Gardens
10901 Old Cutler Road
Coral Gables, Florida 33156, USA
jgordon@fairchildgarden.org

René Guillén
Capatanía de Alto y Bajo Izozog (CABI)
Santa Cruz, Bolivia

Eileen Helmer
International Institute of Tropical Forestry
USDA Forest Service
Calle Ceiba 1201
Jardín Botánico Sur
Río Piedras, Puerto Rico 00926-1119
ehelmer@fs.fed.us

Otto Huber
CoroLab Humboldt, CIET/IVIC
Apartado 21827
Caracas 1020-A, Venezuela
Present address:
Fundación Instituto Botánico de Venezuela
Apartado 2156
Caracas 1010-A, Venezuela
ohuber@mac.com

João André Jarenkow
Departamento de Botânica
Instituto de Biociências
Universidade Federal do Rio Grande do Sul
Av. Bento Gonçalves, 9500
91501-970, Porto Alegre, RS, Brazil

Manoel Cláudio da Silva Júnior
Departamento de Engenharia Florestal
CP 04357, Universidade de Brasília
70919-970, Brasília, DF, Brazil

Timothy J. Killeen
Center for Applied Biodiversity Science
  (CABS) at Conservation International
1919 M Street NW
Washington, D.C. 20036, USA
t.killeen@conservation.org

Bente B. Klitgaard
Department of Botany
Natural History Museum
Cromwell Road
London SW7 5BD, UK
B.Klitgaard@nhm.ac.uk

Matt Lavin
Plant Sciences and Plant Pathology
Montana State University
Bozeman, Montana 59717, USA
mlavin@montana.edu

Gwilym P. Lewis
Herbarium, Royal Botanic Gardens
Kew, Richmond
Surrey TW9 3AB, UK
g.lewis@rbgkew.org.uk

Reynaldo Linares-Palomino
Herbario Forestal MOL
Facultad de Ciencias Forestales
Universidad Nacional Agraria La Molina
Apartado 456
Lima, Peru
pseudobombax@yahoo.co.uk

J. Michael Lock
Royal Botanic Gardens
Kew, Richmond
Surrey TW9 3AB, UK
mike@xyris.fsnet.co.uk

Emily J. Lott
Plant Resources Center
University of Texas at Austin
Austin, Texas 78712, USA
emilyjlott@hotmail.com

Ariel E. Lugo
International Institute of Tropical
  Forestry
USDA Forest Service
Calle Ceiba 1201
Jardín Botánico Sur
Río Piedras, Puerto Rico 00926-1119
alugo@fs.fed.us

Francis E. Mayle
Institute of Geography
School of GeoSciences
University of Edinburgh
Drummond Street, Edinburgh EH8 9XP, UK
Francis.Mayle@ed.ac.uk

Ernesto Medina
International Institute of Tropical
  Forestry
USDA Forest Service
Calle Ceiba 1201
Jardín Botánico Sur
Río Piedras, Puerto Rico 00926-1119
and
Centro de Ecología
Instituto Venezolano de Investigaciones
  Científicas
Caracas, Venezuela
emedina@ivi.ve

Yamama Naciri
Molecular Genetics and Phylogenetics
  Laboratory
Conservatoire et Jardin botaniques de la
  Ville de Genève
Chemin de l'Impératrice 1
CP 60, 1292 Chambésy
Genève, Switzerland
Yamama.Naciri@ville-ge.ch

Paulo Ernane Nogueira
Departamento de Engenharia Florestal
CP 04357, Universidade de Brasília
70919-970, Brasília, DF, Brazil

Ary T. Oliveira-Filho
Departamento de Ciências Florestais
Universidade Federal de Lavras
37200-000, Lavras, MG, Brazil

Marielos Peña-Claros
Instituto Boliviano de Investigaciones
  Forestales (IBIF)
Santa Cruz, Bolivia
mpena@ibifbolivia.org.bo

R. Toby Pennington
Royal Botanic Garden Edinburgh
20a Inverleith Row
Edinburgh EH3 5LR, UK
t.pennington@rbge.org.uk

Darién E. Prado
Cátedra de Botánica Morfológia y Sistemática
Facultad de Ciencias Agrarias, UNR
Casilla de Correo No. 14
S2125ZAA, Zavalla, Argentina
dprado@jcagr.unr.edu.ar

Luciano Paganucci de Queiroz
Universidade Estadual de Feira de
 Santana
Departamento de Ciências Biológicas
km 03 BR 116 Campus
44031-460 Feira de Santana
Bahia, Brazil
lqueiroz@uefs.br

Roberto Quevedo
Proyecto BOLFOR II
Santa Cruz, Bolivia

James A. Ratter
Royal Botanic Garden Edinburgh
20a Inverleith Row
Edinburgh EH3 5LR, UK
j.ratter@rbge.org.uk

Alba Valéria Rezende
Departamento de Engenharia Florestal
CP 04357, Universidade de Brasília
70919-970, Brasília, DF, Brazil

J. Felipe Ribeiro
EMBRAPA, Centro de Pesquisa Agropecuária
 dos Cerrados (CPAC)
BR020 – km 18 CEP 73301
Planaltina, DF, Brazil
felipe@cpac.embrapa.br

Ricarda Riina
Botany Department
University of Wisconsin
132 Birge, 430 Lincoln Drive
Madison, Wisconsin 53706, USA
rgriinaoliva@wisc.edu

Maria Jesus Nogueira Rodal
Departamento de Biologia
Universidade Federal Rural de Pernambuco
R. Dom Manoel de Medeiros s/n
52.171-900, Recife, PE, Brazil

Mario Saldias
Herbario del Oriente
Museo de Historia Natural Noel Kempff
 Mercado
Casilla 2489
Santa Cruz, Bolivia
msaldias@museonoelkempff.org

Brian D. Schrire
Herbarium, Royal Botanic Gardens
Kew, Richmond
Surrey TW9 3AB, UK
B.Schrire@rbgkew.org.uk

Liliana Soria
Departamento de Geografía
Museo de Historia Natural Noel Kempff
 Mercado
Casilla 2489
Santa Cruz, Bolivia
lsoria@museonoelkempff.org

Rodolphe Spichiger
Conservatoire et Jardin botaniques de la Ville
 de Genève
Chemin de l'Impératrice 1
CP 60, CH-1292 Chambésy
Genève, Switzerland
rodolphe.spichiger@cjb.ville-ge.ch

Rodrigo Duno de Stefano
Centro de Investigación Científica de Yucatán
AC, Calle 43 no. 130
Col. Chuburná de Hidalgo
CP 97200 Mérida, Yucatán, Mexico
roduno@cicy.mx

Marc Steininger
Center for Applied Biodiversity Science
 (CABS) at Conservation International
1919 M Street NW
Washington, D.C. 20036, USA
m.steininger@conservation.org

Marisol Toledo
Instituto Boliviano de Investigaciones
 Forestales (IBIF)
Santa Cruz, Bolivia

J. Carlos Trejo-Torres
Ciudadanos del Karso
497 Emiliano Pol
Box 230
San Juan, Puerto Rico 00926-5636
karsensis@yahoo.com.mx

Ynés Uslar
Herbario del Oriente
Museo de Historia Natural Noel Kempff
 Mercado
Casilla 2489
Santa Cruz, Bolivia

Israel Vargas
Herbario del Oriente
Museo de Historia Natural Noel Kempff
 Mercado
Casilla 2489
Santa Cruz, Bolivia

John R.I. Wood
Department of Plant Sciences
University of Oxford
South Parks Road
Oxford OX1 3RB, UK
jriwood@hotmail.com

Table of Contents

1 An Overview of the Plant Diversity, Biogeography and Conservation of Neotropical Savannas and Seasonally Dry Forests

R. Toby Pennington, Gwilym P. Lewis and James A. Ratter

CONTENTS

ABSTRACT

The Neotropics contain globally significant areas of seasonally dry forests and savannas. Seasonally dry tropical forests are found in scattered areas, with the most species-rich in Mexico, but similar levels of diversity may be found elsewhere (e.g. Peru, Bolivia). Levels of floristic similarity between areas are often low, and some areas (e.g. Brazilian caatingas, Peruvian inter-Andean valleys, Mexican Pacific coast) are rich in endemic species, whereas others (e.g. Bolivian chiquitano) are not. By far the largest savannas are the cerrados of central Brazil and the Llanos of Venezuela and Colombia. The cerrados have the highest species richness and endemism, and the woody flora of other areas is often just a subset of cerrado species. The fossil record and dated phylogenetic trees suggest that seasonally dry forests are at least as old as the Miocene, but that savannas dominated by grasses photosynthesizing using the C4 pathway may not have risen to dominance until the late Pliocene. Fossil pollen evidence and models of the distribution of species and vegetation indicate that savanna and seasonally dry tropical forest species changed their ranges during the climatic oscillations during the Pleistocene, but the extent of these shifts is uncertain. Clarification of whether disjunct distributions of species in both vegetations reflect migration through previously continuous formations or long-distance dispersal might be provided by population genetic studies of widespread species. Conservation strategy for seasonally dry forests has failed to take account of floristic patterns across the Neotropics, and some areas rich in endemic species (e.g. Peruvian inter-Andean valleys) lack protection entirely. However, a pragmatic approach, protecting any areas where social and political opportunities permit, is also necessary in these severely impacted ecosystems.

1.1   INTRODUCTION

The chapters of this book arise largely from a plant diversity symposium that formed part of a conference, *Tropical savannas and seasonally dry forests: ecology, environment and development*, held at the Royal Botanic Garden Edinburgh in September 2003. For the general public, and indeed many scientists, forests in Latin America are synonymous with rain forests, especially those of the Amazon basin, and tropical savannas conjure images of the African plains. However, the Neotropics holds globally significant areas of both savanna and dry forest. Miles et al. (in press) estimate that they have 67% (*c.*700,000 km²) of the world's remaining dry forests, and the original extent of neotropical savannas is estimated as three million km² (Huber, 1987). This book aims to provide a synthesis of current knowledge of the plant diversity and geography in savannas and seasonally dry forests, and their conservation status, across their broad neotropical range. This synthesis is necessary because, despite many pleas to direct greater attention to these ecosystems, they continue to receive much less scientific research and conservation resources than are devoted to neotropical rain forests.

A principal aim of this introductory chapter is to place each contribution in context, and in places to fill a few gaps where important topics are not covered elsewhere in the book. We aim to provide a broad synthesis of the book's main topics. These are (i) diversity and geography – where are the centres of species richness and endemism of the floras of neotropical seasonally dry forest and savannas?; (ii) historical biogeography – how and why did this endemism and diversity arise?; and (iii) conservation – are these ecosystems adequately protected, which areas of them should be priorities for conservation, and how can this be best achieved?

1.2 WHAT ARE NEOTROPICAL SAVANNAS AND SEASONALLY DRY FORESTS AND WHERE DO THEY GROW?

In discussing vegetation in seasonal areas of the tropics it is useful to distinguish dry forests and savannas, although their relationships are notoriously complex (Furley, Proctor and Ratter, 1992). Seasonally dry tropical forests (SDTF; Figures 1.1a–d) are essentially tree-dominated ecosystems with a continuous or almost continuous canopy and a ground layer in which grasses are a minor element (Mooney et al., 1995), whilst a xeromorphic, fire-tolerant grass layer is an important component of savannas (Figures 1.1e, f).

1.2.1 SEASONALLY DRY TROPICAL FORESTS

Seasonally dry tropical forest occurs where the rainfall is less than 1600 mm/yr, with a period of at least 5–6 months receiving less than 100 mm (Gentry, 1995; Graham and Dilcher, 1995). The vegetation is mostly deciduous during the dry season (Figure 1.1d), and this increases along a gradient as rainfall declines, although in the driest forests there is a marked increase in evergreen and succulent species (Mooney et al., 1995).

Seasonally dry tropical forests have a smaller stature and lower basal area than tropical rain forests (Murphy and Lugo, 1986), and thorny species are often prominent. Ecological processes are strongly seasonal, and net primary productivity is lower than in rain forests because growth only takes place during the wet season. There is a build-up of leaf litter during the dry season caused by leaf-fall of the majority of trees followed by sunlight penetration to the forest floor and consequent cessation of decomposition in the low relative humidity. Flowering and fruiting phenologies are strongly seasonal, and many species flower synchronously at the transition between the dry and wet seasons whilst the trees are still leafless (Bullock, 1995). In contrast to rain forests, conspicuous flowers and wind-dispersed fruits and seeds are frequent.

We follow Murphy and Lugo (1995) and Lugo et al. (Chapter 15) and use a wide interpretation of SDTF, including formations as diverse as tall forest on moister sites (Figure 1.1a) to cactus scrub on the driest. Many different names are used for the vegetation which we include under this definition (e.g. tropical and subtropical dry forests, caatinga, agreste, mata acatingada, mesotrophic, mesophilous or mesophytic forest, semideciduous or deciduous forest, bosque caducifolio, bosque espinoso; see Murphy and Lugo [1995] and Lugo et al. [Chapter 15] for a fuller discussion), and this has probably served to confuse the links among the forests of different regions, rather than emphasize their similarities. We have encouraged authors to use the term seasonally dry tropical forest in this book that aims to gather information about this vegetation over wide geographical areas.

Seasonally dry tropical forests generally occur on fertile soils with a moderate to high pH and nutrient status and low levels of aluminium. Such soils are very suitable for agriculture (Ratter et al., 1978), which has resulted in enormous forest destruction in many areas (e.g. less than 2% of seasonally dry forests on the Pacific coast of Mesoamerica are still intact; Janzen, 1988), a problem exacerbated by the large human populations in many neotropical dry forest life zones (Murphy and Lugo, 1995).

Miles et al. (in press) estimated that 54.2% of the word's remaining SDTFs are contained in South America, and 12.5% in Central America. The estimate for South America may be too high as these authors include chaco vegetation of Argentina, Paraguay and Bolivia in their definition of SDTF. This is a floristically different, subtropical formation because it receives frost (Spichiger et al., Chapter 8; Pennington et al., 2000). However, it is clear that the Neotropics contains globally significant stands of SDTF.

Seasonally dry tropical forests occur in disjunct patches scattered throughout the Neotropics. These are dealt with in a series of chapters with geographical focus from Florida in the north (Gillespie, Chapter 16) to Bolivia (Killeen et al., Chapter 9) and Paraguay (Spichiger et al., Chapter 8) in the south. Figure 1.2 is a map, modified from Pennington et al. (2000), showing schematically the distribution of neotropical SDTF and savannas.

FIGURE 1.1 (See colour insert following page 208) Neotropical seasonally dry forests and savannas: (a) Inter-Andean SDTF in the Marañon valley, Cajamarca, Peru. The big trees with pale trunks are *Eriotheca peruviana* A. Robyns, and most of the shrubs in the foreground are *Croton* sp. (photo: R. Linares-Palomino). (b) SDTF in Oaxaca, Mexico, with abundant cacti (photo: C. Pendry; reproduced from Pennington et al., 2004). (c,d) The same area of SDTF at Sagarana, Minas Gerais, Brazil, in the wet and dry seasons, illustrating complete deciduousness in the dry season (photos: J. Ratter). (e) Savanna (cerrado) with a scattering of low shrubs and occasional small trees (*campo sujo* = dirty field) in Distrito Federal, Brazil (photo: J. Ratter). (f) Typical savanna (cerrado) vegetation in Mato Grosso, Brazil (photo: S. Bridgewater).

(e)

(f)

FIGURE 1.1 (Continued).

The largest areas of SDTF in South America are found in north-eastern Brazil (the caatingas; Queiroz, Chapter 6), in two areas defined by Prado and Gibbs (1993) as the Misiones (Spichiger et al., Chapter 8) and the Piedmont nuclei (see Wood, Chapter 10, for a treatment of the Bolivian part of this area), and on the Caribbean coasts of Colombia and Venezuela. Other, smaller and more isolated areas of SDTF occur in dry valleys in the Andes in northern Bolivia (Wood, Chapter 10), Peru (Linares-Palomino, Chapter 11; Figure 1.1a), southern Ecuador (Lewis et al., Chapter 12) and Colombia. Other areas of limited extent are in coastal Ecuador and adjacent northern Peru (Linares-Palomino, Chapter 11), the Mato Grosso de Goiás in central Brazil and scattered throughout the Brazilian Cerrado Biome on areas of fertile soils (Ratter et al., 1978; Figure 1.1e,f). In Mesoamerica, seasonally dry forests are concentrated along the Pacific coast from Guanacaste in northern Costa Rica, to just north of the Tropic of Cancer in the Mexican state of Sonora (Lott and Atkinson, Chapter 13; Gordon et al., Chapter 14). Other significant stands are in the Yucatán peninsula, and elsewhere in Mesoamerica smaller, isolated SDTF areas are found in interior dry valleys and on the Pacific coast of Panama. In the Caribbean, the most extensive stands of SDTF are found in Cuba, but they also occur on other islands including Hispaniola, Puerto Rico, Jamaica and the Lesser Antilles (Lugo et al., Chapter 15), and reach their most northern position in Florida (Gillespie, Chapter 16). In all of these areas SDTFs occur within a complex of vegetation types depending on local climatic, soil and topographical conditions.

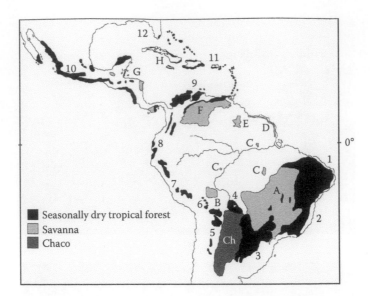

FIGURE 1.2 Schematic distribution of seasonally dry forests and savannas in the Neotropics, highlighting areas mentioned in the text. Seasonally dry forest: 1. caatingas; 2. south-east Brazilian seasonal forests; 3. Misiones nucleus; 4. Chiquitano; 5. Piedmont nucleus; 6. Bolivian inter-Andean valleys; 7. Peruvian and Ecuadorean inter-Andean valleys; 8. Pacific coastal Peru and Ecuador; 9. Caribbean coast of Colombia and Venezuela; 10. Mexico and Central America; 11. Caribbean Islands (small islands coloured black are not necessarily entirely covered by seasonally dry forests); 12. Florida. Savannas: (A) cerrado; (B) Bolivian; (C) Amazonian (smaller areas not represented); (D) coastal (Amapá, Brazil to Guyana); (E) Rio Branco-Rupununi; (F) Llanos; (G) Mexico and Central America; (H) Cuba. Ch: Chaco. Modified after Pennington et al. (2000) and Huber et al. (1987).

1.2.2 SAVANNAS

Savannas are defined by their grass-rich ground layer. They are found under similar or slightly wetter climatic conditions to SDTF but tend to be on poorer soils (Sarmiento, 1992), which may be the reason why savanna trees frequently have sclerophyllous, evergreen leaves (Ratter et al., 1997). By far the most extensive savannas in the Neotropics are found in the Cerrado Biome (sensu Ratter et al., Chapter 2*), which covers some two million km^2 of central Brazil. It contains several vegetation types, but the dominant form of sclerophyllous vegetation is known as cerrado sensu lato (s.l.; Figure 1.1e,f). This grows on dystrophic, acid soils, with low calcium and magnesium availability, and usually with high levels of aluminium (Furley and Ratter, 1988; Ratter et al., 1997). Cerrado soils are always well drained and cerrado vegetation is intolerant of waterlogging. Cerrado s.l. varies from dense grassland with a sparse covering of shrubs and small trees, to an almost closed woodland known as cerradão with a canopy height of 12–15 m (see, for instance, Ratter et al., 1997). These differences in vegetation structure have been related to soil fertility gradients (e.g. Goodland and Pollard, 1973), but other data fail to show this correlation (e.g. Ribeiro, 1983), and Durigan (Chapter 3) presents convincing evidence that fire frequency is a critical factor. Fire is undoubtedly an important factor throughout the biome (Ratter et al., 1997), and the woody flora of the cerrado shows adaptations such as thick, corky bark, xylopodia and the ability to sprout from lateral buds if the growing apex is killed by fire. In fact

---

* Ratter et al., Durigan and Felfili et al. (Chapters 2, 3, 4) use Cerrado 'Biome' to describe the geographic area dominated by Cerrado vegetation, which also contains other vegetation types. Elsewhere in this introduction and other chapters (e.g, Huber, Chapter 5; Killeen et al., Chapter 9), 'biome' describes a physiognomic vegetation type (e.g. SDTF, rain forest).

many cerrado species are so pre-adapted that the replacement of the growing apex is normal; the terminal bud is replaced by an adventitious one for each flush of growth.

Other vegetation types occur within the Cerrado Biome. Gallery forests occur along the rivers, and contain many species which also occur in the rain forests of the Amazon and the Atlantic coast of Brazil (Oliveira-Filho and Ratter, 1995). Wet campos, lacking trees, except the fan palm *Mauritia flexuosa* L.f., frequently occur between cerrados and gallery forest, where there is extreme fluctuation of the water table. In these areas, waterlogging in the wet season excludes woody cerrado species, while in the dry season the soils are too dry to support gallery forest species. Seasonally dry tropical forest occurs on areas of fertile soil, which are often associated with calcareous rocks (Ratter et al., 1978).

Disjunct, isolated areas of cerrado-like vegetation also occur within the Brazilian Amazon rain forest (Eiten, 1972). These occur both to the north of the Amazon river (e.g. in Amapá State and Roraima State) and to the south closer to the cerrados of central Brazil (e.g. Humaitá in Amazonas State) and even close to the Amazon river itself at Alter do Chão (Pará State). These occur on poor, sandy soils (Solbrig, 1993) and with the exception of Alter do Chão and one of the Humaitá sites (Janssen, 1986) are depauperate in numbers of species compared to the cerrados of central Brazil (Sanaiotti, 1996). The southern sites and Alter do Chão have floristic affinities with central Brazil, whilst the northern sites show affinities to the savannas of the Llanos and the Rupununi savannas (Ratter et al., 1996; Sanaiotti, 1996; Sanaiotti et al., 1997). An additional disjunct area of cerrado-like vegetation is found in the Bolivian chiquitano, which, as pointed out by Killeen et al. (1990), has strong affinities with the cerrados of central Brazil. Killeen and Nee (1991) recorded 581 species for this area, a much richer flora than is normal for most outlying cerrado areas, other than those of São Paulo and Rondônia. About 100 of these are typical trees and larger shrubs of the central cerrado area, of which 30 belong to the element characteristic of richer (mesotrophic) soils (e.g. *Callisthene fasiculata* (Spreng.) Mart., and *Dilodendron bipinnatum* Radlk.), demonstrating the prevalence of such soils in the region.

The other large areas of savanna in South America are the Venezuelan and Colombian Llanos (see Huber et al., Chapter 5) and the adjoining savannas in Guyana and Brazil. They also frequently differ ecologically since many are hydrologic savannas, with extreme seasonal fluctuations in water table levels. Huber et al. (Chapter 5) consider the savannas of the Rio Branco-Rupununi region in northernmost Brazil and western Guyana similar to the Llanos because they occupy a comparable landscape mosaic of flooded and non-flooded alluvial plains filled with recent sediments deriving from the surrounding mountains of the Guayana Shield, and are exposed to a similarly strong biseasonal climatic regime. The partly flooded Bolivian savannas of Santa Cruz resemble the Venezuelan Llanos in physiographical and floristic aspects, but have distinct climatic and edaphic characteristics (Huber et al., Chapter 5).

Smaller areas of savanna are present on the Amazonian flanks of the Andes of Peru (Weberbauer, 1945; Scott, 1978; Bridgewater et al., 2002; Pennington et al., 2004) and Bolivia (Wood, Chapter 10). These have received little recent attention, especially in discussions of neotropical savanna biogeography. These small pockets vary from 100 ha to only a few hectares in size (Scott, 1978; R.T. Pennington, pers.obs.). In Peru, they are most extensive in the Gran Pajonal (departments of Ucayali, Junín and Pasco; Scott, 1978), but are also found in the valley of the Río Urubamba, the Tarapoto region (department of San Martín), and in the central Andean Chanchamayo valley (department of Junín). In Bolivia they can be found in the Yungas (R.T. Pennington, pers. obs; Wood, Chapter 10). We suspect that other small pockets of savanna await discovery elsewhere in Amazonian valleys of the Andes.

The small Central American savannas in Belize and adjacent Guatemala, Mexico and Nicaragua are largely hydrologic savannas (Bridgewater et al., 2002). The principal areas of natural savannas in Cuba are also hydrologic (Borhidi, 1991), but with characteristic endemic palm species. In Cuba and Central America pine savannas are also found, which are not seasonally inundated, but which may be the result of degradation of pine woodlands, especially by fire (Borhidi, 1991; Bridgewater et al., 2002). Other areas of the extensive Cuban savannas are considered the result of alteration of the natural forest vegetation, and classified as agricultural by Borhidi (1991). Similarly, in Central America, SDTF areas degraded by cattle ranching are converted to savanna such as the 'Crescentia savannas' described by Bass (2004).

1.3 FLORISTIC COMPOSITION OF NEOTROPICAL SAVANNAS AND SEASONALLY DRY FORESTS

Although not as diverse as rain forests, neotropical savannas and SDTF are species rich (e.g. there are estimates of 10,000 species of vascular plants in the Brazilian Cerrado Biome; Myers et al., 2000). Quantifying and comparing patterns of diversity across the huge geographical range of these ecosystems is a logistic and scientific challenge. Ideally, floristic surveys should be carried out in all areas using uniform methodology, and species identifications made in different areas should be cross checked. In the past 10 years, enormous progress has been made with studies of this type in the Brazilian cerrados (Ratter et al., Chapter 2; Felfili et al., Chapter 4; Durigan et al., Chapter 3). The situation is less satisfactory for SDTF. The only study that compares all neotropical SDTF based upon inventories of uniform methodology is still that of Gentry (1995), comparing only 28 sites. The comparative floristic studies of SDTF presented in this book are generally drawn from surveys based upon a variety of methodologies such as 1-ha forest inventory plots (Killeen et al., Chapter 9), transects (Gillespie, Chapter 16) and general floristic lists (e.g. Oliveira-Filho et al., Chapter 7). In these studies species identifications are often only cross-checked by searching for synonyms because specimens cannot be consulted due to constraint of time. Where herbarium specimens have been consulted and taxonomy checked (e.g. Lewis et al., Chapter 12; Queiroz, Chapter 6; Wood, Chapter 10), the studies focus on specific families rather than the whole flora. Despite these frequent drawbacks, the studies presented here provide a preliminary picture of the patterns of floristic composition of neotropical SDTF that allows the identification of centres of diversity and endemism.

1.3.1 FLORISTIC COMPOSITION OF NEOTROPICAL SEASONALLY DRY FORESTS

Previous reviews (Gentry, 1995; Pennington et al., 2000) and biogeographical studies of widespread SDTF species (Prado and Gibbs, 1993; Pennington et al., 2000) have perhaps overemphasized floristic similarities between the separate areas of neotropical SDTF. New data such as those presented in several chapters here, plus recent studies such as Gillespie et al. (2000) and Trejo and Dirzo (2002) allow a refinement of the floristic picture of neotropical SDTF, and the emerging view is of greater heterogeneity. Although comparable quantitative inventory studies are lacking for many areas, it seems clear that the taxonomic composition of different areas is variable and floristic similarity between areas can be low.

Despite these differences, there are some common factors that should be emphasized. First, Leguminosae is the most species-rich family in all areas with the exception of the Caribbean (Lugo et al., Chapter 15) and Florida (Gillespie, Chapter 16), where Myrtaceae predominate. Second, Cactaceae are often both common and one of the most species-rich families (Figure 1.1b). Third, there are a few woody families that are more abundant in SDTF vegetation than elsewhere, and in this sense are characteristic of this vegetation. These are Capparidaceae, Zygophyllaceae (especially in Central America) and Bombacaceae (Figure 1.1a). Erythroxylaceae are common, but also characteristic of the Brazilian cerrados (Ratter et al., Chapter 2), and *Bursera* (Burseraceae) is very common in Mexican SDTF. Seasonally dry tropical forests usually have a closed canopy, with a sparse ground flora that often contains Bromeliaceae, Compositae, Malvaceae, Araceae, Portulacaceae and Marantaceae, and rather few grasses. The only contribution in this book that examines a selection of families of this herbaceous flora is by Wood (Chapter 10), and this aspect of the flora clearly merits more study.

Gentry (1995), despite lengthy discussion of floristic differences between his study sites, concluded that at the familial level dry forest communities were "remarkably consistent in their taxonomic composition." Examination of the floristic data presented in several chapters in this book (and in fact the discussion in Gentry [1995]) leads to conclusions somewhat at odds with this statement. The order of families by species richness after Leguminosae is variable. The second-place family in Mexico (Lott and Atkinson, Chapter 13) and the Brazilian caatingas (Queiroz, Chapter 6) is Euphorbiaceae, whereas in Peru (Linares-Palomino, Chapter 11) it is Cactaceae and in south-east Brazilian seasonal forests (Oliveira-Filho, Chapter 7), Myrtaceae. The third most species-rich family in the Brazilian caatingas is Cactaceae, in the south-east Brazilian seasonal forests, Rubiaceae, and in Peru, Bignoniaceae. Bignoniaceae were said by Gentry (1995) to dominate the woody floras of SDTF, and their lesser importance in the contributions included in this book is probably accounted for by Gentry's inclusion of lianas in his transect studies. Lianas are a life form unfortunately ignored by many of the authors here. Gentry (1995) emphasized a series of other families such as Rubiaceae and Sapindaceae that were on average most species rich in his samples. The importance of these families is geographically variable. For example, Rubiaceae are often the third most species-rich family in south-east Brazilian forests, but are of much less importance in Peru (Linares-Palomino, Chapter 11).

There has been considerable interest in patterns of β-diversity – how species composition changes with distance – in neotropical rain forests (Pitman et al., 1999; 2001; Condit et al., 2002), and in the Brazilian cerrados (Ratter et al., 1988; Bridgewater et al., 2004; Ratter et al., Chapter 2), but few comparable studies in SDTF. Decay of species similarity with distance is often measured using the Sørensen index (2 × the shared species in two sites divided by the total species richness of both sites). Between Ecuadorean and Peruvian rain forests, separated by c.1400 km, the Sørensen index is c.0.35 (Condit et al., 2002), whilst non-adjacent cerrado floristic provinces in Brazil separated by c.1000 km have a Sørensen similarity of 0.35–0.45 (Bridgewater et al., 2004). The Sørensen similarity between the north-eastern and south-eastern Brazilian seasonal forests (as defined by Oliveira et al., Chapter 7, Figure 7.4), which are separated by c.1500 km, is 0.29. This is slightly lower than the values for cerrado and rain forest, which might reflect the less clear present continuity of this SDTF vegetation between these areas (Oliveira et al., Chapter 7, Figure 7.1). However, the broadly similar values for these areas of cerrado, rain forest and seasonal forest might indicate commonality of processes of both contemporary ecological and historical assembly of these different ecosystems.

In Central America, Gillespie et al. (2000) showed SDTF sites on the Pacific coast of Costa Rica and Nicaragua to have Sørensen similarities of a similar order or somewhat higher (c.0.35–0.50), though these were only separated by c.200 km. However, the Sørensen similarity between other sites was as low as 0.16. Even lower similarity was found by Trejo and Dirzo (2002), who studied 20 Mexican SDTF sites and showed their average Sørensen similarity to be only 0.09. This mirrors data presented here for Peru (Linares-Palomino, Chapter 11) where the floras of inter-Andean valleys separated by short distances are dissimilar. Only 16 species are shared between the Mantaro and Marañon valleys (Sørensen index of 0.14; Linares-Palomino, Chapter 11), which are separated by 400 km, and Bridgewater et al. (2003) reported only a single species shared between forest plots in the Tarapoto region and the Marañon valley, separated by only 150 km. The substantial mountain barriers separating SDTF areas in Mexico and Peru probably explain these low levels of floristic similarity in both countries.

Gentry's (1995) review has also been influential in consideration of SDTF α-diversity. He highlighted the SDTF of western Mexico (94.3 spp. of diameter ≥ 2.5 cm in 0.1 Ha based upon data from the single site of Chamela; see Lott and Atkinson, Chapter 13) and Bolivia (86 spp. ≥ 2.5 cm in 0.1 Ha based upon data from the single site of Quiapaca) as especially species rich, and noted that this pattern of high diversity near both the tropics of Cancer and Capricorn is a significant departure from the pattern in neotropical rain forest where diversity tends to decrease with distance from the Equator. Whilst there is no doubt that SDTF on the Pacific coast of Mexico are amongst the most species rich in the Neotropics, more recent work highlights the need to re-examine Gentry's conclusions carefully.

Killeen et al. (1998) reported 107 tree species from a single (400 ha) site in the Bolivian Chiquitano, corroborating Gentry's conclusions (i.e. of high diversity far from the Equator), but in Chapter 9 Killeen et al. present data from 56 1-ha plots from a far larger area in the same region and record only 155 tree species with a maximum of 48 species of diameter $\geq$ 10 cm in 1 Ha of STDF. They conclude that the region's SDTF may therefore be only moderately diverse. The SDTF of Tarapoto, Peru, situated only 5°S from the Equator, may have diversity approaching those of the Mexican forests. For example, Bridgewater et al. (2003) found 24 species of tree ($\geq$ 10 cm dbh) in only 0.2 ha of SDTF at Tarapoto. However, the forest of their study site was disturbed, which has been shown to reduce species richness of SDTF in Central America by Gillespie et al. (2000) and Gordon et al. (2004). These authors argued that Gentry had not taken this disturbance factor into account, with an implication that sites much further south in Central America may have been equally species rich as Chamela in the state of Jalisco, Mexico. This seems to be corroborated by Trejo and Dirzo (2002) who showed that many other SDTF sites in Pacific coastal Mexico have a similar species richness to the Chamela site used by Gentry (1995). Furthermore, there are some indications that overall species diversity and levels of endemism within Pacific coastal SDTF in Central America may be highest in geographically central areas of these forests, especially in Oaxaca, Mexico. For example, a checklist for Zimatan (Salas-Morales et al., 2003) lists c.20% more species than for Chamela, and Gordon et al. (2004) point to some sites especially rich in range-restricted endemics further east in the Tehuantepec area. A possible partial explanation for this modest gradient in species diversity and the higher diversity in the middle of the Pacific coastal dry forest zone is the mid-domain effect of range geometry (C.E. Hughes, Univ. Oxford, pers. comm., Colwell and Lees, 2000; see Lott and Atkinson, Chapter 13). This predicts that geographically central areas will tend to have more species.

A problem in making uniform comparisons of both α- and ß-diversity over a broad geographical scale in neotropical SDTF is the different methodologies used in various studies. For example, whilst Trejo and Dirzo (2002) and Gillespie et al. (2000) used 0.1-ha plots, their qualifying size for stem diameter was different. In contrast, the data of Oliveira et al. (this volume) and Linares-Palomino (Chapter 11) are a mixture of inventory plots of varying methodology and general floristic lists since the authors wished to use all the relevant data available. A drawback of inventory plots and transects is that they may fail to capture all species growing in the wider surrounding area and therefore overemphasize floristic dissimilarity between sites. Lewis et al. (Chapter 12) combined general legume surveys with plot and transect data in an attempt to overcome this bias. The advantage of plots and transects is that they do capture data on species abundance, which is another area where studies of SDTF lag behind those of rain forest and savannas. To study patterns of diversity in neotropical SDTF more fully, a priority must be to gather inventory data using uniform methodology.

1.3.2 Species Distribution and Endemism in Neotropical Seasonally Dry Forests

From the perspective of conservation, endemism may be more important than diversity (Gentry, 1995). High-diversity forests may be composed of wide-ranging species that merit less conservation focus than local endemics. From the evolutionary perspective, endemics are also interesting. Some areas of dry forest are clearly much richer in endemics than others. The Bolivian chiquitano (only three woody endemics; Killeen et al., Chapter 9) and seasonal forests of the Atlantic coastal forests of Brazil (Oliveira-Filho et al., Chapter 7) are relatively poor in endemic species, whereas some inter-Andean valleys of Peru (Linares-Palomino, Chapter 11), Ecuador (Lewis et al., Chapter 12), Bolivia (Wood, Chapter 10), the Brazilian caatingas (Oliveira-Filho et al., Chapter 7; Queiroz, Chapter 6) and Mexican dry forests (Lott and Atkinson, Chapter 13) show higher endemism. This has considerable implications for both conservation prioritization and inference of history as discussed in the sections below.

Gentry (1982a) suggested that dry forest species tend to be unusually wide-ranging and little prone to local endemism. Wide-ranging dry forest species have also been highlighted in a biogeographical context by Prado and Gibbs (1993) and Pennington et al. (2000). These studies have the fault of not clearly placing these widespread species in the context of the entire flora of a given area of SDTF – separate areas of SDTF might share a few widespread species (which may be used as the marker species for classifying the forest type), but be floristically very dissimilar. This, to some extent, is being rectified by the observations of ß-diversity patterns in SDTF outlined above. The emerging picture is of a flora where widespread species are in a minority in terms of species numbers. For example, of 466 tree species recorded for the caatingas, 215 (46%) are endemic, and only 56 (12%) reach the peripheral chaco forests c.2500 km distant (Oliveira-Filho et al., Chapter 7). This does not mean, however, that the few widespread species must be insignificant in abundance and dominance. The ß-diversity studies of Pitman et al. (1999; 2001) in western Amazonia and Bridgewater et al. (2004) in the cerrados show remarkably similar patterns, with small oligarchies of common species dominating widely separate sites, which nonetheless differ floristically because they do not share a high percentage of more range-restricted species. We are unaware of any studies that have examined this issue in SDTF, though our impression from our own field experience and the literature is that domination by oligarchies of common species over wide areas may not always be the case. In addition, Gillespie et al. (2000) stated that in Costa Rica and Nicaragua, "no species was repeatedly abundant based upon number of stems per 0.1 ha in all fragments of dry forests sites".

1.3.3 FLORISTIC COMPOSITION, SPECIES DISTRIBUTION AND ENDEMISM IN NEOTROPICAL SAVANNAS

The next three chapters of this book (by Ratter et al., Durigan et al., and Felfili et al.) are detailed studies of the Brazilian Cerrado Biome. They are interrelated and deal with distribution patterns of the larger woody species (trees and shrubs) of the savannic element of the biome, the cerrado sensu lato (s.l.).

The Cerrado Biome has a rich flora of 6429 recorded species of native vascular plants (Mendonça et al., 1998), which will certainly be increased by further research, and higher estimates of c.10,000 species have been made (e.g. Myers et al., 2000). It has been identified as one of the world's top 25 biodiversity hotspots and centres of endemism (Myers et al., 2000). Out of the total recorded species, 2880 belong to the savannic cerrado s.l., while the remaining 3549 are native in the gallery forests, SDTF, wetlands, rocky mountain habitats (campo rupestre), etc., of the biome. The flora of the cerrado s.l. consists of c.800 tree and shrub species, of which probably about 35% are endemic, and rather more than 2000 ground species (herbs and subshrubs) for which it has been suggested (e.g. Machado et al., 2004) that the endemic element may be over 70%. In addition to this, many of the non-endemic tree species seem so completely adapted to the pyrophytic habitat of the cerrado (thick, corky bark, sclerophylly etc.) that one feels it is probable they have originated there and only occur in other biomes in a secondary manner as a result of colonization. Although this high endemicity is characteristic of the cerrado s.l., very few species belonging to the other vegetation types of the biome are endemics, apart from the campo rupestre.

Chapter 5 by Huber et al. is devoted to the Venezuelan Llanos, the second largest savanna area in South America, covering 240,000 km² and with a recorded flora of 3200 species of native vascular plants. There are often strong similarities in the appearance of the savanna woodland of the Llanos and the cerrado. However, the α-diversity of cerrado communities is far higher and the levels of endemicity are strikingly different, representing only c.1.1% of the total flora of the Llanos, whereas that of the cerrado is much higher. This difference seems to be due to the very recent establishment of the Llanos on a late Quaternary landscape, giving insufficient time for significant levels of speciation to occur. Whereas the typical vegetation of cerrado s.l. is restricted to well-drained soils, much of the Llanos occurs on seasonally ill-drained areas and here characteristic hydrologic savannas are found. Such areas carry a small number of tree species and are usually dominated by *Curatella americana* L. and *Byrsonima crassifolia* (L.) Kunth. Both of these are enormously widespread and extend from Central America to the south of Brazil. They are both natives of the Cerrado Biome where they form a community typical of damper, ill-drained areas, but *C. americana*, in particular, also thrives in drier habitats (it is a fast-growing, colonizing species and is self-compatible, unlike the majority of cerrado plants for which information is available, which are self-incompatible). The two species dominate a community termed *Curatella americana/ Byrsonima crassifolia* savanna (Milliken and Ratter, 1998) widespread in South American savannas north of the Amazon, and were the two most cosmopolitan species in a survey of representative New World savannas (Lenthall et al., 1999). This community is relatively species poor, particularly in larger woody taxa, and is typical of the Rio Branco savannas of Roraima which extend into the Rupununi savannas of Guyana. The two dominants are often associated with a number of common, widespread larger woody species of the Brazilian cerrado flora, particularly *Bowdichia virgilioides* Kunth, *Byrsonima coccolobifolia* Kunth, *B. verbascifolia* Rich. ex A. Juss., *Casearia sylvestris* Sw., *Erythroxylum suberosum* A. St. Hil., *Himatan-thus articulatus* (Vahl) Woodson, *Palicourea rigida* Kunth, and sometimes *Anacardium occiden-tale* L. and *Antonia ovata* Pohl. The savannas of the Brazilian state of Amapá (Sanaiotti et al., 1997) in the extreme north-east of the country are of this type, but rather richer in characteristic cerrado trees than, for instance, the Rio Branco savannas of Roraima state. Even Central American savannas are related to this type of savanna; they can look just like a typical campo cerrado at a distance but the woody element is composed of *Byrsonima crassifolia*, *Curatella americana*, *Quercus oleoides* Schlect. & Cham. (usually the commonest tree), and the palm *Acoelorrhaphe wrightii* (Griseb.) H. Wendl. ex Becc., together with a handful of other very rare species (Furley and Ratter, 1989). Cuban savannas, however, are floristically different, with a more diverse flora characterized by palm species (e.g. *Sabal parviflora* Becc., *Copernicia* spp.), many of which are endemic (Borhidi, 1991; Bridgewater et al., 2002).

Comprehensive inventories of the savannas on the Amazonian flanks of the Peruvian and Bolivian Andes are lacking. Scott (1978) describes grassland dominated by *Imperata brasiliensis* Trin. as a successional stage to grasslands dominated by *Andropogon* spp., *Axonopus* spp., *Eriochrysis cayanensis* P. Beauv., *Isachne rigens* Trin., *Spilanthes americana* Hieron. and *Panicum polygonatum* Schrad. Woody species are either absent, or present as scattered individuals. This woody flora includes *Astronium fraxinifolium* Schott, *Roupala montana* Aubl., *Physocalymma scaberrimum* Pohl, *Diloden-dron bipinnatum* Raldk., *Luehea paniculata* Mart., *Cybistax antisyphilitica* (Mart.) Mart., *Curatella americana* and *Byrsonima crassifolia* (R.T. Pennington, pers. obs.). These are mostly species char-acteristic of relatively fertile soils in the Brazilian cerrados.

It is clear that much of the woody element of neotropical savannas outside of the Brazilian cerrados comprises widespread cerrado species. How these species have achieved their distributions and reached areas as isolated as some of the Andean valleys is an intriguing biogeographical question that has received far less recent attention than the same question for broadly distributed SDTF species.

1.4 BIOGEOGRAPHICAL HISTORY OF NEOTROPICAL SAVANNAS AND SEASONALLY DRY FORESTS

Key biogeographical questions that have been asked of the neotropical rain forest flora also apply to seasonally dry ecosystems. Here, for neotropical SDTF and savannas we ask:

1. When did they originate?
2. What are their historical links with similar ecosystems on other continents?
3. Has their species diversity originated recently or slowly over time, and what has driven this evolution?
4. How did their component species react to climatic change, especially during recent ice ages?

Evidence from floristic composition, macrofossils, microfossils, molecular phylogenetics and molecular population genetics can be used to answer these questions.

1.4.1 Fossil Evidence

Direct evidence of the minimum age of origin of a biome is most objectively inferred from the fossil record. For example, Burnham and Johnson (2004) examined the macrofossil record for characteristics of neotropical rain forests such as large, entire leaves with drip tips, a high diversity of angiosperm trees and vines, and a suite of characteristic families (e.g. Sapotaceae, Leguminosae, Palmae). We are aware of no similar review of the time of origin of neotropical SDTF and savannas except that of Graham and Dilcher (1995) who examined the Cenozoic microfossil record of tropical dry forest in northern Latin America and the southern USA, but lacked any extensive recently studied macrofossil floras. Not only do dry environments have fewer sites of fossil deposition, but in terms of taxonomic composition at a generic level, SDTF can be similar to rain forests (e.g. dominated by legumes), making clear distinction using taxonomic criteria difficult. This latter problem is exacerbated by the close interdigitation of rain forest and SDTF in some areas (e.g. see Oliveira-Filho et al., Chapter 7 and Killeen et al., Chapter 9).

The microfossil record, especially for pollen, has been used to study the response of neotropical species to the climatic changes of the Pleistocene, with much debate focusing on the Amazon basin, and whether rain forest was fragmented into refugia by the expansion of dry-adapted species during drier, cooler Pleistocene climates. These species may have come from both cerrado and SDTF, although most models of refuge theory (e.g. Colinvaux et al., 1996) only considered cerrado species. Pennington et al. (2000) pointed out that the possibility of historically widespread SDTF species had not been considered by many workers (e.g. Hoorn, 1997; Haberle, 1997), who in seeking evidence for drier Amazonian climates, searched for grass pollen characteristic of cerrado vegetation. This neglects the fact that SDTF, a closed canopy formation with few grasses, can grow right alongside cerrado under the same climatic, but different edaphic, conditions (e.g. Furley and Ratter, 1988). More recent literature has given SDTF species more attention (e.g. Mayle, 2004; Chapter 17).

1.4.2 Floristic Composition

Floristic comparisons provide circumstantial evidence about biogeographical history, given certain assumptions. For example, high species-level similarity between two separate SDTF nuclei (Figure 1.1) implies a recent historical connection between the areas or high levels of dispersal. Conversely, high levels of species endemism may indicate a physically isolated area that received few immigrants over a long enough period (probably millions of years) to allow endemic species to evolve. These species may also be explained as the result of recent explosive evolution (e.g. Richardson et al., 2001).

1.4.3 Molecular Evidence

1.4.3.1 Dated Phylogenies

Recent theoretical developments (e.g. Sanderson, 1997, 2002; Near and Sanderson, 2004; Thorne et al., 1998; Thorne and Kishino, 2002; Lavin, Chapter 19) can place an estimated dimension of absolute time on phylogenetic trees derived from DNA sequence data, and potentially distinguish between recent evolutionary radiations and older events (see Lavin, Chapter 19). If an age can be assigned to a single node in a phylogenetic tree using external evidence, these methods can estimate the ages of all other nodes of that tree. This permits estimation of when specific monophyletic groups originated and when lineages arrived in different geographical areas (e.g. Lavin, Chapter 19). These methods are not without critics (e.g. Heads, 2005), who point out that calibrations generally rely on an inadequate fossil record. These drawbacks must be taken into account, and this can be done by the use of multiple fossil calibrations that can cross-validate (e.g. Near and Sanderson, 2004; Lavin et al., 2005), and also by explicitly biasing the analysis against the favoured hypothesis. This latter approach is used by Lavin (Chapter 19), whose hypothesis is that endemic species in SDTF areas predate the Pleistocene. He biases his analyses towards estimating younger dates, and the fact that the dates calculated still fall outside the Pleistocene gives more confidence that the hypothesis is not falsified.

Like the fossil record, dated phylogenies can allow us to examine both old and more recent events. They can be used to estimate the time of origin of families, genera or groups of species that characterize neotropical savannas and SDTF. The date of origin of these biome-specific groups might be indicative of the origin of the biome itself (for an application of this approach to the age of rain forests, see Davis et al., 2005). They can also be used to date the onset of radiations of species that characterize SDTF and savannas, for example to assess whether this coincides with climatic changes in the Pleistocene (e.g. Pennington et al., 2004; Lavin, Chapter 19).

1.4.3.2 Population Genetics

Molecular population genetics offers powerful tools for elucidating the history of populations within species, and therefore for understanding the more recent demographic history of neotropical savanna and SDTF (Naciri et al., Chapter 18). For example, it can contribute to the debate of whether the disjunct distributions of many SDTF species are remnants of previously more widespread distributions (e.g. Prado and Gibbs, 1993), the result of long-distance dispersal (Mayle, 2004; Chapter 17), or a combination of both. In the simplest cases, if two equal sized populations result from vicariance of a single ancestral population, their genetic diversity should be about equal. In contrast, if a population derives from a few recent long-distance dispersal events from a single source population, its genetic diversity should be a subset of the diversity of this source.

Given the power of molecular population genetics for understanding the demographic history of widespread species, it is remarkable that there are no published studies using these techniques to address this problem for neotropical SDTF or savannas. In fact, there are few such studies for analogous problems in other vegetation types, even in Amazonia where range-wide studies of the genetic variation in geographically widespread species can be used to test the predictions of various hypotheses of diversification such as refuge theory (e.g. Silva and Patton, 1998; Lougheed et al., 1999; Aleixo, 2004). This paucity of studies probably reflects logistical difficulties — these studies need range-wide sampling of many (hundreds of) individuals, and in the Neotropics this generally involves working in several countries and the negotiation of separate collection permits in each.

1.4.4 BIOGEOGRAPHICAL HISTORY OF NEOTROPICAL SEASONALLY DRY FORESTS

1.4.4.1 Deep History

Graham and Dilcher (1995) reported no clear fossil evidence of dry forest in Mesoamerica and the Caribbean basin before the Eocene, and no evidence for widespread SDTF vegetation before the Miocene/Pliocene.

Important recent work by Burnham and co-workers on deposits from the Loja and Cuenca basins in Ecuador gives clear evidence of SDTF ecosystems in this area in the mid-Miocene. The deposits have been dated as 13.0–10.2 Ma (Cuenca) and 10.7–10 Ma (Loja; Hungerbühler et al., 2002). Macrofossils of genera with species mostly characteristic and common in SDTF, such as

Tipuana, Cedrela, Ruprecthia and *Loxopterygium,* have been found in these deposits, consistent with this being a SDTF fossil flora. In the cases of *Loxopterygium* and *Tipuana,* the fossils are described as new species (Burnham, 1995; Burnham and Carranco, 2004). Species of these genera are still important components of SDTF in this area, the adjacent inter-Andean valleys of Peru and the Pacific coast of Ecuador and Peru (Bridgewater et al., 2003; Linares-Palomino et al., 2003; Linares-Palomino, Chapter 11). The exception is *Tipuana,* which is restricted to the Piedmont nucleus of SDTF in southern Bolivia and northern Argentina. However, a close relative of *Tipuana, Maraniona,* has recently been described from the Marañon valley in northern Peru (Hughes et al., 2004).

This suite of Ecuadorean fossils shows remarkable evidence for a habitat similar to present-day SDTF occurring in the same area in the mid-Miocene (Burnham and Carranco, 2004). This long-term stability is the probable explanation for the high level of endemism of the inter-Andean dry valleys of Peru (Linares-Palomino, Chapter 11), Ecuador (Lewis et al., Chapter 12) and possibly Bolivia (Wood, Chapter 10) discussed above. Cladistic biogeographical study of a series of genera with species showing a high level of endemicity in the different areas of neotropical SDTF showed a sister relationship between the Ecuadorean/Peruvian inter-Andean valleys and the Andean Piedmont nucleus of southern Bolivia/northern Argentina (Pennington et al., 2004). This, together with the present-day and fossil distribution of *Tipuana,* is perhaps suggestive of a wider Miocene SDTF in this area that has been disrupted by the Andean orogeny and concomitant climatic change (Hughes, 2005).

Fossils provide evidence of the minimum age of taxa, and therefore of the ecosystems that they inhabited. It is therefore probable that neotropical SDTF is older than the fossil evidence indicates. Additional lines of evidence suggest that this is probably the case. First, palaeoclimatic reconstructions suggest that there was an increasing trend in aridity in global climates from the middle Eocene (reviewed by Willis and McElwain, 2002; see also Scotese, 2002), so the climates required to maintain SDTF would have been available. This is corroborated by the Middle Eocene (45–48 Ma) Green river flora of Colorado-Utah (USA), which is likely to represent a dry subtropical forest (Graham, 1999), implying a much earlier origin for an analogue of SDTF in Laurasia.

1.4.4.2 Inter-Continental Biogeographical Links

A first step in making global biogeographical comparisons for any biome is to identify comparable physiognomic vegetation types on different continents. This is straightforward for rain forests, but less so for SDTF and savanna, where physiognomy can be tremendously variable. Few authors have made intercontinental floristic and physiognomic comparisons of SDTF and savannas (e.g. Aubreville, 1961), though more recently Schrire et al. (2005a,b; see Lewis et al., Chapter 12) have done so in their analyses of the biogeography and evolution of the Leguminosae. Schrire et al. (2005a,b) identified a globally distributed 'Succulent biome', which corresponds approximately to SDTF as defined here.

Seasonally dry tropical forest and savanna vegetation is extensive in Africa and we encouraged a contribution comparing the two vegetations of this continent with those in South America (Lock, Chapter 20). Seasonally dry tropical forests are also represented in tropical Asia, though to a lesser extent (Miles et al., in press). Given the wide geographical separation of these areas, their SDTF floras may have evolved in complete isolation. However, deep in their history, they had much closer physical links — for example, 100 Ma Africa and South America were joined as part of west Gondwana. Subsequently, although South America was an island continent until the closure of the Panama isthmus 3 Ma, it may have been accessible by stepping-stone migration routes, and transoceanic dispersal (Pennington and Dick, 2004). The possible effects of this geological history on the neotropical flora, especially rain forests, have been widely discussed (e.g. Gentry, 1993), but little attention has been paid to SDTF and savannas.

Shared Gondwanan history has been implicated as important for the entire tropical South American flora by Gentry (1982b), who estimated that 90% of its species had been produced as a result of in situ evolution of ancestors isolated after the vicariance of west Gondwana. However, recent dated molecular phylogenies have started to question this view (e.g. Renner et al., 2001; Chanderbali et al., 2001; Davis et al., 2002; Lavin et al., 2004; Weeks et al., 2005), and implicate Tertiary stepping-stone migration routes such as a boreotropical route via tropical Laurasia, or long-distance transoceanic dispersal in the creation of global tropical distributions. Few of these studies have focused upon taxa characteristic of seasonally dry areas, but given the dominance of Leguminosae in neotropical SDTF and savannas, the biogeographical history of this family is of particular relevance.

The strong representation of Leguminosae in both African and neotropical forests had been assumed to result from a shared west Gondwanan history (e.g. Raven and Polhill, 1981). However, a dated molecular phylogeny (Lavin et al., 2004; 2005) implies that this cannot be the case, as most transoceanic disjunctions in the family date to only $c.6$–16 Ma. Despite the imprecision of these molecular biogeographical dating techniques, an error of almost an order of magnitude is necessary to force these dates back to the Cretaceous, and the fossil record of the legumes only stretches back $c.60$ Ma. Lavin et al. (2004) implicated long-distance transoceanic dispersal as the major force shaping the distribution of the legume family. Schrire et al. (2005a,b) emphasized more strongly stepping-stone migration along the margins of the Tethys seaway. Whichever scenario is correct (and they are not mutually exclusive), the legume evidence does not support a role of Gondwana vicariance in the historical assembly of neotropical savannas and SDTF.

This legume evidence is corroborated by a recent biogeographical study of the Burseraceae (Weeks et al., 2005), which is especially relevant to the historical assembly of Mexican SDTF (Becerra, 2005), where *Bursera* is diverse and characteristic (Lott and Atkinson, Chapter 13). The global distribution of Burseraceae had also been considered to have been shaped by Gondwanan vicariance (e.g. Raven and Axelrod, 1974; Gentry, 1982b; Becerra 2003, 2005). Weeks et al. (2005) used a dated phylogeny to show that Burseraceae date to $c.60$ Ma, far younger than the breakup of west Gondwana, and argue that its distribution is much more likely to be explained by boreotropical migration across a north Atlantic land bridge.

1.4.4.3 Quaternary History

Because some areas of neotropical SDTF have existed for at least 10 million years, this raises the question of whether their species diversity has accumulated gradually over this longer time period, or is a more recent phenomenon, perhaps influenced by Pleistocene climatic changes. This debate of the time of origin of tropical species richness has generally been asked of rain forest biota, and has neglected seasonally dry ecosystems. However, recent studies are re-addressing this balance (e.g. Prado and Gibbs, 1993; Pennington et al., 2000; Mayle, 2004, Chapter 17; Lavin, Chapter 19)

Most of these papers have examined the potential effects of Pleistocene climate changes on SDTF species. The very influential paper of Prado and Gibbs (1993; see also Prado, 1991) examined a series of SDTF species whose distributions are disjunct between separated SDTF nuclei (Figure 1.2). Several taxa such as *Anadenanthera colubrina* (Vell.) Brenan var. *cebil* (Griseb.) Altschul have a broad, arc-like distribution, being found in some or all of the disjunct areas of SDTF found from the Brazilian caatingas to the Ecuadorean inter-Andean valleys (see Figure 1.2 in Mayle, Chapter 17). Prado and Gibbs characterized this area as the residual Pleistocenic seasonal formations arc or Pleistocenic arc because they suggested that in times of drier Pleistocene climates, the areas of SDTF it comprises were more widespread and continuous. They suggested that these repeated disjunctions were indicative of a more widespread SDTF formation in drier, cooler Pleistocene climates. This may have even penetrated Amazonia in areas where soils are relatively rich (Pennington et al., 2000). These ideas lead to the concept that the present-day areas of SDTF might be considered as remnants, or refuges of these more widespread glacial formations (Pennington et al., 2000; 2004).

That levels of floristic similarity within south-eastern Brazilian SDTF match those of the cerrados and western Amazonian forests gives some further support for the Pleistocenic arc in

Brazil, Paraguay and Argentina at least, as does Prado's subsequent (2000) recognition of these SDTFs as a phytogeographic unit (Tropical Seasonal Forests Region). However, whether this unit can be considered to extend into Bolivia and Peru is less clear. In the context of entire woody floras, the species found throughout the whole arc are few, and overall floristic similarity between areas such as the Peruvian inter-Andean valleys and Brazilian seasonal forests are low (Linares-Palomino et al., 2003).

Further analysis that is critical of the hypothesis of widespread Pleistocene SDTF comes from fossil pollen evidence in the Bolivian Chiquitano (Mayle et al., 2000; Mayle, 2004; Mayle, Chapter 17). This work shows that *Anandenathera colubrina*, geographically widespread in SDTFs in Brazil, Paraguay, Argentina, Bolivia, Peru and Ecuador, arrived only after the last glaciation in the Chiquitano. Mayle (2004; Chapter 17) infers that the Chiquitano SDTF is a recent formation, and cannot be considered a vestigial remnant of an older, more widespread formation. This is lent support by the remarkably low level of plant species endemism shown by the Chiquitano dry forests. Killeen et al. (Chapter 9) list only three confirmed endemic species of woody plants, which implies that this may be a young formation with little time to have evolved a distinctive flora.

Support for the expansion of the ranges of SDTF species during the last glacial maximum (LGM) comes from models of their distribution based upon their current climatic preferences and assumed climatic conditions at this time (Spichiger et al., Chapter 8). These workers show the distributions of SDTF species such as *Geoffroea spinosa* and *Astronium urundeuva* are modelled to have invaded the Amazon basin at the LGM. This work is lent support by dynamic vegetation model simulations (reviewed by Mayle et al., 2004) that deciduous broadleaf forests covered the southern half of Amazonia at the last LGM. We should caution, however, that the models used do not incorporate edaphic variables, which must be important as SDTF species, particularly *A. urundeuva*, tend to prefer more fertile soils than those underlying most of the *terra firme* (non-flooded) Amazonian forest.

1.4.4.4 Molecular Biogeographical Evidence: (i) Dated Phylogenies

An equation of high species endemism with antiquity might not be correct if the species diversity of SDTF were the result of recent speciation. Pleistocene origin of SDTF species was suggested by Pennington et al. (2000). They highlighted a series of genera (e.g. *Pereskia*, *Loxopterygium*, *Coursetia*), whose species show a high level of endemicity in the separate areas of SDTF (Figure 1.1). Pennington et al. (2000) speculated that these endemic species might be the result of Pleistocene vicariance of the widespread formations of SDTF envisaged by Prado and Gibbs (1993) followed by differentiation and eventual speciation.

Pennington et al. (2004, 2005) dated phylogenies of four clades rich in species endemic to the separate areas of neotropical SDTF to test this hypothesis of Pleistocene speciation. These were *Ruprechtia* (Polygonaceae), robinioid legumes, *Chaetocalyx/Nissolia* (Leguminosae) and *Loxopterygium* (Anacardiaceae). In South America, all these taxa started diversifying at least by the late Miocene/Pliocene, with very few species originating in the Pleistocene. This clearly refuted the hypothesis of recent origin of South American SDTF species. It does not, however, necessarily contradict the model of a previously widespread and now fragmented expanse of SDTF because it could be that populations were isolated for insufficient time for speciation, or that gene flow was sufficient to prevent speciation. In contrast, there was clear evidence for Pleistocene origin of SDTF species of *Coursetia*, *Nissolia* and *Ruprechtia* in Central America.

The picture suggested by the Miocene fossil record from Ecuador of at least some areas of South American SDTF being relatively ancient is given further support by evidence from these molecular phylogenies showing Miocene/Pliocene diversification dates (Pennington et al., 2004, 2005). In this book, Lavin (Chapter 19) examines in detail one of these phylogenies, which includes the SDTF legume genera *Coursetia* and *Poissonia*, and contrasts it with that of the neotropical rain forest legume genus *Inga*. He argues that *Coursetia/Poissonia* are representative of other largely

South American SDTF genera such as *Ruprechtia, Loxopterygium* and *Chaetocalyx/Nissolia*, which have similar dates of divergence and patterns of species endemism (Pennington et al., 2004). *Coursetia* and *Poissonia* have phylogenies that are highly structured geographically, with sister species generally inhabiting the same SDTF area. Lavin considers that this pattern of phylogenetic geographical structure argues for limited historical dispersal between the separate areas of South American SDTF, which is corroborated by the low levels of floristic similarity between them (see above). In contrast, *Inga*, whose 300 species have arisen remarkably rapidly in the past two million years (Lavin, Chapter 19), shows little phylogenetic geographical structure because *Inga* species growing alongside one another in the same area of rain forest are unlikely to be closely related. Lavin explains this by the cumulative effect of greater powers of dispersal, which is corroborated by the fact that many more *Inga* species are widespread than those of *Coursetia/Poissonia*. Lavin argues that this difference in historical dispersal processes between rain forest and dry forest may be general, but to verify this more groups need to be studied, especially in rain forest, where *Inga* may not be typical if older genera predominate (Bermingham and Dick, 2001)

The more recent origins of SDTF species of *Ruprechtia, Coursetia/Poissonia* and *Chaetocalyx /Nissolia* in Central America may also be congruent with the fossil record, which shows that SDTF did not become widespread in this area until the Miocene/Pliocene (Graham and Dilcher, 1995). However, much older patterns of diversification are found in *Bursera*, with the crown group of subgenus *Bursera* dating to 30 Ma, and that of subgenus *Elaphrium* to 37 Ma. We consider the diversification dates of Burseracae calculated by Weeks et al. (2005) to be more accurate than those of other recent studies of *Bursera* (Becerra, 2003, 2005), which erroneously assumed that the transatlantic distribution of Burseracae is the result of Gondwanan vicariance (Dick and Wright, 2005). The legume *Leucaena* is less species rich than *Bursera*, but also has a series of restricted endemics in Mexican SDTF, and shows similar relatively old divergence dates, with a crown group estimated at 10 Ma (Lavin et al., 2004). Another legume, *Ateleia*, shows similar patterns of species endemism in Mexico and Central America, and whilst its molecular phylogeny (Ireland, 2001) has not been subject to formal dating, we suspect that the relatively high divergence of ITS sequences among species is indicative of antiquity.

Bursera, Leucaena and *Ateleia* have their centres of species diversity in Mexico and Central America, with likely origin in these areas or in the wider Laurasian boreotropics (Weeks et al., 2005; Schrire et al., 2005a,b). In contrast, Central American clades of *Ruprechtia, Nissolia* and *Coursetia* have ancestors in South America, and arrived in Central America more recently (*c.*4–9 Ma; Pennington et al., 2004). Hence there are indications of a pattern of historical assembly for Central American SDTF congruent with that proposed for its rain forests (e.g. Gentry, 1982b; Wendt, 1993) — a mixture of more ancient Laurasian lineages with more recent arrivals from South America. We emphasize one important element of this traditional model that requires revision: the idea that species migration from South America occurred only after the closure of the Panama isthmus. The ages of Central American lineages in *Ruprechtia, Nissolia* and *Coursetia* largely predate the closure of the isthmus (*c.*3 Ma), ranging from *c.*4 to 9 Ma.

There is a paucity of dated phylogenetic studies for plants characteristic of Caribbean island SDTF. The one exception is work on the legumes *Pictetia, Poitea* and *Hebestigma* (Lavin et al. 2001, 2003, 2004). All species of these genera with a single exception are endemic to the Greater Antilles. They are shown to have diverged from their continental sister genera 38 Ma (*Hebestigma*), 16 Ma (*Poitea*) and 14 Ma (*Pictetia*), and the diversification in each occurred during the Tertiary. Each is therefore a relatively old radiation, and in the case of the Cuban endemic *Hebestigma*, the divergence date from its sister genus *Lennea* strongly implies that the vicariance of Cuba from Central America underlies its distribution. There has, however, been increasing realization of the importance of dispersal in Caribbean biogeography (de Queiroz, 2004), and this also appears to have been an important factor in the historical assembly of the Caribbean island SDTF flora. Gillespie (Chapter 16) points out that SDTFs in Florida are composed of 90% animal (mostly bird) -dispersed tree species, and that this over-representation of zoochorous taxa is a characteristic of Caribbean SDTF and other oceanic island SDTF worldwide. Caribbean SDTF have few lianas compared to their continental counterparts, which may reflect that dry forest lianas are wind dispersed and unable to disperse long distances over water, and perhaps that many liana seeds may not be able to germinate after long periods in salt water (Gillespie, Chapter 16).

1.4.4.5 Molecular Biogeographical Evidence: (ii) Population Genetics

Naciri et al. (Chapter 18) present the theoretical framework for the population genetic study of geographically widespread SDTF species, and preliminary results from range-wide molecular genetic studies of two such species, *Astronium urundeuva* and *Geoffroea spinosa*. *Astronium urundeuva* is distributed from the Brazilian caatingas to Paraguay and Bolivia and *G. spinosa* is even more widely distributed in SDTF nuclei including the Caribbean coast of Colombia and Venezuela, Ecuador and Peru, Brazil, Paraguay and Bolivia. The differentiation observed within *G. spinosa* for microsatellite markers is higher than in *A. urundeuva*, and some of these markers did not amplify in some *G. spinosa* populations, probably because of mutations in conserved flanking sequences. Both factors imply that the events underlying the disjunct distribution of *G. spinosa* may be more ancient than those that have caused the distribution of *A. urundeuva*. In *A. urundeuva*, chloroplast haplotype data imply a migration from the Misiones to the Piedmont SDTF nucleus. This agrees with Mayle's suggestion (Chapter 17) of recent migration of *Anadenathera colubrina* into the forest of the Bolivian Chiquitano, with the SDTF of the Misiones nucleus being the probable source. The data presented by Naciri et al. are preliminary, but hint that ancient vicariance and recent dispersal have shaped the distributions of different species in different ways.

Naciri et al. demonstrate the power of molecular population genetics for distinguishing, at the intraspecific level, whether widespread, disjunct species distributions reflect vicariance of previously continuous distributions, or recent dispersal, a question that is the subject of many chapters in this book. More such studies on diverse species will be vital to resolve this question.

1.4.5 Biogeographical History of Neotropical Savannas

1.4.5.1 Deep History

The fossil record of the grass family is the obvious starting point for discussion of the earliest appearance of grass-dominated habitats such as savannas. The distinctive pollen of the grasses first appeared in the Palaeocene of South America and Africa between 60 and 55 Ma (Jacobs et al., 1999). However, C4 grasses which dominate dry, open habitats such as neotropical savannas are nested relatively high in the grass phylogenetic tree (Kellogg, 2001), and must have therefore originated later. This is corroborated by a marked increase in the abundance of fossil grass pollen in the mid-Miocene in South America and elsewhere (Jacobs et al., 1999), presumably indicating the establishment of grass-rich habitats. However, the South American fossil record of grazing animals suggests that there may have been a significant grass component in ecosystems from the Oligocene (*c*.27 Ma) based upon the appearance in the fauna of adaptations to grazing. The rise to dominance of C4 grasses can also be tracked by studies of isotope content of herbivore teeth. In South America, C4 grasses were clearly present in ecosystems by 10 Ma, but the present-day proportion of C4 grasses was not reached until *c*.4 Ma. The picture for the rise of C4 grasslands in North and Central America is similar.

While there are more lines of palaeontological evidence for the origin of neotropical savannas, they do not produce an entirely clear picture, especially for the Brazilian cerrados, partly because it seems that virtually none of the fossil deposits (reviewed by Jacobs et al. [1999]) come from this area. It seems that neotropical savanna is unlikely to be older than the mid-Miocene, but the South American herbivore fossil evidence does not rule out an Oligocene origin. Nevertheless, savannas dominated by C4 grasses, similar to those of today, are clearly more recent, establishing fully by the late Pleistocene. One might speculate about the possibility of the cerrado tree flora being older, and derived from a prototypic forest that perhaps resembled the cerradão (see Ratter et al., Chapter 2), with presence of C3 grasses. It seems unlikely that this grass-poor prototypic forest would have been especially fire prone, but most of the contemporary woody flora of neotropical savannas displays fire adaptations associated with the burning of a dry grass-dominated ground layer. This implies that any prototypic woody flora has been considerably modified phenotypically. Furthermore, Bond and Keeley (2005) suggested that fire, rather than grazing herbivores, has been the key evolutionary factor in establishing savannas, and this implies that the rise of more flammable C4 grasses may have been essential for savanna formation.

1.4.5.2 Intercontinental Biogeographical Links

As outlined above under SDTF (Section 1.4.4), there is little evidence to support a role for Gondwanan vicariance in the assembly of the important legume element of the flora of neotropical savannas. We suspect that this will be found to be the case for other elements of the savanna flora, and that many savanna lineages will have arrived in the Neotropics by stepping-stone migration or long-distance dispersal during the Cenozoic. Lavin et al. (2004) and Schrire et al. (2005a,b) speculated that the larger areas of savanna relative to SDTF in the Neotropics and Africa may have formed larger targets for long-distance dispersal events. They also argued that savannas are more prone to invasion by immigrant taxa than SDTF because of lesser water stress, though this may not be reasonable for some neotropical areas such as central Brazil where savanna and SDTF grow side by side and receive the same rainfall. They couple these factors to argue for higher historical rates of successful immigration by long-distance dispersal into savanna areas. Their predictions should be tested by examining the level of geographical structuring within clades endemic to savanna areas using the methodology explained by Lavin (Chapter 19).

1.4.5.3 Quaternary History

The vigorous debate of how glacial climates affected the Amazon rain forest has resulted in much attention to the Quaternary history of neotropical savannas because most authors have assumed that if climates dried in glacial times, then savanna vegetation would have spread into Amazonia (e.g. Haffer, 1969). This is the model of refuge theory tested by several Quaternary palaeontologists (e.g. Colinvaux et al., 1996; 2001; Haberle, 1997; Hoorn, 1997), who have searched for grass pollen as evidence of savanna in Amazonian sediments, and the marine sediments at the mouth of the Amazon river (the Amazon fan). They found no evidence for grass pollen, and concluded that the majority of Amazonia must have remained forested. However, there is evidence for expansion of savanna at the edges of Amazonia during glacial periods both from fossil pollen and vegetation-climate models (reviewed by Mayle et al., 2004; Cowling et al., 2001). Whether savanna expansion was sufficient to provide connections to the isolated Amazonian savannas, the small areas of savanna on the Amazonian flanks of the Andes, and the Llanos and Rupununi savannas is uncertain. We suggest that some glacial expansion of savanna at the periphery of Amazonia makes more plausible the generation of the disjunct distributions shown by tree species characteristic of these areas because it would decrease dispersal distances between the areas. However, for savanna areas within the Amazon rain forest between Porto Velho (Rondonia) and Humaitá (Amazonas), soil carbon isotope analysis does not suggest savanna expansion at the last glacial maximum. If there was no savanna expansion into Amazonia, then long-distance dispersal must be invoked.

Pollen records only skirt the borders of the Cerrado Biome (see Mayle, Chapter 17, Figure 17.1), making inference of the Quaternary history of this area difficult. The most relevant sites are at the edges of the southern cerrados, and are reviewed by both Mayle (Chapter 17) and Durigan (Chapter 3). They show a complex shifting over the past 30,000–40,000 years of cerrado, SDTF and subtropical grassland and forest (e.g. *Araucaria* forests) formations that are currently found only at higher altitudes and further south. Whether climate-induced vicariance of cerrado vegetation elsewhere in the biome during the Quaternary can explain the patterns of species distribution in the biome that underlie the phytogeographical regions delimited by Ratter et al. (1996, 2003, Chapter 2; see also Pennington, 2003) is uncertain.

The history of the entire Venezuelan and Colombian Llanos is very recent. Huber et al. (Chapter 5) invoke the low level of species endemism (only 1.1% of the total flora) and recent alluvial geological history to postulate a late Quaternary origin of the Llanos vegetation.

1.4.5.4 Molecular Biogeographical Evidence: (i) Dated Phylogenies

We are aware of only three molecular phylogenetic studies that include cerrado plants. Perhaps the most comprehensive study is that of tribe Microliceae (Melastomataceae; Fritsch et al., 2004). This group traditionally comprises 15–17 genera and 275–300 species, of which 90% are endemic to the Brazilian cerrado and *campo rupestre* (rocky field) vegetations, the latter a higher elevation habitat found within or at the eastern borders of the Cerrado Biome (see Lock, Chapter 20; Ratter et al., Chapter 2). Fritsch et al.'s study identifies a clade of core Microliceae comprising six genera. The closest relatives of this group are found in mesic habitats, so there is a clear indication of a shift to seasonally dry habitats in the lineage leading to core Microliceae. The majority of diversification in the core Microliceae clade took place in the past c.4 Ma. Similar dates were found by Schilling et al. (2000), who sampled 21 accessions representing 15 species of *Viguiera* (200 species, Asteraceae), including 14 accessions representing nine species from Brazil. The majority of the 34 Brazilian species of the genus occur in the cerrado. Schilling et al. estimated that the divergence among the South American species dates to c.3 Ma.

The genus *Andira* (29 species, Leguminosae) has four species, *A. humilis* Mart. ex Benth., *A vermifuga* (Mart.) Benth., *A. cujabensis* Benth. and *A. cordata* R.T. Penn & H.C. Lima that are cerrado endemics. Phylogenetic studies show that these species have arisen independently with the exception of *A. cujabensis* and *A. cordata*, which are likely to be sister species (Pennington, 2003; Skema, 2003). This implies that *Andira* has moved three times from more humid habitats to the cerrado. This habitat shift into cerrado is dated at c.2 Ma (Skema, 2003; R.T. Pennington, unpubl.).

The evidence from dated phylogenies is remarkably congruent in suggesting recent diversification of cerrado species in the past 4 Ma. This is in turn congruent with the concept that this vegetation cannot have developed fully until the rise of C4 grasses was complete, c.4 Ma (Jacobs, 1999). We do, however, emphasize how limited the evidence is — representing a tiny percentage of the woody flora of the cerrados — and further phylogenetic studies of cerrado endemics may show earlier divergence times, which could add plausibility to the concept that grass-dominated habitats may have arisen earlier in South America than elsewhere.

1.4.5.5 Molecular Biogeographical Evidence: (ii) Population Genetics

Although there are population genetic studies of some neotropical savanna plant species (e.g. Collevatti et al., 2001; Lacerda et al., 2001), we are aware of none that address historical biogeographical questions explicitly. Range-wide studies of widespread savanna species such as those described for SDTF by Naciri et al. (Chapter 18) could be extremely important for elucidating how their distributions have been achieved. For example, such studies might detect if long-distance dispersal is the cause of populations of widespread cerrado species in isolated Amazonian savannas.

Historical population migration at the rain forest–savanna boundary could also be addressed with molecular population genetics, although, in this case, studies should focus on rain forest species that are hypothesized to have recently expanded their distributions. If demographic expansion of rain forest species at the rain forest–savanna ecotone in areas such as Bolivia has been as rapid as the fossil pollen record implies (Mayle et al., 2000; Mayle, Chapter 17), then this should leave a genetic signature, just as rapid population expansion has done in European species that have spread from glacial refugia in southern Europe (e.g. Démesure et al., 1996; Hewitt, 1999, see Naciri et al., Chapter 18).

1.5 CONSERVATION

1.5.1 Savannas

Aspects of deforestation and conservation are considered in the chapters of Ratter et al. (Chapter 2), Durigan (Chapter 3), and Felfili et al. (Chapter 4) in the Brazilian Cerrado Biome, while Huber et al. (Chapter 5) deal with the same subject in the Venezuelan Llanos. Because of its higher diversity and endemism, the cerrado is considered the highest conservation priority at a global scale (e.g. Myers et al., 2000). The situation is critical since the destruction of natural savanna vegetation is enormous, far exceeding both in absolute and relative terms that of the Amazon rain forest. For instance, at least 1.3 million km² of the Brazilian cerrado has been deforested (Machado et al., 2004) representing about 66% of the original two million km² of the biome, while the equivalent figure for the 3.5 million km² of Brazilian Amazonian rain forest is probably little more than 700,000 km², representing approximately 21% of its original area (Ratter et al., Chapter 2). The most extreme example of cerrado destruction is in the southern state of São Paulo, originally a cerrado biodiversity hotspot, but which has lost 88.3% of the cerrado cover existing in 1962, so that now only fragments totalling approximately 2,100 km² remain (Durigan, Chapter 3). In addition, the advance of the agricultural frontier continues relentlessly over the cerrado, a highly mechanized blitzkrieg converting the landscape to endless expanses of pastures planted with exotic grasses to feed millions of cattle and enormous arable areas covered in soya, maize, millet and other crops (Ratter et al., 1997; Chapter 2). Against this background of rapid development, only 2.2% of the total Cerrado Biome is in officially conserved areas (Machado et al., 2004) while in some places there is even less protection, for example in São Paulo state the figure is only 0.5% (Durigan, Chapter 3).

Various initiatives are being taken to address the problems of cerrado conservation and the Brazilian Ministry of the Environment, other Brazilian official bodies, foreign government aid offices, together with a plethora of international and local NGOs, and other interested organizations are planning conservation (Ratter et al., Chapter 2). The focus is to recognize priority areas for conservation both to safeguard hot-spots and to provide an adequate representation of the various environments across this notably heterogenous area. The extensive vegetation surveys carried out by a number of teams over the last 15 or so years have provided data that are proving very useful in this planning. Nevertheless, there will be a great deal of hard work necessary to achieve even a barely moderate success in terms of conservation.

It is also important to stress that selection and initial installation of conservation units will only be the beginning of a difficult process. To maintain the range of physiognomies seen in pyrophytic savanna vegetation it is important to have the correct fire regime. Complete protection from fire can result in an almost impenetrable thicket (Ratter et al., 1988) or at least establishment of a tall, forest-like cerradão (Durigan, Chapter 3). A great deal has been learned about use of fire regimes, from work over many years (e.g. Coutinho, 1990; Miranda et al., 2002), and experience of management techniques for reserves is improving.

Clearly only a relatively small part of the Cerrado Biome could be maintained as reserves, and one of the hopes for conservation must be the promotion of an environmentally friendly, sustainable form of agriculture to be established and maintained in some parts of the region (Ratter et al., Chapter 2). Such a system would be particularly appropriate for large areas of cerrado on sandy soil not suitable for arable cultivation. In addition, the team of Dr Felipe Ribeiro at EMBRAPA/Cerrados has been working for many years to develop strategies for sustainable extractivism of native cerrado plants and is obtaining promising results.

1.5.2 SEASONALLY DRY TROPICAL FORESTS: CONSERVING FRAGMENTS OF FRAGMENTS

The reasons for the relative neglect of SDTF by conservationists are complex. Highlighting a lack of financial resources for conservation may be a factor, but it is simplistic to invoke this in all cases. For example, in the USA, which is comparatively rich, Gillespie (Chapter 16) contrasts the excellent protection of the remaining SDTF in Florida with the parlous state of those in Hawaii. This suggests that the perception of the biological and social importance of SDTF may be paramount. Because many remaining areas of SDTF vegetation are small in area and highly degraded, they may be devalued in countries that possess undisturbed tracts of more species-rich vegetation such as rain forest. Following Murphy and Lugo (1995), we suggest that one fundamental difficulty for the conservation of neotropical SDTF has been the failure to consider it as a single biome, with a consistent name, in contrast to the universally recognized 'rain forest'. This is a longstanding problem. Lugo et al. (Chapter 15) point out that Beard (1955) referred to young secondary dry forests with names like cactus bush, logwood thicket, logwood-acacia bush, leucaena thicket and thorn savanna. More generally, across the Neotropics, SDTF is known by a plethora of Spanish, Portuguese and English names such as bosque seco, bosque deciduo, bosque espinoso, selva baja caducifolia, caatinga, mata seca, dry forest, seasonal forest, and mesophilous forest. The major problem of this tangle of nomenclature has been a failure to consider links between the different areas of neotropical SDTF at a continental scale. This is critical for conservation because the natural distribution of neotropical SDTF is fragmented (Figure 1.2), and many of the separate major areas of SDTF are rich in endemics. If the majority of these species are to be protected in situ, a suite of reserves across the Neotropics will be necessary, and planning these will require a full understanding of patterns of endemism and diversity, which is currently lacking. However, even the incomplete data available allow us to highlight some areas that have high endemism of SDTF species but hardly any protected areas (e.g. the inter-Andean valleys of Ecuador; Lewis et al., Chapter 12) or none whatsoever (e.g. the inter-Andean valleys of Peru; Linares-Palomino, Chapter 11). These areas should be considered conservation priorities at a continental scale, but not all receive the attention that they deserve. For example, the only Peruvian inter-Andean valley listed among the terrestrial ecoregions delimited by the World Wildlife Fund's Conservation Science Program (see http://www.worldwildlife.org/science/ecoregions/neotropic.cfm) is the Marañon valley. Only four neotropical SDTF areas are highlighted as priority ecoregions for conservation – Mexican dry forests, Tumbesian-inter-Andean dry forests, Brazilian Atlantic dry forests and Chiquitano dry forests (Olson and Dinerstein, 2002). Of these, the Chiquitano and Brazilian Atlantic dry forests have relatively few endemic plant species (Killeen et al., Chapter 9; Oliveira et al., Chapter 7) in comparison to other areas that are not listed.

A second geographical level to consider in conservation planning is the selection of reserves within the scattered major areas of neotropical SDTF (Figure 1.2). This is necessary because with the single exception of the Bolivian Chiquitano, the forests in each area have been decimated and reduced to small, fragmented areas. Conservation will therefore be principally of fragments within fragments. Ideally, conservation planning within each major SDTF area should take into account its patterns of species richness and endemism. For example, Queiroz (Chapter 6) discusses how only one of the centres of endemism of Leguminosae in the caatingas contains a well-protected Brazilian permanent protection reserve, and makes recommendations for reserves in other areas of endemism.

Some authors consider that the spectre of climatic change must also be taken into account in conservation planning (e.g. Bush, 2002; Durigan, Chapter 3). Our understanding of the Quaternary history of neotropical SDTF and savanna shows that species' ranges can change remarkably quickly. Allowing for migration across ecotones into other habitats may therefore be essential in the long term. This would require the protection of preferably large, intact SDTF areas and other biomes surrounding them. Given the heavily impacted nature of most neotropical SDTF, few areas would fulfil these criteria, though the Bolivian Chiquitano (Killeen et al., Chapter 9) is a potential candidate.

The severe fragmentation typical of SDTF means that conservation should consider the wider agro-ecosystem outside of any protected fragments (Boshier et al., 2004). This may contain SDTF species, and their seed and pollen may be able to traverse the spaces between reserves easily (e.g. White et al., 2002). Such species are, effectively, not restricted to, nor dependent upon reserves. Other species that require low disturbance and are restricted to reserves may have very small population sizes that would provoke concerns for their long-term viability.

An attribute of SDTF that may help contribute to their future conservation in the face of high levels of human disturbance is their ecological resilience. Lugo et al. (Chapter 15) discuss how a diversity of life forms, resistance to wind, high proportion of root biomass, high soil carbon and nutrient accumulation below ground, the ability of most tree species to resprout and high nutrient-use efficiency make Antillean SDTF robust in the face of disturbance. This may apply particularly to hurricane-adapted Caribbean SDTF, but Josse and Balslev (1994) suggest that because Ecuadorean SDTF can regenerate quickly, even highly degraded areas are worth protecting. The Brazilian SDTF are similarly robust ecosystems, as anyone will realize who has seen the regeneration of their native trees such as aroeira (*Astronium urundeuva*; J.A. Ratter, pers. obs.). Lugo et al. (Chapter 15) discuss successful rehabilitation of degraded dry forests in Puerto Rico. They highlight the need to remove both grazing animals and fire, and to replant native species if none remain as seed sources. Although alien species are present in these Antillean dry forests, they can facilitate the establishment of native species in degraded stands. Perhaps because of the harsh natural conditions of these forests, aliens never appear to dominate mature stands of forest, and Lugo et al. (Chapter 15) speculate that the SDTF of the future in the Antilles may combine a small percentage of aliens with native species. This rehabilitation of SDTF is, however, unlikely to be universally possible throughout the Neotropics. In some areas of steep slopes in the Andes (e.g. the Huancabamba region of Peru), soil erosion after the removal of SDTF cover appears so severe that re-establishment of forest seems improbable (R.T. Pennington, pers. obs.). In the context of rehabilitation of SDTF, it is also worth noting that in many areas, SDTF was destroyed so long ago that we do not have a clear picture of the structure and floristic composition of the undisturbed ecosystem (Gordon et al., Chapter 14). In this case, perhaps the best that can be done is to create a forest dominated by native species, which conserves both them and the ecological processes of a SDTF (J. Gordon, pers. comm.).

It is depressing that neotropical SDTF have been virtually entirely destroyed except in the Bolivian Chiquitano, which was estimated as 85% intact in 2001 (Killeen et al, Chapter 9). A much-quoted statistic is that only 2% of Mesoamerican dry forests are intact (Janzen, 1988), and the situation is similar elsewhere. For example, only 3.2% of the vegetation of the caatinga biome is unaltered (Queiroz, Chapter 6). Given the urgent need to protect the few remaining areas of SDTF, it might be argued that there is no time to perfect our biological knowledge of these ecosystems in order to plan ideal conservation units. Gordon et al. (Chapter 14) discuss conservation of SDTF in Central America and argue for a pragmatic approach to reserve selection that may proceed without detailed prior biological assessments. They highlight two successful protected SDTF areas in Mexico and Honduras that were designated because of a complex mixture of socio-economic, and political reasons. Some authors would contend that this is flawed conservation planning (e.g. Pressey, 1994), but Gordon et al. (Chapter 14) suggest that this is inevitable given the social and political complexities in these areas. Because of the precarious state of the few remaining fragments of neotropical SDTF and savanna, we agree absolutely with Gordon et al. that protecting what can be conserved today is as important as deciding what should be conserved tomorrow.

ACKNOWLEDGEMENTS

We thank Jamie Gordon, Robyn Burnham, Chris Dick, Sam Bridgewater and Matt Lavin for constructive criticism, and Reynaldo Linares for assistance with the map.

REFERENCES

Aleixo, A., Historical diversification of a terra-firme bird superspecies: a phylogenetic perspective on the role of different hypotheses of Amazonian diversification, *Evolution*, 58, 1303, 2004.

Aubreville, A., Étude des principales formations végétales du Brésil, Centre Technique Forestier Tropical, Nugent-sur-Marne, France, 1961.

Bass, J., Incidental agroforestry in Honduras: the jícaro tree (*Crescentia* spp.) and pasture land use, *J. Latin American Geog.*, 3, 67, 2004.

Beard, J., The classification of tropical American vegetation types, *Ecology* 36, 89, 1953.

Becerra, J.X., Synchronous adaptation in an ancient case of herbivory, *Proc. Natl. Acad. Sci. USA*, 100, 12804, 2003.

Becerra, J.X., Timing the origin and expansion of the Mexican tropical dry forest, *Proc. Natl. Acad. Sci. USA*, 102, 10919, 2005.

Bermingham, E. and Dick, C., The *Inga* — newcomer or museum antiquity? *Science*, 293, 2214, 2001.

Bond, W.J. and Keeley, J.E., Fire as a global 'herbivore': the ecology and evolution of flammable ecosystems, *Trends Ecol. Evol.*, 20, 387, 2005.

Borhidi, A., *Phytogeography and vegetation ecology of Cuba*, Akadémiai Kiadó Budapest, Hungary, 1991.

Boshier, D.H. et al., Prospects for circa situm tree conservation in Mesoamerican dry forest agro-ecosystems, in *Biodiversity Conservation in Costa Rica, Learning the Lessons in a Seasonal Dry Forest*, G.W. Frankie et al., Eds, University of California Press, Berkeley, USA, 2004, 210.

Bridgewater, S. et al., Vegetation classification and floristics of the savannas and associated wetlands of the Rio Bravo Conservation and Management Area, Belize. *Edinb. J. Bot.* 59, 421, 2002.

Bridgewater, S. et al., A preliminary floristic and phytogeographic analysis of the woody flora of seasonally dry forests in northern Peru, *Candollea*, 58, 129, 2003.

Bridgewater, S., Ratter, J.A., and Ribeiro J.F., Biogeographic patterns, β-diversity and dominance in the Cerrado Biome of Brazil, *Biodiv. Cons.*, 13, 2295, 2004.

Burnham, R.J., A new species of winged fruit from the Miocene of Ecuador: *Tipuana ecuatoriana* (Leguminosae), *Amer. J. Bot.*. 82, 1599, 1995.

Burnham, R.J. and Carranco, N., Miocene winged fruits of *Loxopterygium* (Anacardiaceae) from the Ecuadorean Andes, *Amer. J. Bot,* 91, 1767, 2004.

Burnham, R.J. and Johnson, K.R., South American palaeobotany and the origins of neotropical rain forests, *Phil. Trans. R. Soc. Lond. B.*, 359, 1623, 2004.

Bush, M.B., Distributional change and conservation on the Andean flank: a palaeoecological perspective, *Global Ecol. Biogeogr.*, 11, 463, 2002.

Chanderbali, A.S., Van der Werff, H., and Renner, S.S., Phylogeny and historial biogeography of Lauraceae: evidence from the chloroplast and nuclear genomes, *Ann. Missouri Bot. Gard.*, 88, 104, 2001.

Colinvaux, P.A. et al., A long pollen record from lowland Amazonia: forest and cooling in glacial times, *Science,* 274, 85, 1996.

Colinvaux, P.A. et al., A paradigm to be discarded: Geological and paleoecological data falsify the Haffer & Prance refuge hypothesis of Amazonian speciation, *Amazoniana*, 16, 609, 2001.

Collevatti, R.G., Grattapaglia, D., and Hay, J.D., Population genetic structure of the endangered tropical tree species *Caryocar brasiliense*, based on variability at microsatellite loci, *Mol Ecol.*, 10, 349, 2001.

Colwell, R.K. and Lees, D.C., The mid-domain effect: geometric constraints on the geography of species richness, *Trends Ecol. Evol.*, 15, 70, 2000.

Condit, R. et al., Beta-diversity in tropical forest trees, *Science*, 295, 666, 2002.

Cowling, S.A., Maslin, M.A., and Sykes, M.T., Paleovegetation simulations of lowland Amazonia and implications for Neotropical allopatry and speciation, *Quat. Res.*, 55, 140, 2001.

Davis, C.C. et al., Laurasian migration explains Gondwanan disjunctions: evidence from Malpighiaceae, *Proc. Natl. Acad. Sci. USA*, 99, 6833, 2002.

Davis, C.C. et al., Explosive radiation of Malpighiales supports a mid-Cretaceous origin of modern tropical rain forests, *Amer Nat.* 162, E36, 2005.

Démesure, B., Comps, B., and Petit, R.J., Chloroplast DNA phylogeography of the common Beech (*Fagus sylvatica* L.) in Europe, *Evolution*, 50, 2515, 1996.

Dick, C.W. and Wright, S.J., Tropical mountain cradles of dry forest diversity, *Proc. Natl. Acad. Sci. USA*, 102, 10757, 2005.

Eiten, G., The cerrado vegetation of Brazil, *Bot. Rev.*, 38, 201, 1972.

Fritsch, P.W. et al., Phylogeny and circumscription of the near-endemic Brazilian tribe Microlicieae (Melastomataceae), *Amer. J. Bot.*, 91, 1105, 2004.

Furley, P.A. and Ratter, J.A., Soil resources and plant communities of the central Brazilian cerrado and their development, *J. Biogeogr.,* 15, 97, 1988.

Furley, P.A. and Ratter, J.A., Further observations on the nature of savanna vegetation and soils in Belize, in *Advances in Environmental and Biogeographical Research in Belize*, Furley, P.A., Ed, Biogeographical Monographs No. 3, Biogeographical Research Group, University of Edinburgh, Edinburgh, UK, 1989, 1.

Furley, P.A., Proctor, J., and Ratter, J.A., Eds, *Nature and Dynamics of Forest-Savanna Boundaries*, Chapman & Hall, London, 1992.

Gentry, A.H., Phytogeographic patterns as evidence for a Chocó refuge, in *Biological Diversification in the Tropics*, Prance, G.T., Ed, Columbia University Press, New York, 1982a, 112.

Gentry, A.H., Neotropical floristic diversity: phytogeographical connections between Central and South America: Pleistocene climatic fluctuations or an accident of the Andean orogeny, *Ann. Missouri Bot. Gard.*, 69, 557, 1982b.

Gentry, A. H., Diversity and floristic composition of lowland tropical forest in Africa and South America, in *Biological Relationships Between Africa and South America*, Goldblatt, P., Ed, New Haven, Yale University Press, 1993, 500.

Gentry, A.H., Diversity and floristic composition of neotropical dry forests, in *Seasonally Dry Tropical Forests*, Bullock, S.H., Mooney, H.A. and Medina, E., Eds, Cambridge University Press, Cambridge, 1995, 146.

Gillespie, T.W., Grijalva, A., and Farris, C.N., Diversity, composition, and structure of tropical dry forests in Central America, *Plant Ecology*, 147, 37, 2000.

Goodland, R.J. and Pollard, R., The Brazilian cerrado vegetation: a fertility gradient. *J. Ecol.,* 61, 219, 1973.

Gordon, J.E. et al., Assessing landscapes: a case study of tree and shrub diversity in the seasonally dry forests of Oaxaca, Mexico and southern Honduras, *Biol. Cons.*, 117, 429, 2004.

Graham, A. *Late Cretaceous and Cenozoic History of North American Vegetation*, Oxford University Press, Oxford, 1999.

Graham, A. and Dilcher, D., The Cenozoic record of tropical dry forest in northern Latin America and the southern United States, in *Seasonally Dry Tropical Forests*, Bullock, S.H., Mooney, H.A., and Medina, E., Eds, Cambridge University Press, Cambridge, 1995, 124.

Haberle, S., Upper Quaternary vegetation and climate history of the Amazon Basin: correlating marine and terrestrial pollen records, *Proceedings of the Ocean Drilling Program, Scientific Results*, 155, 381, 1997.

Haffer, J., Speciation in Amazonian forest birds, *Science,* 165, 131, 1969.

Heads, M., Dating nodes on molecular phylogenies: a critique of molecular biogeography, *Cladistics,* 21, 62, 2005.

Hewitt, G., Post-glacial re-colonization of European biota, *Biol. J. Linn. Soc.*, 68, 87, 1999.

Hoorn, C., Palynology of the Pleistocene glacial/interglacial cycles of the Amazon fan (holes 940A, 944A, and 946A), *Proceedings of the Ocean Drilling Program, Scientific Results,* 397, 1997.

Huber, O., Neotropical savannas: their flora and vegetation, *Trends Ecol. Evol.* 2, 67, 1987.

Hughes, C.E., Four new legumes in forty-eight hours, *Oxford Plant Systematics*, 12, 6, 2005.

Hughes, C.E. et al., *Maraniona*, a new dalbergioid legume genus (Leguminosae Papilionoideae) from Peru, *Syst. Bot.*, 29, 366, 2004.

Hungerbühler, D. et al., Neogene stratigraphy and Andean geodynamics of southern Ecuador, *Earth Science Reviews*, 57, 75, 2002.

Ireland, H., The taxonomy and systematics of *Ateleia* and *Cyathostegia* (Leguminosae-Swartzieae), PhD thesis, University of Reading, 2001.

Jacobs, B.F., Kingston, J.D., and Jacobs, L.L., The origin of grass-dominated ecosystems, *Ann. Missouri Bot. Gard.*, 86, 590, 1999.

Janssen, A., *Flora und vegetation der savanna von Humaitá und ihre Standortbedingungen*, Dissertationes Botanicae, Band 93, J. Cramer, Berlin, Stuttgart, Germany, 1986.

Janzen, D., Tropical dry forests, in *Biodiversity*, Wilson, E.O. and Peter, F.M., Eds, National Academy Press, Washington, DC, 1988, 130.

Josse, C. and Balsler, H., The composition and structure of a semideciduous forest in western Ecuador, *Nordic J. Bot.*, 14, 425, 1994.

Kellogg, E., Evolutionary history of the grasses, *Plant Physiol.*, 125, 1198, 2001.

Killeen, T.J., Louman, B.T., and Grimwood. La ecologia paisajística de la region de Concepción y Lomerio en la provincia Ñuflo de Chávez, Santa Cruz, Bolivia, *Ecol. Bolivia*, 16, 1, 1990.

Killeen, T.J. and Nee, M., Catalogo de las plantas sabaneras de Concepción, Depto. Santa Cruz, Bolivia, *Ecol. Bolivia*, 17, 53, 1991.

Killeen, T.J. et al., Diversity, composition, and structure of a tropical semideciduous forest in the Chiquitanía region of Santa Cruz, Bolivia, *J. Tropical Ecology,* 14, 803, 1998.

Lacerda, D.R. et al., Genetic diversity and structure of natural populations of *Plathymenia reticulata* (Mimosoideae), a tropical tree from the Brazilian Cerrado, *Mol. Ecol.,* 1143, 2001.

Lavin, M., Herendeen, P.S., and Wojciechowski, M.F., Evolutionary rates analysis of Leguminosae implicates a rapid diversification of the major family lineages immediately following an Early Tertiary emergence, *Syst. Biol.,* 54, 530, 2005.

Lavin, M. et al., Identifying Tertiary radiations of Fabaceae in the Greater Antilles: alternatives to cladistic vicariance analysis, *Int. J. Plant Sci.,* 162 (6 supplement), S53, 2001.

Lavin, M. et al., Phylogeny of robinioid legumes (Fabaceae) revisited: *Coursetia* and *Gliricidia* recircumscribed, and a biogeographical appraisal of the Caribbean endemics, *Syst. Bot.,* 28, 387, 2003.

Lavin, M. et al., Metacommunity processes rather than continental tectonic history better explain geographically structured phylogenies in legumes, *Phil. Trans. R. Soc., Lond. B.,* 359, 150, 2004.

Lenthall, J.C., Bridgewater, S., and Furley, P.A., A phytogeographic analysis of the woody elements of New World savannas, *Edinb. J. Bot.,* 56, 293, 1999.

Linares-Palomino, R., Pennington, R.T., and Bridgewater, S., The phytogeography of seasonally dry tropical forests in Equatorial Pacific South America, *Candollea,* 58, 473, 2003.

Lougheed, S.C. et al., Ridges and rivers: a test of competing hypotheses of Amazonian diversification using a dart-poison frog (*Epipedobates femoralis*), *Proc. R. Soc. Lond. B. Biol. Sci.,* 266, 1829, 1999.

Machado, R.B. et al., *Estimativas de perda da área do Cerrado brasileiro*, unpublished technical report, Conservação Internacional, Brasília, DF, Brazil, 2004.

Mayle, F.E., Assessment of the Neotropical dry forest refugia hypothesis in the light of palaeoecological data and vegetation simulations, *J. Quaternary Sci.,* 19, 713, 2004.

Mayle, F.E. et al., Responses of Amazonian ecosystems to climatic and atmospheric carbon dioxide changes since the last glacial maximum, *Phil. Trans. R. Soc. Lond. B,* 359, 499, 2004.

Mayle, F.E., Burbidge, R., and Killeen, T.J., Millennial-scale dynamics of southern Amazonian rain forests, *Science,* 290, 2291, 2000.

de Mendonça, R.C. et al., Flora vascular do cerrado, in *Cerrado: Ambiente e Flora*, Sano, S.M. and de Almeida, S.P., Eds, EMBRAPA, Planaltina, DF, Brazil, 1998, 289.

Miles, L. et al., A global overview of the conservation status of tropical dry forests, *J. Biogeog.*, in press.

Milliken, W. and Ratter, J.A., The vegetation of the Ilha de Maracá, in *Maracá: the Biodiversity and Environment of an Amazonian Rainforest,* Milliken, W. and Ratter, J.A., Eds, John Wiley, Chichester, UK, 1998, 71.

Miranda, H.S., Bustamente, M.M.C. and Miranda, A.C., The Fire Factor, in *The Cerrados of Brazil*, Oliveira, P.S. and Marquis, R.J., Eds, Colombia University Press, New York, 2002, 51.

Mooney, H.A., Bullock, S.H., and Medina, E., Introduction, in *Seasonally Dry Tropical Forests*, Bullock, S.H., Mooney, H.A., and Medina, E., Eds, Cambridge University Press, Cambridge, 1995, 1.

Murphy, P. and Lugo, A.E., Ecology of tropical dry forest, *Annu. Rev. Ecol. Syst.,* 17, 67, 1986.

Murphy, P. and Lugo, A.E., Dry forests of Central America and the Caribbean, in *Seasonally Dry Tropical Forests*, Bullock, S.H., Mooney, H.A., and Medina, E., Eds, Cambridge University Press, Cambridge, 1995, 9.

Myers, N. et al., Biodiversity hotspots for conservation priorities, *Nature*, 403, 853, 2000.

Near, T.J. and Sanderson, M.J., Assessing the quality of molecular divergence time estimates by fossil calibrations and fossil-based model selection, *Phil. Trans. R. Soc., Lond. B.,* 359, 1477, 2004.

Oliveira-Filho, A.T. and Ratter, J.A., A study of the origin of central Brazilian forests by the analysis of the plant species distributions, *Edinb. J. Bot.* 52, 141, 1995.

Olson, D.M and Dinerstein, E., The global 200: Priority ecoregions for global conservation. *Ann. Missouri Bot. Gard.* 89, 199, 2002.

Pennington, R.T., A monograph of *Andira* (Leguminosae-Papilionoideae), *Syst. Bot. Monog.* 64, 2003.

Pennington, R.T. and Dick, C.W., The role of immigrants in the assembly of the Amazonian tree flora, *Phil. Trans. R. Soc. Lond. B,* 359, 1611, 2004.

Pennington, R.T., Prado, D.E., and Pendry, C.A., Neotropical seasonally dry forests and Quaternary vegetation changes, *J. Biogeogr.,* 27, 261, 2000.

Pennington, R. T. et al., Historical climate change and speciation: neotropical seasonally dry forest plants show patterns of both Tertiary and Quaternary diversification. *Phil. Trans. R. Soc. Lond B,* 359, 515, 2004.

Pennington, R.T. et al., Climate change and speciation, in *Tropical Forests and Global Atmospheric Change,* Malhi, Y. and Phillips, O.L., Eds, Oxford University Press, Oxford, 2005, 199.

Pitman, N.C.A. et al., Tree species distributions in an upper Amazonian forest, *Ecology,* 80, 2651, 1999.

Pitman, N.C.A. et al., Dominance and distribution of tree species in upper Amazonian terra firme forests, *Ecology,* 82, 2101, 2001.

Prado, D.E., *A critical evaluation of the floristic links between chaco and caatingas vegetation in South America,* unpublished PhD thesis, University of St Andrews, St Andrews, 1991.

Prado, D.E., Seasonally dry forests of tropical South America: from forgotten ecosystems to a new phytogeographic unit, *Edinb. J. Bot.,* 57, 437, 2000.

Prado, D.E. and Gibbs, P.E., Patterns of species distribution in the dry seasonal forests of South America, *Ann. Missouri Bot. Gard.,* 80, 902, 1993.

Pressey, R.L., *Ad Hoc* Reservations: Forward or backward steps in developing representative reserve systems? *Conservation Biology,* 8, 662, 1994.

de Queiroz, A., The resurrection of oceanic dispersal in historical biogeography, *Trends Ecol. Evol.,* 20, 68, 2004.

Ratter, J.A., Ribeiro, J.F., and Bridgewater, S., The Brazilian cerrado vegetation and threats to its biodiversity, *Ann. Bot.,* 80, 223, 1997.

Ratter, J.A. et al., Observations on forests of some mesotrophic soils in central Brazil, *Rev. Bras. Bot.,* 1, 47, 1978.

Ratter, J.A. et al., Observations on the woody vegetation types in the Pantanal and at Corumbá, Brazil, *Notes R. Bot. Gdn Edinb.,* 45, 503, 1988.

Ratter, J.A. et al., Analysis of the floristic composition of the Brazilian cerrado vegetation II. Comparison of the woody vegetation of 98 areas, *Edinb. J. Bot.,* 53, 153, 1996.

Ratter, J.A. et al., Analysis of the floristic composition of the Brazilian cerrado vegetation III. Comparison of the woody vegetation of 376 areas, *Edinb. J. Bot.,* 60, 57, 2003.

Raven, P.H. and Polhill, R.M., Biogeography of the Leguminosae, in *Advances in Legume Systematics, Part 1,* Polhill, R.M. and Raven, P.H., Eds, Royal Botanic Gardens, Kew, 1981, 27.

Renner, S.S., Clausing, G., and Meyer, K., Historical biogeography of Melastomataceae: the roles of Tertiary migration and long-distance dispersal, *Amer. J. Bot.,* 88, 1290, 2001.

Ribeiro, J.F., *Comparação da concentração de nutrientes na vegetação arbórea de um cerrado e um cerradão no Distrito Federal, Brasil,* MSc thesis, University of Brasília, Brazil, 1983.

Richardson, J.A. et al., Recent and rapid diversification of a species-rich genus of neotropical trees, *Science,* 293, 2242, 2001.

Salas-Morales, S.H., Saynes-Vasquez, A., and Schibli, L., Flora de la costa de Oaxaca, Mexico: Lista floristica de la region de Zimatan, *Bol. Soc. Bot. Mex.,* 72, 21, 2003.

Sanaiotti, T.M., *The woody flora and soils of seven Brazilian Amazonian dry savanna areas,* PhD thesis, University of Stirling, Scotland, 1996.

Sanaiotti, T.M., Bridgewater, S., and Ratter, J.A., A floristic study of the savanna vegetation of the state of Amapá, Brazil, and suggestions for its conservation, *Bol. Mus. Para Emílio Goeldi, Sér. Bot.,* 13, 3, 1997.

Sanderson, M.J., A nonparametric approach to estimating divergence times in the absence of rate constancy, *Mol. Biol. Evol.,* 14, 1218, 1997.

Sanderson, M.J., Estimating absolute rates of molecular evolution and divergence times: a penalized likelihood approach, *Mol. Biol. Evol.,* 19, 101, 2002.

Sarmiento, G., A conceptual model relating environmental factors and vegetation formations in the lowlands of tropical South America, in *Nature and Dynamics of Forest-Savanna Boundaries,* Furley, P.A., Proctor, J., and Ratter, J.A., Eds, Chapman & Hall, London, 1992, 583.

Schilling, E.E. et al., Brazilian species of *Viguiera* (Asteraceae) exhibit low levels of ITS sequence variation, *Edinb. J. Bot.* 57, 323, 2000.

Schrire, B.D., Lavin, M., and Lewis, G.P. Global distribution patterns of the Leguminosae: insights from recent phylogenies, *Biol. Skr.,* 55, 375, 2005a.

Schrire, B.D., Lewis, G.P., and Lavin, M., Biogeography of the Leguminosae, in *Legumes of the World*, Lewis, G. et al., Eds, Royal Botanic Gardens, Kew, 2005b, 21.

Scotese, C.R., Atlas of earth history, volume 1, paleogeography, PALEOMAP project, Arlington, Texas, (website: http://www.scotese.com), 2002.

Scott, G.A.J., *Grassland development in the Gran Pajonal of eastern Peru. A study of soil-vegetation nutrient systems*, Hawaii Monographs in Geography, No. 1. University of Hawaii at Manoa, Department of Geography, Honolulu, 1978.

da Silva, M.N.F. and Patton, J.L., Molecular phylogeography and the evolution and conservation of Amazonian mammals, *Mol. Ecol.*, 7, 475, 1998.

Skema, C., Phylogeny and biogeography of *Andira*, MSc thesis, University of Edinburgh and Royal Botanic Garden Edinburgh, 2003.

Solbrig, O.T., Ecological constraints to savanna land use, in *The World's Savannas. Economic Driving Forces, Ecological Constraints and Policy Options for Sustainable Land Use,* Young, M.D. and Solbrig, O.T., Eds, MAB series 12, Parthenon Publications, London, 1993, 21.

Thorne, J.L. and Kishino, H., Divergence time and evolutionary rate estimation with multilocus data, *Syst. Biol.*, 51, 689, 2002.

Thorne, J.L., Kishino, H., and Painter, I.S., Estimating the rate of evolution and the rate of molecular evolution, *Mol. Biol. Evol.*, 15, 1647, 1998.

Trejo, I. and Dirzo, R., Floristic diversity of Mexican seasonally dry tropical forests, *Biodiv. Cons.*, 11, 2063, 2002.

Weberbauer, A., *El mundo vegetal de los Andes Peruanos*, Estac. Exper. Agric. La Molina, Edit. Lume, Lima, 1945.

Weeks, A., Daly, D.C., and Simpson, B.B., The phylogenetic history and biogeography of the frankincense and myrrh family (Burseraceae) based upon nuclear and chloroplast sequence, *Mol. Phylogenetics Evolution*, 35, 85, 2005.

Wendt, T., Composition, floristic affinities and origins of the Mexican Atlantic slope rain forests, in *Biological Diversity of Mexico, Origins and Distribution,* Ramamoorthy, T.P. et al., Eds, Oxford University Press, Oxford, 1993, 595.

White, G.M., Bowshier, D.H., and Powell, W., Increased pollen flow counteracts fragmentation in a tropical dry forest: an example from *Swietenia humilis* Zucc. *Proc. Natl Acad. Sci. USA*, 99, 2038, 2002.

Willis, K.J. and McElwain, J.C., *The Evolution of Plants*, Oxford University Press, Oxford, 2002.

2 Biodiversity Patterns of the Woody Vegetation of the Brazilian Cerrado

James A. Ratter, Samuel Bridgewater and J. Felipe Ribeiro

CONTENTS

ABSTRACT

The Cerrado Biome is one of the world's principal centres of biodiversity and has 6429 recorded native species of vascular plants, although according to estimates the total may be as high as 10,500. Tree and large shrub species of the savannic element (i.e. cerrado *sensu lato*) total *c.*700–1000, of which the great majority are rare: only 300 species occurred at ≥2.5% of 316 sites surveyed in the main core cerrado area and its southern and western outliers. An oligarchy of 116 species dominates the woody flora, while a total of *c.*340 species provides the vast majority of the vegetation cover. The cerrado woody flora probably evolved from an ancestral cerradão of endemic species to which were added a large number of species from neighbouring biomes. Multivariate analyses of the results of 376 floristic surveys produced six geographical groups: southern (principally São Paulo state); central and south-eastern, and central-western — both subdivisions of the main Planalto core area; far-western; north and north-eastern; and disjunct Amazonian. In addition, another group is linked to the presence of mesotrophic soils and is scattered across a large area of the cerrado region. The explosion of modern highly mechanized agriculture during the last *c.* 35 years has destroyed the natural vegetation of at least 1.3 million km² representing about 66% of the biome's area, and continues to expand. Major conservation initiatives to protect biodiversity hotspots and provide a representation of the geographical variation across the biome are urgent.

2.1 INTRODUCTION

The aim of this chapter is to discuss the results obtained by our team, and compare and integrate them with those of other workers to give a picture of the patterns of floristic diversity of woody cerrado vegetation. During the last 20 or so years the increase of data available to make such a study has been enormous. The first paper we published on the subject (Ratter and Dargie, 1992) reported a comparison of 26 areas of cerrado, representing all the records available to us in 1986–87, while the second (Ratter et al., 1996) compared 98 areas for which information existed in 1994. This was followed by an intensive period of targeted research on our part and by other workers, and the publication of Ratter et al. (2003) reported the results of the survey of 376 areas throughout the Cerrado Biome. This activity has continued and now results for over 450 floristic vegetation surveys are available for comparison and analysis. Much of the present interest in the cerrado has been stimulated by the realization that the biome is a world centre of biodiversity (Dias, 1992; Fonseca et al., 2000; Myers et al., 2000) and that it is endangered by the expansion of modern agriculture, already having lost at least 66% of its original area (Alho and Martins, 1995; Ratter et al., 1997; Cavalcanti, 1999; Cavalcanti and Joly, 2002; Machado et al., 2004).

A great deal of the information has been provided by a number of major projects. The Conservação e Manejo da Biodiversidade do Bioma Cerrado (CMBBC) project based on collaboration of Embrapa Cerrados, the University of Brasília, and the Royal Botanic Garden Edinburgh, with funding from the UK Department for International Development (DFID), has carried out 170 surveys in the northern and central-western parts of the Cerrado Biome (Ratter et al. 2000, 2001, 2003). Miranda (1997) has made a comprehensive survey of the Amazonian savannas of Roraima, including the inventory of 45 sites. The Biogeografia dos Cerrados project team based in the University of Brasília has made many surveys in the extensive Chapadas Pratinha and dos Veadeiros in Central Brazil (see, for example, Felfili and Silva Junior, 1993; Felfili et al., 1997, 2004 and Ch. 4), while the group of Brandão and associates has worked over a great area of southern and central Minas Gerais. In addition, Durigan et al. (2003b) have studied 86 areas as part of the Conservation Feasibility of the Cerrado Remnants in São Paulo State project, financed by the Biota Programme. As discussed in our previous works, studies similar to ours have been carried out by Dr Alberto Jorge F. de Castro of the Federal University of Piauí (Castro, 1994a,b; Castro and Martins, 1999; Castro et al., 1998, 1999) and provide essential data for comparison.

2.2 CERRADO — ENVIRONMENT AND PHYSIOGNOMY

The Brazilian Cerrado Biome originally covered an area of some 2 million km² of central Brazil (Figure 2.1), representing about 23% of the land surface of the country. In terms of area, it is exceeded by only one other vegetation formation in Brazil, the Amazonian forest covering approximately 3.5 million km². The dominant vegetation of the biome is a tree savanna termed *cerrado*, but more open forms of scrubby or grassy savanna are also very important. The cerrado region extends from the southern margin of the Amazonian forest to outlying areas in the southern states of São Paulo and Paraná, occupying more than 20° of latitude and an altitudinal range from

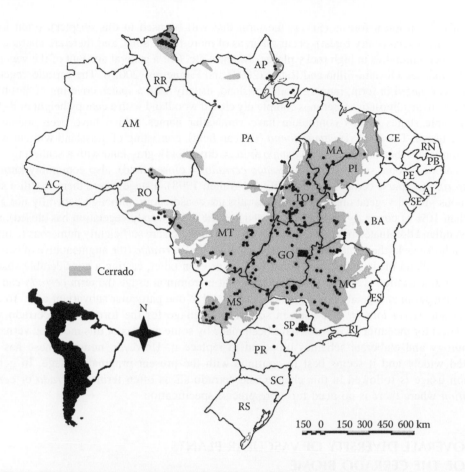

FIGURE 2.1 Map showing the distribution of cerrado vegetation in Brazil. The dots and rectangle (Federal District) represent the sites compared in the CMBBC study (Ratter et al., 2003). Letters are state abbreviations; those referred to in the text are: AM, Amazonas; AP, Amapá; BA, Bahia; CE, Ceará; GO, Goiás; MA, Maranhão; MG, Minas Gerais; MS, Mato Grosso do Sul; MT, Mato Grosso; PA, Pará; PI, Piauí; PR, Paraná; RO, Rondônia; RR, Roraima; SP, São Paulo; TO, Tocantins. Reprinted with permission of Cambridge University Press; Ratter, J.A., Bridgewater, S. and Ribeiro, J.F. 2003. Analysis of the floristic composition of the Brazilian cerrado vegetation. III: Comparison of the woody vegetation of 376 areas. *Edinburgh J. Bot.* 60: 57–109.

sea-level to 1800 m; about 700,000 km² of the total area of cerrado vegetation is within the southern Amazon basin. In addition, there are areas of cerrado in Bolivia and Paraguay, while related, but species-poor, savanna vegetation such as the Roraima and Rupununi savannas and the Venezuelan Llanos occur north of the Amazon.

The continuous area of the Cerrado Biome in central Brazil is often called the core area, although usage of this term is sometimes restricted to the Planalto, the extensive high plateau lands of central Brazil.

The cerrado climate is typical of the rather moister savanna regions of the world, with an average precipitation for over 90% of the area of 800–2000 mm and a very strong dry season during the southern winter (approximately April–September), while average annual temperatures are 18–28°C (Dias, 1992). The soils of most of the area are dystrophic with low pH and availability of calcium and magnesium, and high aluminium content (Lopes and Cox, 1977; Furley and Ratter, 1998), most are oxisols (ferralitic soils).

The typical vegetation landscape within the Cerrado Biome consists of cerrado on the well-drained interfluves with gallery forests along the watercourses. In addition, deciduous and semideciduous seasonally dry tropical forests (SDTF, the term that will be used in this chapter), often locally known as *mata seca* (= dry forest), occur on areas of more fertile soils, and there are characteristic vegetation communities in high rocky places, swamps, etc. (for a general account of the vegetation of the biome, see Oliveira-Filho and Ratter (2002) and Filgueiras (2002)). The cerrado vegetation itself is very varied in form, ranging from grassland, usually with a sparse covering of shrubs and small trees, to an almost or sometimes completely closed woodland with a canopy height of 8–13 m. Recognizable stages in this continuum have vernacular names, which have been adapted for scientific usage. These are (i) *campo limpo* (= clean field), consisting of grassland without woody vegetation larger than subshrubs; (ii) *campo sujo* (= dirty field), grassland with a scattering of low shrubs and occasional small trees; (iii) *campo cerrado* (= closed field), also commonly known as *cerrado ralo* (= sparse cerrado, see Ribeiro and Walter, 1998), a stage where there is still a strong continuous ground vegetation but trees and shrubs are conspicuous, although generally not giving more than 10% of cover; (iv) *cerrado* (= closed, i.e. the taller woody vegetation has closed), a low open woodland dominated by trees frequently 3–8 m tall which are sufficiently numerous to impede travel on horseback (the vernacular name refers to this); (iv) *cerradão* (the augmentative of cerrado), a dense woodland made up of trees, often 8–13 m or even taller, casting a considerable shade so that ground vegetation is sparse. It is confusing that in common usage the term *cerrado* can refer to Brazilian savanna vegetation in its generic sense and to one particular subvariant of it. To avoid this, cerrado *sensu lato* (*s.l.*) is used in scientific language for the former, and cerrado *sensu stricto* (*s.s.*) for the latter. Criticisms have been made by some authors of this modified vernacular terminology and elaborate schemes proposed to replace it. However, none of these has been accepted widely and it seems best to continue with the present pragmatic usage. In general, common usage is followed in this chapter and cerrado s.l. is often termed *cerrado* or *cerrado vegetation* where there is no need for more precise specification.

2.3 OVERALL DIVERSITY OF VASCULAR PLANTS OF THE CERRADO BIOME

The Cerrado Biome is one of the world's principal centres of biodiversity, ranked by Myers et al. (2000) as among 25 global hotspots of absolute importance for conservation. It is difficult to provide accurate figures, but Dias (1992) estimated that it contained a total of 160,000 species of plants, animals and fungi. It is probably the world's overall richest area for parasitic organisms causing plant galls, while recent work by Dr José Dianese of the University of Brasília has demonstrated the occurrence of at least three or four specific fungal parasites for each species of the cerrado higher plants he investigated (Dianese et al., 1993, and Dianese, pers. comm.). The establishment of such abundant host-parasite interactions must reflect a very long evolutionary history.

Dias (1992) has estimated the total number of native species of vascular plants in the biome, including all vegetation types (cerrado s.l., riverine forest, dry forest, hill grassland, swamp, etc.) as 10,500, while Myers et al. (2000) and Simon and Proença (2000) both estimate 10,000. Mendonça et al. (1998) give an overall checklist, including habitat data, for the same category of 6429 recorded species; which will certainly be increased by new discoveries. Figures for species numbers are summarized in Table 2.1.

The number of species belonging to the savannic element, i.e. cerrado s.l., is obviously much less than that for all the vegetation types of the biome. An analysis of the checklist of Mendonça et al. (1998) gives a total of 2,880 species for cerrado s.l. of which only approximately 477 (17%) are listed as trees. Thus the ground layer of herbs and low shrubs contribute by far the greater part of higher species diversity. However, paradoxically, the great majority of studies so far made have concentrated on the larger woody component of the vegetation. The reason for this lies in the ease with which surveys can be conducted. It is possible for an experienced person to recognize and inventory, even in vegetative condition, the few hundred species of cerrado trees

---

TABLE 2.1
Total Number of Species of Native Vascular Plants
in the Cerrado Biome

Entire Biome (All Vegetation Types: Cerrado s.l., Forest, Wet Habitats, etc.)

6,429 (recorded)	Mendonça et al. (1998)
10,500 (estimate)	Dias (1992)
10,000 (estimate)	Myers et al. (2000)
10,000 (estimate)	Simon and Proença (2000)

Cerrado Sensu Lato

2,880 (recorded)	Mendonça et al. (1998)
2–7,000 (estimate)	Castro et al. (1999)

Trees and Large Shrubs of Cerrado Sensu Lato

774	Rizzini (1963) and Heringer et al. (1977)
477 + 296* (=773)	Mendonça et al. (1998)
1,000–2,000 (estimate)	Castro et al. (1999)
973 (ident. with confidence)	Castro et al. (1999)
951** (incl. disjunct Amazonian savannas)	Ratter et al. (2003)
914*** (core cerrado area and southern outliers only)	Ratter et al. (2003)

* 477 are listed as tree species, but a further 296 listed by these authors as cerrado shrubs or forest
  trees have been regarded as 'cerrado' trees by other authors
** Of which 334 (35%) occurred at only a single site locality (unicates) out of 376 areas surveyed.
*** Of which 309 (34%) were unicates in 316 areas surveyed.

---

and large shrubs normally encountered, whereas it would be enormously time-consuming to do the same thing for perhaps five times the number of ground-layer species. Furthermore, the dried-up, hay-like condition of the herbaceous vegetation makes its survey during the dry season even more difficult. Thus rapid surveys to compare cerrado vegetation over large areas, such as that of the Conservation and Management of the Biodiversity of the Brazilian Cerrado Vegetation Project (CMBBC) (Ratter et al., 2003), are, of necessity, limited to the larger woody element. The flora of cerrado trees and large shrubs is therefore much better known than the floristically richer ground layer.

2.3.1 Diversity of the Larger Woody Species of the Cerrado Sensu Lato

Since it is often difficult to distinguish between trees and large shrubs in the cerrado, the two must be considered together. Most workers agree fairly well on the definition of this element. The criterion used in our surveys for inclusion of a species in the tree/large shrub category is that it should have the capacity to reach at least 1.5 m tall and 3 cm basal diameter and have a perennial woody stem. This means that small, slender shrubs with shoots of short duration (hemixyles or geoxyles — an important growth form in the cerrado) are not included. The situation is complicated by some species having different growth forms in different areas. For instance *Caryocar brasiliense** grows as a low hemixyle in the southern cerrado islands of São Paulo State but is a thick-trunked large tree often reaching 15 m in Central Brazil, *Byrsonima basiloba* is a low hemixyle

---

* Authorities for species mentioned in the text are given in Appendices 1, 2 and 3.

in the Federal District but a small tree reaching *c.*7 m on the Rio Araguaia drainage of Mato Grosso State, while *Brosimum gaudichaudii* is a low slender shrub in the Federal District but sometimes reaches 7 m or more in other localities. We shall term trees and large shrubs as larger woody species in this account.

A useful base list of larger woody species was produced by Rizzini (1963) and this was later added to by his collaborative group (Heringer et al., 1977). These authors record 774 species belonging to 261 genera, of which they regarded 336 species (43%) as endemic to the cerrado. Mendonça et al. (1998) list 477 species of trees of the cerrado s.l., but scrutiny of their overall checklist reveals at least 296 species classified as cerrado shrubs or forest trees that have been included as larger woody species of cerrado s.l. by other authors. Addition of these extra species gives a total of 773. A recent study by Castro et al. (1999) gives 973 species and 337 genera identified with confidence and in addition mentions a large number of records of undetermined or partially determined taxonomic entities. These authors suggest that the total larger woody flora of cerrado s.l. may be 1000–2000 species. Ratter et al. (2003) recorded a total of 951 species in 376 sites across the whole Cerrado Biome, including 56 in Brazilian areas of savanna north of the River Amazon. If the latter are excluded, since they are species poor and very different from the main core cerrado area, 914 species are recorded in the remaining 316 sites. Table 2.1 summarizes records and estimates of species numbers made by various authors.

2.3.1.1 Most Important Families and Genera

The most important ten families in terms of species numbers according to Heringer et al. (1977) are given in Table 2.2 and compared with figures from Mendonça et al. (1998) and Ratter et al. (2003). To make figures comparable, those of Mendonça et al. have been increased to include some species regarded as larger woody species of cerrado in our studies but only as forest trees or smaller cerrado shrubs by those authors. The figures from Ratter et al. (2003) are for the commonest 300 species, and do not include the 614 species occurring at less than 2.5% of surveys; they are therefore not distorted by the large numbers of extremely rare species, many of which are 'casuals' (i.e. not true natives of the cerrado). A similar treatment for genera is given in Table 2.3. The very extensive data given by Castro et al. (1999) have not been included in these tables since they are not strictly comparable. This is largely because these authors often included plants smaller than those accepted

TABLE 2.2
Number of Species of the Largest Families of Trees and Large Shrubs of the Cerrado, in Order of Heringer et al. (1977)

	Heringer et al. (1977)	Mendonça et al. (1998)	Ratter et al. (2003) (in commonest 300 spp.)
Leguminosae (all 3 subfamilies)	153	91	45
Malpighiaceae	46	14	9
Myrtaceae	43	30	23
Melastomataceae	32	18	13
Rubiaceae	30	21	16
Apocynaceae	29	20	9
Annonaceae	27	16	6
Bignoniaceae	27	12	10
Vochysiaceae	23	20	16
Palmae	21	14	4

TABLE 2.3
Number of Species of the Largest Genera of Trees and Large Shrubs of the Cerrado, in Order of Heringer et al. (1977)

	Heringer et al. (1977)	Mendonça et al. (1998)	Ratter et al. (2003) (in commonest 300 spp.)
Byrsonima (Malphigiaceae)	22	9	8
Myrcia (Myrtaceae)	18	10	14
Kielmeyera (Guttiferae)	16	8	5
Miconia (Melastomataceae)	15	13	10
Eugenia (Myrtaceae)	14	6	4
Aspidosperma (Apocynaceae)	13	14	6
Mimosa (Leguminosae)	12	7	2
Vochysia (Vochysiaceae)	12	8	7
Annona (Annonaceae)	11	5	3

by other workers as belonging to the larger woody category (e.g. *Annona warmingiana* Mello-Silva and Pirani [= *A. pygmaea* (Warm.) Warm., nom. illeg.], and *Pterandra pyroidea* Adr. Juss.) and thus their figures are almost always considerably higher. However, their list is an extremely valuable data-source of cerrado floristics. In many cerrado areas the vegetation is dominated by Vochysiaceae because of the abundance of the three species of pau-terra (*Qualea grandiflora, Q. parviflora* and *Q. multiflora*, the first two of which were the commonest species in the study of Ratter et al., 2003). The importance of Vochysiaceae in the cerrados of the Triângulo Mineiro, Minas Gerais State, was also stressed by Goodland (1970) and Goodland and Ferri (1979).

2.3.2 ALPHA DIVERSITY OF CERRADO WOODY VEGETATION

Diversity of woody species in a community at a single site (α-diversity) varies greatly across the cerrado region. The number of species ranges in the 376 surveys recorded in Ratter et al. (2003) from 193 at Assis, São Paulo State (Durigan et al., 1999), to less than 10 in many savannas north of the River Amazon. It is rare, however, to find over 100 species in any community, other than in sites of very large area and/or where intensive studies have been conducted over a long period, thus allowing extreme rarities to be encountered. Since methods of collecting data have differed considerably, particularly in size of areas studied, it is possible only to give some very general observations. However, the following results obtained by the CMBBC project (Ratter et al., 2003) using the same rapid survey method for 170 sites are fairly comparable: 28–55 (average 47) species for Bahia (14 sites), 65–91 (average 79) for Goiás (19 sites), 33–70 (average 54) for northern Minas Gerais (11 sites), 19–79 (average 49) for Maranhão (13 sites), 41–106 (average 65) for Mato Grosso (23 sites), 28–76 (average 58) for Mato Grosso do Sul (31 sites), 30–55 (average 41) for southern Piauí (5 sites), 23–97 (average 72) for Tocantins (43 sites), 21–63 (average 42) for Rondônia (10 sites), and 20–44 (average 32) for Ceará (2 sites). Particularly high species richness was noted in surveys in the Araguaia and Tocantins drainage regions of the states of Goiás, Tocantins and Mato Grosso, and in the Xingu drainage of Mato Grosso. There are also many records of high α-diversity in the São Paulo cerrados where many very detailed surveys have been carried out.

Diversity is generally lower in areas with richer soils, where there is dominance of mesotrophic cerradão with characteristic calcicolous indicator species such as *Callisthene fasciculata, Magonia pubescens, Terminalia argentea, Luehea paniculata*, etc., (Ratter et al., 1977; Furley and Ratter, 1988). Such sites occur throughout a great part of the cerrado region wherever mesotrophic soils occur, but are particularly common in Mato Grosso do Sul and Rondônia.

The number of congeneric species growing in a small area of cerrado can be surprisingly high. For instance, at Fazenda Água Limpa, the ecological reserve of the University of Brasília, one can find six species of *Byrsonima* and *Miconia*, and five species of *Erythroxylum* and *Kielmeyera* side by side in the same community. There seems to be no interspecific hybridization in such communities and records of hybrids are extremely rare in cerrado. This is interesting since recent development has produced vast areas of disturbed habitats in the Cerrado Biome, and in other parts of the world interspecific hybridization has often been recorded as prevalent in such places, e.g. in *Ceanothus* and *Quercus* in North America and in *Eucalyptus* in Australia. Some information relevant to this subject is available from meticulous studies on breeding systems of the cerrado flora carried out over the last 30 years (see Oliveira and Gibbs, 2002). In one of these, Barros (1989) experimented with interspecific hybridization in a number of genera. Her attempts to produce interspecific hybrids amongst cerrado species of *Erythroxylum*, *Kielmeyera* and *Tabebuia* failed completely, although nearly 1000 interspecific pollinations were made and there was a high level of success in intraspecific pollinations. The situation was nearly the same amongst five species of *Diplusodon* where a total of four fruits were produced as a result of 1174 interspecific pollinations, although again success from intraspecific pollinations was very high (68.0–89.8% fruit-set according to species). Fluorescence microscopy of pollen-tube development showed irregularities associated with interspecific incompatibility, and these were often similar to the reactions given to self-pollen (all species studied were strongly self-incompatible with the exception of partial self-compatibility in *Erythroxylum campestre* and two species of *Diplusodon*). One might speculate that the strong barriers to interspecific fertilization in this representative group of unrelated cerrado genera may be the result of the powerful systems of self-incompatibilty also demonstrated in them by Barros (1989, 1996, 1998). This may have a wider application in cerrado communities since work on the breeding systems of many cerrado native species (see Gibbs, 1990; Oliveira and Gibbs, 2002) has shown that they are almost all obligate outbreeders, and that barriers to selfing in the form of self-incompatibility are common. Thus, if such a link exists between strong self-incompatibility systems and barriers to interspecific hybridization, one should expect the latter to be widespread in the cerrado flora. Extrapolating from this, one might speculate further and view the cerrado flora as one rich in congeneric species probably isolated from each other and maintained genetically by strong barriers to interspecific hybridization. Repeated changes in distribution of the biome, with fragmentation and isolation of populations in refuge pockets during the drastic climatic variations of the Quaternary could perhaps account for the establishment of such barriers.

2.3.2.1 Comparative Frequency of Woody Species of Cerrado Sensu Lato

Data published in Ratter et al. (2003) covering the entire Cerrado Biome, but excluding savanna areas north of the River Amazon and two at Humaitá, Amazonas, are used as the principal basis for this section*. A total of 914 species was recorded in 316 surveyed sites distributed throughout this vast area (Figure 2.1), but of these only 300 species occur at eight or more

---

* The data-base used consisted of the results of 316 surveys carried out by about 40 research groups. Methods of collecting data differed considerably, particularly in the size of areas studied, but nevertheless we consider the floristic lists provide a reasonable basis for comparison. The *sine qua non* of surveys for floristic comparison is that they give complete species lists of the communities being studied, and works were only included if they seemed adequate in this respect. Methods used for our surveys are discussed in Ratter et al. (2003).

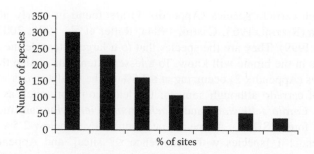

FIGURE 2.2 Species occurrence vs. percentage of sites in the cerrado core area (e.g. 38 species occur at ≥50% of sites, while 300 species occur at ≥2.5%). Reprinted with permission of Cambridge University Press; Ratter, J.A., Bridgewater, S., and Ribeiro, J.F. 2003. Analysis of the floristic composition of the Brazilian cerrado vegetation. III: Comparison of the woody vegetation of 376 areas. *Edinburgh J. Bot.* 60: 57–109.

sites (i.e. ≥2.5% of the total), while the remaining 614, including 309 (34% of the total) present at only a single site, are very rare. Figure 2.2 shows percentage bands for the number of species occurring between ≥50% and ≥2.5% of the 316 sites. Only 39 species occur at ≥50% of sites (see Appendix 2) and these can be considered the most typical of cerrado as a whole, although in parts regional abundance can make other species very prominent (e.g. *Acosmium subelegans* and *Pipto-carpha rotundifolia* in the south and *Parkia platycephala* and *Hirtella ciliata* in the north-east). The 300 species occurring at ≥2.5% of the sites represent the most common and widespread woody species of at least 75% of the cerrado region. To these we added approximately 40 species relatively common in São Paulo state but apparently rare or absent elsewhere (Durigan, pers. comm.) to produce a group of about 340 species that overwhelmingly dominates the woody vegetation of the Cerrado Biome (Ratter et al., 2003, and see Appendix 3). This concept was later refined to produce a smaller suite of 121 species forming a dominant oligarchy (Bridgewater et al., 2004) and we have now further reduced this to the 116 species listed in Appendix 1. The latter suite was derived from the 107 species occurring at ≥20% of the 316 sites plus another nine we regard as particularly important. An analysis of the data from 26 published phytosocio-logical surveys across the cerrado region shows that this suite contributes a high proportion of total species in each area (66% on average) and also 75% on average of the Importance Value Index (Bridgewater et al., 2004).

It is interesting to compare our conclusion that the woody vegetation of the cerrado is dominated by a relatively small oligarchy of species with recent results from the Amazonian forest. Pitman et al. (2001) have studied dominance and distribution of tree species in great areas of *terra firme* forests in Ecuador and Peru. They found oligarchies of 150 species dominating thousands of square kilometres of forest in each country, although the total species diversity was much greater because of the huge number of rare species present. This clearly bears a strong resemblance to the situation across the Cerrado Biome.

As shown in the previous paragraphs, the pattern of diversity of the woody cerrado vegetation consists of a relatively moderate number of common, widely distributed species, and a vast number of rarities. In terms of biomass (basal area) and importance value index, the contribution of these common species outweighs even the frequency of their occurrence — when they are present they are usually abundant and of relatively large size. This pattern of diversity is common to many vegetation types (e.g. Amazonian forest, see above) and other biological systems. It is because of this that someone knowing a couple of hundred tree species, or even less, can make a reasonable job of a cerrado or forest survey. On the other hand, floristic surveys would be a formidable challenge to one's taxonomic memory if the 914 woody species we recorded in the cerrado core area and southern outliers were equally common and evenly distributed!

The 116 oligarch cerrado species (Appendix 1) are found in nearly all compiled lists of cerrado woody flora (Rizzini, 1963; Castro, 1994a; Ratter et al., 1996, 2003; Mendonça et al., 1998; Castro et al., 1999). They are the species that to a large extent define the cerrado s.l. and which most workers in the biome will know. To a lesser extent this is also the case for the most common 300 species (Appendix 2) occurring at ≤2.5% of the total surveys where all the species are characteristic of cerrado although some of the least common, such as *Apuleia leiocarpa*, *Casearia rupestris*, *Cordia sellowiana* and *Machaerium scleroxylon*, are more at home in seasonally dry tropical forest (SDTF). At species occurrence levels below 2.5% of sites (see Ratter et al. [2003], Appendix 1 [species with occurrence ≤7 sites], and Appendix 2 [unicates]) a considerable number of the records are rarities much more characteristic of the SDTF or other vegetation types (e.g. *Albizia niopoides, Attalea speciosa* ['babaçu'], *Calycophyllum multiflorum* and *Cedrela fissilis*). The concept of endemic and non-endemic accessory species in the cerrado is discussed in following sections.

We conclude that contrary to the views of some workers, the larger woody element of the core area of cerrado is really rather well known. This is illustrated by the fact that during our long experience of work covering a large part of the biome we have only recorded four apparently undescribed species — a *Trischidium* (Leguminosae, Papilionoideae), a *Callisthene* (Vochysiaceae), a *Kielmeyera* (Guttiferae) and a *Mezilaurus* (Lauraceae). However, there is little doubt that many species of herbs and subshrubs remain to be discovered.

2.3.3 ORIGIN OF THE PRESENT WOODY CERRADO VEGETATION

It would be appropriate to consider the origin of the woody vegetation of the cerrado before discussing endemic species, non-endemics and geographical elements. Much of the information and ideas come from the work of Rizzini and his collaborators and are based on their long experience and that of earlier workers such as Lund and Loefgren.

Rizzini postulated that the major progenitor of the present more open forms of cerrado was cerradão which he called *floresta xeromorfa* (Rizzini and Heringer, 1962; Rizzini, 1963), a term he later abandoned (Rizzini, 1979). He considered that this cerradão represented an ancient xeromorphic forest, the endemic element of which had originally evolved *in loco* from Atlantic and Amazonian forest floras during the period from the late Eocene to the Miocene. This evolution had produced vicariant species in the cerradão showing characters related to heliomorphic adaptation (e.g. sclerophylly). Other elements from neighbouring vegetation formations (principally the Atlantic and Amazonian forests) had later been added to the endemic core to produce the modern cerradão.

Before the great anthropic pressure of more recent years there is strong evidence to indicate that tall cerradão (xeromorphic forest) with virtually no ground layer was a dominant constituent of the landscape of the Cerrado Biome. Lund (fide Rizzini, 1979) commented on its prevalence in his travels across Central Brazil in 1833–1835, while Loefgren (1898) considered it 'the second category of primitive forests of São Paulo State', and Warming (1892) deduced its former widespread occurrence (although it was no longer important in the region of Lagoa Santa, Minas Gerais state, where he worked in the 1860s). Rizzini's attention focussed particularly on cerradão as a result of the construction of the Belo Horizonte-Brasília highway in the early 1960s giving access to large undisturbed tracts of this vegetation (that were rapidly destroyed in the next few years). One of us (J.A.R.) was similarly fascinated by seeing areas of cerradão in newly opened expanses of eastern Mato Grosso state in the 1960s. Rizzini, and all the others who observed it, stressed that cerradão was a true forest formation, hence his use of the term *floresta xeromorfa*, and similarly of *mata de terceira classe* (third-class forest) recorded by Waibel (1948) in Goiás state. They also emphasized that the cerradão was the primitive base vegetation from which more open forms of cerrado had originated, rather than merely an exuberant form of cerrado s.s. (Heringer et al., 1977).

The latter idea, regarding cerrado s.s. as the 'true' climax, was quite prevalent in the 1960s–1970s and earlier, when it covered huge extensions of the Cerrado Biome due to the fire regimes practiced in that period.

The concept of cerradão as the climax vegetation of cerrado s.l. is certainly still true for great areas of the biome. Examples of its re-establishment in areas where sparser forms of cerrado have been protected from fire and other human disturbance can be seen in a number of places, such as São Paulo state (Ratter et al., 1988; Durigan, 2003b and Chapter 3; Durigan and Ratter, in press) and in the Federal District (Ratter, 1980, 1986). The work of Durigan (Chapter 3) and Durigan and Ratter (in press) communicates data of surveys conducted by the Biota Programme project Conservation Feasibility of the Cerrado Remnants in São Paulo State, and shows that as a result of protection the most common cover of the ever-diminishing cerrado remnants of the state have changed from more open forms to cerradão during the last 40 years. However, even with complete protection from fire, succession to produce tall, dense cerradão cannot occur where soils are unfavourable and here the climax may be a sparse, low, cerrado scrub.

Rizzini was strongly influenced in reaching conclusions on the relatively recent changes in cerrado vegetation by the work of French investigators studying savannas in Africa, particularly that of Aubréville who visited Brazil in the late 1950s. As a result of observation and thorough consideration of relevant literature, he made a convincing exposition of what we call here the Rizzini-Lund-Aubréville theory of the origin of the Afro-Brazilian savannas, called the Teoria de Lund-Aubréville by Rizzini himself (1979). Basically this consists of the continual frequent repetition of man-made fires damaging native woodlands (cerradões, tree savannas and SDTF) to such an extent that a more open savanna vegetation of mixed floristic origin (e.g. campo cerrado) is produced. The process had already been recognized by a number of earlier workers, (e.g. Lund, 1843, and Ab'Saber and Junior, 1951 [fide Rizzini, 1979]). Eiten (1972), in his monumental review of cerrado vegetation, also describes exactly this type of deterioration of dense, tall, forms of cerrado under frequent burning regimes.

As already mentioned, Rizzini and his co-workers regarded the flora of the larger woody vegetation of the present-day cerrado as composed of an exclusive endemic element consisting of species they termed '*peculiares*' native to the ancient cerradão (or mata xeromorfa) to which has been added a large number of '*acessórias*' (accessory species). They defined the accessory species as those which occur preferentially in other vegetation types but are found in a secondary or sporadic manner in cerrado, although sometimes becoming common there. In a differential analysis of 774 cerrado woody species they regarded only 336 species (43%) as *peculiares* (the endemic component), while the other 438 (56%) were *acessórias* (Heringer et al., 1977). The latter are principally derived from Atlantic forest (which provided the largest number of species), Amazonian forest, and the semideciduous and deciduous dry forests (SDTF) occurring on mesotrophic soils in the Cerrado Biome. In our experience (Ratter et al., 1973, 1977; Furley and Ratter, 1988) the last is a particularly important element and often dominates the community as a component of mesotrophic cerradão where more fertile soils occur. In the next section we endeavour to identify non-endemic accessory species in our data-base.

Most of the theories of the evolution of cerrado vegetation proposed by the Rizzini school seem very sound, but their idea that the endemic element was originally derived from Atlantic and Amazonian forests seems questionable. Other authors consider the cerrado more ancient than the forests (e.g. Cole [1986], basing her arguments on geomorphology). In support of this, molecular studies in *Inga* species of the Amazonian forest indicate that they are of remarkably recent origin (Richardson et al., 2001), suggesting perhaps that the evolution of the Amazonian forest is much less ancient than previously considered. If the cerrado were viewed as ancestral to the forests then the concept of non-endemic accessory species should be considered in the opposite direction, i.e. they are native cerrado species which also occur as accessories in forest, not the reverse.

2.3.3.1 Nature and Numbers of Endemic and Non-Endemic (Accessory) Species

We have attempted to distinguish here endemic (*peculiares*) and non-endemic species (*acessórias*) in our CMBBC data-base (Ratter et al., 2003), identifying such species according to (a) the publications of Rizzini (1963, 1979), Heringer et al. (1977), Oliveira-Filho and Fontes (2000), and Méio et al. (2003), (b) information from Oliveira-Filho (pers. comm.), and/or (c) our own experience. However, we prefer to use the term *non-endemic* in place of accessory, since the latter infers that the species plays only a secondary role, whereas it has been applied to many important, well-adapted, native cerrado species simply because they also occur in other biomes.

The characteristics of non-endemic species are, however, very variable. They encompass the following range:

a. Typical species of the cerrado with all the usual pyrophytic and associated adaptations (thick corky bark, sclerophylly, etc.) only classified as accessories by the Rizzini school because they also occur in other vegetation formations, e.g. *Curatella americana, Byrsonima coccolobifolia, Pouteria ramiflora*, etc. As pointed out by Heringer et al. (1977), such species are often an important or even dominant element of cerrado vegetation. They are typical natives of the vegetation and certainly should not be considered as second-class members of the cerrado flora because they are not strictly endemic.

b. Species characteristic of cerradão and common to various forest types, generally showing a lower resistance to fire than species typical of more open forms of cerrado, e.g. *Protium heptaphyllum, Siphoneugena densiflora, Virola sebifera*, etc. Such species have a more mesophytic appearance than the more strongly adapted pyrophytes. Some of them are geographically widespread, occurring in various vegetation types, and have been classed by Oliveira-Filho and Fontes (2000) as 'supertramps'. In some cases their adaptability is astonishing, for instance *Tapirira guianensis* will thrive in habitats ranging from cerradão on well-drained soils to swampy riverine forest. In the latter habitat it develops buttress roots, but whether adaptation to such different habitats involves ecotypic differentiation is not known.

c. Those least adapted to the conditions of more open cerrado, such as *Amaioua guianensis*, are often seral and represent late stages in the succession to forest, or simply 'casuals' resulting from a seed deposited in the wrong habitat managing to establish itself and survive. Everybody with enough field experience has seen examples of this, e.g. small trees of *Apuleia leiocarpa* in cerrado — although, in fact, it seems commoner to find young cerrado trees without hope of a long-term future sprouting in forest rather than vice-versa.

Appendix 2 lists the commonest 300 species from the CMBBC data-base (those found in ≥2.5% of the sites studied) in order of frequency of occurrence, and designates those recognized as non-endemics, while Table 2.4 gives the number and percentage of occurrence of the latter. In the group of 39 commonest species occurring at ≥50% of sites (see first section of Appendix 2) no less than 20 (51.3%) have been ranked as non-endemics. However, with the possible exception of *Astronium fraxinifolium*, all of these belong to category (a) above, i.e. species fully adapted to the rigours of more open forms of cerrado. *Astronium fraxinifolium* is a marginal case; it is a weakly calcicolous species very common in mesotrophic cerradões and SDTF, and probably less fire resistant than the others.

Carrying the analysis on to the 107 species recorded at ≥20% of areas studied, the number of non-endemics has now risen to 62, representing 57.9% of the total. Many of the extra 33 species are now

TABLE 2.4
Number of Non-Endemic Species in Percentage Bands Occurring in the Most Common 300 Species (Those Recorded at ≥8, i.e. ≥2.5% of 316 Sites) in Ratter et al., 2003

% Sites	Total spp.	Non-Endemic spp.	% Non-Endemic spp.
≥50%	39	20	51.3%
≥20%	107	62	57.9%
≥10%	160	104	65.0%
≥5%	231	144	62.3%
≥2.5%	300	196	65.3%

cerradão elements belonging to category (b) above, rather than typical of the sparser forms of cerrado. Of these *Aspidosperma subincanum, Dilodendron bipinnatum, Guazuma ulmifolia, Guettarda viburnoides, Luehea paniculata* and *Myracrodruon urundeuva** are calcicolous species found in mesotrophic cerradões and also in the SDTF of the Cerrado Biome, while *Emmotum nitens, Matayba guianensis, Protium heptaphyllum, Simarouba versicolor,* and *Siparuna guianensis* are plants of dystrophic cerradão and the related dystrophic forests. *Agonandra brasiliensis, Anacardium occidentale, Copaifera langsdorffii, Dalbergia miscolobium, Pouteria torta, Pterodon polygalaeflorus, P. pubescens* and *Stryphodendron adstringens* are category (a) non-endemics, most of which can also flourish in forest (in the case of *C. langsdorffii, Pouteria torta* and *Pterodon polygalaeflorus,* often in SDTF).

Extending the analysis to the 300 species occurring at ≥2.5% of the areas studied (≥8 sites out of 316) through intermediate points at ≥10% and ≥5% shows a corresponding rise in number of non-endemics (Table 2.4). At ≥10% (≥32 sites) the total number of species recorded is 160 and the number of non-endemics is 104 (≥65%). These new non-endemics include important SDTF species associated with mesotrophic soils (*Acrocomia aculeata, Anadenanthera colubrina, Coussarea hydrangeaefolia, Jacaranda cuspidifolia, Platypodium elegans, Rhamnidium elaeocarpum* and *Zanthoxylum riedelianum*) and a north/north-eastern element represented by *Caryocar cuneatum, Hirtella ciliata* and *Parkia platycephala,* as well as dystrophic forest/cerradão species (e.g. *Callisthene major* and *Zanthoxylum rhoifolium*) and adventives (*Cecropia pachystachya* and *Solanum lycocarpum*). At ≥5% (≥16 sites) 231 species are recorded and the number of non-endemics has risen to 144 (62.3%). The non-endemic newcomers include the SDTF species *Cordia glabrata, Luehea speciosa, Sterculia striata, Tabebuia impetiginosa, T. roseo-alba, Vitex polygama* and the adventive *Apeiba tibourbou,* while the north/north-eastern group is represented by *Vochysia gardneri.* At ≥2.5% (≥8 sites) 300 species are recorded with 196 (65.3%) non-endemics. The SDTF element is represented by *Alibertia verrucosa, Casearia rupestris, Combretum leprosum, Cordia trichotoma, Enterolobium contortisiliquum, Luehea candicans, Machaerium hirtum, M. scleroxylon, Priogymnanthus hasslerianus* and the adventive *Trema micrantha,* while many of the other species are characteristic of dystrophic forests. A few of the newcomers, such as *Symplocos rhamnifolia, Eremanthus goyazensis* and *Myrcia uberavensis* seem to be type (a) non-endemics.

When the analysis is continued to species occurring at less than 2.5% of the surveys (<8 sites) the great majority are probably extremely rare non-endemics, but we are usually too unfamiliar with such species to be able to designate them to a particular element.

It is interesting to compare our figures of endemic and non-endemic accessory species with those given by Heringer et al. (1977) which includes data from earlier studies (Rizzini, 1963, 1971). Those works gave a total of 774 species, while Ratter et al. (2003) has 914 for the core cerrado area and

---

* Considered by some authors as *Astronium urundeuva* (e.g. Chapter 18).

southern outliers. Heringer et al. (1977) regarded 438 (56%) of their species as non-endemic accessories while we have considered 196 (65.3%) of the commonest 300 species as belonging to this category. At least 54 (27.5%) and probably many more of the non-endemics we recognized were of type (a), i.e. fully adapted to harsh cerrado environments.

2.3.3.2   Origin of Non-Endemic (Accessory) Species

The non-endemic (accessory) species of cerrado s.l. are derived from the following sources according to Heringer et al. (1977): Atlantic forest, Amazonian forest, SDTF ('mata seca', i.e. the deciduous and semideciduous forests of richer more calcareous soils in the Cerrado Biome), the cerrado campestrine (or ground) flora and 'xerophyllous species of wide distribution'. We have attempted to mark non-endemic species in our CMBBC data base (Appendix 2) according to these categories, with the exception of campestrine which we found a difficult concept to understand. Many species are simply designated as non-endemics when they have been recognized as such by other authors but we do not have reliable information on their distribution. Similarly, others are just marked as forest when we are not sure of the nature of their forest origin. The total number of non-endemics we have recognized is 196 (65.3%) out of the 300 commonest species, and of these at least 54% are associated with Atlantic forest, 24% with SDTF, and 8% with Amazonian forest. However, many of these species are associated with more than one forest type. The figures for non-endemics shared with SDTF are fairly accurate, whereas those for the Amazonian element are very approximate and probably underestimated. In comparison, Heringer et al. (1977) considered 300 (39%) of a total cerrado woody flora of 774 as belonging to the Atlantic and Amazonian forest elements.

Two other recent works furnish important information on the affinities of the cerrado flora:

1. Oliveira-Filho and Fontes (2000) investigated the floristic links of both Atlantic rain- and semideciduous forests to the cerrado and to Amazonian forests, as well as comparing the relationships of many forms of Atlantic forest. In a comprehensive and meticulous study they compared 125 areas of Atlantic forest containing a total of 2532 species with 98 areas of cerrado containing 528 species (using data for the latter from Ratter et al., 1996) and 22 areas of terra firme (upland) Amazonian rain forest (data from Oliveira-Filho and Ratter, 1995). They showed that the cerrado shared a large proportion (55%, 281 species) of their flora with the Atlantic forests, and that the cerrado flora was much more closely related to Atlantic semideciduous forests than to Atlantic rain forests. The strongest association of cerrado was with western montane and submontane semideciduous forests in which 38 of the 48 preferential species selected by TWINSPAN (i.e. 79%) are characteristic of cerrado s.l. Of the species associated with major groups of Atlantic forest formations based on TWINSPAN, 22 out of 57 in the eastern low-altitude semideciduous forests, 33 of 48 in the western montane and submontane semideciduous forests, and 35 of 57 in the 'supertramp' category (i.e. of widespread distribution) are also typical of cerrado.

2. In a study to investigate the influence of Amazonian and Atlantic forest on the flora of cerrado, Méio et al. (2003) compared 12 cerrado surveys at distances varying from 93–1166 km and 37–1055 km from the margins of the Amazonian forest and Atlantic forest respectively. A total of 290 tree and shrub species were recorded in these surveys of which 41% were found only in cerrado and are probably endemic, while 59% also occurred in Atlantic and/or Amazonian forests. The contribution of Atlantic forest was much greater (44.8%) than for Amazonian (1.4%), with the remaining 12.8% found in both of these forest biomes. The proportion of species with centres of distribution in Atlantic and Amazonian forest showed a slight decrease towards the centre of the cerrado core area. For the former, the distance explained only 30% of variation in proportion using a polynomial model to fit the data, and for the latter a significant linear model explained 78% of variation.

We conclude that all studies considered here strongly support the already existing knowledge of a close relationship of the Atlantic forest and Cerrado Biomes.

2.4 PHYTOGEOGRAPHIC PATTERNS OF THE CERRADO BIOME

A number of studies carried out over the last 15 or so years have attempted to elucidate distribution patterns of cerrado vegetation, e.g. Ratter and Dargie (1992), Castro (1994a), Ratter et al. (1996, 2003), Felfili et al. (2004 and Chapter 4), Castro and Martins (1999), Durigan et al. (2003a), Bridgewater et al. (2004), and Durigan (Chapter 3). In addition, Oliveira-Filho and Ratter (1995) have published data on the distribution of tree species of the gallery forests of the biome.

Our study, published in Ratter et al. (2003), has compared floristic data from a total of 376 areas ranging from the states of Amapá and Roraima, north of the River Amazon, to Paraná in the south (Figure 2.1). A total of 316 of these areas lay in the core cerrado area and its southern outliers in São Paulo and Paraná, and western outliers in Rondônia. Of these, 170 were studied by targeted rapid surveys of the Conservation and Management of the Biodiversity of the Brazilian Cerrado Biome (CMBBC) project designed to produce data from previously floristically little-known areas (particularly in Mato Grosso do Sul, Tocantins and Rondônia states (Ratter et al., 2000, 2001, 2003); while 42 of the others were surveys from our previous work (see Ratter et al., 1996, 2003). The rest came from reliable published work or theses of other workers. The overall matrix, consisting of 376 areas and 951 species was analysed using two techniques of multivariate analysis which we had found particularly appropriate in our previous work: (a) a divisive hierarchical classification by Two-Way Indicator Species Analysis (TWINSPAN) (Hill, 1979), and (b) an agglomerative hierarchical classification by UPGMA (Unweighted Pair-Groups Method using Arithmetic Averages), using the Sørensen Coefficient of Community (cc) as a measure of similarity (Kent and Coker, 1992).

The results produced by both methods of multivariate analysis were in accord. They produced geographical groups discussed in detail in Ratter et al. (2003) and provided the basis for the map shown in Figure 2.3. The following were recognized:

Southern sites. These form one of the most distinctive groups recognized and consist of all 18 São Paulo sites, the single site recorded for Paraná and three in the south of Minas Gerais.

Central and south-eastern sites. These are made up of all 13 Federal District sites, one in neighbouring Goiás and 51 from Minas Gerais, mostly in the southern and central parts of the state.

Northern and north-eastern sites. These are sites from Bahia, Ceará, the extreme north of Minas Gerais, Maranhão, Piauí, Tocantins and one in Pará very close to the Tocantins border.

Central-western sites. This group is made up of a huge swathe of about 100 sites running across the states of Mato Grosso do Sul, Mato Grosso, Tocantins and into Pará close to the Tocantins border. It is characterized by the presence of many mesotrophic indicator species related to the presence of richer cerrado soils.

Widespread sites of strongly mesotrophic character. This group occurs in Ceará, Goiás, Mato Grosso, Piauí and Tocantins, but is particularly well represented in Mato Grosso do Sul where the original dominant vegetation of large areas was mesotrophic cerradão with *Terminalia argentea* as the most characteristic tree. The group is linked to the occurrence of mesotrophic soils which are often of very scattered occurrence. Because of this it cannot be mapped satisfactorily but much of its range falls within that of the large central-western group.

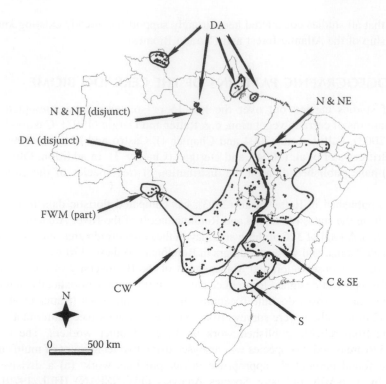

FIGURE 2.3 Floristic regions within the Cerrado Biome of Brazil. C & SE, central and south-eastern; CW, central-western; DA, disjunct Amazonian savannas; FWM, far-western mesotrophic sites; N & NE, north and north-eastern; S, southern. Reprinted with permission of Cambridge University Press; Ratter, J.A., Bridgewater, S., and Ribeiro, J.F. 2003. Analysis of the floristic composition of the Brazilian cerrado vegetation. III: Comparison of the woody vegetation of 376 areas. *Edinburgh J. Bot.* 60: 57–109.

Far-western mesotrophic sites. This is a small set of 10 sites occurring on mesotrophic soils in Rondônia, Mato Grosso do Sul and in the Mato Grosso Pantanal near Poconé. The sites in Rondônia are on the widespread 'solo chocolate', common in that state and so-named because of the presence of a shallow, mineral-rich, chocolate-brown pan. Clearly the dominant factor in determining the characteristics of this floristic group is the presence of mesotrophic soils.

Disjunct Amazonian sites. These floristically poor disjunct savannas, mainly situated north of the River Amazon are the group most strongly distinguished from all the others by both methods of multivariate analysis. One disjunct site south of the Amazon at Humaitá, Amazonas, also belongs to this group, while two (at Alter do Chão, Pará, and Humaitá, Amazonas) are more species-rich and floristically allied to the core area cerrados. These Amazonian savannas seldom have more than 20 tree species, and those in Roraima usually have much less. They are closely related to the Guayanan and other savannas of northern South America (see Huber et al., Chapter 5, and Pennington et al., Chapter 1).

In a further study made by our team, Bridgewater et al. (2004) used the same data-base and investigated the patterns of β-diversity of the geographic groups recognized above (with the exception of 'widespread sites of strongly mesotrophic character'). The species lists for each group were compared using Sørensen similarity indices (c.c.) (Figure 2.4) and showed that the most similar pair were the central and south-eastern and the central-western with a very high similarity index of 0.625. The group most dissimilar from the others was the disjunct Amazonian sites which showed indices of 0.256, 0.268 and 0.274 when compared with the southern, central and south-eastern, and central-western groups

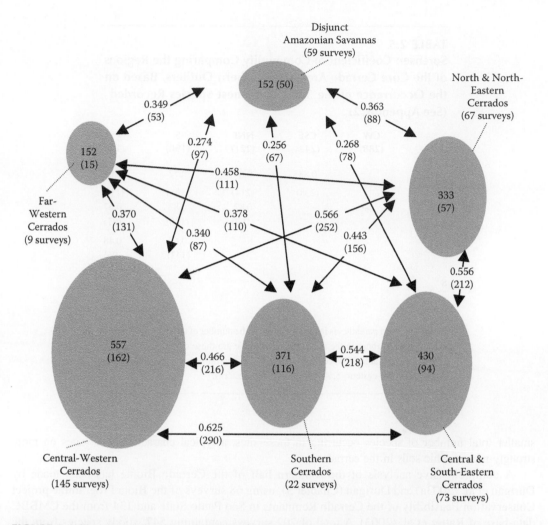

FIGURE 2.4 Floristic relationships of cerrado phytogeographic regions. Figures in ovals are the number of species recorded in each region, with those restricted to it in parentheses. The figures between ovals are the number of species in common and the Sørensen similarity index. The size of oval indicates the species richness of each province. Reprinted with permission of Kluwer Academic Publishers; Bridgewater, S., Ratter, J.A., and Ribeiro, J.F. 2004. Biogeographic patterns, ß-diversity and dominance in the Cerrado Biome of Brazil. *Biodiversity and Conservation* 13: 2295–2318.

respectively. This conforms exactly to the results of both methods of multivariate analysis which showed that the strongest division was between the disjunct Amazonian group and the rest. The analysis showed that 494 species (more than half of the 951 species recorded) are found in only one of the six geographical groups compared in this study, while only 37 (3.9%) are recorded from all groups. However, this apparent high degree of heterogeneity is not so surprising when one considers that over 620 species of a total 951 occur at less than 2.5% of 376 sites surveyed and this includes 334 (35%) found at only a single locality (unicates). Really the figures demonstrate the abundance of extremely rare accessory and casual species rather than anything else. More significant figures are provided by a Sørensen analysis of the commonest 300 species (see Appendix 2) and these are given in Table 2.5 for the core cerrado area and its outliers (with disjunct Amazonian savannas excluded). Again the pattern agrees with that coming from the multivariate analyses and the central and south-eastern and central-western pair now have the remarkably high Sørensen similarity index of 0.88. The lower indices in comparisons involving the far-western mesotrophic sites are explained by the

TABLE 2.5
Sørensen Coefficient of Community Comparing the Regions of the Core Cerrado Area and Southern Outliers, Based on the Occurrence of the 300 Commonest Species Recorded (See Appendix 2)

	CW (288)	CSE (248)	NNE (221)	S (190)	FW (116)
CW	—	0.88 (238)	0.84 (214)	0.77 (185)	0.57 (115)
CSE		—	0.82 (192)	0.82 (180)	0.57 (87)
NNE			—	0.69 (143)	0.48 (96)
S				—	0.54 (82)
FW					—

Figures given in parentheses in the headings are the number of these species present in each region; those in the main part of the table are those common to the pairs of regions compared (CW, central-western; CSE, central and south-eastern; NNE, north and north-eastern; S, southern; FW, far-western).

smaller total number of species occurring in these sites, a typical characteristic of sites on more strongly mesotrophic soils in the cerrado.

A comprehensive analysis of the southern half of the Cerrado Biome has been made by Durigan et al. (2003a) and Durigan (Chapter 3), using 68 surveys of the Biota Programme project Conservation Feasibility of the Cerrado Remnants in São Paulo State and 134 from the CMBBC data-base of Ratter et al. (2003). A total of 202 surveys containing 547 woody species from the states of São Paulo, Paraná, Mato Grosso do Sul, Minas Gerais and Goiás were compared using three techniques of multivariate analysis: TWINSPAN, UPGMA and Detrended Correspondence Analysis (DCA). Two distinct geographic groups were distinguished in São Paulo state: the almost endemic Paulista and the more widespread Interestadual. The former contains 70 sites from São Paulo, one from Paraná, and one in Minas Gerais lying close to the São Paulo border. It subdivides into four groups, one of which is physiognomically particularly characteristic, consisting of 26 sites of cerradão and cerradão/forest ecotone in the west of São Paulo. The Interestadual group consists of more open forms of cerrado and divides into various subgroups, although the 13 sites from São Paulo, 20 from the basin of the Rio Paraná, Mato Grosso do Sul, and 38 from western Minas Gerais remain together after the third level of TWINSPAN division. The three other subgroups of Interestadual at the third level of TWINSPAN division consist of (i) 24 sites from the east of Minas Gerais, (ii) 14 Mato Grosso do Sul sites from the drainage basin of the Rio Paraguai and a single Minas Gerais site, and (iii) 20 sites from Goiás. One feature demonstrated by the analysis, and most particularly by DCA, was the placing of sites in groups determined by their hydrographic basins: those of the Rio Paraná, Rio Paraguai, Rio Paraíba do Sul, Rio São Francisco, Rio Tocantins, and Rio Araguaia. It was extremely rare for a site to be misclassified and placed in the wrong river basin. An interesting exception is the placing of the northern Minas Gerais sites at Januária and Sagarana amongst those of the Rio Paraguai basin, when in fact they are on the drainage of the Rio São Francisco. However, both these sites are on the margin of the north-eastern caatinga vegetation and a large disjunct patch of this vegetation also occurs in the drainage of the Rio Paraguai, producing similar floristic influences in the cerrado of both areas — hence the geographical misclassification. This study demonstrates the great importance of hydographic basins in determining patterns of vegetation distribution; other influencing factors discussed by the authors will be considered later.

As discussed in Ratter et al. (2003), our observations on the phytogeography of the Cerrado Biome agree quite well with those of Castro (1994a,b) and Castro and Martins (1999). These authors recognize the following three 'supercentres of biodiversity' within the Cerrado Biome on the basis of the analysis of 145 surveys:

a. São Paulo state (where two subdivisions, SP1 and SP2, were recognized with cerradão and campo cerrado respectively as the dominant components).
b. Planalto central with four subdivisions (PC1, PC2, PC3 and the Pantanal).
c. A northeastern group.

These more or less correspond in our classification to: (a) the southern sites; (b) the central and south-eastern, the central-western and to some extent the widely spread mesotrophic sites; and (c) our northern and north-eastern sites. The southern sites form a very distinct group in our analysis (Ratter et al., 1996, 2003) and those of Castro (1994a,b) and Castro and Martins (1999). Their separation into two physiognomically differentiated subgroups characterized by presence of cerradão or more open forms of cerrado respectively corresponds exactly with the observations of Durigan (Chapter 3) and Durigan et al. (2003a,b). The cerradão subgroup (SP1) is the endemic 'Paulista' group of Durigan's work and shows strong separation from areas external to the state. Our data-base (Ratter et al., 2003) did not include any sites from this exclusively São Paulo element.

Castro (1994a,b) and Castro and Martins (1999) correlate their 'supercentres of biodiversity' and subdivisions with climatic factors of the environment. They point out that across the cerrado region soil hydric deficiency increases in a south-east to Planalto central to northeast direction, as does mean temperature, and suggest that species distributions can be correlated with this trend. They state that two climatic barriers cut across the region: occurrence of frosts to the south of 20°S and of severe droughts to the north and east of 15°S 45°W. In addition, altitude with its effects on temperature and other aspects of the climate reinforces the separation of their super-centres.

Durigan et al. (2003) also stress the importance of climatic factors, particularly temperature, precipitation and the length of the dry season. They regard this as the main factor responsible for differentiation of the two floristic groups occurring in São Paulo state. The more open, lower cerrados of the Interestadual group, which spreads into southern Minas Gerais and eastern Mato Grosso do Sul, are characteristic of a hotter climate with a longer dry season, whereas the endemic Paulista cerradão group occupies areas of colder climate with a shorter dry season. Soil type is also regarded as an important secondary factor.

The conclusions of these authors on the importance of climate are undoubtedly very valid but soil type (mesotrophic vs. dystrophic) can also be the major factor in determining the floristics and structure of cerrado communities (Ratter et al., 1977; Furley and Ratter, 1988; Ratter and Dargie, 1992). It should also be remembered that in addition to present-day environmental factors the vegetation pattern of the Cerrado Biome must reflect the major dynamic changes which are known to have occurred repeatedly during the Quaternary period, and much further research is necessary to correlate observations with new data emerging on this subject.

Our scheme (Ratter et al., 2003) recognizes six geographical groups within the vast cerrado region, although some show considerable overlap (so much so that it was not practical to place the 'widely spread sites of strong mesotrophic character' on the map) and, in addition, a very distinct disjunct Amazonian Savanna group. Over most of the area, we are, of course, describing regional variation of a continuum, although the floristic heterogeneity of different parts of this continuum can be great. In general, these geographical groups could be recognized in the field by the few people familiar with the whole of the cerrado region. However, it is difficult to see differences in the floristics of the continuum over short distances, other than when strong differences in soil-type or topography occur. But occasionally our team has noticed differences on road journeys, and remarked on 'the flora becoming like that of the Federal District here' (this was at Arinos, Minas Gerais, when we had just passed from the north and north-eastern group to the central and south-eastern), or 'this reminds me of Xavantina', or 'quite a strong southern element here.' The clues to the differences generally lie in the presence of characteristic marker species, for example *Hirtella ciliata*, *Caryocar cuneatum* and *Parkia platycephala* in northern and north-eastern sites, *Vochysia gardneri*, *Platonia insignis* Mart. and *Martiodendron mediterraneum* (Mart. ex Benth.) Koeppen in the north-east (the latter two rather unusual in appearance for trees of the cerrado), and *Acosmium subelegans*, *Anadenanthera peregrina*, *Campomanesia adamantium*, *Erythroxylum cuneifolium*, *Gochnatia* species and many Lauraceae in southern sites.

2.5 BIODIVERSITY PATTERNS, CONCLUSION AND SUMMARY

The Cerrado Biome is one of the largest in terms of area in South America and is amongst the world's richest in species and endemicity. The total number of native vascular plant species so far recorded for all vegetation types of the biome (cerrado s.l., forest, wet habitats, etc.) is 6429 (Mendonça et al., 1998), but is estimated to be considerably greater than this. The number of species recorded for cerrado s.l., the savannic element of the vegetation, is 2880, but again this will certainly rise with further botanical exploration. This figure is made up of 700 to 900 larger woody species and some 2000 species in the ground layer. About 35% of the commonest 300 species of the former are endemic to the biome, as are the great majority of the ground species. The tree and large shrub species consist of a group of rather more than 100 relatively common and mainly widespread species and an abundance of less common and rare ones. In surveys across the whole core area of the biome and its southern and western outliers in São Paulo and Rondônia, 914 species were recorded at 316 sites, but of these only 39 occurred at ≥50% of sites, 107 at ≥20% and 300 at ≥2.5% (see Appendix 2), while the remainder were extremely rare, including 309 species recorded at only a single site. We have selected suites of 116 and c.340 species to represent groups dominating the woody flora of cerrado at higher and lower levels (Appendices 1, 2 and 3). Analysis of 26 phytosociological surveys across the cerrado region shows that the former suite constitutes an 'oligarchy' contributing an average 66% of species and 75% of Importance Value Index (Bridgewater et al., 2004). Comparable patterns of oligarchy have been described by Pitman et al. (2001) in large areas of Amazonian *terra firme* forest in Ecuador and Peru. The second group of c.340 species represents the flora that provides the dominant overall woody vegetation cover of the cerrado s.l. throughout the greater part of the biome.

Cerradão is the climax vegetation of cerrado s.l. in more favourable habitats. There is abundant evidence to show that this physiognomy was widespread before the occurrence of recent high levels of anthropic influence, particularly the practice of frequent burning. In many protected areas cerradão is now rapidly re-establishing itself as the dominant form of cerrado. The Rizzini-Lund-Aubréville theory of the origin of cerrado vegetation postulates modern cerrado as derived from an ancient sclerophyllous cerradão (or mata xeromorfa) of endemic species to which a large number of non-endemic species, principally from adjacent biomes, became associated. It also regards modern sparse forms of cerrado (e.g. campo cerrado, campo sujo, etc.) as having been derived from previous cerradões and dry forests by long-term frequent repetition of man-made fires and other disturbance. A number of other early workers reached similar conclusions independently.

Various studies have been directed at discovering the proportion of endemics and non-endemic species in the woody cerrado flora. The number of the latter is surprisingly high, reaching 65% of the commonest 300 large woody species (see Appendix 2), and most of these are also found in the Atlantic forest and/or the dry forest. However, many of these species are equally well adapted to the cerrado environment so that accessory (as they have frequently been designated in the past, indicating that they are more at home in a different vegetation type) is a misnomer and they are better described as non-endemics, the terminology we have used in this chapter. Such well-adapted non-endemics have been designated by us as type (a). Table 2.4 shows the percentage of non-endemic species derived from analysis of our data-base and further data are given in Appendix 2.

Phytogeographic studies of the Cerrado Biome based on the multivariate analysis of hundreds of floristic surveys of the woody vegetation (Castro, 1994a; Ratter et al., 1996, 2003; Castro and Martins, 1999; Durigan et al. 2003a; Bridgewater et al., 2004; Durigan, Chapter 3) have revealed significant distribution patterns (see Figure 2.3). These patterns can be related to climate and soil type but most probably also reflect major changes occurring during the Quaternary and earlier geological periods.

2.6 DESTRUCTION OF THE CERRADO AND THE CONSERVATION SITUATION

The present state and conservation situation of the Cerrado Biome is an enormous problem that can only here be considered very briefly. Readers wishing to obtain more information should consult works such as Dias (1992), Alho and Martins (1995), Ratter et al. (1997), Cavalcanti (1999), Busch-bacher (2000), Cavalcanti and Joly (2002), Klink and Moreira (2002) and Machado et al. (2004).

Prior to about 35 years ago agricultural activity in the cerrado *sensu lato* was confined largely to low-density cattle raising in natural vegetation. This provided a low-yielding, environmentally friendly system that could be sustained indefinitely. Crop production (rice, beans, maize etc.) was confined to the extremely fertile soils of the mesotrophic SDTF which yielded profitably year after year without fertilizer. However, a considerable conservation problem was already occurring here, for instance much of the Mato Grosso de Goiás, a huge forest area of $c.40,000$ km^2 stretching from Anapolis and Goiânia right to the western borders of Goiás state, had already been devastated. In more accessible areas cerrado trees were cropped for charcoal production for the steel industry, but this was probably a sustainable system with timber harvested according to a regular cycle.

All this changed with the introduction of modern mechanized agriculture, bringing with it techniques for the rapid clearing of extensive landscapes, preparation involving heavy applica-tion of fertilizers to counteract the low pH and toxic levels of aluminium inherent in the cerrado soil, use of huge rotating sprinklers to irrigate arable crops, and advanced harvesting systems, etc. This process became established with the help of agricultural subsidies and loans, generous tax incentives, and other attractions for the development of a gigantic capital-intensive agri-business. Over the last 30 or so years it has continued to grow at a remarkable rate so that the cerrado has been transformed into Brazil's major producer and exporter of important cash crops. The principal crop is soya bean which in 2002 yielded 27 million tonnes from 9.4 million hectares of land; to give an idea of the rate of production increase, the figure seven years earlier was approximately 10.4 million tonnes from 4.2 million hectares (figures from Machado et al., 2004). Other important arable crops are maize (5.5 million tonnes in 1995 [Klink and Moreira, 2002]), cotton, millet, sorghum and sunflower, but in the last decade production of traditional staples such as mandioca, beans and rice has declined. However, by far the greatest land-use area (estimated as 67% of the total cleared) is occupied by short-lived pastures planted with

African grasses and carrying the great majority of at least 51 million cattle raised in the cerrado (Klink and Moreira, 2002).

Accurate observations of deforestation in the Cerrado Biome by analysis of remote sensing imagery have recently been made by Machado et al. (2004). These authors consider that 34% of the original area still remains in more or less natural form, although estimates of some other workers are considerably more pessimistic. This means that 66% of the biome (i.e. 1.32 million km^2) has already been converted to predominantly anthropic landscape, representing slightly more than the combined area of Belgium, Denmark, France, Germany, the Netherlands, Switzerland and the United Kingdom (1,301,489 km^2), or nearly twice that of the state of Texas (695,673 km^2), or more than the area of the two large Brazilian states of Mato Grosso and Mato Grosso do Sul together (1,231,549 km^2). Furthermore, the process continues at an estimated rate of 1.1% of the total original cerrado area per year (Machado et al., 2004). The land devoted to cultivation of soya alone (94,000 km^2) is larger than the area of Portugal (88,940 km^2)! The area of deforestation of the cerrado is nearly twice that which has occurred in the Brazilian Amazonian forest, and in terms of percentage of the original area represents approximately 66%, in contrast to about 21% of the 3.5 million km^2 of the forest. But whereas the destruction of the Amazonian forest has rightly received enormous publicity, the even worse plight of the Cerrado Biome has been largely neglected by the world media despite its status as a global megadiversity hot-spot.

This drastic transformation of the land use of the cerrado during a relatively short time period has caused immense environmental and social problems. The onslaught of an extreme form of modern mechanized agriculture and the resultant environmental destruction has made conservation planning an urgent necessity, particularly since agricultural expansion continues at a more or less unabated rate. At present, conservation areas within the Cerrado Biome cover only 2.2% of its original area (Machado et al., 2004) and many more are clearly needed to give an adequate representation of the heterogeneous biodiversity present. Various bodies from the Brazilian Ministry of the Environment (Ministério do Meio Ambiente), state governments and major NGOs to a host of minor organizations are attempting to address this challenge. For instance, a large workshop, Ações Prioritárias da Biodiversidade de Cerrado e Pantanal, took place in Brasília in 1998. It was sponsored and organized by no less than 14 organizations including the Ministério do Meio Ambiente, Conservation International, CNPq (Brazilian government department for organization and support of research), the World Bank and GEF. It brought together about 250 specialists covering all aspects of biological conservation in the cerrado whose brief was to identify priority areas for protection throughout the biome, and a total of 17 locations for major conservation units were selected. Since the workshop two of these outstanding priority areas have been decreed: the Peruaçu valley in the north of the state of Minas Gerais and the Serra da Bodoquena in the state of Mato Grosso do Sul (see Cavalcanti [1999] and Cavalcanti and Joly [2002] for further information). The Nature Conservancy do Brasil (TNC) is also carrying out a similar exercise state by state and has already held its workshop for Mato Grosso and Mato Grosso do Sul in late 2002, when a representative selection of priority areas was identified. WWF Brasil, for its part, has been very active in securing the large Chapada dos Veadeiros reserve in Goiás and the great Grande Sertão Veredas reserve in north Minas Gerais where FUNATURA, a dynamic Brasília NGO, played a vital role. Surveys carried out by the CMBBC and Biogeografia dos Cerrados projects (see Introduction) have provided important data for decisions leading to proposals for the selection of these reserves. An outstanding example of enthusiasm and effectiveness has been shown by a young group led by Marcelo Lima (nicknamed 'Cegonha') of the University of Brasília, Andre Lima of the Instituto Socio Ambiental and Alexandre Bonesso Sampiao, also of the University of Brasília, who founded the NGO Pequi (Pesquisa e Conservação do Cerrado). Shortly after its foundation in 2000 the group were leading biodiversity surveys instrumental in the establishment of the huge Jalapão reserve in Tocantins State, and attracting funds from international NGOs and other organizations to mount large successful workshops on rapid survey methods and other relevant subjects.

An aim of the Ações Prioritárias workshop in Brasília was to protect 10% of the original area in conservation units, which would expand the existing area fivefold, and this was accepted as a target by the Brazilian Ministry of the Environment. This certainly would be a most desirable goal, and we can only hope that it comes close to being achieved.

An important factor which should mitigate cerrado destruction is that Brazilian law requires 20% of any agricultural land to be maintained under natural vegetation, and if this were better observed by landowners most of the problems of the extinction of the cerrado would be avoided. Clearly a more rigorous enforcement of this requirement would help the conservation situation enormously — the legislation is already in force to protect one fifth of the biome but it requires enforcement!

Another approach to conservation is one which promotes sustainable, non-destructive use of the natural habitat: essentially a refinement of the best traditional systems used in the cerrado before the introduction of modern intensive agriculture. Programa Fazenda Trijunção is being developed as a model for such a system on a large fazenda (ranch) at the junction of the borders of the states of Bahia, Goiás and Minas Gerais by three environmentally dedicated partners, Srs Theodoro Machado, Neuber dos Santos and José Roberto Marinho. The project is meticulously planned and even involves development of the traditional, locally adapted, Caracu cattle to improve yield on poor native pasture. It is hoped that it may prove a rational, profitable, alternative long-term use of cerrado areas, as opposed to the present short-term, massively destructive systems.

Lastly it should be mentioned that the Cerrado stands alone among the major Brazilian biomes in not being recognized in the constitution as a national heritage, a status accorded to the Amazon, Atlantic Rain Forest, Pantanal and coastal areas. Efforts are now being made to amend the constitution to afford the cerrado such recognition. If successful they should at long last remove the biome's 'poor relative' status and allow it to compete on equal terms for conservational resources.

ACKNOWLEDGEMENTS

We wish to thank Dr Ary T. Oliveira-Filho, Federal University of Lavras, Minas Gerais, Brazil for discussions and provision of data on the status and distribution of non-endemic ('accessory') species of the cerrado, and Dr Giselda Durigan, Instituto Florestal, Assis, São Paulo, Brazil for information and advice on the São Paulo cerrados. We also wish to acknowledge Dr Mariluza Araújo Granja e Barros for kindly allowing us to use unpublished data from her PhD thesis on the breeding systems of cerrado plants.

The greater part of the data on which the chapter is based were obtained during the years 1993–1996 by the Biodiversity of the Cerrados project, financed by the Baring Foundation and the European Community (financial contribution B92/4-3040/9304), and continued from 1996–2000 as the Conservation and Management of the Biodiversity of the Cerrado Biome (CMBBC) Project funded by the British Department for International Development (contract no. CNTR 95 5574). EMBRAPA/CERRADOS provided generous counterpart funding, logistic support (including transport) and staff time for these projects, while the University of Brasília also provided facilities (including those of their herbarium), transport and staff participation. A large number of people contributed to these projects over the years and detailed acknowledgements are given in Ratter et al. (2003).

REFERENCES

Ab'Saber, A.N. and Junior, M.C., Contribuição ao estudo do Sudoeste Goiano, *Bol. Geográfico*, 9, 123, 1951.

Alho, C.J.R. and Martins, E. de S., *Bit by bit the cerrado loses space*, WWF, Brasília, DF, Brazil, 1995.

Barros, M.A.G., *Studies on the pollination biology and breeding systems of some genera with sympatric species in the Brazilian cerrados*, PhD thesis, University of St. Andrews, UK, 1989.

Barros, M.A.G., Biologia reproductiva e polinização de espécies simpátricas de *Diplusodon*, *Acta Bot. Mex.*, 37, 11, 1996.

Barros, M.A.G., Sistemas reprodutivas e polinização em espécies simpátricas de *Erythroxylum* R.Br. (Erythroxylaceae) do Brasil, *Revista Brasil. Bot*, 21, 159, 1998.

Bridgewater, S., Ratter, J.A., and Ribeiro, J.F., Biogeographic patterns, β-diversity and dominance in the Cerrado Biome of Brazil, *Biodiversity and Conservation*, 13, 2295, 2004.

Buschbacher, R. (co-ord.), *Expansão Agrícola e Perda da Biodiversidade no Cerrado: origens históricas e o papel do comércio Internacional*, WWF Brasil, Série Técnica 8, Brasilia, DF, Brazil, 2000.

Castro, A.A.J.F., *Comparação florístico-geográfica (Brasil) e fitossociológica (Piauí-São Paulo) de amostras de cerrado*, PhD thesis, Universidade Estadual de Campinas, SP, Brazil, 1994a.

Castro, A.A.J.F., Comparação florística de espécies de cerrado, *Silvicultura (Brazil)*, 15 (58), 16, 1994b.

Castro, A.A.J.F. and Martins, F.R., Cerrados do Brasil e do Nordeste: caracterização, área de ocupação e considerações sobre a sua fitodiversidade, *Pesq. Foco, São Luís*, 7 (9), 147, 1999.

Castro, A.A.J.F., Martins, F.R., and Fernandes, A.G., The woody flora of the cerrado vegetation in the state of Piauí, northeastern Brazil, *Edinburgh J. Bot.*, 55, 455, 1998.

Castro, A.A.J.F. et al., How rich is the flora of Brazilian cerrados? *Ann. Missouri Bot. Gard.* 86, 192, 1999.

Cavalcanti, R.B. (Scientific co-ordinator), *Ações Prioritárias para Conservação da Biodiversidade do Cerrado e Pantanal* (Ministério de Meio Ambiente, Funatura, Conservation Internacional, Fundação Biodiversitas, Universidade de Brasília), Brasília, DF, Brazil, 1999.

Cavalcanti, R.B. and Joly, C.A., Biodiversity and conservation priorities in the cerrado region, in *The Cerrados of Brazil*, Oliveira, P.S. and Marquis, R.J., Eds, Columbia University Press, New York, USA, 2002, 351.

Cole, M.M., *The Savanas: biogeography and geobotany*, Academic Press, London, UK, 1986.

Dianese, J.C., Buricá, P., and Hennen, J.F., *Batistopsora gen. nov.* and new *Phakopsora, Ravenelia, Cerotelium*, and *Skierska* species from the Brazilian Cerrado, *Fitopatol. Brasil*, 18, 436, 1993.

Dias, B.F. de S., Cerrados: uma caracterização, in *Alternativas de desenvolvimento dos cerrados: manejo e conservação dos recursos naturais renováveis*, Dias, B.F. de S., Co-ord., FUNATURA-IBAMA, Brasília, DF, Brazil, 1992, 11.

Durigan, G. and Ratter, J.A., Notes on successional changes in cerrado and cerrado-forest ecotonal vegetation in western São Paulo State, *Edinburgh J. Bot.*, in press.

Durigan, G. et al., Inventário florístico do cerrado na estação ecológica de Assis, SP, *Hoehnea*, 26, 149, 1999.

Durigan, G. et al., Padrões fitogeográficos do cerrado paulista sob uma perspectiva regional, *Hoehnea*, 30, 39, 2003a.

Durigan, G. et al., The vegetation of priority areas for cerrado conservation in São Paulo State, Brazil, *Edinburgh J. Bot.* 60, 217, 2003b.

Eiten, G., The cerrado vegetation of Brazil, *Bot. Rev. (Lancaster)*, 38, 201, 1972.

Felfili, J.M. and Silva Junior, M.C., A comparative study of cerrado (*sensu stricto*) vegetation in Central Brazil, *J. Trop. Ecol.*, 9, 277, 1993.

Felfili, J.M. et al., Comparação do cerrado (*sensu stricto*) nas Chapadas Pratinha e dos Veadeiros, in *Contribuição ao conhecimento ecológico do cerrado (trabalhos selecionados do 3° Congresso de Ecologia do Brasil, 1996)*, Leite, L.L. and Saito, C.H., Eds, Universidade de Brasilia, DF, Brazil, 1997, 6.

Felfili, J.M. et al., Diversity, floristic and structural patterns of cerrado vegetation in Central Brasil, *Plant Ecology*, 175, 37, 2004.

Filgueiras, T.S., Herbaceous plant communities, in *The Cerrados of Brazil*, Oliveira, P.S. and Marquis, R.J., Eds, Columbia University Press, New York, USA, 2002, 121.

da Fonseca, G.A.B. et al., Eds, *Hotspots, Earth's biologically richest and most endangered terrestrial ecoregions*, Conservation International, Chicago, Illinois, USA, 2000, 148.

Furley, P.A. and Ratter, J.A., Soil resources and plant communities of the Central Brazilian cerrado and their development, *J. Biogeogr.*, 15, 97, 1988.

Gibbs, P., Self-incompatibility in flowering plants: a neotropical perspective, *Revista Brasil. Bot.*, 13, 125, 1990.

Goodland, R.J.A., Plants of the cerrado vegetation of Brazil, *Phytologia*, 20, 57, 1970.

Goodland, R.J.A. and Ferri, M.G., *Ecologia do Cerrado*, Editora Univ. São Paulo, São Paulo, Brazil, 1979.

Heringer, E.P. et al., A flora do cerrado, in *IV Simpósio sobre o cerrado*, Ferri, M.G., Ed, Editora Univ. São Paulo, São Paulo, Brazil, 1977, 211.

Hill, M.O., *TWINSPAN — a FORTRAN program for arranging multivariate data in an ordered two-way* table *by classification of individuals and attributes*, Cornell University, Ithaca, NY, USA, 1979.

Kent, M. and Coker, P., *Vegetation description and analysis : a practical approach*, Belhaven Press, London, UK, 1992.

Klink, C.A. and Moreira, A.G., Past and current human occupation, and land use, in *The Cerrados of Brazil*, Oliveira, P.S. and Marquis, R.J., Eds, Columbia University Press, New York, USA, 2002, 69.

Loefgren, A., Ensaio para uma distribuição dos vegetaes no diversos grupos florísticos no Estado de São Paulo, Ed. 2, *Bol. Commiss. Geogr. Estado São Paulo*, no. 11, São Paulo, Brazil, 1898.

Lopes, A.S. and Cox, F.R., A survey of the fertility status of surface soils under 'cerrado' vegetation in Brazil, *J. Soil. Sci. Soc. Amer.*, 41, 742, 1977.

Lund, P.W., Blik paa Brasiliens Dyreverden, Kongel. *Danske Vidensk. Selsk. Skr.*, Ser. 1, 1, 1843.

Machado, R.B. et al., *Estimativas de perda da área do Cerrado brasileiro*, unpublished technical report, Conservação Internacional, Brasilia, DF, Brazil, 2004. http:www.conservation.org.br/arquivos/Relat Dermatam Cerrado.pdf.

Méio, B.B. et al., Influência da flora das florestas Amazônica e Atlântica na vegetação do cerrado sensu stricto, *Revista Brasil. Bot.*, 26, 437, 2003.

de Mendonça, R.C. et al., Flora vascular do cerrado, in *Cerrado: ambiente e flora*, Sano, S.M. and Almeida, S.P. de, Eds, EMBRAPA, Planaltina, DF, Brazil, 1998, 289.

Miranda, I.S., *Flora, fisionomia e estrutura das savanas de Roraima, Brazil*, PhD thesis, INPA, Manaus, AM, Brazil, 1997.

Myers, N. et al., Biodiversity hotspots for conservation priorities, *Nature*, 403, 853, 2000.

Oliveira, P.E. and Gibbs, P.E., Pollination and reproductive biology in cerrado plant communities, in *The Cerrados of Brazil*, Oliveira, P.S. and Marquis, R.J., Eds, Columbia University Press, New York, USA, 2002, 329.

Oliveira-Filho, A.T. and Fontes, M.A., Patterns of floristic differentiation among Atlantic forests in southeastern Brazil and the influence of climate, *Biotropica*, 32, 793, 2000.

Oliveira-Filho, A.T. and Ratter, J.A., A study of the origin of central Brazilian forests by the analysis of plant species distribution patterns, *Edinburgh J. Bot.*, 52, 141, 1995.

Oliveira-Filho, A.T. and Ratter, J.A., Vegetation physiognomies and woody flora of the Cerrado Biome, in *The Cerrados of Brazil*, Oliveira, P.S. and Marquis, R.J., Eds, Columbia University Press, New York, USA, 2002, 91.

Pitman, N.C.A. et al., Dominance and distribution of tree species in Upper Amazonian terra firme forests, *Ecology*, 82, 2101, 2001.

Ratter, J.A., Notes on the vegetation of Fazenda Água Limpa (Brasília, DF, Brazil), Royal Botanic Garden Edinburgh, Edinburgh, UK, 1980.

Ratter, J.A., *Notas sobre a vegetação da Fazenda Água Limpa (Brasilia, DF, Brasil)*. Textos universitários No. 003, Editora UnB, Brasília, 1986.

Ratter, J.A. and Dargie, T.C.D., An analysis of the floristic composition of 26 cerrado areas in Brazil, *Edinburgh J. Bot.*, 49, 235, 1992.

Ratter, J.A., Bridgewater, S., and Ribeiro, J.F., Espécies lenhosas da fitofisionomia cerrado sentido amplo em 170 localidades do bioma cerrado, *Bol. Herb. Ezechias Heringer*, 7, 5, 2001.

Ratter, J.A., Bridgewater, S., and Ribeiro, J.F., Analysis of the floristic composition of the Brazilian cerrado vegetation III. Comparison of the woody vegetation of 376 areas, *Edinburgh J. Bot.*, 60, 57, 2003.

Ratter, J.A., Ribeiro, J.F., and Bridgewater, S., The Brazilian cerrado vegetation and threats to its biodiversity. *Ann. Bot. (Oxford)*, 80, 223, 1997.

Ratter, J.A. et al., Observations on the vegetation of northeastern Mato Grosso: 1. The woody vegetation types of the Xavantina-Cachimbo Expedition Area, *Phil. Trans. Roy. Soc. London B*, 226, 449, 1973.

Ratter, J.A. et al., Observações adicionais sobre o cerradão de solo mesotrófico no Brasil Central, in *IV Simposio sobre o cerrado*, Ferri, M.G., Ed, Editora Univ. São Paulo, São Paulo, Brazil, 1977, 303.

Ratter, J.A. et al., Floristic composition and community structure of a southern cerrado area in Brazil, *Notes Roy. Bot. Gard. Edinburgh*, 45, 137, 1988.

Ratter, J.A. et al., Analysis of the floristic composition of the Brazilian cerrado vegetation II. Comparison of the woody vegetation of 98 areas, *Edinburgh J. Bot.,* 53, 153, 1996.

Ratter, J.A. et al., Estudo preliminar da distribuição das espécies lenhosas da fitofisionomia cerrado sentido restrito nos estados compreendidos pelo Bioma Cerrado, *Bol. Herb. Ezechias Heringer,* 5, 5, 2000.

Ribeiro, J.F. and Walter, B.M.T., Fitofisionomias do Bioma Cerrado, in *Cerrado: Ambiente e Flora,* Sano, S.M. and Almeida, S.P., Eds, EMBRAPA, Planaltina, DF, Brazil, 1988, 87.

Richardson, J.E. et al., Rapid diversification of a species-rich genus of neotropical rain forest trees, *Science,* 293, 2242, 2001.

Rizzini, C.T., A flora do cerrado – análise florística das savanas centrais, in *Simposio sobre o cerrado,* Ferri, M.G., Ed, Editora Univ. São Paulo, São Paulo, Brazil, 1963, 125.

Rizzini, C.T., Árvores e arbustos do cerrado, *Rodriguésia,* 38, 63, 1971.

Rizzini, C.T., *Tratado de fitogeografia do Brasil, 2º volume — aspectos sociológicos e florísticos,* Editora Univ. São Paulo, São Paulo, Brazil, 1979.

Rizzini, C.T. and Heringer, E.P., *Preliminares acêrca das formações vegetais e do reflorestamento no Brasil Central,* Ministério da Agricultura, Rio de Janeiro, Brazil, 1962.

Simon, M.F. and Proença, C., Phytogeographical patterns of *Mimosa* (Mimosoideae, Leguminosae) in the Cerrado Biome of Brazil: an indicator genus of high-altitude centers of endemism? *Biol. Conservation,* 96, 279, 2000.

Waibel, L., Vegetation and land use in the planalto central of Brazil, *Geogr. Rev. (New York),* 38, 529, 1948.

Warming, E., Lagoa Santa: Et bidrag til den biologiska planegeografi, *Kongel. Danske Vidensk. Selsk. Naturvidensk,* 6, 153, 1892.

APPENDIX 1

The 116 dominant woody species of the cerrado flora. Those in bold are typical of mesotrophic (more fertile) soils.

Acosmium dasycarpum (Vog.) Yakovlev

A. subelegans (Mohl.) Yakovlev

Aegiphila lhotskiana Cham.

Agonandra brasiliensis Miers

Alibertia edulis (L.Rich) A.Rich

A. obtusa K. Schum.

Anacardium occidentale L.

Andira cuiabensis Benth.

A. vermifuga Mart.

Annona coriacea Mart.

A. crassiflora Mart.

Aspidosperma macrocarpon Mart.

A. nobile Müll. Arg.

A. subincanum Mart.

A. tomentosum Mart.

Astronium fraxinifolium Schott

Austroplenckia populnea (Reissek) Lundell

Bauhinia rufa (Bong.) Stcud.

Bowdichia virgilioides Kunth

Buchenavia tomentosa Eichler

Brosimum gaudichaudii Trécul

Byrsonima coccolobifolia Kunth

B. crassa Nied.(syn. *B. pachyphylla* A. Juss.)

B. intermedia A. Juss.

B. verbascifolia Rich. ex A. Juss.

Callisthene fasciculata (Spreng.) Mart.

Caryocar brasiliense Cambess.

C. cuneatum Wittm.

Casearia sylvestris Sw.

Connarus suberosus Planch.

Copaifera langsdorffii Desf.

Couepia grandiflora (Mart.) Benth.

Curatella americana L.

Cybistax antisyphilitica Mart.

Dalbergia miscolobium Benth. (=*D. violacea* (Vog.) Malme)

Davilla elliptica A. St. Hil.

Dilodendron bipinnatum Radlk.

Dimorphandra mollis Benth. (incl. *D. gardnerianum* Tul.)

Diospyros hispida DC.

Dipteryx alata Vogel

Duguetia furfuracea (A. St. Hil.) Benth. & Hook. f.

Emmotum nitens (Benth.) Miers

Enterolobium gummiferum (Mart.) J. Macbr.

Eriotheca gracilipes (Schum.) A. Robyns

E. pubescens (Mart. & Zucc.) Schott. & Endl.

Erythroxylum suberosum A. St. Hil.

E. tortuosum Mart.

Eugenia dysenterica DC.

Guapira noxia (Netto) Lundell

Guazuma ulmifolia Lam.

Guettarda viburnoides Cham. & Schltdl.

Hancornia speciosa Gomez (incl. *H. pubescens* Nees & Mart.)

Heteropterys byrsonimifolia A. Juss.

Himatanthus obovatus (Müll. Arg.) Woodson

Hirtella ciliata Mart. ex Zucc.

Hymenaea stigonocarpa Mart. ex Hayne

Kielmeyera coriacea (Spreng.) Mart.

K. rubriflora Cambess.

K. speciosa A. St. Hil.

Lafoensia pacari A. St. Hil. (incl. *L. densiflora* Pohl)

Luehea paniculata Mart.

Machaerium acutifolium Vogel

M. opacum Vogel

Magonia pubescens A. St. Hil.

Maprounea guianensis Aubl.

Miconia albicans (Sw.) Triana

M. ferruginata DC.

Mouriri elliptica Mart.

M. pusa Gardner

Myracrodruon urundeuva Fr. Allem.

Neea theifera Oerst.

Ouratea hexasperma (A. St. Hil.) Benth.

O. spectabilis (Mart.) Engl.

Palicourea rigida Kunth

Piptocarpha rotundifolia (Less.) Baker

Plathymenia reticulata Benth.

Pouteria ramiflora (Mart.) Radlk.

P. torta (Mart.) Radlk.

Protium heptaphyllum (Aubl.) Marchal

Pseudobombax longiflorum (Mart. & Zucc.) A. Robyns

P. tomentosum (Mart. & Zucc.) A. Robyns

Psidium myrsinoides O. Berg

Pterodon polygalaeflorus Benth.

P. pubescens Benth.

Qualea grandiflora Mart.

Q. multiflora Mart.

Q. parviflora Mart.

Rapanea guianensis Kuntze

Roupala montana Aubl.

Rourea induta Planch.

Rudgea viburnoides (Cham.) Benth.

Salacia crassifolia (Mart.) Peyr.

Salvertia convallariodora A. St. Hil.

Schefflera macrocarpum (Cham. & Schltdl.) DC. Frodin

Sclerolobium aureum (Tul.) Benth.

S. paniculatum Vogel

Simarouba versicolor A. St. Hil.

Siparuna guianensis Aubl.

Strychnos pseudoquina A. St. Hil.

Stryphnodendron adstringens (Mart.) Cov.

S. obovatum Benth.

Styrax ferrugineus Nees & Mart.

Syagrus comosa (Mart.) Mart.

S. flexuosa (Mart.) Becc.

Tabebuia aurea (Manso) Benth. & Hooker f. ex S. Moore

T. ochracea (Cham.) Standl.

Tapirira guianensis Aubl.

***Terminalia argentea* Mart. & Zucc.**

T. fagifolia Mart. & Zucc.

Tocoyena formosa (Cham. & Schltdl.) K. Schum.

Vatairea macrocarpa (Benth.) Ducke

Vernonia ferruginea Less.

Virola sebifera Aubl.

V. rufa (C. K. Spreng.) Mart.

Xylopia aromatica Lam.

Zeyheria montana Mart.

APPENDIX 2

Woody species occurring at ≥2.5% (≥8 sites) of the 316 core area and related outlier sites from the CMBBC data-base (see Ratter, Bridgewater, and Ribeiro, 2003). They are given in percentage bands, and the total species number given for each band is its own total added to that of all previous bands. Total 300 species.

Information on occurrence in Atlantic forest is from Oliveira-Filho and Fontes (2000), and Oliveira-Filho (pers. comm.). Other information on non-endemic 'accessory' species was derived from Rizzini (1963), Heringer et al. (1977) and personal observations. Records for Amazonian occurrence are probably very incomplete.

Code: NE, non-endemic ('accessory') species; a, type (a) non-endemic sp (see p. 42); Am, Amazonian forest; Atl, Atlantic forest; C, colonizer; F, forest; MS, SDTF and mesotrophic cerradão species; S, supertramp species (Oliveira-Filho and Fontes, 2000); X, xerophyllous species of wide distribution (Heringer et al., 1977). Where no code is given species are probably cerrado endemics.

I. Species at ≥ 50% of sites (total = 39)	No. sites	% of sites	Habitat code
Qualea grandiflora Mart.	274	87	
Qualea parviflora Mart.	251	79	
Bowdichia virgilioides Kunth	249	79	NE, a
Dimorphandra mollis Benth. (incl. D. gardnerianum Tul.)	238	75	
Lafoensia pacari A. St. Hil. (incl. L. densiflora Pohl)	238	75	NE, F (Atl), a
Connarus suberosus Planch.	237	75	
Hymenaea stigonocarpa Mart. ex Hayne	236	75	
Kielmeyera coriacea (Spreng.) Mart.	227	72	
Tabebuia aurea (Manso) Benth. & Hook. f. ex S. Moore	217	68	
Tabebuia ochracea (Cham.) Standl.	213	67	
Byrsonima coccolobifolia Kunth	211	67	NE, F, a
Pouteria ramiflora (Mart.) Radlk.	210	66	NE, F (Atl, Am), a
Casearia sylvestris Sw.	207	65	NE, F, X, a
Roupala montana Aubl.	202	64	NE, F (Atl), a
Acosmium dasycarpum (Vogel) Yakovl.	200	63	NE, F (Atl), a
Curatella americana L.	200	63	NE, C, F, X, a
Erythroxylum suberosum A. St. Hil.	199	63	NE, a
Caryocar brasiliense Cambess.	198	63	
Brosimum gaudichaudii Trec.	195	62	NE, a
Byrsonima crassa Nied. (syn. B. pachyphylla Adr. Juss.)	195	62	
Himatanthus obovatus (Müll. Arg.) Woods	192	61	
Vatairea macrocarpa (Benth.) Ducke	190	60	
Machaerium acutifolium Vogel	189	60	NE, C, F (Atl), a
Tocoyena formosa (Cham. & Schlecht.) K. Schum.	188	59	NE, F (Atl), a
Davilla elliptica A. St. Hil.	186	59	
Diospyros hispida DC.	184	58	NE, F (Atl), a
Salvertia convallariodora A. St. Hil.	182	58	
Xylopia aromatica Lam.	179	57	NE, C, F (Atl, Am)
Sclerolobium aureum (Tul.) Benth.	178	56	
Astronium fraxinifolium Schott	177	56	NE, F (Atl, MS)
Annona coriacea Mart.	173	55	
Ouratea hexasperma (A. St. Hil.) Baill.	172	54	NE, F (Atl), a
Plathymenia reticulata Benth.	172	54	NE, F, a
Hancornia speciosa Nees & Mart. (incl. H. pubescens Gomes)	170	54	NE, a
Aspidosperma tomentosum Mart.	165	52	
Qualea multiflora Mart.	164	52	NE, F, a

I. Species at ≥ 50% of sites (cont.) (total = 39)	No. sites	% of sites	Habitat code
Byrsonima verbascifolia (Rich. ex) Adr. Juss.	161	51	
Eriotheca gracilipes (Schum.) Robyns	159	50	
Sclerolobium paniculatum Vog.	158	50	NE, F (Atl, Am), a

II. Species at ≥ 40% of sites (total = 53)	No. sites	% of sites	Habitat code
Simarouba versicolor A. St. Hil.	153	48	NE, F
Miconia albicans (Sw.) Triana	150	47	NE, F (Atl), a
Pseudobombax longiflorum (Mart. & Zucc.) Robyns	149	47	
Andira vermifuga Mart.	148	47	
Annona crassiflora Mart.	148	47	
Copaifera langsdorfii Desf.	147	46	NE, F (Atl, S)
Magonia pubescens A. St. Hil.	146	46	NE, F (MS)
Dalbergia miscolobium Benth. (syn. *D. violacea* (Vog.) Malme)	142	45	NE, F (Atl), a
Emmotum nitens (Benth.) Miers	141	45	NE, F (Atl)
Strychnos pseudoquina A. St. Hil.	137	43	
Terminalia argentea Mart. & Zucc.	136	43	NE, F (Atl, MS)
Couepia grandiflora (Mart. & Zucc.) Benth.	131	41	
Palicourea rigida Kunth	129	41	
Agonandra brasiliensis Miers	127	40	NE, F (Atl), a

III. Species at ≥ 30% of sites (total = 75)	No. sites	% of sites	Habitat code
Rourea induta Planch.	126	39.9	
Piptocarpha rotundifolia (Less.) Bak.	125	39	
Vochysia rufa (Spr.) Mart.	124	39	
Aspidosperma macrocarpon Mart.	123	39	
Eugenia dysenterica DC.	121	38	
Enterolobium gummiferum (Mart.) J. Macbride	119	38	
Neea theifera Oerst.	116	37	
Alibertia edulis (L.Rich) A. Rich (incl. *A. lanceolata*)	114	36	NE, F (Atl)
Anacardium occidentale L.	113	36	NE, F (Atl), a
Stryphnodendron obovatum Benth.	110	35	
Salacia crassifolia (Mart.) G. Don	108	34	
Tapirira guianensis Aubl.	105	33	NE, F (Atl, S, Am)
Zeyheria montana Mart.	103	33	
Dipteryx alata Vogel	101	32	NE, F (MS)
Luehea paniculata Mart.	101	32	NE, F (Atl, MS)
Matayba guianensis Aubl.	100	32	NE, F (Atl, S, Am), X
Erythroxylum tortuosum Mart.	99	31	NE, a
Machaerium opacum Vogel	99	31	
Protium heptaphyllum (Aubl.) March.	98	31	NE, F (Atl, S, Am), X
Cybistax antisyphilitica Mart.	96	30	NE, F (Atl), a
Stryphnodendron adstringens (Mart.) Cov.	96	30	NE, a
Mouriri elliptica Mart.	95	30	

IV. Species at ≥ 20% of sites (total = 107)	No. sites	% of sites	Habitat code
Styrax ferrugineus Nees & Mart.	94	29.7	NE, F (Atl), a
Pterodon pubescens (Benth.) Benth.	92	29	NE, F, a
Pouteria torta (Mart.) Radlk.	91	29	NE, F (Atl, S, Am, MS), a
Andira cuiabensis Benth.	90	28	
Duguetia furfuracea (A. St. Hil.) Benth. & Hook. f.	90	28	

IV. Species at ≥ 20% of sites (cont.) (total = 107)	No. sites	% of sites	Habitat code
Vernonia ferruginea Less.	90	28	NE, C, F (Atl)
Aegiphila lhotskyana Cham.	86	27	NE, C, F (Atl)
Erythroxylum deciduum A. St. Hil.	85	27	NE, F (Atl), a
Guazuma ulmifolia Lam.	85	27	NE, F (Atl, S, MS), X
Heteropterys byrsonimifolia Adr. Juss.	84	27	NE, F (Atl), a
Guettarda viburnoides Cham. & Schlecht.	82	26	NE, F (MS), X
Syagrus comosa (Mart.) Mart.	82	26	NE, F (Atl), a
Buchenavia tomentosa Eichl.	81	26	NE, a
Ouratea spectabilis (Mart.) Engl.	81	26	
Austroplenckia populnea (Reiss) Lund	80	25	NE, F (Atl), a
Aspidosperma nobile Müll. Arg.	77	24	
Myracrodruon urundeuva Fr. Allem.	77	24	NE, F (MS)
Psidium myrsinoides O. Berg	76	24	
Callisthene fasciculata (Spr.) Mart.	74	23	NE, F (MS)
Pseudobombax tomentosum (Mart. & Zucc.) Robyns	74	23	NE, F (Atl, MS)
Siparuna guianensis Aubl.	72	23	NE, F (Atl)
Terminalia fagifolia Mart. & Zucc.	72	23	NE, F (Atl), a
Aspidosperma subincanum Mart.	70	22	NE, F (MS)
Bauhinia rufa (Bong.) Steud.	69	22	
Pterodon polygalaeflorus Benth.	68	21	NE, F (MS), a
Rudgea viburnoides (Cham.) Benth.	68	21	NE, F (Atl), a
Syagrus flexuosa (Mart.) Becc.	68	21	NE, F (Atl), a
Dilodendron bipinnatum Radlk.	67	21	NE, F (MS)
Schefflera macrocarpa (Cham. & Schltdl.) D.C.Frodin	66	21	NE, F (Atl), a
Rapanea guianensis Aubl.	65	21	NE, F (S)
Acosmium subelegans (Mohl.) Yakovl.	64	20	
Alibertia obtusa Cham.	64	20	

V. Species at ≥ 10% of sites (total = 160)	No. sites	% of sites	Habitat code
Guapira noxia Netto var. *noxia*	63	19.9	NE, F (Atl), a
Maprounea guianensis Aubl.	63	19.9	NE, F (Atl, Am)
Mouriri pusa Gardn.	63	19.9	
Byrsonima intermedia Adr. Juss.	62	19.9	
Eriotheca pubescens (Mart. & Zucc.) Schott. & Endl.	61	19	
Hirtella glandulosa Spreng.	60	19	NE, F, X, a
Rhamnidium elaeocarpum Reiss.	58	18	NE, F (MS)
Copaifera martii Hayne	57	18	NE
Ferdinandusa elliptica Pohl	56	18	NE
Kielmeyera rubriflora Cambess.	56	18	NE
Virola sebifera Aubl.	55	17	NE, F (Atl), X
Caryocar cuneatum Wittm.	54	17	
Acrocomia aculeata (Jacq.) Lodd. ex Mart.	53	17	NE, F (Atl, MS)
Copaifera malmei Harms	51	16	
Licania humilis Cham. & Schlecht.	50	16	
Hirtella ciliata Mart. & Zucc.	49	15	NE, a
Myrcia tomentosa (Aubl.) DC.	49	15	NE, F (Atl)
Ouratea castaneaefolia Engl.	49	15	NE, F (Atl)
Zanthoxylum rhoifolium Lam.	49	15	NE, F (Atl), S
Anadenanthera peregrina (L.) Speg.	48	15	NE, F (Atl, Am), X
Guapira graciliflora (Mart. ex J.A. Schmidt) Lundell	48	15	NE, F (Atl), a
Physocallyma scaberimmum Pohl	48	15	NE, F (MS)
Solanum lycocarpum St. Hil.	47	15	NE, C

V. Species at ≥ 10% of sites (cont.) (total = 160)	No. sites	% of sites	Habitat code
Coussarea hydrangeaefolia Benth. & Hook.f.	46	15	NE, F (MS)
Jacaranda brasiliana Pers.	46	15	NE, a
Cecropia pachystachya Tréc.	45	14	NE, C, F (Atl, Am)
Platypodium elegans Vogel	44	14	NE, F (Atl, MS), X
Byrsonima basiloba Adr. Juss.	43	14	
Styrax camporum Pohl	43	14	NE, F (Atl), a
Vochysia elliptica (Spr.) Mart.	43	14	NE, a
Antonia ovata Pohl	41	13	NE, F (Am)
Vochysia tucanorum (Spr.) Mart.	41	13	NE, F (Atl)
Alibertia sessilis (Cham.) Schum.	40	13	NE, F (MS)
Kielmeyera lathrophyton Saddi	40	13	NE, F (Atl), a
Anadenanthera colubrina (Vell.) Brenan var. *cebil* (Griseb.) Altschul	39	12	NE, F (MS)
Byrsonima crassifolia (L.) Kunth (syn. *B. fagifolia* Nied.)	39	12	NE
Guapira noxia (Netto) Lundell var. *psammophila* (Mart. ex J. A . Schmidt) ined.	39	12	
Vochysia cinnamomea Pohl	39	12	
Diospyros sericea DC.	38	12	NE, F
Casearia grandiflora Cambess.	37	12	NE, F
Didymopanax vinosum (Cham. & Schlecht.) March.	37	12	NE, a
Tabebuia serratifolia (Vahl) Nicholson	37	12	NE, F (Atl, S), X
Hymenaea courbaril L. var. *stilbocarpa* (Haync) Lee & Langenheim	36	11	NE, F (Atl, S), X
Vochysia thyrsoidea Pohl	36	11	NE, a
Jacaranda cuspidifolia Mart.	35	11	NE, F (MS)
Kielmeyera speciosa St. Hil.	34	11	
Miconia ferruginata DC.	34	11	
Parkia platycephala Benth.	34	11	NE, a
Callisthene major Mart.	33	10	NE, F (Atl)
Eugenia aurata O. Berg	33	10	NE, F (Atl)
Terminalia glabrescens Mart.	33	10	NE, F (Atl)
Campomanesia pubescens O. Berg	32	10	NE, F (Atl)
Zanthoxylum riedelianum Engl.	32	10	NE, F (Atl, MS)

VI. Species at ≥ 5% of sites (total = 231)	No. sites	% of sites	Habitat code
Diptychandra aurantiaca Tul. (syn. *D. glabra* Benth.)	31	9.8	
Erythoxylum daphnites Mart.	31	9.8	NE, F (Atl)
Eschweilera nana (O. Berg) Miers	31	9.8	
Heisteria ovata Benth.	31	9.8	NE, a
Myrcia sellowiana O. Berg	31	9.8	NE, F (Atl), a
Vochysia gardneri Warm.	31	9.8	
Coccoloba mollis Casar.	30	9	NE, F (Atl)
Tabebuia roseoalba (Ridley) Sandw.	30	9	NE, F (Atl, MS)
Vellozia squamata Pohl	30	9	
Diospyros burchellii Hiern	29	9	
Cordia glabrata (Mart.) A. DC.	28	9	NE, F (MS)
Eremanthus glomerulatus Less.	28	9	
Hyptidendron canum (Pohl ex Benth.) Harley (=*Hyptis cana* Pohl ex Benth.)	28	9	
Stryphnodendron polyphyllum Benth.	28	9	NE
Xylopia sericea St. Hil.	28	9	NE, F (Atl)
Mezilaurus crassiramea (Meisn.) Taub.	27	8	
Cochlospermum vitifolium (Willd.) Spreng.	26	8	

VI. Species at ≥ 5% of sites (cont.) (total = 231)	No. sites	% of sites	Habitat code
Combretum mellifluum Eichler	26	8	
Eriotheca parvifolia (Mart. & Zucc.) A. Robyns	26	8	
Miconia rubiginosa (Bonpl.) DC.	26	8	NE, C?, F (Atl)
Miconia stenostachya DC.	26	8	
Myrcia albo-tomentosa Cambess.	26	8	
Myrcia rostrata DC.	26	8	NE, F (Atl, S)
Qualea dichotoma (Mart.) Warm.	26	8	NE, F (Atl)
Annona tomentosa R.E. Fr.	25	8	
Euplassa inaequalis (Pohl) Engl.	25	8	NE
Cardiopetalum calophyllum Schltdl.	24	8	NE
Stryphnodendron coriaceum Benth.	24	8	
Lithraea molleoides (Vell.) Engl.	23	7	NE, X
Luehea grandiflora Mart.	23	7	NE, F (MS), X
Myrcia lingua (O. Berg) Mattos	22	7	
Peltogyne confertiflora (Hayne) Benth.	22	7	NE, F, a
Pera glabrata (Schott.) Baill.	22	7	NE, F (Atl, S)
Salacia elliptica G. Don	22	7	NE, F (Atl)
Banisteriopsis latifolia (Adr. Juss.) Cuatrec.	21	7	
Blepharocalyx salicifolius (Kunth) O. Berg (=*B. suaveolens* (Cambess.) Bur.)	21	7	NE, F
Myrcia rorida Kiaersk.	21	7	
Myrcia splendens (Sw.) DC.	21	7	NE, F (Atl)
Psidium warmingianum Kiaersk.	21	7	
Sterculia striata St. Hil. & Naud.	21	7	NE, F (MS)
Vochysia haenkeana Mart.	21	7	NE, X
Chomelia ribesioides Benth.	20	6	
Miconia pohliana Cogn.	20	6	
Rapanea umbellata (Mart.) Mez	20	6	NE, F (Atl)
Alibertia elliptica (Cham.) K. Sch.	19	6	NE, F (Atl)
Callisthene sp. nov. (R7228)	19	6	
Cenostigma macrophyllum Tul.	19	6	C
Didymopanax distractiflorum Harms	19	6	
Eugenia bimarginata DC.	19	6	
Kielmeyera sp. nov.? (R7954)	19	6	
Ocotea pulchella Mart.	19	6	NE, F
Pseudobombax marginatum (A.St.-Hil., A.Juss. & Camb.) Robyns	19	6	NE, F (MS), a
Schinus terebinthifolius Raddi	19	6	NE, F
Tabebuia impetiginosa (Mart. ex A. DC.) Standl.	19	6	NE, F (MS)
Andira cordata Arroyo ex R.T.Pennington	18	6	
Cybianthus detergens Mart.	18	6	NE, F (MS)
Helicteres brevispira A . St. Hil.	18	6	NE, F (MS)
Jacaranda caroba (Vell.) DC.	18	6	NE
Mimosa laticifera Rizzini & Mattos	18	6	
Myrcia variabilis DC.	18	6	NE, F (Atl)
Protium ovatum Engl.	18	6	
Senna rugosa (G. Don) Irwin & Barneby	18	6	NE
Apeiba tibourbou Aubl.	17	5	NE, C, F (Atl, MS), X
Butia archeri (Glassman) Glassman (= *B. leiospatha* (Mart.) Becc.)	17	5	
Miconia fallax DC.	17	5	NE, F (Atl)
Bredemeyera floribunda Willd.	16	5	
Cnidoscolus vitifolia (L.) Pohl	16	5	NE, C

VI. Species at ≥ 5% of sites (cont.) (total = 231)	No. sites	% of sites	Habitat code
Mabea fistulifera Mart.	16	5	NE, C, F (Atl, S)
Miconia ligustroides (DC.) Naud.	16	5	NE, F (Atl)
Mimosa clausennii Benth.	16	5	NE, F (Atl)
Vitex polygama Cham.	16	5	NE, F (MS)

VII. Species at ≥2.5% of sites (total = 300)	No. sites	% of sites	Habitat code
Byrsonima sericea DC.	15	4.7	NE, F (Atl)
Licania gardneri (Hook. f.) Fritsch	15	4.7	NE, F, S
Luehea divaricata Mart.	15	4.7	NE, F (Atl, S)
Ocotea minarum (Nees) Mez	15	4.7	NE, F (Atl), a
Aspidosperma multiflorum A. DC.	14	4	NE, F (Atl), a
Bauhinia brevipes Vogel	14	4	
Priogymnanthus hasslerianus (Chodat) P.S. Green (syn. *Linociera hassleriana* (Chodat) Hassler)	14	4	NE, F (MS)
Senna silvestris (Vell.) Irwin & Barneby	14	4	NE, C, F (Atl)
Vismia glaziovii Ruhl.	14	4	
Cordia trichotoma (Vell.) Arrab. ex Steud.	13	4	NE, F (Atl, MS)
Enterolobium contortisiliquum (Vell.) Morong	13	4	NE, F (Atl, MS)
Psidium pohlianum O. Berg	13	4	
Rapanea ferruginea (Ruiz. & Pav.) Mez	13	4	NE
Tapura amazonica Poepp. & Endl.	13	4	NE, F (Am)
Alchornea schomburgkii Klotzsch	12	4	NE, F (Am)
Bauhinia ungulata L.	12	4	
Licania sclerophylla (Mart.ex Hook.f.) Fritsch	12	4	NE, F (Am)
Machaerium hirtum (Vell.) Stellfeld	12	4	NE, F (MS)
Myrcia lanuginosa O. Berg	12	4	
Myrcia pallens DC.	12	4	NE, F (Atl)
Psidium guinense Sw.	12	4	NE, F (Am)
Symplocos rhamnifolia A. DC.	12	4	NE, X, a
Virola subsessilis Warb.	12	4	
Bauhinia pulchella Benth.	11	3	NE, F (Atl)
Byrsonima cydoniifolia Adr. Juss.(syn. *B. orbignyana* Adr. Juss.)	11	3	
Casearia arborea Urb.	11	3	NE, F
Chomelia obtusa Cham. & Schlecht.	11	3	NE, F (Atl)
Chrysophyllum marginatum Radlk.	11	3	NE
Cupania vernalis Cambess.	11	3	NE, F (Atl)
Eremanthus goyazensis Sch. Bip.	11	3	
Luehea candicans Mart.	11	3	NE, X, MS
Ximenia americana L.	11	3	NE
Acacia paniculata Willd.	10	3	NE, F (MS)
Bauhinia forficata Link.	10	3	NE, F (MS)
Casearia rupestris Eichl.	10	3	NE, F (MS)
Erythroxylum cuneifolium Poepp. ex O.E. Schulz	10	3	NE, F (Atl)
Eugenia punicifolia (Kunth) DC. (syn. *E. polyphylla* O. Berg)	10	3	NE, F (Atl)
Gochnatia barrossii Cabrera	10	3	
Machaerium villosum Vogel	10	3	NE, F
Miconia sellowiana Naud.	10	3	NE, F (Atl)
Qualea cordata Spreng.	10	3	NE
Sapium haematospermum Müll. Arg. (syn. *S. longifolium* (Müll. Arg.) Huber)	10	3	NE, F (Atl)
Apuleia leiocarpa Macbride	9	3	NE, F (Atl)
Aspidosperma parvifolium A. DC.	9	3	NE, F (Atl, S)

VII. Species at ≥2.5% of sites (cont.) (total = 300)	No. sites	% of sites	Habitat code
Erythroxylum ambiguum Peyr.	9	3	NE, F
Himatanthus articulatus (Vahl) R.E. Woodson	9	3	
Manihot tripartita Müll. Arg.	9	3	
Miconia burchellii Triana	9	3	NE, C, F (Atl)
Miconia macrothyrsa Benth.	9	3	NE, F
Myrcia lasiantha DC.	9	3	
Myrcia uberavensis O. Berg	9	3	
Siphoneugena densiflora O. Berg	9	3	NE, F
Vitex cymosa Bert.	9	3	NE, F (Atl)
Alibertia concolor (Cham.) K. Sch.	8	2.5	NE, F (Atl)
Alibertia macrophylla K. Sch.	8	2.5	NE, F (Atl)
Alibertia verrucosa S. Moore	8	2.5	NE, F (MS)
Baccharis dracunculifolia DC.	8	2.5	NE, C
Celtis pubescens (Kunth) Spreng.	8	2.5	NE, F (MS)
Chamaecrista orbiculata (Benth.) H.S.Irwin & R.C.Barneby	8	2.5	
Chomelia pohliana Müll. Arg.	8	2.5	NE, F (Atl)
Combretum leprosum Mart.	8	2.5	NE, F (MS)
Cordia sellowiana Cham.	8	2.5	NE, F (Atl/S)
Hirtella gracilipes (Hook. f.) Prance	8	2.5	NE, F (Atl)
Luetzelburgia praecox (Harms) Harms	8	2.5	
Macairea radula DC.	8	2.5	
Machaerium scleroxylon Tul.	8	2.5	NE, F (Atl, MS)
Myrcia canescens O. Berg.	8	2.5	
Myrcia multiflora DC.	8	2.5	NE, F (Atl, Am)
Trema micrantha (L.) Blume	8	2.5	NE, C, F (Atl/S, MS)

APPENDIX 3

The following 37 species are rare and occur in less than 2.5% of the surveys in the CMBBC database or are absent from it. However, they are common in São Paulo state (Durigan, pers. comm.) and on these grounds have been included here.

Acacia polyphylla DC.
Actinostemon conceptionis (Chodat & Hassler) Pax & K. Hoffm.
Alchornea triplinervia Müll. Arg.
Amaiouea guianensis Aubl.
Aspidosperma cylindrocarpum Müll. Arg.
Byrsonima coriacea DC.
Calyptranthes concinna DC.
Campomanesia adamantium (Cambess.) O. Berg
Casearia lasiophylla Eichler
Croton floribundus Spreng.
Erythroxylum pelleterianum A.St.Hil.
Eugenia livida O. Berg
E. pluriflora Mart.
E. pyriformis Cambess.
Gochnatia polymorpha DC.
Guapira opposita (Vell.) Reitz
Helietta apiculata Benth.
Lacistema hasslerianum Chodat
Machaerium brasiliense Vogel
Matayba elaeagnoides Radlk.
Maytenus robusta Reissek
Miconia langsdorffii Cogn.
Myrcia bella Cambess.
M. fallax (Rich.) DC.
M. venulosa DC.
Myrciaria ciliolata Cambess.
Nectandra cuspidata Nees & Mart.
Ocotea corymbosa (Meisn.) Mez
O. velloziana (Meisn.) Mez
O. velutina Mart.
Pera obovata Baill.
Persea pyrifolia Nees & Mart.
Syagrus romanzoffiana (Cham.) Glassman
Symplocos pubescens Klotzsch ex Benth.
Tabernaemontana hystrix Steud.
Vernonia diffusa Less.
Zeyheria tuberculosa (Vell.) Bureau ex Verlot

3 Observations on the Southern Cerrados and their Relationship with the Core Area

Giselda Durigan

CONTENTS

ABSTRACT

The cerrado vegetation of São Paulo and neighbouring states occurs under more extreme environmental conditions, especially lower temperatures (occasionally with severe frosts), and a shorter dry season than the core area of the biome. As a consequence, these southern cerrados form a distinct phytogeographical region, which can be locally divided into two separate floristic (and physiognomic) groups: (1) areas covered by cerradão (associated with more fertile soils) and (2) areas covered by sparser vegetation, e.g. cerrado *sensu stricto* (associated with very poor and sandy soils). The southern cerrado areas are bordered by the Atlantic forest, which exerts a strong influence on their flora. In recent years, a rapid physiognomic and floristic change has taken place at the southern limits of the Cerrado Biome, especially in São Paulo state: strong protection against fire has caused open forms of cerrado vegetation to change into closed cerradão. The southern cerrados contain some of the floristically richest sites in the cerrado domain and require special protection by conservation strategies.

3.1 GEOGRAPHY, ENVIRONMENT, AND LATE QUATERNARY HISTORY

The Cerrado Biome covers an area of 2,000,000 km², from 3°N to 24°S, extending from the Amazon basin, in the north, to São Paulo (SP) and Paraná states, in the south, with an altitudinal range from sea level to 1800 m (Ratter et al., 1997). It has been subject to many changes over the Quaternary period and it is relevant to consider these since they are still in progress and particularly important in the periphery of the biome, despite being largely obscured by the overwhelming destructive effect of human activities.

The vegetation of central Brazil has been a shifting mosaic of tropical and subtropical savanna and forest over the last 40,000 to 30,000 years (Ledru, 1993; Salgado-Labouriau et al., 1997). Before that, temperatures were probably too cold and dry to support cerrado species, and the region was occupied by subtropical grasslands (Ledru et al., 1996; Behling and Lichte, 1997). Since then, cerrado has occupied varying proportions of the region, correlated with major changes in climate. It has occupied larger areas in the past, when it probably expanded north-westwards into the Amazon region and southwards over the region presently occupied by semideciduous forest in south-eastern Brazil. In the late Pleistocene, between 30,000 and 23,000 years BP, the climate was moist in this region, and cooler than at present and the vegetation was similar to today's cerrado and campo de altitude (subtropical grassland of high altitudes, occurring nowadays generally above 1000–1500 m). Later, between 17,000 and 11,000 years BP, the climate became drier and was still cooler than at present, and although forest became more common, it was found only along rivers and in valleys, and the dominant vegetation was still grassland. At the beginning of the Holocene, between 11,000 and 5,500 years BP, the climate became warmer, but still drier and cooler than at present, and with a long dry season, probably of about 4–5 months. This was a period of cerrado expansion (Behling, 1998). Approximately 5,500 years BP the climate became semihumid warm tropical, like today, and modern vegetation began to develop, with development of forest and a more diverse cerrado (Behling, 1995). At this point the cerrado decreased in area in southern and northern Brazil, with the advance of evergreen and semideciduous forests (Ledru, 1993; Behling, 1998). The remaining cerrado is considered to be a relict of the era of warmer and drier climate in these southern and northern peripheral regions of the biome.

The present cerrado vegetation is found in a range of climatic types across the whole biome, from tropical with short dry season (Am, Köppen system) and tropical with winter dry season (Aw) to subtropical without dry season (Cfb) (Camargo, 1963). The average rainfall ranges from 800 to 2000 mm, and average annual temperatures from 18°C to 28°C (Dias, 1992). Extremes of minimum temperature across the biome vary from 18°C to 4°C, the latter in parts of São Paulo state (Eiten, 1972). The dry season can last from one to five months, usually from April to September. In the southernmost cerrados, however, some regions, like the Paranapanema watershed, have no dry period at all in some years.

The cerrado domain is dominated by latosols (Lopes, 1984; Adámoli et al., 1987; Reatto et al., 1998), but a number of different soil types and a wide range of chemical and physical characteristics occur. In general, cerrado soils are dystrophic, acid, and with high levels of aluminium (Furley and Ratter, 1988).

High floristic heterogeneity and distinct phytogeographical patterns occur across the vast area of the Cerrado Biome, and this can probably be correlated to a large extent with the wide variation of the environmental conditions occurring within it (Castro, 1994a,b; Castro and Martins, 1999; Ratter et al., 1996, 2003; Bridgewater et al., 2004).

3.2 THE SOUTHERN CERRADOS

The so-called southern cerrados (Ratter et al., 2003 and Chapter 2) originally covered an area of some 50,000 km^2 (2.5% of the total Cerrado Biome) in the states of São Paulo, and a few parts of Paraná and Minas Gerais (Figure 3.1), a region with the wettest and coldest climate in the cerrado domain (mean annual temperature ranging from 20–24ºC, and average annual rainfall of about 1450 mm). In São Paulo state, they more or less fall into two physiognomically distinct groups, one comprising more open forms and the other cerradão (cerrado woodland)*. These groups were also recognized in the works of Castro (1994a,b) and Castro and Martins (1999).

---

* For a description of the physiognomic variation of cerrado see Chapter 2, Section 2.2.

FIGURE 3.1 The location of the southern cerrados province in the cerrado domain.

The cerrado as a whole is surrounded by very distinct neighbouring biomes: the semi-arid caatinga (to the north-east), the wet pantanal (to the south-west), the moist and hot Amazonian forest (to the north-west) and the somewhat drier and cooler Atlantic forest (to the south-east). The cerrado flora is influenced by all these surrounding ecosystems and characteristic ecotonal vegetation occurs. Elements derived from the neighbouring biomes become non-endemic, so-called accessory species in the cerrado flora and this process has consequently produced contrasting regional variation in its flora (see Méio et al., 2003; and Ratter et al., Chapter 2).

The southern cerrados are bordered by forms of the Atlantic forest, mostly by its seasonal dry semideciduous (inland Atlantic) subtype, but also by mixed ombrophilous (evergreen, with *Araucaria*) forest in Paraná state and at high altitudes in São Paulo and south Minas Gerais states, or dense ombrophilous (evergreen broadleaf) forest in the Paraíba valley, eastern São Paulo state. In a recent study Oliveira-Filho and Fontes (2000) have demonstrated a close association between cerrado and Atlantic dry semideciduous forests, with many non-endemic accessory species in common.

3.3 FLORA OF THE SOUTHERN CERRADOS

For reasons discussed in Ratter et al. (2003 and Chapter 2) floristic inventories of cerrado have usually been restricted to larger woody species (trees and large shrubs) and this is the case for the data given here. A number of authors have given numbers (both recorded and estimated) for the total of such species, and these are summarized in Chapter 2, Table 2.1. We recorded 383 tree/large shrub species in 86 sites surveyed in São Paulo state (Durigan et al., 2003b) and the 54 most frequent (occurring at ≥50% of the sites) are listed in Table 3.1. This reflects the high diversity occurring in the state since Ratter et al. (2003) record a total of 914 species for the whole core Cerrado Biome and its southern outliers, and our figure of 383 species means that more than 40% occur in São Paulo state, although it represents less than 2% of the total cerrado area. All 39 of the commonest large woody species recorded as occurring in ≥50% of sites sampled across the Cerrado Biome (Ratter et al., Chapter 2, Appendix 2) are found in the southern cerrados. However, only 14 of them occur amongst

TABLE 3.1
Woody Species (Trees and Large Shrubs) Recorded in 50% or more of 80 Cerrado Sites in São Paulo State*

Species	Frequency (% of the Sites)
Casearia sylvestris Sw.	90
Byrsonima intermedia A. Juss.	88
Copaifera langsdorffii Desf.	87
***Gochnatia barrosii* Cabrera**	**86**
Tabebuia ochracea (Cham.) Standl.	85
Siparuna guianensis Aubl.	81
Machaerium acutifolium Vogel	80
Platypodium elegans Vogel	79
Roupala montana Aubl.	79
Stryphnodendron obovatum Benth.	78
Miconia albicans (Sw.) Triana	78
Syagrus romanzoffiana (Cham.) Glassman	77
Bauhinia rufa (Bong.) Steud.	76
Tapirira guianensis Aubl.	76
Xylopia aromatica (Lam.) Mart.	74
Solanum paniculatum L.	72
Terminalia glabrescens Mart.	72
Vochysia tucanorum (C.K. Spreng.) Mart.	71
Baccharis dracunculifolia DC.	71
Luehea grandiflora Mart.	70
Acosmium subelegans (Mohl.) Yakovlev	70
Didymopanax vinosum (Cham. & Schltdl.) March.	67
Anadenanthera peregrina (L.) var. *falcata* (Benth.) Altschul	66
***Gochnatia polymorpha* (Less.) Cabrera**	**66**
***Ocotea corymbosa* (Meisn.) Mez**	**66**
Protium heptaphyllum (Aubl.) Marchand	65
Aegiphila lhotskiana Cham.	65
Cecropia pachystachya Trécul	64
Brosimum gaudichaudii Trécul	63
Bredemeyera floribunda Willd.	63
***Matayba elaeagnoides* Radlk.**	**63**
***Tabernaemontana hystrix* (Steud.) DC.**	**62**
Qualea grandiflora Mart.	60
Dimorphandra mollis Benth.	60
Miconia stenostachya DC.	59
Duguetia furfuracea (A. St.-Hil.) Benth. & Hook. f.	59
Myrcia albotomentosa Cambess.	59
Styrax camporum Pohl	59
Zanthoxylum rhoifolium Lam.	58
Caryocar brasiliense Cambess.	58
Tocoyena formosa (Cham. & Schltdl.) K. Schum.	57
Annona coriacea Mart.	57
***Campomanesia adamantium* (Cambess.) O. Berg**	**57**
***Ouratea spectabilis* (Mart.) Engl.**	56

TABLE 3.1

Woody Species (Trees and Large Shrubs) Recorded in 50% or more of 80 Cerrado Sites in São Paulo State* (Continued)

Species	Frequency (% of the Sites)
Acacia polyphylla DC.	55
Eugenia aurata O. Berg	55
Pouteria ramiflora (Mart.) Radlk.	52
Erythroxylum cuneifolium Poepp. Ex O. E. Schulz	52
Lacistema hasslerianum Chodat	52
Qualea multiflora Mart.	51
Diospyros hispida DC.	51
Machaerium brasiliense Vogel	**51**
Rapanea umbellata (Mart. ex DC.) Mez	51
Luehea candicans Mart.	50

*Species thought to be endemic to São Paulo State are in bold.
Derived from Durigan et al., 2003b.

the 54 species recorded at ≥50% of our São Paulo sites (Table 3.1). This difference probably indicates the usual distributional heterogeneity encountered from site to site across the Cerrado Biome. As one would expect, the similarity decreases when the less common species are compared: there is 84% co-occurrence of the 116 species considered the dominant woody oligarchy of the cerrado flora by Ratter et al. (Chapter 2, Appendix 1), and 66% when the 300 species recorded in ≥2.5% of the 316 sites analysed are considered (Chapter 2, Appendix 2).

There are a considerable number of endemic and near-endemic species in the southern cerrados. Some occur in typical cerrado vegetation — the cerrado sensu stricto (e.g. *Campomanesia adamantium* (Cambess.) O. Berg, *Eugenia livida* O. Berg, *Eugenia pyriformis* Cambess. and *Myrcia bella* Cambess.), some are more frequent in cerradão (cerrado woodland) (e.g. *Gochnatia barrosii* Cabrera, *Myrcia venulosa* DC. and *Symplocos pubescens* Klotzsch ex Benth.) and a number also occur in the seasonal semideciduous forest (such as *Gochnatia polymorpha* (Less.) Cabrera, *Maytenus robusta* Reissek, *Ocotea corymbosa* (Meisn.) Mez, *Nectandra cuspidata* Nees & Mart. and *Pera obovata* Baill.).

Even some species indicating richer (mesotrophic) soils (Furley and Ratter, 1988; Ratter et al., 2003) occur in São Paulo state, although the so-called mesotrophic cerradão was never recorded there.

A much larger region of south-eastern Brazil was covered by savanna-type vegetation in the early Holocene than in the late Holocene and, according to Behling (2002), especially after 1000 carbon-14 yr BP, cerrado was replaced by semideciduous forest in this region due to the shorter annual dry season and wetter climatic conditions. This process of cerrado invasion by Atlantic species is probably still occurring and explains the high proportion of non-endemic forest accessory species in the cerrado, especially in the southern region of the cerrado domain. It also explains the occurrence of relict typical cerrado trees in the Atlantic forest community, as also observed by Ratter et al. (1978) and Ratter (1992) in the Amazon forest.

The most widespread woody species in São Paulo state cerrados (occurring in more than 80% of the sites surveyed by Durigan et al. (2003b) and listed in Table 3.1), which can be considered characteristic of the southern cerrados, are in decreasing order: *Casearia sylvestris*, *Byrsonima intermedia, Copaifera langsdorffii, Gochnatia barrosii, Tabebuia ochracea* and *Siparuna guianensis*. Except for *Byrsonima intermedia* and *Gochnatia barrosii*, which are cerrado endemics ('peculiares' and 'campestres', according to Rizzini, 1963, 1971 and see Chapter 2), the others also occur in the Atlantic forest ('acessórias' or 'silvestres', according to Rizzini, 1963, 1971) and other vegetation formations; in fact they are 'supertramps' which occur in many vegetation types (see Chapter 2).

As recognized by various workers (Castro, 1994a,b; Castro and Martins, 1999; Durigan et al., 2003b; Ratter et al., 2003), the São Paulo cerrados are very distinct. In the multivariate analyses of Ratter et al. (2003), the southern cerrados consisting of all the São Paulo sites together with one from Paraná and three from the very south of Minas Gerais, split off as a natural group by TWINSPAN and were classed together as a similar integrated group by UPGMA (Unweighted Pair-Groups Method using Arithmetic averages) using the Sørensen Coefficient of Community (c.c.) as a measure of similarity (Kent and Coker, 1992). Durigan et al. (2003b) in a major study comparing 202 sites from the southern states by three methods of multivariate analysis (TWINSPAN, DCA and UPGMA) recognized two distinct geographical floristic groups in São Paulo correlated with physiognomy and environmental conditions: the almost endemic 'Paulista' including cerradão sites from the west of the state, and the 'Interestadual' (Interstate) which extends into the state of Paraná, and the south of Minas Gerais and Mato Grosso do Sul. These two groups are mentioned in a number of places in this chapter.

Bridgewater et al. (2004) compared the floristic relationships of six phytogeographic provinces identified by Ratter et al. (2003) using the Sørensen Index for comparison. This work demonstrated, not unexpectedly, that the southern cerrados were mostly closely related to their neighbour to the north, the central and south-eastern group (central-west Minas Gerais state, east of Goiás state and the Federal District) with a Sørensen similarity index of 0.544. The second most similar group was the central-western (covering a huge area of central and west Goiás, south-western Tocantins, most of Mato Grosso do Sul, the cerrado area of Mato Grosso and extending into Rondônia and Pará) with a Sørensen similarity index of 0.466.

3.4 FACTORS DETERMINING VEGETATION STRUCTURE (PHYSIOGNOMY) AND FLORISTIC COMPOSITION

Durigan et al. (2003a,b), analysed the distribution of species in São Paulo and neighbouring states and identified geographical and environmental features as the most important factors correlated with floristic patterns. Floristic similarity was highest among sites located in the same watershed, and was related across the region to the duration of the dry season, the temperature and the vegetational physiognomy present, the last of which is closely related to soil chemical and physical properties.

In the southern cerrados, areas covered by two major physiognomies can be recognized, each with its own distinct floristic composition: 1) cerradão, in the west of the region (associated with more fertile soils) and 2) cerrado sensu stricto in the eastern part of the region (associated with very poor, sandy soils).

Recent studies in São Paulo state (Toppa, 2004) have demonstrated a strong correlation between cerrado biomass, floristic composition and soil-water capacity (sand/clay proportion), the last of which was shown to be a more important factor than fertility, acidity or aluminium saturation of the soil. This is a revival of the ideas of more than a hundred years ago (Warming, 1892), when water availability was considered to be the main constraint to the occurrence of forest vegetation in the cerrado domain.

More recently, fire frequency and soil chemical properties have been considered the main factors conditioning cerrado distribution, structure and floristic composition, water availability being disregarded, since forests and cerrado can occur under the same annual rainfall. However, in a region where the dry season can last for up to five months, the soil-water capacity, depending on the proportion of sand and clay, plays an important role in determining water availability. In addition to the chemical differences in the soil, the variation in soil-water capacity may explain the existing mosaic of distinct physiognomies of cerrado vegetation and forest patches under the same homogeneous weather conditions.

3.5 FROST AS A FACTOR IN THE SOUTHERN CERRADOS

Cold weather, especially the occurrence of frosts, has been reported as an important causative factor of the distinct floristic pattern in the southern cerrados or, sometimes, as the factor limiting the occurrence of cerrado vegetation south of its present limits. Eiten (1972, 1990) asserts that cerrado only occurs where frosts are absent or extremely rare, giving way to grasslands under cooler conditions with more frequent frosts. I consider that infrequent frosts do not cause important changes in the structure of cerrado vegetation, although they are associated with some differences in species composition. Other authors also regard frosts as possibly important factors influencing cerrado species distribution (Silberbauer-Gottsberger et al., 1977; Filgueiras, 1989; Toledo Filho, 1984; Vuono et al., 1982; Ribeiro and Walter, 1998; Castro and Martins, 1999).

Brando and Durigan (2004), monitoring the changes in cerrado vegetation after frosts in São Paulo state, concluded that cerrado species are capable of occupying and colonizing frost-susceptible sites. They observed that a very severe frost did not eliminate even a single species from the community, and although the structure of the vegetation changed for a short time it recovered within a year. However, frost tolerance is extremely variable among species, and in some of the more susceptible (e.g. *Caryocar brasiliense*, *Xylopia aromatica*, and *Vochysia tucanorum*) reproductive processes can be interrupted, thus affecting community dynamics. According to these authors, the phytogeographical patterns of the southern cerrado area are influenced by frosts, favouring occurrence of typical frost-resistant species (e.g. *Campomanesia adamantium*, *Erythroxylum cuneifolium*, *Gochnatia polymorpha*, *Machaerium brasiliense* and *Myrcia multiflora*). They also consider that sites where frequent frosts occur maintain more open forms of cerrado vegetation, even if both water and nutrient availability could support denser vegetation.

3.6 THE PHYSIOGNOMIES OF THE SOUTHERN CERRADOS

The most widespread physiognomy in the southern cerrados is at present the cerradão, which has a continuous or near-continuous arboreal stratum and high biomass, being essentially a dry forest (termed Floresta Xeromorfa by Rizzini, 1963). It was recorded as the dominant vegetation in 70% of the cerrado sites surveyed by Durigan et al. (2003a) in São Paulo state.

This is in sharp contrast to the situation in the cerrado core area and it was not the case 40 years ago, when Chiarini and Coelho (1969) mapped the cerrado vegetation of the state. These authors surveyed the vegetation of cerrado sensu lato using aerial photographs taken in 1962 and determined the area covered by each physiognomy. At that time, cerrado sensu stricto, with a fair amount of structural variation (including low arboreal, dense scrub, and savanna forms) was the most extensive vegetation type (75%), followed by 16% of the more open campo cerrado, and only 9% of cerradão. Individual areas covered by sparse forms of cerrado in 1962 had become dense cerradão when resurveyed in 2000 (Durigan et al., 2003a). Such rapid closing of more open forms of cerrado to produce dense cerradão are generally associated with protection from fire and have been described in three localities in São Paulo state, Assis (Durigan et al., 1987), Emas (Goodland and Ferri, 1979) and Angatuba (Ratter et al., 1988), and also in the Federal District (Ratter, 1992). This indicates a seral tendency towards cerradão for most of the southern cerrados, except in some areas where soils are so poor or perhaps water availability so low that the open cerrado represents the maximum of biomass production that can be supported, or where there is still pressure of fire.

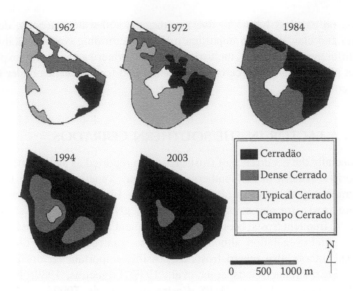

FIGURE 3.2 Changes in cerrado physiognomy during 41 years, observed in 180 ha of cerrado vegetation protected from fire at Assis State Forest, São Paulo State, Brazil.

Figure 3.2 maps the closing of sparser forms of cerrado to form cerradão on a 180-ha protected area at Assis, SP, over 41 years.

There is good evidence that before the advent of frequent man-made fires the denser forest forms of cerrado (i.e. cerradão) were more common and occupied a much larger area than they do today, representing in fact the dominant climax of cerrado vegetation (Warming, 1892; Eiten, 1972; Rizzini, 1979; Furley and Ratter, 1988; Ratter et al., Chapter 2). Surveys in São Paulo state in the nineteenth century (Löfgren, 1890) showed large areas of cerrado, and especially cerradão all over the state. Protection in reserves is thus now recreating the climax form of cerrado sensu lato once more, after it had previously declined to such an extent due to anthropic pressure that it was considered in many areas an exceptional rarity.

The suppression of fires in western São Paulo state in recent decades is probably the main causative factor for the recent changes in cerrado vegetation. No evidence of recent fires at all has been recorded by Durigan et al. (2004) in 79% of 86 cerrado remnants surveyed in São Paulo state. The change in physiognomy is being accompanied by one in floristic composition. Some typical heliophyte cerrado species are disappearing while cerradão and forest species have increased their populations in transitional zones, succeeding and engulfing the patches of cerrado sensu stricto.

3.7 THE IMPORTANCE OF THE SOUTHERN CERRADOS

The Cerrado Biome has a diverse flora, both at local and regional levels, and species composition usually varies greatly between areas (Ratter and Dargie, 1992; Ratter et al., 1996, 2003), giving high alpha-, beta- and gamma-diversities. Alpha-diversity of trees and large shrubs may be low, less than ten species per hectare in some isolated Amazonian sites, but it can often be high in the core area and southern outliers with more than 100 species of trees and shrubs growing together. Some of the richest sites are located in São Paulo state, considered by Ratter et al. (2003) as a particular diversity hotspot, together with some central-western areas.

The southern cerrados, especially in São Paulo state, have also been reported as one of the three biodiversity supercenters in the biome by Castro and Martins (1999), with the others also occurring in transition zones between cerrado and forest vegetation.

According to recent predictions of species extinction risk from climate change (Thomas et al., 2004), the rich, colder and wetter cerrado areas in the south may be the last refuges for cerrado species in the future and must receive priority for cerrado conservation strategy at a national scale. São Paulo state could be the best place in a future warmer and drier climatic scenario for *in situ* conservation of cerrado, and could even be used for *ex situ* conservation by transplanting species that are today's northern endemics and under warmer conditions may not be able to survive there. Conservation strategies must be focused on the southern cerrados and their transition areas with the Atlantic forest in an attempt to rectify the present extreme fragmentation of the existing vegetation remnants and provide reserves of a genetically viable size.

3.8   DEFORESTATION AND CONSERVATION

Accurate data for cerrado deforestation are available for São Paulo state, but are more difficult to access for the neighbouring areas in the north of Paraná and the south of Minas Gerais included in the southern cerrados (Ratter et al., 2003). The following account is therefore based entirely on São Paulo, but it can probably be assumed that the situation is very similar in the relatively small areas outside the state.

There still remained a little more than 2100 km^2 of the original cerrado cover of São Paulo state in 2001 (the latest year for which data are available), comprised of thousands of fragments of which only 42 were larger than 400 ha and none reached 10,000 ha (Kronka et al., 2005). This remnant was the result of deforestation of 88.3% of the cerrado cover existing in the state in 1962, which in four decades has been converted principally to the following uses (in descending order of importance): sugar cane, pastures, arable crops, citrus and plantation forestry (principally eucalypts and pines) (Kronka et al., 2005). Only about 0.5% of the original cerrado area is now protected in conservation units and all the scattered fragments suffer the problems of isolation, fire-risk, and invasion by alien species (about 70% of them have been invaded by African pasture grasses, according to Durigan et al., 2004).

In response to the rapid deforestation of the São Paulo cerrado a workshop was held in 1995 to promote conservation and sustainable use. This was the Bases para a Conservação e Uso Sustentável das Áreas de Cerrado do Estado de São Paulo, organized by the State Environmental Secretariat, which resulted in the recommendation of priority conservation areas and identification of the principal threats to the biome. However, in the decade since then not a single new unit has been created to protect the cerrado in the areas indicated, although some existing conservation units have been enlarged. There have also been no changes in legislation to stop or reduce the rate of deforestation. Thus, it is probable that in the future the São Paulo cerrados will occupy only the established protected areas (c.0.5% of the original area) and the 20% legal reserves (20% of agricultural land that should legally be maintained under natural vegetation; see Ratter et al., Chapter 2). Judging from other states, many of the latter may be in poor condition and far below their minimum stipulated area. One line of salvation might be that, in theory, Brazilian law obliges restoration of deforested legal reserves (i.e. owners have to reinstate native vegetation to bring their properties up to the 20% requirement). However, although the techniques of replanting native forest have been developed, less is known about those necessary for the recuperation of the cerrado and therefore owners may not attempt such an enterprise. Only a very small protected area is likely to represent the southern cerrados in the future: one hopes that it will be sufficient to maintain the gene pool and characteristics of this unique biodiversity hotspot.

ACKNOWLEDGEMENTS

I wish to thank the invitation by the organizers of the International Conference on Tropical Savannas and Seasonally Dry Forests, Edinburgh, 2003, where the contents of this chapter were first presented, and to the Fundação de Amparo à Pesquisa do Estado de São Paulo (FAPESP), for supporting my participation in the meeting.

REFERENCES

Adámoli, J. et al., Caracterização da região dos cerrados, in *Solos dos cerrados: tecnologias e estratégias de manejo*, Goedert, W. J. Ed, Nobel, São Paulo, 1987, 33.

Behling, H., A high resolution Holocene pollen record from Lago do Pires, SE Brazil: vegetation, climate and fire history, *J. Paleolimnol.*, 14, 253, 1995.

Behling, H., Late Quaternary vegetational and climatic changes in Brazil, *Review of Palaeobotany and Palynology*, 99(2), 143, 1998.

Behling, H., South and southeast Brazilian grasslands during Late Quaternary times: a synthesis. *Palaeogeogr. Palaeoclimatol. Palaeoecol*, 177, 19, 2002.

Behling, H. and Lichte, M., Evidence of dry and cold climatic conditions at glacial times in tropical southeastern Brazil, *Quat. Res.* 48, 348, 1997.

Brando, P.M. and Durigan, G., Changes in cerrado vegetation after disturbance by frost (São Paulo State, Brazil), *Plant Ecology*, 175, 205, 2004.

Bridgewater, S., Ratter, J.A., and Ribeiro, J.F., Biogeographic patterns, ß-diversity and dominance in the Cerrado Biome of Brazil, *Biodiversity and Conservation*, 13, 2295, 2004.

Camargo, A. P., Clima do Cerrado, in Simpósio sobre o cerrado, Ferri, M.G., Ed, Editora. Univ. São Paulo, São Paulo, Brazil, 1963, 94.

Castro, A.A.J.F., *Comparação florístico geográfica (Brasil) e fitossociológica (Piauí – São Paulo) de amostras de cerrado*, PhD thesis, Universidade Estadual de Campinas, São Paulo, Brazil. 1994a.

Castro, A.A.J.F., Comparação florística de espécies do cerrado. *Silvicultura* 15(58), 16, 1994b.

Castro, A.A.J.F. and Martins, F.R., Cerrados do Brasil e do nordeste: caracterização, área de ocupação e considerações sobre a sua fitodiversidade, *Pesquisa em Foco* 7(9), 147, 1999.

Chiarini, J.V. and Coelho, A.G.S., *Cobertura vegetal natural e áreas reflorestadas do Estado de São Paulo. Secretaria da Agricultura do Estado de São Paulo*, Instituto Agronômico (Boletim 193), Campinas, Brazil, 1969.

Dias, B.F.S., Cerrados: uma caracterização, in *Alternativas de desenvolvimento do cerrado: manejo e conservação dos recursos naturais renováveis*, Dias, B.F.S, Coord., FUNATURA — IBAMA, Brasília, Brazil, 1992, 11.

Durigan, G. et al., Fitossociologia e evolução da densidade da vegetação de cerrado em Assis, SP, *Bol. Técn. Inst. Flor.* 41, 59, 1987.

Durigan, G. et al., Padrões fitogeográficos do cerrado paulista sob uma perspectiva regional, *Hoehnea*, 30, 39, 2003a.

Durigan, G. et al., The vegetation of priority areas for cerrado conservation in São Paulo State, Brazil, *Edinburgh J. Bot.*, 60, 217, 2003b.

Durigan, G. et al., A vegetação dos remanescentes de cerrado no Estado de São Paulo, in *Viabilidade da conservação do cerrado no Estado de São Paulo*, Bitencourt, M. D. and Mendonça, R., Eds, Anablume, São Paulo, 2004, 29.

Eiten, G., The cerrado vegetation of Brazil, *Botanical Reviews*, 38, 201, 1972.

Eiten, G., Vegetação do cerrado, in *Cerrado caracterização e perspectivas*, 2nd ed., Pinto, M.N., coord., UnB/SEMATEC, Brasília, Brazil, 1990, 17.

Filgueiras, T.S., Efeito de uma geada sobre a flora do cerrado na Reserva Ecológica do IBGE, DF, Brasil, *Cadernos de Geociências*, 2, 67, 1989.

Furley, P.A. and Ratter, J.A., Soil resources and plant communities of the central Brazilian cerrado and their development, *J. Biogeogr.*, 15, 97, 1988.

Goodland, R. and Ferri, M.G., *Ecologia do cerrado*, Editora Itatiaia, Belo Horizonte and EDUSP, São Paulo, 1979.

Kent, M. and Coker, P., *Vegetation description and analysis: a practical approach*, Belhaven Press, London, 1992.

Kronka, J.F. et al., *Inventário Florestal da vegetação natural do Estado de São Paulo*, Secretaria do Meio Ambiente/Instituto Florestal/Imprensa Oficial, São Paulo, Brazil, 2005, 33.

Ledru, M. P., Late Quaternary environmental and climatic changes in central Brazil, *Quat. Res.* 39, 90, 1993.

Ledru, M. P. et al., The last 50,000 years in the Neotropics (southern Brazil): evolution of vegetation and climate, *Palaeogeogr. Palaeoclimatol. Palaeoecol.* 123, 239, 1996.

Löfgren, A., *Contribuições para a botânica paulista. Região campestre*, Bol. Comiss. Geogr. Estado de São Paulo, 5, São Paulo, 1890.

Lopes, A.S., *Solos sob o cerrado: características, propriedades, manejo*, 2nd ed., Potafos, Piracicaba, 1984.

Méio, B.B. et al., Influência da flora das florestas Amazônica e Atlântica na vegetação do cerrado *sensu stricto*, *Revista Brasil. Bot.* 26, 437, 2003.

Oliveira-Filho, A.T. and Fontes, M.A., Patterns of floristic differentiation among Atlantic forests in southeastern Brazil and the influence of climate. *Biotropica*, 32, 793, 2000.

Ratter, J.A., Transitions between cerrado and forest vegetation in Brazil, in *Nature and dynamics of forest – savanna boundaries*, Furley, P.A., Proctor, J., and Ratter, J.A., Eds, Chapman and Hall, London, 1992, 417.

Ratter, J.A. and Dargie, T.C.D., An analysis of the floristic composition of 26 cerrado areas in Brazil, *Edinburgh J. Bot.* 49, 235, 1992.

Ratter, J.A., Bridgewater, S., and Ribeiro, J.F., Analysis of the floristic composition of the Brazilian cerrado vegetation. III. Comparison of the woody vegetation of 376 areas, *Edinburgh J. Bot.* 60, 57, 2003.

Ratter, J.A., Ribeiro, J.F., and Bridgewater, S., The Brazilian cerrado vegetation and threats to its biodiversity, *Ann. Bot. (Oxford)* 80, 223, 1997.

Ratter, J.A. et al., Observations on the vegetation of northeastern Mato Grosso II. Forests and soils of the Rio Suiá-Missu area. *Proc. R. Soc. Lond.* 203, 191, 1978.

Ratter, J.A. et al., Floristic composition and community structure of a southern cerrado area in Brazil, *Notes Roy. Bot. Gard. Edinburgh* 45, 137, 1988.

Ratter, J.A. et al., Analysis of the floristic composition of the Brazilian cerrado vegetation. II. Comparison of the woody vegetation of 98 areas, *Edinburgh J. Bot.*, 53, 153, 1996.

Reatto, A., Correia, J.R., and Spera, S.T., Solos do bioma cerrado: aspectos pedológicos, in *Cerrado: ambiente e flora*, Sano, S.M. and Almeida, S.P., Eds, EMBRAPA-CPAC, Planaltina, Brazil, 1998, 47.

Ribeiro, J.F. and Walter, B.M.T., Fitofisionomias do bioma cerrado, in *Cerrado: ambiente e flora*, Sano, S.M. and Almeida, S.P., Eds, EMBRAPA-CPAC, Planaltina, Brazil, 1998, 89.

Rizzini, C. T., A flora do cerrado — análise florística das savanas centrais, in *Simpósio sobre o cerrado*, Ferri, M.G., Ed, Editora Univ. São Paulo, São Paulo, Brazil, 1963, 125.

Rizzini, C.T., Árvores e arbustos do cerrado, *Rodriguesia*, 38, 63, 1971.

Rizzini, C.T., *Tratado de Fitogeografia do Brasil*, 2°. *volume – aspectos sociológicos e florísticos*, São Paulo, Brazil, Editora Univ. São Paulo, 1979.

Salgado-Labouriau, M.L. et al., Late Quaternary vegetational and climatic changes in cerrado and palm swamp from central Brazil, *Palaeogeogr. Palaeoclimatol. Palaeoecol*, 128, 215, 1997.

Silberbauer-Gottsberger, I., Morawetz, W., and Gottsberger, G., Frost damage of cerrado plants in Botucatu, Brazil, as related to the geographical distribution of species, *Biotropica* 9, 253, 1977.

Thomas, C.D. et al., Extinction risk from climate change, *Nature*, 427, 145, 2004.

Toledo Filho, D.V., *Composição florística e estrutura fitossociológica da vegetação do cerrado no município de Luís Antonio (SP)*, MSc Dissertation, Campinas, Instituto de Biologia da Universidade Estadual de Campinas, Brazil. 1984, 24.

Toppa, R.H., *Estrutura e diversidade florística das diferentes fisionomias de cerrado e suas correlações com o solo na Estação Ecológica de Jataí, Luiz Antônio, SP*, PhD thesis, Departamento de Ecologia e Recursos Naturais, Universidade Federal de São Carlos, 2004.

Vuono, Y.S. et al., Efeitos biológicos da geada na vegetação do cerrado, in Congresso Nacional Sobre Essências Nativas, 1, Campos do Jordão, *Anais Instituto Florestal: Silvicultura em São Paulo*, 16A(1), 545, 1982.

Warming, E., Lagoa Santa: Et bidrag til den biologiska plantegeografi, *K. Dansk. Vidensk. Selsk. Skr.* 6, 153, 1892.

4 Phytogeography of Cerrado Sensu Stricto and Land System Zoning in Central Brazil

Jeanine Maria Felfili, Maria Cristina Felfili, Christopher William Fagg, Alba Valéria Rezende, Paulo Ernane Nogueira and Manoel Cláudio da Silva Júnior

CONTENTS

ABSTRACT

The dominant vegetation of the Cerrado Biome of central Brazil is composed of mosaics of savanna woodland and grasslands. This study deals with only one of the main savanna woodland physiognomies, cerrado sensu stricto. Floristic and structural patterns of the woody flora of cerrado s.s. are studied over an area of 10° latitude and 8° longitude. Data from 220 (20 × 50 m) plots containing 22,306 stems of woody plants ≥5 cm diameter at 30 cm above ground level were compared to determine phytogeographical patterns. Two main phytogeographical units were detected, the highlands in the central area, and the lowlands in the north-north-eastern borders of the biome. β-diversity was high especially due to the variation in density of species between sites. The floristic and structural variations fitted well with the pattern of land system zoning of the Cerrado Biome based on variation in climate, landscape and soil types. Sites grouped in the same physiographical unit as defined by Cochrane et al. (1985) showed a higher similarity in the presence of preferential species related to altitude, latitude and soil types.

4.1 INTRODUCTION

Phytogeographical patterns within the same biome are especially related to physical factors such as soils, relief and topography. These factors were used in a land system zoning of central Brazil by Cochrane et al. (1985) who identified a total of 70 land systems within 25 physiographical units in the Cerrado Biome. They identified, characterized and mapped land systems defined as an area, or a group of areas, with a recurrent pattern of climate, landscape, soils and vegetation physiognomies. Land systems were grouped into regional units based on dominant landform and vegetation called physiographical units. Several different physiognomies (structural forms) of savanna (Eiten, 1972), and forest formations (Felfili and Silva Júnior, 1992) can be related to physical features within the biome (Haridasan et al., 1997; Haridasan, 2001). A wide variation in floristic composition within the same physiognomy is also related to environmental conditions (Felfili et al., 1994, 1997, 2004b; Filgueiras et al., 1998), suggesting the need for conservation and management planning based on phytogeography and the analysis of diversity. It is necessary to consider the recurrent patterns of disjunct landscapes in this vast biome covering over 20° of latitude where sites far from each other can be more environmentally similar than sites nearby (Felfili et al., 2004a).

The Cerrado Biome of two million km^2 occupies central Brazil almost entirely and has as adjacent neighbours the Amazonian and Atlantic humid forests, the dry forests and scrub of the north-eastern caatinga and the subtropical vegetation of southern Brazil. Altitude varies from sea level to over 1000 m, precipitation from 800 mm to 1800 mm, and soils form a mosaic composed of 21 major types dominated by latossols (40%) and sandy quartz soils (15.2%) (Reatto et al., 1998). The cerrado is amongst the richest and most threatened of the world's ecosystems, identified as a biodiversity hot spot (Mittermeier et al. 1999), based on species richness and endemism. Several sources suggest that the biome has already lost about 70% of its natural vegetation (Mittermeier et al., 1999; UNESCO, 2000; and Ratter et al., Chapter 2) and we are close to the eleventh hour for planning its conservation and sustainable management.

Floristic diversity is very high, with 6429 vascular species recorded for the biome (Mendonça et al., 1998) and various higher estimates given in the literature (see Chapter 2, Table 2.1). Endemism is high, estimated at c.44% of the vascular flora by Silva and Bates (2002). Of the 6,429 species listed by Mendonça et al. (1998), c.2900 occur in the cerrado sensu lato and of those c.800 are larger woody species. Cerrado sensu lato is a savanna formation ranging from open grassland through forms with varying tree cover to quite dense xeromorphic woodland (Eiten, 1972, and see Ratter et al., Chapter 2, for classification of cerrado structural forms). This vegetation formerly covered most of central Brazil and is easily cleared by mechanized agriculture for soya bean and other grain production, and for planted pastures.

The flora is rich and diverse and varies in a mosaic pattern of different communities (Castro, 1994; Felfili et al., 1994, 2004a; Durigan et al., 2003; Ratter et al., 2003). Altitude and soil types have been suggested as the main determinants of the floristic composition and structure of the cerrado (Felfili and Silva Júnior, 1993; Felfili et al., 1994, 1997, 2004a). The connection with the large hydrographical basins of Araguaia-Tocantins (belonging to the Amazonian realm), São Francisco and Paraná rivers also appear to influence the phytogeographical patterns of the cerrado (Felfili, 1998; see also Durigan, Chapter 3), as has been demonstrated for the gallery forests of the region (Oliveira-Filho and Ratter, 1995).

The exceptionally fast rate of destruction of the Cerrado Biome means that botanical data to be applied in conservation have to be obtained rapidly, and the only practical way to do that is to survey the woody vegetation (Ratter et al., 2000, and Chapter 2). Fortunately results obtained by Filgueiras et al. (1998) indicate a similarity in patterns of β-diversity of ground (herbaceous and low shrub) and woody vegetation, so hopefully data from the woody vegetation will relate to the whole vascular flora.

Our approach has been to study and compare patterns of the α and β diversity of the woody flora over a large area containing several land systems. To detect diversity and abundance patterns over five physiographical units in central Brazil we have carried out widespread surveys in one particular physiognomy, the cerrado sensu stricto (s.s.), a particularly abundant and characteristic vegetation from with tree cover varying from 10 to 60%.

4.2 METHODS

4.2.1 SITES, SAMPLING AND ENVIRONMENTAL FEATURES

A total of 22 selected sites in five physiographical units (Cochrane et al., 1985) were chosen for the comparison of woody cerrado sensu stricto. These five physiographical units include 11 land systems, as seen in Table 4.1.

The study sites were selected over an area of 10° of latitude and 8° of longitude. Six sites were in the Pratinha physiographical unit (Felfili et al., 1994), five in the Veadeiros (Felfili et al., 1997), four in São Francisco (Felfili and Silva Júnior, 2001), three in the Complex Xavantina (Felfili et al., 2002) and two in the Paranã valley.

All sites are classified as having Aw climate according to Köppen's system. Location and other physical characteristics (Cochrane et al., 1985) are given in Figure 4.1 and Table 4.1.

4.2.2 DATA COLLECTION AND ANALYSES

The study sites were chosen at the extreme and central parts of each physiographical unit and also to cover the recognized land systems occurring within each unit. Maps, satellite images, aerial photos and preliminary field surveys were used to select sites still covered by native vegetation without evidence of disturbance.

In each physiographical unit most study sites were 100 km or more distant from each other, but in Pratinha, the first physiographical unit to be assessed, three sites in conservation areas within the Federal District were only 50 km apart. The overall distance between the plots at Patrocínio-MG in Pratinha and those in the Xavantina complex was over 1500 km, showing the size of the total area studied.

The data were obtained by sampling 10 (20 × 50 m) plots randomly placed at each study site. No less than 220 of these 1000 m^2 plots were surveyed covering a total area of 22 ha and containing 22,306 stems reaching qualifying size. All woody stems ≥5 cm diameter at 30 cm above ground level were measured with a caliper, and if the stem had an irregular shape, two measures, at right angles to each other, were taken and an average used. Trees were identified to species and their heights were measured. Lianas were not included, since they are rare in the cerrado and do not reach the minimum diameter for sampling. The definition of individual plants is sometimes difficult in cerrado since many species can have vegetative reproduction from the root system. However, most aerial stems stand alone, but in the few cases when multiple or bifurcated stems were found they were considered as a single individual for calculation of density and summed together for calculation of basal area (in any case, the normal procedure for calculating basal area).

Voucher specimens were collected and were deposited in the IBGE herbarium, Brasília. Species lists were published in Felfili et al. (1994), Mendonça et al. (1998) and Felfili and Silva Júnior (2001).

Community composition and phytosociology were evaluated at each site. Shannon's diversity index and Pielou's evenness index were used to evaluate α-diversity, which is related to the number of species and the distribution of individuals per species in a community. β-diversity, related to the differences in species composition and abundance between sites (Kent and Coker, 1992), was evaluated by similarity indices and also by multivariate analyses. Density, basal area averages, and standard deviation at 95% probability level per site were calculated to evaluate the community structure.

Shannon's diversity index (nats./individual) varies from zero to positive values depending on the chosen logarithmic scale but rarely surpasses five. The less even the distribution of the number of individuals per species, the closer to zero becomes the Pielou evenness index (Margurran, 1988). The Sørensen similarity index, based on presence-absence of species, and the Czekanowski index,

TABLE 4.1
Latitude (S), Longitude (W), Altitude (m), Mean Monthly Precipitation (mm) and Soil Classes at the Sites Studied in Central Brazil

Physiographical Unit/Site/Land System (LS) (Cochrane et al. 1985)	Latitude (S)	Longitude (W)	Altitude (m)	Average Precipitation (mm)	Soils
Chapada Pratinha					
EE de Águas Emendadas (EEAE)-DF (ls 1 = Pratinha Highlands)	15°31′–15° 35′S	47°32′–47°37′	1100	1552	
APA Gama-Cabeça de Veado (APA)-DF (ls 1 = Pratinha Highlands)	15°52′–15°59′	47°50′–47°58′	1100	1552	
PARNA Brasília Distrito Federal — PNB-DF (ls 1 = Pratinha Highlands)	15°37′–15°45′	47°54′–47°59′	1100	1552	Latossols
Silvânia-GO (ls 1 = Pratinha Highlands)	16°30′–16°50′	48°30′–48°46′	1050	1552	
Paracatu-MG (ls 2 = Eroded Surface Pratinha Lowlands)	17°00′–17°20′	46°45′–47°07′	900	1438	
Patrocínio-Ibiá-Pratinha-MG (ls 2 = Eroded Surface Pratinha Lowlands)	18°47′–19°45′	46°25′–47°09′	950	1438	
Chapada dos Veadeiros					
Vila Propício-GO. (ls 1 = Tocantins Highlands - Bc 16A)	15°16′–15°26′	48°40′–49°04′	750–1100	1200–1800	
Alto Paraíso de Goiás-GO. (ls 1 = Tocantins Highlands - Bc 16A)	14°–14°10′	47°20′–47°58′	1200	1500–1750	
PARNA Chapada dos Veadeiros-GO. (ls 2 = Tocantins High peneplains)	13°50′–14°12′	47°24′–47°58′	620–1650	1500	Latossols, Sandy Quartz, Cambissols
Serra da Mesa –Minaçu-GO. (ls 3 = Tocantins Highlands - Bc 18A)	13°34′–13°50′	48°10′–48°22′	450–1110	1500	
Serra Negra - Minaçu-GO. (ls 3 = Tocantins Highlands - Bc 18A)	13°58′–14°05′	48°10′–48°24′	450–1110	1500	
Vale do Paranã					
Alvorada do Norte-Simolândia- Buritinópolis-GO. (ls 18 = Nativity Highlands)	14°23′–14°31′	46°27′–46°34′	450–750	1300	
Iaciara-Guarani de Goiás-Posse-GO. (ls 110 = Eroded Espigão Mestre Sandy Plateaux)	14°01′–14°15′	46°05′–46°11′	650	1300	Latossols, Sandy Quartz, Cambissols
Campos Belos-São Domingos-GO. (ls 110 = Eroded Espigão Mestre Sandy Plateaux)	13°02′–13°35′	46°19′–46°38′	650	1300	

(continued)

TABLE 4.1
Latitude (S), Longitude (W), Altitude (m), Mean Monthly Precipitation (mm) and Soil Classes at the Sites Studied in Central Brazil (Continued)

Physiographical Unit/Site/Land System (LS) (Cochrane et al. 1985)	Latitude (S)	Longitude (W)	Altitude (m)	Average Precipitation (mm)	Soils
Damianópolis-Mambaí (ls 110 = Eroded Espigão Mestre Sandy Plateaux)	14°26–14°36′	46°05–46°14′	650	1300	
Espigão Mestre do São Francisco					
Correntina – BA (ls 1 = Espigão Mestre Sandy Plateaux)	13°31′–13°32′	45°22–45°25′	586	1086	
São Desidério – BA (ls 1 = Espigão Mestre Sandy Plateaux)	12°35′–12°46′	45°34′–45°48′	695–775	1121	Latossols, Sandy Quartz
PARNA Grande Sertão Veredas - Formoso-MG. (ls 1 = Espigão Mestre Sandy Plateaux)	15°10′–15° 21′	45°45′–46°00′	635–850	1185	
Formosa do Rio Preto – BA (ls 1 = Espigão Mestre Sandy Plateaux)	11°06′–11°12′	45°18′–45°35′	550	1006	
Xavantina Complex					
Nova Xavantina-MT. (ls 33 = Xavantina Complex)	14°45′–15°45′	52°00′–52°15′	450–500	1600	
Canarana – MT. (ls 28 = Xavantina Complex)	13°15′–14°00′	51°50′–53°10′	375–400	1600	Latossols, Cambissols
Água Boa – MT. (ls 31 = Xavantina Complex)	13°50′–14°30′	52°0′–52°45′	450–500	1600	

also known as percent similarity index, and based on the number of individuals per species (Kent and Coker, 1992), were used for the comparison between sites.

The complete data-set was also classified by TWINSPAN (Kent and Coker, 1992). The variable density (number of trees) of species per hectare for each site was used for the construction of the matrix for the multivariate analyses with cut levels of 0, 2, 5 and 10 for TWINSPAN classification.

4.3 RESULTS AND DISCUSSION

4.3.1 SPECIES RICHNESS AND ALPHA-DIVERSITY

Richness varied from 18 to 35 species per 1000 m² with the broadest confidence interval around 20% (Table 4.2), suggesting that these values are representative of the species richness per area for cerrado sensu stricto in central Brazil.

Shannon's indices (H′) per 1000 m² varied from 2.2 to 3.2 nats./ind with most values around 3 and Pielou's (J) varied from 0.77 to 0.90 (Table 4.2) indicating that in this large region of cerrado sensu stricto α-diversity is high, and shows a consistent pattern, which should be considered in conservation and management planning.

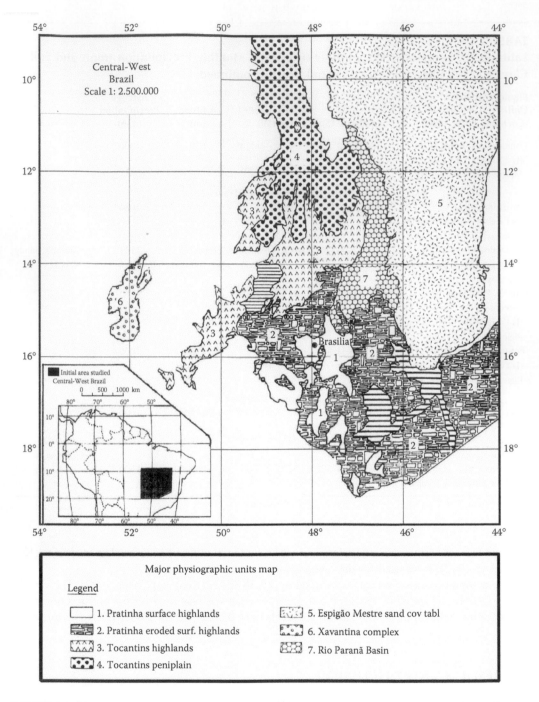

FIGURE 4.1 Physiographical units sampled in this study according to Cochrane et al. (1985). 1, 2. Chapada Pratinha; 3, 4, Chapada dos Veadeiros; 5, Espigão Mestre; 6, Xavantina Complex; 7, Rio Paranã Basin.

Average values per hectare for 15 of the 22 sites studied (Felfili and Silva Júnior, 2001, Felfili et al., 2004a) showed that most sites had around 60 woody species per hectare, varying from 55 to 97, and a Shannon diversity index from 3.04 to 3.73 with most values around 3.5 nats./ind, a range similar to those found in tropical humid forests (Felfili, 1995; Silva Júnior et al., 2001), which places the cerrado sensu stricto as one of the richest savanna physiognomies worldwide.

TABLE 4.2
Species Richness and Diversity (Average and Confidence Intervals (CI)
for the Woody Flora Including All Individuals ≥5 cm Diameter at 30 cm
Above Ground Level, Based on Samples of 10 Units of 1000 m²

Study Sites	Richness AV.	CI	Shannon's Div. Index (Nats/Ind) AV.	CI	Pielou's Evenness Index
Chapada Pratinha					
E.E Águas Emendadas–DF	33	1.98	3.1	0.05	0.88
APA Gama-Cabeça de Veado–DF	35	3.06	3.2	0.08	0.90
PARNA Brasília–DF	24	2.81	2.8	0.10	0.89
Silvânia-GO	29	2.47	2.8	0.10	0.83
Paracatu-MG	19	2.84	2.5	0.18	0.83
Patrocínio-Ibiá-Pratinha-MG	21	2.05	2.6	0.11	0.84
Chapada dos Veadeiros					
Vila Propício-GO.	28	4.34	2.9	0.21	0.87
Alto Paraíso de Goiás-GO.	18	2.91	2.2	0.31	0.77
PARNA Chapada dos Veadeiros-GO.	24	5.15	2.6	0.34	0.82
Serra da Mesa -Minaçu-GO.	25	3.38	2.7	0.30	0.83
Serra Negra – Minaçu-GO.	20	2.69	2.5	0.17	0.83
Vale do Paranã					
Alvorada do Norte-Simolândia- Buritinópolis-GO.	18	2.08	2.4	0.14	0.83
Iaciara-Guarani de Goiás-Posse-GO.	23	3.67	2.6	0.24	0.83
Campos Belos-São Domingos-GO.	21	2.90	2.5	0.17	0.81
Damianópolis-Mambaí -GO.	22	2.48	2.6	0.14	0.86
Espigão Mestre do São Francisco					
Correntina-BA.	22	3.35	2.8	0.18	0.90
São Desidério-BA.	23	2.70	2.7	0.11	0.87
PARNA Grande Sertão Veredas-MG	22	3.12	2.6	0.12	0.86
Formosa do Rio Preto-BA	20	3.36	2.6	0.18	0.90
Xavantina Complex					
Nova Xavantina-MT	31	4.69	2.9	0.20	0.85
Canarana-MT	31	4.18	3.0	0.19	0.87
Água Boa-MT	22	2.98	2.5	0.23	0.82

Species richness per 1000 m² plot as well as diversity (Table 4.2) were higher at most sites in the Pratinha plateau in the core area of the Cerrado Biome, but were also high in the transition zones of cerrado/Amazonian forests of the Xavantina complex; both those sites were on latossols. Lower richness and diversity were found on cambissols (Veadeiros, Paranã) and on sandy quartz soils (São Francisco). Richness was at a similar level at all sites in São Francisco, a large and uniform physiographical unit, where the most distant sites, Correntina and Formosa, are more than 500 km apart. However, it was most variable in the Veadeiros, a physiographical unit containing three land systems, where the most distant sites, Serra da Mesa and Vila Propício were c.200 km apart. These results support the findings of Felfili and Silva Júnior (2001) and Felfili et al., (2004b)

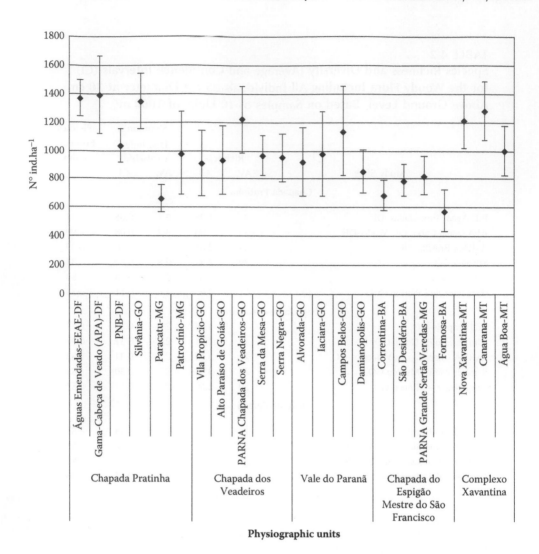

FIGURE 4.2 Density and confidence intervals for the 220 sampled (20 × 50 m) plots, totalling 22 hectares of cerrado sensu stricto, in five physiographical units in central Brazil. Bars represent confidence intervals and central diamonds the average number of individuals per hectare within each physiographical unit. PARNA-Parque Nacional.

that richness and structure of the vegetation conform well with the land system zoning proposed by Cochrane et al. (1985) for central Brazil and allow the cerrado of the central region to be distinguished from the other groupings as suggested by Ratter et al. (2003).

4.3.2 COMMUNITY STRUCTURE

Eighteen of the 22 sites studied contained between 800 and 1400 indiv./ha (Figure 4.2), indicating that this is a representative range for cerrado sensu stricto. Densities over 1000 stems per hectare were found in the physiographical units where most plots were on latossols, while those on cambissols and sandy quartz soils were below this limit. Nunes et al. (2002) analysed 100 plots of similar size in cerrado sensu stricto in the Federal District finding a confidence interval of 1042,8 ± 35,0 ind./ha, well within the range found in this study for cerrado sensu stricto on latossols.

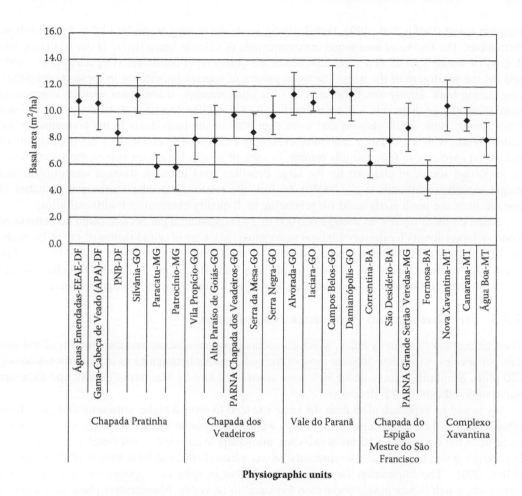

FIGURE 4.3 Basal area and confidence intervals for the 220 sampled plots. Bars represent confidence intervals and the central diamonds the average basal area per hectare within each physiographical unit. PARNA = Parque Nacional

The São Francisco sites showed the lowest density of all the physiographical units, ranging from 577 to 825 indiv./ha. The highest values were found in Pratinha, ranging from 664 to 1395 indiv./ha, followed by sites in Xavantina. These results are within the limits found in other studies of cerrado sensu stricto sites studied under a similar methodology (Felfili and Silva Júnior, 1993; Felfili et al., 1994; Rossi et al., 1998; Andrade et al., 2002; Assunção and Felfili, 2004).

Over 80% of our sites had basal areas from 7 to 12 m²/ha (Figure 4.3). The highest values occurred in Paranã, Pratinha and the Xavantina complex. The Paranã valley is a physiographical unit where the vegetation is a mosaic of cerrado and seasonal dry forests on limestone outcrops, and the cerrado occurs mostly on cambissols of higher soil fertility (Cochrane et al., 1985) which may explain why larger trees occur there. The other sites with higher basal area were on latossols in Pratinha and in the Xavantina complex, while the lowest values occurred at the São Francisco sites on sandy quartz soils and also in the lowlands of the Chapada Pratinha. The last group of sites are fragments surrounded by farms and dense urban settlements and certainly suffer from uncontrolled fires, and possibly had some larger trees felled in the past, although no recent evidence of such activity was found during the sampling.

The basal area of the cerrado sensu stricto we studied is about half of the value found in cerradão (a dense woodland with a mixed flora of cerrado and forest species) and a third of that found in forest (Felfili et al., 1994; Felfili, 1995), while the density is similar to that found in those formations. The low basal area found in communities of cerrado sensu stricto is due to (a) the low density of woody species in a vegetation where the grassy layer dominates (Filgueiras et al. 1998) and (b) the small size of the trees. The main pattern of species importance in these communities, considering both density and basal area, is of a small number of about ten species, comprising more than 50% of these values in each site (Felfili et al., 2001). Diameter distributions (Felfili and Silva Júnior, 1988, 2001) showed that more than 80% of the individuals in the communities are under 10 cm, with a few larger individuals reaching *c*.50 cm. The majority of the populations of individual species also followed this pattern. In spite of its low basal area values, the vegetation is an important source of charcoal for the large Brazilian steel industry, through uncontrolled and illegal exploitation (Felfili et al., 2004b). In fact, the relatively slender trunks and branches of cerrado trees are particularly good for producing high quality charcoal in traditional kilns.

Based on these figures, an average of 1000 to 1400 individuals per hectare could be considered a normal target density for conservation and management planning of cerrado sensu stricto on latossols, and 400 to 1000 for the other soil types, while basal areas generally should vary from 7 to 12 m^2/ha. Knowledge of these values will be useful in planning harvesting limits for wild fruit collection and timber production for firewood and charcoal, as well as for the recovery of degraded land.

4.3.3 BETA-DIVERSITY AND PHYTOGEOGRAPHICAL PATTERNS

The Czekanowsky similarity index, which considers presence-absence and also number of individuals per species, was below 50% for most of the comparisons between plots in a matrix containing 220 plots. Similarity indices under 50% were considered low by Margurran (1988), and therefore our results indicate high β-diversity.

As would be expected, plots from the same site tend to show a higher similarity than plots from other localities; there is also a greater similarity within land systems and within physiographical units. Even when the same species occurs at two sites, the density in each varies and therefore the difference in population size can produce low similarity between sites (Felfili and Silva Júnior, 1993; Felfili and Felfili, 2001). The implication for conservation is that in spite of a species being present in a conservation unit it may have a population too small to be viable. Management plans for economic exploitation should also consider the changes in population size over the area and focus on the use of multiple species. Densities per species are generally low and variable from place to place, making the exploitation of just one or a few species very unlikely to bring sustainable profits.

TWINSPAN classification (Figure 4.4), indicates that the cerrado sensu stricto, in general, is a vegetation continuum with more or less characteristic communities occurring in mosaics varying along the geographical gradient correlated with land systems and physiographical units. The analysis grouped most plots by site, land system and physiographical unit; this was especially clear in the highlands of the Chapada Pratinha, the Xavantina complex, Paranã and a large part of the Espigão Mestre do São Francisco and the Veadeiros land systems.

TWINSPAN classification using density of species as a variable, detected two phytogeographical groups: (A) Highlands in the central area and (B) Lowlands in the north-northeastern borders of the biome. In group A most sites are on latossols, while in group B most sites are on cambissols and sandy quartz soils and some in the Xavantina complex on latossols. Soil classes were also regarded by Haridasan (2001) as one of the main differentiating factors within cerrado sensu stricto in central Brazil. Groupings with altitude as a main gradient were found by Felfili and Silva Júnior (1993) in the Chapada Pratinha. Group B contained the sites in the cerrado/caatinga (São Francisco) ecotones and cerrado/Amazonian forest ecotone (Xavantina). Castro (1994), Ratter et al. (1996, 2003), and Castro (1994) also correlated phytogeographical patterns with latitudinal gradients in the cerrado sensu lato.

TWINSPAN classification indicates preferential species and their density levels for each division (Table 4.3). The higher density levels for preferential species varies from two to 20 stems/ha

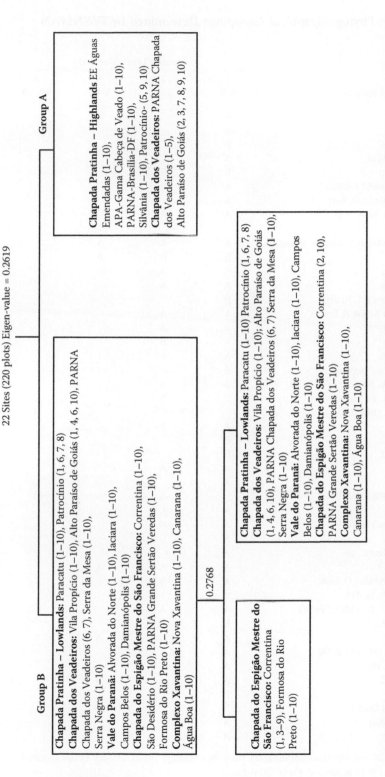

FIGURE 4.4 TWINSPAN classification of a matrix of 291 species × 220 plots (20 × 50 m) sampled in cerrado sensu stricto in central Brazil. PARNA = Parque Nacional, APA = Área de Proteção Ambiental, EE = Estação Ecológica. (Values between brackets are number of plots per study site at each side of the division).

TABLE 4.3
Preferential Species by Phytogeographical Groupings Determined by TWINSPAN Classification

Preferential Species	Density Levels (ind/ha)			
	1 0–2	2 2–5	3 5–10	4 10–20
Anacardium occidentale L.	B			
Andira vermifuga Mart.	B			
Annona coriacea Mart.	B			
Aspidosperma macrocarpon Mart.	A			
Astronium fraxinifolium Schott	B	B		
Austroplenckia populnea (Reissek) Lundell	A			
Blepharocalyx salicifolius (Kunth) O. Berg.	A			
Byrsonima verbascifolia Rich. ex A. Juss.	A	A		
Caryocar brasiliense Cambess.			A	
Chamaecrista claussenii (Benth.) H.S. Irwin & Barneby	A	A	A	
Curatella americana L.	B	B	B	
Dalbergia miscolobium Benth.	A	A		
Eremanthus glomerulatus Less.	A	A		
Eriotheca gracilipes (K. Schum.) A. Robyns	B			
E. pubescens (Mart. & Zucc.) Schott & Endl.	A	A		
Erythroxylum tortuosum Mart.	A	A		
Eugenia dysenterica DC.		B		
Guapira noxia (Netto) Lundell	A	A		
Hancornia speciosa Gomes var. *speciosa*	A			
Heteropterys byrsonimifolia A. Juss.	A	A		
Hymenaea stigonocarpa Mart. ex Hayne		B		
Kielmeyera coriacea (Spreng.) Mart.				A
K. speciosa A. St. Hil.	A	A		
Machaerium acutifolium Vogel	B			
Magonia pubescens A. St.-Hil.	B			
Miconia ferruginata DC.	A	A		
M. pohliana Cogn.	A			
Ouratea hexasperma (A. St.-Hil.) Baill.		A	A	A
Palicourea rigida Kunth	A	A		
Piptocarpha rotundifolia (Less.) Baker	A	A		
Psidium myrsinoides O. Berg	B	B		
Myrsine guianensis (Aubl.) Kuntz.	A			
Roupala montana Aubl.	A	A	A	
Salacia crassiflora (Mart. ex Shult.) G. Don	A	A		
Salvertia convallariaeodora A. St. Hil.	B			
Schefflera macrocarpa (Cham. & Schltdl.) Frodin	A	A		
Sclerolobium aureum (Tul.) Benth.	B			
S. paniculatum Vogel	A	A	A	
Stryphnodendron adstringens (Mart.) Coville	A	A		
Strychnos pseudoquina A. St. Hil.	A			
Styrax ferrugineus Nees & Mart.	A	A	A	
Terminalia argentea Mart.	B			
Vatairea macrocarpa (Benth.) Ducke	B			
Vochysia elliptica (Spreng.) Mart.	A			
V. thyrsoidea Pohl	A	A		

(A) Highlands of Central Brazil and (B) Lowlands/Ecotones North-Northeastern Brazil. Density Levels (n/ha): 1 = 0–2 ind/ha, 2 = 25 ind/ha, 3 = 5–10 ind/ha and 4 = 10–20 ind/ha

TABLE 4.4
Generalist Species Present in Both the Highlands of Brazil and the Lowlands and Ecotones of North-Northeastern Brazil

Generalist Species	Density Levels (nd/ha)			
	1 0–2	2 2–5	3 5–10	4 10–20
Acosmium dasycarpum (Vog.) Yakovl.	x			
Annona crassifolia Mart.	x	x		
Aspidosperma tomentosum Mart.	x	x		
Bowdichia virgilioides Kunth	x	x		
Byrsonima coccolobifolia (Spreng.) Kunth	x	x		
B. crassa Nied.	x	x		
* *Caryocar brasiliense* Cambess.	x	x		
Connarus suberosus Planch.	x	x		
Davilla elliptica A. St. Hil.	x	x		
Dimorphandra mollis Benth.	x	x		
Erythroxylum deciduum A. St. Hil.	x			
Erythroxylum suberosum A. St. Hil.	x	x		
+ *Eugenia dysenterica* Mart. ex DC.	x			
+ *Hymenaea stigonocarpa* Mart.	x			
Kielmeyera coriacea (Spreng.) Mart.	x	x		x
Lafoensia pacari A. St. Hil.	x	x		
Machaerium opacum Vogel	x			
* *Ouratea hexasperma* (A. St. Hil.) Baill.	x			
Pouteria ramiflora (Mart.) Radlk.	x	x	x	
Pterodon pubescens (Benth.) Benth.	x			
Qualea grandiflora Mart.	x	x	x	x
Q. multiflora Mart.	x	x		
Q. parviflora Mart.	x	x	x	x
Tabebuia aurea (Manso) Benth. & Hook.f. ex S. Moore	x			
Tabebuia ochracea (Cham.) Standl.	x			
Vochysia rufa Mart.	x	x		

(*) Preferential at higher densities in the highlands; (+) Preferential at higher densities in the lowlands.

Density levels: 1 = (0–2 ind/ha), 2 = (2–5 ind/ha), 3 = (5–10 ind/ha) and 4 = (10–20 ind/ha)

showing the absence of dominance in this vegetation where most species have low density levels. This means that in any site the most abundant species may have, at the most, just 20 stems per hectare, which, together with the low stature of cerrado trees, implies a low timber production per species per hectare.

In the highlands of central Brazil (group A), see Table 4.3, *Kielmeyera coriacea** and *Ouratea hexasperma* with 10–20 stems/ha and *Caryocar brasiliense, Chamaecrista claussenii, Roupala montana, Sclerolobium paniculatum* and *Styrax ferrugineus* with 5–10 stems/ha were the preferential species, while *Curatella americana* (5–10 stems/ha), plus *Astronium fraxinifolium, Eugenia dysenterica, Hymenaea stigonocarpa* and *Psidium myrsinoides* (with 2–5 stems/ha) were preferential for the north-northeastern lowlands and ecotones (group B). Non-preferential (generalist) species (Table 4.4), such as *Qualea grandiflora* and *Qualea parviflora* (10–20 stems/ha), *Kielmeyera*

* Authorities for species given in Table 4.3

coriacea and *Pouteria ramiflora* (5–10 stems/ha) should be species used for recuperation of degraded land, at planting densities based on our field observations. All non-preferential species recorded by Ratter et al. (2003) as important widespread species can be regarded as generalists at a regional level.

Some species, such as *Caryocar brasiliense*, *Kielmeyera coriacea* and *Ouratea hexasperma*, were preferential in the highlands (Group A) at higher densities but generalists at lower densities, suggesting better growth conditions for them in the highlands. The opposite occurred with *Eugenia dysenterica* and *Hymenaea stigonocarpa* which were preferential at higher densities in the lowlands (Group B).

TWINSPAN classification of the cerrado at the third level division agreed with the zoning of the cerrado into physiographical units by Cochrane et al. (1985). The subdivision of the units into land systems by these authors corresponded with the floristic and structural classification produced by TWINSPAN for all systems except the Alto Paraíso-Vila Propício subset of Veadeiros, which did not fit so well into their classification.

4.4 FINAL CONSIDERATIONS

Cerrado sensu stricto is a rich and diverse vegetation with a high β-diversity, especially due to the variation in density of species between sites. Even when there is a large number of species in common between sites, the population size of individual species varies considerably.

The floristic and structural characteristics of vegetation communities conform well with the land system zoning of the Cerrado Biome based on variation in climate, landscape and soil (Cochrane et al., 1985). Land-system zoning should be used as a guide for establishing conservation strategies at a regional level to maximize the protection of biodiversity.

Our quantitative analyses, based on a standardized methodology, confirm the findings of previous authors based on floristic data (e.g. Castro, 1994; Ratter et al., 2003) and those from our previous work using the same methodology but based on a smaller sample size (Felfili and Silva Júnior, 1993, 2001; Felfili et al., 1994, 1997, 2004b).

Two main phytogeographical units could be detected by the vegetation classification, the highlands in the central area, and the lowlands in the north-north-eastern borders of the biome. Further studies on direct gradient analyses should be performed to investigate the main factors determining these groupings.

ACKNOWLEDGEMENTS

We are grateful to the Brazilian Ministry of Environment-FNMA (Projeto Biogeografia do bioma Cerrado), to the British Department for International Development (Conservation and Management of the Biodiversity of the Brazilian Cerrado Biome Project) and to the Brazilian Research Council (CNPq) for support of the projects Biogeography of the Cerrado Biome and Conservation and Management of the Biodiversity of the Cerrado Biome. We also wish to thank our field and herbarium teams, especially Roberta Mendonça, curator of the IBGE herbarium, and Newton Rodrigucs and Edson Cardoso from the University of Brasília. Special thanks also to Dr James Alexander Ratter from the Royal Botanic Garden Edinburgh and to an anonymous reviewer for reviewing the manuscript.

REFERENCES

Andrade, L.A.Z., Felfili, J.M., and Violatti, L., Fitossociologia de uma área de cerrado denso na RECOR-IBGE, *Acta Bot. Brasil*, 16, 2, 225, 2002.
Assunção, S.L. and Felfili, J.M., Fitossociologia de um fragmento de cerrado sensu stricto na APA do Paranoá, DF, Brasil., *Acta Bot. Brasil*, 18, 4, 903, 2004.

Castro, A.A.J.F., *Comparação florístico-geográfica (Brasil) e fitossociológica (Piauí-São Paulo) de amostras de cerrado*, PhD thesis, UNICAMP, Campinas, Brasil, 1994.

Cochrane, T.T. et al., *Land in Tropical América*, CIAT/EMBRAPA- CPAC, Cali, Colombia, 1985.

Durigan, G. et al., Padrões fitogeográficos do cerrado paulista sob uma perspectiva regional, *Hoehnea*, 30, 1, 39, 2003.

Eiten, G., The cerrado vegetation of Brazil, *Bot. Rev. Lancaster*, 38, 201, 1972.

Felfili, J.M., Diversity, structure and dynamics of a gallery forest in central Brazil, *Vegetatio* 117, 1, 1995.

Felfili, J.M., Águas Emendadas no contexto fitogeográfico do cerrado, in *Seminário Sobre Unidades de Conservação*, SEMATEC, Linha Gráfica Editora, Brasília, DF, Brazil, 1998, 71.

Felfili, M.C. and Felfili, J.M., Diversidade alfa e beta no cerrado sensu stricto da Chapada Pratinha, Brasil, *Acta Bot. Brasil*, 15, 2, 243. 2001.

Felfili, J.M. and Silva Júnior, M.C., Distribuição dos diâmetros numa faixa de cerrado na Fazenda Água Limpa (FAL) em Brasília, DF, Brazil, *Acta Bot. Brasil*, 2, 85, 1988.

Felfili, J.M. and Silva Júnior, M.C., Floristic composition, phytosociology and comparison of cerrado and gallery forests at Fazenda Água Limpa, Federal District, Brazil, in *Nature and Dynamics of Forest-Savanna Boundaries*, Furley, P.A., Proctor, J.A., and Ratter, J.A., Eds, Chapman and Hall, London. 1992, 393.

Felfili, J.M. and Silva Júnior, M.C., A comparative study of cerrado (sensu stricto) vegetation in Central Brazil, *J. Trop. Ecol.*, 9, 277, 1993.

Felfili, J.M. and Silva Júnior, M.C., *Biogeografia do Bioma Cerrado, Estudo Fitofisionômico na Chapada do Espigão Mestre do São Francisco*, UnB, Brasília, DF, Brazil, 2001, 152.

Felfili, J.M. et al., Projeto biogeografia do bioma cerrado: vegetação e solos. *Cadernos de geociências do IBGE*, 12, 75, 1994.

Felfili, J.M. et al., Comparação florística e fitossociológica do cerrado nas Chapadas Pratinha e dos Veadeiros, in *Contribuição ao Conhecimento Ecológico do Cerrado*, Leite, L. and Saito, C., Eds, UnB, Brasília, DF, Brazil, 1997, 6.

Felfili, J.M. et al., Composição florística e fitossociologia do cerrado sentido restrito no Município de Água Boa-MT, *Acta Bot. Brasil*, 16, 1, 103, 2002.

Felfili, J.M. et al., Potencial econômico da Biodiversidade do Cerrado: estádio atual e possibilidades de manejo sustentável dos recursos da flora, in *Cerrado: Ecologia e Caracterização*, Aguiar, L.M.S and Camargo, A.J.A., Eds, EMBRAPA, Brasília, DF, Brazil, 2004a, 177.

Felfili, J.M. et al., Diversity, floristic and structural patterns of cerrado vegetation in Central Brasil, *Pl. Ecol.* 175, 37, 2004b.

Filgueiras, T.S. et al., Floristic and structural comparison of cerrado sensu stricto vegetation in Central Brasil, in *Measuring and Monitoring Forest Biological Diversity*, Dallmeier, F. and Comiskey, J.A., Eds, Midsomer Norton, UK, 1998, 2, 633.

Haridasan, M., Solos, in *Biogeografia do Bioma Cerrado. Estudo Fitofisionômico na Chapada do Espigão Mestre do São Francisco*, Felfili, J.M. and Silva Júnior, M.C., Eds, UnB, Brasília, DF, Brazil, 2001, 12.

Haridasan, M. et al., Gradient analysis of soil properties and phytosociological parameters of some gallery forests in the Chapada dos Veadeiros, in *Proceedings of the International Symposium on Assessment and Monitoring of Forests in Tropical Dry Regions*, UnB, Brasília, DF, Brazil, 1997, 259.

Kent, M. and Coker, P., *Vegetation Description and Analysis*, Belhaven Press, London, 1992, 363.

Margurran, A.E., *Ecological Diversity and its Measurement*, Chapman and Hall, London, 1988, 179.

Mendonça, R. et al., Flora vascular do Cerrado, in *Cerrado: Ambiente e Flora*, Sano, S. and Almeida, S., Eds, EMBRAPA-CPAC, Planaltina, DF, Brazil, 1998, 287.

Mittermeier, R.A., Myers, N., and Mittermeier, C.G., *Hotspots: Earth's Biologically Richest and Most Endangered Terrestrial Ecoregions*, Conservation International-CEMEX, New York, 1999, 430.

Nunes, R.V. et al., Intervalo de classe para abundância, dominância e freqüência do componente arbóreo do Cerrado sentido restrito no Distrito Federal, *Árvore*, 26, 2, 173, 2002.

Oliveira-Filho, A.T. and Ratter, J.A., A study of the origin of central Brazilian forests by the analysis of plant species distribution patterns, *Edinburgh J. Bot.* 52, 141, 1995.

Ratter, J.A., Bridgewater, S., and Ribeiro, J.F., Analysis of the floristic composition of the Brazilian cerrado vegetation III. Comparison of the woody vegetation of 376 areas, *Edinburgh J. Bot.*, 60, 57, 2003.

Ratter, J.A. et al., Analysis of the floristic composition of the Brazilian cerrado vegetation II. Comparison of the woody vegetation of 98 areas. *Edinburgh J. Bot.*, 53, 153, 1996.

Ratter, J.A. et al., Estudo preliminar da distribuição das espécies lenhosas da fitofisionomia cerrado sentido restrito nos estados compreendidos pelo bioma cerrado, *Bol. Herb. Ezechias Paulo Heringer*, 5, 5, 2000.

Reatto, A., Correia, J.R., and Spera, S.T., Solos do Bioma Cerrado, in *Cerrado: Ambiente e Flora*, Sano, S. and Almeida, S., Eds, EMBRAPA-CPAC, Planaltina, DF, Brazil, 1998, 47.

Rossi, C.V., Silva Júnior, M.C., and Santos, C.E.N., Fitossociologia do estrato arbóreo do Cerrado (sensu stricto) no Parque Ecológico Norte, Brasília, DF, *Bol. Herb. Ezechias Paulo Heringer*, 2, 49, 1998.

Silva, J.M.C. and Bates, J.M., Biogeographic patterns and conservation in the South American Cerrado: a tropical savanna hotspot, *BioScience*, 52, 3, 225, 2002.

Silva Júnior, M.C. et al., Análise da flora arbórea de Matas de Galeria no Distrito Federal: 21 levantamentos, in *Cerrado: Caracterização e Recuperação de Matas de Galeria*, Ribeiro, J.F., Fonseca, C.E.L. da, and Sousa-Silva, J.C., Eds, EMBRAPA-CPAC, Planaltina, DF, Brazil, 2001, 143.

UNESCO, *Vegetação no Distrito Federal: tempo e espaço: uma avaliação multitemporal da perda da cobertura vegetal no DF e da diversidade florística das áreas nucleares da Reserva da Biosfera do Cerrado – Fase I*, Brasília, DF, Brazil, 2000, 29.

5 Flora and Vegetation of the Venezuelan Llanos: A Review

Otto Huber, Rodrigo Duno de Stefano, Gerardo Aymard and Ricarda Riina

CONTENTS

ABSTRACT

The Venezuelan Llanos, a well-defined ecoregion in northern South America, occupy an area of approximately 240,000 km² lying between 7° and 10°N. These wide lowland plains consist mainly of Quaternary alluvial sediments covered by a mosaic of savannas, gallery forests and dry to semideciduous forests. The climate is markedly macrothermic (>24°C) and tropophilous, with a strong alternation between one rainy and one dry season, and with an average rainfall between 800 and 2200 mm/year. A recent thorough inventory of vascular plants collected in this region during the past two centuries has yielded a surprisingly high number of 3219 species in 1117 genera and 190 families, including 127 species of ferns, 860 monocotyledons and 2232 dicotyledons. The ten most important families account for almost 42% of the total flora, of which the grasses and sedges alone comprise *c.*450 species; the Leguminosae is the most diverse family (*c.*350 species) of the dicotyledons, followed by the Rubiaceae, Asteraceae, Euphorbiaceae and Melastomataceae. Despite the large area and the great number of taxa, the level of endemism of the Llanos is low (approximately only 1% of the total flora), probably because of the very young (Quaternary) alluvial landscape and the consequently short evolutionary time available for speciation, together with the lack of major geographical barriers across the entire region. The highest species diversity is concentrated in the semideciduous forest types of the western Llanos and in the extensive gallery forests, but other typical Llanos families, such as grasses and sedges, have their greatest diversity in the open savanna and shrubland communities. β-diversity in the extensive grassland ecosystems has never been studied for the entire region, but it is suspected to be extraordinarily high, as a result of the relatively complex pattern of soil and climatic parameters in each of the six or seven major landscape types recognized in the region.

5.1 INTRODUCTION

Savannas in northern South America have been known to represent an important landscape element for more than 200 years. The first geographical and botanical descriptions of the wide interior savannas of Venezuela, called *Llanos* ('plains'), were made by Alexander von Humboldt in 1818–1819, who visited them twice in 1800 with his travelling companion Aimé Bonpland. Humboldt compared these extremely hot, level plains to an 'endless sea covered by sargasso grass or pelagic algae', where an ever-vanishing horizon produces a strange sensation of infinity in the deepness of the soul. He also noticed that these 'steppes' are extremely monotonous and have little attraction from the scientific point of view when compared with the much more diverse forests of the American tropics.

Since then, many explorers and scientists have visited these remote regions, steadily accumulating an increasing amount of knowledge on their natural and human history. Today, a large body of publications exists on the flora, vegetation and plant ecology of the Venezuelan Llanos (see Huber, 1974), and it appears that the uniformity and paucity of plants and plant communities mentioned by Humboldt should be interpreted as a result of the limited knowledge that he and his colleague Bonpland were able to gather during the few days they spent in the Llanos.

This chapter aims to provide an update on recent advances in floristic knowledge of one of the major neotropical savanna regions. The imminent publication of the *Catálogo anotado e ilustrado de la Flora Vascular de los Llanos de Venezuela* gives evidence of another serious effort to cover progressively all major biomes of Venezuela; as a matter of fact, its appearance coincides with the conclusion of one of the most important floristic projects of the Neotropics, the monumental *Flora of the Venezuelan Guayana* by Steyermark et al. (1995–2005, in nine volumes).

Unfortunately, however, even such enormous extensions as the Llanos in Venezuela and adjacent Colombia do not escape from a rapidly increasing colonization and subsequent population pressure. Intensification and expansion of large industrial-scale farms and logging exploitation, extensive tree plantations and progressive exploitation of petroleum are causing the rapid disappearance of natural ecosystems with their specialized plant communities. So much so, that the savannas of the Venezuelan Llanos and the Brazilian cerrados are today among the most threatened biomes in tropical South America. Therefore the publication of the plant catalogue of the Venezuelan Llanos should also represent a last-minute effort to document the diversified flora of this extensive, characteristic landscape of the Venezuelan interior.

5.2 PHYSICAL CHARACTERISTICS OF THE VENEZUELAN LLANOS

5.2.1 GEOMORPHOLOGY

The Venezuelan Llanos are a well-defined ecoregion in northern South America occupying an area of approximately 240,000 km². These extensive plains (*Llanos*), located roughly at 7–10°N and 62–72°W, are mostly flat or undulating lowlands (with few elevations 300 m a.s.l.). They are limited

FIGURE 5.1 Location of the Llanos region in Venezuela (map by R. Schargel).

to the north by the Coastal Cordillera mountain system (of Upper Tertiary origin), to the south by the ancient (Precambrian) and massive mountain system of the Guayana Shield, to the west by the Andean Cordillera of mid-Tertiary age and to the east by the alluvial Quaternary deposits of the Orinoco Delta (Figure 5.1).

The Llanos themselves extend along a huge, south-west–north-east-oriented geosyncline occurring between the basement of the Guayana Shield and the Coastal Cordillera. As a consequence of the orogenic processes caused by the Andean uplifting, this enormous depression, originally flooded by marine waters, has been filled up since the Upper Tertiary and during the Quaternary with sediments that now dominate the superficial geology of the region. As in the Brazilian Pantanal do Mato Grosso, this has produced vast areas lacking stones in the soil.

Furthermore, during mid- and late-Tertiary times, the Llanos depression was subject to differential tectonic processes. Its central and eastern sectors have suffered from a moderate uplift causing subsequent pronounced erosion activities in the area, whereas in the western sector subsidence prevailed, which was then followed by intense sediment accumulation.

As a result of this mainly Tertiary and Quaternary geological and tectonic history, four principal geomorphological landscapes determining soils, vegetation and fauna can be identified in the Llanos region (Figure 5.2).

1. Alluvial plains (*planicies aluviales*): plains with low slopes (usually < 1%) produced mainly by overflow of the rivers with consequent periodic accumulation of sediments; they are most evident in the west and the extreme north-east Llanos.

2. Eolian plains (*planicies eólicas*): plains covered by sand and silt, which were deposited by wind during the Pleistocene; in the Llanos, eolian plains with stabilized Pleistocenic dunes more than 10 m high are distinguished from completely flat eolian plains covered by silt strata. These plains extend principally from the southern central Llanos to the lower Arauca-Cinaruco basins in the west Llanos.

3. High plains or plateaux (*altiplanicies*): these are plains formed by flat or slightly undulated extensions where the rivers are flowing in valleys or fluvial incisions more than 10 m deep (in some cases >150 m deep). They consist of flat mesas elevated by tectonic movements and are most frequent and visible in the east Llanos.

4. Hills and denuded surfaces (*colinas y superficies de denudación*): when the erosion has dismantled a high plain, hills less than 300 m high remain, including low mesa remnants and colluvial glacis; this landscape can be seen mainly in the central Llanos.

Geomorphological Landscapes

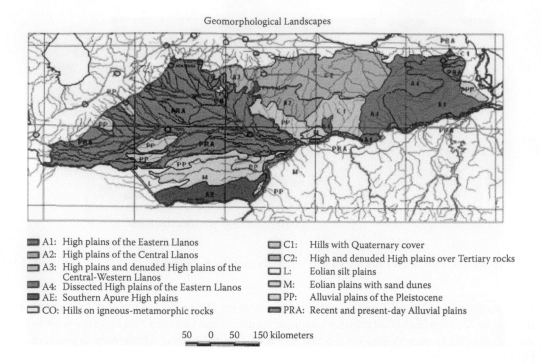

A1:	High plains of the Eastern Llanos		C1:	Hills with Quaternary cover
A2:	High plains of the Central Llanos		C2:	High and denuded High plains over Tertiary rocks
A3:	High plains and denuded High plains of the Central-Western Llanos		L:	Eolian silt plains
A4:	Dissected High plains of the Eastern Llanos		M:	Eolian plains with sand dunes
AE:	Southern Apure High plains		PP:	Alluvial plains of the Pleistocene
CO:	Hills on igneous-metamorphic rocks		PRA:	Recent and present-day Alluvial plains

50   0   50   150 kilometers

FIGURE 5.2 (See colour insert following page 208) Geomorphological subdivision of the Llanos region in Venezuela (map by R. Schargel).

Several detailed geomorphological studies on Llanos landscapes in Venezuela have been produced in the past few decades by the governmental agencies Comisión de Planificación Nacional de los Recursos Hidráulicos (COPLANARH) of the Ministry of Public Works (MOP) and the Ministry of the Environment (Ministerio del Ambiente y de los Recursos Naturales Renovables, MARNR); the most important results were published by Berroterán (1985, 1988), Comerma and Luque (1971), Montes et al. (1987), Morales (1978), Schargel (2003), Schargel and Aymard (1993) and Zinck and Stagno (1966). Goosen (1971) gives detailed geomorphological information on the Colombian Llanos region.

5.2.2 Palaeo-Ecology

Much has been debated about the palaeo-ecological evolution of the landscape types in the Llanos region, including the associated history of the plant cover. Although there is general agreement that the area was subject to several prolonged marine transgressions during the late Tertiary and the Quaternary (see Díaz de Gamero, 1996), the hypothesized climatic oscillations accompanying them are still little understood. The discussions about climatically induced alternations between arboreal (i.e. evergreen to dry forest) and herbaceous ecosystems (savanna) in the region are largely extrapolated from data found in the surrounding mountains or regions, a fact that underlines the lack of concrete palynological field data from the Llanos region proper. Some of the publications dealing with this subject were produced by Behling and Hooghiemstra (1998, 2001), Gentry (1982), Graham and Dilcher (1995), Meave and Kellman (1996), Pennington et al. (2000, 2004), Roa-Morales (1979), Rull (1998), Salgado-Labouriau (1980), Schubert (1988) and Tricart (1985).

5.2.3 Climate

Because Venezuela is located in the American tropics, all vegetation types growing there are characteristic of a tropical nature, not only with reference to the thermic regime, but especially concerning the prevailing annual energy and radiation cycles and budgets, the planetary wind regime, etc. Furthermore, because the upper altitudinal level of the Llanos region is 250 m a.s.l., it is obvious that all plant cover included in it belongs to the macrothermic lowland regime (i.e. mean annual temperatures always >24°C).

On the other hand, the entire Llanos region of Venezuela (and also of adjacent Colombia) is subject to a strongly seasonal (tropophilous) climate, with a single dry season extending between November and April–May and a single rainy season between April–May and October. This biseasonal climatic regime is caused by the alternation of the north-east trade winds blowing from the Atlantic Ocean over north-east South America, and the intertropical convergence zone (ITCZ) which annually oscillates north and south of the Equator, penetrating into Venezuela during the months from April to November (Sánchez, 1960). The relatively moist north-eastern trade winds coming from the Atlantic discharge their humidity along the coastal area of the continent, and then continue over the Llanos as dry winds until meeting with the eastern slopes of the Andean Cordillera; therefore, during the prevalence of this wind regime (i.e. during November–December to April–May), virtually no rain falls over the wide Llanos plains. This rainless period of the year is locally called *verano* (summer), although that is geographically not correct, because the Llanos are located in the northern hemisphere. Generally each May, the sky becomes cloudy as a result of the arrival from the south of the ITCZ with its moisture-laden local climatic regimes, which overwhelm the north-east trade winds. Then torrential rains begin, and extensive flooding occurs in the lower-lying parts of the Llanos, such as in the states of Apure and southern Guárico. Normally, more than 80% of the annual rainfall falls during the 5–6 months of rainy season (*invierno*, winter).

According to Köppen's classification (1936), the Llanos are dominated by two climate types: Am (tropical monsoon climate type) in the south-western region of Apure state and in the eastern Llanos reaching the Orinoco Delta, and Aw (tropical savanna climate type) in all the remaining area.

The temperature regime prevailing in these tropical American lowlands is macrothermic, with pronounced hot mean annual temperatures ranging between 26°C and 28°C; monthly average maxima are between 34°C and 37°C, but absolute maxima of more than 40°C in the shade during a typically cloudless winter day have been reported from various places. Because of the near-Equatorial position of the region, the temperature oscillations are markedly diurnal (average daily range 10–12°C) but almost imperceptible on an annual basis (average annual range 1–2°C); likewise, the length of daylight varies only between 11.5 and 12.5 h during summer and winter times, therefore offering relatively uniform radiation balances to plant cover throughout the year (Vareschi and Huber, 1971).

In contrast to this relative thermic uniformity, the rainfall of the Llanos region shows a clearly alternating (tropophilous) regime characterized by very pronounced differences during the various months of the year. As can be seen in the climadiagrams of representative Llanos stations and in the climatogram of Calabozo (Figure 5.3), both the annual and the monthly average rainfall differ markedly; monthly rainfall at Calabozo, in the centre of the Llanos, varies between 0 mm and almost 500 mm, while the annual rainfall amount may vary by more than 100%, from 850 mm to 1800 mm. Obviously, there are also noticeable local variations of rainfall between the various sections of the Llanos, but, generally speaking, one can observe an increasing pluviometric gradient from east to west, i.e. from the Atlantic coast to the Andean piedmont, and a similar increasing gradient running from north to south (Figure 5.4).

Relative humidity of the air is directly related to the rainfall regime; during the dry season, it can drop to 20–25%, increasing rapidly after the first rainfalls in May and then reaching levels between 70% and 98% during most of the rainy period. In the same context, potential evapotranspiration is highest during the dry season and lowest in the wet season.

The annual regime of winds is strongly influenced by the planetary wind currents prevailing in this near-Equatorial region. During the dry season, the north-east trade winds blow with remarkable constancy and moderate strength, but during the rainy season local cyclonic and anticyclonic

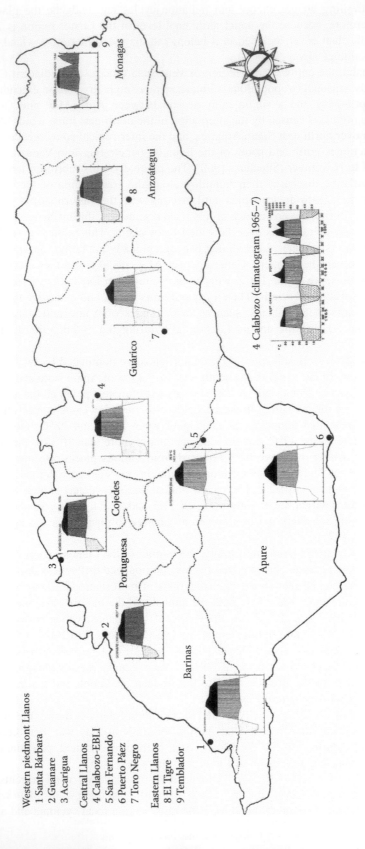

FIGURE 5.3 Climadiagrams of selected stations of the Venezuelan Llanos (see Walter and Medina, 1971). Inset: climatogram of 3 years at the station of Calabozo (Guárico state, central High Llanos). (After Vareschi, V. and Huber, O., *Bol. Soc. Venez. Ci. Nat.*, 119–120, 50, 1971. With permission.)

Western piedmont Llanos
1 Santa Bárbara
2 Guanare
3 Acarigua

Central Llanos
4 Calabozo-EBLI
5 San Fernando
6 Puerto Páez
7 Toro Negro

Eastern Llanos
8 El Tigre
9 Temblador

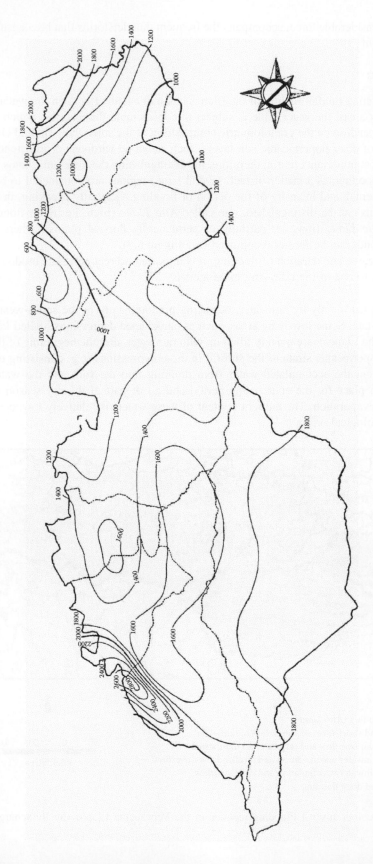

FIGURE 5.4 Isohyets of the Venezuelan Llanos region, based on data from the Ministry of the Environment (Ministerio del Ambiente y de los Recursos Naturales).

winds, often of considerable force, accompany the frequent thunderstorms that break usually during afternoon and night.

5.2.4 FLOODING

The strong, alternating rainfall regime of the Llanos has immediate effects on the water regime of the soils. In many areas, the water table is subject to considerable fluctuations, which may reach 10 m or more, depending on the granulometric composition of the soils, their silt and clay content, and the presence of other impermeable soil layers, such as buried hardpans or ironstone cuirasses.

One of the most important criteria, therefore, for distinguishing the different Llanos landscapes is the degree of flooding; as a matter of fact, several larger regions are considered to be wetlands according to the length and intensity of the period of flooding. Generally speaking, in the Llanos region two sections can be distinguished, one called *Alto Llano* (high, i.e. never-flooded plains) and the other *Bajo Llano* (low, i.e. partially or permanently flooded plains), where a variable spectrum of wetlands can be further recognized (see Figure 5.5).

Besides the degree and duration of flooding, it is important to recognize the mode of flooding, which can belong to one of the following three models.

1. Areas flooded mainly by rainwater, occurring predominantly in the south-west Llanos of Apure state. Some low-lying areas, such as pan-shaped depressions (called *laguna* or *estero* in the Llanos), are quickly filled up with rainwater after the beginning of the rainy season. Impermeable strata of the subsoil in these depressions (e.g. consisting of silt or clay) prevent the accumulated water from draining into the soil, and the water body remains in place for the entire rainy period and all or part of the dry season until its complete evaporation. The nutrient content of these waters is relatively low, because of its meteorological origin.

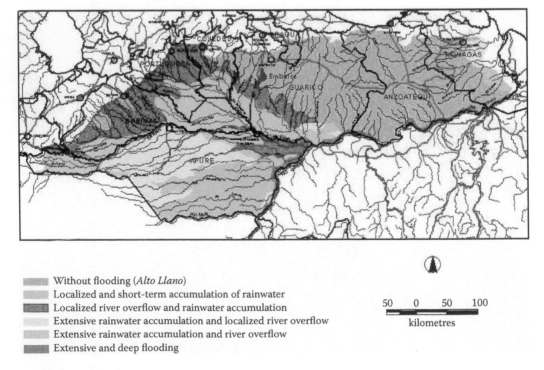

Without flooding (*Alto Llano*)
Localized and short-term accumulation of rainwater
Localized river overflow and rainwater accumulation
Extensive rainwater accumulation and localized river overflow
Extensive rainwater accumulation and river overflow
Extensive and deep flooding

50   0   50   100
kilometres

FIGURE 5.5 (See colour insert.) Flooding regimes in the Venezuelan Llanos (by R. Schargel and J.G. Quintero).

2. Flooding caused by river overflow, which is predominant in the central Llanos (the states of Portuguesa, Guárico and Anzoátegui) and also in large parts of the south-western Llanos of the state of Apure. After the first couple of weeks of the rainy season, many torrential streams descending from the steep eastern Andean slopes reach the western piedmont Llanos; on their way eastwards to the Atlantic Ocean, they normally overflow from their relatively flat beds, causing extensive flooding of the adjacent riverine and interfluvial areas and forming so-called continental deltaic floodplains. These floods are normally of shorter duration, but the muddy waters charged with a high sedimentary load derived from the weathering mountain slopes provide important nutrient inputs to the soils of these floodplains.

3. Areas flooded by the combination of rainwater and river flooding, found in most parts of the Bajo Llano.

Evidently, the soil and associated vegetation types of the Llanos are strongly influenced by the flooding regime of the terrain. Because water is one of the main limiting factors in large parts of the Llanos during the dry season, the presence of permanent or semipermanent water bodies is of great importance for livestock production or other agricultural activities. Attempts have been made since prehistoric times to regulate the availability of soil water (Denevan and Zucchi, 1978); more recently, a large government programme called Módulos de Apure, consisting of the construction of a complex network of dykes, locks and flooded fields, has been implemented in a large wetland area of north-western Apure state.

5.2.5 SOILS

As mentioned in the section on geomorphology, the apparently uniform Venezuelan Llanos plains consist of a mosaic of at least four different types of landscape (alluvial, eolian, high plains and hills, and other denuded surfaces). Depending on the geological origin of the parent material, as well as the prevailing flooding regime, a variety of soil types has developed in each of these broad geomorphic units. According to soil and landscape studies made in several parts of the Llanos (e.g. by Montes and San José, 1995; Berroterán, 1998; Schargel, 2003), the predominant soil families in this region are Ultisols, Inceptisols, Vertisols and Alfisols.

Ultisols are widespread in almost all high plains of the Venezuelan Llanos; these generally acidic soils have moderate to high levels of clay, but low contents of organic matter and low base saturation. Sometimes ferruginous concretions (iron hardpan and plinthite) may be present within the first 2 m of depth. Many Ultisols of the high plains have a very low cation exchange capacity and also sandy surface horizons, which make them extremely infertile. The only fertile soils of this landscape type, such as Entisols and Inceptisols, are generally found in the valley bottoms dissecting the high plains. The extensive hill lands of the central Llanos are covered mostly by Alfisols and Vertisols deriving from the weathering of the underlying sedimentary rocks.

The soils of the eolian plains with dunes are often sandy Entisols with pH below 5.5 and low cation exchange capacity; at the base of the sand dunes, where temporary flooding may occur during the rainy season, the predominant soil type is Ultisol with hardpan at <1.5 m of depth. Extensive wetlands or floodplains can usually be found in the aeolian plains with silt deposits, and Ultisols and Oxisols are again the most frequent soil types.

Finally, the widespread alluvial plains of the Venezuelan Llanos are typical floodplains with a characteristic altitudinal sequence of land forms consisting of (from top to bottom) the elevated river bank or levee (*banco*), followed by a slightly descending basin (*bajío*), and ending in a lower-lying pond (*estero* or *laguna*). The most important soil types of these geomorphic units are Inceptisols in all three land forms, but on banks Entisols and Mollisols may also be found frequently, as well as Vertisols in the lower-lying basins and temporary ponds. Frequently, the water-saturated soils form on their surface a characteristic microtopography of small channels alternating with low elevations, locally called *tatucos*. Many of these recent alluvial plains are located near the piedmont of the Andean Cordillera, where extensive flooded forests were situated until their clear-cutting for agricultural expansion during the past decades.

Somewhat older (Pleistocenic) alluvial plains are widely scattered over the Llanos, especially along the northern bank of the Orinoco river and in certain regions of the states of Barinas and Apure. Ultisols are, again, one of the more important soil types of these plains, but Alfisols can also be found under forest cover near the Andean piedmont of Barinas.

In some parts of the western Llanos less than 100 m a.s.l., soils with high levels of exchangeable sodium are found; these sites — possibly relics of small ponds evaporated during Pleistocenic climatic oscillations — are often covered by large colonies of the characteristic Llanos palm *Copernicia tectorum* (Kunth) Mart. (García-Miragaya et al., 1990).

5.2.6 COMPARATIVE REMARKS

Even from this short description of the physical parameters of the Venezuelan Llanos ecoregion, it is evident that this second-largest neotropical savanna complex differs notably from many other savanna areas in South America. The Brazilian cerrados are located mostly on the well-drained hilly or low- to medium-elevated montane topography of the Brazilian Shield (see Eiten, 1972 and Chapter 2). Because of a much denser vegetation cover consisting largely of inundation forests, the extensive floodplains of the Pantanal may not classify as a true neotropical savanna ecoregion, but rather as a wide swamp woodland in the sense of Beard (1953) alternating with open lagoons and drainage channels. The partly flooded Bolivian savannas of Santa Cruz in the upper Beni and Mamoré basins resemble the Venezuelan Llanos in some aspects (mainly physiographical and floristic), but they are much smaller in extent and show quite distinct climatic and edaphic characteristics because of their much more pronounced continentality (Beck, 1980). Perhaps the shrub savannas and palm groves of the Rio Branco–Rupununi region in northernmost Brazil and western Guyana may be considered the most similar ecoregion to the Llanos, because they occupy a comparable landscape mosaic of flooded and non-flooded alluvial plains filled with recent sediments deriving from the surrounding mountains of the Guayana Shield, and are exposed to a similarly strong biseasonal climatic regime.

In a forthcoming paper, a more thorough comparative analysis of the principal environmental and floristic characteristics of these major South American savanna regions will be presented.

5.3 FLORA AND VEGETATION

5.3.1 HISTORY OF BOTANICAL EXPLORATION

Pehr Loefling, a young Swedish disciple of Linnaeus, was probably the first botanist to set foot on the Venezuelan Llanos; he was appointed to accompany the Royal Spanish Expedición de Límites commissioned to demarcate the frontiers between the Portuguese and the Spanish colonies in the northern Amazon basin. During 1755, Loefling crossed the Llanos from (Nueva) Barcelona in the north to the lower Orinoco, making numerous botanical collections in the savannas and forests along this traverse. The classical neotropical savanna shrub genus *Curatella* was published posthumously by Linnaeus in Loefling's name (von Linné, 1758), as Loefling had died in 1756 of malaria a few months after having reached the lower Orinoco river. Unfortunately, Loefling's historical herbarium and abundant notes on medicinal and ethnobotanical plants have never been found (Castroviejo, 1989), but the accurate descriptions written by him, together with the excellent drawings made by his assistants Juan de Diós Castel and Bruno Salvador Carmona, allowed Linnaeus to publish numerous new genera and species from these hitherto unexplored regions of the New World (Pelayo and Puig-Samper, 1992).

Over the following 250 years, numerous botanists and naturalists have travelled and collected in almost all parts of the wide Llanos plains, accumulating a large number of botanical specimens now lodged in many herbaria across the world. Among the classical collections, those of Alexander von Humboldt and Aimé Bonpland are probably the first botanical specimens from the Llanos to reach European herbaria; they were collected during Humboldt and Bonpland's travel to the upper Orinoco in 1800, during which they had to cross the Llanos twice using two different routes.

Among the many botanists and naturalists visiting the Llanos in the nineteenth and twentieth centuries, only a few will be mentioned here. In 1840–1841, the German botanist G.F.E. Otto explored the region between Cumaná and the Orinoco river (Pittier, 1978). J.G. Myers, a British naturalist travelling from Caracas to Georgetown, probably wrote the first modern scientific publication on the flora and vegetation of the Llanos (Myers, 1933). In the same period, the Swiss botanist Henri Pittier, resident in Venezuela since 1920, had already started to dedicate much time and effort to the botanical exploration of several areas of the Llanos in the periods 1923–1927 and 1940–1945. He published not only the first ecological map of Venezuela in 1920, but also a landmark study of one of the most interesting sections of the eastern Llanos, the Mesa de Guanipa (Pittier, 1948). One of Pittier's disciples, Francisco Tamayo, continued the botanical inventories in many areas of the Llanos during innumerable field trips made between approximately 1950 and 1975; he also published a first detailed classification of the savanna types found in the Llanos region (Tamayo 1958, 1964, 1972).

Another important Venezuelan botanist, Leandro Aristeguieta, besides making numerous collections in many different parts of the Llanos, published the first florula of a savanna area; this was located in the upper Llanos near Calabozo, which had been included in the famous *Estación Biológica de los Llanos* (Aristeguieta, 1966). At the same time, the plant ecologist Volkmar Vareschi carried out the first ecological experiments and ecophysiological measurements at the same biological station, concentrating especially on the role of fire in savanna ecosystems (Vareschi, 1960, 1962). Julian A. Steyermark, the most prolific botanist ever living in Venezuela, contributed enormously to the floristic knowledge of the semideciduous forests in the western Llanos, making several intensive collecting expeditions to the various forest reserves of Ticoporo, Caparo and San Camilo, often accompanied by botanists from the Faculty of Forestry of the Universidad de los Andes in Mérida. Steyermark's collections have been included in a checklist of plant species of the Caparo forest reserve published later by Hernández and Guevara (1994). Unfortunately, these forests have now nearly disappeared as a consequence of intensive timber exploitation and the agricultural colonization process started in the 1950s with the construction of the Pan-American Highway across this area (Crist, 1956; Veillon, 1976).

Mauricio Ramia is one of the main researchers in the ecology and floristics of the open, treeless southern Llanos, especially in the state of Apure. He published a most useful pocketbook on the flora of the Llanos savannas (1974), and his classification of inundated savannas of the Llanos was the first attempt to combine floristic, hydrological and edaphic criteria for this purpose (Ramia, 1959, 1967). Later, Guillermo Sarmiento and Maximina Monasterio developed a similar, hydrologically based classification covering the entire neotropical savannas (Sarmiento, 1984).

Of the numerous botanists working in the Llanos more recently, a few deserve special mention because of their substantial contribution to herbaria by means of their extensive collections: Gerrit Davidse and Angel González, who made the largest and by far the most complete set of collections in Apure state and in several other little-explored savanna areas of the Llanos; Báltazar Trujillo and Francisco Delascio, the first collecting widely in the Llanos for the Agricultural Herbarium of the Universidad Central de Venezuela in Maracay, and the second for the purpose of the establishment of a regional herbarium of the Sociedad de Ciencias Naturales La Salle in San Carlos, Cojedes state, in the north-west Llanos; and lastly Basil Stergios, †Francisco Ortega and Gerardo Aymard, who are the founders of the largest herbarium (PORT) existing today in the Venezuelan Llanos region at the Universidad Nacional Experimental de los Llanos Ezéquiel Zamora in Guanare, Portuguesa state — they also began publishing a local flora of the state (Stergios, 1984).

    Table 5.1 contains the names and areas of activity of the main research stations existing today in the Venezuelan Llanos.

TABLE 5.1
Main Institutions Dedicated to Botanical and Ecological Research in the Llanos

Name and Location	Area of Research	Main Researchers	Comments
Estación Biológica de los Llanos, Calabozo (Guárico)	North-central Llanos	Aristeguieta, Blydenstein, Eden, Foldats, Huber, Medina, Monasterio, Montes, San José, Sarmiento, Tamayo, Vareschi, etc.	Belongs to the Sociedad Venezolana de Ciencias Naturales
Universidad del Oriente, Maturín (Monagas)[a]	North-east Llanos	Lárez, Russell	Herbarium UOJ
Universidad de Los Andes, Mérida (Mérida)[a]	Forest reserves Ticoporo and Caparo, and experimental forest Caimital, Barinas (west Llanos)	Arends, Finol, Guevara, Marcano Berti, Rodríguez	Herbarium MER
	Savannas of Barinas in west Llanos	Sarmiento, Silva	
Universidad Nacional Experimental de los Llanos 'Ezéquiel Zamora,' Guanare (Portuguesa)[a]	North-west, west and south-west Llanos	Aymard, Licata, Ortega, Stergios, etc.	Herbarium PORT
Universidad Central de Venezuela, Caracas (Distrito Federal), Maracay (Aragua)[a]	Módulos de Apure, south-west Llanos; savannas of Santa Rita, south-central Llanos	Berroterán, González, Ponce, Ramia, Susach, Velásquez	Herbaria VEN, MYF (Caracas)
	Savannas of Guárico, central-south Llanos	Badillo, Benítez, Cárdenas, De Martino, Rodríguez, Trujillo	Herbarium MY (Maracay)
Universidad Simón Rodríguez, La Iguana (Guárico)	South-west Llanos	Chacón, Pérez	Experimental research station
Fundación La Salle, San Carlos (Cojedes)[a]	North-west Llanos	Delascio, Reyes-López	Herbarium CAR
Jardín Botánico del Orinoco, Ciudad Bolívar (Bolívar)[a]	South-east Llanos	Delascio, Díaz	Herbarium GUYN

[a] With active herbarium.

Source: Huber, O. et al., *Estado actual del conocimiento de la flora de Venezuela*, Documentos Técnicos de la Estrategia Nacional de Diversidad Biológica no. 1, MARNR–Fundación Instituto Botánico de Venezuela, Caracas, 1998, 153. With permission.

TABLE 5.2
Main Components of the Flora of the Venezuelan Llanos (Preliminary Version)

Plant Group	Number of Families	Number of Genera	Number of Species	Number of Species Endemic to: Llanos	Number of Species Endemic to: Venezuela	Number of Introduced Species
Dicotyledons	133	819	2232	23	79	93
Monocotyledons	35	253	860	1	0	20
Pteridophytes	22	45	127	1	—	1
Total	190	1117	3219	—	—	114

5.3.2 FLORA OF THE VENEZUELAN LLANOS

Considering the huge extension of the Llanos region (almost 240,000 km^2) and its location in Equatorial America, one might expect to find there a rich and diversified flora, similar to that of the adjacent Amazon lowlands. Nevertheless, for a long time the floristic richness of the Llanos was reputed to be relatively low in number and uniform in its distribution. It is only with the conclusion of a detailed and accurate inventory of most botanical collections made in the region during the past two centuries that a more concrete view of the composition and subregional distribution of the flora can be made; this will be presented here in its first approximation. Table 5.2 offers an overview of the presently known composition of the flora of the Venezuelan Llanos region.

The Llanos region covers approximately one-quarter of Venezuela and contains approximately one-fifth of its total vascular plant flora of 15,353 species, according to Huber et al. (1998). Curiously, the Llanos flora represents also roughly one-fifth of the total angiosperm flora of the country but only one-tenth of its fern flora; this difference is probably a result of the unfavourable climatic conditions in the Llanos for the growth of a more diversified fern flora and of the large extension of floodplains.

The ten most important families in the Llanos flora account for approximately 42% of the total flora (see Table 5.3), of which the grasses and sedges together sum c.450 species. The Leguminosae (Fabaceae s.l.) are the most diverse of the dicotyledons (c.350 species), followed by the Rubiaceae, Asteraceae, Euphorbiaceae and Melastomataceae. Orchidaceae are the third most diverse family of the monocotyledons.

The level of endemism in the Llanos flora is low (approximately only 1.1% of the total flora), despite the large area and the great number of taxa present. This is probably because of the very young alluvial landscape and the consequently short evolutionary time available for speciation, together with the lack of major geographical barriers across the entire region. So far, only one fern, 12 monocotyledons and 23 dicotyledonous taxa have been recognized as being endemic to this region. Surprisingly, although grasses are by far the largest family of the Llanos flora (altogether 274 species), no endemic species are reported from there and only a few species of wider neotropical distribution are restricted to the Llanos region in Venezuela. But there are at least three endemic species in the family Cyperaceae (*Calyptrocarya delascioi* Davidse & Kral, *Eleocharis venezuelensis* S. González & Reznicek and *Rhynchospora papillosa* W.W. Thomas). Among the dicotyledons, the Rubiaceae have the most endemic species (*Borreria aristeguietaeana* Steyerm., *Chomelia ramiae* Steyerm. and *Coccocypselum apurense* Steyerm.).

The highest species diversity is concentrated in the semideciduous to evergreen forest types of the western Llanos and in the gallery forests (Aymard, 2003); in contrast, certain typical Llanos families, such as grasses and sedges, have their greatest diversity in the open savanna and shrubland communities.

TABLE 5.3
The 10 Most Important Families of Angiosperms
in the Llanos Flora

	No. of spp.	Percentage of Total no. of spp.
Leguminosae *sensu lato*[a]	360	11.2
Poaceae	274	8.5
Papilionoideae	204	6.3
Cyperaceae	175	5.4
Rubiaceae	132	4.1
Asteraceae	106	3.3
Orchidaceae	102	3.2
Euphorbiaceae	95	2.9
Melastomataceae	90	2.8
Mimosoideae	83	2.6
Caesalpinioideae	73	2.3
Total[b]	1334	41.4

[a] Incl. all three subfamilies
[b] Excl. Leguminosae *sensu lato*

It should also be mentioned that the Llanos plains, as traditionally the most important and extensive agricultural area of Venezuela, have received during the past three centuries numerous introductions of plants from many other regions of the world, which today may cover even larger areas than those occupied by the original savannas. This is the case not only for the extensive sugar cane, corn, rice and sorghum plantations established during the past 50 years, but also for the even more extensive and rapidly increasing pine plantations established during the past decade, which should provide the basic material for a huge programme of paper production. Also, some exotic grass species introduced for livestock production, such as *Urochloa* (*Brachiaria*) *decumbens* (Stapf) R.D. Webster, *Hyparrhenia rufa* (Nees) Stapf and *Imperata brasiliensis* Trin., have escaped from cultivation and are now present in almost all savanna types.

After the publication of the *Catálogo de la Flora Vascular de los Llanos de Venezuela*, it will be possible to make reliable comparisons between the floristic components of different savanna types of the Llanos and other neotropical savanna regions, such as the Brazilian cerrados and the Bolivian savannas. Such studies have not yet been carried out, but they will certainly help to evaluate the relationships between the great savanna regions of the Neotropics, as well as those of the floristic elements of which they themselves are composed. At present, it is assumed that the Venezuelan savannas have a relatively rich and diverse stock of grass species but a low diversity in the woody component, where Caribbean phytogeographical elements predominate, especially among the Leguminosae, Bignoniaceae, Anacardiaceae, Simaroubaceae, Capparaceae and Burseraceae. In fact, some phytogeographers (e.g. Huber et al., 1998) consider the Venezuelan Llanos as the southernmost province of the Caribbean region, since several characteristic elements of that region reach their southern limits there. However, only much more detailed quantitative as well as qualitative floristic analyses of each of the major savanna communities of the Llanos will allow us to understand their present phytogeographical affinities, which in turn are the basis for meaningful conjectures about their geographical and, eventually, palaeo-ecological connections.

5.3.3 VEGETATION OF THE VENEZUELAN LLANOS

Although generally known as one of the characteristic neotropical savanna regions, the Venezuelan Llanos harbour a great variety of different ecosystems with a correspondingly large diversity of plant communities and vegetation types. Unfortunately, a complete inventory of the vegetation types of the Llanos has not yet been written; instead, only a few regional vegetation studies have been made using different classification schemes and methods. The most important of such classifications of the savannas of the Llanos have been produced by Tamayo (1964), Ramia (1967) and Sarmiento (1984), but probably Beard's classification of 1953 is still the one with the widest degree of acceptance at a worldwide level.

In their vegetation map of Venezuela, Huber and Alarcón (1988) included all major vegetation types found in the Llanos under the hierarchical classification scheme given in Table 5.4. The mainly physiognomic categories employed in this classification should be complemented with detailed floristic field data, such as those elaborated by Susach (1989) for the south-central Llanos. Applying the phytosociological methodology of Braun-Blanquet to a large area of temporarily flooded savannas, he elaborated a sophisticated classification of vegetation units that are strongly correlated with the edaphic and hydrological properties of the substrate. Previously, similar studies had been made only at an experimental level by Velásquez (1965) in the dry savannas of the high plains near Calabozo (Guárico state) and, more recently, by Castroviejo and López (1985) in the floodplains of the Llanos of Apure.

In view of the lack of a generally accepted classification of the many different savanna types found in the Llanos, other authors prefer to apply classifications based on key characters of the Llanos landscape, such as hydrology (e.g. Ramia, 1967; Sarmiento, 1984) or the nutrient status of the soils (e.g. Schargel, 2003). Obviously, a truly coherent and practical classification should include as many criteria as possible in order to be sufficiently flexible to take into account the different environmental and biotic factors influencing the nature of each savanna type. It is our view that a sound phytosociological inventory carried out across the entire range of the Llanos savannas in both Venezuela and Colombia will provide the most useful background for establishing a coherent classification of neotropical savannas. Considering the pioneering works made in this field by van Donselaar (1965, 1969) in the savannas of the Guianas, together with the results of Susach, Castroviejo and López in the Venezuelan Llanos, by Janssen (1986) in the Amazonian campos of Humaitá, and by Beck (1980, 1984) and other workers in the northern Bolivian savanna floodplains of Beni, it appears that meaningful comparisons of savanna plant communities on a phytosociological basis should be possible and therefore should be stimulated. Sadly, an unknown number of unique, native savanna plant communities have probably already disappeared as a consequence of the widely progressing transformation of natural savanna landscape into monotonous agro-industrial fields in the Brazilian cerrado, the Bolivian campos and the Venezuelan Llanos.

In conclusion, the Venezuelan Llanos are apparently floristically and ecologically well known and explored, but this wide and important region is still lacking a detailed vegetation map with a coherent classification of all the Llanos plant communities. This information gap is a great obstacle to the implementation of appropriate land management assigning equal values to criteria for social and economic development, as well as to the conservation of unique and biologically significant features of this magnificent and diverse region of the Neotropics.

5.3.3.1 Major Vegetation Types

The natural vegetation of a large majority of the Venezuelan Llanos region consists of a complex mosaic of herbaceous ecosystems, especially savannas. The next most important biome of these wide plains is that of the forests, ranging from small areas of evergreen (ombrophilous) lowland forests to entirely deciduous (tropophilous) low forests, and to riverine forests of the gallery type.

TABLE 5.4

Classification of the Main Vegetation Types in the Venezuelan Llanos

Region, Subregion and Sector	Predominant Vegetation Type		
	Forest	Shrubland	Savanna
B Lowland plains[a]			
B.2 Llanos[b]			
B.21 Western Llanos	Semideciduous to evergreen premontane forests	Degraded woodlands (matorrales)	Temporarily flooded savanna with palms (*Attalea butyracea* (L. f.) Wess. Boer)
B.22 Central High Llanos	Deciduous tropophilous forests; deciduous gallery forests; deciduous forest islands (matas)	Degraded woodlands (matorrales)	Dry *Trachypogon* savannas • with trees • with shrubs • open Dry *Paspalum* or *Axonopus* savannas • with trees • with shrubs • open Temporarily flooded savanna with *Copernicia tectorum* (Kunth) Mart. palms (palmar llanero)
B.23 Central Low Llanos	Deciduous tropophilous forests; deciduous gallery forests; deciduous, temporarily flooded forest islands	Temporarily flooded shrublands with *Curatella americana* L. (chaparrales); temporarily flooded shrublands with *Acosmium nitens* (Vogel) Yakovlev (congriales)	Temporarily flooded *Paspalum* and/or *Mesosetum* savannas • with trees • with *Acrocomia aculeata* (Jacq.) Mart. palms (corozales)
B.24 South-western Llanos (Llanos de Apure)	Subevergreen gallery forests; evergreen flooded forest islands with *Caraipa llanorum* Cuatrec. and *C. densifolia* Mart. (saladillales)	Temporarily flooded shrublands with *Acosmium nitens* (congriales)	Flooded *Leersia, Hymenachne, Paspalum* savannas; temporarily flooded *Paspalum* and/or *Mesosetum* savannas with *Mauritia flexuosa* L. f. palms (morichales)
B.25 Unare depression	Deciduous tropophilous forests	Degraded woodlands (matorrales)	—
B.26 Eastern high plains (mesas orientales)	Deciduous gallery forests	Evergreen shrublands in depressions	Dry *Trachypogon* savannas • with trees • with shrubs • open
B.27 Eastern Llanos	Evergreen gallery forest with *Mauritia flexuosa* (morichales)	Evergreen shrublands in depressions	Dry, open *Trachypogon* savannas

[a] Less than 250 m a.s.l., macrothermic, >24°C.

[b] An area with a mainly tropophilous climatic regime.

Source: after Huber, O. and Alarcón, C., *Mapa de vegetación de Venezuela, con base en criterios fisiográfico-florísticos*, 1:2.000.000, MARNR–the Nature Conservancy, Caracas, 1988. With permission.

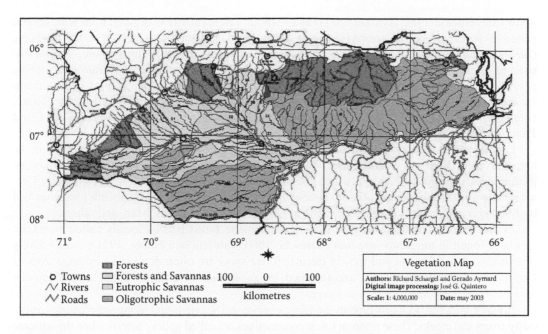

FIGURE 5.6 (See colour insert) General vegetation map of the Venezuelan Llanos (by R. Schargel and G. Aymard).

In contrast to these two widespread Llanos biomes, shrublands are only found in some smaller areas, as are pioneer communities, which are restricted to rock outcrops or sandy beaches in river areas.

In the following sections, some of the more important or widespread vegetation types will be briefly characterized (see also Figure 5.6).

5.3.3.2 Forests

The largest continuous forest cover in the Llanos region used to extend along the eastern piedmont of the Andean Cordillera, in the states of Táchira, Apure, Barinas, Portuguesa and Cojedes (from south-west to north-east). This truly premontane belt of forests, locally designated as the *pie de monte* or *selvas llaneras* (forested plains), was originally between 50 km and 80 km wide, and its existence is clearly conditioned by the permanent input of groundwater and nutrients descending from the adjacent mountain slopes. As a consequence, several forest types of this area are flooded during 6–8 months of the year.

The predominant forest type growing on the drier or little-inundated alluvial plains of the piedmont (Zinck and Stagno, 1966) is a medium tall (25–30 m) forest, semideciduous or deciduous, with usually more than 100 tree species, many of them of valuable commercial timber. Dominant species are *Croton gossypiifolius* Vahl, *Brownea macrophylla* Mast., *Banara guianensis* Aubl., *Pachira quinata* (Jacq.) W.S. Alverson, *Brosimum alicastrum* Sw., *Lonchocarpus hedyosmus* Miq. and *Chrysophyllum argenteum* Jacq. subsp. *auratum* (Miq.) Penn.

Other important forest types are found somewhat more to the east of the piedmont region, on frequently inundated floodplains (Finol, 1976); usually they are medium to tall (up to 40 m), semideciduous to subevergreen forests, often with three well-defined tree layers, and are rich in palms but with a relatively open understorey. The dominant tree species are *Swietenia macrophylla* King, *Cedrela odorata* L. and *Pachira quinata*, accompanied by large stands of the palm *Attalea butyracea* (L. f.) Wess. Boer.

The most interesting forests in terms of structure and floristic composition, however, are those growing in the San Camilo forest reserve located in western Apure state. The forests of this area are clearly more diverse than those found in the Caparo and Ticoporo forest reserves farther to the north (Steyermark, 1982; Aymard, 2003) and harbour a rich, mostly evergreen flora with several endemic species (e.g. *Forsteronia apurensis* Markgr., *Odontocarya steyermarkii* Barneby, *Ouratea apurensis* Sastre and *Ouratea pseudomarahuacensis* Sastre). Furthermore, several floristic elements of the Amazon and Guayana regions can often be found in these transition forests, pointing towards strong connections with the Llanos region in the recent past (e.g. *Licania latifolia* Hook. f., *Abarema laeta* (Benth.) Barneby & Grimes, *Caraipa punctulata* Ducke, *Guatteria cardoniana* R. E. Fr., *G. schomburgkiana* Mart., *Roucheria columbiana* Hallier f. and *Pouteria bangii* (Rusby) Penn.).

Dry, deciduous forests of relatively low structure (10–20 m) but dense understorey grow in the high and central Llanos of Guárico state, where rainfall is less abundant than along the eastern Andean piedmont (Aristeguieta, 1968a); their tree communities are usually rich in legumes (especially *Caesalpinia mollis* (Kunth) Spreng., *Cassia moschata* Kunth and *Acacia glomerosa* Benth.) accompanied by *Bourreria cumanensis* (Loefl.) O.E. Schulz, *Luehea candida* (DC.) C. Mart. and others.

In the same region of the High Llanos, characteristic forest islands locally called *matas* are regularly found in the shrub savannas (Vareschi, 1960; Vareschi and Huber, 1971); these entirely deciduous, up to 25 m tall and mostly circular forest spots are often dominated by huge trees of *Copaifera officinalis* (Jacq.) L., accompanied by *Jacaranda obtusifolia* Bonpl. and *Connarus venezuelanus* Baill. var. *orinocensis* Forero.

Another widespread forest type in the Llanos region is the one found along watercourses, especially rivers and creeks. These riparian forest communities are called gallery forests when the adjacent vegetation types are non-forest plant communities such as savannas or semidesertic scrub. Most of the gallery forests of the Llanos are on slightly elevated river banks (levees) and are predominantly evergreen or semi-evergreen. The most important tree species in gallery forests of the central Llanos are *Vochysia venezuelana* Stafleu, *Symmeria paniculata* Benth. and *Licania apetala* (E. Mey.) Fritsch, together with *Palicourea croceoides* Ham. in the understorey. In gallery forests of the south-west Apure state, various members of the Guayanan-Amazonian genus *Campsiandra* are found, together with other species of similar floristic affinity, such as *Laetia suaveolens* (Poepp.) Benth., *Licania wurdackii* Prance and *Spathanthus bicolor* Ducke, and the palms *Leopoldinia pulchra* Mart. and *Mauritiella aculeata* (Kunth) Burret (Aymard, 2003).

There are also various palm communities, sometimes forming dense, forest-like vegetation, that are typical Llanos elements. The dense palm stands formed by *Mauritia flexuosa* L. f. together with other trees and shrubs are called *morichal* and occur widely in all Llanos landscapes but especially in riparian forests of the eastern, south-central and south-western Llanos (Aristeguieta, 1968b; González, 1987). Other dense palm stands are formed by *Attalea butyracea* in flooded terrains of the western Llanos forests, whereas *Copernicia tectorum*, the classical *palma llanera*, is found in great numbers mixed within deciduous forests covering large floodplains in the central-northern Llanos.

Lastly, a peculiar forest type growing only along the Guayana–Llanos border in Venezuela and Colombia is represented by dense and often quite extensive, but almost pure, stands of *Caraipa llanorum* Cuatrec. and *C. densifolia* Mart.; these form more or less extensive forest islands, locally called *saladillales*, occurring mainly in low depressions that are flooded for at least 8–10 months every year.

5.3.3.3 Savannas

Savannas are defined as tropical herbaceous ecosystems in which a more or less continuous grass cover formed mainly by C4 grass species represents the dominant ecological element; shrubs or trees may be present or not, but they never form a closed canopy (Huber and Riina, 1997). The reference to C4 grasses indicates that savannas are typically lowland or upland vegetation types (*c.*0–1500 m a.s.l.), thus differing ecologically and floristically from tropical high mountain (e.g. *puna*) or extratropical grasslands (e.g. steppes, prairies or *pampas*), which are essentially dominated by C3 grass species.

Savanna types of the Venezuelan-Colombian Llanos, the second largest savanna region in the American tropics, are naturally very diverse, both in their physiognomic aspect and in their floristic composition (Blydenstein, 1967; Sarmiento, 1984). Although the presence of a common 'neotropical floristic matrix' (Huber, 1987) at the generic level may well be recognizable in almost all Llanos savannas (indicated mainly by the important presence of *Trachypogon*, *Axonopus*, *Panicum* and *Mesosetum* in the herbaceous element and *Curatella*, *Bowdichia*, *Byrsonima* and *Roupala* in the ligneous element), there is a wide array of physiognomically and floristically different savanna types depending on regional and local mesoclimatic, edaphic and hydrological variations.

From the floristic point of view, the overwhelming majority of Llanos savanna types belong to *Trachypogon*-dominated grass communities, reunited by van Donselaar (1965, 1969) in the phytosociological order Trachypogonetalia. The principal species, *Trachypogon spicatus* (L. f.) Kuntze, has been reported in numerous inventories from dry to inundated savannas (e.g. in Blydenstein, 1963; Monasterio and Sarmiento, 1968; and San José et al., 1985), although very rare in the latter, where it is commonly replaced by other species of the genera *Axonopus*, *Panicum*, *Paspalum*, *Mesosetum*, *Andropogon* or *Schizachyrium*. It is only in more permanently flooded savannas, such as those of the Apure floodplains, that other grasses such as *Paspalum fasciculatum* Flügge, *Hymenachne amplexicaulis* (Rudge) Nees and *Leersia hexandra* Sw. (Ramia, 1959) predominate; the fact that these grass species belong to the C3 photosynthetic pathway, together with the great importance of sedges in these flooded savannas, emphasizes the different evolutionary strategies that obviously have been developed by various members of the herbaceous savanna element.

In addition to the floristic criterion, the physiognomic aspect of the savannas also provides useful characteristics for their classification. For instance, the complete absence of woody plants emerging from the grass stratum is a remarkable character of large, open savanna tracts in the states of south-west Apure and of Monagas, although the dominant grass species are quite different from each other in the two areas. Much more common, however, are savannas with some kind of woody plants present. One can find all transitions from low to tall shrub savannas, from isolated to dense shrub savannas, from savannas with trees, isolated or in groups, or palms as their most visible components, to savannas with trees and/or palms and/or shrubs in varying proportions. The well-known physiognomic gradient from the Brazilian cerrados, *campo limpo-campo sujo-campo cerrado* (see Chapter 2), is therefore perfectly recognizable also in the Llanos savannas, where these same categories are called, respectively, *sabana abierta* or *lisa*, *sabana arbustiva* and *sabana arbolada*.

When the woody element is particularly conspicuous in the savanna, the Llanos residents distinguish *sabana con mata*, savanna with forest islands; *sabana con palmas*, with scattered palms, usually of the genera *Copernicia* or *Acrocomia*; *palmar llanero*, savanna with large palm stands of *Copernicia tectorum*; *corozal*, savanna with large palm stands of *Acrocomia aculeata* (Jacq.) Mart.; and morichal, savanna with large palm stands of *Mauritia flexuosa*. The Brazilian *cerradão*, a dense arboreal vegetation type, is not considered to be a *sabana* in the Llanos, but rather a *matorral* (scrub) or *monte seco* (dry low forest).

Other useful parameters for the classification of savanna types in the Llanos are the degree of inundation (e.g. non-seasonal, seasonal or hyperseasonal; Sarmiento, 1984) and the nutrient status of the soil (e.g. eutrophic or oligotrophic; Schargel, 2003). Because both refer to environmental conditions and not to the savanna vegetation proper, we suggest they should be employed as complementary characters of each savanna type or for upper hierarchical classification levels.

So far, however, nobody has made a coherent inventory and classification of all savanna types found in the Venezuelan Llanos; therefore, it is still difficult to make meaningful and objective comparisons of the Llanos savannas with the other large neotropical savanna areas in Brazil and Bolivia, or with other tropical savanna regions of the world.

Although not savannas in the strict sense, the diverse and widespread aquatic vegetation of the floodplains, lagoons and rivers of the Llanos should also be mentioned here. More than 200 aquatic plant species have been inventoried by Rial (2001) in water bodies of central Apure state, the most important belonging to the families Pontederiaceae (*Eichhornia*), Alismataceae (*Echinodorus* and *Sagittaria*), Nymphaeaceae (*Nymphaea*), Mayacaceae (*Mayaca*) and Najadaceae (*Najas*). They either grow isolated in the water or form small groups or even massed communities, often appearing as true floating meadows. Certain areas of the inundated Llanos contain very typical aquatic ecosystems, such as the esteros, which may be flooded for six months by more than 2 m of water; during that period one can travel by boat for hours through an intriguing labyrinth of channels and observe the rich floating vegetation, as well as the many palms and tree and shrub islands with dense flocks of red and white birds living in their crowns.

5.3.3.4 Shrublands and Matorrales

Natural (or primary, undisturbed) ecosystems dominated by shrubs are called shrublands or scrub; in some cases, when the woody component is relatively open and consists mainly of small trees, the vegetation type may also be classified as a woodland, which then represents a transitional situation towards the true forest, where the crown cover should always be sufficiently closed to avoid permanent direct sunlight at the ground level. In Venezuela, shrublands are normally found in mountain environments, such as the Andes or the Guayana Shield, where they often represent some of the most extraordinary plant communities, such as the Andean *páramos* or the Guayanan sclerophyllous *Bonnetia* scrub.

Natural shrublands of the Venezuelan Llanos are characteristically of two types: *chaparral* and *congrial*. The first is a dense grass savanna with equally dense stands of *Curatella americana* L. up to 6 m tall; this most typical neotropical savanna bush or tree, with its contorted trunks and very hard, rugged and rough leaves, forming completely open crowns, is called *chaparro* or *chaparro sabanero* (sandpaper tree). It grows in virtually pure stands in some dry, stony areas near the Orinoco river, but it has also been observed in depressions near the same river, which are flooded by 1–1.5 m of water during the rainy season. The Venezuelan chaparral is quite different from the various types of Mexican-Californian chaparrales, with which it shares only a remote physiognomic resemblance but no floristic elements (see Huber and Riina, 2003).

The congrial (*congrio*) is another type of shrubland in which more or less dense colonies of *Acosmium nitens* (Vogel) Yakovlev occupy extensive depressions in the Orinoco floodplains of the states of Guárico and Apure. This same shrub is otherwise found occasionally in the riparian vegetation of many rivers and creeks of the Llanos drainage.

Finally, the so-called *farallones* form a very restricted Llanos ecosystem with interesting shrub communities occupying 5 m to over 50 m deep valleys and clefts located mostly along the southern border of the high plains or mesas of the central Llanos. In this rugged terrain of continuously eroding, steep hill slopes and *barrancas*, dense scrub alternates with spots of relatively open shrub savanna on more level terrain or *moriche* palm communities along the watercourses; the dominant shrub species in these rather inaccessible places have been discovered only recently and consist of *Humiria balsamifera* (Aubl.) J. St. Hil., *Genipa americana* L. and a still unidentified species of *Ternstroemia*.

The term *matorral*, widely used in Latin America and designating a great variety of mostly shrubby or low forest vegetation types, is applied in modern Venezuelan vegetation science to woody secondary or non-climax ecosystems. It includes degraded forests from which almost all trees have been extracted, leaving behind a dense, often thorny scrub or woodland of up to 5–8 m tall, or shrubby anthropogenic ecosystems following overgrazing or excessive burning, dominated by legume-invaded plant communities or natural shrubby regeneration phases of destroyed forests. Such matorrales are very common in many places of the Llanos colonized for several centuries by European cattle ranchers. In the north, along the piedmont region of the Coastal Cordillera, a broad belt of matorrales extends into and fringes the Llanos plains themselves; it is dominated by *Bourreria cumanensis* and numerous fabaceous thorny shrubs and treelets of the genera *Acacia*, *Prosopis*, *Cercidium* and *Mimosa*, together with Capparidaceae, Lamiaceae and Bignoniaceae. This is obviously a heavily degraded phase of the otherwise predominant dry deciduous forests of the area, originally dominated by *Lonchocarpus*, *Pterocarpus*, *Cordia* and *Sterculia* but destroyed as a consequence of severe overexploitation for fuel wood during the past three or four centuries. Also, many forest islands in the interior regions of the High Llanos have been degraded, but the floristic composition of the resulting matorrales varies greatly from one place to another.

5.3.3.5 Pioneer Vegetation

True pioneer plant communities in the Llanos include either rock outcrops or sandbanks in or near the rivers of the floodplains. The former are confined to the border area of the Llanos with the Guayana Shield in Apure state and in some places on the northern shore of the lower Orinoco river, for example in Cabruta. The saxicolous vegetation growing on huge and very ancient granitic boulders (locally called *lajas*) is similar to that found profusely on lajas located in the Venezuelan Guayana: tiny herbs of *Portulaca*, *Borreria* and *Bulbostylis*; colonies of terrestrial bromeliads of the genera *Pitcairnia* and *Ananas*; shrubs of *Acanthella*; lianas of *Mandevilla*; and palms such as *Syagrus orinocensis* (Spruce) Burret are all members of the rich Guayanan laja flora, which have also been found in the south-western Llanos of Apure state.

On sandy beaches of riverbeds or on banks, small psammophilous plant communities formed by grasses, sedges and a few subshrubs grow during the dry season, when the low water level allows them to complete their annual life cycle quickly. The more permanent woody plants of these temporal habitats belong mainly to rheophytic genera such as *Ludwigia* and *Tessaria*, which are able to survive long periods of the year underwater.

5.4 CONSERVATION AND OUTLOOK

Among all major Venezuelan natural regions, the Llanos have the lowest number of national parks (see Table 5.5). In fact, only three parks have been declared in the region during the past 20 years, and although their area may seem sufficiently large, especially by European standards, their real status of protection is still relatively low (Estaba, 1998). This is because of the lack of specific management plans and the low enforcement capacity of the official authorities (Instituto Nacional de Parques Nacionales, INPARQUES) in these remote and scarcely populated areas of the country's interior.

Botanical and ecological study of these protected areas is also sparse; only two studies have been published on the flora and vegetation of the National Park Aguaro-Guariquito (Montes et al., 1987; Susach, 1989) and a third, on the ecology of the inundated savanna types, is in preparation. In the National Park Capanaparo-Cinaruco, intensive botanical collections have been made, but the results have not yet been published. This lack of solid scientific baseline information is a common situation in many Venezuelan national parks and represents a serious obstacle to the elaboration of sound and applicable management plans.

Three natural monuments are usually cited for the Llanos region, but they are all located outside the Llanos proper and consist essentially of atypical Llanos landscape features, such as calcareous mountains or other montane environments in the transition zones between the Coastal Cordillera and the northern Llanos piedmont.

TABLE 5.5

Main Areas under Special Administration (*Áreas Bajo Regimen de Administración Especial*) in the Venezuelan Llanos

Name	Location (State)	Area (ha)	Main Landscape Type(s)	Degree of Anthropic Impact Present
National Parks (when declared)		1,230,118		
Aguaro–Guariquito (1974)	Central Llanos (Guárico)	585,750	High plains, valleys, river floodplains	Moderate to high
Cinaruco–Capanaparo ('Santos Luzardo') (1988)	South-west Llanos (Apure)	584,368	Alluvial floodplains, eolian dunes	Low
Río Viejo–San Camilo (1992–3)	South-west Llanos (Apure)	60,000	Alluvial floodplains	Low
Protected Areas		17,431		
Zona Protectora Tortuga Arrau	Orinoco river (Apure and Bolívar)	17,431	—	Moderate
Forest Reserves		945,026		
Turén	North-west Llanos (Portuguesa)	116,400	Alluvial plains	Extremely high
Ticoporo	West Llanos (Barinas)	187,156	Piedmont, alluvial floodplains	Extremely high
Caparo	West Llanos (Barinas)	174,370	Piedmont, alluvial floodplains	High
San Camilo	South-west Llanos (Apure)	97,100	Alluvial floodplains	Moderate to high
Guarapiche	North-east Llanos (Monagas)	370,000	Alluvial floodplains	Moderate to high

Source: after Weidmann, K. et al., *Parques nacionales de Venezuela*, Oscar Todtmann Editores, Caracas, 2003, 256. With permission.

The four forest reserves originally existing in the Llanos region are located along the eastern Andean piedmont. Far from being protected, these areas are designed for so-called sustainable exploitation, but the results are invariably more or less accelerated disappearance of native vegetation caused by excessive exploitation, agricultural invasions, or both. Thus, of these four forest reserves, at time of writing two (F.R. Turén and F.R. Ticoporo) have already disappeared entirely, while the remaining two (F.R. Caparo and F.R. San Camilo) are almost reaching the same situation. A fifth forest reserve cited in Table 5.5, F.R. Guarapiche, is actually located on the north-eastern border of the Llanos and is mostly in the adjacent ecoregion of the Orinoco river delta floodplains.

The four principal human activities affecting the Llanos ecosystems are as follows.

1. Farming (traditional since the sixteenth century), consisting mainly of both extensive and intensive cattle ranching and the exploitation of caymans.
2. Agro-industry (since 1950), devoted to the cultivation of soya beans, rice, sugar cane and corn maize and with irrigated fields for the production of vegetables.
3. Agro-forestry (recently at an experimental scale).
4. Forestry (since the beginning of the twentieth century), consisting principally of logging natural forests (mainly forest reserves), cellulose production from *Pinus caribaea* Morelet plantations, and silicon production from plantations of *Eucalyptus* sp.

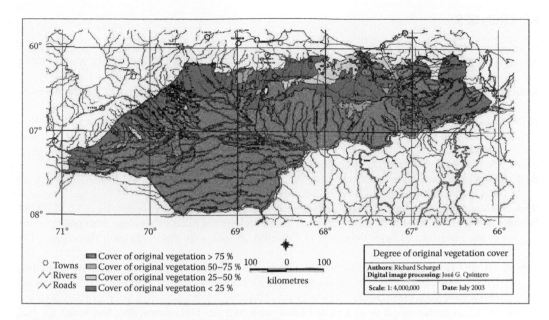

FIGURE 5.7 (See colour insert) Proportion of original vegetation remaining in the Venezuelan Llanos (by R. Schargel).

Each of these activities has increased considerably during the recent past, pushing the agricultural frontier continuously further south and leaving every year less land for remaining natural ecosystems and their native flora and vegetation (see Figure 5.7). This expansion process, well known from the Brazilian cerrado region, is advancing steadily and is probably leading towards the disappearance of the entire biome during the next one or two decades.

In an attempt to counteract this situation, the Venezuelan non-governmental conservation organization Fundación para la Defensa de la Naturaleza (FUDENA) applied to the Global Environmental Facility and obtained a grant in 1999 for the elaboration of a diagnostic inventory of natural resources, human settlements, activities and future trends in the Llanos ecoregion. The publication of the *Catálogo anotado e ilustrado de la Flora Vascular de los Llanos de Venezuela* mentioned in this chapter is one of the main outcomes of the first phase of this project. In a second, subsequent phase, FUDENA will be engaged in the elaboration of a regional management plan for the sustainable use and conservation of the main biomes and ecosystems of the Llanos ecoregion. It is hoped that this plan will be completed in time for rapid implementation of its recommendations, leading to survival of at least some of the most important remaining ecosystems of this once apparently endless, but now so highly threatened, landscape of the Venezuelan interior.

ACKNOWLEDGEMENTS

The authors wish to thank Angel Fernández, Alfred Zinck, Richard Schargel and Rafael García for their helpful comments and provision of detailed information on different subjects relevant to this chapter; Laurie Fajardo for researching and obtaining the new meteorological data and Richard Schargel for the accurate elaboration of the set of electronic maps.

O. Huber wishes to thank the organizers of the *Plant Diversity* symposium of the *Tropical Savannas and Seasonally Dry Forests* international conference (Edinburgh, September 2003) for their kind invitation, and especially Toby Pennington, Jim Ratter and Peter Furley for their friendly collaboration in all phases of the conference. He also thanks the Royal Society of Edinburgh for the generous support of his participation in the conference by providing funds for travel, accommodation, etc. Jim Ratter, Toby Pennington and an anonymous reviewer made useful suggestions that improved the manuscript.

REFERENCES

Aristeguieta, L., Flórula de la Estación Biológica de los Llanos, *Bol. Soc. Venez. Ci. Nat.*, 110, 228, 1966.

Aristeguieta, L., El bosque caducifolio seco de los Llanos altos centrales, *Bol. Soc. Venez. Ci. Nat.*, 113–114, 395, 1968a.

Aristeguieta, L., Consideraciones sobre la flora de los morichales llaneros al norte del Orinoco, *Acta Bot. Venez.*, 3, 19, 1968b.

Aymard, G., Bosques de los Llanos de Venezuela: consideraciones generales sobre su estructura y composición florística, in *Tierras Llaneras de Venezuela*, Hétier, J.M. and López F.R., Eds, SC-77, IRD-CIDIAT, Mérida, 2003, 19.

Beard, J.S., The savanna vegetation of northern tropical America, *Ecol. Monogr.*, 23, 149, 1953.

Beck, S.G., Vegetationsökologische Grundlagen der Viehwirtschaft in den Überschwemmungs-Savannen des Rio Yacumá (Departamento Beni, Bolivien), *Diss. Bot.*, 80, 1, 1980.

Beck, S., Comunidades vegetales de las sabanas inundables del NE de Bolivia, *Phytocoenologia*, 12, 321, 1984.

Behling, H. and Hooghiemstra, H., Late Quaternary palaeoecology and palaeoclimatology from pollen records of the savannas of Llanos Orientales in Colombia, *Palaeogeogr. Palaeoclimatol. Palaeoecol.*, 139, 251, 1998.

Behling, H. and Hooghiemstra, H., Neotropical savanna environments in space and time: late Quaternary interhemispheric comparisons, in *Interhemispheric Climate Linkages*, Markgraf, V., Ed, Academic Press, New York, 2001, 307.

Berroterán, J.L., Geomorfología de un área de Llanos bajos centrales venezolanos, *Bol. Soc. Venez. Ci. Nat.*, 143, 31, 1985.

Berroterán, J.L., Paisajes ecológicos de sabanas en Llanos altos centrales de Venezuela, *Ecotropicos*, 1, 92, 1988.

Berroterán, J.L., Relationships between floristic composition, physiognomy, biodiversity, and soils of the ecological systems of the central high Llanos of Venezuela, in *Forest Biodiversity in North, Central and South America, and the Caribbean (Research and Monitoring)*, Man and the Biosphere Series, vol. 21, Dallmeier, F. and Comiskey, J.A., Eds, Parthenon, New York, 1998, 481.

Blydenstein, J., La sabana de *Trachypogon* del alto Llano. Estudio ecológico de la región alrededor de Calabozo, Estado Guárico, *Bol. Soc. Venez. Ci. Nat.*, 102, 139, 1963.

Blydenstein, J., Tropical savanna vegetation of the Llanos of Colombia, *Ecology*, 48, 1, 1967.

Castroviejo, S., Spanish floristic exploration in America, in *Tropical Forests (Botanical Dynamics, Speciation and Diversity)*, Holm-Nielsen, L.B., Nielsen, I.C., and Balslev, H., Eds, Academic Press, London, 1989, 347.

Castroviejo, S. and López, G., Estudio y descripción de las comunidades vegetales del 'Hato El Frío' en los Llanos de Venezuela, *Mem. Soc. Ci. Nat. La Salle*, 124, 79, 1985.

Comerma, J.A. and Luque, O., Los principales suelos y paisajes del estado Apure, *Agron. Trop.*, 21, 379, 1971.

Crist, R., Along the Llanos-Andes border in Venezuela: then and now, *Geogr. Rev. (New York)*, 46, 187, 1956.

Denevan, W.M. and Zucchi, A., Ridged-field excavations in the central Orinoco Llanos, Venezuela, in *Advances in Andean Archaeology*, Browman, D.L., Ed, Mouton, the Hague, 1978, 235.

Díaz de Gamero, M.L., The changing course of the Orinoco river during the Neogene: a review, *Palaeogeogr. Palaeoclimatol. Palaeoecol.*, 123, 385, 1996.

van Donselaar, J., An ecological and phytogeographic study of northern Surinam savannas, in *The Vegetation of Suriname, vol. 4*, Lanjouw, J. and Versteegh, P.J.D., Eds, van Eedenfonds, Amsterdam, 1965, 163.

van Donselaar, J., Observations on savanna vegetation-types in the Guianas, *Vegetatio*, 17, 271, 1969.

Eiten, G., The cerrado vegetation of Brazil, *Bot. Rev.*, 38, 201, 1972.

Estaba, R.M., Parques nacionales y monumentos naturales de Venezuela: un esfuerzo para la memoria colectiva, *Terra*, 14, 21, 1998.

Finol, H., Estudio fitosociológico de las unidades 2 y 3 de la Reserva Forestal de Caparo, Estado Barinas, *Acta Bot. Venez.*, 11, 17, 1976.

García-Miragaya, J. et al., Chemical properties of soils where palm trees grow in Venezuela, *Commun. Soil Sci. Pl. Anal.*, 21, 337, 1990.

Gentry, A., Neotropical floristic diversity: phytogeographical connections between Central and South America, Pleistocene climatic fluctuations, or an accident of the Andean orogeny? *Ann. Missouri Bot. Gard.*, 69, 557, 1982.

González, V., *Los Morichales de los Llanos Orientales. Un Enfoque Ecológico*, Ediciones Corpoven, Caracas, 1987, 56.

Goosen, D., *Physiography and Soils of the Llanos Orientales, Colombia*, Publications Series B, no. 64, International Institute for Aerial Survey and Earth Sciences, Enschede, 1971, 199.

Graham, A. and Dilcher, D., The Cenozoic record of tropical dry forest in northern Latin America and the southern United States, in *Seasonally Dry Tropical Forests*, Bullock, S.H., Mooney, H.A., and Medina, E., Eds, Cambridge University Press, Cambridge, 1995, 124.

Hernández P.C. and Guevara G., J.R., Especies vegetales de la Unidad I de la Reserva Forestal de Caparo, *Cuaderno Comodato ULA-MARNR* 23, 1, 1994.

Huber, O., *The Neotropical Savannas. Select Bibliography on their Plant Ecology and Phytogeography*, Instituto Italo Latino-Americano, Roma, 1974, 855.

Huber, O., Neotropical savannas: their flora and vegetation, *Trends Ecol. Evol.*, 2, 67, 1987.

Huber, O. and Alarcón, C., *Mapa de Vegetación de Venezuela, con Base en Criterios Fisiográfico-Florísticos, 1:2.000.000*, MARNR–the Nature Conservancy, Caracas, 1988.

Huber, O. and Riina, R., Eds, *Glosario Fitoecológico de las Américas. Vol. 1: América del Sur: Países Hispanoparlantes*, Ediciones Tamandúa, Caracas, 1997, 500.

Huber, O. and Riina, R., Eds, *Glosario Fitoecológico de las Américas. Vol. 2: México, América Central e Islas del Caribe: Países Hispanoparlantes*, Ediciones UNESCO, Paris, 2003, 474.

Huber, O. et al., *Estado actual del conocimiento de la flora de Venezuela*, documentos técnicos de la Estrategia Nacional de Diversidad Biológica, no. 1, MARNR–Fundación Instituto Botánico de Venezuela, Caracas, 1998, 153.

von Humboldt, A., *Personal Narrative of Travels to the Equinoctial Regions of the New Continent, During the Years 1799–1804*, Williams, H.M., Tr, H.G. Bohn Brothers Press, London, 1818–19, 230.

Janssen, A., Flora und Vegetation der Savannen von Humaitá und ihre Standortbedingungen, *Diss. Bot.*, 93, 1986, 324.

Köppen, W., Das geographische System der Klimate, in *Handbuch der Klimatologie*, Bd. 1, Köppen, W. and Geiger, R., Eds, Gebrüder Bornträger, Berlin, 1936, 44.

von Linné, C., *Petri Loefling, Iter Hispanicum eller resa til Spanska Länderna uti Europa och America etc.*, Salvius, Stockholm, 1758.

Meave, J. and Kellman, M., Maintenance of rain forest diversity in riparian forests of tropical savannas: implications for species conservation during Pleistocene drought, *J. Biogeogr.*, 21, 121, 1996.

Monasterio, M. and Sarmiento, G., Análisis ecológico y fitosociológico de la sabana de la Estación Biológica de los Llanos, *Bol. Soc. Venez. Ci. Nat.*, 113-114, 477, 1968.

Montes, R. and San José, J., Vegetation and soil analysis of topo-sequences in the Orinoco Llanos, *Flora*, 190, 1, 1995.

Montes A.R. et al., Paisajes-vegetación e hidrografía del Parque Nacional Aguaro-Guariquito, Estado Guárico, *Bol. Soc. Venez. Ci. Nat.*, 144, 73, 1987.

Morales, F., *El Alto Llano, Estudio de su Geografía Física*, Universidad Central de Venezuela, Caracas, 1978, 185.

Myers, J.G., Notes on the vegetation of the Venezuelan Llanos, *J. Ecol.*, 21, 335, 1933.

Pelayo, F. and Puig-Samper, M.A., *La Obra Científica de Löfling en Venezuela*, Cuadernos Lagoven, Caracas, 1992, 163.

Pennington, R.T., Prado, D.E., and Pendry, C.A., Neotropical seasonally dry forests and Quaternary vegetation changes, *J. Biogeogr.*, 27, 261, 2000.

Pennington, R.T. et al., Historical climate change and speciation: neotropical seasonally dry forest plants show patterns of both Tertiary and Quaternary diversification, *Philos. Trans., Ser. B*, 359, 515, 2004.

Pittier, H., *Esbozo de las Formaciones Vegetales de Venezuela con una Breve Reseña de los Productos Naturales y Agrícolas*, Litografía del Comercio, Caracas, 1920, 44, 1.

Pittier, H., La Mesa de Guanipa. Ensayo de fitogeografía, in *Trabajos Escogidos*, Ministerio de Agricultura y Cría, Caracas, 1948, 195.

Pittier, H., *Manual de las Plantas Usuales de Venezuela y su Suplemento*, Fundación Eugenio Mendoza, Caracas, 1978, 620.

Ramia, M., *Las Sabanas de Apure*, Ministerio de Agricultura y Cría, Caracas, 1959, 134.

Ramia, M., Tipos de sabanas en los Llanos de Venezuela, *Bol. Soc. Venez. Ci. Nat.*, 112, 264, 1967.

Ramia, M., *Plantas de las Sabanas Llaneras*, Monte Ávila Editores, Caracas, 1974, 287.

Rial, A., *Plantas acuáticas de los Llanos inundables del Orinoco, estado Apure. Contribución taxonómica y ecológica*, PhD thesis, Universidad de Sevilla, Seville, 2001, 560.

Roa-Morales, P., Estudio de los médanos de los Llanos Centrales de Venezuela: evidencias de un clima desértico, *Acta Biol. Venez.*, 10, 19, 1979.

Rull, V., Biogeographical and evolutionary considerations of *Mauritia* (Arecaceae), based on palynological evidence, *Rev. Paleobot. Palynol.*, 100, 109, 1998.

Salgado-Labouriau, M.L., A pollen diagram of the Pleistocene–Holocene of Lake Valencia, Venezuela, *Rev. Paleobot. Palynol.*, 30, 297, 1980.

San José, J.J. et al., Bio-production of *Trachypogon* savannas in a latitudinal cross-section of the Orinoco Llanos, Venezuela, *Acta Ecol./Ecol. Gen.*, 6, 25, 1985.

Sánchez, J.M., Aspectos meteorológicos del Llano, *Bol. Soc. Venez. Ci. Nat.* 97, 323, 1960.

Sarmiento, G., *The Ecology of Neotropical Savannas*, Harvard University Press, Cambridge, 1984, 235.

Schargel, R., Geomorfología y suelos de los Llanos Venezolanos, in *Tierras Llaneras de Venezuela*, Hetier, J.M. and López Falcón, R., Eds, Serie Suelos y Clima SC-77, CIDIAT, Mérida, 2003, 89 and 487.

Schargel, R. and Aymard, G., Observaciones sobre suelos y vegetación en la llanura eólica limosa entre los Ríos Capanaparo y Riecito, Estado Apure, Venezuela, *BioLlania*, 9, 119, 1993.

Schubert, C., Climatic changes during the last glacial maximum in northern South America and the Caribbean. A review, *Interciencia*, 13, 128, 1988.

Stergios, B., Flora de la Mesa de Cavacas. I. Introducción, *BioLlania*, 1, 1, 1984.

Steyermark, J.A., Relationships of some Venezuelan forest refuges with lowland tropical floras, in *Biological Diversification in the Tropics*, Prance G.T., Ed, Columbia University Press, New York, 1982, 182.

Steyermark, J.A., Berry, P.E., and Holst, B.K., Eds, *Flora of the Venezuelan Guayana*, vols 1–9, Missouri Botanical Garden Press, St Louis, 1995–2005.

Susach C.F., Caracterización y clasificación fitosociológica de la vegetación de sabanas del sector oriental de los Llanos centrales bajos venezolanos, *Acta Biol. Venez.*, 12, 1, 1989.

Tamayo, F., Notas explicativas del ensayo del mapa fitogeográfico de Venezuela (1955), *Rev. For. Venez.*, 1, 7, 1958.

Tamayo, F., *Ensayo de Clasificación de Sabanas de Venezuela*, Universidad Central de Venezuela, Caracas, 1964, 63.

Tamayo, F., *Los Llanos de Venezuela*, vols 1 and 2, Monte Ávila Editores, Caracas, 1972, 123 and 152.

Tricart, J., Evidence of upper Pleistocene dry climates in northern South America, in *Environmental Change and Tropical Geomorphology*, Douglas, I. and Spencer, T., Eds, G. Allen and Unwin, London, 1985, 197.

Vareschi, V., La Estación Biológica de los Llanos de la Sociedad Venezolana de Ciencias Naturales y su tarea, *Bol. Soc. Venez. Ci. Nat.*, 96, 107, 1960.

Vareschi, V., La quema como factor ecológico en los Llanos, *Bol. Soc. Venez. Ci. Nat.*, 101, 9, 1962.

Vareschi, V. and Huber, O., La radiación solar y las estaciones anuales de los Llanos de Venezuela, *Bol. Soc. Venez. Ci. Nat.*, 119–120, 50, 1971.

Veillon, J.P., Las deforestaciones en los Llanos Occidentales de Venezuela desde 1959 hasta 1975, in *Conservación de los Bosques Húmedos de Venezuela*, Hamilton, L., Ed, Sierra Club, Consejo de Bienestar Rural, Caracas, 1976, 97.

Velásquez, J., Estudio fitosociológico acerca de los pastizales de las sabanas de Calabozo, Estado Guárico, *Bol. Soc. Venez. Ci. Nat.*, 109, 59, 1965.

Walter, H. and Medina, E., Caracterización climática de Venezuela sobre la base de climadiagramas de estaciones particulares, *Bol. Soc. Venez. Ci. Nat.*, 119–120, 212, 1971.

Weidmann, K. et al., *Parques Nacionales de Venezuela*, Oscar Todtmann Editores, Caracas, 2003, 256.

Zinck, A. and Stagno, P., *Estudio Edafológico de la Zona Sto Domingo-Pagüey, Estado Barinas*, Ministerio de Obras Públicas, División de Edafología, Guanare, 1966, 304.

6 The Brazilian Caatinga: Phytogeographical Patterns Inferred from Distribution Data of the Leguminosae

Luciano Paganucci de Queiroz

CONTENTS

ABSTRACT

The caatinga represents the largest and most isolated of the South American dry forests, occupying more than 850,000 km² in the semi-arid region of north-eastern Brazil. Leguminosae is the best represented family in the caatinga, with 293 species in 77 genera, and comprises almost one-third of the total plant diversity there. The geographical distribution patterns of its taxa define seven major centres of endemism. Similarity analyses combined with these patterns reveal two major floristic groups with distinct composition and phenological traits. One group occupies *c.*70% of a mostly continuous surface of soils primarily derived from crystalline basement rocks. The other group occurs on disjunct and sandy sedimentary surfaces. The data support a scenario wherein the sedimentary areas became dissected during a huge process of pediplanation during the Tertiary period, which promoted both the vicariance of its flora and opened the way to the expansion of the flora related to the seasonally dry tropical forests.

6.1 INTRODUCTION

The caatinga is the dominant vegetation form in the semi-arid area of north-eastern Brazil. It covers more than 850,000 km², from *c*.02°50′S at its northern limits in the states of Ceará and Rio Grande do Norte to *c*.17°20′S in northern Minas Gerais state. Caatinga occurs under a prevailing semi-arid climate with a high evapotranspiration potential (1500–2000 mm/year) and a low precipitation (300–1000 mm/year) that is usually concentrated within 3–5 months (Sampaio, 1995). The caatinga may be characterized as a low forest composed mostly of small trees and shrubs, frequently having twisted trunks and thorns, with small leaves that are deciduous in the dry season. Succulent plants of the family Cactaceae are common, and there is an ephemeral herbaceous layer present only during the short rainy season.

Prado (1991) and Prado and Gibbs (1993) proposed that the flora of the dry areas of South America may represent a relict of a wider seasonally dry forest biota that reached its maximum expansion during the driest phases of the Pleistocene epoch. The present-day distribution of this flora forms an arc running from the caatinga (the caatinga nucleus in north-east Brazil) southwards, through the semideciduous forests of south-eastern Brazil, to the confluence of the Parana and Paraguay rivers in north-eastern Argentina (the Paranaense or Misiones nucleus), and then turning northwards to south-eastern Bolivia and north-western Argentina (the Piedmont nucleus). These three major areas were considered part of a new phytogeographical unit of South America called the neotropical seasonally dry tropical forests (SDTFs) (Pennington et al., 2000; Prado, 2000). These dry forests also occur in smaller areas in the dry Andean valleys of Bolivia and Peru, coastal Ecuador and the Guajira region of the Caribbean coast of Colombia and Venezuela.

The caatinga constitutes the largest and most isolated nucleus of the SDTF. Historically, it has been considered as having a low number of species, and as poor in endemism at both the generic and the specific levels (Rizzini, 1963, 1979; Andrade-Lima, 1982). Following this view, it was proposed that the caatinga does not have an autochthonous flora, and that most of its elements were derived from the chaco and the Atlantic rain forest (Rizzini, 1963, 1979; Andrade-Lima, 1982).

Other authors raised doubts about the floristic relationships of some types of vegetation often included within the caatinga, such as the carrasco, which is frequently interpreted as being related to the cerrado vegetation (Fernandes, 1996; Prado, 2003). Prado and Gibbs (1993) rejected the hypothesis of an ancient link between the caatinga and the chaco, based on their analysis of distribution patterns of many woody plant species. More recently, Harley (1996), Giulietti et al. (2002) and Prado (2003) came out against the predominant view of the caatinga as having both a low floristic diversity and a low degree of endemism. They demonstrated that the caatinga contains a surprisingly high number of endemic taxa, including at least 18 genera and 318 species (Giulietti et al., 2002). These data suggest that the caatinga displays the most diverse and divergent biota of any other nucleus of the SDTF, a position that is supported by data from some groups of animals, particularly lizards (Rodrigues, 1996, 2003) and bees (Zanella and Martins, 2003).

Although often viewed as a single unit, a number of questions still arise concerning the caatinga. What are its limits? What is its floristic composition? How should its natural vegetation units be classified? How have different environmental factors affected diversity patterns? Is the caatinga a natural unit or the product of intensive human activity during four centuries of occupation? A careful reading of published works concerning the caatinga reveals a fundamental lack of agreement on all such matters. As such, these questions might benefit from a phytogeographical approach that seeks to recognize a hierarchical classification of its floristic composition in order to establish a relationship between vegetation and possible controlling factors (such as climate, soils and other environmental determinants), and that takes into account the putative palaeo-environmental history of the region.

Most of the phytogeographical studies carried out from the 1950s to the 1970s in areas of caatinga were sporadic surveys that employed no explicit methodologies for comparing different areas (e.g. Andrade-Lima, 1954, 1971, 1977). More recently, the use of multivariate analysis to compare species lists has produced more objective results and has allowed the analysis of a larger volume of data (e.g. Rodal and Nascimento, 2002; Alcoforado-Filho et al., 2003). These studies based on quantitative techniques have some predictive value concerning the distribution of biota and its putative determining factors. Groups derived from such analyses could stimulate hypotheses concerning past historical links among different areas, as an initial step towards a historical biogeographical approach. Unfortunately, these works did not advance interpretations of possible historical events that could have contributed to the present-day distribution of this vegetation.

In this chapter, some of these questions are considered in the light of the distribution data of the family Leguminosae in the caatinga. This family is particularly species-rich in the dry lands (the 'succulent biome' of Schrire et al., 2005), and metacommunity rather than vicariance processes are being proposed to explain its distribution (Lavin et al., 2004). As in other seasonally dry forest areas, Leguminosae is the most diverse plant family in the caatinga. An examination of the published literature allowed Queiroz (2002) to estimate a total of 256 species of legumes for the caatinga, of which 139 are possibly endemic to it. These numbers (256 and 139) correspond to approximately one-third of the total number of plant species in the caatinga, and one-third of the total number of endemic caatinga species surveyed by Giulietti et al. (2002), respectively. Queiroz (2002) also concluded that most of these endemic taxa are not widely distributed throughout the caatinga, but are more narrowly distributed within the area of the biome. Because of its high number of taxa and the distribution patterns of its species, the Leguminosae family seems to represent a model group for examining biogeographical hypotheses concerning the caatinga.

This chapter therefore aims to discuss the limits of the caatinga and to establish biogeographical hypotheses that help to understand floristic relationships of its biota by identifying centres of endemism and distribution patterns, using the Leguminosae as a model.

6.2 MATERIAL AND METHODS

Data concerning plant distribution were mostly derived from three primary sources:

1. published floristic surveys of areas in north-eastern Brazil (cited in Table 6.1);
2. taxonomic monographs of groups of the Leguminosae (cited in Table 6.2); and
3. almost 20,000 specimens of the Leguminosae surveyed in the ALCB, CEPEC, EAC, HRB, HUEFS, IPA, K and PEUFR herbaria.

Data from areas of caatinga in the state of Bahia were taken from Lewis (1987) and from unpublished floristic inventories prepared principally by researchers from the Feira de Santana State University (Table 6.1). The inclusion of these unpublished data from Bahia was critical, for this state contains the greatest portion of caatinga and yet is more poorly surveyed in comparison with caatinga areas located further north. In fact, not a single Bahian site was cited in the principal published revisions dealing with floristics (Sampaio, 1995; Rodal and Sampaio, 2002; Prado, 2003) and comparative floristic composition (e.g. Araújo et al., 1998; Alcoforado-Filho et al., 2003).

Distribution patterns for taxa were defined by their limits of distribution. Dot maps were prepared and analysed, taking into consideration data concerning the physical environment of the area occupied by the taxon, especially the underlying geology, the soil and the local geomorphology. These data were acquired directly from the area descriptions accompanying floristic surveys, and were complemented where necessary by data available on the website of the Brazilian Institute of Geography and Statistics (http://www.ibge.gov.br).

TABLE 6.1
Areas Selected for Similarity Analysis

Site[a]	Abbreviation[b]	Coordinates	Vegetation Type and Physiognomy	Topography and Substrate	Reference(s)
Abaíra, BA (carrasco)	ABAI	13°17'S, 41°46'W	Ecotone caatinga-cerrado	Hilly, mostly quartzite	Zappi et al. (2003)
Barreiras, BA	BARR	11°59'S, 45°34'W	Open cerrado	Cental Brazilian plateau, dystrophic soil	Felfili and Silva (2001), L.P. Queiroz (unpublished data)
Bezerros, PE	BEZE	08°12'S, 35°49'W	Brejo forest	Hilly, mostly eutrophic soil	Sales et al. (1998), Santos (2002)
Biological Station of Canudos, BA	CANU	09°58'S, 39°06'W	Dense to open, mostly shrubby caatinga	Hilly, deep dystrophic sand	Queiroz (2004b)
Bom Jesus da Lapa, BA	BJLP	13°24'S, 43°21'W	Tall caatinga forest	Calcareous soil	Andrade-Lima (1977), L.P. Queiroz (unpublished data)
Bonito, PE	BONI	08°29'S, 35°41'W	Brejo forest	Hilly, mostly eutrophic soil	Sales et al. (1998), Santos (2002)
Brejo da Madre de Deus, PE	BJMD	08°09'S, 36°22'W	Brejo forest	Hilly, mostly eutrophic soil	Lyra (1982), Sales et al. (1998), Santos (2002)
Buíque, PE	BUIQ	08°39'S, 38°41'W	Mostly shrubby, dense caatinga	Sedimentary tablelands, deep dystrophic sand	Rodal et al. (1998)
Campo Alegre de Lurdes, BA	CPAL	09°30'S, 43°05'W	Tall caatinga forest	Hilly, eutrophic soil	Queiroz (2004b)
Cariris, PB	CARI	07°28'S, 36°53'W	Low caatinga forest	Hilly, mostly granite	Gomes (1979), Lima (2004)
Caruaru, PE	CARU	08°11'S, 36°01'W	Caatinga forest	Undulated, eutrophic soil	Alcoforado-Filho et al. (2003)
Casa Nova, BA	CSNV	09°30'S, 41°12'W	Patchy scrub on sandy dunes	Dune, deep dystrophic sand	Queiroz (2004b)
Catolés, BA (campo rupestre)	CTCR	13°17'S, 41°46'W	Campo rupestre	Hilly, mostly quartzite exposed rocks	Zappi et al. (2003)
Catolés, BA (cerrado)	CTCE	13°17'S, 41°46'W	Cerrado	Upland tableland, deep dystrophic sand	Zappi et al. (2003)
Catolés, BA (mountain forest)	CTMT	13°17'S, 41°46'W	Upland evergreen forest	Hilly, mostly eutrophic soil	Zappi et al. (2003)
Correntina, BA	CORR	13°23'S, 44°34'W	Dense shrubby cerrado	Central Brazilian plateau, deep dystrophic sand	Felfili and Silva (2001), L.P. Queiroz (unpublished data)
Dois Irmãos, PE	DOIR	08°04'S, 34°57'W	Atlantic rain forest	Coastal plain, eutrophic soil	Guedes (1992)
Floresta, PE	FLOR	08°35'S, 38°02'W	Brejo forest	Hilly, mostly eutrophic soil	Sales et al. (1998), Santos (2002)
Grão-Mogol, MG (campo rupestre)	GMCR	16°35'S, 42°52'W	Campo rupestre	Hilly, mostly quartzite exposed rocks	Queiroz (2004a)

Location	Code	Coordinates	Vegetation type	Description	Reference
Grão-Mogol, MG (cerrado)	GMCE	16°35'S, 42°52'W	Cerrado	Upland tableland, deep dystrophic sand	Queiroz (2004a)
Ibiraba, BA	IBIR	10°48'S, 42°50'W	Patchy, dense, non-deciduous scrub	Stabilized sandy dunes	Rocha et al. (2004)
Ilhéus, BA	ILHE	14°47'S, 39°10'W	Atlantic rain forest	Coastal plain, eutrophic soil	Mori et al. (1983)
Ipirá, BA	IPIR	12°08'S, 40°00'W	Low caatinga forest	Undulated, mostly eutrophic	L.P. Queiroz (unpublished data)
Itiúba, BA	ITIU	10°21'S, 39°36'W	Tree caatinga on a sloped topography	Hilly, mostly eutrophic soil	Queiroz (2004b)
Jaburuna, CE	IBIA	03°54'S, 40°59'W	Carrasco	Elevated plateau with deep sandy soil	Araújo et al. (1999)
Januária, MG	JANU	15°28'S, 44°23'W	Seasonally dry forests	Calcareous soil derived from limestone	Ratter et al. (1978)
Jataúba, PE	JATA	08°02'S, 36°28'W	Brejo forest	Hilly, mostly eutrophic soil	Moura (1997)
Maracás, BA	MARA	13°23'S, 43°21'W	Tall caatinga forest	Elevated plateau, eutrophic soil	Andrade-Lima (1971), L.P. Queiroz (unpublished data)
Milagres, BA	MILA	12°53'S, 39°49'W	Low caatinga forest	Pediments of granitic inselbergs, soil?	França et al. (1997)
Mucugê, BA (campo rupestre)	MUCR	12°59'S, 41°20'W	Campo rupestre	Hilly, mostly quartzite exposed rocks	Harley and Simmons (1986)
Mucugê, BA (cerrado)	MUGE	12°59'S, 41°20'W	Cerrado	Upland tableland, deep dystrophic sand	Harley and Simmons (1986)
Novo Oriente, CE	NVOR	05°28'S, 40°52'W	Carrasco	Deep sandy dystrophic soils	Araújo et al. (1998)
Ouricuri, PE	OURI	07°57'S, 39°38'W	Mostly open, shrubby caatinga	Seasonally flooded depression, mostly sand	Silva (1985)
Pesqueira, PE	PESQ	08°21'S, 36°41'W	Brejo forest	Hilly, mostly eutrophic soil	Correia (1996)
Raso da Catarina, BA	RSCT	09°31'S, 38°46'W	Dense mostly shrubby caatinga	Sedimentary tableland, deep dystrophic sand	Guedes (1985); Queiroz (2004b); L.P. Queiroz, F.P. Bandeira and M.L.S. Guedes (in prep.)
Remanso, BA	REMA	09°33'S, 42°05'W	Open shrubby caatinga	Seasonally flooded depression, dystrophic sand	Queiroz (2004b)
São Raimundo Nonato, PI	SRNN	07°54'S, 42°35'W	Mostly shrubby caatinga with low trees	Seasonally flooded depression	Emperaire (1983), L.P. Queiroz (unpublished data)

(continued)

TABLE 6.1
Areas Selected for Similarity Analysis (Continued)

Site[a]	Abbreviation[b]	Coordinates	Vegetation Type and Physiognomy	Topography and Substrate	Reference(s)
São Vicente Ferrer, PE	SVFE	07°35'S, 35°30'W	Brejo forest	Hilly, mostly eutrophic soil	Sales et al. (1998), Santos (2002)
Seridó Ecological Station, RN	SERI	06°35'S, 37°15'W	Open low caatinga	Hilly, mostly shallow gravelled soil	Camacho (2001)
Serra da Capivara, PI	SACA	08°26'S, 42°19'W	Caatinga: shrubby deciduous scrub	Deep sandy dystrophic soils	Lemos and Rodal (2002)
Serra da Jibóia, BA	SAJI	12°51'S, 39°28'W	Atlantic rain forest	Hilly, eutrophic soil	Carvalho-Sobrinho and Queiroz (2005)
Triunfo, PE	TRIU	07°51'S, 38°04'W	Brejo forest	Hilly, mostly eutrophic soil	Ferraz et al. (1998), Sales et al. (1998), Santos (2002)
Xingó, SE	XING	09°35'S, 37°35'W	Open shrubby caatinga	Slightly undulated on granitic substrate	Fonseca (1991)

[a] BA, Bahia; CE, Ceará; MG, Minas Gerais; PB, Paraíba; PE, Pernambuco; PI, Piauí; RN, Rio Grande do Norte; SE, Sergipe.

[b] Used in Figure 6.2.

TABLE 6.2
Genera of the Leguminosae with Their Respective Numbers of Native Species and Infraspecific Taxa in the Caatinga and with Relevant References for Biogeographical Data

Genera	Number of Species	Number of Additional Infraspecific Taxa	Number of Taxa	Reference(s)
		Caesalpinioideae		
Apuleia	2	—	2	—
Bauhinia	16	—	16	Vaz (1979, 2001)
Caesalpinia	8	3	11	Lewis (1998)
Cassia	1	—	1	Irwin and Barneby (1982)
Cenostigma	1	—	1	—
Chamaecrista	26	9	35	Irwin and Barneby (1978, 1982)
Copaifera	4	1	5	—
Diptychandra	1	—	1	Lima et al. (1990)
Goniorrhachis	1	—	1	—
Guibourtia	1	—	1	Barneby (1996)
Hymenaea	4	1	5	Lee and Langenhein (1975)
Martiodendron	1	—	1	—
Melanoxylon	1	—	1	—
Parkinsonia	1	—	1	—
Peltogyne	1	—	1	Silva (1976)
Peltophorum	1	—	1	—
Poeppigia	1	—	1	—
Pterogyne	1	—	1	—
Senna	19	6	25	Irwin and Barneby (1982)
		Mimosoideae		
Acacia	11	—	11	—
Albizia	2	—	2	Barneby and Grimes (1996)
Anadenathera	2	1	3	Altschul (1964)
Blanchetiodendron	1	—	1	Barneby and Grimes (1996)
Calliandra	12	1	13	Barneby (1998)
Chloroleucon	3	—	3	Barneby and Grimes (1996)
Desmanthus	1	—	1	—
Enterolobium	2	—	2	Mesquita (1990)
Leucochloron	1	—	1	Barneby and Grimes (1996)
Mimosa	37	4	41	Barneby (1991)
Neptunia	2	—	2	—
Parapiptadenia	2	—	2	Lima and Lima (1984)
Pithecellobium	1	—	1	Barneby and Grimes (1997)
Piptadenia	7	—	7	—
Plathmenia	1	—	1	Warwick and Lewis (2003)
Prosopis	1	—	1	—
Pseudopiptadenia	3	—	3	Lewis and Lima (1990)
Samanea	1	—	1	Barneby and Grimes (1996)
Zapoteca	2	—	2	—
		Papilionoideae		
Acosmium	2	—	2	—
Aeschynomene	9	1	10	—
Amburana	1	—	1	—

(continued)

TABLE 6.2
Genera of the Leguminosae with Their Respective Numbers of Native Species and Infraspecific Taxa in the Caatinga and with Relevant References for Biogeographical Data (Continued)

Genera	Number of Species	Number of Additional Infraspecific Taxa	Number of Taxa	Reference(s)
Andira	2	—	2	Pennington (2003)
Arachis	4	—	4	Krapovickas and Gregory (1994)
Canavalia	2	—	2	—
Centrolobium	1	—	1	—
Centrosema	6	—	6	Williams and Clements (1990)
Chaetocalyx	2	—	2	—
Coursetia	2	—	2	Lavin (1988)
Cratylia	2	—	2	—
Crotalaria	3	—	3	—
Dalbergia	2	—	2	Carvalho (1985)
Desmodium	2	—	2	—
Dioclea	4	—	4	—
Discolobium	1	—	1	—
Erythrina	1	—	1	—
Galactia	2	—	2	—
Geoffroea	1	—	1	Ireland and Pennington (1999)
Indigofera	3	—	3	—
Lonchocarpus	5	—	5	Tozzi (1985)
Luetzelburgia	3	—	3	—
Machaerium	5	1	6	—
Macroptilium	6	—	6	Fevereiro (1986)
Mysanthus	1	—	1	Lewis and Delgado (1994)
Periandra	1	—	1	—
Platymiscium	2	—	2	Klitgaard (1995)
Platypodium	1	—	1	—
Poecilanthe	3	—	3	—
Poiretia	1	—	1	—
Pterocarpus	3	—	3	—
Rhynchosia	2	—	2	—
Riedeliella	3	—	3	Lima and Vaz (1984)
Sesbania	1	—	1	Monteiro (1994)
Stylosanthes	7	—	7	Ferreira and Costa (1979)
Tephrosia	2	1	3	—
Trischidium	2	—	2	Ireland (2001)
Zollernia	1	—	1	—
Zornia	9	—	9	Mohlenbrock (1961)
Total	**293**	**—**	**322**	**—**

In order to analyse possible correlations among the distribution patterns of different taxa within the caatinga boundary, the ecoregion limits proposed by Velloso et al. (2002) were used. These authors applied the concept of ecoregions (sensu Bailey, 1998) to the caatinga biome, recognizing eight major units that were defined by key biotic and abiotic factors regulating the structure and function of the natural communities (Figure 6.1). These ecoregions were treated as an initial hypothesis for natural phytogeographical units of lower hierarchical level, and were tested by examining the congruence of distribution patterns of the Leguminosae. A taxon was considered

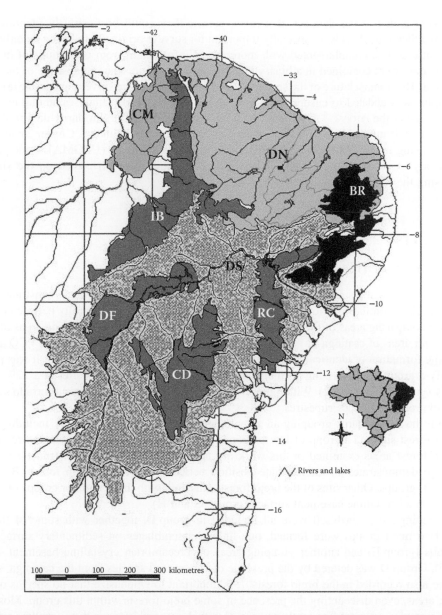

FIGURE 6.1 Limits of the morphoclimatic caatinga domain based on the 800 mm/year isohyet and the ecoregions proposed by Velloso et al. (2002). BR, Borborema; CD, Chapada Diamantina complex; CM, Campo Maior complex; DF, dunes of the São Francisco; DN, northern Sertaneja depression; DS, southern Sertaneja depression; IB, Ibiapaba–Araripe plateaux; RC, Raso da Catarina.

endemic if it was found in only one of these ecoregions. Taxa occurring in two or more contiguous ecoregions were considered widely distributed, and those found in two or three non-contiguous ecoregions were considered disjunct within the caatinga.

Studies of floristic similarity were carried out using 43 areas selected as representing the major biomes of north-eastern Brazil: Atlantic rain forest (*mata atlântica*), savannas (*cerrados*), upland rocky fields (*campos rupestres*) and caatinga. In the case of the caatinga, areas that were different based on physiognomy, type of substrate and occurrence in different geomorphological units were included (Table 6.1). An additional criterion for selection of an area was the availability of voucher specimens to check identification. In addition to the species cited in the published checklists, other species were added to the data-set of a given area if they could be positively identified from herbarium collections. This was especially important for surveys originally focusing on only woody species that could be complemented with more recent information from collections of herbs and vines. The data were combined in a binary matrix based on the presence or absence of the species in each area. The nomenclature of the species cited, both in the floristic surveys and in the taxonomic monographs, was updated to ensure taxonomic uniformity. Similarity analyses among areas were carried out using the FITOPAC 1.1 software program (Shepherd, 1995). Similarity among the areas was calculated from the binary matrix using the Sørensen (Dice) index. Cluster analysis was performed using unweighted pair group method with arithmetic mean (UPGMA). Principal coordinates analyses (PCO) using Gower distance measurements were performed on the similarity matrix using the same software.

6.3 RESULTS

6.3.1 ANALYSIS OF FLORISTIC SIMILARITY

The analysis of floristic similarity performed on the areas cited in Table 6.1 demonstrated low to medium levels of similarity among the areas. The Sørensen index was generally between 0.20 and 0.67 when comparing areas of the same biome, but reaching zero when comparing areas of campo rupestre with areas of caatinga or Atlantic rain forest. As both the UPGMA and the PCO analyses support the formation of identical groups, we have presented here only the former. It was possible to identify a group encompassing all the areas of cerrado and campos rupestres at a 22% level of similarity (group A, Figure 6.2). Within this class, the analysis further separates a cerrado subgroup from a subgroup of campos rupestres.

Within the second large grouping an assemblage of moist forests (group B), including a high mountain forest site and a group of low mountain forest areas (*brejo* forests), can be recognized. The brejo forest areas examined in this work did not group together; the easternmost (group C) assemblage demonstrates an intermediate position between the Atlantic rain forest (B) and the caatinga (D) groups. Other sites of the brejo forests clustered with the two major groups of caatinga occurring on a crystalline basement surface (groups G and H).

The caatinga sites analysed were all included in group D, together with some of the brejo forests. Two main groups were formed, one linking assemblages on sedimentary surfaces with sandy soils (group E) and another grouping areas on Precambrian crystalline basement surfaces (group F). Group D was defined by the presence of widespread species from the caatinga, some of which are also recorded in the brejo forests. It is important to emphasize that no species occurred in all the areas, even discounting the presence of some brejo forests within this group. Most of the diagnostic species are more frequently associated with the assemblages on the crystalline substrates (group F) and are more occasional in sandy environments (group E), for example, *Acacia langsdorffii* Benth., *Anadenanthera colubrina* (Vell.) Brenan var. *cebil* (Griseb.) Altschul, *Caesalpinia ferrea* Mart. ex Tul., *C. pyramidalis* Tul., *Senna macranthera* (Collad.) H.S. Irwin & Barneby and *S. spectabilis* (DC.) H.S. Irwin & Barneby. In contrast, *Piptadenia moniliformis* Benth. is common in sandy areas and is only occasional on crystalline surfaces. Species ubiquitous in both assemblages in the caatinga were *Amburana cearensis* (Allemão) A.C. Sm., *Chloroleucon foliolosum* (Benth.) G.P. Lewis, *Mimosa ophthalmocentra* Mart. ex Benth., *M. tenuiflora* (Willd.) Poir., *Piptadenia stipulacea* (Benth.) Ducke and *Senna rizzinii* H.S. Irwin & Barneby. Thus there are relatively few species of Leguminosae that could be considered characteristic of the entire caatinga, although some weedy species, taking advantage of anthropogenic environments, occur scattered across the caatinga, such as *Centrosema brasilianum* (L.) Benth., *Chaetocalyx scandens* (L.) Urb. var. *pubescens* (DC.) Rudd, *Crotalaria holosericea* Nees & Mart., *Indigofera suffruticosa* Mill., *Macroptilium lathyroides* (L.) Urb., *Tephrosia purpurea* (L.) Pers. and *Zornia brasiliensis* Vog.

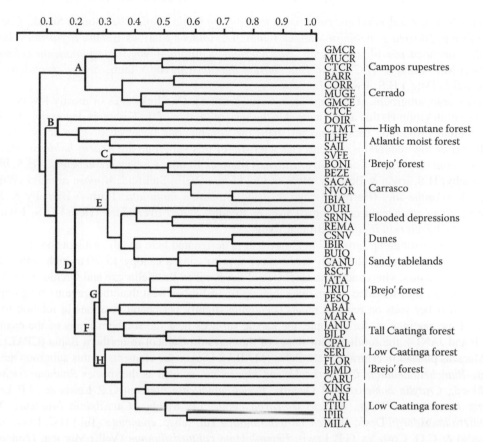

FIGURE 6.2 Dendrogram showing the similarity between 43 areas of north-eastern Brazil, representing the major vegetation types based on the presence of species of Leguminosae. See Table 6.1 for definitions of abbreviations.

Group E is associated with sites on dystrophic quartzose sandy soils. These areas form disjunct patches of sedimentary surfaces in the Ibiapaba plateau (IBIA and NVOR in Figure 6.2), the Serra da Capivara in Southern Piauí (SACA), the Tucano-Jatobá sedimentary basin (BUIQ, CANU and RSCT), the dunes in the middle course of the São Francisco river valley (CSNV and IBIR) and depressed areas subject to seasonal flooding (OURI, REMA and SRNN). It is a well-defined group with 13 exclusive species: *Aeschynomene martii* Benth., *Trischidium molle* (Benth.) H.E. Ireland, *Caesalpinia microphylla* Mart. ex Tul., *Calliandra depauperata* Benth., *Calliandra macrocalyx* Harms, *Cenostigma gardnerianum* Tul., *Cratylia mollis* Mart. ex Benth., *Dioclea marginata* Benth., *Galactia remansoana* Harms, *Hymenaea eriogyne* Benth., *H. velutina* Ducke, *Lonchocarpus araripensis* Benth. and *Zornia echinocarpa* (Moric.) Benth. None of these species occur in all the sites of group E, but many of them appear disjunctly in two or more of these areas (refer to the section on distribution patterns).

Group F comprises typical areas of caatinga together with some brejo forest areas. This group includes assemblages occurring on mostly shallow and rocky soils derived from the crystalline bedrock, but it also includes some areas on deeper and richer, often calcareous, soils derived from limestone formations. This group is almost entirely situated on the wide undulated pediplains of the Sertaneja depression. The vegetation in these areas corresponds to typical caatinga, with a low to tall open canopy, often spiny woody plants with caducous foliage, ephemeral herbs, and succulents, mainly of the Cactaceae. This group is characterized by the presence of *Albizia polycephala*

(Benth.) Killip, *Anadenanthera colubrina* var. *cebil*, *Bauhinia cheilantha* (Bong.) Steud., *Caesal-pinia ferrea*, *Dioclea grandiflora* Mart. ex Benth., *D. violacea* Mart. ex Benth., *Erythrina velutina* Willd., *Lonchocarpus obtusus* Benth., *Machaerium acutifolium* Vog., *Parapiptadenia zehntneri* (Harms) M.P. Lima & H.C. Lima, *Peltophorum dubium* (Spreng.) Taub., *Senna spectabilis* and *S. splendida* (Vog.) H.S. Irwin & Barneby.

Two large subgroups of caatinga areas were identified. One links sites of mostly low caatinga forest (within group H) that are all located in the eastern section of the Sertaneja depression. Besides their mutual proximity, these areas all share the conspicuous presence of granitic or gneissic inselbergs or low mountains. Species characteristic of this subgroup are *Acacia bahiensis* Benth., *A. farnesiana* (L.) Willd., *Canavalia brasiliensis* Mart. ex Benth., *Chamaecrista belemii* (H.S. Irwin & Barneby) H.S. Irwin & Barneby, *Machaerium hirtum* (Vell.) Stellfeld, *Mimosa arenosa* (Willd.) Poir., *Poecilanthe ulei* (Harms) Arroyo & Rudd, *Samanea inopinata* (Harms) Barneby & J.W. Grimes, *Vigna peduncularis* (Kunth) Fawc. & Rendle, *Senna aversiflora* (Herb.) H.S. Irwin & Barneby and *Zollernia ilicifolia* (Brongn.) Vog.

The second subgroup (group G in Figure 6.2) is designated here as tall caatinga forest, because it brings together areas with mostly forest species, possessing a canopy 15–20 m high, and is also composed of lianas, vines and some epiphytes of the families Bromeliaceae and Orchidaceae. Most of the caatinga sites included within this subgroup occur further west than the previous subgroup, in areas with richer soils on residual reliefs occupying slightly peripheral positions in relation to the limits of the caatinga as, for example, the karstic surfaces near the southern limits of the caatinga (BJLP and JANU), the south-eastern slopes of the Serra do Caracol in northern Bahia (CPAL) and the Maracás plateau in central-eastern Bahia (MARA). Diagnostic elements of this subgroup are the genera *Blanchetiodendron*, *Goniorrhachis* and *Mysanthus*, as well as the species *Bauhinia trichose-pala* ined., *Cratylia bahiensis* L.P. Queiroz, *Pseudopiptadenia brenanii* G.P. Lewis & M.P. Lima, *Acacia monacantha* Willd., *Apuleia leiocarpa* (Vog.) J.F. Macbr., *Caesalpinia bracteosa* Tul., *Copaifera langsdorffii* Desf., *Diptychandra aurantiaca* Tul. subsp. *epunctata* (Tul.) H.C. Lima, A.M. Carvalho & C.G. Costa ex G.P. Lewis, *Enterolobium contortisiliquum* (Vell.) Morong, *Hymenaea martiana* Hayne, *Piptadenia viridiflora* (Kunth) Benth., *Plathymenia reticulata* Benth., *Pseudopipta-denia contorta* (DC.) G.P. Lewis & M.P. Lima, *Pterogyne nitens* Tul. and *Pterocarpus villosus* (Mart. ex Benth.) Benth.

6.3.2 PATTERNS OF GEOGRAPHICAL DISTRIBUTION

The main distribution patterns are summarized in Tables 6.3 and 6.4. It was possible to define patterns for 274 of the 322 caatinga taxa (at or below the species level). There is only fragmentary information for the remaining taxa, which does not allow the definition of their distribution patterns.

6.3.2.1 Wide Neotropical or Pantropical Pattern

Almost 20% of the taxa showed a wide neotropical or, less commonly, pantropical pattern. Most of them are weedy herbs of the genera *Chamaecrista*, *Senna*, *Mimosa* and *Centrosema*.

6.3.2.2 Species of the Caatinga Extending to the Atlantic Rain Forest or to the Cerrado

A low proportion of taxa occur in the caatinga range eastwards to the Atlantic rain forest (*c.*7%) or south-westwards to the central Brazilian cerrado (*c.*8.5%). Examples of species extending to the Atlantic rain forest are *Acacia martiusiana* (Steud.) Burkart, *Bauhinia outimouta* Aubl., *Melanoxy-lon brauna* Schott, *Piptadenia adiantoides* (Spreng.) J.F. Macbr. and *Pseudopiptadenia contorta*. Species found in the caatinga and in the cerrado are exemplified by *Lonchocarpus obtusus*, *Lue-tzelburgia auriculata* (Allemão) Ducke, *Mimosa pithecolobioides* Benth. and *Periandra coccinea* (Schrad.) Benth.

TABLE 6.3
Patterns of Geographical Distribution of the Leguminosae of the Caatinga

Pattern	Taxa[b]	
	n	%
Pantropical	17	6.20
Widely distributed in the Neotropics	39	14.23
Disjunct among the caatinga and other areas of the Pleistocenic arc	29	10.58
Disjunct between the caatinga and the chaco	1	0.36
Distributed in the caatinga and in eastern Brazil	21	7.66
Distributed in the caatinga and in central Brazil	23	8.39
Endemic to the caatinga	144	52.55
Total	274	100.00

[a]It was not possible to define a pattern for 18 taxa (c.6.2% of the total taxa).
[b]Taxa comprise species and infraspecific taxa.

6.3.2.3 Taxa Disjunct among the SDTF Nuclei

About 11% of the taxa showed a distribution pattern associated with the SDTF that fits the Pleistocenic arc pattern proposed by Prado and Gibbs (1993). These plants display a disjunct distribution in two or more nuclei of the forests cited: caatinga, Paranaense, piedmont and Andean dry valleys, referred to by Pennington et al. (2000) and Prado (2000). Examples of species characteristic of this pattern are *Anadenanthera colubrina*, *Amburana ceurensis*, *Caesalpinia ferrea*, *Enterolobium contortisiliquum*, *Geoffroea spinosa* Jacq., *Mimosa tenuiflora*, *Piptadenia viridiflora*, *Poeppigia procera* C. Presl and *Pseudopiptadenia contorta* (most of these taxa are mapped in Prado and Gibbs, 1993; Pennington et al., 2000; Prado, 2000). It is interesting to note that these species usually occur in the caatinga in areas with forest formations, and mostly on rich Latosols.

TABLE 6.4
Patterns of Geographical Distribution of Leguminosae
Taxa Endemic to the Caatinga

Pattern	Taxa[a]	
	n	%
Widely distributed in the caatinga	27	18.75
Restricted to the northern depression	13	9.03
Restricted to the southern depression	57	39.58
Restricted to the Ibiapaba–Araripe complex	3	2.08
Restricted to the dunes of the São Francisco	15	10.42
Restricted to the Raso da Catarina	6	4.17
Restricted to the Borborema plateau	3	2.08
Restricted to the Campo Maior complex	0	0.00
Restricted to the caatingas of the Chapada Diamantina	7	4.86
Species disjunct between two or three areas	13	9.03
Total	144	100.00

[a] Taxa comprise species and infraspecific taxa.

6.3.2.4   Species Disjunct with the Chaco

Only *Prosopis ruscifolia* Griseb. occurs disjunctly in both the caatinga and the chaco. It represents a typical chaquean species (Prado, 1991) and is known in the caatinga from only two plants in western Pernambuco, collected only twice, first by Luetzelburg and then by Andrade-Lima after a gap of more than 50 years. Prado (2003) proposed that this plant was accidentally introduced by imported cattle and therefore not a true case of disjunction. It is puzzling, however, that it has a local indigenous name, *juncumarim* (Andrade-Lima, in sched.), in an area where indigenous populations have long been absent, thus raising the possibility that it was familiar to the native people at an earlier time. For this reason, we cannot discard the possibility that it represents a true disjunct species, or that it was perhaps introduced in the past by migrating indigenous peoples.

6.3.2.5   Taxa Endemic to the Caatinga

Most of the taxa examined (about 52%) are endemic to the caatinga. The analysis of the limits of distribution of these taxa revealed coincident patterns that could be correlated with the geomorphology and the different substrates within the caatinga domain. As stated earlier, the geographical distribution patterns were checked against the units proposed by Velloso et al. (2002). Not a single endemic taxon has yet been found for the Campo Maior complex ecoregion. This area is located almost entirely within central-northern Piauí state, and covers *c.*41,000 km^2 of flat lowlands (altitudes 50–200 m) with periodically flooded Oxisols. The vegetation is a complex transition between caatinga and cerrado, with elements of babaçu palm forests. Almost nothing is known about the botany of this area. The few species of Leguminosae recorded (12) are mostly widespread in the caatinga, as in the cases of *Amburana cearensis* and *Hymenaea courbaril* L. It is still premature to proffer conclusions about the floristic relationships of this area based on such sparse data. Floristic surveys are urgently needed and could well reveal interesting novelties for the herbaceous flora.

Some endemic species, such as *Calliandra leptopoda* Benth., *Dioclea grandiflora* and *Senna martiana* (Benth.) H.S. Irwin & Barneby, occur scattered across the caatinga domain, usually associated with temporary water bodies (Figure 6.3). However, most of the species have a narrower range, permitting the definition of the following distribution patterns.

6.3.2.5.1   Species Endemic to the Northern Sertaneja Depression (Figure 6.4)

Eighty taxa (species and infraspecific taxa) of Leguminosae have been recorded from this area, of which 13 are endemic to this region, including *Caesalpinia gardneriana* Benth., *Chamaecrista duckeana* (P. Bezerra & Afr. Fern.) H.S. Irwin & Barneby, *Chamaecrista tenuisepala* (Benth.) H.S. Irwin & Barneby, *Senna trachypus* (Benth.) H.S. Irwin & Barneby, *Arachis dardani* Krapov. & W.C. Greg. and *Zornia afranioi* R. Vanni.

6.3.2.5.2   Species Endemic to the Borborema Plateau (Figure 6.4)

Only 18 species of Leguminosae have been recorded here, but this may more reflect the sparse botanical coverage of this area. Some possible endemic legumes are *Mimosa borboremae* Harms, *Mimosa paraibana* Barneby (Barneby, 1991) and *Chamaecrista trichopoda* (Benth.) Britton & Killip.

6.3.2.5.3   Species Endemic to the Southern Sertaneja Depression (Figure 6.5)

From this region, 202 taxa have been recorded. Of these, 57 taxa are restricted to this unit, including three monotypic genera: *Blanchetiodendron*, *Goniorrhachis* and a new genus currently being described (L.P. Queiroz, G.P. Lewis, and M.F. Wojciechowski, in prep.). Data on the

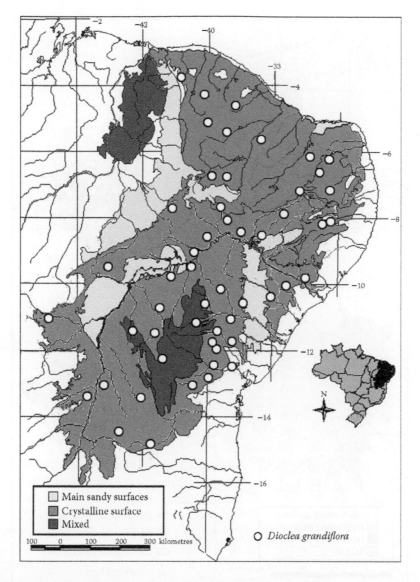

FIGURE 6.3 Distribution of *Dioclea grandiflora* Mart. ex Benth., illustrating its widespread distribution pattern on the crystalline basement surface.

distribution patterns of the Leguminosae permit a finer subdivision of this region into the following, more homogeneous areas.

- Tall caatinga forest on karstic surfaces near Januária-Bom Jesus da Lapa and Irecê, growing on rich calcareous soils derived from limestone. Taxa characteristic of this subtype are *Anadenanthera colubrina* var. *cebil*, *Apuleia leiocarpa*, *Bauhinia trichosepala*, *Caesalpinia pluviosa* DC. var. *sanfranciscana* G.P. Lewis, *Cratylia bahiensis*, *Goniorrhachis marginata* Taub. ex Glaz., *Mimosa pithecolobioides* and *Piptadenia viridiflora*.
- Medium to low caatinga forests of eastern Bahia and northern Minas Gerais states, occurring mostly in the lowland region between the Chapada Diamantina and the eastern limits of the caatinga. Diagnostic species of this subtype are *Acacia kallunkiae* J.W. Grimes & Barneby, *Acacia santosii* G.P. Lewis, *Leucochloron limae* Barneby & J.W. Grimes and *Zapoteca filipes* (Benth.) H.M. Hern.

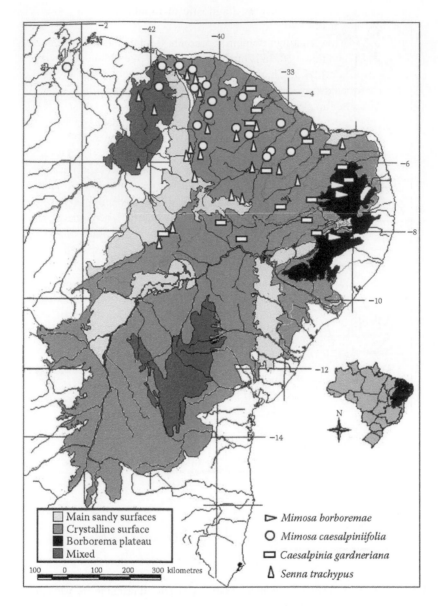

FIGURE 6.4 Distribution of taxa endemic to the northern Sertaneja depression and the Borborema plateau in north-eastern Brazil.

6.3.2.5.4   Species Endemic to the Dunes of the São Francisco River (Figure 6.6)

Thirty-five taxa of the Leguminosae have been recorded from this dune area, of which half (15 taxa) are endemic. These include the trees *Pterocarpus monophyllus* Klitgaard, L.P. Queiroz & G.P. Lewis, *Calliandra macrocalyx* Harms var. *aucta* Barneby and *Luetzelburgia bahiensis* Yakovlev; the liana *Dioclea marginata*; and the prostate herbs and subshrubs *Mimosa xiquexiquensis* Barneby, *M. setuligera* Harms, *Zornia harmsiana* Standl. and *Zornia ulei* Harms. New species of *Rhynchosia* and *Aeschynomene* are being described from this region (L.P. Queiroz, in prep.).

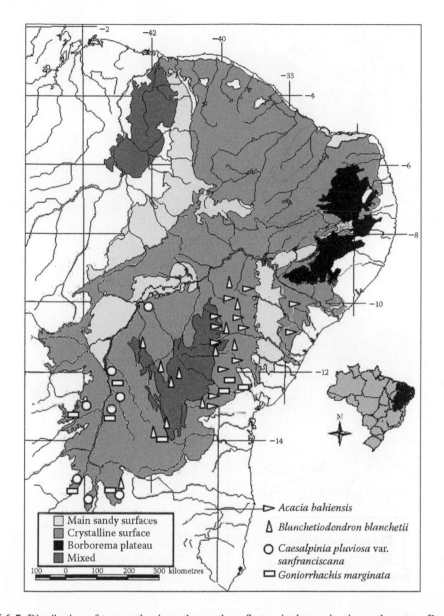

FIGURE 6.5 Distribution of taxa endemic to the southern Sertaneja depression in north-eastern Brazil.

6.3.2.5.5   Species Endemic to the Raso da Catarina (Figure 6.6)

Forty-three species of Leguminosae have been recorded from this area, six of which may be considered endemic (although they may slightly extend their range southwards or eastwards where the sandy tableland penetrates into areas with moister conditions). Such species include *Calliandra aeschynomenoides* Benth., *C. squarrosa* Benth., *Chamaecrista swainsonii* (Benth.) H.S. Irwin & Barneby, *Chloroleucon extortum* Barneby & J.W. Grimes, *Dioclea lasiophylla* Mart. ex Benth. and *Zornia echinocarpa. Mimosa lewisii* Barneby is centred on Raso da Catarina but reaches the coastal restinga, as well as the lower slopes of the Chapada Diamantina.

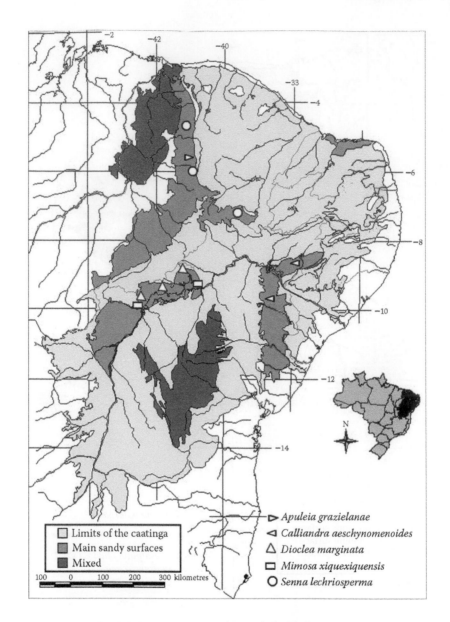

FIGURE 6.6 Distribution of taxa endemic to the three major nuclei of sedimentary surfaces: the Ibiapaba–Araripe plateaux, the dunes of the middle São Francisco river and the Raso da Catarina.

6.3.2.5.6 Species Endemic to the Ibiapaba and Araripe Plateaux (Figure 6.6)

Thirty-four species of legumes are known from the Ibiapaba plateau, six of which are possibly endemic: *Apuleia grazielanae* Afr. Fern., *Senna lechriosperma* H.S. Irwin & Barneby, *Calliandra fernandesii* Barneby, *Mimosa poculata* Barneby, *Trischidium decipiens* (R.S. Cowan) H.E. Ireland and *Aeschynomene matosii* Afr. Fern.

The species from Araripe plateau are typical cerrado plants, such as *Dimorphandra gardneriana* Tul., *Parkia platycephala* Benth., *Cratylia argentea* (Desv.) Kuntze and *Hymenaea stigonocarpa* Hayne. No species of Leguminosae has yet been identified as being endemic to the Araripe plateau.

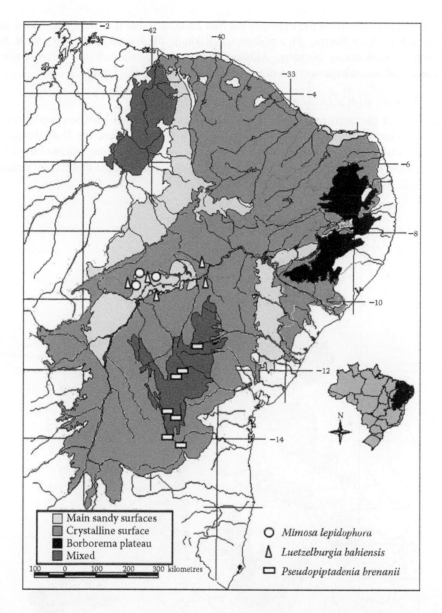

FIGURE 6.7 Distribution of taxa endemic to the caatingas of the Chapada Diamantina mountain range and the seasonally flooded depressions of south-western Piauí, northern Bahia and western Pernambuco.

6.3.2.5.7 Species Endemic to the Caatingas of the Chapada Diamantina (Figure 6.7)

The Leguminosae are represented in this unit by 43 recorded taxa, most of them extending to the surrounding areas of the southern Sertaneja depression. Seven species can be considered as endemic to these caatingas, including *Calliandra pilgeriana* Harms, *Crotalaria bahiaensis* Windler & S.G. Skinner, *Mimosa irrigua* Barneby, *M. morroensis* Barneby and *Pseudopiptadenia brenanii*.

6.3.2.5.8 Species Endemic to Seasonally Flooded Depressions (Figure 6.7)

This encompasses a single area whose flora is related to the sandy pediments that extend from south-eastern Piauí to western Pernambuco and northern Bahia on alluvial sands. This flora has some elements encountered on the dunes of the São Francisco, but it is rich in endemic taxa such as *Mimosa ulbrichiana* Harms, *M. lepidophora* Rizzini, *M. hirsuticaulis* Harms, *M. hortensis* Barneby and *M. nothopteris* Barneby. Although not endemic, *Caesalpinia microphylla*, *Senna martiana* and *Calliandra depauperata* are very characteristic of this region.

6.3.2.5.9   Species with a Disjunct Distribution

Fifty-seven taxa of the caatinga (*c*.18% of the total taxa) are disjunct between two or more areas. The most conspicuous disjunction involves three widely separate areas, the Ibiapaba plateau, the Raso da Catarina and the dunes of the São Francisco river, a pattern shown by 13 species (Figure 6.8). Most of these species are restricted to the caatinga and occur in different combinations

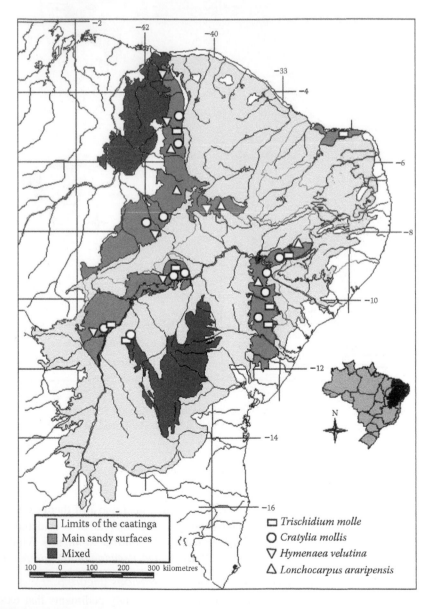

FIGURE 6.8 Distribution of disjunct taxa among the major nuclei of the sedimentary surfaces of the caatinga.

of areas. *Cratylia mollis* and *Trischidium molle* occur in all three areas, as is the case for *Piptadenia moniliformis* and *Chamaecrista repens* (Vog.) H.S. Irwin & Barneby var. *multijuga* (Benth.) H.S. Irwin & Barneby, but unlike the two first taxa cited, the two latter are not restricted only to the caatinga. Other species occur in two of these three areas, especially in the Ibiapaba area and the dunes of the São Francisco river, such as *Senna gardneri* (Benth.) H.S. Irwin & Barneby, *Hymenaea eriogyne*, *H. velutina* and *Mimosa verrucosa* Benth. (this last species very slightly extending westwards to areas of cerrado). Species occurring in the Ibiapaba plateau and in the Raso da Catarina, but not recorded from the dunes of the São Francisco river, are *Copaifera cearensis* Ducke var. *arenicola* Ducke and *Lonchocarpus araripensis*. *Mimosa modesta* Mart. var. *ursinoides* (Harms) Barneby, *Calliandra squarrosa* and *Aeschynomene martii* occur in the dunes of the São Francisco river and in the Raso da Catarina, but they have not been recorded for the Ibiapaba plateau.

Another group of species demonstrated a disjunct distribution pattern involving areas of caatinga forest in central-southern Bahia as well as areas ranging from northern Paraíba, southern Ceará and north-eastern Piauí. Habitat has been poorly recorded on material coming from these last two states, but some of these species seem to occur in the piedmont of residual landforms. Some examples of this type of distribution are seen with *Hymenaea martiana*, *Calliandra spinosa* Ducke and *Chloroleucon foliolosum*.

6.4 DISCUSSION

6.4.1 DIVERSITY OF THE LEGUMINOSAE AND LINKS WITH OTHER FLORAS

The family Leguminosae is represented in the caatinga by 293 species belonging to 77 genera (Table 6.2; L.P. Queiroz, in prep.). With the addition of the infraspecific taxa for polytypic species, the family increases to 322 taxa at or below the rank of species (Table 6.2). The best-represented genera are *Mimosa* (37 species, 41 taxa), *Chamaecrista* (26, 35), *Senna* (19, 25), *Bauhinia* (16, 16), *Calliandra* (12, 13), *Acacia* (11, 11), *Caesalpinia* (8, 11), *Aeschynomene* (9, 10) and *Zornia* (9, 9). Eight genera are represented by five to seven species, 29 genera by two to four species and the remaining 31 genera by a single species each.

The genera most speciose in the caatinga are relatively large and widespread in the Neotropics, mostly from savanna and dry vegetation regions. However, a closer inspection of the species present in the caatinga reveals some taxonomic biases. In the genus *Mimosa*, 26 of the 37 species belong to three series recognized by Barneby (1991): *Leiocarpae* (10 of a total of 27 species in the series), *Bimucronatae* (5 of 9) and *Cordistipulae* (11 of 13). Discounting weedy species, these plants account for 81% of the species of *Mimosa* in the caatinga. A similar situation can be found with *Calliandra*, with all but one species (*C. leptopoda*) belonging to the section *Androcallis* (Barneby, 1998). No species from the section *Calliandra* has yet been recorded for the caatinga, although this group is represented by 40 species in the mountains of the Chapada Diamantina (Souza, 2001), a mountain range embedded within the caatinga region. This suggests that the diversity observed in these genera resulted from in situ diversifications, reflecting the isolation of the caatinga from other similar vegatation forms.

Chamaecrista is represented mainly by shrubby species of the section *Absus*, subsection *Absoideae* (12 species). This group is conspicuously present in planaltine Brazil, mainly in the cerrado and campo rupestre vegetations. Of the remaining species, eight could be considered weeds, mostly belonging to section *Chamaecrista*. In *Caesalpinia*, the caatinga species belong to groups strongly connected with dry sites: eight out of nine species are included in the *Poincianella-Erythrostemon* group (Lewis, 1998), while *C. ferrea* belongs to the *Libidibia* group and is also associated with dry forests in South America.

Three genera are endemic to the caatinga, and each is represented by a single species. *Blanchetiodendron* was segregated from *Albizia* by Barneby and Grimes (1996). *Goniorrhachis* was described at the end of the nineteenth century (Taubert, 1892). The third genus, belonging to the tribe *Brongniartieae*, is currently being described (L.P. Queiroz, G.P. Lewis and M.F. Wojciechowski, in prep.). *Mysanthus*, a monospecific genus segregated from *Phaseolus* by Lewis and Delgado (1994), has been considered endemic to the caatinga (Prado, 2000). It is known only from the caatinga, except for two collections in the state of São Paulo, far removed from the caatinga area. These two collections, however, are from areas where species from different parts of Brazil were introduced in the past, and *Mysanthus* may not be native to São Paulo.

The view that the caatinga has a biota with few endemics, and that its flora represents an impoverished composition in relation to that of the chaco, the cerrado or the Atlantic rain forest, has prevailed for a long time (Rizzini, 1963, 1979; Andrade-Lima, 1982). Prado (1991) convincingly demonstrated the separation of the caatinga flora from that of the chaco. In comparing the legumes of the caatinga with those of the Cerrado Biome (checklist compiled by Kirkbride, 1984, with additions from the list published by Ratter et al., 2003), we found a lower number of legume genera and species in the caatinga (77 genera and 293 species) than in the cerrado (85 genera and 551 species). The larger number of taxa from the cerrado may be explained by the greater topographical diversity in the region where it occurs and the explosive speciation seen in higher altitudes, especially in the central Brazilian plateau and the Espinhaço range in Bahia and Minas Gerais. In these areas, genera typical of savanna habitats are represented by a large number of species, such as *Chamaecrista* (116 recorded species), *Mimosa* (47), *Senna* (19) and *Bauhinia* (15) (all these are among the most diverse genera in the caatinga). There are also genera poorly represented, or absent, from the caatinga, such as *Eriosema* (21), *Crotalaria* (17), *Stryphnodendron* (9), *Desmodium* (8), *Camptosema* (8), *Sclerolobium* (4) and *Pterodon* (2 or 3).

It must be noted that almost half of the genera, and only *c*.9% of the species, are common to both the caatinga and Cerrado Biomes, thus demonstrating great differences between these two floristic areas. This becomes even more evident when considering that some of the genera common to both these biomes represent taxa of the SDTF that occur in patches of forests growing on mesotrophic soils within the range of the cerrado (Ratter et al., 1978; Oliveira-Filho and Ratter, 1995). These taxa include *Apuleia*, *Caesalpinia*, *Diptychandra*, *Martiodendron*, *Pterogyne* and *Riedeliella*, which are recorded from only a few sites among the 376 areas of cerrado compiled by Ratter et al. (2003).

Unfortunately, there is no available total compilation for the entire complement of legumes from the Atlantic rain forest that could be compared with the data presented here for the caatinga. Nonetheless, the low number of taxa of the caatinga extending eastwards (Table 6.3), as well as data derived from floristic surveys in Atlantic rain forest sites (e.g. Mori et al., 1983; Guedes, 1992; Guedes-Bruni and Lima, 1994; Thomas et al., 1998), favour an interpretation of two quite distinct floras, at least with respect to the Leguminosae.

Brejo forests are usually considered fragments of a more ancient and more widespread rain forest that penetrated inland from the coast, possibly linking with the Amazonian forest. The present study is not conclusive with respect to the relationships among the floras of the brejo forests, or between them and the caatinga. In fact, there is evidence that these brejo forests are not a homogeneous unit (Sales et al., 1998; Santos, 2002). It is interesting to note that a parsimony analysis of endemicity using areas of brejo forests reveals that the western sites surrounded by drier landscapes compose a clade, while those found more eastwards, and embedded in areas subjected to moister conditions, split from a basal grade that includes the rain forest floras of the Amazon and the Atlantic rain forest (Santos, 2002). The hierarchical structure found in the parsimony analysis of Santos (2002) is similar to that found here using multivariate techniques (Figure 6.2).

Unfortunately, Santos (2002) did not include areas of caatinga in examining the influence of the caatinga flora on brejo forest composition. In the present work, we find indications that the surrounding caatinga could in fact influence brejo composition, but this may also reflect the imprecise delimitation of the surveyed floras, which possibly included elements of neighbouring vegetation types (see Sales et al., 1998). The clustering of different sites of brejo forests with one or another group of caatinga forests could therefore reflect this contamination of the matrix, rather than express true relationships.

Nevertheless, the impressive number of endemic taxa of the legumes supports the view of the caatinga as a distinct floristic province. Prado (2003) listed 14 genera of angiosperms as being endemic to the caatinga, while Giulietti et al. (2002) cited 18. At the species level, Prado (1991) determined that 42% of the flowering plant species of the caatinga were endemic (183 of the 437 species surveyed), while Giulietti et al. (2002) cited 318 endemic species. These figures strongly demonstrate the distinctiveness of the flora of the caatinga in relation to that of the neighbouring biomes.

Another question could be raised concerning the relationship of the caatinga flora with that of the SDTF. Prado (2000) listed 11 genera of angiosperms endemic to the SDTF that occur in more than one of its major nuclei. All but one of these (*Perianthomega*, Bignoniaceae) occur in the caatinga and include species that are usually dominant elements in caatinga forests. It must be taken into consideration, however, that these genera, common to the caatinga and the SDTF, are found almost exclusively in areas of the crystalline basement group (group F in Figure 6.2), particularly in tall forest formations, a fact that has consequences for the interpretation of the history of this vegetation (see Section 6.4.3).

6.4.2 Centres of Endemism

The major patterns of endemism are partially congruent with the limits of the ecoregions proposed for the caatinga biome (Velloso et al., 2002). These could be considered as putative centres of endemism for the caatinga, based on the family Leguminosae.

Two of the major centres of endemism are related to Precambrian terrains in the wide Sertaneja depression. The number of species of Leguminosae decreases northwards, but a marked transition in the floristic composition seems to occur in the middle of Pernambuco state, close to the southern slopes of the Araripe and Borborema plateaux that delimit two extensive depressed areas on the crystalline basement surface.

The northern Sertaneja depression includes mostly lowlands (50–500 m alt.). There are no large perennial rivers there, and the soils are mostly shallow and rocky, derived from the Precambrian crystalline bedrock. In addition to the endemic species of legumes, other characteristic elements of the flora include the genera *Auxemma* (Boraginaceae, two species) and *Hydrothrix* (Pontederiaceae). The latter is endemic to seasonally flooded areas in the states of Ceará and Paraíba (Harley, 1996; Velloso et al., 2002). Some other conspicuous species in the northern depression just extend into areas of cerrado, such as *Luetzelburgia auriculata* and *Martiodendron mediterraneum* (Mart. ex Benth.) Köppen. This latter species is known from a single collection in central-northern Bahia but has not been recollected there since the nineteenth century. *Mimosa caesalpiniifolia* Benth. is another species characteristic of this region. Its range extends slightly westwards to Maranhão state; records from Manaus (Amazonas) and Ilhéus (southern Bahia) are almost certainly from plants that escaped from cultivation.

Data from Leguminosae do not justify considering the Borborema plateau as an independent centre of endemism. Although it presents a quite distinct landscape, with a series of plateaux and low mountains with altitudes ranging from 150 m to 650 m, with some peaks reaching 1000 m, the flora appears to be a subset of that of the northern depression. It must be pointed out, however, that most of the original vegetation has been removed, and that one possibly isolated genus of Scrophulariaceae, *Ameroglossum*, was recently described and is known from rocky surfaces on the southern Borborema plateau (Fisch et al., 1999).

The southern Sertaneja depression area is the largest of the caatinga ecoregions, occupying an area of *c.*374,000 km², and ranges from central-northern Pernambuco and central-eastern Piauí southwards to northern Minas Gerais state. The typical landscape there is that of a wide, low plain. However, unlike the northern depression it retains more significant residual elevations and mountain ranges, and contains some perennial watercourses. It likewise occupies the crystalline basement surface, but it possesses deeper soils than those found in the northern depression. This area concentrates most of the remnants of the tall caatinga forests, as well as most of the endemic genera and species found on the crystalline basement surface, such as the monospecific genera *Haptocarpus* (Capparaceae), *Piriadacus* (Bignoniaceae), *Holoregmia* (Martyniaceae), *Gorceixia* (Compositae) and *Anamaria* (Scrophulariaceae; Harley, 1996; Velloso et al., 2002; R.M. Harley, Royal Botanic Gardens, Kew, Richmond, pers. comm.), besides the legumes *Blanchetiodendron* and *Goniorrhachis*. *Argyrovernonia* (Compositae) comprises two species and is also possibly endemic to this region.

Four major centres of endemism related to sandy sedimentary surfaces were identified. A relatively high number of endemic taxa occur in each area, but they can be treated as a single unit, owing to the many examples of taxa occurring disjunctly among the four centres.

The dunes of the São Francisco river centre include an area of *c.*36,000 km^2 in the mid São Francisco river valley. This region is characterized by extensive deposits of dystrophic quartzose sands. The vegetation is mainly shrubby and occurs in patches of low trees and shrubs. Part of the soil remains exposed, and the presence of *Bromelia antiacantha* Bertol. (Bromeliaceae) and *Tacinga inamoena* (K. Schum.) N.P. Taylor & Stuppy (Cactaceae; Rocha et al., 2004) is notable there. In addition to the high levels of endemism for the Leguminosae, other elements of the flora seem to be endemic to this region, such as the monospecific genus *Glischrothamnus* (Molluginaceae), a possibly new genus of Bignoniaceae (A.H. Gentry, in sched.), *Croton paludosus* Müll. Arg. (Euphorbiaceae) and a new species of *Eugenia* (Myrtaceae, M.L. Kawasaki, Field Museum, Chicago, pers. comm.). Rocha et al. (2004) pointed out that most of the animals surveyed in this region are endemic: for example, four species of amphisbaenians, 16 species of lizards, eight species of snakes and one amphibian species (Rodrigues, 1996; Rocha, 1998; Rodrigues and Juncá, 2002; Rodrigues, 2003). These figures are even more impressive in light of the fact that 37% of all lizards and amphisbaenians known from the caatinga are endemic to these dunes (Rodrigues, 2003).

The Raso da Catarina centre is confined to the area of the Tucano-Jatobá sedimentary basin, most of which is included in the north-eastern portion of Bahia, but also extends northwards to central-southern Pernambuco as a broad tableland. It occupies an area of *c.*31,000 km^2, with mostly deep and dystrophic sandy soils. The vegetation is mostly shrubby and is very dense. Besides the examples of the Leguminosae already cited, other interesting floristic elements include many taxa of Malpighiaceae, such as the genera *Ptilochaeta*, a genus with three or four species possibly endemic to the SDTF, and disjunct between the caatinga and Paraguay, Argentina and Bolivia; *Barnebya*, with only two species, one endemic to this region and a second occurring disjunctly in Rio de Janeiro; and *Macvaughia*, a monotypic genus originally thought to be endemic to this region but recently collected on the São Francisco dunes. The reptiles *Amphisbaena arenaria* Vanzolini and *Tropidurus cocorobensis* Rodrigues seem to be restricted to this unit, as well as a new species of the rodent genus *Dasyprocta* (Velloso et al., 2002). The blue macaw *Anodorhynchus leari* Bonaparte (Aves: Psittacidae) is found only in this region, but this may be related to hunting pressure.

The Ibiapaba-Araripe centre encompasses three landscape features: the Chapada do Ibiapaba, a north–south mountain massif at the border between the states of Ceará and Piauí; a set of low mountains and tablelands in central-southern Piauí; and the Chapada do Araripe, which runs east-west at the border between the states of Ceará and Pernambuco. The Ibiapaba plateau consists primarily of sandy soils of low fertility, and the vegetation there is mainly composed of a dense shrubby scrub or a low forest type named *carrasco* (see Figueiredo, 1986; Araújo et al., 1998; Araújo and Martins, 1999). The Araripe plateau hosts a mosaic of areas of carrasco, on deep sandy soils, as well as cerrado, on mostly acidic Latosols. There has been no analysis of the species distribution of this area that could identify endemic taxa beyond the six species of Leguminosae cited earlier (see under *Species endemic to the Ibiapaba and Araripe plateaux*). Velloso et al. (2002), however, noted that *Antilophia bokermanni* (Aves: Pipridae) is possibly endemic to this area, as well as *Hyptidendron amethystoides* (Benth.) Harley (Labiatae; R.M. Harley, pers. comm.).

The seasonally flooded sandy depressions of south-eastern Piauí, northern Bahia and western Pernambuco can be viewed as another centre of endemism, as indicated by the distribution patterns of the Leguminosae and by similarity analyses. Velloso et al. (2002) included this area in the southern Sertaneja depression, but many taxa of Leguminosae seem to be exclusive to it (see Section 6.3.2.5.8). The flora is mostly composed of small annual herbs that grow rapidly and flower at the start of the rainy season. Besides the Leguminosae, *Fraunhofera* (Celastraceae) seems to be restricted to this area, while *Apterokarpos* (Anacardiaceae) occurs disjunctly here and on the Ibiabapa plateau. Both these genera are monospecific.

An additional centre of endemism is located in the caatinga of the Chapada Diamantina range. Both these genera are monospecific. The Chapada Diamantina is the largest mountain range in north-eastern Brazil, occupying *c.*50,000 km², with peaks reaching slightly above 2000 m. It hosts a mosaic of different vegetation types associated with a large array of soil types, topography and hydrological conditions. Climatic conditions there are usually more humid than those found in the surrounding caatinga area (Harley, 1995; Giulietti et al., 1996). The most characteristic types of vegetation found in this region are campos rupestres (literally, rocky fields) and open forms of cerrado (locally termed *gerais*). Caatinga vegetation occurs mainly in valleys and on the western slopes of the main mountains that have drier conditions than the upper slopes. These caatinga areas range from dense stands of shrubs with a few low trees in areas with poor soils, to tall caatinga forests in areas with richer soils. The latter are usually dominated by *Pseudopiptadenia brenanii* and *Piptadenia moniliformis*. The genus *Mysanthus*, which was discussed earlier, could represent an example of endemism to the Chapada Diamantina caatinga at the generic level. Examples of endemic genera of other families include *Raylea* (Malvaceae; Cristóbal, 1981) and *Heteranthia* (Scrophulariaceae; Harley, 1996).

6.4.3 THE MAIN PHYTOGEOGRAPHICAL UNITS AND THEIR IMPLICATIONS FOR THE RECONSTRUCTION OF THE HISTORY OF THE CAATINGA

The results presented and discussed in this chapter strongly support the view that the caatinga comprises two separate biotas, one associated with soils derived from the crystalline basement surface, and the other with sandy sedimentary surfaces.

The first phytogeographical unit encompasses areas on the crystalline basement surface and covers approximately 622,000 km² (73% of the total area of the caatinga biome). It corresponds to the most typical form of the caatinga, as described by many authors (e.g. Andrade-Lima, 1954, 1981). The physiognomy of this vegetation type varies along a gradient that runs from tall forests (with canopies reaching *c.*15–20 m), to low forests (with a dense to sparse tree layer associated with a usually conspicuous shrubby layer), to open shrub communities (on areas with large rock outcrops and gravelly soils). Both the trees and the shrubs usually have contorted trunks and small leaves, and frequently bear thorns. Cacti and terrestrial bromeliads are important floristic elements, notably in the more open communities, decreasing in frequency in the more arborescent vegetation.

Phenological cycles are strongly seasonal and are governed by rainfall, as in other dry forests (Bullock, 1995). The most conspicuous feature of the vegetation is the deciduous character of most of the trees and shrubs during the dry season. It is this feature that provides the origin of the word caatinga (literally, whitish forest in the indigenous Tupi language). The proportion of species that keep their leaves in the dry season ranges from 26% (Machado et al., 1997) to almost zero in areas where only *Ziziphus joazeiro* Mart. (Rhamnaceae) could be considered an evergreen (Oliveira et al., 1998). In terms of their reproductive phenology, Machado et al. (1997) demonstrated that most of the woody species flower at the beginning of the rainy season, a pattern observed in other dry forest sites (Bullock and Solís, 1990; Guevara de Lampe et al., 1992; Bullock, 1995).

The flora of the crystalline basement surfaces is clearly related to the general SDTF flora, and all the genera endemic to the SDTF flora that occur in the caatinga (and in other areas of these forests; Prado, 2000) are found in this group. Besides the leguminous genera *Amburana*, *Apuleia* and *Pterogyne*, other genera that link the flora of the crystalline areas with the SDTF include *Myracrodruon* (Anacardiaceae), *Patagonula* (Boraginaceae), *Quiabentia* (Cactaceae) and *Balfourodendron* (Rutaceae). Virtually all the species of Leguminosae occurring in the caatinga and in other areas of the SDTF belong to areas overlying crystalline soils. The exceptions amount to only a few species that occur scattered throughout all of the caatinga, mostly associated with permanent or intermittent water bodies, such as *Geoffroea spinosa*, *Albizia inundata* (Mart.) Barneby & J.W. Grimes, *Discolobium hirtum* Benth. and *Mimosa hexandra* Micheli.

Andrade-Lima (1981) presented a classification of the vegetation of the caatinga based mainly on its floristic composition, although taking into consideration physiognomic and ecological aspects of the vegetation. He recognized six major units, further divided into 12 subtypes. Prado (2003) followed the same conceptual framework, recognizing the same six major units but adding a 13th subtype. Unfortunately, neither of these authors mapped the proposed phytogeographical zones. Four of the main units proposed by Andrade-Lima (1981) belong to the phytogeographical unit recognized here as caatinga on crystalline basement surfaces. The four main units are unit I (tall caatinga forests), II (median to low caatinga forests), IV (tall to low open shrubby caatinga) and V (open, shrubby caatinga). The patterns of distribution of the Leguminosae only partially fit these proposed units, allowing the recognition of a subunit of tall caatinga forest on richer soils (unit I of Andrade-Lima) and a subunit combining the remaining units of this author. Sampaio et al. (1981) analysed the correlation between variations of some attributes of the plant community and abiotic factors. They found a positive correlation between the plant height and a soil depth gradient. The number of plants per hectare demonstrated a negative correlation with this same gradient. Thus part of the observed variation in the plant community structure may be a reflection of minor local soil conditions (affecting mostly plant height and frequencies) rather than distinguishing the different plant communities, as proposed by Andrade-Lima (1981) and Prado (2003).

The second floristic unit accepted here corresponds to the vegetation overlying sandy sedimentary areas. This covers more than 136,000 km² (*c.*16% of the entire area of the caatinga) and encompasses mainly three disjunct areas: the Ibiapaba and Araripe plateaux, the tableland surfaces of the Tucano-Jatobá sedimentary basins, and the dunes of the mid São Francisco river valley. Other smaller areas occur in north-eastern Rio Grande do Norte state, in the Chapada do Apodi, and scattered across the area of the caatinga biome (mostly associated with residual elevations). Although comprising discontinuous patches of vegetation, this unit may be defined by the species that are disjunct among these same patches, suggesting a past floristic link between the floras of these areas. In addition to the Leguminosae, plants from other families show similar patterns of disjunctions among these three sandy areas, such as *Harpochilus neesianus* Mart. ex Nees (Acanthaceae), *Godmania dardanoi* (J.C. Gomes) A.H. Gentry (Bignoniaceae; Gentry, 1992), *Pilosocereus tuberculatus* (Werderm.) Byles & G.D. Rowley, *Tacinga inamoena* (Cactaceae; Taylor and Zappi, 2004), *Pavonia glazioviana* Gürke (Malvaceae) and *Jatropha mutabilis* (Pohl) Baill. (Euphorbiaceae; Rocha et al., 2004).

Phenological data presented by Rocha et al. (2004) from an area of continental dunes showed that vegetative and reproductive cycles are not strongly affected by rainfall distribution. They determined that budding and leaf drop, and floral development and anthesis, as well as fruit production and dispersion were not synchronized between species, and that at least 50% of the individuals produced leaves throughout the year. This is in sharp contrast with the marked pattern of leaf fall and strongly synchronous phenological patterns found in the dry forests, including caatinga overlying crystalline basement surfaces.

The relationship between the floras of sandy areas of caatinga is a controversial matter and often overlooked. Some authors consider these areas as belonging to the caatinga, mostly because they used criteria related to the physical environment and physiognomy of the vegetation (e.g. Egler, 1951; Andrade-Lima, 1978, 1981). On the other hand, other authors have excluded the carrasco vegetation of the Ibiapaba plateau from the caatinga, emphasizing its floristic singularity while considering the vegetation overlying other sandy areas as part of the caatinga, and failing to recognize the many floristic links between them (Araújo et al., 1998; Prado, 2003).

Unit III in the classification of Andrade-Lima (1981) corresponds to one of the three main nuclei of the vegetation overlying sandy areas. This author also failed to perceive the relationships between these areas. Rodal and Sampaio (2002) suggest that unit III of Andrade-Lima could be expanded to embrace the continental dunes and upland plateaux. In analysing the relationship of the flora of one site of the dunes of the São Francisco river, Rocha et al. (2004) highlighted the fact that many of the non-endemic species of these dunes occur in other sandy areas. These two last papers are the first recognition of a distinct flora in these sandy areas.

It is intriguing to ask why the clear differences between these biotas remained unrecognized for such a long time. Possible causes could be the strong physiognomic similarity caused by convergence driven by the harsh environment. Another possible explanation may reside in the scales used in most phytogeographical studies, often focussing on a continental scale and viewing the entire caatinga as an analytical unity (e.g. Pennington et al., 2000), or sometimes focusing on a very regional scale, which tended to exaggerate minor variations and produce a plethora of designations for different vegetation types (e.g. Egler, 1951; Rodal and Sampaio, 2002). The lack of selection of an appropriate scale to deal with regional biogeography may have prevented the major revisions from recognizing the distinct nature of these biotas (Andrade-Lima, 1981; Sampaio, 1995; Rodal and Sampaio, 2002; Prado, 2003). Additionally, the absence of a solid taxonomic framework for most of the important taxa of the caatinga, plus an almost complete absence of published data concerning the floristic composition of the caatinga of the state of Bahia, surely delayed the full comprehension of this flora. Nevertheless, the elucidation of these two different biotas, previously obscured under the broadly applied name of caatinga, could permit new and different interpretations of the phytogeographical history of this area.

A great deal of information supports the view that a widespread process of pediplanation took place in the region during the Neogene and lower Quaternary (Ab'Saber, 1974). This process uncovered the Precambrian crystalline bedrock, and left as remnants the present small mountain ranges, inselbergs and significant areas of sedimentary surfaces that today occur as elevated plateaux or tablelands, and represent remnants of the older surface that covered most of the region during the upper Cretaceous and lower Tertiary (Tricart, 1961; Ab'Saber, 1974). Fossil records can often provide direct evidence of past biotas, but unfortunately the fossil record for caatinga plants is very sparse (see review in Burnham and Graham, 1999). This is because most of its surface resulted from erosional processes, and the arid conditions precluded fossilization in most of the area. As such, except for very rare instances, indirect evidence must be used to infer the local biogeographical history.

One of the few physical records of the past comes from recent studies of dune field sediments in the São Francisco river valley dated with carbon-14 techniques. The dating suggests that these dunes were active during the Tertiary, through the Pleistocene, and possibly until c.1200 years BP in the Holocene (Oliveira et al., 1999; Colinvaux et al., 2001). These findings appear to reinforce the suggestion of Ab'Saber (1974) that the palaeoclimatic conditions in the area of the caatinga have been subject to drought events at least since the Neogene. Indirect evidence of the remote age of a xeromorphic vegetation, in the area now occupied by the caatinga, comes from phylogeographical studies of small forest-dwelling mammals undertaken by Costa (2003). This author found that most of the divergence among Amazonian and Atlantic rain forest lineages of six genera of didelphoid marsupials and three genera of sigmodontine rodents occurred prior to the Pleistocene. This may indicate a division of the two major South American rain forests at the end of the Tertiary, and it is reasonable to assume that, by that time, the region was already occupied by vegetation adapted to dry conditions.

With respect to the Quaternary, recent evidence argues that the caatingas did not undergo the intense climatic changes that affected other parts of South America (Cailleux and Tricart, 1959). The origin of the dune fields in the mid São Francisco river valley is frequently cited as evidence of a long dry phase. Ab'Saber (1974) and Tricart (1985) argue that these massive sand deposits probably originated during the driest phases of the Quaternary, when the São Francisco river may have dried up in mid course. The sandy sediments left during this drying were subsequently remodelled as dune fields by the wind. On the other hand, studies of the palynological profile of the sediments of the Icatu river valley indicated that, in the last 11,000 years, this region may have experienced at least three periods with humid climatic conditions, alternating with two dry periods, including the present one, whose onset could date from 4200 years BP (Oliveira et al., 1999). Studies of marine and lacustre sedimentation patterns carried out in north-eastern Brazil suggested that, during the last glacial period, the Heinrich events known from the North Atlantic region represented phases of increased humidity and precipitation in the caatinga zone (Jennerjahn et al., 2004).

Indirect evidence of climatic fluctuation during the Quaternary was recently provided by de Vivo and Carmignotto (2004). These authors attributed the loss of large-sized South American mammal lineages to the decrease in the extent of areas with open vegetation, linked with the more humid conditions of the Holocene climatic optimum. This seems to be reinforced by fossils of two primate genera, *Caipora* and *Protopithecus*, from a limestone cave in an area of caatinga in Bahia state (Cartelle and Hartwig, 1996; Hartwig and Cartelle, 1996). These are the largest known living or extinct neotropical monkeys, and de Vivo (2002) argues that these large atelid monkeys could only have inhabited a major mesic forest. Disjunct distribution of forest-dwelling mammals was also taken by de Vivo (2002) as evidence of the existence of a mesic forest in the area presently occupied by the caatinga. Costa (2003) dated the origin of some of the mammalian disjunctions between the two largest South American rain forests to the Pleistocene, and postulated the existence of an ancient corridor of moist tropical forest linking the Amazonian and Atlantic rain forests whose relicts can be seen as brejo forests.

On the other hand, caution must be used in proposing a direct relationship between the Quaternary glacial periods with drier climatic conditions and, conversely, a humid climate under interglacial periods. In the case of the area now occupied by the caatinga, the limited palynological evidence available suggests the occurrence of humid conditions under both warmer and colder environments at the end of the Pleistocene (Oliveira et al., 1999; Behling et al., 2000; Jennerjahn et al., 2004). As such, the alternation between dry and humid periods would not necessarily be related to the glacial cycles of the Pleistocene, and other factors that could influence the climatic changes in north-eastern Brazil would need to be proposed. Tectonic events are generally overlooked, although they may influence our interpretation of causal factors of the patterns observed today. For example, the origin of the dunes along the São Francisco river is usually considered to be the result of a severe dry period (see earlier), but Lima (2002) pointed out that the São Francisco river experienced a drastic change of direction in its mid course approximately 12 million years BP, diverting from a northern flow to the sea to its present eastern course. This change in course may have resulted in the exposure of sediments left behind in parts of its ancient riverbed (C.C.U. Lima, Universidade Estadual de Feira de Santana, pers. comm.).

In summary, the geological and (more limited) palaeontological evidence favours the view of alternating dry and wet periods from the Neogene until the Holocene in the region presently occupied by the caatinga. One of the geological consequences was the exposure of large areas of Precambrian crystalline basement surfaces in most of the area of the caatinga, following a huge process of pediplanation.

Combining available data concerning palaeogeography and present-day plant distribution patterns, especially the disjunct distribution of taxa among the sedimentary sandy areas, we may propose a scenario in which the Cretaceous and early Tertiary sedimentary surfaces (widespread in north-eastern Brazil by the Paleogene) became dissected, and their present configuration arose as a consequence of pediplanation processes. As a result, different kinds of substrates (crystalline granites, gneisses and schists) became exposed, creating conditions for the invasion of a new biota, that of the SDTF. This implies that the biota of the sedimentary areas must represent a more ancient flora and fauna in the dry areas of north-eastern Brazil, while that of the SDTF is a relatively more recent arrival.

The model proposed here combines the present-day geomorphological configuration of the extensive and mostly continuous crystalline surface with discontinuous areas of sedimentary surfaces. As such, it would also provide an explanation of the disjunct distribution of the species inhabiting sandy areas, as being a consequence of the events promoting vicariance within a formerly continuous area. The model is also compatible with the relatively high proportion of endemics in these sandy areas, because they are presently isolated from one another, in accordance with an allopatric speciation model. It is interesting to note that an intensive study of the herpetofauna of the caatinga undertaken by Rodrigues (2003) concluded that almost all cases of endemism for reptiles and amphibians were associated with sandy soils, and that the disjunct patterns found in some species of lizards (such as *Tropidurus cocorobensis*, and the fossorial and saxicolous genus *Calyptommatus*, for example), could be understood only if we assume that the sandy areas were much more extensive in the past (Rodrigues, 2003).

In terms of the vegetation on the crystalline basement surfaces, the model proposed here advocates the more recent arrival of a flora related to the SDTF, probably close to the conclusion of the main events that resulted in the pediplanation by the end of the Tertiary period. Pennington et al. (2004), in a study based on the evolutionary rate of the nuclear internal transcribed spacer in different SDTF genera, demonstrated that the speciation of these dry forest taxa apparently took place mostly during the Miocene-Pliocene. These authors therefore reject the widespread hypothesis that most speciation occurred allopatrically by the separation of these forests during the Pleistocene, as postulated by Prado and Gibbs (1993) and Pennington et al. (2000). Thus many SDTF species may have existed since the Neogene and have occupied the pediplane areas during the drier periods at the end of the Tertiary and the Pleistocene. The caatinga is presently the most isolated area of the SDTF, and its isolation helps to explain why it displays the greatest taxonomic diversity and a greater number of endemic taxa than any other area of the SDTF in South America (even discounting the figures presented by Prado, 1991, and Giulietti et al., 2002, for species associated with the sandy sedimentary surfaces). One cannot reject, however, the possibility of multiple invasions of SDTF taxa during the driest phases of the Pleistocene, when the SDTF could have undergone its greatest expansion.

6.4.4 IMPLICATIONS FOR CONSERVATION

The conservation status of the caatinga region is extremely heterogeneous, both in terms of the distribution and types of conservation areas already established, and in light of the centres of endemism encountered in the present work. Brazil recognizes many different kinds of conservation reserves (see descriptions available online at http://www.ibama.gov.br), and they have been grouped into three general categories according to the degree of protection they can confer on local biodiversity: permanent protection areas, which forbid any use of natural resources by local populations; sustainable use areas, which allow for human use of the natural resources; and private reserves, which confer varying degrees of protection and are usually administered on a family basis. It is assumed that the permanent protection areas offer more effective protection than sustainable use reserves, and that both are potentially less ephemeral than private reserves.

Data compiled by Velloso et al. (2002) indicate that all the protected lands in the caatinga compose less than 5% of the total area of that biome (Table 6.5). Additionally, only about 1% of this protected land is included within areas that can be considered permanent protection reserves. The great majority of the conservation units are sustainable use zones (8.8% of the total surface). The fact that only 1% of the caatinga is included within nominally permanent protection reserves is worrisome, for a large fraction of the original vegetation of this biome has already been transformed to pasture or is being used for subsistence farming. Data from the Brazilian Ministry of the Environment indicate that only 3.2% of the caatinga biome can now be considered as unaltered, and even this small percentage is dispersed into a tenuous archipelago (Brazilian Ministry of the Environment, 2002).

Perspectives for the conservation of the biodiversity of the caatinga biome are even more alarming when the distribution of its conservation units are considered in terms of the centres of endemism that have been determined for the Leguminosae. The largest fraction of permanent protection areas is located in the geomorphological sedimentary zones. Only one of the centres of endemism can be considered as being reasonably well represented in any permanent protection reserve

TABLE 6.5
Summary of the Conservation Areas in Each Ecoregion of the Caatinga Biome

Ecoregion	Biota	Total Conserved Area		Permanent Protection Areas		Sustainable Use Areas		Private Reserves		Total Area (km²)
		ha	%	ha	%	ha	%	ha	%	
Campo Maior	Mixed	168,829	4.08	7700	0.19	159,255	3.84	1874	0.05	41,420
Ibiapaba–Araripe	Sedimentary	3,143,166	45.22	602,974	8.67	2,534,921	36.47	5271	0.08	69,510
Northern depression	Crystalline	348,250	1.68	25,286	0.12	314,205	1.52	8759	0.04	206,700
Borborema	Crystalline	1440	0.03	0	0.00	0	0.00	1440	0.03	41,940
Southern depression	Crystalline	178,940	4.79	0	0.00	162,970	4.36	15,970	0.43	373,900
Dunes	Sedimentary	1,085,000	30.00	0	0.00	1,085,000	30.00	0	0.00	36,170
Chapada Diamantina	Mixed	169,034	3.34	158,000	3.12	11,034	0.22	0	0.00	50,610
Raso da Catarina	Sedimentary	107,353	3.49	100,872	3.28	1321	0.04	5160	0.17	30,800
Total	—	5,202,012	6.11	894,832	1.05	7,452,412	8.76	38,474	0.05	851,050

Source: Velloso et al., 2002

(the Ibiapaba-Araripe complex in the states of Ceará and Piauí has 8.7% of its area within a permanent protection area, as well as an additional 36.5% within a sustainable use zone). In the region of the Raso de Catarina, there is a national reserve that protects approximately 3.5% of its area. It would be desirable to establish other permanent preservation zones to the north of this region, near Buíque and the valley of Ipojuca in the state of Pernambuco, as these areas have been designated as extremely important zones of biodiversity (Brazilian Ministry of the Environment, 2002).

The most serious situation within the sedimentary zones, however, is encountered on the sand dunes of the São Francisco river. Despite the fact that this region contains the greatest number of endemic species of Leguminosae, as well as a significant number of endemic species belonging to other groups of plants and animals, no permanent protection areas have yet been set up. There is one large sustainable use reserve covering 30% of the sand dunes, but it includes many small villages and farms. Fortunately, it is a region with a very low population density (fewer than five inhabitants per square kilometer; Superintendência de Estudos Econômicos e Sociais da Bahia, 2005), and this has reduced human impact on the natural vegetation. Nevertheless, the creation of permanent protection reserves near Ibiraba, Casa Nova and Pilão Arcado should be considered a priority, as this area retains large extensions of intact native vegetation.

The crystalline bedrock geomorphological zone has the greatest human population density within the caatinga, and almost all the vegetation there has been altered through clearing for cattle raising and for subsistence agriculture. Unfortunately, the number of established conservation units is still lower than that found in the sedimentary zones.

There are no full protection reserves in the southern Sertaneja depression zone, and sustainable use areas and private reserves occupy less than 5% of the total area. The largest areas of reasonably well-preserved vegetation in this zone are located near the south-western border of the caatinga biome, between southern Bahia state (between the municipalities of Bom Jesus da Lapa and Santa Maria da Vitória) and northern Minas Gerais state (in the region of Peruaçu–Jaíba). This should be considered a priority region for conservation planning. Other important areas in the southern Sertaneja depression include Senhor do Bonfim and Maracás, in Bahia state, and the Serra Negra region in Pernambuco (Brazilian Ministry of the Environment, 2002).

Similarly, in the northern Sertaneja depression only 0.12% of its surface is included within permanent protection reserves at Aiuaba (Ceará) and Seridó (Rio Grande do Norte) regions. Urgent protection measures, which include setting up reserves, are needed in Jaguaribe and Chapada do Apodi, the Baturité range and the region near Quixadá in the states of Ceará and Rio Grande do Norte, as well as the region near Cariri, in Paraíba state.

6.5 CONCLUSIONS

Patterns of geographical distribution of taxa of the Leguminosae, as well as the similarity analysis of areas representing different types of vegetation of north-eastern Brazil (based on the presence of species of this family), all point to the recognition of two different biotas covered by a generic application of the term caatinga. One biota is associated with the crystalline basement surfaces that cover most of the caatinga's region, and demonstrates floristic relationships with the SDTF. The second biota is associated with sandy sedimentary surfaces and, although it occupies geomorphologically diverse areas, it retains a strong floristic unity, with a relatively high proportion of taxa occurring disjunctly among its main nuclei.

These two larger floras can be divided into seven major centres of endemism recognized here. Two of these centres are associated with the crystalline basement surface:

1. the northern Sertaneja depression ranging northwards from the borders of Pernambuco and Ceará states, including the ecoregion of the Borborema plateau; and
2. the southern Sertaneja depression, ranging southwards from the southern limits of the northern Sertaneja depression to northern Minas Gerais.

Four other centres are associated with the sandy sedimentary surfaces:

3. the sedimentary tablelands of the Tucano-Jatobá basin;
4. the dunes of the mid São Francisco river valley;
5. the Ibiapaba–Araripe plateaux; and
6. the seasonally flooded depressions of south-eastern Piauí, western Pernambuco and northern Bahia.

The seventh centre, the caatingas of the Chapada Diamantina mountain range, is not clearly related to any of the two major floras cited here.

The data presented here, interpreted using available palaeogeographical information, favour a model that considers that plants of the sandy areas compose most of the original flora of the dry areas of north-eastern Brazil. They were replaced (on most of this surface) during the late Tertiary and early Quaternary, when processes of pediplanation exposed the Precambrian crystalline bedrock, allowing for the establishment of species typical of the SDTF.

The hypotheses presented here could quite possibly be tested using molecular phylogenetic techniques (see Lavin, Chapter 19; Pennington et al., Chapter 1:), which could estimate the age of diversification of the characteristics of the two major floristic groups, as well as raise phylogeographical hypotheses about past movements of key taxa. Other lines of investigation, such as population genetics, could perhaps be used to determine if SDTF taxa invaded the crystalline basement surfaces on more than one occasion (e.g. see Naciri-Graven et al., Chapter 18).

Floristic relationships between the sedimentary areas of the caatinga and other South American vegetation types remain to be examined. Based on the data presented here, we recommend that phytogeographical studies of South America must not assume that the flora of the dry region of north-eastern Brazil is a single unit, but that it is divided into two major floristic units, as recognized here.

ACKNOWLEDGEMENTS

The author would like to thank Dr Gwilym Lewis for sharing data concerning the Leguminosae; Dr Ray Harley, Dr Ana Giulietti, Dr Freddy Bravo, Dr Toby Pennington, Dr Gwilym Lewis and an anonymous reviewer for their critical reading of the manuscript and useful suggestions; Dr Carlos C. Uchôa Lima for his help with the geological literature; Gil I. Ximenes and Dr Flora Juncá for their help with relevant bibliography on animals; Eduardo L. Borba for his assistance with multivariate analysis; Francisco Haroldo F. do Nascimento for his assistance with herbaria information; and Roy Funch and Ray Harley for improving the English text. Fieldwork was partially supported by the Instituto do Milênio do Semi-árido (Ministério de Ciência e Tecnologia–CNPq).

This research was supported by grants from the CNPq (process 350715/2000-9 and 470221/2001-1) and Fapesb (Prodoc).

REFERENCES

Ab'Saber, A.N., O domínio morfoclimático semi-árido das caatingas brasileiras, *Geomorfologia*, 43, 1, 1974.
Alcoforado-Filho, F.G., Sampaio, E.V.S.B., and Rodal, M.J.N., Florística e fitossociologia de um remanescente de vegetação caducifólia espinhosa arbórea em Caruaru, Pernambuco, *Acta Bot. Bras.*, 17, 287, 2003.
von Altschul, S. R., A taxonomic study of the genus *Anadenanthera*, *Contr. Gray Herb.*, 193, 1, 1964.
Andrade-Lima, D., *Contribution to the Study of the Flora of Pernambuco, Brazil*, Universidade Rural de Pernambuco, Recife, 1954.
Andrade-Lima, D., Vegetação da área Jaguaquara-Maracás, Bahia. *Ci. Cult.*, 23, 317, 1971.
Andrade-Lima, D., Flora de áreas erodidas de calcário Bambuí, em Bom Jesus da Lapa, Bahia, *Rev. Bras. Biol.*, 37, 179, 1977.

Andrade-Lima, D., As formações vegetais da bacia do Parnaíba, in *Bacia do Parnaíba: Aspectos Fisiográficos*, Lins, R.C., Ed, Instituto de Pesquisas Sociais, Recife, 1978.

Andrade-Lima, D., The caatinga dominium, *Rev. Bras. Bot.*, 4, 149, 1981.

Andrade-Lima, D., Present day forest refuges in northeastern Brazil, in *Biological Diversification in the Tropics*, Prance, G.T., Ed, Columbia University Press, New York, 1982, 245.

Araújo, F.S. and Martins, F.R., Fisionomia e organização da vegetação do carrasco no planalto da Ibiapaba, estado do Ceará, *Acta Bot. Bras.*, 13, 1, 1999.

Araújo, F.S., Martins, F.R., and Shepherd, G.J., Variações estruturais e florísticas do carrasco no planalto da Ibiapaba, estado do Ceará, *Rev. Bras. Biol.*, 59, 663, 1999.

Araújo, F.S. et al., Composição florística da vegetação de carrasco, Novo Oriente, CE, *Rev. Bras. Bot.*, 21, 105, 1998.

Bailey, R.G., *Ecoregions: the Ecosystem Geography of the Oceans and Continents*, Springer-Verlag, New York, 1998.

Barneby, R.C., Sensitivae censitae, a description of the genus *Mimosa* L. (Mimosaceae) in the New World, *Mem. New York Bot. Gard.*, 65, 1, 1991.

Barneby, R.C., Neotropical Fabales at NY: asides and oversights, *Brittonia*, 48, 174, 1996.

Barneby, R.C., Silk tree, guanacaste, monkey's earring — a generic system for the synandrous Mimosaceae of the Americas: part III, *Calliandra*, *Mem. New York Bot. Gard.*, 74, 3, 1998.

Barneby, R.C. and Grimes, J.W., Silk tree, guanacaste, monkey's earring — a generic system for the synandrous Mimosaceae of the Americas: part I. *Abarema*, *Albizia*, and allies, *Mem. New York Bot. Gard.*, 74, 1, 1996.

Barneby, R.C. and Grimes, J.W., Silk tree, guanacaste, monkey's earring — a generic system for the synandrous Mimosaceae of the Americas: part II. *Pithecellobium*, *Cojoba*, and *Zygia*, *Mem. New York Bot. Gard.*, 74, 2, 1, 1997.

Behling, H. et al., Late Quaternary vegetational and climate dynamics in northeastern Brazil, inferences from marine core GeoB 3104-1, *Quat. Sci. Rev.*, 19, 981, 2000.

Brazilian Ministry of the Environment, *Avaliação e Ações Prioritárias para a Conservação da Biodiversidade da Caatinga, Brasília*, Brazilian Ministry of the Environment, Brasília, 2002, 36.

Bullock, S.H., Plant reproduction in neotropical dry forests, in *Seasonally Dry Tropical Forests*, Bullock, S.H., Mooney, H.A., and Medina, E., Eds, Cambridge University Press, New York, 1995, 277.

Bullock, S.H. and Solís M.J.A., Phenology of canopy trees of a tropical deciduous forest in Mexico, *Biotropica*, 22, 22, 1990.

Burnham, R.J. and Graham, A., The history of neotropical vegetation: new developments and status, *Ann. Missouri Bot. Gard.*, 86, 546, 1999.

Cailleux, A. and Tricart, J., Zonas fitogeográficas e morfoclimáticas do Quaternário no Brasil, *Not. Geomorf.*, 2, 12, 1959.

Camacho, R.G.V., *Estudo fitofisiográfico da caatinga do Seridó — Estação Ecológica do Seridó, RN*, PhD thesis, Universidade de São Paulo, São Paulo, 2001.

Cartelle, C. and Hartwig, W.C., A new extinct primate among the Pleistocene megafauna of Bahia, Brazil, *Proc. Natl. Acad. Sci. USA*, 93, 6405, 1996.

Carvalho, A.M., *Systematics studies of the genus* Dalbergia *L. f. in Brazil*, PhD thesis, University of Reading, Reading, 1985.

Carvalho-Sobrinho, J.G. and Queiroz, L.P., Composição florística de um fragmento de mata atlântica na Serra da Jibóia, Santa Terezinha, Bahia, Brasil, *Sitientibus Sér. Ci. Biol.*, 5, 20, 2005.

Colinvaux, P.A. et al., A paradigm to be discarded: geological and palaeoecological data falsify the Haffer & Prance refuge hypothesis of Amazonian speciation, *Amazoniana*, 16, 609, 2001.

Correia, M.S., *Estrutura da vegetação da mata serrana de um brejo de altitude em Pesqueira — PE*, MSc thesis, Universidade Federal de Pernambuco, Recife, 1996.

Costa, L.P., The historical bridge between the Amazon and the Atlantic forest of Brazil: a study of molecular phylogeography with small mammals, *J. Biogeogr.*, 30, 71, 2003.

Cristóbal, C.L., *Rayleya*, nueva Sterculiaceae de Bahia — Brasil, *Bonplandia*, 5, 45, 1981.

Egler, W.A., Contribuição ao estudo da caatinga pernambucana, *Rev. Bras. Geogr.*, 13, 577, 1951.

Emperaire, L., Végétation de l'État du Piauí (Brésil), *C. R. Seances Soc. Biogeogr.*, 60, 151, 1983.

Felfili, J.M. and Silva, M.C., Jr, *Biogeografia do Bioma Cerrado: Estudo Fisionômico na Chapada do Espigão Mestre do São Francisco*, Universidade de Brasília, Brasília, 2001.

Fernandes, A., Fitogeografia do semi-árido, in *Anais da IV Reunião Especial da SBPC*, Universidade Estadual de Feira de Santana, Feira de Santana, 1996, 215.

Ferraz, E.M.N. et al., Composição florística de vegetação de caatinga e brejo de altitude na região do vale do Pajeú, Pernambuco, *Rev. Bras. Bot.*, 21, 7, 1998.

Ferreira, M.B. and Costa, N.M.S., *O Gênero* Stylosanthes *no Brasil*, EPAMIG, Belo Horizonte, 1979, 107.

Fevereiro, V.P.B., *Macroptilium* (Bentham) Urban do Brasil (Leguminosae-Faboideae-Phaseoleae-Phaseolinae), *Arq. Jard. Bot. Rio de Janeiro*, 28, 109, 1986.

Figueiredo, M.A., Vegetação, in *Atlas do Ceará*, SUDEC, Fortaleza, 1986, 24.

Fisch, E., Vogel, S., and Lopes, A.V., *Ameroglossum*, a new monotypic genus of Scrophulariaceae-Scrophularioideae from Brazil, *Feddes Repert.*, 110, 529, 1999.

Fonseca, M.R., *Análise da vegetação arbustiva-arbórea da caatinga hiperxerófila do noroeste do estado de Sergipe*, PhD thesis, Universidade Estadual de Campinas, Campinas, 1991.

França, F., de Melo, E., and Santos, C.C., Flora de inselbergs da região de Milagres, Bahia, Brasil: I. Caracterização da vegetação de lista de espécies de dois inselbergs, *Sitientibus*, 17, 171, 1997.

Gentry, A.H., Bignoniaceae part II (Tecomeae), in *Fl. Neotrop.*, 25, 1, 1992.

Giulietti, A.M., Queiroz, L.P., and Harley, R.M., Vegetação e flora da Chapada Diamantina, Bahia, in *Anais da IV Reunião Especial da SBPC*, Sociedade Brasileira para o Progresso de Ciência, Feira de Santana, 1996, 144.

Giulietti, A.M. et al., Espécies endêmicas da caatinga, in *Vegetação e Flora da Caatinga*, Sampaio, E.V.S.B. et al., Eds, Associação Plantas do Nordeste, Recife, 2002, 103.

Gomes, M.A.F., *Padrões de caatinga nos Cariris Velhos, Paraíba*, MSc thesis, Universidade Federal Rural de Pernambuco, Recife, 1979.

Guedes, M.L.S., *Estudo florístico e fitossociológico de um trecho da reserva ecológica da Mata dos Dois Irmãos, Recife — Pernambuco*, MSc thesis, Universidade Federal Rural de Pernambuco, Recife, 1992.

Guedes, R.R., Lista preliminar das angiospermas ocorrentes no Raso da Catarina e arredores, Bahia, *Rodriguésia*, 37, 5, 1985.

Guedes-Bruni, R.R. and Lima, M.P.M., Abordagem geográfica, fitofisionômica, florística e taxonômica de Reserva Ecológica de Macaé de Cima, in *Reserva Ecológica de Macaé de Cima, Nova Friburgo — RJ: Aspectos Florísticos das Plantas Vasculares*, Guedes-Bruni, R.R. and Lima, M.P.M., Eds, Jardim Botânico do Rio de Janeiro, Rio de Janeiro, 1994.

Guevara de Lampe, M.C. et al., Seasonal flowering and fruiting patterns in tropical semi-arid vegetation in northeastern Venezuela, *Biotropica*, 24, 64 1992.

Harley, R.M., Introduction, in *Flora of the Pico das Almas, Chapada Diamantina, Bahia, Brazil*, Stannard, B.L., Ed, Royal Botanic Gardens, Kew, Richmond, 1995, 1.

Harley, R.M., Examples of endemism and phytogeographical elements in the caatinga flora, in *Anais da IV Reunão Especial da SBPC*, Sociedade Brasileira para o Progresso de Ciência, Feira de Santana, 1996, 219.

Harley, R.M. and Simmons, N.A., *Florula of Mucugê, Chapada Diamantina — Bahia, Brazil*, Royal Botanic Gardens, Kew, Richmond, 1986, 228.

Hartwig, W.C. and Cartelle, C., A complete skeleton of the giant South American primate *Prothopithecus*, *Nature*, 301, 307, 1996.

Ireland, H.E., *The taxonomy and systematics of* Ateleia *and* Cyathostegia *(Leguminosae — Swartzieae)*, PhD thesis, University of Reading, Reading, 2001.

Ireland, H.E. and Pennington, R.T., A revision of *Geoffroea* (Leguminosae–Papilionoideae), *Edinburgh J. Bot.*, 56, 329, 1999.

Irwin, H.S. and Barneby, R.C., Monographic studies in *Cassia* (Leguminosae, Caesalpinioideae). III. Sections *Absus* and *Grimaldia*, *Mem. New York Bot. Gard.*, 30, 1, 1978.

Irwin, H.S. and Barneby, R.C., The American Cassiinae. A synoptical revision of Leguminosae tribe Cassieae subtribe Cassiinae in the New World, *Mem. New York Bot. Gard.*, 35, 1, 1982.

Jennerjahn, T.C. et al., Asynchronous terrestrial and marine signals of climate change during Heinrich event, *Science*, 306, 2236, 2004.

Kirkbride, J.H., Jr, Legumes of the cerrado, *Pesq. Agropec. Bras.*, 19, 23, 1984.

Klitgaard, B.B., *Systematics of* Platymiscium *(Leguminosae: Papilionoideae: Dalbergieae): taxonomy, morphology, ontogeny, and phylogeny*, PhD thesis, University of Aarhus, Aarhus, 1995.

Krapovickas, A. and Gregory, W.C., Taxonomia del género *Arachis* (Leguminosae), *Bonplandia*, 8, 1, 1994.

Lavin, M., Systematics of *Coursetia* (Leguminosae-Papilionoideae), *Syst. Bot. Monogr.*, 21, 1, 1988.

Lavin, M. et al., Metacommunity process rather than continental tectonic history better explains geographically structured phylogenies in legumes, *Philos. Trans., Ser. B*, 359, 1509, 2004.

Lee, Y.-T. and Langenhein, J.H., Systematics of the genus *Hymenaea* L. (Leg. Caesalpinioideae, Detarieae), *Univ. California Publ. Bot.*, 69, 1, 1975.

Lemos, J.R. and Rodal, M.J.N., Fitossociologia do componente lenhoso de um trecho de vegetação de caatinga no Parque Nacional da Serra da Capivara, Piauí, Brasil, *Acta Bot. Bras.*, 16, 23, 2002.

Lewis, G.P., *Legumes of Bahia*, Royal Botanic Gardens, Kew, Richmond, 1987.

Lewis, G.P., Caesalpinia, *A Revision of the* Poincianella-Erythrostemon *Group*, Royal Botanic Gardens, Kew, Richmond, 1998.

Lewis, G.P. and Delgado S.,A., *Mysanthus*, a new genus in tribe Phaseoleae (Leguminosae: Papilionoideae) from Brazil, *Kew Bull.*, 49, 343, 1994.

Lewis, G.P. and Lima, M.P.M., *Pseudopiptadenia* Rauschert no Brasil (Leguminosae — Mimosoideae), *Arq. Jard. Bot. Rio de Janeiro*, 30, 43, 1990.

Lima, C.C.U., *Sedimentologia e aspectos neotectônicos do grupo barreiras no litoral sul do estado da Bahia*, PhD thesis, Universidade Federal da Bahia, Salvador, 2002.

Lima, H.C. and Vaz, A.M.S.F., Revisão taxonômica do gênero *Riedeliella* Harms (Leguminosae — Faboideae), *Rodriguésia*, 36, 9, 1984.

Lima, H.C., Carvalho, A.M., and Costa, C.G., Estudo taxonômico do genero *Diptychandra* Tulasne (Leguminosae: Caesalpinioideae), in *35th Brazilian Botanical Congress, Annals*, 1990, 175.

Lima, I.B., *Levantamento florístico da Reserva Particular do Patrimônio Natural Fazenda Almas, São José dos Cordeiros — PB*, monograph, Universidade Federal da Paraíba, João Pessoa, 2004.

Lima, M.P.M. and Lima, H.C., *Parapiptadenia* Brenan (Leg. Mim.) – estudo taxonômico das espécies brasileiras, *Rodriguésia*, 36, 23, 1984.

Lyra, A.L.R.T., *A condição de 'brejo': efeito do relevo na vegetação de duas áreas no município do Brejo da Madre de Deus — PE*, MSc thesis, Universidade Federal Rural de Pernambuco, Recife, 1982.

Machado, I.C.S., Barros, L.M., and Sampaio, E.V.S.B., Phenology of the caatinga species at Serra Talhada, PE, north-eastern Brazil, *Biotropica*, 29, 57, 1997.

Mesquita, A.L., *Revisão taxonômica do gênero* Enterolobium *Mart. (Mimosoideae) para a região Neotropical*, MSc thesis, Universidade Federal Rural de Pernambuco, Recife, 1990.

Mohlenbrock, R.H., A monograph of the leguminous genus *Zornia*, *Webbia*, 16, 1, 1961.

Monteiro, R., The species of *Sesbania* Scop. (Leguminosae) in Brazil, *Arq. Biol. Tecnol.*, 37, 309, 1994.

Mori, S.A. et al., Southern Bahian moist forests, *Bot. Rev. (Lancaster)*, 49, 155, 1983.

Moura, F.B.P., *Fitossociologia de uma mata serrana semidecídua no brejo de Janaúba, Pernambuco, Brasil*, MSc thesis, Universidade Federal de Pernambuco, Recife, 1997.

Oliveira, J.G.B. et al., Observações preliminares de fenologia de plantas de caatinga na Estação Ecológica de Aiuaba, Ceará, Esam, Mossoró, *Coleção Mossoroense série B*, 538, 1, 1998.

Oliveira, P.E., Barreto, A.M.F., and Suguio, K., Late Pleistocene/Holocene climatic and vegetational history of the Brazilian caatinga: the fossil dunes of the middle São Francisco river, *Palaeogeogr. Palaeoclimatol. Palaeoecol.*, 152, 319, 1999.

Oliveira-Filho, A. and Ratter, J.A., A study of the origin of central Brazilian forests by the analysis of plant species distribution patterns, *Edinburgh J. Bot.*, 52, 141, 1995.

Pennington, R.T., A monograph of *Andira* (Leguminosae-Papilionoideae), *Syst. Bot. Monogr.*, 64, 1, 2003.

Pennington, R.T., Prado, D.E., and Pendry, C.A., Neotropical seasonally dry forests and Quaternary vegetation changes, *J. Biogeogr.*, 27, 261, 2000.

Pennington, R.T. et al., Historical climate change and speciation: neotropical seasonally dry forest plants show patterns of both Tertiary and Quaternary diversification. *Philos. Trans., Ser. B*, 359, 515, 2004.

Prado, D.E., *A critical evaluation of the floristic links between chaco and caatinga vegetation in South America*, PhD thesis, University of Saint Andrews, Saint Andrews, 1991.

Prado, D.E., Seasonally dry forests of tropical South America: from forgotten ecosystems to a new phytogeographical unity, *Edinburgh J. Bot.*, 57, 437, 2000.

Prado, D.E., As caatingas do Brasil, in *Ecologia e Conservação da Caatinga*, Leal, I.R., Tabarelli, M., and Silva, J.M.C., Eds, Universidade Federal de Pernambuco, Recife, 2003, 3.

Prado, D.E. and Gibbs, P.E., Patterns of species distributions in the dry seasonal forest of South America, *Ann. Missouri Bot. Gard.*, 80, 902, 1993.

Queiroz, L.P., Distribuição de espécies de Leguminosae na caatinga, in *Vegetação e Flora da Caatinga*, Sampaio, E.V.S.B. et al., Eds, Associação Plantas do Nordeste, Recife, 2002, 141.

Queiroz, L.P., Leguminosae, in *Flora de Grão-Mogol, Minas Gerais*, Pirani, J.R., Ed, *Bol. Bot. Univ. São Paulo*, 22, 213, 2004a.

Queiroz, L.P., *Biodiversidade da família Leguminosae na caatinga da Bahia: florística, biogeografia e disseminação*, technical report, Universidade Estadual de Feira de Santana, Feira de Santana, 2004b.

Ratter, J.A. et al., Observations on forests of some mesotrophic soils in central Brazil, *Rev. Bras. Bot.*, 1, 47, 1978.

Ratter, J.A., Bridgewater, S., and Ribeiro, J.F., Analysis of the floristic composition of the Brazilian cerrado vegetation III. Comparision of the woody vegetation of 376 areas, *Edinburgh J. Bot.*, 60, 57, 2003.

Rizzini, C.T., Nota prévia sobre a divisão fitogeográfica do Brasil, *Rev. Bras. Geogr.*, 25, 3, 1963.

Rizzini, C.T., *Tratado de Fitogeografia do Brasil*, Hucitec-Universidade de São Paulo, São Paulo, 1979.

Rocha, P.L.B., *Uso e partição de recursos pelas espécies de lagartos das dunas do rio São Francisco, Bahia (Squamata)*, PhD thesis, Universidade de São Paulo, São Paulo, 1998.

Rocha, P.L.B., Queiroz, L.P., and Pirani, J.R., Plant species and habitat structure in a sand dune field in the Brazilian caatinga: a homogeneous habitat harbouring an endemic biota, *Rev. Bras. Bot.*, 27, 739, 2004.

Rodal, M.J.N. and do Nascimento, L.M., Levantamento florístico de uma floresta serrana da reserva biológica de Serra Negra, microrregião de Itaparica, Pernambuco, Brasil, *Acta Bot. Bras.*, 16, 481, 2002.

Rodal, M.J.N. and Sampaio, E.V.S.B., A vegetação do bioma caatinga, in *Vegetação e Flora da Caatinga*, Sampaio, E.V.S.B. et al., Eds, Associação Plantas do Nordeste, Recife, 2002, 11.

Rodal, M.J.N. et al., Fitossociologia do componente lenhoso de um refúgio vegetacional no município de Buíque, Pernambuco, *Rev. Bras. Biol.*, 58, 517, 1998.

Rodrigues, M.T.U., Lizards, snakes and amphisbaenians from the Quaternary and dunes of the middle Rio São Francisco: Bahia: Brazil, *J. Herpet.*, 30, 513, 1996.

Rodrigues, M.T.U., Herpetofauna da caatinga, in *Ecologia e Conservação da Caatinga*, Leal, I.R., Tabarelli, M., and Silva, J.M.C., Eds, Universidade Federal de Pernambuco, Recife, 2003, 181.

Rodrigues, M.T.U. and Juncá, F.A., Herpetofauna of the Quaternary sand dunes of the middle Rio São Francisco: Bahia: Brasil. VII. *Tiphlops amoipira* sp. nov, a possible relative of *Typhlops yonenagae* (Sepentes, Typhlopidae), *Pap. Avulsos Zool. S. Paulo*, 42, 325, 2002.

Sales, M.F., Mayo, S.J., and Rodal, M.J.N., *Plantas Vasculares das Florestas Serranas de Pernambuco*, Universidade Federal Rural de Pernambuco, Recife, 1998, 130.

Sampaio, E.V.S.B., Overview of the Brazilian caatinga, in *Seasonally Dry Tropical Forests*, Bullock, S.H., Mooney, H.A., and Medina, E., Eds, Cambridge University Press, New York, 1995, 35.

Sampaio, E.V.S.B., Andrade-Lima, D.A., and Gomes, M.A.F., O gradiente vegetacional das caatingas e áreas anexas, *Rev. Bras. Bot.*, 4, 27, 1981.

Santos, A.M.M., *Distribuição de plantas lenhosas e relações históricas entre a Amazônia, a floresta Atlântica Costeira e os brejos de altitude do nordeste Brasileiro*, MSc thesis, Universidade Federal de Pernambuco, Recife, 2002.

Schrire, B.D., Lavin, M., and Lewis, G.P., Global distribution patterns of the Leguminosae: insights from recent phylogenies, *Biologiske Skrifter*, 55, 375, 2005.

Shepherd, G. J., *FITOPAC 1*, Universidade Estadual de Campinas, Campinas, 1995.

Silva, G.C., *Flora e vegetação das depressões inundáveis da região de Ouricuri — PE*, MSc thesis, Universidade Federal Rural de Pernambuco, Recife, 1985.

Silva, M.F., Revisão taxonômica do gênero *Peltogyne* Vog. (Leguminosae — Caesalpinioideae), *Acta Amazonica*, 6, 1, 1976.

Souza, E.R., *Aspectos taxonômicos e biogeográficos do gênero Calliandra Benth. (Leguminosae–Mimosoideae) na Chapada Diamantina, Bahia, Brasil*, MSc thesis, Universidade Estadual Feira de Santana, Feira de Santana, 2001.

Superintendência de Estudos Econômicos e Sociais da Bahia, http://www.sei.ba.gov.br, 2005.

Taubert, P., Leguminosae novae v. minus cognitae austro-americanae, *Flora*, 75, 68, 1892.

Taylor, N.P. and Zappi, D.C., *Cacti of Eastern Brazil*, Royal Botanic Gardens, Kew, Richmond, 2004.

Thomas, W.W. et al., Plant endemism in two forests in southern Bahia, Brazil, *Biodivers. Conservation*, 7, 311, 1998.

Tozzi, A.M.G.A., *Estudos taxonômicos nos gêneros Lonchocarpus Kunth e Deguelia Aubl. no Brasil*, PhD thesis, Universidade Estadual de Campinas, Campinas, 1985.

Tricart, J., As zonas morfoclimáticas do nordeste brasileiro, *Not. Geomorf.*, 3, 17, 1961.

Tricart, J., Evidence of upper Pleistocene dry climates in northern South America, in *Environmental Change and Tropical Geomorphology*, Douglas, I. and Spencer, T., Eds, Allen & Unwin, London, 1985, 197.

Vaz, A.M.S.F., Considerações sobre a taxonomia do gênero *Bauhinia* L. sect. *Tylotea* Vogel (Leguminosae – Caesalpinioideae) no Brasil, *Rodriguésia*, 31, 127, 1979.

Vaz, A.M.S.F., *Taxonomia de* Bauhinia *sect.* Pauletia *(Leguminosae: Caesalpinioideae: Cercideae) no Brasil*, PhD thesis, Universidade Estadual de Campinas, Campinas, 2001.

Velloso, A.L. et al., *Ecorregiões: Propostas para o Bioma Caatinga*, Associação Plantas do Nordeste–Nature Conservancy do Brasil, Recife, 2002.

de Vivo, M., Mammalian evidence of historical ecological change in the caatinga semiarid vegetation of northeastern Brazil, *J. Comp. Biol.*, 2, 65, 2002.

de Vivo, M. and Carmignotto, A.P., Holocene vegetation change and the mammal faunas of South America and Africa, *J. Biogeogr.*, 31, 943, 2004.

Warwick, M.C. and Lewis, G.P., Revision of *Plathymenia* (Leguminosae — Mimosoideae), *Edinburgh J. Bot.*, 60, 2, 111, 2003.

Williams, R.J. and Clements, R.J., Taxonomy of *Centrosema*, in *Centrosema, Biology, Agronomy, and Utilization*, Centro Internacional de Agricultura Tropical, Cali, 1990, 29.

Zanella, F.C.V. and Martins, C.F., Abelhas da caatinga: biogeografia, ecologia e conservação, in *Ecologia e Conservação da Caatinga*, Leal, I.R., Tabarelli, M., and Silva, J.M.C., Eds, Universidade Federal de Pernambuco, Recife, 2003, 75.

Zappi, D.C. et al., Lista das plantas vasculares de Catolés, Chapada Diamantina, Bahia, Brasil, *Bol. Bot. Univ. São Paulo*, 21, 345, 2003.

7 Floristic Relationships of Seasonally Dry Forests of Eastern South America Based on Tree Species Distribution Patterns

Ary T. Oliveira-Filho, João André Jarenkow and Maria Jesus Nogueira Rodal

CONTENTS

ABSTRACT

The tree flora of seasonally dry tropical forests (SDTF) of eastern tropical and subtropical South America was investigated according to two main aspects: (a) the variations in floristic composition were analysed in terms of geographical and climatic variables by performing multivariate analyses on 532 existing floristic checklists; and (b) the links among different seasonally dry forest formations, Amazonian forests and cerrados (woody savannas) were assessed. Analyses were performed at the species, genus and family levels. There was a strong spatial pattern in tree species distribution that only receded and allowed clearer climate-related patterns to arise when either the geographical range was restricted or data were treated at the genus and family levels. Consistent floristic differences occurred between rain and seasonal forests, although these were obscured by strong regional similarities which made the two forest types from the same region closer to each other floristically than they were to their equivalents in different regions. Atlantic rain and seasonal forests were floristically closer to each other than to Amazonian rain forests but north-east rain and seasonal forests were both closer to Amazonian rain forests than each other, though only at the generic and familial levels. Atlantic seasonal forests also share a variable proportion of species with caatingas, cerrados and the chaco, and may represent a transition to these open formations. Increasing periods of water shortage, with increases in soil fertility and temperature are characteristic of a transition from semideciduous to deciduous forests and then to the semi-arid formations, either caatingas (tropical) or chaco forests (subtropical), while increasing fire frequency and decreasing soil fertility lead from seasonal forests to either cerrados (tropical) or southern campos (subtropical). The SDTF vegetation of eastern South America may be classified into three floristic nuclei: caatinga, chaco and Atlantic forest (sensu latissimo). Only the last, however, should be linked consistently to the residual Pleistocenic dry seasonal flora (RPDS). Caatinga and chaco represent the extremes of floristic dissimilarity among the three nuclei, also corresponding to the warm–dry and warm–cool climatic extremes, respectively. In contrast to the caatinga and chaco nuclei, the Atlantic SDTF nucleus is poor in endemic species and is actually a floristic bridge connecting the two drier nuclei to rain forests. Additionally, there are few grounds to recognize the Atlantic nucleus flora as a clearly distinct species assemblage, since there is a striking variation in species composition found throughout its wide geographical range. Nevertheless, there is a group of wide-range species that are found in most regions of the Atlantic nucleus, some of which are also part of the species blend of the caatinga and chaco floras, though the latter plays a much smaller part. We propose that it is precisely this small fraction of the Atlantic nucleus flora that should be identified with the RPDS vegetation.

7.1 INTRODUCTION

Neotropical seasonally dry tropical forests, or SDTF, are presently an increasing focus of attention because of both their very threatened status and poorly studied flora and ecology, and this is striking when compared to traditional flag ecosystems like rain forests and savannas (Mooney et al., 1995). They occur where annual rainfall is less than 1600 mm and more than 5–6 months receive less than 100 mm (Graham and Dilcher, 1995) and therefore include a diverse array of vegetation formations, from tall semideciduous forests to thorny woodland with succulents (Murphy and Lugo, 1995). Despite all this variation, Pennington et al. (2000) argue that the concept of SDTF should exclude fire-related formations, such as savannas and cerrados, and the non-tropical chaco forests.

The distribution of SDTF in South America forms an arc with the ends positioned at the caatinga domain of north-eastern Brazil and the Caribbean coast of Colombia and Venezuela and a long curved route connecting the ends through the seasonal forests of the Atlantic forest domain, the patches of seasonal forests of the cerrado domain, and the seasonal forests of the Andean piedmont, inter-Andean valleys, Pacific coast and Caribbean coast. Prado (1991) and Prado and Gibbs (1993) suggested the hypothesis that this arc is a relic of a much wider distribution of SDTF in South America reached during the Pleistocene glacial maxima. They based their model on the present distribution of what they called residual Pleistocenic dry seasonal (RPDS) flora. Since then a number of studies have analysed species distribution patterns of this flora in order to assess the validity of the RPDS arc hypothesis in different geographical contexts (e.g. Pennington et al., 2000, 2004; Prado, 2000; Bridgewater et al., 2003; Linares-Palomino et al., 2003; Spichiger et al., 2004). The assessment of floristic links among species assemblages of SDTF areas scattered over the putative RPDS arc has proved a useful tool to elucidate patterns of historical vegetation change. Linares-Palomino et al. (2003) performed a detailed phytogeographical analysis of SDTF areas of Pacific South America, i.e. the western section of the RPDS Arc, and found three main groups with a considerable dissimilarity among them. In the present contribution, we perform a similar analysis of seasonally dry forest areas of the eastern section of the RPDS arc. As we dealt with both the tropical and subtropical regions of eastern South America we incorporated seasonally dry subtropical forests into the SDTF concept.

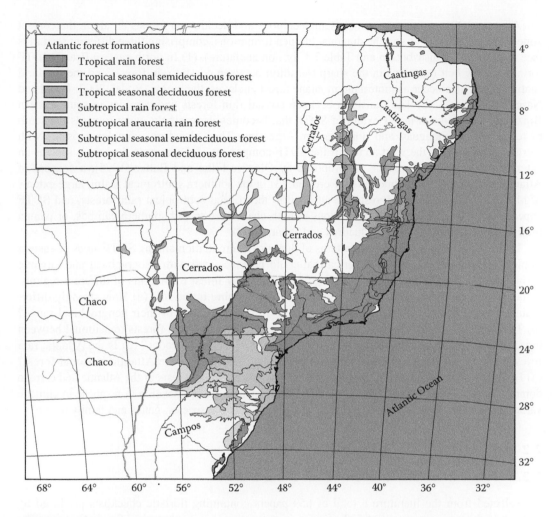

FIGURE 7.1 (See colour insert following page 208) Map of eastern South America showing the distribution of the predominant vegetation formations of the South American Atlantic forest domain. Caatingas, cerrados, chaco and campos are the adjacent domains that make up the "diagonal of open formations".

The geographical range of the SDTF areas analysed in the present study is large enough to include four vast vegetation domains: Atlantic forest, caatinga, cerrado and chaco (Figure 7.1). The Atlantic forest domain stretches for >3300 km along the eastern Brazilian coast between the latitudes of 6°S and 30°S and makes up the second largest tropical moist forest area of South America, exceeded only by the vast Amazonian domain. The two forest domains are separated by the so-called diagonal of open formations, a corridor of seasonally dry formations that includes another three domains: the caatingas (mostly tropical thorny woodlands), cerrado (mostly woody savannas), and the chaco (mostly subtropical thorny woodlands). Each domain contains its particular SDTF formations. The now widely accepted concept of Atlantic forests (sensu lato) attaches seasonal forests, the Atlantic SDTF, to the coastal rain forests, formerly considered as the true (sensu stricto) Atlantic forests (Oliveira-Filho and Fontes, 2000; Galindo-Leal and Câmara, 2003). Caatingas and carrascos (tropical deciduous scrubs) are both SDTF and make up the predominant vegetation cover of the caatinga domain (See Chapter 6). SDTF formations are also an important component of the cerrado domain where they occur as forest patches on more fertile soils and on the freely drained slopes of gallery forests (Oliveira-Filho and Ratter, 1995, 2002). In the chaco domain, SDTF occur on peripheral areas and in some internal forest patches (Prado, 2000).

Atlantic SDTF occur as seasonal (semideciduous and deciduous) forests all along the contact zone between rain forests and the diagonal of open formations, comprising three different scenarios (see Figure 7.1 for distribution and Table 7.1 for nomenclature). (1) In north-eastern Brazil, SDTF form a narrow belt (<50 km) in the sharp transition between coastal rain forests and the semiarid caatingas, but also occur as hinterland montane forest enclaves, the brejos (Rodal, 2002; Rodal and Nascimento, 2002). (2) The transition between coastal rain forests and cerrados in south-eastern Brazil involves a much larger extent of SDTF that becomes increasingly wider towards the south to reach eastern Paraguay and north-eastern Argentina. They also form complex mosaics with cerrado vegetation to the west so that if the SDTF component of the cerrado domain is seen as an extension of the Atlantic SDTF, as proposed by Oliveira-Filho and Ratter (1995), a concept of Atlantic forests sensu latissimo must be created. (3) In the southern subtropical realm, large extents of hinterland araucaria rain forests are attached to the coastal subtropical rain forests, and SDTF appear in the west and south as transitions to both chaco forests and southern campos, or pampa prairies (Spichiger et al., 1995; Quadros and Pillar, 2002).

In the present study we sought patterns of floristic differentiation among SDTF areas of eastern tropical and subtropical South America that could be associated with geographical and climatic variables, and assessed the floristic links of seasonally dry forest formations of different regions, Amazonian forests and cerrados. We addressed the following questions: (a) How strongly differentiated are Atlantic seasonal and rain forests in different sections of their geographical range? (b) To what extent is the tree species composition of Atlantic seasonal forests transitional between those of rain forests and open formations, such as caatingas and cerrados? (c) Is the Atlantic rain forest flora closer to that of Amazonian rain forests or to that of the Atlantic seasonal forest? (d) How strong are the floristic links among caatingas, the seasonal forests of the Atlantic and cerrado domains, and the chaco forests? (e) Does SDTF flora change its composition in response to climatic variations? and (f) How are the above questions answered at the species, genus and family levels?

7.2 METHODS

7.2.1 PREPARATION AND REVISION OF THE DATABASES

We selected from the literature a total of 659 papers containing floristic checklists produced by surveys of the tree component of 532 areas of eastern tropical and subtropical South America. The geographical range included the Atlantic forest, caatinga, cerrado and chaco domains (Figure 7.1). Vegetation formations included seasonally dry tropical forests (SDTF), which are the focus of the present study, as well as tropical and subtropical rain forests, and subtropical araucaria rain forests (see Table 7.1). SDTF formations (Figures 7.2 and 7.3) are a broad category that contains tropical and subtropical seasonal forests (both deciduous and semideciduous) as well as caatingas and carrascos. Mesotrophic cerradões were treated as SDTF-cerrado transition.

Individual areas were defined arbitrarily within a maximum range of 20 km width and 400 m elevation, and thus included sections of large continuous forest tracts (e.g. Tiradentes), assemblages of forest fragments (e.g. Santa Maria) and nearby areas at different altitudes (e.g. Lençóis and Palmeiras). We obtained the following geographical information for each area: latitude and longitude at the centre of the area, median altitude, and shortest distance from the ocean. We also obtained the annual and monthly means for the temperature and rainfall of each area or the nearest meteorological stations. When the source of the checklist did not provide the climatic records, they were obtained from DNMet (1992) and from governmental websites (http://masrv54.agricultura.gov.br/rna; http://www.inmet.gov.br/climatologia). Some areas required interpolation and/or standard correction for altitude (Thornthwaite, 1948).

We entered the information from the 532 areas on to spreadsheets using Microsoft Excel 2002 in order to produce two databases. The first consisted of basic information about each area including locality, forest classification (see below), geographical and climatic variables, and literature sources. The second database was a matrix of tree species presence in the 532 areas plus three additional

TABLE 7.1

Nomenclature Used in the Present Chapter for Vegetation Classification of Eastern Tropical and Subtropical South America

Vegetation Formations	Altitudinal Belt		Regions	Main Formations		
Tropical rain forests	lowland	low altitude	Northeast, East and Southeast	Atlantic rain forests (Atlantic forests *sensu stricto*)		Rain forests
	submontane					
	lower montane	high altitude				
	upper montane					
Subtropical rain forests	lowland	low altitude	South			
	submontane					
	lower montane	high altitude				
	upper montane					
Subtropical araucaria rain forests	lower montane	high altitude	South and Southeast			
	upper montane					
Tropical seasonal semideciduous forests	lowland	low altitude	NorthEast, East, Southeast and Central-West	Atlantic seasonal forests (NE/E/SW) and Central-western seasonal forests (CW)	Atlantic forests (Atlantic forests *sensu latissimo*; *sensu lato* excludes CW and SW)	SDTF – Seasonally dry tropical forests
	submontane					
	lower montane	high altitude				
	upper montane					
Tropical seasonal deciduous forests	lowland	low altitude	Northeast, East, Southeast and Central-West			
	submontane					
	lower montane	high altitude				
Subtropical seasonal semideciduous forests	lowland	low altitude	South	Atlantic seasonal forests (S)		
	submontane					
	lower montane	high altitude				
Subtropical seasonal deciduous forests	lowland	low altitude	South and Southwest	Atlantic seasonal forests (S) and Peripheral Chaco seasonal forests (SW)		
	submontane					
	lower montane	high altitude				
	upper montane					
Caatingas (tropical thorny woodlands)	lowland and submontane	low altitude	Northeast and East			
Carrascos (tropical deciduous scrubs)						
Cerrados (*sensu lato*: open savannas to forests, or cerradões)	lowland to lower montane	low-high altitude	Central-West			
Chaco (subtropical thorny woodland)	lowland to lower montane	low altitude	Southwest			
Southern campos or pampa prairies	lowland	low altitude	South			

FIGURE 7.2 Seasonally dry tropical forests (SDTF) of eastern South America: (A) caatinga in São Raimundo Nonato, Piauí; (B) submontane tropical seasonal deciduous forest in the Serra das Confusões, Piauí; (C–F) submontane tropical seasonal deciduous forest in Três Marias, Minas Gerais in the dry (C and D) and wet (E and F) seasons (Image credits: F. Filetto [A and B] and M. A. Fontes [C–F]).

checklists that we included to compare them to Amazonian rain forests, cerrados (s.l.) and chaco forests. The first of these combined the flora of Reserva Ducke (Ribeiro et al., 1999) with the 22 checklists of Amazonian rain forests compiled by Oliveira-Filho and Ratter (1994) and contained 2190 tree species. The second contained 528 species present in 98 areas of cerrado (Ratter et al., 1996) and the third 183 chaco species listed by Prado (1991), Lewis et al. (1994) and Spichiger et al. (1995).

Before reaching its final form, the information contained in the species database underwent a detailed revision to check all species names cited in the checklists for growth form, synonymy and geographical distribution. Only species capable of growing to trees or treelets, i.e. producing a free-standing woody stem ≥3 m in stature, were maintained in the database. The task involved consultation of 387 published revisions of families and genera, 32 specialists of various institutions and four

FIGURE 7.3 Seasonally dry tropical forests (SDTF) of eastern tropical and subtropical South America: (A) lower montane tropical semideciduous forest in Itambé do Mato Dentro, Minas Gerais; (B) submontane tropical seasonal semideciduous forest in the Chapada dos Guimarães, Mato Grosso; (C) lowland subtropical seasonal semideciduous forest in Praia do Tigre, Rio Grande do Sul; (D) lowland subtropical seasonal deciduous forest in Cachoeira do Sul, Rio Grande do Sul (Image credits: A.T. Oliveira-Filho [A, B], J. A. Jarenkow [D] and J. C. Budke [E]).

websites (http://www.cnip.org.br; http://www.ipni.org/index.html, 2003-2005; http://www.mobot.org/W3T/Search/vast.html; http://sciweb.nybg.org/science2/hcol/sebc/index.asp). When these sources referred to herbarium specimens unequivocally collected in any of the 532 areas, the species was added to the database. The final database contained 6598 species, 976 genera, and 128 families. The species classification into families followed the Angiosperm Phylogeny Group II (APG, 2003).

7.2.2 VEGETATION CLASSIFICATION

We extend here the vegetation classification proposed by Oliveira-Filho and Fontes (2000) for southeast Brazil to include a much wider geographical range as well as additional vegetation formations (Table 7.1). This extended classification was based on exploratory multivariate analyses of both floristic and climatic data (ongoing studies). We defined the top classification level by combining main thermoclimate (either tropical or subtropical) and rainfall seasonality (rainy, seasonally rainy and semi-arid) and established the limit between the two thermoclimates at the latitudes of 25°30'S and 26°30'S for rain and seasonal forests, respectively. Areas of the chaco domain and araucaria rain forests were all subtropical; areas of the cerrado and caatinga domains were all tropical. We classified areas with tropical climates as rain forests, seasonal forests/cerrados and caatingas/carrasco in which the dry season lasts for up to 30 days (rainy), ≥30–160 days (seasonally rainy) and ≥160 days (semi-arid), respectively.

In subtropical climates, we classified the areas as chaco forests/peripheral chaco seasonal forests where the dry season lasts for ≥30 days (semi-arid) and areas below this limit as either rain forests or seasonal forests/campos in which the difference in mean monthly temperatures between the coolest and warmest months is up to 10°C or ≥10–15°C, respectively (rainfall seasonality secondary).

Seasonal deciduous and semideciduous forests of both tropical and subtropical climates are commonly distinguished by the amount of leaf-fall during the slow-growth season (Veloso et al., 1991). Except for the degree of deciduousness, it is often difficult to tell them apart, particularly in central Brazil where they commonly form continua determined by local variations of soil moisture and fertility (Oliveira-Filho and Ratter, 2002). Therefore, in most cases we opted to trust the authors' experience to classify seasonal forests as either deciduous or semideciduous. Savannas, i.e. cerrado (sensu lato) and campo are also a very important component of the vegetation in seasonal climates but we included only the mesotrophic cerradão (plural, cerradões) in the analyses because of its forest-like physiognomy. Again, the authors' experience was trusted to distinguish this vegetation from seasonal forests. Another distinction was made between subtropical araucaria rain forests and other rain forests, based on their geographical location in the inner highlands and the conspicuous presence of emergent trees of *Araucaria angustifolia* (Bert.) O.Kuntze. Areas of caatinga and carrasco were distinguished by elevation and substrate, the former occurring on dissected lowlands and the latter on sandy plateaus (see Queiroz, Chapter 6).

We also used the exploratory multivariate analyses of floristic data to classify the above vegetation formations according to altitude and geographical region. We defined elevation ranges as follows. For latitudes <16°S: lowland, <400 m; submontane, 400 – <800 m; lower montane, 800 – <1200 m; upper montane, ≥1200 m. For latitudes between 16° and <23°30'S: lowland, <300 m; submontane, 300 – <700 m; lower montane, 700 – <1100 m; upper montane, ≥1100 m. For latitudes between 23°30' and <32°S, lowland, <200 m; submontane, 200 – <600 m; lower montane, 600 – <1000 m; upper montane, ≥1000 m. The geographical regions recognized were north-east, east, south-east, south, central-west and south-west. The resulting classification categories are given in Table 7.1 and the geographical distribution of the 532 areas are shown in Figure 7.4. Limited space does not allow us to list the areas, neither to provide here their description and source references. We intend to make this information available in a forthcoming publication.

7.2.3 MULTIVARIATE ANALYSES

We used detrended correspondence analysis, DCA (Hill and Gauch, 1980), processed by the program PC-ORD 4.0 (McCune and Mefford, 1999) to seek main species distribution gradients across 341 SDTF areas. We removed rain forests because the patterns within this vegetation type were not the focus of the present work. An additional DCA was performed with the 243 areas of tropical seasonal forests (subtropical and semiarid formations excluded) to seek more detailed patterns within the group. Two other DCA were performed separately for the areas of caatinga and subtropical seasonal forests. We chose DCA coupled to *a posteriori* interpretation of ordination results because we aimed at patterns dictated solely by the species without the interference of environmental variables, as occurs with joint analyses such as CCA (Kent and Coker, 1992). We used two interpretation tools: the previous vegetation classification of the areas and 13 geographical and climatic (hereafter geo-climatic) variables. They were both plotted (*a posteriori*) on the DCA diagrams as symbols and arrows, respectively. The geo-climatic variables were latitude, longitude, median altitude, distance from the ocean, mean annual temperature, mean monthly temperatures in the warmest and coolest months, mean temperature range obtained from the difference between the two previous variables, mean annual rainfall, mean monthly rainfall of the dry (June–August) and rainy (December–February) seasons, rainfall distribution ratio obtained from the proportion between the two previous variables and mean duration of the dry season obtained from the number of days of water shortage given by Walter diagrams (Walter, 1985). We also obtained the Pearson correlation coefficients between the geo-climatic variables and the ordination scores of the areas in each DCA axis.

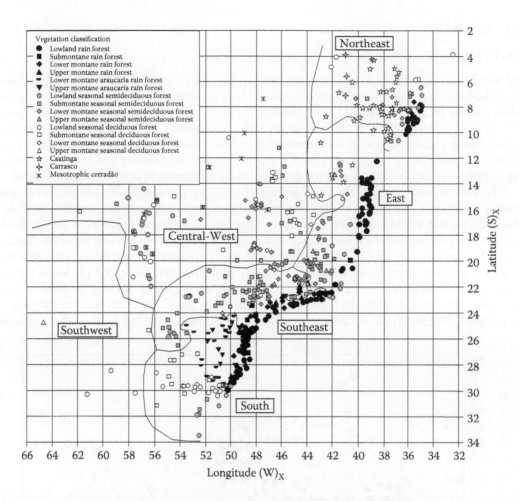

FIGURE 7.4 Geographical coordinates diagram showing the location and vegetation classification of the 532 areas used in the analyses and the six geographical regions. Forest areas situated in the north-east, east, south-east and central-west regions are classified as tropical and those in the south and south-west are subtropical.

7.2.4 CONDENSED FLORISTIC DATA

Because both previous and present multivariate analyses demonstrated that the vegetation classification system adopted was highly consistent, we eventually condensed the floristic information contained in the database by lumping together the species records within main vegetation formations (Table 7.1). Atlantic rain forests as well as the cerrado, chaco and Amazonian rain forest checklists were incorporated here. We also merged lowland and submontane categories as low altitude and lower and upper montane categories as high altitude. The resulting lumped matrix of binary data of species presence in the main vegetation formations was used to produce two additional matrices, for genera and families. The generic and familial matrices were both quantitative, as they consisted of the number of species per genus or family, respectively, in each main vegetation formation. We performed cluster analyses of the condensed matrices using the program PC-ORD 4.0. Cluster analyses used Jaccard's floristic similarity for species and relative squared Euclidian distances for genera and families (number of species as abundance data); the linkage method was group average (Kent and Coker, 1992). We also used the condensed data to perform a direct quantitative assessment of the floristic links between the vegetation formations by plotting the number of shared and exclusive species in Venn diagrams. The most frequent species, and the richest genera and families of main vegetation formations were extracted from the matrices.

7.3 RESULTS

7.3.1 MULTIVARIATE ANALYSES

The ordination diagrams yielded by DCA are shown in Figures 7.5 and 7.6 for the two assemblages of vegetation areas, seasonally dry tropical forests (SDTF) and tropical seasonal forests. Their eigenvalues are first DCA, 0.688 (axis 1) and 0.394 (axis 2); second DCA, 0.400 (axis 1) and 0.475 (axis 2). According to ter Braak (1995), these eigenvalues are relatively high (>0.3), indicating considerable species turnover along the gradients summarized in the first two axes. In addition, most DCA axes produce a number of high values of Pearson correlation coefficients between geo-climatic variables and ordination scores (Table 7.2) giving consistency to the interpretation of the emerging patterns.

The first ordination axis in the DCA for SDTF is chiefly correlated with latitude, minimum monthly temperature, duration of the dry season, annual temperature and annual rainfall (Table 7.2). This indicates that the data structure summarized by the first axis primarily reflects a geographical gradient based on latitude which corresponds to a major climatic gradient towards the south characterized by decreasing temperatures and duration of the dry season and increasing total rainfall. Longitude and distance to the ocean are more strongly correlated with the second DCA axis but no climatic variable accompanies this gradient. The areas of caatinga and carrasco are found at the right side of the diagram associated with latitudes near the Equator, longer dry seasons and higher temperatures (Figure 7.5). No distinction is made between north-east and east caatingas but the three

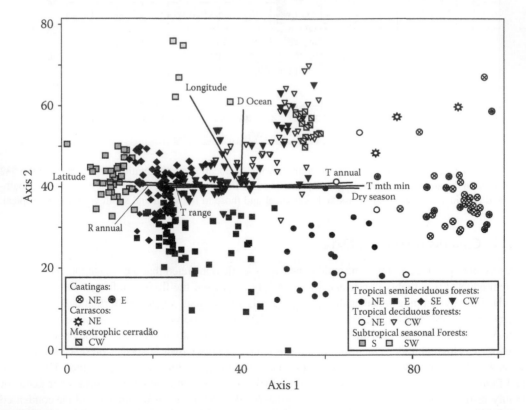

FIGURE 7.5 Diagram yielded by detrended correspondence analysis (DCA) showing the ordination of 341 areas of seasonally dry tropical forests (SDTF) of eastern South America on the first two DCA axes, based on the presence of 3018 tree species. The areas are classified according to main vegetation formation and geographical region. The centred straight lines show the correlation between axes and geoclimatic variables (only those with r > 0.3 with at least one axis are shown): T = temperature, Mth = monthly, Min = minimum, R = rainfall, D = distance.

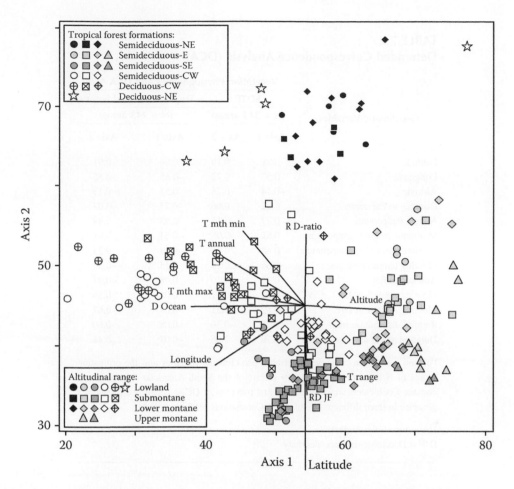

FIGURE 7.6 Diagram yielded by detrended correspondence analysis (DCA) showing the ordination of 243 areas of tropical seasonal forests of the South American Atlantic forest domain on the first two DCA axes, based on the presence of 2680 tree species. The areas are classified according to forest formation, geographical region and altitudinal range. The centred straight lines show the correlation between axes and geoclimatic variables (only those with r > 0.3 with at least one axis are shown): T = temperature, Mth = monthly, Min = minimum, Max = maximum, D = distance, R = rainfall, DJF = December-January-February, D-Ratio = distribution ratio.

carrasco areas are displaced to the top on the second axis together with three areas of caatinga (Serra da Capivara, São José do Piauí and Ibiraba) which differ from other caatingas in their sandy substrate, as do the carrascos (see Chapter 6). Towards the left side of the diagram, caatingas and carrascos are followed by areas of tropical seasonal forests of the north-east and central-west, discriminated at the bottom and top halves of the diagram, respectively. For the north-east areas, deciduous forests come first followed by semideciduous forests. For the central-west, however, deciduous and semidec- iduous forests are not distinguished from each other and only mesotrophic cerradões form a consistent clump. Along the latitudinal sequence of the first DCA axis, north-east tropical seasonal forests are followed by those of the east and then south-east regions. This sequence ends at the left side of the diagram where the areas of subtropical seasonal forests that correspond to the extremes of higher latitudes, lower temperatures and shorter dry seasons are situated. In addition, the second axis discriminates, at the top, the four south-west areas of peripheral chaco seasonal forests.

The DCA performed for tropical seasonal forests shows additional patterns linked to altitude that do not arise when other seasonally dry forests are included. The first axis is primarily correlated with

TABLE 7.2
Detrended Correspondence Analysis (DCA)

| Geoclimatic Variables | Vegetation Physiognomies and DCA Axes | | | |
| | SDTF (N = 341 areas) | | Tropical Seasonal Forests (N = 243 areas) | |
	Axis 1	Axis 2	Axis 1	Axis 2
Latitude	−0.88	0.19	0.08	−0.91
Longitude	−0.52	0.72	−0.65	−0.55
Altitude	−0.14	−0.26	0.58	−0.13
Distance to the ocean	0.12	0.66	−0.73	−0.07
Annual temperature	0.77	0.15	−0.63	0.49
Minimum monthly temperature	0.81	0.08	−0.54	0.61
Maximum monthly temperature	0.48	0.26	−0.59	0.23
Monthly temperature range	−0.66	0.14	0.15	−0.62
Annual rainfall	−0.72	0.10	0.08	−0.37
Monthly rainfall in JJA	−0.27	−0.17	0.05	0.39
Monthly rainfall in DJF	−0.52	0.21	0.07	**−0.67**
Rainfall distribution ratio	0.08	−0.30	0.08	0.60
Duration of the dry season	0.79	0.06	−0.07	0.41

Pearson correlation coefficients between geo-climatic variables and the ordination scores of N areas of seasonally dry forests of the South American Atlantic forest domain. Coefficients are given for the first two axes of DCAs performed for species presence in three different sets of areas. Correlations >0.5 are in bold.

JJA = June–July–August
DJF = December–January–February

distance to the ocean, longitude, temperatures (annual, minimum and maximum) and altitude, while the second axis is more strongly correlated with latitude, monthly rainfall in December–January–February (DJF), rainfall distribution ratio and monthly temperature range (Table 7.2). This suggests that the first axis primarily reflects a geographical gradient based on penetration into the continental interior together with decreasing altitude, both of which also correspond to a climatic gradient characterized mainly by increasing temperatures. To a great extent, the second axis repeats patterns already shown by the first axis of the previous DCA, so that all north-east seasonal forests are strongly discriminated at the top of the diagram (Figure 7.6). As opposed to the others, north-east areas show stronger correlations with decreasing latitude, rainfall in DJF and temperature range and with increasing rainfall distribution ratio. The first ordination axis, however, discriminates north-east areas of deciduous and semideciduous forests to the left and right sides of the diagram, respectively, with the single exception of the oceanic island of Fernando de Noronha, located at the top right corner. At the bottom half of the diagram the first ordination axis discriminates central-west seasonal forests from those of the south-east and east regions in such a way that two concurrent geographical gradients were distinguished, one related to longitude and distance from the ocean and the other with altitude, both involving decreasing temperatures. The areas at the extreme left of the diagram are the westernmost low-altitude seasonal forests of Mato Grosso and Bolivia, while those at the extreme right are mostly the eastern high-altitude seasonal forests of Bahia and Minas Gerais. As in the previous DCA, central-west deciduous and semideciduous forests are not discriminated from each other.

Additional DCAs were performed separately for the areas of caatinga and subtropical seasonal forests, but no relevant additional patterns arose. Deciduous and semideciduous subtropical forests of the south were not discriminated amongst themselves and only altitude showed a weak correlation with the floristic patterns.

7.3.2 Analyses of Condensed Floristic Information

As they were extracted from floristic checklists for specific forest areas, the condensed information must be regarded as a means of assessing the floristic links between the main forest formations quantitatively and not as a register of actual figures for number of species, either total or in common.

The classification dendrograms (Figure 7.7) show different patterns for each of the three taxonomic levels. A clear general trend arising from the species dendrogram is that regional patterns are stronger than vegetation formation patterns. Four main geographical groups are discriminated. The first contains all vegetation formations of the north-east region, including tropical rain and seasonal forests, caatingas and carrascos. The second, and largest, group contains tropical rain and seasonal forests of four regions (east, south-east, central-west and south) plus cerrados and cerradões. The third main group is composed of the Amazonian rain forests and the fourth, and most distinct, by chaco forests and peripheral chaco seasonal forests. The north-east main group is split into two subgroups, the first containing moister formations (tropical rain forests and seasonal semideciduous forests) and the second drier formations (tropical seasonal deciduous forests, caatingas and carrascos). The second main group is split into five subgroups, all of clear geographical nature: (a) tropical rain and seasonal semideciduous forests of the east and south-east; (b) tropical seasonal semideciduous and deciduous forests of the central-west; (c) subtropical low-altitude seasonal forests and rain forests of the south, plus subtropical high-altitude araucaria rain forests of both the south and south-east; (d) cerrados (sensu lato) and mesotrophic cerradões; and (e) subtropical high-altitude seasonal forests of the south.

The dendrogram for genera shows different main groups and an increased role of vegetation formation over regional patterns. The first main group contains tropical low-altitude rain forests and semideciduous forests of the north-east and Amazonian regions. The second main group contains two subgroups: (a) tropical seasonal semideciduous forests of the east, south-east and central-west, plus tropical rain forests of the east, and (b) tropical rain forests of the south-east, subtropical rain forests of the south and subtropical araucaria rain forests of both the south and south-east. The third, fourth and fifth main groups contain, respectively, subtropical seasonal forests of the south, tropical seasonal deciduous forests of the north-east and central-west, and cerrados and cerradões. The last two main groups contain caatingas, carrascos, chaco forests and peripheral chaco seasonal forests.

The dendrogram for families goes a step further in generating groups with a strong vegetation formation character. The semi-arid formations, namely the chaco forests, caatingas and carrascos, form a clump that merges, at the subsequent level, with a group that includes tropical and subtropical seasonal deciduous forests and mesotrophic cerradões. An oddity of this group is the presence of a side subgroup containing low-altitude rain forests and semideciduous forests of the north-east and Amazonian regions. Tropical seasonal semideciduous forests predominate in another main group that also includes the cerrado and tropical rain forests of the east and north-east. The following main group contains tropical and subtropical rain forests and subtropical araucaria rain forests, and the last main group contains subtropical seasonal forests of the south.

The tree floras represented in the rain and seasonal forest checklists are similar in species richness: 3009 and 2903 species, respectively. On the other hand, the number of rain forest checklists, 191, is considerably smaller than that of seasonal forests, 285, therefore suggesting that the species richness of the latter may actually be lower. In fact, the species/area curves of the two vegetation formations (Figure 7.8) demonstrate that, at any number of areas, the mean cumulative number of species is much higher in rain than in seasonal forests. The two formations also share a high proportion of tree species, 1814 out of 4098, or 44.3%, but both also have a considerable number of putative endemics, 1195 (29.2%) and 1089 (26.6%) for rain and seasonal forests, respectively.

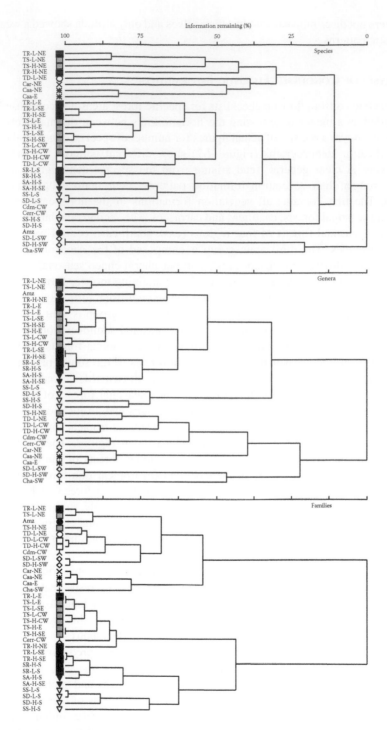

FIGURE 7.7 Dendrograms produced by group averaging of Jaccard's floristic similarity for species and relative squared Euclidian distances for genera and families of the tree flora 23 areas of eastern Amazonian rain forests (Amz), 39 areas of caatinga (Caa), 3 areas carrasco (Car), 376 areas of cerrado (Cerr), 11 areas of mesotrophic cerradão (Cdm), 5 areas of chaco forests (Cha) and 479 areas of Atlantic forests *s.l.* merged into 28 main forest formations abbreviated as follows: the first set of letters stands for either tropical (T) or subtropical (S) forests and for rain (R), araucaria rain (A), seasonal semideciduous (S) and seasonal deciduous (D) forests; the middle letters stand for low (L) and high (H) altitude; the last letters stand for north-east (NE), east (E), south-east (SE), central-west (CW), south (S) and south-west (SW) regions. Scale in dendrograms expresses the remaining information after clustering.

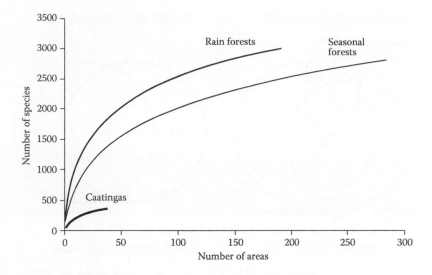

FIGURE 7.8 Mean cumulative number of species in areas of rain forests and seasonally dry tropical forests of eastern South America with increasing number of areas.

The Venn diagrams in Figure 7.9 show the relationship of the tree flora of rain and seasonal forests in different geographical regions. The number of species of both formations is smaller in the north-east and south and larger in the east and south-east. The non-Atlantic seasonal forests have higher numbers of species in the central-west and lower in the southwest. Seasonal deciduous forests of the north-east, despite sharing many species with regional rain and semideciduous forests, have their own group of putative endemics. The proportions of species shared by Atlantic rain and seasonal forests are very similar in all regions: 20.4% in the north-east, 21.0% in the east, 22.0% in the south-east, and 18.8% in the south. The species proportions in rain and seasonal forests, however, show opposing trends from the north-east to the south, and are, respectively, 31.8% and 47.8% in the north-east, 39.2% and 39.8% in the east, 48.3% and 29.7% in the south-east, and 58.5% and 22.7% in the south. Subtropical araucaria rain forests share high proportions of species with both rain and seasonal forests in both the south and south-east regions. However, they also contain their group of putative endemics, particularly in the south. The central-west seasonal forests share 76.2% of their species with Atlantic tropical rain and seasonal forests (north-east, east and south-east), though 60.6% are present in both formations, 15.6% in seasonal forests only, and none in rain forests only.

The seasonal forests of all six geographical regions contain 2903 species, of which only 40 are registered in all regions, 81 in five regions, 257 in four, 414 in three, 630 in two, and 1481 in one. These putative endemic species are in higher proportion in the floras of the east (619; 35.0%) and north-east (241; 32.1%), followed by the south-west (68; 27.1%), central-west (310; 24.4%), south (66; 15.7%) and south-east (177; 14.9%). The relationships among seasonal forests of adjacent geographical regions are shown in the left-side Venn diagrams of Figure 7.10. The three regions of Atlantic tropical seasonal forests share a small number of species, 273 out of 2062 (13.2%), but adjacent regions share larger proportions: 24.8% between the north-east and east and 29.0% between the east and south-east. Subtropical seasonal forests share a high number of species with the tropical seasonal forests of the south-east: 76.3% and 50.2% for the south and south-west, respectively. The latter also share 59.8% of their species with the tropical seasonal forests of the central-west and 67.7% with both the south-east and central-west.

Caatingas are considerably poorer in tree species than are rain- and seasonal forests (Figure 7.8). The relationships between caatingas and adjacent vegetation formations are shown in the right-side

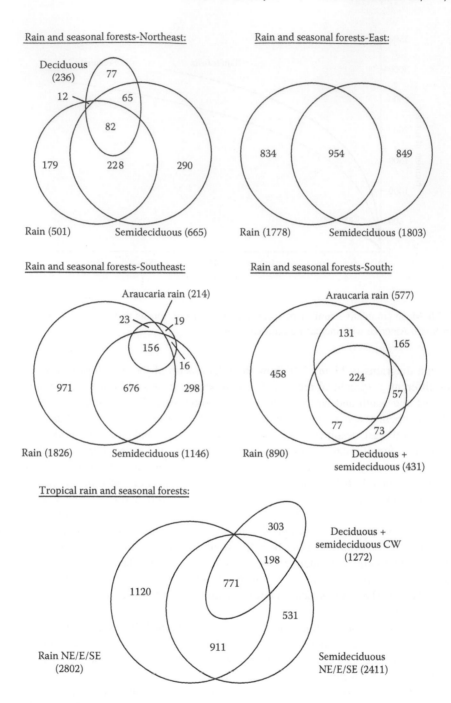

FIGURE 7.9 Venn diagrams extracted from the checklists showing the number of tree species shared by rain and seasonal forests in different geographical regions of eastern South America.

Venn diagrams of Figure 7.10. A high proportion of their 466 species is shared with adjacent seasonal forests, 61.2%, but this also leaves 38.8% of putative endemics. The proportion of shared species is higher with the north-east seasonal forests (49.4%) than with the central-west (39.3%). The proportion shared with cerrados is much smaller, 17.6%, and most of this is also shared with central-west seasonal forests. The number of species shared with chaco forests is very small, only five. In fact, both semi-arid formations, caatingas and chaco forests, share more species with the central-west seasonal forests

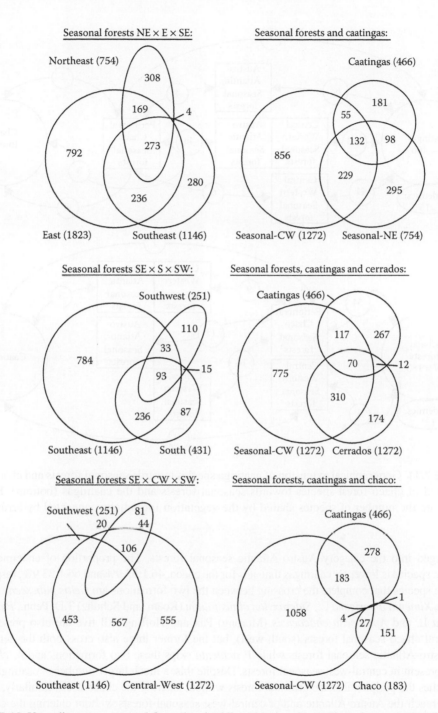

FIGURE 7.10 Venn diagrams extracted from the checklists showing the number of tree species shared by seasonal forests in different geographical regions of eastern South America (left side), and by seasonal forests, caatingas, cerrados and chaco forests (right side).

than between themselves. The geographical range of the tree flora of those two formations is illustrated in the two flow diagrams of Figure 7.11. Lowland seasonal deciduous forests and carrascos of the north-east are excluded because they have a strong floristic identity with the caatingas and are not in the route between the caatinga and chaco domains. Seasonal forests of the east, south-east and south

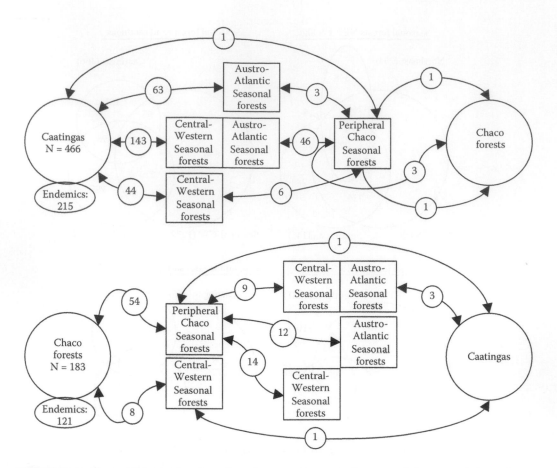

FIGURE 7.11 Geographical extension of caatinga species towards seasonal forests and chaco forests (top), and of chaco forest species towards seasonal forests and the caatingas (bottom). Encircled figures are the number of species shared by the vegetation formations connected by arrows.

are merged into the category Austro-Atlantic seasonal forests. The proportion of endemic to non-endemic species is lower for caatingas than it is for the chaco, 46.1:53.9% and 66.1:33.9%, respectively. The five species that complete the crossing between the two formations are *Celtis pubescens* (Jacquin) Sargent, *Ximenia americana* L., *Sideroxylon obtusifolium* (Roem. and Schultz) T.D. Penn., *Parkinsonia aculeata* L. and *Aporosella chacoensis* (Morong) Pax and Hoffmg. All five are also present in the peripheral chaco seasonal forests (south-west), but the former three also cross both the central-west and Austro-Atlantic seasonal forests while *P. aculeata* skips these two formations and *A. chacoensis* is also present in central-west seasonal forests. Despite this, a much larger number of caatinga species, 55, reaches the peripheral chaco seasonal forests without entering the chaco itself. Similarly, 42 chaco species reach the Austro-Atlantic and/or central-west seasonal forests without entering the caatingas.

The most species-rich genera and families of each main vegetation formation are given in Tables 7.3 to 6, and the most frequent species of the same formations are provided in the Appendix. Some genera rank high in most main Atlantic seasonal forest formations, e.g. *Eugenia*, *Myrcia*, *Ocotea* and *Miconia* (Table 7.3). Some trends can be observed with increasing altitude: the relative importance decreases for some genera such as *Eugenia* (except in the north-east), *Inga* and *Ficus*, and increases for others such as *Miconia* and *Tibouchina* (east and south-east), *Ilex* and *Solanum*. Subtropical seasonal forests of the south are similar to tropical seasonal forests in their generic profile, but the south-west subtropical seasonal forests, chaco forests and the caatingas have very

TABLE 7.3
Genera with the Highest Number of Species (S) in the Tree Flora of Tropical Seasonal Forests of the South American Atlantic Forest Domain Classified into Four Geographical Regions and Two Altitudinal Ranges

North-East Region				East Region				South-East Region				Central-West Region			
Low Altitude (N = 13)		High Altitude (N = 11)		Low Altitude (N = 29)		High Altitude (N = 26)		Low Altitude (N = 47)		High Altitude (N = 35)		Low Altitude (N = 74)		High Altitude (N = 23)	
	S		S		S		S		S		S		S		S
	542		420		1317		1193		848		911		1129		624
Eugenia	16	Eugenia	17	Eugenia	49	Miconia	43	Eugenia	37	Miconia	36	Miconia	31	Miconia	28
Myrcia	12	Erythroxylum	15	Ocotea	32	Myrcia	37	Ocotea	27	Ocotea	26	Eugenia	26	Myrcia	18
Cordia	9	Myrcia	12	Miconia	32	Ocotea	29	Miconia	24	Eugenia	26	Myrcia	23	Machaerium	13
Erythroxylum	9	Cordia	9	Myrcia	30	Eugenia	29	Myrcia	18	Myrcia	18	Machaerium	20	Eugenia	12
Senna	9	Senna	9	Machaerium	23	Tibouchina	18	Ficus	16	Solanum	16	Ficus	18	Aspidosperma	11
Inga	9	Ocotea	9	Inga	22	Machaerium	17	Solanum	15	Ficus	15	Aspidosperma	16	Ocotea	11
Miconia	9	Croton	9	Ficus	22	Ilex	16	Machaerium	13	Machaerium	14	Inga	15	Ficus	11
Psidium	9	Ouratea	9	Tabebuia	15	Erythroxylum	16	Nectandra	12	Piper	14	Bauhinia	14	Nectandra	10
Aspidosperma	8	Zanthoxylum	8	Pouteria	15	Solanum	16	Erythroxylum	11	Ilex	12	Ocotea	14	Casearia	9
Ficus	8	Maytenus	8	Guatteria	14	Inga	15	Inga	11	Nectandra	11	Erythroxylum	13	Symplocos	9
Casearia	8	Acacia	8	Psychotria	14	Maytenus	14	Tabebuia	10	Tibouchina	11	Byrsonima	12	Ilex	8
Pouteria	8	Byrsonima	8	Solanum	14	Psychotria	14	Piper	10	Erythroxylum	10	Casearia	12	Inga	8
Tabebuia	7	Helicteres	7	Cordia	13	Casearia	13	Senna	9	Inga	10	Nectandra	11	Maytenus	7
Bauhinia	7	Casearia	7	Maytenus	13	Guatteria	13	Psychotria	9	Casearia	10	Trichilia	11	Campomanesia	7
Ocotea	7	Solanum	7	Casearia	13	Psidium	13	Zanthoxylum	9	Leandra	9	Psidium	11	Vochysia	7
Mimosa	6	Aspidosperma	6	Aspidosperma	12	Cinnamomum	12	Maytenus	8	Trichilia	9	Cordia	10	Byrsonima	6
Guapira	6	Cyathea	6	Trichilia	12	Nectandra	12	Croton	8	Myrsine	9	Maytenus	10	Trichilia	6
Coccoloba	6	Inga	6	Rudgea	12	Ouratea	12	Bauhinia	8	Psychotria	9	Senna	10	Psidium	6
Zanthoxylum	6	Psidium	6	Erythroxylum	11	Cordia	11	Myrsine	8	Aspidosperma	8	Acacia	10	Guapira	6
Tabernaemontana	5	Psychotria	6	Campomanesia	11	Croton	11	Pouteria	8	Tabebuia	8	Zanthoxylum	10	Tabebuia	5
Licania	5	Capparis	5	Psidium	11	Byrsonima	11	Cestrum	8	Cordia	8	Tabebuia	8	Licania	5
Clusia	5	Pilosocereus	5	Croton	10	Ficus	10	Styrax	8	Croton	8	Dalbergia	8	Dalbergia	5
Croton	5	Clusia	5	Swartzia	10	Campomanesia	10	Aspidosperma	7	Mollinedia	8	Cupania	8	Alibertia	5
Copaifera	5	Bauhinia	5	Ilex	9	Cupania	9	Mollinedia	7	Gomidesia	8	Vochysia	8	Styrax	5
Swartzia	5	Caesalpinia	5	Nectandra	9	Vochysia	9	Casearia	7	Symplocos	8	Capparis	7	Calyptranthes	5

N = Number of Areas

TABLE 7.4

Genera with the Highest Number of Species (S) in the Tree Flora of Subtropical Seasonal Forests of the South American Atlantic Forest Domain, Chaco Forests, and Caatingas

Subtropical Seasonal Forests							
South Region (N = 37)	S 542	South-West Region (N = 5)	S 420	Chaco Forests (N = 39)	S 1193	Caatingas (N = 6)	S 1317
Eugenia	18	Acacia	9	Prosopis	19	Croton	14
Ocotea	11	Schinus	7	Acacia	9	Mimosa	13
Myrsine	8	Aspidosperma	6	Lycium	8	Senna	11
Tabebuia	7	Tabebuia	5	Echinopsis	7	Erythroxylum	10
Erythroxylum	7	Zanthoxylum	5	Opuntia	7	Bauhinia	9
Ficus	7	Chloroleucon	4	Jatropha	7	Manihot	8
Myrcia	7	Ficus	4	Senna	6	Eugenia	8
Solanum	7	Eugenia	4	Capparis	5	Aspidosperma	7
Ilex	5	Schinopsis	3	Bougainvillea	5	Cordia	7
Maytenus	5	Tecoma	3	Bulnesia	5	Pilosocereus	7
Sebastiania	5	Maytenus	3	Schinopsis	4	Acacia	7
Machaerium	5	Bauhinia	3	Cereus	4	Helicteres	7
Trichilia	5	Mimosa	3	Harrisia	4	Tabebuia	6
Myrciaria	5	Ceiba	3	Maytenus	4	Maytenus	6
Psychotria	5	Luehea	3	Caesalpinia	4	Caesalpinia	6
Zanthoxylum	5	Cedrela	3	Mimosa	4	Psidium	6
Cestrum	5	Trichilia	3	Aloysia	4	Zanthoxylum	5
Schinus	4	Myrsine	3	Myrcianthes	3	Capparis	4
Cordia	4	Myracrodruon	2	Ruprechtia	3	Pereskia	4
Lonchocarpus	4	Ilex	2	Condalia	3	Cnidoscolus	4
Inga	4	Ruprechtia	2	Zanthoxylum	3	Hymenaea	4
Nectandra	4	Prosopis	2	Quiabentia	2	Chloroleucon	4
Miconia	4	Ziziphus	2	Ceiba	2	Guapira	4
Gomidesia	4	Capparis	2	Aspidosperma	2	Ruprechtia	3
Symplocos	4	Carica	2	Berberis	2	Facheiroa	3

N = Number of Areas

particular sets of species-richest genera (Table 7.4). Among the families, Legominosae (Fabaceae) is top in all vegetation formations except the subtropical seasonal forests where it switches places with Myrtaceae (Tables 7.5 and 7.6). In all other formations, Myrtaceae ranks second, except in the south-west subtropical seasonal forests (3rd), chaco forests (16th), and caatingas (5th). Other families ranking high among tropical and subtropical seasonal forests (except in the south-west) are Rubiaceae, Melastomataceae and Lauraceae. Families showing increasing importance at higher altitudes are Asteraceae, Melastomataceae (except in the north-east) and Lauraceae (though unchanged in the east and south-east). Euphorbiaceae are particularly important in most formations but rank higher in the north-east and central-west low-latitude seasonal forests, as well as in the caatingas and chaco forests, which are also distinguished by the high ranking of Cactaceae.

7.4 DISCUSSION

An overall pattern emerging from the floristic analyses of the vegetation formations of eastern South America was the strong influence of distance on tree species distribution. This influence only receded and allowed clear climate-related patterns to be discerned when either the geographical range considered

TABLE 7.5

Families with the Highest Number of Species (S) in the Tree Flora of Tropical Seasonal Forests of the South American Atlantic Forest Domain Classified into Four Geographical Regions and Two Altitudinal Ranges

| North-East Region | | | | East Region | | | | South-East Region | | | | Central-West Region | | | |
Low Altitude (N = 13)	S 542	High Altitude (N = 11)	S 420	Low Altitude (N = 29)	S 1317	High Altitude (N = 26)	S 1193	Low Altitude (N = 47)	S 848	High Altitude (N = 35)	S 911	Low Altitude (N = 74)	S 1129	High Altitude (N = 23)	S 624
Fab	113	Fab	81	Fab	192	Fab	151	Fab	111	Fab	108	Fab	210	Fab	93
Myrt	48	Myrt	44	Myrt	140	Myrt	137	Myrt	101	Myrt	96	Myrt	95	Myrt	62
Rubi	28	Rubi	22	Rubi	83	Melastomat	73	Rubi	55	Melastomat	59	Rubi	58	Rubi	37
Euphorbi	19	Euphorbi	17	Laur	68	Laur	65	Laur	49	Laur	52	Euphorbi	46	Melastomat	34
Apocyn	18	Erythroxyl	15	Melastomat	48	Rubi	58	Melastomat	34	Rubi	44	Melastomat	39	Laur	29
Malv	17	Rut	12	Annon	44	Aster	51	Euphorbi	31	Aster	33	Malv	36	Annon	20
Sapot	16	Malv	11	Euphorbi	44	Euphorbi	38	Solan	29	Euphorbi	32	Annon	32	Malv	17
Annon	15	Laur	10	Mor	35	Annon	32	Rut	24	Solan	27	Laur	32	Mor	16
Mor	14	Solan	10	Sapot	31	Solan	27	Mor	22	Mor	21	Mor	27	Vochysi	15
Clusi	13	Boragin	9	Solan	29	Clusi	24	Annon	18	Malv	20	Sapind	27	Salic	14
Bignoni	11	Malpighi	9	Aster	27	Sapind	24	Bignoni	17	Annon	19	Rut	26	Apocyn	13
Chrysobalan	11	Anacardi	8	Bignoni	27	Malv	21	Malv	16	Rut	19	Apocyn	25	Arec	13
Salic	11	Cact	8	Malv	26	Rut	20	Salic	16	Salic	17	Salic	21	Bignoni	12
Boragin	10	Clusi	8	Sapind	25	Apocyn	19	Sapind	16	Bignoni	15	Nyctagin	18	Celastr	12
Melastomat	10	Salic	8	Arec	24	Chrysobalan	19	Aster	15	Vochysi	15	Meli	18	Clusi	12
Sapind	10	Apocyn	7	Clusi	24	Mor	19	Myrsin	14	Meli	14	Chrysobalan	17	Euphorbi	12
Anacardi	9	Aster	7	Rut	24	Bignoni	18	Meli	12	Piper	14	Arec	17	Meli	12
Combret	9	Bignoni	7	Apocyn	23	Salic	18	Anacardi	11	Sapind	11	Bignoni	16	Chrysobalan	11
Erythroxyl	9	Ochn	7	Celastr	19	Vochysi	18	Celastr	11	Myrsin	11	Vochysi	16	Sapind	11
Laur	9	Celastr	6	Chrysobalan	19	Celastr	17	Erythroxyl	11	Apocyn	12	Celastr	16	Aster	10
Rut	9	Chrysobalan	6	Meli	19	Aquifoli	16	Sapot	11	Aquifoli	12	Aster	16	Myrsin	10
Arec	9	Meli	6	Salic	19	Erythroxyl	16	Apocyn	10	Celastr	12	Combret	15	Nyctagin	10
Nyctagin	7	Melastomat	6	Anacardi	16	Sapot	16	Lami	10	Clusi	10	Malpighi	15	Anacardi	9
Polygon	7	Mor	6	Nyctagin	14	Malpighi	14	Piper	10	Erythroxyl	10	Sapot	14	Symploc	9
Lecythid	6	Sapind	5	Boragin	13	Meli	13	Vochysi	10	Cyathe	9	Boragin	13	Aquifoli	8

Suffix '-aceae' omitted from all families; N = Number of Areas

TABLE 7.6
Families with the Highest Number of Species (S) in the Tree Flora of Subtropical Seasonal Forests of the South American Atlantic Forest Domain, Chaco Forests and Caatingas

Subtropical Seasonal Forests							
South Region (N = 37)	S 542	South-West Region (N = 5)	S 420	Chaco Forests (N = 39)	S 1193	Caatingas (N = 6)	S 1317
Myrt	60	Fab	55	Fab	63	Fab	106
Fab	49	Anacardi	14	Cact	29	Euphorbi	36
Laur	21	Myrt	14	Solan	13	Cact	24
Rubi	18	Malv	13	Euphorbi	11	Malv	19
Solan	18	Sapind	12	Zygophyll	9	Myrt	16
Euphorbi	17	Bignoni	10	Anacardi	7	Apocyn	11
Bignoni	11	Apocyn	8	Nyctagin	7	Bignoni	10
Meli	11	Euphorbi	8	Rhamn	6	Erythroxyl	10
Mor	11	Rut	7	Brassic	5	Boragin	9
Aster	10	Meli	6	Celastr	5	Rubi	9
Salic	10	Mor	6	Malv	5	Rut	8
Rut	9	Salic	6	Verben	5	Celastr	7
Myrsin	8	Arec	5	Arec	4	Combret	7
Anacardi	7	Celastr	5	Bignoni	4	Annon	6
Erythroxyl	7	Nyctagin	5	Cannab	4	Nyctagin	6
Malv	7	Sapot	5	Myrt	4	Polygon	6
Melastomat	7	Aster	4	Rut	4	Sapind	6
Sapind	7	Cannab	4	Sapind	4	Arec	5
Urtic	7	Laur	4	Sapot	4	Brassic	5
Celastr	6	Polygon	4	Apocyn	3	Anacardi	4
Sapot	6	Rubi	4	Malpighi	3	Rhamn	4
Apocyn	5	Brassic	3	Mor	3	Salic	4
Aquifoli	5	Rhamn	3	Polygon	3	Solan	4
Arec	5	Myrsin	3	Santal	2	Aster	3
Boragin	5	Phytolacc	3	Aster	2	Malpighi	3

Suffix '-aceae' omitted from all families; N = Number of Areas

was restricted or data were treated at generic and familial levels. Likewise, geographically restricted analyses of Atlantic forest sections, such as those performed for south-east Brazil by Oliveira-Filho and Fontes (2000) and north-east Brazil by Ferraz et al. (2004), could clearly detect species patterns primarily related to the climate. However, analyses performed for wider geographical ranges, such as the Amazon, could best detect patterns related to climate and vegetation formations when dealing with genera and families rather than species (ter Steege et al., 2000; Oliveira and Nelson, 2001). The geographical proximity among different vegetation formations within the same region and evolution through adaptive radiation into adjacent habitats could explain much of the strong effect of distance found in species patterns throughout the geographical range analysed here. On the other hand, the patterns related to climate and vegetation formations found for genera and families strongly suggest that climatic variables, particularly temperature and water availability, have had a long influence on the evolution and speciation of tree taxa in eastern South America. This is not a surprise since water and temperature are the chief factors determining the distribution of most world vegetation formations, and the history of the climate of eastern South America during the Quaternary shows dramatic shifts in both temperature and rainfall regime (Salgado-Labouriau et al., 1997; Behling, 1998; Ledru et al., 1998; Oliveira et al., 1999).

One important result of the above-mentioned geographical pattern is that there is greater similarity in species composition between Atlantic rain and seasonal forests of the same region than between either seasonal or rain forests of disjunct regions, although this holds true only when east and south-east are merged. In the same region the tree flora of seasonal forests is much less diverse than that of the rain forests, and is probably composed of species able to cope with relatively longer dry seasons. Tree species diversity in tropical forests is often correlated with water consumption and energy uptake, resources that are partitioned among species and limit their number in forest communities (Hugget, 1995). Water shortage probably plays the chief role in reducing species richness of seasonal forests compared to rain forests, and even more so of semi-arid formations such as chaco forests, caatingas and carrascos. Moreover, the structure of seasonal forests is also less complex, therefore favouring a reduced number of understory species compared to rain forests (Gentry and Emmons, 1987).

The intimate relationship between the two floras within each geographical region supports the wider definition of Atlantic forests to include both rain and seasonal forests as physiognomic and floristic expressions of a single great vegetation domain (Oliveira-Filho and Fontes, 2000; Galindo-Leal and Câmara, 2003). For all regions but the north-east, there is a greater floristic similarity at all three taxonomic levels between Atlantic rain and seasonal forests than between either of these and Amazonian rain forests. The exceptions are the north-east rain and seasonal semideciduous forests, both closer to Amazonian rain forests though only at the generic and familial levels. As the coastal north-east is climatically and geographically closer to the Amazon, a stronger past link could have existed through the so-called north-east bridge (Bigarella et al. 1975; Mori et al., 1981; Andrade-Lima, 1982; Cavalcanti and Tabarelli, 2004). Nevertheless, as shown by the present results, this alleged link also includes rain and seasonal forests and, for that reason, there is little floristic ground for viewing Atlantic rain forests as being closer to Amazonian rain forests than to their adjacent seasonal forests.

In all four Atlantic regions, seasonal forests and their rain forest neighbours share a similar proportion of the total species count (*c*.20%) and are both poorer in species in the north-east and south and much richer in the east and south-east. Seasonal forests of the central-west are also comparatively rich. An inspection of the distribution map (Figure 7.1) helps us understand this. Of all regions, the north-east has the smallest forest area and also lacks the highly rugged relief of other regions, the latter being a feature that may boost species richness through increased environmental heterogeneity. In addition, the region may have lost much of its primitive species richness because it was the first to go through mass deforestation, beginning in the sixteenth century. It is the least known, and most threatened and reduced of all Atlantic forests, now covering only 3.76% of its original area (Silva and Tabarelli, 2000, 2001). Towards the south, Atlantic seasonal forests expand increasingly more into the continental interior until reaching Mato Grosso do Sul and eastern Paraguay, so that they cover a wide area with pronounced variation in relief and climate (Oliveira-Filho and Fontes, 2000). In addition, seasonal forests also spread towards the west into the whole of the cerrado domain as galleries and forest patches that are found as far as in the Bolivian Chiquitanía (see Kileen et al., Chapter 9). The high environmental heterogeneity of this large area, combined with the complex contact with the cerrado, certainly explains the high species richness of the east, south-east and central-west tropical seasonal forests. The comparatively lower species richness of the subtropical seasonal forests of the south may also be explained by their comparatively smaller area and modest relief, but these forests are also at the southernmost range of Atlantic forests where extreme low temperatures in winter coupled with frosts may have already selected the small proportion of tree species able to cope with this climatic harshness (Rambo, 1980; Leite, 2002; Jurinitz and Jarenkow, 2003).

Surprisingly the proportion of seasonal forest species shared with rain forests remains more or less constant throughout the geographical range, despite the increase in species numbers of rain forest from north-east and north to south-east and south. This is brought about by an increase in the percentage of seasonal forest species occurring in rain forest from 51% and

52.9% in the north-east and east, respectively, to 74% and 83% in the south-east and south, thus counterbalancing the increase in rain forest endemics and maintaining the proportion. Unlike the situation in the other regions, the number of species in the north-east seasonal forests actually surpasses that of rain forests. The main contrast between the two pairs of northern and southern regions is that the former (north-east and east) have warmer temperatures and a much more pronounced variation in rainfall totals and seasonality than the latter (south-east and south) since they are adjacent to the caatinga domain. This is probably why north-east seasonal forests are the only ones to show a clear distinction between semideciduous and deciduous formations. The wider rainfall gradient is correlated with a relatively rapid transition from rain- to semideciduous and deciduous forests, and from those to caatingas, and thus because of ecotonal effects increasing the species richness of seasonal forests relative to their 'purer' rain forest partners. This explains, for example, why Cactaceae and Euphorbiaceae are so important in the flora of montane semide-ciduous forests, the so-called brejo forests, which occur as hinterland forest islands on mountains surrounded by lowland caatingas, and inevitably share a number of species with the latter (Rodal, 2002; Pôrto et al., 2004). In addition to this, the assemblage of north-east seasonal forests includes those influenced by other neighbouring formations, such as the coastal sandy *restingas* (e.g. at Fernando de Noronha and Natal) and the cerrado (e.g. at Araripe, Campo Maior and Sete Cidades) that may also boost their species richness (Farias and Castro, 2004). Similar effects may have occurred in the eastern region which also combines floristic interactions with both caatingas and cerrados, in addition to the effects of the rugged relief of the Espinhaço mountain range and the Chapada Diamantina (Guedes and Orge, 1998; Zappi et al., 2003).

The central-west seasonal forests also have strong floristic links with both the cerrados and Atlantic forests and share a considerable number of species. In fact, one could extract a continuum in tree species distribution determined by rainfall seasonality starting at the east and south-east Atlantic rain and seasonal forests, and extending towards the central-west to reach its seasonal forests and, lastly, the cerradões and cerrados, as already suggested by Leitão-Filho (1987). However, this is an oversimplified view of the floristic gradient because it is now known that, under seasonal climates, more important factors are involved in determining the forest–cerrado transition, and fire frequency and soil fertility and moisture play the chief role here (Oliveira-Filho and Ratter, 2002). As a result, it is not uncommon to find in the central-west two or more of those formations on a single slope (Furley and Ratter, 1988; Furley et al., 1988; Ratter et al., 1978). The complex mosaic of vegetation formations of the region and the species interchange among them probably explain why the analyses failed to discriminate deciduous from semideciduous forests floristically. Although the two formations do form a continuum, it is usually easy to tell at least their extremes apart in the field on the basis of physiognomy and floristic composition (Oliveira-Filho and Ratter, 2002). For that reason, particular attention should be paid to such nuances in the preparation of checklists for the region.

Towards the south and south-east, the declining temperatures and related vapor pressure curtail the water deficit gradient and this probably favors a stronger floristic relationship between rain and seasonal forests expressed by the much higher proportions of shared species. Extremes of low temperatures may be important determinants of tree species distribution. Occasional frosts have been mentioned by Oliveira-Filho et al. (1994) as an important factor limiting species distribution both in relation to higher elevations and latitudes in south and south-east Brazil. Resistance to frosts was suggested as a key factor determining the special nature of the chaco flora, together with their saline to alkaline soils (Pennington et al., 2000). The influence of latitude and altitude on climate, however, is far more complex than simply that of temperature and frosts. Increasing latitude also means increasing year-round variation of the daily sunlight period. Rising elevation also decreases atmospheric pressure, increases solar radiation, accelerates wind movement, promotes greater cloudiness and boosts rainfall (Jones, 1992). For tropical forests, rainfall seasonality is apparently more important than annual rainfall in determining presence of rain or seasonal forests, and the occurrence of at least a

30-day dry season produces effects which can clearly be shown on a vegetation map (IBGE, 1993). For subtropical forests, however, temperature range prevails over rainfall seasonality in separating rain and seasonal forests, probably because the contrast of low winter and high summer temperatures plays an additional role in forest deciduousness (Holdridge et al., 1971). Low temperatures alone, without the strong annual oscillation, are not associated with the presence of subtropical seasonal forests, and other formations appear, in particular araucaria rain forests, in the hinterland highlands, and upper montane rain forests (cloud forests) on the mountain ridges near the coast (Roderjan et al., 2002). Semi-arid formations such as chaco forests are found only where very strong rainfall seasonality occurs in subtropical climates, but under these conditions forests also give way to open grasslands (campos or pampas) in many areas of the south, and this is probably linked to the past history of fire and grazing (Behling, 1995, 1997; Quadros and Pillar, 2002).

Tree species composition of seasonal forests is highly influenced by altitude and associated temperatures, a well-known fact for mountain vegetation worldwide (Hugget, 1995). Because most mountain ranges and plateaus in our area of study are concentrated in the east and the lowlands of the Paraguay river basin lie in the west, the seasonal forest gradient related to decreasing altitude and increasing temperatures is highly coincident with increasing distance from the ocean. For this reason, one might speculate that this gradient was another primarily related to distance. Nevertheless, altitude-related gradients have already been detected for Atlantic rain and seasonal forests at more regional scales by Oliveira-Filho and Fontes (2000) and Ferraz et al. (2004) in south-east and north-east Brazil, respectively, and by Salis et al. (1995), Torres et al. (1997) and Scudeller et al. (2001) in the state of São Paulo. Moreover, some floristic patterns found with increasing altitude also coincided with those cited by the above authors and by Gentry (1995) for Andean and Central American forests. Among these, are the increasing contribution to the tree flora of Melastomataceae (particularly *Miconia* and *Tibouchina*), Solanaceae (*Solanum*), Lauraceae (*Ocotea* and *Nectandra*), Aquifoliaceae (*Ilex*) and Asteraceae, and the decrease of *Eugenia* and *Ficus* correlated with an increase in altitude. A detailed treatment of genera and species diagnostic of montane Atlantic forests is given by Oliveira-Filho and Fontes (2000).

It is now accepted that the caatinga domain represents the largest, most isolated and species-rich nucleus of the SDTF and that its flora is made up of a blend of endemic and wide-range species (Giulietti et al., 2002; Prado, 2003; Chapter 6). The strongest internal floristic dichotomy of the semi-arid vegetation of the caatinga is that linked to soils derived from either crystalline base rock or sandy deposits (Araújo et al., 1998; Rodal and Sampaio, 2002; Rocha et al., 2004). The present analyses largely support these findings, which are considered by Queiroz in Chapter 6. He hypothesizes that the vast proportion of caatinga non-endemics results from post-Tertiary migration of wide-range SDTF species into the region as a result of the progressive retreat of sandy deposits and exposure of the crystalline bedrock. In fact, most non-endemic caatinga species are found throughout the Austro-Atlantic and central-west seasonal forests and many reach the peripheral chaco seasonal forests without entering the chaco itself, as already emphasized by Prado (1991) and Prado and Gibbs (1993) to demonstrate the strong differentiation of the chaco and caatinga floras. Interestingly, our analyses also suggest that a similar pattern may occur in non-endemic chaco species that are also found in Austro-Atlantic and/or central-west seasonal forests but do not enter the caatingas. Despite this similarity there is also an important difference in that most non-endemic chaco species show a more limited distribution outside the chaco domain and do not reach as far as the caatinga periphery. Thus, chaco and caatinga are well-defined floristic nuclei with very weak relationships between their floras at both specific and generic levels. Only at the family level do the two floras show a stronger link, thus suggesting that if a common proto-flora did exist it must have been in the very remote past. Both floras also show floristic connections with the Atlantic and central-west seasonal forests but this is probably mainly the result of both species interchange in transitional areas and expansion of wide-range SDTF species.

7.5 CONCLUSION

We propose here that one can best describe Atlantic seasonal forests as a section of a complex floristic gradient extending from evergreen forests to semideciduous and deciduous forests (the SDTF section), and then running in a partial blending of floras to open formations, such as cerrados and campos or, alternatively, to caatingas and chaco forests. This gradient is chiefly related to decreasing water availability through either increasing rainfall seasonality and/or decreasing soil moisture content, but there is also a strong interference of temperature gradients along the latitudinal and altitudinal ranges, and of variations in soil fertility and fire frequency. The flora of the Atlantic seasonal forests occurs mostly in the section of the tropical gradient corresponding to annual periods of water shortage between 30 and 160 days, and in the section of the subtropical range where periods of water shortage are below 30 days but year-round monthly temperature oscillation is above 10°C. Increasing periods of water shortage, soil fertility and temperature range normally lead from semideciduous to deciduous forests and then to the semi-arid formations, either caatingas (tropical) or chaco forests (subtropical), while increasing fire frequency and decreasing soil fertility frequently lead from seasonal forests to either cerrados (tropical) or southern campos (subtropical). For this reason we suggest here that the definition of SDTF should be reshaped to include both cerrados and the chaco.

In conclusion, we believe that the best view of the SDTF vegetation of eastern South America is that of three floristic nuclei: caatinga, chaco and Atlantic forest (sensu latissimo). Only the latter, however, should be linked consistently to the residual Pleistocenic dry seasonal (RPDS) flora. Caatinga and chaco form the extremes of floristic dissimilarity among the three SDTF nuclei, also corresponding to the warm–dry and warm–cool climatic extremes, respectively. In contrast, the Atlantic SDTF nucleus is poor in endemic species and is actually a floristic bridge connecting the two drier nuclei to rain forests. Additionally, there is little evidence to describe the Atlantic nucleus flora as a clearly distinct species assemblage, as are those of the caatinga and chaco nuclei, because of the striking variation in species composition found throughout the vast geographical extent of Atlantic seasonal forests. Nevertheless, there is a group of wide-range species that is found in most regions of the Atlantic nucleus, some of which are also part of the species blend of the caatinga and chaco floras, though involving the latter to a much lesser extent. We believe that, at least in eastern South America, it is precisely this small fraction of the Atlantic nucleus flora that should be identified with the RPDS vegetation. To clarify the past history of neotropical SDTF, we propose that the focus should now be shifted to the investigation of the distribution patterns of those species and the past history of their populations in different locations of their geographical range.

ACKNOWLEDGEMENTS

The first author thanks the CNPq and the Royal Society of Edinburgh for the financial support to the present study and the Royal Botanic Garden Edinburgh for warmly hosting him once more. We were helped during the taxonomic revision of the database by the following: Marcos Sobral (Myrtaceae) and João Renato Stehmann (Solanaceae), both from the Federal University of Minas Gerais; Haroldo Lima (Fabaceae), Alexandre Quinet (Lauraceae), José Fernando Baumgratz (Melastomataceae) and Angela Studart da Fonseca Vaz (*Bauhinia*) from the Rio de Janeiro Botanic Garden; Maria Célia Vianna (*Vochysia*) from the Alberto Castellanos Herbarium; José Rubens Pirani (Simaroubaceae, Picramniaceae and Rutaceae) and Pedro Fiaschi (Araliaceae) from the University of São Paulo; Maria Lúcia Kawazaki (Myrtaceae) and Inês Cordeiro (Euphorbiaceae) from the São Paulo Botanic Institute; Washington Marcondes-Ferreira (*Aspidosperma*) from the State University of Campinas; Germano Guarim Neto (*Cupania*) from the Federal University of Mato Grosso; and Toby Pennington and Maureen Warwick (Fabaceae) from the Royal Botanic Garden Edinburgh. We also thank Toby Pennington, James Ratter and an anonymous reviewer for their critical and constructive reading of the first draft.

REFERENCES

Andrade-Lima, D., Present-day forest refuges in northeastern Brazil, in *Biological Diversification in the Tropics*, Prance, G.T., Ed, Plenum Press, New York, 1982, 220.

APG, An update of the Angiosperm Phylogeny Group classification for the orders and families of flowering plants: APG II, *Bot. J. Linn. Soc.*, 141, 399, 2003.

Araújo, F.S. et al., Composição florística da vegetação de carrasco, Novo Oriente, CE, *Rev. Brasil. Bot.*, 21, 105, 1998.

Behling, H., Investigations into the Late Pleistocene and Holocene history of vegetation and climate in Santa Catarina (S. Brazil), *Veget. Hist. Archaeo.*, 4, 127, 1995.

Behling, H., Late Quaternary vegetation, climate and fire history in the Araucaria forest and campos region florm Serra Campos Gerais (Paraná) S. Brazil., *Rev. Palaeobot. Palyn.* 97, 109, 1997.

Behling, H., 1998. Late Quaternary vegetational and climatic changes in Brazil, *Rev. Palaeobot. Palyn.*, 99, 143, 1998.

Bigarella, J.J., Andrade-Lima, D., and Riehs, P.J., Considerações a respeito das mudanças paleoambientais na distribuição de algumas espécies vegetais e animais no Brasil, *Anais Acad. Brasil. Ciênc.*, 47, 411, 1975.

ter Braak, C.J.F., Ordination, in *Data Analysis in Community and Landscape Ecology*, Jongman, R.H.G., ter Braak, C.J.F., and van Tongeren, O.F.R., Eds, Cambridge University Press, Cambridge, 1995, 91.

Bridgewater, S. et al., A preliminary floristic and phytogeographic analysis of the woody flora of seasonally dry forests in northern Peru, *Candollea*, 58, 129, 2003.

Cavalcanti, D. and Tabarelli, M., Distribuição das plantas Amazônico-Nordestinas no Centro de Endemismo Pernambuco: brejos de altitude vs. florestas de terras baixas, in *Brejos de Altitude em Pernambuco e Paraíba, História Natural, Ecologia e Conservação*, Pôrto, K.C., Cabral, J.J.P., and Tabarelli, M., Eds, Ministério do Meio Ambiente, Brasília, 2004, 285.

DNMet, *Normais climatológicas (1961–1990)*. Departamento Nacional de Meteorologia, Ministério da Agricultura, Brasília, 1992.

Farias, R.R.S. and Castro, A.A.J.F., Fitossociologia da vegetação do Complexo de Campo Maior, Campo Maior, PI, Brasil, *Acta Bot. Brasil.*, 18, 949, 2004.

Ferraz, E.M.N., Araújo, E.L., and Silva, S.I., Floristic similarities between lowland and montane areas of Atlantic Coastal Forest in Northeastern Brazil, *Plant Ecol.* 174, 59, 2004.

Furley, P.A. and Ratter, J.A., Soil resources and plant communities of the central Brazilian cerrado and their development, *J. Biogeogr.*, 15, 97, 1988.

Furley, P.A., Ratter, J.A., and Gifford, D.R., Observations on the vegetation of eastern Mato Grosso, Brazil. III. The woody vegetation and soils of the Morro de Fumaça, Torixoreu, *Proc. Roy. Soc. London B*, 235, 259, 1988.

Galindo-Leal, C. and Câmara, I.G., Atlantic Forest hotspot status: an overview, in *The Atlantic Forest of South America*, Galindo-Leal, C. and Câmara, I.G., Eds, Center for Applied Biodiversity Science, Washington, 2003, 3.

Gentry, A.H., Patterns of diversity and floristic composition, in *Neotropical Montane Forests, Biodiversity and Conservation of Neotropical Montane Forests, Neotropical Montane Forest Biodiversity and Conservation Symposium, 1*, Churchill, S.P. et al., Eds, The New York Botanical Garden, New York, 1995, 103.

Gentry, A.H. and Emmons, L.H., Geographical variation in fertility, phenology and composition of the understory of Neotropical forests, *Biotropica*, 19, 216, 1987.

Giulietti, A.M. et al., Espécies endêmicas da Caatinga, in *Vegetação e Flora da Caatinga*, Sampaio, E.V.S.B. et al., Eds, Associação Plantas do Nordeste, Recife, 2002, 103.

Graham, A. and Dilcher, D., The Cenozoic record of tropical dry forest in northern Latin America and the southern United States, in *Seasonally Dry Tropical Forests*, Mooney, H.A., Bullock, S.H., and Medina, E., Eds, Cambridge University Press, Cambridge, 1995, 124.

Guedes, M.L.S. and Orge, M.D.R., *Checklist das espécies vasculares de Morro do Pai Inácio (Palmeiras) e Serra da Chapadinha (Lençóis). Chapada Diamantina, Bahia*, Instituto de Biologia da Universidade Federal da Bahia, Salvador, 1998.

Hill, M.O. and Gauch, H.G., Detrended correspondence analysis: an improved ordination technique, *Vegetatio*, 42, 47, 1980.

Holdridge, L.R. et al., *Forest Environments in Tropical Life Zones, A Pilot Study*, Pergamon, Oxford, 1971.

Hugget, R.J., *Geoecology, An Evolutionary Approach*, Routledge, London, 1995.

IBGE, *Mapa de Vegetação do Brasil*, Instituto Brasileiro de Geografia e Estatística, Rio de Janeiro, 1993.

Jones, H.G., *Plants and Microclimate: A Quantitative Approach to Environmental Plant Physiology*, 2, Cambridge University Press, Cambridge, 1992.

Jurinitz, C.F. and Jarenkow, J.A., Estrutura do componente arbóreo de uma floresta estacional na Serra do Sudeste, Rio Grande do Sul, Brasil, *Rev. Brasil. Bot.*, 26, 475, 2003.

Kent, M. and Coker, P., *Vegetation Description and Analysis, A Practical Approach*, Belhaven Press, London, 1992.

Ledru, M.P., Salgado-Labouriau, M.L., and Lorscheiter. M.L., Vegetation dynamics in southern and central Brazil during the last 10,000 yr. B.P., *Rev. Palaeobot. Palyn.* 99, 131, 1998.

Leitão-Filho, H.F., Considerações sobre a florística de florestas tropicais e sub-tropicais do Brasil, *Rev. IPEF* 35, 41, 1987.

Leite, P.F., Contribuição ao conhecimento fitoecológico do Sul do Brasil, *Ciênc. Amb.*, 24, 51, 2002.

Lewis, J.P., Pire, E.F., and Vesprini, J.L., The mixed dense forest of the Southern Chaco. Contribution to the study of flora and vegetation of the Chaco. VIII, *Candollea*, 49, 159, 1994.

Linares-Palomino, R., Pennington, R.T., and Bridgewater, S., The phytogeography of the seasonally dry tropical forests in Equatorial South America, *Candollea*, 58, 473, 2003.

McCune, B. and Mefford, M.J., *PC-ORD version 4. 0., Multivariate Analysis of Ecological Data, Users Guide*, MjM Software Design, Glaneden Beach, 1999.

Mooney, H.A., Bullock, S.H., and Medina, E., Introduction, in *Seasonally Dry Tropical Forests*, Mooney, H.A., Bullock, S.H., and Medina, E., Eds, Cambridge University Press, Cambridge, 1995, 1.

Mori, S.A., Boom, B.M., and Prance, G.T., Distribution patterns and conservation of eastern Brazilian coastal forest tree species, *Brittonia*, 33, 233, 1981.

Murphy, P.G. and Lugo, A.E., Dry forests of Central America and the Caribbean, in *Seasonally Dry Tropical Forests*, Mooney, H.A., Bullock, S.H., and Medina, E., Eds, Cambridge University Press, Cambridge, 1995, 146.

Oliveira, A.A. and Nelson, B.W., Floristic relationships of terra firme forests in the Brazilian Amazon, *Forest Ecol. Manag.*, 146, 169, 2001.

Oliveira, P.E., Barreto, A.M.F., and Suguio, K., Late Pleistocene/Holocene climatic and vegetational history of the Brazilian Caatinga: the fossil dunes of the middle São Francisco river, *Palaeogeogr. Palaeoclimat. Palaeocecol.*, 152, 319, 1999.

Oliveira-Filho, A.T. and Fontes M.A.L., Patterns of floristic differentiation among Atlantic forests in southeastern Brazil, and the influence of climate, *Biotropica*, 32, 793, 2000.

Oliveira-Filho, A.T. and Ratter, J.A., *Database: Woody Flora of 106 Forest Areas of Eastern Tropical South America*, Royal Botanic Garden Edinburgh, Edinburgh, 1994.

Oliveira-Filho, A.T. and Ratter, J.A., A study of the origin of central Brazilian forests by the analysis of plant species distribution patterns, *Edinburgh J. Bot.*, 52, 141, 1995.

Oliveira-Filho, A.T. and Ratter, J.A., Vegetation physiognomies and woody flora of the Cerrado Biome, in *The Cerrados of Brazil: Ecology and Natural History of a Neotropical Savanna*, Oliveira, P.S. and Marquis, R.J., Eds, Columbia University Press, New York, 2002, 91.

Oliveira-Filho, A.T. et al., Comparison of the woody flora and soils of six areas of montane semi-deciduous forest in southern Minas Gerais, Brazil, *Edinburgh J. Bot.* 51, 355, 1994.

Pennington, R.T., Prado, D.E., and Pendry, C.A., Neotropical seasonally dry forests and Quaternary vegetation changes, *J. Biogeogr.*, 27, 261, 2000.

Pennington, R.T. et al., Historical climate change and speciation: neotropical seasonally dry forest plants show patterns of both Tertiary and Quaternary diversification, *J. Biogeogr.*, 359, 515, 2004.

Pôrto, K.C., Cabral, J.J.P., and Tabarelli, M., *Brejos de Altitude em Pernambuco e Paraíba, História Natural, Ecologia e Conservação*, Ministério do Meio Ambiente, Brasília, 2004.

Prado, D.E., *A critical evaluation of the floristic links between Chaco and Caatingas vegetation in South America*, PhD thesis, University of St Andrews, St Andrews, 1991.

Prado, D.E., Seasonally dry forests of tropical South America: from forgotten ecosystems to a new phytogeographic unit, *Edinburgh J. Bot.*, 57, 437, 2000.

Prado, D.E., As Caatingas do Brasil, in *Ecologia e Conservação da Caatinga*, Leal, I.R., Tabarelli, M., and Silva, J.M.C., Eds, Editora Universidade Federal de Pernambuco, Recife, 2003, 3.

Prado, D.E. and Gibbs, P.E., Patterns of species distribution in the dry seasonal forests of South America, *Ann. Mo. Bot. Gard.*, 80, 902, 1993.

Quadros, F.L.F. and Pillar, V.P., Transições floresta–campo no Rio Grande do Sul, *Ciênc. Amb.*, 24, 109, 2002.

Rambo, B., A mata pluvial do Alto Uruguai, *Roessléria* 3, 101, 1980.

Ratter, J.A. et al., Observations on the vegetation of northeastern Mato Grosso, II. Forests and soils of the Rio Suiá-Missu area, *Proc. R. Soc. London B*, 203, 191, 1978.

Ratter, J.A. et al., Analysis of the floristic composition of the Brazilian cerrado vegetation, *Edinburgh J. Bot.*, 53, 153, 1996.

Ribeiro, J.E.L.S. et al., *Flora da Reserva Ducke: Guia de Identificação das Plantas Vasculares de uma Floresta de Terra-Firme na Amazônia Central*, INPA-DFID, Manaus, 1999.

Rocha, P.L.B., Queiroz, L.P., and Pirani, J.R., Plant species and habitat structure in a sand dune field in the Brazilian Caatinga: a homogeneous habitat harbouring an endemic biota, *Rev. Brasil. Bot.*, 27, 739, 2004.

Rodal, M.J.N., Montane forests in Northeast Brazil: a phytogeographical approach, *Bot. Jarh. Syst.*, 124, 1, 2002.

Rodal, M.J.N. and Nascimento, L.M., Levantamento florístico de uma floresta serrana da Reserva Biológica de Serra Negra, microrregião de Itaparica, Pernambuco, Brasil, *Acta Bot. Brasil.*, 16, 481, 2002.

Rodal, M.J.N. and Sampaio, E.V.S.B., A vegetação do Bioma Caatinga, in *Vegetação e Flora da Caatinga*, Sampaio, E.V.S.B. et al., Eds, Associação Plantas do Nordeste, Recife, 2002, 11.

Roderjan, C.V. et al., As unidades fitogeográficas do Estado do Paraná, *Ciênc. Amb.*, 24, 75, 2002.

Salgado-Labouriau, M.L. et al., Late Quaternary vegetational and climatic changes in cerrado and palm swamp from central Brazil, *Palaeogeogr. Palaeoclimat. Palaeocecol.*, 128, 215, 1997.

Salis, S.M., Shepherd, G.J., and Joly, C.A., Floristic comparison of mesophytic semi-deciduous forests of the interior of the state of São Paulo, southeast Brazil, *Vegetatio*, 119, 155, 1995.

Scudeller V.V., Martins F.R., and Shepherd G.J., Distribution and abundance of arboreal species in the Atlantic ombrophilous dense forest in Southeastern Brazil, *Plant Ecol.*, 152, 185, 2001.

Silva, J.M.C. and Tabarelli, M., Tree species impoverishment and the future flora of the Atlantic Forest of northeast Brazil, *Nature*, 404, 72, 2000.

Silva, J.M.C. and Tabarelli, M., The future of Atlantic forest in Northeastern Brazil, *Conserv. Biol.* 15, 819, 2001.

Spichiger, R. et al., Origin, affinities and diversity hotspots of the Paraguayan dendrofloras, *Candollea*, 50, 515, 1995.

Spichiger, R., Calange, C., and Bise, B., Geographical zonation in the Neotropics of tree species characteristic of the Paraguay-Paraná Basin, *J. Biogeogr.* 31, 1489, 2004.

ter Steege, H. et al., An analysis of the floristic composition and diversity of Amazonian forests including those of the Guiana Shield, *J. Trop. Ecol.*, 16, 801, 2000.

Thornthwaite, C.W., An approach toward a rational classification of climate, *Geogr. Rev.*, 38, 55, 1948.

Torres, R.B., Martins, F.R., and Gouvea, L.S.K., Climate, soil and tree flora relationships in forests in the state of São Paulo, southeastern Brazil, *Revta. Brasil. Bot.*, 20, 41, 1997.

Veloso, H.P., Rangel Filho, A.L.R., and Lima, J.C.A., *Classificação da Vegetação Brasileira, Adaptada a um Sistema Universal*, Instituto Brasileiro de Geografia e Estatística, Rio de Janeiro, 1991.

Walter, H., *Vegetation of the Earth and Ecological Systems of the Geo-Biosphere*, 3rd ed., Springer-Verlag, Berlin, 1985.

Zappi, D.C. et al., Lista das plantas vasculares de Catolés, Chapada Diamantina, Bahia, Brasil, *Bol. Bot. Univ. São Paulo*, 21, 345, 2003.

APPENDIX: MOST FREQUENT SPECIES (≥70% OF CHECKLISTS) IN THE TREE FLORA OF SELECTED SDTF FORMATIONS OF EASTERN SOUTH AMERICA

Low altitude tropical seasonal forests — North-east region: *Abarema cochliacarpos, Acacia polyphylla, Albizia pedicellaris, A. polycephala, Allophylus edulis, Alseis pickelii, Anacardium occidentale, Andira fraxinifolia, A. nitida, Apeiba tibourbou, Apuleia leiocarpa, Aspidosperma pyrifolium, Astronium fraxinifolium, Bowdichia virgilioides, Brosimum gaudichaudii, B. guianense, Buchenavia capitata, Byrsonima sericea, Caesalpinia ferrea, Campomanesia aromatica, C. dichotoma, Capparis flexuosa, Casearia sylvestris, Cecropia pachystachya, C. palmata, Cereus jamacaru, Chamaecrista apoucouita, C. ensiformis, Chrysophyllum rufum, Clusia nemorosa, Coccoloba* *alnifolia, C. cordifolia, Cordia trichotoma, Coutarea hexandra, Cupania revoluta, Curatella americana, Enterolobium contortisiliquum, Erythrina velutina, Erythroxylum citrifolium, Eschweilera ovata, Eugenia florida, E. punicifolia, E. uniflora, Guapira noxia, G. opposita, G. pernambucensis, Guarea guidonia, Guazuma ulmifolia, Guettarda platypoda, Himatanthus phagedaenicus, Hirtella ciliata, H. racemosa, Hymenaea courbaril, H. rubriflora, Inga capitata, I. ingoides, I. laurina, I. thibaudiana, Lecythis pisonis, Luehea ochrophylla, L. paniculata, Manihot epruinosa, Manilkara salzmannii, Maytenus distichophylla, M. erythroxylon, Miconia albicans, Myrcia multiflora, M. sylvatica, M. tomentosa, Myrsine guianensis, Ocotea notata, Ouratea hexasperma, Palicourea crocea, Pera glabrata, Pogonophora schomburgkiana, Pouteria grandiflora, Pradosia lactescens, Protium heptaphyllum, Psidium oligospermum, Pterocarpus rohrii, Rauvolfia ligustrina, Sacoglottis mattogrossensis, Schefflera morototoni, Spondias mombin, Strychnos parvifolia, Stryphnodendron pulcherrimum, Swartzia pickelii, Tabebuia impetiginosa, T. roseo-alba, T. serratifolia, Talisia esculenta, Tapirira guianensis, Thyrsodium spruceanum, Trema micrantha, Vismia guianensis, Vitex triflora, Ximenia americana, Ziziphus joazeiro, Zollernia latifolia.*

High altitude tropical seasonal forests — North-east region: *Acacia polyphylla, A. riparia, A. tenuifolia, Albizia polycephala, Allophylus edulis, Anadenanthera colubrina, Bowdichia virgilioides, Buchenavia capitata, Byrsonima sericea, Caesalpinia ferrea, Campomanesia aromatica, Capparis flexuosa, Casearia sylvestris, Ceiba glaziovii, Celtis iguanaea, Clusia nemorosa, Copaifera langsdorffii, Cordia trichotoma, Coutarea hexandra, Croton rhamnifolius, Cupania revoluta, Cyathea microdonta, Enterolobium contortisiliquum, Erythroxylum citrifolium, Eugenia punicifolia, Guapira laxiflora, G. opposita, Guazuma ulmifolia, Guettarda sericea, Hymenaea courbaril, Machaerium hirtum, Manilkara rufula, Maprounea guianensis, Maytenus obtusifolia, Miconia albicans, Myrcia fallax, M. multiflora, M. sylvatica, M. tomentosa, Myroxylon peruiferum, Myrsine guianensis, Ocotea duckei, Piptadenia stipulacea, Platymiscium floribundum, Prockia crucis, Psidium guineense, Randia nitida, Roupala cearensis, Ruprechtia laxiflora, Sapium glandulosum, Schoepfia brasiliensis, Senna macranthera, S. spectabilis, S. splendida, Tabebuia impetiginosa, T. serratifolia, Talisia esculenta, Vitex rufescens, Zanthoxylum rhoifolium.*

Low altitude tropical seasonal forests — East region: *Acacia polyphylla, Aegiphila sellowiana, Albizia polycephala, Alchornea glandulosa, Allophylus edulis, Amaioua guianensis, Anadenanthera colubrina, Andira fraxinifolia, Aparisthmium cordatum, Apuleia leiocarpa, Aspidosperma parvifolium, Astrocaryum aculeatissimum, Astronium graveolens, Bathysa nicholsonii, Bauhinia fusco-nervis, Brosimum guianense, B. lactescens, Byrsonima sericea, Cabralea canjerana, Carpotroche brasiliensis, Casearia sylvestris, C. ulmifolia, Cassia ferruginea, Cecropia glaziovii, C. hololeuca, Cedrela fissilis, Copaifera langsdorffii, Croton urucurana, Cyathea delgadii, Dalbergia nigra, Endlicheria paniculata, Erythrina verna, Erythroxylum pelleterianum, E. pulchrum, Eugenia florida, Euterpe edulis, Ficus gomelleira, Gallesia integrifolia, Guapira opposita, Guarea macrophylla, Guatteria australis, G. villosissima, Guettarda uruguensis, Himatanthus lancifolius, Hortia arborea, Hymenolobium janeirense, Inga capitata, I. vera, Joannesia princeps, Lacistema pubescens, Lecythis lurida, L. pisonis, Luehea divaricata, L. grandiflora, Mabea fistulifera, Machaerium brasiliense, M. hirtum, M. stipitatum, Maclura tinctoria, Maprounea guianensis, Melanoxylon brauna, Miconia cinnamomifolia, Myrcia fallax, M. rufula, Myrciaria floribunda, Nectandra oppositifolia, Ocotea dispersa, Pera glabrata, Piptadenia gonoacantha, Plathymenia reticulata, Platymiscium floribundum, Platypodium elegans, Pogonophora schomburgkiana, Pourouma guianensis, Protium warmingianum, Pseudobombax grandiflorum, Pseudopiptadenia contorta, Pterocarpus rohrii, Pterygota brasiliensis, Rollinia laurifolia, Senna macranthera, S. multijuga, Siparuna guianensis, Sorocea guilleminiana, Sparattosperma leucanthum, Stryphnodendron pulcherrimum, Swartzia acutifolia, S. myrtifolia, Syagrus romanzoffiana, Tabebuia serratifolia, Tabernaemontana hystrix, Tapirira guianensis, Trichilia lepidota, T. pallida, Urbanodendron verrucosum, Virola bicuhyba, Vismia guianensis, Xylopia brasiliensis, Xylopia sericea, Zanthoxylum rhoifolium.*

High altitude tropical seasonal forests — East region: *Aegiphila sellowiana, Alchornea triplinervia, Amaioua guianensis, Anadenanthera colubrina, Andira fraxinifolia, Apuleia leiocarpa, Aspidosperma discolor, A. olivaceum, Astronium graveolens, Bauhinia longifolia, Blepharocalyx salicifolius, Bowdichia virgilioides, Byrsonima sericea, Cabralea canjerana, Campomanesia xanthocarpa, Casearia arborea, C. decandra, C. obliqua, C. sylvestris, Cassia ferruginea, Cecropia glaziovii, C. hololeuca, C. pachystachya, Cedrela fissilis, Celtis iguanaea, Chrysophyllum gonocarpum, Clethra scabra, Copaifera langsdorffii, Cordia sellowiana, Croton floribundus, C. urucurana, Cupania paniculata, C. vernalis, Cyathea corcovadensis, C. delgadii, C. phalerata, Dalbergia frutescens, Dictyoloma vandellianum, Eugenia florida, Gochnatia polymorpha, Guapira opposita, Guarea macrophylla, Guatteria australis, G. sellowiana, G. villosissima, Guazuma ulmifolia, Hyptidendron asperrimum, Inga laurina, I. marginata, I. sessilis, Kielmeyera lathrophyton, Lamanonia ternata, Leandra melastomoides, Luehea divaricata, Machaerium brasiliense, M. hirtum, M. nictitans, M. villosum, Maprounea guianensis, Matayba elaeagnoides, Maytenus salicifolia, Miconia cinnamomifolia, M. ligustroides, M. pepericarpa, Myrcia detergens, M. fallax, M. guianensis, M. rostrata, M. tomentosa, Myrsine coriacea, M. umbellata, Nectandra lanceolata, N. oppositifolia, Ocotea corymbosa, O. odorifera, O. spixiana, Pera glabrata, Platypodium elegans, Protium heptaphyllum, Prunus myrtifolia, Psychotria vellosiana, Rollinia laurifolia, Rollinia sylvatica, Roupala brasiliensis, Sapium glandulosum, Sclerolobium rugosum, Senna macranthera, S. multijuga, Siparuna guianensis, Siphoneugena densiflora, Sorocea guilleminiana, Tabebuia serratifolia, Tapirira guianensis, T. obtusa, Terminalia glabrescens, Tibouchina candolleana, Trichilia pallida, Vitex polygama, Vochysia tucanorum, Zanthoxylum rhoifolium.*

Low altitude tropical seasonal forests — South-east region: *Acacia polyphylla, Actinostemon klotzschii, Aegiphila sellowiana, Albizia niopoides, Alchornea glandulosa, A. triplinervia, Allophylus edulis, Aloysia virgata, Annona cacans, Apuleia leiocarpa, Aralia warmingiana, Aspidosperma polyneuron, Astronium graveolens, Balfourodendron riedelianum, Bastardiopsis densiflora, Cabralea canjerana, Campomanesia guazumifolia, C. xanthocarpa, Cariniana estrellensis, Casearia gossypiosperma, C. sylvestris, Cecropia pachystachya, Cedrela fissilis, Ceiba speciosa, Celtis iguanaea, Chrysophyllum gonocarpum, C. marginatum, Colubrina glandulosa, Copaifera langsdorffii, Cordia ecalyculata, C. trichotoma, Croton floribundus, Cupania vernalis, Dalbergia frutescens, Dendropanax cuneatus, Diatenopteryx sorbifolia, Endlicheria paniculata, Enterolobium contortisiliquum, Esenbeckia febrifuga, Eugenia florida, E. involucrata, Euterpe edulis, Gallesia integrifolia, Guapira opposita, Guarea guidonia, G. kunthiana, G. macrophylla, Guatteria australis, Gymnanthes concolor, Heliocarpus americanus, Holocalyx balansae, Inga marginata, I. striata, I. vera, Ixora venulosa, Jacaranda micrantha, Jacaratia spinosa, Lonchocarpus cultratus, L. muehlbergianus, Luehea divaricata, Machaerium hirtum, M. nictitans, M. paraguariense, M. stipitatum, Maclura tinctoria, Matayba elaeagnoides, Metrodorea nigra, Myrcia multiflora, Myrciaria floribunda, Myrocarpus frondosus, Myrsine umbellata, Nectandra megapotamica, Ocotea diospyrifolia, O. puberula, O. pulchella, Parapiptadenia rigida, Patagonula americana, Peltophorum dubium, Piper amalago, Prunus myrtifolia, Rollinia emarginata, R. sylvatica, Roupala brasiliensis, Schefflera morototoni, Sebastiania commersoniana, Seguieria langsdorffii, Sorocea bonplandii, Syagrus romanzoffiana, Tabernaemontana catharinensis, Terminalia triflora, Trema micrantha, Trichilia catigua, T. clausseni, T. elegans, T. pallida, Vitex megapotamica, Zanthoxylum caribaeum, Z. fagara, Z. rhoifolium, Z. riedelianum.*

High altitude tropical seasonal forests — South-east region: *Aegiphila sellowiana, Albizia polycephala, Alchornea triplinervia, Amaioua guianensis, Andira fraxinifolia, Annona cacans, Aspidosperma olivaceum, Byrsonima laxiflora, Cabralea canjerana, Calyptranthes clusiifolia, Campomanesia guazumifolia, Cariniana estrellensis, Casearia decandra, C. lasiophylla, C. obliqua, C. sylvestris, Cecropia glaziovii, C. pachystachya, Cedrela fissilis, Chrysophyllum marginatum, Cinnamomum glaziovii, Clethra scabra, Copaifera langsdorffii, Cordia sellowiana, Croton floribundus, C. verrucosus, Cryptocarya aschersoniana, Cupania vernalis, Cyathea delgadii, C. phalerata, Dalbergia* *villosa, Daphnopsis brasiliensis, D. fasciculata, Dendropanax cuneatus, Endlicheria paniculata, Eugenia florida, Gomidesia affinis, Guapira opposita, Guarea macrophylla, Guatteria australis, Gymnanthes concolor, Heisteria silvianii, Hyeronima ferruginea, Inga striata, Ixora warmingii, Jacaranda macrantha, Lamanonia ternata, Leucochloron incuriale, Lithraea molleoides, Luehea divaricata, L. grandiflora, Machaerium hirtum, M. nictitans, M. stipitatum, M. villosum, Maclura tinctoria, Matayba elaeagnoides, Miconia cinnamomifolia, Mollinedia widgrenii, Myrcia fallax, M. rostrata, Myrciaria floribunda, Myrsine coriacea, M. umbellata, Nectandra grandiflora, N. oppositifolia, Ocotea corymbosa, O. diospyrifolia, O. odorifera, O. pulchella, Pera glabrata, Persea pyrifolia, Piptocarpha macropoda, Platycyamus regnellii, Platypodium elegans, Protium widgrenii, Prunus myrtifolia, Psychotria vellosiana, Rollinia dolabripetala, R. laurifolia, R. sylvatica, Roupala brasiliensis, Sapium glandulosum, Schinus terebinthifolius, Sclerolobium rugosum, Solanum pseudoquina, Sorocea bonplandii, Syagrus romanzoffiana, Tabebuia serratifolia, Tapirira guianensis, T. obtusa, Ternstroemia brasiliensis, Trichilia emarginata, T. pallida, Vernonanthura diffusa, Vismia brasiliensis, Vitex polygama, Vochysia tucanorum, Xylopia brasiliensis, Zanthoxylum rhoifolium.*

Low altitude tropical seasonal forests — Central-west region: *Acacia polyphylla, Acrocomia aculeata, Albizia niopoides, Alibertia concolor, Anadenanthera colubrina, A. peregrina, Apeiba tibourbou, Apuleia leiocarpa, Aspidosperma cylindrocarpon, A. olivaceum, A. pyrifolium, A. subincanum, Astronium fraxinifolium, Attalea phalerata, Bauhinia longifolia, Bowdichia virgilioides, Cabralea canjerana, Callisthene fasciculata, C. major, Calophyllum brasiliense, Cariniana estrellensis, Casearia gossypiosperma, C. rupestris, C. sylvestris, Cecropia pachystachya, Cedrela fissilis, Ceiba speciosa, Celtis iguanaea, Chrysophyllum gonocarpum, Combretum leprosum, Copaifera langsdorffii, Cordia glabrata, C. trichotoma, Coutarea hexandra, Cupania vernalis, Dilodendron bipinnatum, Diospyros hispida, D. sericea, Enterolobium contortisiliquum, Eugenia florida, Genipa americana, Guapira opposita, Guarea guidonia, Guazuma ulmifolia, Guettarda uruguensis, Hymenaea courbaril, Inga laurina, I. marginata, I. vera, Jacaranda cuspidifolia, Licania apetala, Luehea divaricata, L. paniculata, Machaerium hirtum, M. stipitatum, M. villosum, Maclura tinctoria, Magonia pubescens, Matayba guianensis, Micropholis venulosa, Myracrodruon urundeuva, Myrcia tomentosa, Myrciaria floribunda, Plathymenia reticulata, Platypodium elegans, Pouteria gardneri, Protium heptaphyllum, P. spruceanum, Pseudobombax tomentosum, Psidium guineense, Pterogyne nitens, Qualea multiflora, Randia nitida, Rhamnidium elaeocarpum, Rollinia emarginata, Salacia elliptica, Sapium glandulosum, Sclerolobium paniculatum, Simira sampaioana, Siparuna guianensis, Sorocea guilleminiana, Spondias mombin, Sterculia striata, Sweetia fruticosa, Tabebuia impetiginosa, T. roseo-alba, T. serratifolia, Talisia esculenta, Tapirira guianensis, Terminalia argentea, T. glabrescens, Tocoyena formosa, Trichilia catigua, T. clausseni, T. elegans, T. pallida, Triplaris gardneriana, Unonopsis lindmanii, Vitex cymosa, Zanthoxylum rhoifolium.*

High altitude tropical seasonal forests — Central-west region: *Aegiphila sellowiana, Alibertia edulis, Amaioua guianensis, Anadenanthera colubrina, Apuleia leiocarpa, Aspidosperma cylindrocarpon, A. australe, A. subincanum, Astronium fraxinifolium, Bauhinia longifolia, Cabralea canjerana, Callisthene major, Calophyllum brasiliense, Cardiopetalum calophyllum, Cariniana estrellensis, Casearia sylvestris, Cecropia pachystachya, Cedrela fissilis, Cheiloclinium cognatum, Chrysophyllum marginatum, Copaifera langsdorffii, Cordia sellowiana, C. trichotoma, Cryptocarya aschersoniana, Cupania vernalis, Dendropanax cuneatus, Diospyros hispida, Emmotum nitens, Endlicheria paniculata, Eugenia florida, Euplassa inaequalis, Faramea cyanea, Ferdinandusa speciosa, Gomidesia fenzliana, Guarea guidonia, G. macrophylla, Guatteria sellowiana, Guazuma ulmifolia, Guettarda uruguensis, Hedyosmum brasiliense, Hirtella glandulosa, H. gracilipes, Hyeronima alchorneoides, Hymenaea courbaril, Inga alba, I. laurina, I. vera, Ixora warmingii, Lamanonia ternata, Licania apetala, Luehea divaricata, Machaerium acutifolium, Maprounea guianensis, Matayba guianensis, Mauritia flexuosa, Miconia chamissois, Micropholis venulosa, Mouriri glazioviana, Myrcia rostrata, M. tomentosa, Myrsine guianensis, M. umbellata, Nectandra cissiflora, Ocotea corymbosa, O. spixiana, Ormosia fastigiata, Ouratea castaneifolia, Pera glabrata,*

Piptadenia gonoacantha, Piptocarpha macropoda, Platypodium elegans, Pouteria gardneri, P. ramiflora, Protium heptaphyllum, P. spruceanum, Prunus myrtifolia, Pseudolmedia laevigata, Qualea dichotoma, Q. multiflora, Richeria grandis, Roupala brasiliensis, Schefflera morototoni, Sclerolobium paniculatum, Siparuna guianensis, Siphoneugena densiflora, Styrax camporum, Symplocos nitens, Tabebuia serratifolia, Talauma ovata, Tapirira guianensis, Terminalia argentea, T. glabrescens, Trichilia catigua, Virola sebifera, Vitex polygama, Vochysia tucanorum, Xylopia aromatica, Xylopia emarginata, X. sericea, Zanthoxylum rhoifolium.

Subtropical seasonal forests — South region: *Aiouea saligna, Alchornea triplinervia, Allophylus edulis, A. guaraniticus, Apuleia leiocarpa, Banara parviflora, B. tomentosa, Blepharocalyx salicifolius, Cabralea canjerana, Calyptranthes concinna, Campomanesia xanthocarpa, Casearia decandra, C. sylvestris, Cedrela fissilis, Celtis iguanaea, Chrysophyllum gonocarpum, C. marginatum, Citronella paniculata, Cordia ecalyculata, C. trichotoma, Cupania vernalis, Dalbergia frutescens, Daphnopsis racemosa, Dasyphyllum spinescens, Diospyros inconstans, Enterolobium contortisiliquum, Erythrina crista-galli, Erythroxylum argentinum, Eugenia hyemalis, E. involucrata, E. opaca, E. ramboi, E. rostrifolia, E. uniflora, Ficus insipida, F. luschnathiana, F. organensis, Gomidesia palustris, Guapira opposita, Guettarda uruguensis, Gymnanthes concolor, Helietta apiculata, Ilex brevicuspis, Inga marginata, I. vera, Jacaranda micrantha, Lithraea brasiliensis, Lonchocarpus nitidus, Luehea divaricata, Machaerium stipitatum, Matayba elaeagnoides, Maytenus ilicifolia, Myrcianthes pungens, Myrciaria tenella, Myrocarpus frondosus, Myrsine coriacea, M. loefgrenii, M. lorentziana, M. umbellata, Nectandra lanceolata, N. megapotamica, Ocotea puberula, O. pulchella, Parapiptadenia rigida, Patagonula americana, Phytolacca dioica, Pilocarpus pennatifolius, Pisonia zapallo, Pouteria gardneriana, P. salicifolia, Prunus myrtifolia, P. subcoriacea, Psidium cattleianum, Quillaja brasiliensis, Randia nitida, Rollinia emarginata, R. sylvatica, Ruprechtia laxiflora, Sapium glandulosum, Schefflera morototoni, Schinus terebinthifolius, Sebastiania brasiliensis, S. commersoniana, Seguieria americana, Solanum granuloso-leprosum, S. pseudoquina, S. sanctaecatharinae, Sorocea bonplandii, Strychnos brasiliensis, Styrax leprosus, Syagrus romanzoffiana, Terminalia australis, Trema micrantha, Trichilia clausseni, T. elegans, Urera baccifera, Vitex megapotamica, Xylosma pseudosalzmanii, Zanthoxylum fagara, Z. rhoifolium.*

Subtropical seasonal forests — South-west region: *Acacia albicorticata, A. caven, A. praecox, Acanthosyris falcata, Achatocarpus praecox, Allophylus edulis, Amburana cearensis, Anadenanthera colubrina, Aralia angelicifolia, Aspidosperma olivaceum, A quebracho-blanco, Caesalpinia paraguariensis, Calycophyllum multiflorum, Capparis retusa, Carica quercifolia, Casearia sylvestris, Celtis pubescens, Chloroleucon tenuiflorum, Chrysophyllum gonocarpum, C. marginatum, Cochlospermum tetraporum, Cordia trichotoma, Crataeva tapia, Cupania vernalis, Diplokeleba floribunda, Enterolobium contortisiliquum, Erythrina falcata, Eugenia uniflora, Geoffroea decorticans, G. striata, Gleditsia amorphoides, Holocalyx balansae, Maclura tinctoria, Myracrodruon balansae, M. urundeuva, Myrcianthes cisplatensis, M. pungens, Myrsine laetevirens, Parkinsonia aculeata, Patagonula americana, Peltophorum dubium, Phyllostylon rhamnoides, Phytolacca dioica, Pilocarpus pennatifolius, Pisonia aculeata, P. zapallo, Pouteria gardneriana, Prosopis nigra, Pterogyne nitens, Rollinia emarginata, Ruprechtia laxiflora, Sapindus saponaria, Sapium haematospermum, Schinopsis brasiliensis, Schinus polygamus, Scutia buxifolia, Sebastiania brasiliensis, Sideroxylon obtusifolium, Solanum granuloso-leprosum, Syagrus romanzoffiana, Tabebuia heptaphylla, T. impetiginosa, Tabernaemontana catharinensis, Terminalia triflora, Tipuana tipu, Ximenia americana, Zanthoxylum fagara, Z. petiolare, Z. rhoifolium, Ziziphus mistol.*

Chaco forests: *Acacia aroma, A. caven, A. curvifructa, A. furcatispina, A. praecox, A. tucumanensis, Acanthosyris falcata, Achatocarpus praecox, Allophylus edulis, Anadenanthera colubrina, Aporosella chacoensis, Aralia angelicifolia, Aspidosperma quebracho-blanco, A triternatum, Athyana weinmanniifolia, Bougainvillea campanulata, B. praecox, Bulnesia bonariensis, Caesalpinia paraguariensis, Capparis atamisquea, C. retusa, C. salicifolia, C. insignis, Cereus stenogonus, Chrysophyllum marginatum, Copernicia alba, Cupania vernalis, Enterolobium contortisiliquum, Eugenia* *uniflora, Geoffroea decorticans, Jacaratia corumbensis, Maytenus scutioides, M. spinosa, M. vitis-idaea, Mimosa castanoclada, M. chacoensis, M. detinens, M. glutinosa, Mimozyganthus carinatus, Myrcianthes cisplatensis, M. pungens, Myrsine laetevirens, Parkinsonia praecox, Patagonula americana, Pereskia sacharosa, Phyllostylon rhamnoides, Pisonia zapallo, Prosopis affinis, P. alpataco, P. elata, P. fiebrigii, P. kuntzei, P. nigra, P. nuda, P. rojasiana, P. ruscifolia, P. sericantha, P. torquata, Quiabentia chacoensis, Ruprechtia apetala, R. laxiflora, R. triflora, Sapium haematospermum, Schinopsis balansae, S. cornuta, S. heterophylla, S. quebracho-colorado, Schinus polygamus, Scutia buxifolia, Sesbania virgata, Sideroxylon obtusifolium, Stetsonia coryne, Tabebuia impetiginosa, T nodosa, Tessaria dodoneifolia, T integrifolia, Thevetia bicornuta, Trema micrantha, Trithrinax schizophylla, Ximenia americana, Zanthoxylum coco, Z. petiolare, Ziziphus mistol.*

Caatingas: *Acacia langsdorffii, A. paniculata, A. polyphylla, Allophylus quercifolius, Amburana cearensis, Anadenanthera colubrina, Annona spinescens, Aspidosperma pyrifolium, Auxemma glazioviana, A. oncocalyx, Balfourodendron molle, Bauhinia acuruana, B. cheilantha, B. pentandra, Bocoa mollis, Brasiliopuntia brasiliensis, Byrsonima gardneriana, Caesalpinia bracteosa, C. ferrea, C. microphylla, C. pyramidalis, Capparis flexuosa, C. jacobinae, C. yco, Ceiba glaziovii, Cereus albicaulis, C. jamacaru, Chloroleucon acacioides, C. foliolosum, C. mangense, Cnidoscolus bahianus, C. obtusifolius, C. quercifolius, Cochlospermum vitifolium, Combretum leprosum, Commiphora leptophloeos, Cordia leucocephala, Coutarea hexandra, Croton rhamnifolius, C. sonderianus, Dalbergia catingicola, D. cearensis, Erythrina velutina, Erythroxylum revolutum, Eugenia punicifolia, E. tapacumensis, Fraunhoffera multiflora, Geoffroea spinosa, Guapira laxa, Jatropha mollissima, J. mutabilis, Manihot dichotoma, M. glaziovii, Maytenus rigida, Mimosa arenosa, M. caesalpinifolia, M. gemmulata, M. malacocentra, M. tenuiflora, Myracrodruon urundeuva, Parapiptadenia zehntneri, Pilosocereus gounellei, P. pachycladus, P. tuberculatus, Piptadenia obliqua, P. stipulacea, Pithecellobium diversifolium, Pseudobombax simplicifolium, Rollinia leptopetala, Sapium argutum, Schinopsis brasiliensis, Senna acuruensis, S. spectabilis, Senna splendida, Sideroxylon obtusifolium, Spondias tuberosa, Tabebuia impetiginosa, Tacinga inamoena, T. palmadora, Thiloa glaucocarpa, Zanthoxylum stelligerum, Ziziphus joazeiro.*

SDTF Supertramp species (present in ≥100 checklists): *Acacia polyphylla, Acrocomia aculeata, Aegiphila sellowiana, Alibertia concolor, Allophylus edulis, Aloysia virgata, Anadenanthera colubrina, Andira fraxinifolia, Apuleia leiocarpa, Aspidosperma olivaceum, A. pyrifolium, Astronium fraxinifolium, Bauhinia forficata, Bowdichia virgilioides, Brosimum gaudichaudii, Cabralea canjerana, Campomanesia xanthocarpa, Casearia decandra, C. sylvestris, Cecropia pachystachya, Cedrela fissilis, Ceiba speciosa, Celtis iguanaea, C. pubescens, Chrysophyllum gonocarpum, C. marginatum, Copaifera langsdorffii, Cordia trichotoma, Coutarea hexandra, Cupania vernalis, Dalbergia frutescens, Diospyros inconstans, Endlicheria paniculata, Enterolobium contortisiliquum, Eugenia florida, E. punicifolia, E. uniflora, Garcinia gardneriana, Guapira opposita, Guarea guidonia, Guazuma ulmifolia, Guettarda uruguensis, Gymnanthes concolor, Hymenaea courbaril, Inga marginata, I. vera, Lithraea molleoides, Lonchocarpus campestris, Luehea divaricata, L. grandiflora, Machaerium acutifolium, M. hirtum, M. stipitatum, Maclura tinctoria, Maprounea guianensis, Matayba elaeagnoides, M. guianensis, Maytenus ilicifolia, Miconia albicans, Myracrodruon urundeuva, Myrcia guianensis, M. multiflora, M. rostrata, M. tomentosa, Myroxylon peruiferum, Peltophorum dubium, Pera glabrata, Piper amalago, Pisonia zapallo, Platypodium elegans, Prockia crucis, Protium heptaphyllum, Prunus myrtifolia, Pterogyne nitens, Randia nitida, Rollinia emarginata, R. sylvatica, Roupala brasiliensis, Ruprechtia laxiflora, Sapium glandulosum, Schefflera morototoni, Sebastiania brasiliensis, Sideroxylon obtusifolium, Siparuna guianensis, Solanum granuloso-leprosum, Sweetia fruticosa, Syagrus oleracea, S. romanzoffiana, Tabebuia impetiginosa, T. serratifolia, Tapirira guianensis, Terminalia fagifolia, Trema micrantha, Trichilia catigua, T. clausseni, T. elegans, Urera baccifera, Zanthoxylum fagara, Z. petiolare, Z. rhoifolium.*

8 Biogeography of the Forests of the Paraguay-Paraná Basin

*Rodolphe Spichiger, Bastian Bise, Clément Calenge
and Cyrille Chatelain*

CONTENTS

ABSTRACT

The biogeography of the neotropical flora since the last glacial maximum is poorly understood. In this study, two biomes, the Chacoan and the Paranean, were investigated. Three homogenous communities are identified: the chaco seco, the chaco seco with a psammophilous facies (generally characteristic of sandy soils) and the Paranean semideciduous forests. Three ecotonal communities are also defined: the chaco húmedo, the Paranean ecotone intermingled with chaco elements, and the Paranean ecotone with cerrado elements. In order to describe the affinities of these communities with other South American vegetation types, we describe three distribution patterns and gradients between the Chacoan pole, the Colombian pole and the Paranean pole. Integrating Mueller's concept of dispersal centres in a model based on the distribution patterns of selected tree species, we can discuss several hypotheses of floristic history, and especially the proposition that present-day Paranean forest is a remnant of a residual Pleistocenic seasonally dry forest flora.

8.1 INTRODUCTION

The distribution pattern of the present Paraguayan floras is correlated with specific ecological, edaphic and climatic trends (Spichiger et al., 1995). Indeed, gradients in climatic and edaphic moisture availability from the Río Paraná to the Andean piedmont explain the division of the Paraguayan territory into two vegetation areas separated in the centre by the Río Paraguay (Figure 8.1).

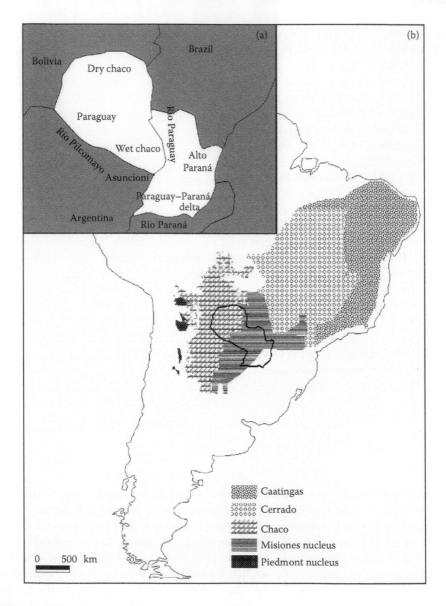

FIGURE 8.1 (a) Map of Paraguay; the main rivers and climatic areas are indicated. (b) Map of South America, showing the main biogeographical areas cited in the text.

The Chacoan vegetation, composed of xeromorphic forests and thickets, extends westwards from the bank of the Río Paraguay to the Bolivian border, growing on alkaline soil with loamy texture. Low annual rainfall (400–1000 mm) characterizes this area (Spichiger et al., 1991). The Paranean semideciduous forest located to the east, between the Río Paraguay and the Río Paraná, grows on acid soil with clay-like texture. A higher annual rainfall level (1500–2000 mm) characterizes this area (Olivera-Filho and Fontes, 2000). Furthermore, a third vegetation type, the cerrado savanna-like vegetation, is found in northern Paraguay. Its soil is acid, like Paranean soil, but its texture is sandy. A low annual rainfall level (750–1250 mm) characterizes this area (Prado, 2003).

The Paraguayan territory can be considered as a huge transition area where various vegetation types, floras and faunas meet (Prado, 1993b). Bernardi (1984) showed that the Paraná basin is connected with the Amazon basin, the Andes and the south of the continent through the large rivers, such as the Río Paraná, the Río Pilcomayo and the Río Paraguay, and some of their tributaries serve as migration routes for modern floras and faunas. Thus he regarded the area as one with a low level of isolation and endemic speciation.

A central issue in biogeography is the identification of variation in vegetation composition in response to climate change. Palaeo-environmental data on the functioning of ecosystems and on the dynamics of the plant communities during the late Quaternary in South America were obtained from palaeoclimatic and vegetation studies (for review, see Spichiger et al., 2004). These studies aim to elucidate the processes that have produced the large vegetation formations encountered in South America today.

Two main hypotheses are commonly advanced, based on the present distribution of plant species and on palynological studies (for a review, see Spichiger et al., 2004). According to the first hypothesis, the Wisconsin period (80,000–10,000 years BP, corresponding to the last major ice age) was characterized by a regression of the forests and a spread of the tropical and subtropical open formations in South America.

The second hypothesis is the residual Pleistocenic seasonally dry forest (RPSD) model (Prado and Gibbs, 1993; Pennington et al., 2000; Prado, 2000). According to this, during the last glacial maximum (LGM) seasonally dry tropical forests with an intermingling of rain forest and montane taxa were confined to the wettest regions, in places now occupied by (semi-) evergreen neotropical forests. The remnants of this once much more extensive mesophilous forest today form a circum-Amazonian arc passing through the Paraguay-Paraná-Uruguay basins (Prado and Gibbs, 1993).

The aims of this chapter are to analyse the floristic communities of the Paraguay-Paraná basin, based on selected tree species; to describe their affinities with the other vegetation types in South America; and to discuss hypotheses on the floristic history of the region by modelling distribution patterns. For that purpose, we try to answer the following questions.

- What do the distributions of selected predominant tree species tell us about the floristic heterogeneity of the Paraguay-Paraná basin?
- What do such distributions tell us about the floristic affinities between the Paraguay-Paraná basin and the other areas of South America?
- What do present-day and modelled distributions tell us about the floristic history of South America in general and the ecotonal status of the Paraguay-Paraná basin in particular?

8.2 FLORISTIC HETEROGENEITY OF THE PARAGUAY-PARANÁ BASIN

8.2.1 METHOD

The Paraguay-Paraná basin is located in the centre of South America, at the confluence of major flora types (Bernardi, 1984). The consequence is a floristic heterogeneity (Spichiger et al., 1995). In this study, we worked on the distribution of 39 common species encountered in Paraguay (Table 8.1), using data from herbarium specimens as our source. We chose these species (which are all trees,

TABLE 8.1
List of the 39 Tree Species Used in the Study (the First 12 Were Used for Construction of the Similarity Distribution Model)

Species Name	Abbreviation[a]	Number of Records			Major Affinity Poles	Polycentric or Monocentric?
		Paraguay	South America	Total		
Anadenanthera colubrina var. *cebil* (Griseb.) Altschul	Ana.col.ceb	8	115	123	Colombian to São Francisco pole	Polycentric
Araucaria angustifolia (Bertol.) Kuntze	—	—	—	418	Paranean pole	Monocentric
Aspidosperma quebracho-blanco Schltdl.	Asp.que	28	0	28	Chacoan pole	Monocentric
Astronium urundeuva (Allemão & M. Allemão) Engl.	Ast.uru	10	56	66	São Francisco to Chacoan pole	Polycentric
Cedrela fissilis Vell.	Ced.fis	9	88	97	Colombian to Paranean pole	Polycentric
Duguetia furfuracea (A. St-Hill.) Benth. & Hook. f.	Dug.fur	0	88	88	São Francisco pole	Monocentric
Geoffroea spinosa Jacq.	Geo.spr	11	41	52	Colombian to São Francisco pole	Polycentric
Maclura tinctoria (L.) D. Don ex Steud.	Mac.tin	20	159	179	Colombian to Paranean pole	Polycentric
Nectandra megapotamica (Spreng.) Mez	Nec.meg	44	47	91	Paranean pole	Monocentric
Phyllostylon rhamnoides (J. Poiss.) Taub.	Phy.rha	0	28	28	São Francisco pole	Monocentric
Schinopsis balansae Engl.	Sch.bal	35	24	59	Chacoan pole	Monocentric
Schinopsis quebracho-colorado (Schltdl.) F.A. Barkley & E. Mey.	Sch.que	26	28	54	Chacoan pole	Monocentric
Acacia praecox Griseb.	Aca.pra	25	0	—	—	—
Astronium fraxinifolium Schott	Ast.fra	9	73	—	—	—
Balfourodendron riedelianum (Engl.) Engl.	Bal.rie	32	15	—	—	—
Bulnesia sarmientoi Griseb.	Bul.sar	9	0	—	—	—
Cabralea canjerana (Vell.) Mart.	Cab.can	0	48	—	—	—
Calycophyllum multiflorum Griseb.	Cal.mul	38	27	—	—	—

(continued)

TABLE 8.1
List of the 39 Tree Species Used in the Study (the First 12 Were Used for Construction of the Similarity Distribution Model) (Continued)

Species Name	Abbreviation[a]	Number of Records			Major Affinity Poles	Polycentric or Monocentric?
		Paraguay	South America	Total		
Caryocar brasiliense Cambess.	Car.bra	0	34	—	—	—
Capparis retusa Griseb.	Cap.ret	37	0	—	—	—
Capparis speciosa Griseb.	Cap.spe	13	0	—	—	—
Cercidium praecox (Ruiz & Pav.) Harms	Cer.pra	16	0	—	—	—
Chrysophyllum gonocarpum (Mart. & Eichler) Engl.	Chr.gon	17	54	—	—	—
Chrysophyllum marginatum (Hook. & Arn.) Radlk.	Chr.mar	36	60	—	—	—
Cochlospermum regium (Schrank) Pilg.	Coc.reg	13	34	—	—	—
Diplokeleba floribunda N.E. Br.	Dip.flo	37	11	—	—	—
Geoffroea decorticans (Hook. & Arn.) Burkart	Geo.dec	9	65	—	—	—
Jacaranda cuspidifolia Mart.	Jac.cus	11	38	—	—	—
Patagonula americana L.	Pat.ame	21	25	—	—	—
Peltophorum dubium (Spreng.) Taub.	Pel.dub	11	40	—	—	—
Pradosia brevipes (Pierre) T. D. Penn.	Pra.bre	0	14	—	—	—
Prosopis alba Griseb.	Pro.alb	11	48	—	—	—
Prosopis nigra (Griseb.) Hieron.	Pro.nig	14	33	—	—	—
Ruprechtia triflora Griseb.	Rup.tri	0	9	—	—	—
Sorocea bonplandii (Baill.) W.C. Burger, Lanj. & Wess. Boer	Sor.bon	27	39	—	—	—
Tabebuia heptaphylla (Vell.) Toledo	Tab.hep	18	38	—	—	—
Tabebuia nodosa (Griseb.) Griseb.	Tab.nod	19	59	—	—	—
Trichilia elegans A. Juss.	Tri.ele	21	49	—	—	—
Xylopia aromatica (Lam.) Mart.	Xyl.aro	0	18	—	—	—

[a] Includes abbreviations used in Figure 8.2.

except for one geoxylic tree (hemixyle) because they are predominant and characteristic of the main Paraguayan habitats. Therefore, the geographical distribution of these species should reflect the spatial variation in floristic composition. In the rest of this chapter, we will use *occurrence* to mean the attested presence of a species at a specified location.

Because our aim is to discriminate these species according to geographical distribution, we used as our spatial discriminant analysis the discriminant analysis of the eigenvectors of the neighbourhood operator (DAENO). This is a multivariate analysis that assigns scores to the occurrences so that the geographical zonation of the area is maximized. The principle of this method is as follows. If n is the number of occurrences registered, one can calculate from the occurrence pattern the $n \times n$ matrix V, named the neighbourhood operator, indicating the neighbouring relationship between the occurrences. At the intersection of the ith row and of the jth column, this matrix contains 1 if the ith occurrence is a neighbour of the jth occurrence, and 0 otherwise. The neighbourhood relationships were here computed by the Delaunay triangulation (Dale, 1999). Then, a matrix S is computed:

$$S = 1/m \, D_n - 1/m \, V,$$

where m is equal to the number of pairs of neighbours (the sum of all values in V), and D_n is a diagonal matrix.

$$D_n = \mathrm{Diag}(V1_n),$$

with 1_n the n vector of 1. Therefore, at the intersection of the row i and of the column i, D_n contains the number of neighbours of the occurrence i. The eigenvectors of S assign scores to each occurrence, so that the score autocorrelation is as high as possible for the study area. These scores can be used to describe the spatial position of each occurrence relative to the others (Thioulouse et al., 1995). Méot et al. (1993) have recommended their use in spatial analyses in place of polynomial functions of geographical coordinates, as they take into account a larger part of the spatial variation. The discriminant analysis of these eigenvectors by the factor species has the following properties (Calenge et al., 2005): the percentage of the spatial variation of the occurrence scores on the first axes explained by the factor species is maximized and the new axes computed by the analysis are uncorrelated.

In other words, two species with a similar spatial distribution have a similar average score, and two species with very different spatial distributions have very different average scores. In this study, we carried out the DAENO at two scales. We first analysed the geographical zonation of the focus species in Paraguay, and then we analysed distributions of the same species at the scale of the whole continent.

8.2.2 RESULTS

Two biomes are clearly separated by the analysis, i.e. the chaco (western Paraguay) and the paraná (eastern Paraguay), defined by their floristic composition (Figure 8.2). From the east to the west, the vegetation does not change continuously but is structured in sharply defined communities. A few species are encountered in both areas (e.g. *Jacaranda cuspidifolia* Mart., *Astronium urundeuva* (Allemão & M. Allemão) Engl. or *Calycophyllum multiflorum* Griseb.), but most species are exclusively chacoan or paranean.

The DAENO of occurrences reveals six vegetation communities in Paraguay, distributed in the two biomes and separated by one ecotone (Figure 8.2). The chaco is composed of three communities (Figure 8.2b): the typical facies of the xeromorphic forests of the chaco at the extreme west (A2); the psammophilous facies (related to sandy soils), with *Schinopsis quebracho-colorado* (Schltdl.) F.A. Barkley & E. Mey, (A1); and the wet chaco on the west side of the Río Paraguay (A0).

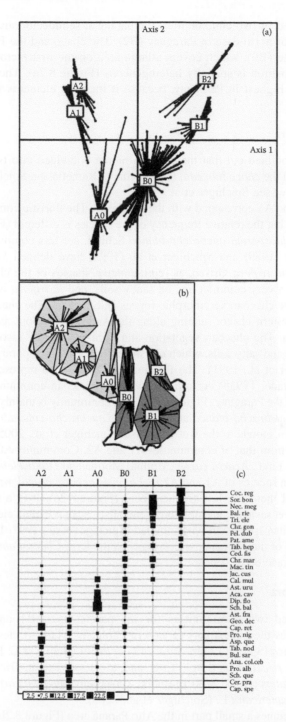

FIGURE 8.2 (a) Typology of the occurrences on the first factorial plane. Six vegetation communities (A0–2, B0–2) with a rather homogeneous floristic composition were visually defined. Each group is identified by a star connecting all the occurrences of the group to its barycentre on the factorial plane. (b) Geographical position of the six vegetation communities defined by discriminant analysis of the eigenvectors of the neighbourhood operator (DAENO). Each community is identified by a star connecting the occurrences to the geographical barycentre of the group. The contour polygon of each group is also displayed (light grey polygons correspond to the Chacoan biome, and dark grey polygons indicate the Paranean biome). (c) Species composition of the six groups of occurrences defined by DAENO (see Table 8.1 for species abbreviations). The importance of a species in a given community is represented by a black square. The square size is proportional to the percentage of the total number of occurrences of the community that is represented by this species.

The Paraná is composed of two communities: the paranean semideciduous forests (B1) and the forest-cerrado mosaic of north-eastern Paraguay (B2). The chaco and the Paraná formations are separated by an ecotone (B0), which covers a large area on the first factorial plane, indicating that the floristic composition is spatially heterogeneous (Figure 8.2a). The floristic diversity of this community is the highest in Paraguay, because it includes elements from the two biomes (Figure 8.2c).

8.2.2.1   Chacoan Flora

Previous studies have pointed out that the chacoan biome is divided into two groups: the *chaco seco*, i.e. dry chaco, and the *chaco húmedo*, i.e. wet chaco (Ramella and Spichiger, 1989; Spichiger et al., 1995; for a review, see Spichiger et al., 2005).

Communities A2 and A1 correspond with the dry chaco. The floristic composition of these two communities is similar, but the relative frequency of the species is different (Figure 8.2c). *Capparis retusa* Griseb. and *Aspidosperma quebracho-blanco* Schltdl. are less common in community A1. Ramella and Spichiger (1989) and Spichiger et al. (1991) have defined *Schinopsis quebracho-colorado* and *Ruprechtia triflora* Griseb. as representative species of the chaco seco. *Schinopsis quebracho-colorado* is a very common tree on sandy soils, and *R. triflora* is a shrub contributing to the understorey of the chacoan xeromorphic forests and thickets. The chaco seco community is centred in the north-western chaco, running along the Andean piedmont and reaching to almost the centre of Argentina. The chacoan vegetation and flora are closely associated with loamy or clay-like loam, alkaline and salty soils, which can suffer from either temporary aridity or temporary waterlogging (Spichiger et al., 1991). The Paraguayan Gran Chaco represented by xeromorphic forests and thickets (Prado, 1993b) corresponds with our chacoan community A2, located at the extreme western part of the Paraguay (Figure 8.2b). This community is mainly made up of *Capparis retusa*, *Aspidosperma quebracho-blanco* and *Schinopsis quebracho-colorado* (Figure 8.2c).

Community A0 corresponds to the wet chaco (see Spichiger et al., 2005). Its species composition is very different from that of communities A1 and A2. Community A0 is mainly composed of *Schinopsis balansae* Engl., *Acacia caven* (Molina) Molina and *Diplokeleba floribunda* N.E. Br., and also of the common species of A1 and A2 (*Schinopsis quebracho-colorado* and *Aspidosperma quebracho-blanco*), but these are less frequent here (Figure 8.2c). It is a transition area where species of dry chaco, wet savanna, mesophilous generalists, scarce Paraná elements and some pan-American or cosmopolitan species are intermingled (Spichiger et al., 1991, 1995, 2005). This area is a southern extension of the Pantanal, which explains the term *Chaco-Pantanal*, which has been used for this region (Prado et al., 1992).

8.2.2.2   Paranean Flora

Our analysis identified two communities for the paraná area: B1 and B2 (Figure 8.2b). Community B1 occupies the largest part of eastern Paraguay. Main species encountered include *Nectandra megapotamica* (Spreng.) Mez, *Sorocea bonplandii* (Baill.) W.C. Burger, Lanj. & Wess. Boer, *Balfourodendron riedelianum* (Engl.) Engl. and, to a lesser extent, *Trichilia elegans* A. Juss., *Chrysophyllum marginatum* (Hook. & Arn.) Radlk., *Chrysophyllum gonocarpum* (Mart. & Eichler) Engl. and *Patagonula americana* L. (Spichiger et al., 2005).

Community B2 occupies a small part in the Alto Paraná area (Figure 8.2b), but it covers a large area on the first factorial plane, indicating that the floristic composition is spatially heterogeneous (Figure 8.2a). Only 15 species compose its floristic diversity (Figure 8.2c). *Cochlospermum regium* (Schrank) Pilg. and *Jacaranda cuspidifolia*, typical species of cerrado (Spichiger et al., 1991), are common, and are intermingled in this area with typical species of Paraná, such as *Nectandra megapotamica*, *Balfourodendron riedelianum* and *Sorocea bonplandii* (Figure 8.2c). The large area on the first factorial plane and the common presence of both cerrado and Paranean species indicate that the community B2 is an ecotone between the cerrado of southern Brazil and the Paranean flora.

8.3 FLORISTIC AFFINITIES OF THE PARAGUAY-PARANÁ BASIN WITH OTHER PARTS OF SOUTH AMERICA

8.3.1 RESULTS

On a wider scale, the continental analysis revealed that the Paraguay basin vegetation is distributed along South American vegetation gradients. Each gradient represents the floristic variation between two extreme points, termed *poles* in the rest of this chapter. Four poles have been identified by the analysis (Spichiger et al., 2004). Three of them are present in Paraguay: the São Francisco, the paranean and the chacoan poles (Figure 8.3). This confirms that Paraguay may be viewed as a huge ecotone in South America, at the intersection of the chacoan, the São Francisco and the paranean poles (Spichiger et al., 2004).

> Every species possesses, or used to possess, at least one dispersal centre that was its centre of origin. During the evolution of the taxon, however, the centre of origin and the centre of dispersal can become widely separated from each other.

(Müller, 1973)

Using Müller's concept of dispersal centres (Müller, 1973; Spichiger et al., 2004), some of our species are monocentric (related to only one pole or dispersal centre), whereas most of them are polycentric, distributed in several poles (Table 8.1).

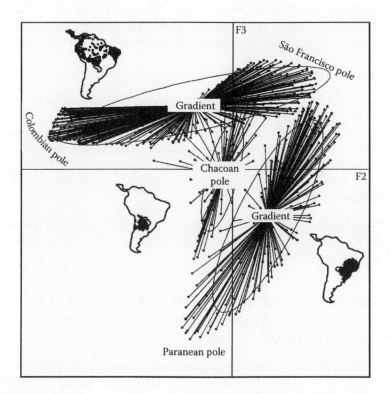

FIGURE 8.3 Factorial map of the occurrence scores on the second and third axes of the discriminant analysis of the eigenvectors of the neighbourhood operator. Categorization of the tree occurrences in four classes according to their position in the three-dimensional space defined by the first three axes: Chacoan, Colombian, São Francisco and Paraná poles. (From Spichiger, R., Calenge, C., and Bise B., *J. Biogeogr.*, 31, 1489, 2004. With permission.)

8.3.1.1 The Chacoan Pole

The chaco is a biogeographical area that has been extensively studied (Ramella and Spichiger, 1989; Spichiger et al., 1991; Prado, 1993a,b; Spichiger et al., 2005). It forms the xeromorphic vegetation of the plains of northern Argentina, western Paraguay and south-eastern Bolivia, and the extreme western edge of Mato Grosso do Sul state in Brazil (Prado, 1993a; Spichiger et al., 2005).

8.3.1.2 The Colombian Pole

The Colombian pole and the Paranean pole have many species in common (Spichiger et al., 2004). This may explain the curved shape of the scatter plot on the factorial plane 2-3 (Figure 8.3). The most characteristic species are *Cedrela fissilis* Vell., *Trichilia elegans*, *Cabralea canjerana* (Vell.) Mart. and *Maclura tinctoria* (L.) D. Don ex Steud., which are widespread in both (semi-) evergreen forests, including gallery forests, and seasonally dry forests. They are mostly generalists, reaching their southern range in the Paraguay basin.

8.3.1.3 The São Francisco Pole

This pole corresponds to the biogeographical caatingas area (Cabrera and Willink, 1973; Prado, 2003). It is located in north-east Brazil, occupying the Rio São Francisco basin. Residual Pleistocenic seasonally dry flora elements (Prado and Gibbs, 1993; Pennington et al., 2000), for example *Astronium urundeuva*, *Anadenanthera colubrina* var. *cebil* (Griseb.) Altschul and *Peltophorum dubium* (Spreng.) Taub., and cerrado-related species, for example *Astronium fraxinifolium* Schott, *Jacaranda cuspidifolia*, *Cochlospermum regium* and *Duguetia furfuracea* (A. St-Hill.) Benth. & Hook. f., are characteristic. The Francisco area has a drier climate than the Paranean and Colombian areas, and is occupied by a mosaic of savannas and gallery forests (Eskuche, 1982).

8.3.1.4 The Paranean Pole

Some monocentric species (*Sorocea bonplandii* and *Nectandra megapotamica*) are only present in the Paraná pole. Moreover, many polycentric elements are strongly related to this dispersal centre: *Balfourodendron riedelianum*, *Chrysophyllum gonocarpum*, *Chrysophyllum marginatum*, *Peltophorum dubium*, *Tabebuia heptaphylla* (Vell.) Toledo, *Astronium urundeuva*, *Xylopia aromatica* (Lam.) Mart., *Peltophorum dubium* and *Patagonula americana* (Spichiger et al., 2004).

8.3.1.5 A Multipolar Species

Finally, *Geoffroea spinosa* Jacq. is related to three poles: Colombian, São Francisco and chacoan (the chaco húmedo). This species is characterized by one of the widest ecological amplitudes among the analysed key species (Spichiger et al., 2004).

8.4 FLORISTIC HISTORY OF SOUTH AMERICA IN GENERAL, AND THE ECOTONAL STATUS OF THE PARAGUAY-PARANÁ BASIN IN PARTICULAR

It is difficult to draw conclusions about past climates from present-day distributions of plant species. The range of a species may change markedly over a 10,000-year period (Webb, 1992), even for species incapable of rapid spread. However, the analysis of present-day distributions of unrelated species may help to evaluate existing hypotheses and formulate new ones (see review in Spichiger et al., 2004). The method is based on the comparison of present species distribution with climatological and palynological data (Markgraf and Bradbury, 1982; Markgraf, 1991; Behling, 1993, 1995; Colinvaux et al., 1996a,b; Behling, 1997a,b; Colinvaux, 1997; Behling, 2002). Several authors have described modelling techniques to investigate the relationships between tree species and their environment (Woodward, 1987; Carpenter et al., 1993; Kutzbach et al., 1998; Guisan and Zimmermann, 2000; Hirzel et al., 2002; Overton et al., 2002; Hirzel and Arlettaz, 2003; Dirnböck and Dullinger, 2004). A model predicting the potential distribution of each species according to environmental and climatic variables would allow testing of the effect of various climatic scenarios. In this study, we have modelled the past and present potential distribution of some species, and we compare these modelled distributions with actual present-day distributions (Spichiger et al., 2004).

8.4.1 DATA COLLECTION

We studied the occurrence of 12 tree species encountered in Paraguay (Table 8.1). Our choice was based on species that are both commonly found in South America and frequently collected. These trees are well represented in the four major 'affinity poles' (the São Francisco, Paranean, Colombian and Chacoan poles). Some are related to only one affinity pole and so are monocentric; others are distributed in several and so are polycentric (Table 8.1). Their distribution data were compiled from the literature (Meyer and Barkley, 1973; Pennington, 1990; Prado, 1991; Totzia, 1992; Golte, 1993; Rohwer, 1993; Ireland and Pennington, 1999; Berg, 2001) and from the Geneva herbarium and the Missouri Botanical Garden (TROPICOS) databases. We digitized the data on a map of South America, using ARCVIEW GIS (Environmental Systems Research Institute, 1996) and processed them with Microsoft ACCESS software.

Some authors (Hugget, 1995; Torres et al., 1997) have demonstrated that tree species distributions in semideciduous forest are influenced by temperature, precipitation and altitude. Six climatic maps were chosen, corresponding to these three factors (Table 8.2). The climatic data, in 0.5° resolution, were sourced from the Intergovernmental Panel on Climate Change (2003), the International Institute for Applied System Analyses (2003) and the International Research Institute for Climate Prediction (2003). They were used to determine the potential distributions of the 12 tree species.

8.4.2 DIFFERENCES BETWEEN PRESENT-DAY AND POTENTIAL DISTRIBUTIONS

8.4.2.1 Methods and Results

The spatial models used in this section are based on the relationship between environmental parameters and the distribution of selected species. They are not based on statistical assumptions and can be applied even to small data-sets. The procedure used here is described by Skov & Borchsenius

---

TABLE 8.2
List of the Six Quantitative Climate Variables Used in the Study

Theme	Unit	Minimum	Maximum	Reference[a]
Mean precipitation per month	Mean/grid (mm)	0	205.5	IPCC
Mean no. of wet days per year	Mean/grid (days)	12.5	260.6	IPCC
Minimum temperature in July	Mean/grid (°C)	−1	25	IIASA climate
No. of frost days in July	No. of frost days/grid (days)	0	30	IRI
Mean yearly temperature	Mean/grid (°C)	−27	30	IIASA climate
Altitude	Maximum/grid (m)	1	8000	GTOPO30

[a] IPCC, Intergovernmental Panel on Climate Change; IIASA, International Institute for Applied Systems Analyses; IRI, International Research Institute for Climate Prediction; GTOPO30, US Geological Survey Global Digital Elevation Data at 30 Arc Second Resolution.

(1997) and Skov (1999). Distribution modelling is a one-step procedure in which a potential map is constructed based on climatic parameters. The application developed by Skov (2000) for ARC VIEW SPATIAL ANALYST (Environmental Systems Research Institute, 1996) produces similarity distribution maps, which give potential distributions based on point to point similarity (Carpenter et al., 1993). These distribution models use a point to point similarity metric to quantify the similarity between two sites (Carpenter et al., 1993). This point to point similarity metric is continuous and varies between 1, when the environment at a given site corresponds completely to a known locality, and 0, when the environment at a given site does not correspond to a known locality. It is worth noting that the point to point similarity metric values do not represent probability (Skov, 2000). A value of 0.90 indicates that the environmental conditions deviate less than 10% from the known range but do not imply a 90% likelihood of finding the species at a given site.

The continuous surface of the potential distribution was converted to a binary potential distribution map by choosing a suitable cut-off level. In our study, only the cells with a point to point similarity metric superior to 0.95 were considered.

Between potential and present-day distributions, two situations occurred: one of equilibrium and one of disequilibrium. The former was met when potential and present-day distributions match almost perfectly. In contrast, the latter occurred when the present-day distribution represents a subset of the potential distribution. This situation of disequilibrium could be explained by an under-collection of the species, but this is unlikely, as we selected common species that are frequently collected. We consider that palaeoclimatic factors could better explain this discrepancy.

A situation of equilibrium means a low ecological amplitude for the studied species. Indeed, all the six climatic factors used in potential distribution restrict this species to its present-day distribution. On the other hand, a situation of disequilibrium means a wide ecological amplitude for the studied species. In fact, no climatic factors restrict this species to its present-day distribution; it potentially has the capacity to grow in other places.

8.4.2.1.1 Monocentric Species in Equilibrium Situation

The present-day distribution of *Araucaria angustifolia* (Bertol.) Kuntze and *Nectandra megapotamica*, representative of the Paranean flora, corresponds to their potential distribution, as is the case for *Schinopsis balansae* and *S. quebracho-colorado*, representative of Chacoan flora (Figure 8.4). Concerning the actual and potential distribution of *Schinopsis* species, *S. balansae* is found eastwards and *S. quebracho-colorado* westwards. This confirms that *S. balansae* is a predominant tree of the chaco húmedo on temporarily waterlogged Chacoan soils, but is also found on other sandier and drier substrates (Spichiger et al., 1995), and that *S. quebracho-colorado* is representative of the chaco seco, with its potential distribution reaching almost to the centre of Argentina.

8.4.2.1.2 Polycentric Species in Disequilibrium Situation

Four species are polycentric (Figure 8.5). They show a potential continuous neotropical distribution (*Cedrela fissilis* and *Maclura tinctoria*) or a potential discontinuous circum-Amazonian distribution (*Astronium urundeuva* and *Geoffroea spinosa*). All of them show a potential distribution much wider than their present-day distribution. With their potential continuous distribution, *Cedrela fissilis* and *Maclura tinctoria* are defined as generalist, because they could grow everywhere in the neotropical region.

8.4.2.2 Discussion

All our monocentric species are in a situation of equilibrium and show a low ecological amplitude, which explains their restricted distribution. Zoologists have explained the distribution of some birds and butterflies by a refuge of Quaternary forest in the Paraná basin (Brown, 1982; Cracraft, 1985). The present-day distribution of *Araucaria angustifolia* and *Nectandra megapotamica* could

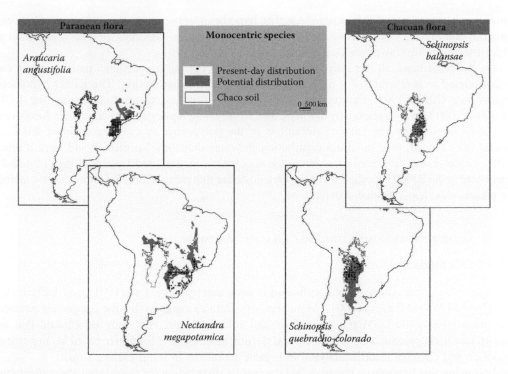

FIGURE 8.4 Potential and present-day distributions of two Paranean and two Chacoan species in an equilibrium situation.

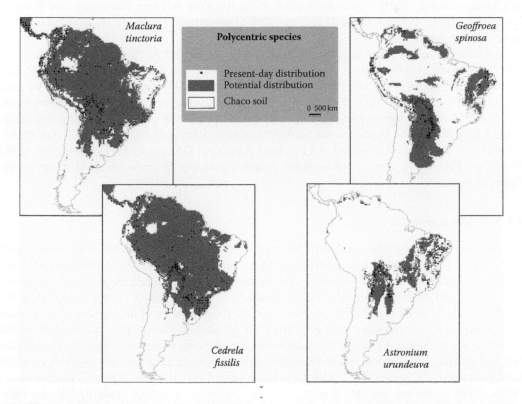

FIGURE 8.5 Potential and present-day distributions of four polycentric species in a disequilibrium situation.

represent the remnants of this refuge. Another hypothesis agrees with the congruence between potential and present-day distribution for the Chacoan species: the Chacoan area is accepted as a Pleistocenic refuge, as postulated by Iriondo and Garcia (1993).

On the other hand, all our polycentric species have a present-day disjunct pattern, which can be considered as a remnant of a once much more expanded distribution. The RPSD hypothesis (Prado and Gibbs, 1993; Pennington et al., 2000) — which states that during the LGM (25,000–15,000 Myr BP) a seasonally dry flora, and not savannas, replaced (semi-)evergreen Amazonian forests — implies that the palaeodistributions of the polycentric species are different from the present one. Indeed, their potential distribution indicates that these species should have a much wider present-day distribution, considering the six climatic factors used for modelling (Table 8.2). According to the RPSD hypothesis, we can then postulate that palaeoclimatic factors explain mainly the present-day, less-extended distributions.

8.4.3 DISTRIBUTIONS DURING THE LAST GLACIAL MAXIMUM

8.4.3.1 Methods and Results

Pleistocenic drier periods have been postulated in particular by Absy et al. (1991), van der Hammen and Absy (1994) and Behling (2001). More precisely, Behling suggests that the completely different vegetation type at the LGM reflects a drier and colder climate. His study established that the temperature was approximately 5°C lower and the precipitation 30% lower. He based his hypothesis on pollen and charcoal records analysed in organic sediments in south-eastern Brazil.

Following this hypothesis, we modelled the species distribution for the rainfall and temperature data during the LGM. Our first step was to create climatic layers on the basis of present-day data — mean precipitation per month and mean minimum temperature from the Intergovernmental Panel on Climate Change (2003) — using a request in the ArcVIEW software (Environmental Systems Research Institute, 1996). A range of 5°C and 30% of precipitation were cut off from each cell of these climatic layers.

A similarity model (Skov and Borchsenius, 1997) was applied to the new layers. The use of this rather crude model indicated three regions (Figure 8.6). The first corresponds with the past potential distributions of the polycentric species (*Astronium urundeuva*, *Cedrela fissilis*, *Duguetia furfuracea*, *Geoffroea spinosa* and *Maclura tinctoria*) and the Paranean species (*Araucaria angustifolia* and *Nectandra megapotamica*). The second is a compilation of past potential distributions of the Chacoan species (*Aspidosperma quebracho-blanco*, *Schinopsis balansae* and *S. quebracho-colorado*). A third region illustrates the Paranean hypothesis (Markgraf, 1991; Pennington et al., 2000), i.e. Patagonian steppic and montane elements invading the area occupied today by the Paranean forest.

8.4.3.2 Discussion

Our modelled patterns corroborate the RPSD flora hypothesis (Prado and Gibbs, 1993; Pennington et al., 2000), which suggests that generalist elements (e.g. *Cedrela fissilis* and *Maclura tinctoria*), cerrado elements (e.g. *Astronium urundeuva* and *Duguetia furfuracea*) and Paranean species (*Araucaria angustifolia* and *Nectandra megapotamica*) invaded the Amazon basin during the LGM at the expense of the Amazonian elements. To understand the limiting factor of the monocentric Paranean species invasion, we modelled their past potential distribution based only on temperature and withdrawing one degree at a time. Their distribution pattern moves northwards when the temperature falls 3°C. We may also suppose that the Paraná basin was occupied by steppic and montane elements during a period 3°C cooler than today (see also Ledru, 1993; Behling, 1995; Colinvaux et al., 1996a, 2000; Colinvaux and De Oliveira, 2001). Furthermore, a floristic study by Oliveira-Filho

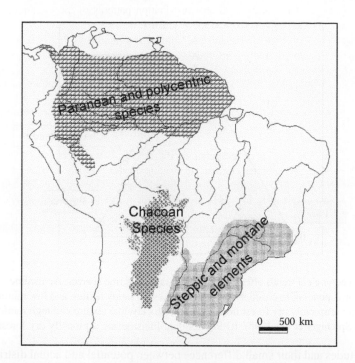

FIGURE 8.6 Past (last glacial maximum) potential distributions of polycentric, Paranean and Chacoan species, illustrated with Patagonian steppic and montane elements (Markgraf, 1991; Pennington et al., 2000). The Amazon basin is invaded by Paranean and polycentric species, and the Paraná basin by steppic and montane elements.

and Fontes (2000) shows that montane taxa (*Clethra, Clusia, Drimys, Hedyosmum, Podocarpus, Prunus* and *Weinmannia*) are present today in the Paraná area.

The past potential distributions of Chacoan species (*Aspidosperma quebracho-blanco, Schinopsis balansae* and *S. quebracho-colorado*) show a slight spread northwards. Indeed, these species are drought- and frost-tolerant (Pennington et al., 2000), and so are scarcely sensitive to LGM climatic changes. Considering their adaptation to salty soils (Spichiger et al., 1991), they could have remained in the north of the Chaco during the LGM, where they are still present today.

CONCLUSION

The distribution patterns of the present Paraguayan floras as explained in the introduction are correlated with specific ecological, edaphic and climatic trends (Spichiger et al., 1995), which are summarized in Figure 8.7. The triangle represents the Paraguay-Paraná basin, which is a huge ecotone where São Francisco, peri-Amazonian generalist, Chacoan and Paranean species compete. The A pole is related to alkaline, salty, loamy soils and low rainfall (chaco seco); the B pole to acid, sandy soils and also low rainfall (RPSD flora); and the C pole to clay-like or sandy–clay-like, acid soils and high rainfall (Paraná). The chaco húmedo is found in the middle of the axis formed by poles A–C.

The monocentric species (*Araucaria angustifolia, Aspidosperma quebracho-blanco, Nectandra megapotamica, Schinopsis balansae* and *S. quebracho-colorado*) are present along the axis from pole A to pole C. They are characteristic of Chacoan and Paranean poles. Because of their narrow

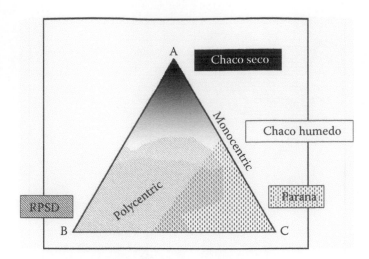

FIGURE 8.7 Main ecological trends affecting present Paraguayan tree species. A, alkaline soil, loamy texture and low rainfall, corresponding to chaco seco flora. B, acid soil, sandy texture and low rainfall, corresponding residual Pleistocenic seasonally dry forest flora. C, acid soil, clay-like texture and high rainfall, corresponding to Paranean flora (Spichiger et al., 1995). RPSD, residual Pleistocenic seasonally dry forest.

ecological amplitudes and their small differences between potential and actual distribution patterns, they can be considered as endemic to the region, or at least as the oldest components of the Paraguay-Paraná ecotone. Salty soil and high rainfall are the two factors that probably best explain the past and present-day distribution of, respectively, Chacoan and Paranean species.

The central position of the polycentric species in the triangle (Figure 8.7) reflects their opportunist behaviour resulting from their wide ecological plasticity. The modelled distributions, and ecological amplitudes, of the polycentric generalist species growing in the Paraguay-Paraná ecotone (*Cedrela fissilis* and *Maclura tinctoria*) show that they should be able to colonize a wider range, corroborating the RPDS flora hypothesis of Pennington and Prado (Pennington et al., 2000). Furthermore, according to the hypotheses of drier climate during the LGM (Absy et al., 1991; van der Hammen and Absy, 1994; Behling, 2001), the low rainfall of pole B (Spichiger et al., 1995) agrees with the spread of the characteristic species of RPSD flora that we studied (*Astronium urundeuva* and *Geoffroea spinosa*).

During the LGM, the Paraná flora, perhaps originating from the Atlantic area, may have invaded the Amazon basin through the São Francisco pole and the gallery forests of the cerrado. Furthermore, the South Amazonian arch pattern of distribution (e.g. for *Anadenanthera colubrina* var. *cebil*) shows that bridges exist, or have existed, between the Paraná basin and the so-called Tucuman-Bolivian forests of the piedmont area. We can infer that extrazonal semideciduous forests in the Chaco (gallery forests and montane forests) connect and have connected the Andean piedmont with the Paraná basin. These connections reinforce the hypothesis of Oliveira-Filho and Ratter (1995), which supports the spread of Paranean semideciduous forests by penetration into the cerrado province via the basalt-derived fertile soils of western Minas Gerais and southern Goías. The ecotone B2 (Figure 8.2a) between Paranean elements and typical species of cerrado agree with the results of Oliveira-Filho and Fontes (2000), which shows that Paranean flora and cerrado have strong links. These authors indicate that cerrados share a much larger proportion (55% of their total species) with the Atlantic semideciduous forests than with the Amazonian forests (20%). Furthermore, a study has demonstrated the coexistence of montane and Paranean forest elements within a part of the present-day cerrado region between 17,000 and 13,000 years BP (Ledru, 1993). All these lines of evidence confirm our model, i.e. that the Paranean flora invaded the Amazonian basin at the LGM through the cerrado (Figure 8.6).

COLOUR FIGURE 1.1 Neotropical seasonally dry forests and savannas: (a) Inter-Andean SDTF in the Marañon valley, Cajamarca, Peru. The big trees with pale trunks are *Eriotheca peruviana* A. Robyns, and most of the shrubs in the foreground are *Croton* sp. (photo: R. Linares-Palomino). (b) SDTF in Oaxaca, Mexico, with abundant cacti (photo: C. Pendry; reproduced from Pennington et al., 2004). (c,d) The same area of SDTF at Sagarana, Minas Gerais, Brazil, in the wet and dry seasons, illustrating complete deciduousness in the dry season (photos: J. Ratter). (e) Savanna (cerrado) with a scattering of low shrubs and occasional small trees (*campo sujo* = dirty field) in Distrito Federal, Brazil (photo: J. Ratter). (f) Typical savanna (cerrado) vegetation in Mato Grosso, Brazil (photo: S. Bridgewater).

(e)

(f)

COLOUR FIGURE 1.1 (Continued).

Geomorphological Landscapes

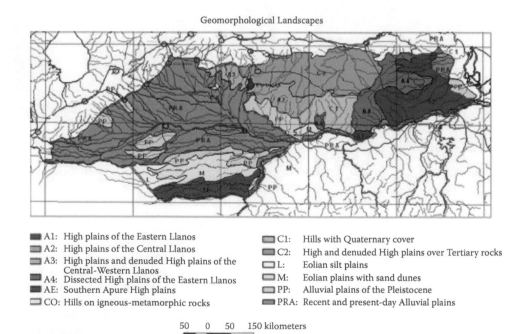

- ■ A1: High plains of the Eastern Llanos
- ▨ A2: High plains of the Central Llanos
- ▨ A3: High plains and denuded High plains of the Central-Western Llanos
- ■ A4: Dissected High plains of the Eastern Llanos
- ■ AE: Southern Apure High plains
- ▢ CO: Hills on igneous-metamorphic rocks

- ▢ C1: Hills with Quaternary cover
- ▢ C2: High and denuded High plains over Tertiary rocks
- ▢ L: Eolian silt plains
- ▢ M: Eolian plains with sand dunes
- ▢ PP: Alluvial plains of the Pleistocene
- ▭ PRA: Recent and present-day Alluvial plains

50   0   50   150 kilometers

COLOUR FIGURE 5.2 Geomorphological subdivision of the Llanos region in Venezuela (map by R. Schargel).

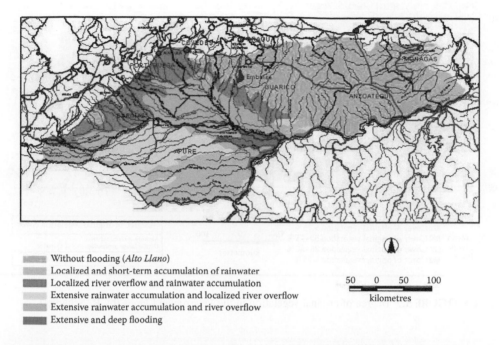

Without flooding (*Alto Llano*)
Localized and short-term accumulation of rainwater
Localized river overflow and rainwater accumulation
Extensive rainwater accumulation and localized river overflow
Extensive rainwater accumulation and river overflow
Extensive and deep flooding

50   0   50   100
kilometres

COLOUR FIGURE 5.5 Flooding regimes in the Venezuelan Llanos (by R. Schargel and J.G. Quintero).

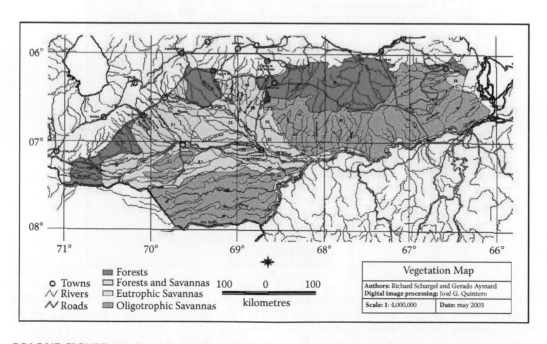

o Towns
∧ Rivers
∧ Roads

Forests
Forests and Savannas
Eutrophic Savannas
Oligotrophic Savannas

100   0   100
kilometres

Vegetation Map	
Authors: Richard Schargel and Gerado Aymard	
Digital image processing: José G. Quintero	
Scale: 1: 4,000,000	**Date:** may 2003

COLOUR FIGURE 5.6 General vegetation map of the Venezuelan Llanos (by R. Schargel and G. Aymard).

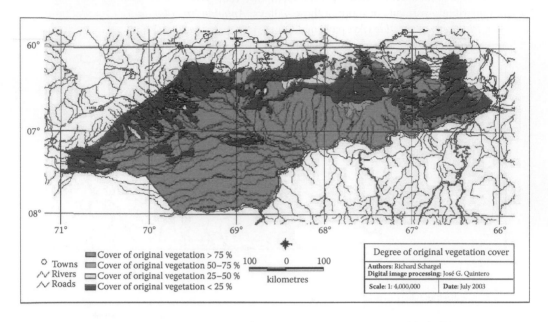

COLOUR FIGURE 5.7 Degree of original vegetation in the Venezuelan Llanos (by R. Schargel).

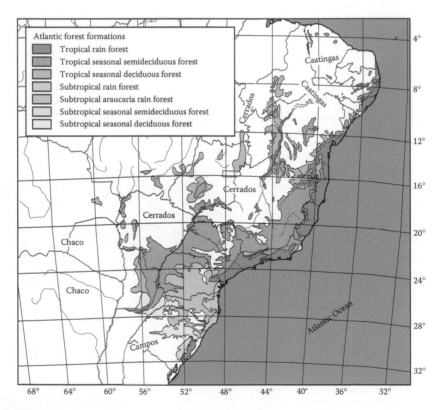

COLOUR FIGURE 7.1 Map of eastern South America showing the distribution of the predominant vegetation formations of the South American Atlantic forest domain. Caatingas, cerrados, chaco and campos are the adjacent domains that make up the "diagonal of open formations".

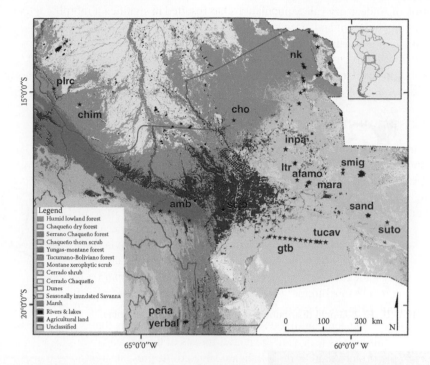

COLOUR FIGURE 9.2 Land cover of eastern Bolivia showing the position of the plot localities in relation to the major vegetation types in the region, which also correspond roughly to existing ecoregional classifications. Land-use change (red) is current to 2002; plot codes as in Table 9.1.

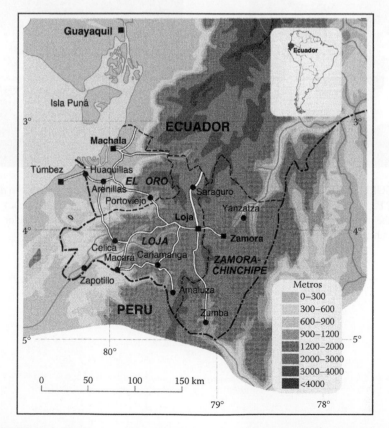

COLOUR FIGURE 12.1 The three southern provinces of Ecuador, El Oro, Loja and Zamora Chinchipe and the location of Ecuador in South America. Reproduced from Botánica Austroecuatoriana with permission from H. Balslev.

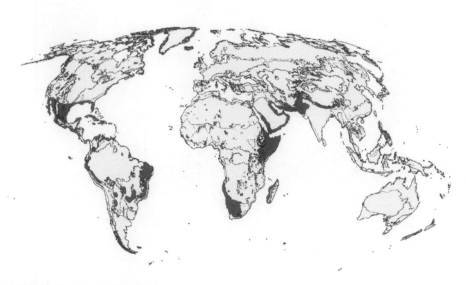

COLOUR FIGURE 12.2 Global distribution of the legume Succulent biome.

COLOUR FIGURE 12.6 *Clitoria brachystegia* Benth. (Leguminosae: Papilionoideae): a narrowly restricted endemic from the bosque petrificado at Puyango on the Loja-El Oro boundary. (Photo: G.P. Lewis.)

COLOUR FIGURE 12.8 *Calliandra tumbeziana* J.F. Macbr. (Leguminosae: Mimosoideae) from the dry spiny forests of south-west Ecuador. (Photo: G.P. Lewis.)

COLOUR FIGURE 12.10 *Pithecellobium excelsum* (Kunth) Mart. ex Benth. (Leguminosae: Mimosoideae): restricted to the SDTF of Ecuador and northern Peru. (Photo: G.P. Lewis.)

COLOUR FIGURE 12.11 *Machaerium millei* Standl. (Leguminosae: Papilionoideae) restricted to southern Ecuador and northern Peru. (Photo: G.P. Lewis.)

COLOUR FIGURE 12.12 *Caesalpinia cassioides* Willd. (Leguminosae: Caesalpinioideae) from the inter-Andean valleys of southern Colombia, southern Ecuador and northern Peru. (Photo: G.P. Lewis.)

COLOUR FIGURE 12.14 *Zapoteca caracasana* (Jacq.) H. Hern. subsp. *weberbaueri* (Harms) H. Hern. (Leguminosae: Mimosoideae): from SDTF in Colombia, Ecuador and Peru. (Photo: G.P. Lewis.)

Semideciduous forest (dry and dry-moist)
Evergreen and seasonal evergreen forests (moist, wet and rain)
Forested wetlands
Emergent wetlands
Pasture and agriculture
Urban and barren
Sand and rock
Water

COLOUR FIGURE 15.3 Distribution of woody vegetation formations typical of dry and dry-to-moist forest formations of Puerto Rico (Helmer et al., 2002). Dry forest formations include woodlands, shrublands and forests that are drought deciduous or semi-deciduous. Unlike most other maps of Puerto Rican forest formations, this map recognizes that dry forest types dominate karst substrate along the north-western coast of the island, which receives less rainfall than inland sites. Based on field observation, Helmer et al. (2002) delineated this area with maps of total annual precipitation and geology, assuming that over karst substrate, where annual rainfall is less than 1500 mm, drier forest formations dominate on limestone hills.

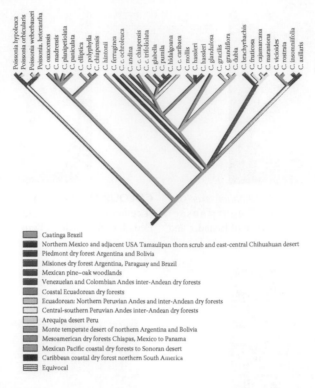

COLOUR FIGURE 19.4 Taxon-area cladogram of *Coursetia*, showing the 14 areas assigned to terminal taxa and that are optimized on a tree topology derived from the Bayesian consensus phylogeny. Duplicate accessions of species from the same geographical setting were omitted from this analysis. Narrow areas of endemism were assigned to the terminal taxa, and much geographical structure was thus lost. For example, the genus *Poissonia* (the outgroup) is confined to the southern Andes, even though each of the four species of this genus is coded for a separate area of endemism. Similarly, the two lineages of *Coursetia hassleri* Chodat occur in adjacent dry forest patches, the Piedmont and Misiones, in South America.

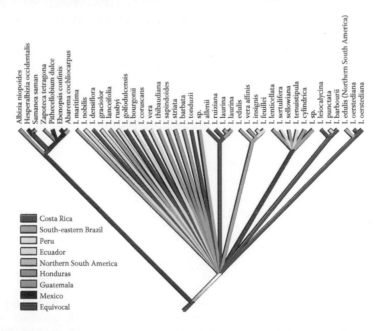

COLOUR FIGURE 19.5 Taxon-area cladogram of *Inga*, showing the eight areas assigned to terminal taxa and that are optimized on a tree topology derived from the Bayesian consensus phylogeny. Duplicate accessions of species from the same geographical setting were included in this analysis. Also, broad geographical areas were assigned to each terminal taxon, which represents the general region from which a particular species was sampled. This area assignment in *Inga* was intentional so as to bias in favour of geographical phylogenetic structure.

REFERENCES

Absy, M.L. et al., Mise en évidence de quatre phases d'ouverture de la forêt dense dans le sud-est de l'Amazonie au cours des 60000 dernières années. Première comparaison avec d'autres régions tropicales, *Compt. Rend. Acad. Sci. Paris, Sér. Gén. Vie Sci.*, 312, 673, 1991.

Behling, H., Untersuchungen zur spätpleistozänen und holozänen Vegetations- und Klimageschichte der tropischen Küstenwälder und der Araukarienwälder in Santa Catarina (Südbrasilien), *Diss. Bot.*, 206, 1, 1993.

Behling, H., A high resolution Holocene pollen record from Lago do Pires, SE Brazil: vegetation, climate and fire history, *J. Paleolimnol.*, 14, 253, 1995.

Behling, H., Late Quaternary vegetation, climate and fire history from the tropical mountain region of Morro de Itapeva, SE Brazil, *Palaeogeogr. Palaeoclimatol. Palaeoecol.*, 129, 407, 1997a.

Behling, H., Late Quaternary vegetation, climate and fire history in the Araucaria forest and campos region from Serra Campos Gerais (Paraná), S Brazil, *Rev. Paleobot. Palynol.*, 97, 109, 1997b.

Behling, H., South and southeast Brazilian grassland during late Quaternary times: a synthesis, *Palaeogeogr. Palaeoclimatol. Palaeoecol.*, 177, 19, 2001.

Behling, H., Carbon storage increases by major forest ecosystems in tropical south America since the Last Glacial Maximum and the early Holocene, *Glob. Planet. Change*, 33, 107, 2002.

Berg, C.C., Moreae, Artocarpeae, and Dorstenia (Moraceae), *Fl. Neotrop.*, 83, 346, 2001.

Bernardi, L., Contribución a la dendrología Paraguaya. Primera parte. Apocynaceae – Bombacaeceae – Euphorbiaceae – Flacourtiaceae – Mimosoideae – Caesalpinioideae – Papilionatae, *Boissiera*, 35, 341, 1984.

Brown, K.S., Paleoecology and regional patterns of evolution in forest butterflies, in *Biological Diversification in the Tropics*, Prance, G.T., Ed, Columbia University Press, New York, 1982, 255.

Cabrera, A.L. and Willink, A., *Biogeogragrafía de América Latina*, Secretaria General OEA, Washington, 1973.

Calenge, C. et al. Discriminant analysis of the spatial distribution of plant species occurrences:1. Theoretical aspects, *Candollea* 60, 563, 2005.

Carpenter, G., Gillison, A.N., and Winter, J., DOMAIN: a flexible modelling procedure for mapping potential distributions of plants and animals, *Biodivers. Conservation*, 2, 667, 1993.

Colinvaux, P.A., *The Ice-Age Amazon and the Problem of Diversity*, NOW–Huygenslezing, The Hague, 1997.

Colinvaux, P.A. and De Oliveira, P.E, Amazon plant diversity and climate through the Cenozoic, *Palaeogeogr. Palaeoclimatol. Palaeoecol.*, 166, 51, 2001.

Colinvaux, P.A. et al., A long pollen record from lowland Amazonia: forest and cooling in glacial times, *Science*, 274, 85, 1996a.

Colinvaux, P.A. et al., Temperature depression in the lowland tropics in glacial times, *Clim. Change*, 32, 19, 1996b.

Colinvaux, P.A., De Oliveira, P.E., and Bush, M.B., Amazonian and neotropical plant communities on glacial time-scales: the failure of the aridity and refuge hypotheses, *Quat. Sci. Rev.*, 19, 141, 2000.

Cracraft, J., Historical biogeography and patterns of differentiation within the South American avifauna: areas of endemism, in *Neotropical Ornithology, Ornithological Monographs, 36*, Buckley, P.A. et al., Eds, American Ornithologists' Union, Washington, 1985, 49.

Dale, M.R.T., *Spatial Pattern Analysis in Plant Ecology*, Cambridge University Press, Cambridge, 1999.

Dirnböck, T. and Dullinger, S., Habitat distribution models, spatial autocorrelation, functional traits and dispersal capacity of alpine plant species, *J. Veg. Sci.*, 15, 77, 2004.

Environmental Systems Research Institute, *Using ArcVIEW GIS. The Geographic Information System for Everyone*, ESRI, Redlands, 1996.

Eskuche, U., Struktur und Wirkungsgefüge eines subtropischen Waldes Südamerikas, in *Bericht über da Internationale Symposium der Internationalen Vereinigung Vegetationskunde, Struktur und Dynamik von Wäldern*, Dierschke, H., Ed, J. Cramer, Vaduz, 1982, 49.

Golte, W., *Araucaria: Verbreitung und Standortansprüche einer Coniferengattung in Vergleichender sicht*, Franz Steiner, Stuttgart, 1993, 20.

Guisan, A. and Zimmermann, N.E., Predictive habitat distribution models in ecology, *Ecol. Modelling*, 135, 147, 2000.

van der Hammen, T. and Absy, M.L., Amazonia during the last glacial, *Palaeogeogr. Palaeoclimatol. Palaeoecol.*, 109, 247, 1994.

Hirzel, A.H. and Arlettaz, R., Modeling habitat suitability for complex species distributions by environmental-distance geometric, *Environm. Managem.*, 32, 614, 2003.

Hirzel, A.H. et al., Ecological-niche factor analysis: how to compute habitat-suitability maps without absence data? *Ecology*, 83, 2027, 2002.

Hugget, R.J., *Geoecology, an Evolutionary Approach*, Routledge, London, 1995.

Intergovernmental Panel on Climate Change, *The IPCC Data Distribution Centre*, http://ipcc-ddc.cru.uea.ac.uk/ 2003.

International Institute for Applied System Analyses, *IIASA Science for Global Insight*, http://www.iiasa.ac.at/, 2003.

International Research Institute for Climate Prediction, *IRI Linking Science to Society*, http://iri.columbia.edu/, 2003.

Ireland, H. and Pennington, R.T., A revision of *Geoffroea* (Leguminosae–Papilionoideae), *Edinburgh J. Bot.*, 56, 329, 1999.

Iriondo, M.H. and Garcia, N.O., Climatic variations in the Argentine plains during the last 18,000 years, *Palaeogeogr. Palaeoclimatol. Palaeoecol.*, 101, 209, 1993.

Kutzbach, J. et al., Climate and biome simulations for the past 21,000 years, *Quat. Sci. Rev.*, 17, 473, 1998.

Ledru, M.P., Late Quaternary environmental and climatic changes in Central Brazil, *Quat. Res.*, 39, 90, 1993.

Markgraf, V., Younger *Dryas* in southern South America? *Boreas (Oslo)*, 20, 63, 1991.

Markgraf, V. and Bradbury, J.P., Holocene climatic history of South America, *Striae*, 16, 40, 1982.

Méot, A., Chessel, D., and Sabatier, R., Opérateurs de voisinage et analyse des données spatio-temporelles, in *Biométrie et Environnement*, Lebreton, J.D. and Asselain, B., Eds, Masson, Paris, 1993, 45.

Meyer, T. and Barkley, F.A., Revision del genero *Schinopsis* (Anacardiaceae), *Lilloa*, 33, 207, 1973.

Müller, P., *The Dispersal Centres of Terrestrial Vertebrates in the Neotropical Realm: A Study in the Evolution of the Neotropical Biota and its Native Landscapes*, Junk, The Hague, 1973.

Oliveira-Filho, A.T. and Fontes, A.L., Patterns of floristic differentiation among Atlantic forests in Southeastern Brazil and the influence of climate, *Biotropica*, 32, 793, 2000.

Oliveira-Filho, A.T. and Ratter, J.A., A study of the origin of central Brazilian forests by the analysis of plant species distribution patterns, *Edinburgh J. Bot.*, 52, 141, 1995.

Overton, J.M. et al., Information pyramids for informed ecosystem management, biodiversity and conservation, *Biodivers. Conservation*, 11, 2093, 2002.

Pennington, R.T., Prado, D.E., and Pendry C.A., Neotropical seasonally dry forests and Quaternary vegetation changes, *J. Biogeogr.*, 27, 261, 2000.

Pennington, T.D., Sapotaceae, *Fl. Neotrop.*, 52, 770, 1990.

Prado, D.E., *A critical evaluation of the floristic links between chaco and caatingas vegetation in South America*, PhD thesis, University of St Andrews, St Andrews, 1991.

Prado, D.E., What is the Gran Chaco vegetation in South America? I. A review. Contribution to the study of the flora and vegetation of the Chaco, *Candollea*, 48, 145, 1993a.

Prado, D.E., What is the Gran Chaco vegetation in South America? II. A redefinition. Contribution to the study of the flora and vegetation of the Chaco, *Candollea*, 48, 615, 1993b.

Prado, D.E., Seasonally dry forests of tropical South America: from forgotten ecosystems to a new phytogeographic unit, *Edinburgh J. Bot.*, 57, 437, 2000.

Prado, D.E., As caatingas da Americá do Sul, in *Ecologia e Conservação da Caatinga*, Leal, I.R., Tabarelli, M., and Cardoso da Silva, J.M., Eds, Universidade Federal de Pernambuco, Recife, 2003, 3.

Prado, D.E. and Gibbs, P.E., Patterns of species distributions in the dry seasonal forests of South America, *Ann. Missouri Bot. Gard.*, 80, 902, 1993.

Prado, D.E. et al., The Chaco–Pantanal transition in southern Mato Grosso, Brazil, in *Nature and Dynamics of Forest–Savanna Boundaries*, Furley, P.A., Proctor, J., and Ratter, J.A., Eds, Chapman and Hall, London, 1992, 451.

Ramella, R. and Spichiger, R., Interpretación preliminar del medio físico y de la vegetación del Chaco boreal. Contribución al estudio de la flora y de la vegetación del Chaco, *Candollea*, 44, 639, 1989.

Rohwer, J.G., Lauraceae: Nectandra, *Fl. Neotrop.*, 60, 332, 1993.

Skov, F., Spatial modelling for tropical biodiversity assessment, in *Scan GIS'99 – Proceedings from the 7th Scandinavian Research Conference on Geographical Information Science*, Stubkjaer E. and Hansen, H.S., Eds, Aalborg Universitetsforlag, Aalborg, 1999, 203.

Skov, F., Potential plant distribution mapping based on climatic similarity, *Taxon*, 49, 503, 2000.

Skov, F. and Borchsenius, F., Predicting plant species distribution patterns using simple climatic parameters: a case study of Ecuadorian palms, *Ecography*, 20, 347, 1997.

Spichiger, R. et al., Proposición de leyenda para la cartografía de las formaciones vegetales del Chaco paraguayo. Contribución al estudio de la flora y de la vegetación del Chaco, *Candollea*, 46, 541, 1991.

Spichiger, R. et al., Origin, affinities and diversity hot spots of the Paraguayan dendrofloras, *Candollea*, 50, 515, 1995.

Spichiger, R., Calenge, C., and Bise B., The geographical zonation in the Neotropics of three species characteristic of the Paraguay–Paraná Basin. *J. Biogeogr.*, 31, 1489, 2004.

Spichiger, R., Calenge, C., and Bise, B., Discriminant analysis of the spatial distribution of plant species occurrences: 2. Distribution of major tree communities in Paraguay, *Candollea* 60, 577, 2005.

Thioulouse, J., Chessel, D., and Champely. S., Multivariate analysis of spatial patterns: a unified approach to local and global structures, *Environm. Ecol. Stat.*, 2, 1, 1995

Torres, R.B., Martins, F.R., and Gouvea, L.S.K., Climate, soil, and tree flora relationships in forests in the state of São Paulo, southeastern Brazil, *Rev. Brasil. Bot.*, 20, 41, 1997.

Totzia, C.A., A re-evalution of the genus *Phyllostyllon* (Ulmaceae), *Sida*, 15, 264, 1992.

Webb, T. III., Past changes in vegetation and climate: lessons for the future, in *Global Warming and Biological Diversity*, Peters, R.L. and Lovejoy, T.E., Eds, Yale University Press, New Haven, 1992, 59.

Woodward, F.I., *Climate and Plant Distribution*, Cambridge University Press, Cambridge, 1987.

9 The Chiquitano Dry Forest, the Transition between Humid and Dry Forest in Eastern Lowland Bolivia

Timothy J. Killeen, Ezequial Chavez, Marielos Peña-Claros, Marisol Toledo, Luzmila Arroyo, Judith Caballero, Lisete Correa, René Guillén, Roberto Quevedo, Mario Saldias, Liliana Soria, Ynés Uslar, Israel Vargas and Marc Steininger

CONTENTS

ABSTRACT

The floristic similarities of 118 permanent plots established in eastern Bolivia were compared using detrended correspondence analysis (DCA) and canonical correspondence analysis (CCA). The Chiquitano dry forest is characterized by a north to south floristic gradient that intergrades with the Amazon flora to the north and with the Gran Chaco flora to the south. However, the forest situated between these two biomes is composed of taxa which are neither Chacoan nor Amazonian, but are a local variant of the seasonal dry tropical forest that is found in other regions of the Neotropics. Species composition varied along both a latitudinal and longitudinal gradient reflecting the floristic differences in the humid forests of southwest Amazon on the Andean piedmont and those of south-central Amazon on the Brazilian shield. The Chiquitano dry forest tree flora, as documented by these plot studies, was not particularly heterogeneous, although forests over calcareous rocks have a different floristic composition when compared with those found on soils derived from granitic rocks. The origin of the floristic gradient is discussed in light of recent discoveries in paleoecology and future climate change.

9.1 INTRODUCTION

The forests of eastern lowland Bolivia are situated across a climatic transition zone between the humid evergreen forests of the Amazon and the deciduous thorn-scrub vegetation of the Gran Chaco. This transition has traditionally been divided into different forest types, or more recently as ecoregions, based on easily observable differences in forest structure, degree of deciduousness and floristic composition (Beck et al., 1993). The most recent classifications (Ibisch and Merida, 2003; Navarro and Maldonado, 2002) recognize several different Amazonian forest types: inundated, pre-Andean, Beni-Santa Cruz, the Chiquitano forest, the Gran Chaco dry forest and serrano-chaco forest. The region also contains numerous savanna habitats which are essentially western outliers of the cerrado Biome of central Brazil (Killeen and Nee, 1991; Mostacedo and Killeen, 1997; Killeen, 1997; Ibisch et al., 2002) or are similar to the inundated savannas of the Gran Pantanal which borders this region to the south-east.

The climatic transition is characterized by a north–south precipitation gradient with mean annual precipitation falling from 1500 to 500 mm (Figure 9.1). However, there is an anomalous non-latitudinal precipitation gradient associated with a topographical feature known as the Elbow of the Andes, where prevailing winds associated with the South American low-level jet create a super humid precipitation zone where mean annual precipitation exceeds 6000 mm per year (Marengo et al., 2004). The entire region is characterized by seasonality that also varies, with the number of months with mean precipitation less than 100 mm increasing from three months in the north to more than seven months at the Paraguayan border. No part of the Bolivian lowlands experiences frost, although occasional southern cold fronts can lower temperatures to below 10°C during the austral winter.

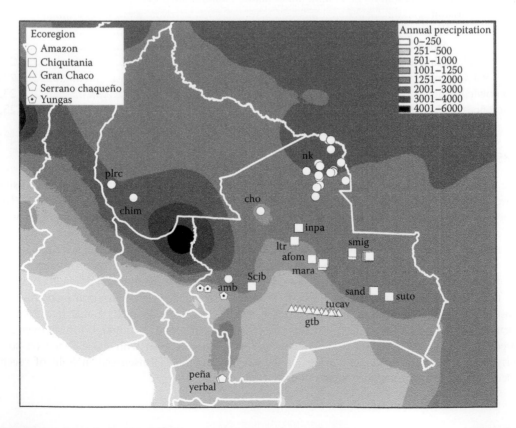

FIGURE 9.1 Precipitation map of Bolivia showing the positions of the plot localities; plot codes as in Table 9.1.

Lowland Bolivia is also characterized by prominent differences in geomorphology and geological history: it is bordered on the west by the relatively young landscapes of the Andean piedmont and to the east by the ancient rocks of the Brazilian Shield. Between these two regions is the Chaco-Beni Plain, a flat plain composed of Quaternary sediments that has been deposited by the numerous rivers originating in the Andes. In the south, where the Amazon is separated from the watershed of the River Plate these alluvial sediments give way to older sedimentary rocks dating from the Cretaceous (Suarez-Soruco, 2000). Soils in the high-precipitation zones are typically acidic, while to the south they tend to the alkaline (Navarro et al., 1998). The piedmont and adjacent plains are characterized by deep soils, which tend to be poorly drained and seasonally waterlogged in the north and well drained and sandy to the south. The landscape of the Brazilian Shield is characterized by rolling hills with low superficial soil, most of which are derived from gneiss or other granitic rocks, although scattered throughout the area are circumscribed areas with metamorphic rock and on the southern edge of the shield several small mountain ranges are composed of calcareous rocks (Litherland, 1984). The variety of landscapes and soils provides a considerable degree of edaphic variability which also influences the distribution of plant species. Ratter et al. (1978) documented that many species characteristic of deciduous dry forest are restricted to mesotrophic soils, while Killeen et al. (2001) demonstrated that edaphic variability was one of the principal drivers of beta diversity in north-east Bolivia at the southern edge of the Amazon forest.

Recent palaeoecological research has documented recent changes in the floristic composition of the forests of the region, with evidence of large-scale latitudinal shifts in the vegetation (Mayle et al., 2000; Burbridge et al., 2004). Humid species characteristic of the Amazon have expanded their ranges to the south, while many species that are today dominant in certain areas of eastern lowland Bolivia were absent from those very same localities just a few thousand years ago (Mayle et al., 2004).

The Chiquitano dry forest is a term used to describe a complex of forest communities that occur across this climatic transition (Killeen et al. 1998), representing what is probably the largest extant patch of what is now broadly recognized as the neotropical seasonal dry tropical forest complex (Prado and Gibbs, 1993; Prado, 2000; Pennington et al., 2004). The Chiquitano dry forest ranges from completely deciduous in the south to semideciduous in the north, while the degree of deciduousness in the intervening areas is highly variable depending on the amount of precipitation that falls within any given year at any given place. At the midpoint of what is considered to be its latitudinal extent, it is a closed-canopy forest about 15 m tall with the largest trees reaching up to 35 m in height and 120 cm in diameter (Killeen et al., 1998). Many, if not all, species are fire adapted with thick corky bark. Most species are relatively slow growing with very high wood densities. The largest trees (*Schinopsis brasilensis* Engl.) are estimated to be more than 500 years old (Dauber et al., 2003).

The derivation of the name comes from the geographical region of Bolivia that is more or less concomitant with this forest type. Known as Chiquitania, it is also the land of the Chiquitano ethnic group which incorporates a number of related indigenous peoples of the Guaraní linguistic heritage (Gott, 1993). The potential economic value of this forest is large because of the very high density wood that characterizes many of the most common species, particularly *Tabebuia impetiginosa* (Mart. ex DC) Standl., *Machaerium scleroxylon* Tul., *Astronium urundeuva* (Alemão) Engl. and *Schinopsis brasilensis* Engl. High transport costs and subsequent isolation from international markets has largely been responsible for conserving this forest ecosystem until the present decade. Unfortunately, areas adjacent to Chiquitania in Bolivia have experienced considerably greater rates of deforestation (Steininger et al., 2001a,b; Pacheco and Mertens, 2004) and the future conservation of the Chiquitano dry forest is very much in doubt. The biggest threat comes from cattle ranching activities and mechanized agriculture, which are both technically feasible and economically attractive investments over the short-term.

In this chapter, we compare the composition of trees in 118 permanent 1-ha forest plots established at different localities in the eastern lowland and evaluate differences in the context of

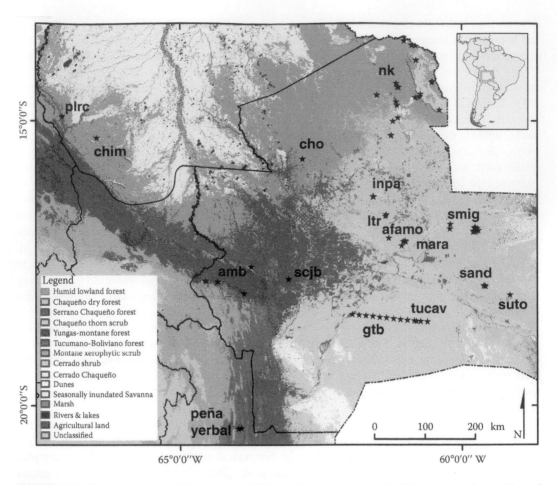

FIGURE 9.2 (See colour insert following page 208) Land cover of eastern Bolivia showing the position of the plot localities in relation to the major vegetation types in the region, which also correspond roughly to existing ecoregional classifications. Land-use change is current to 2002; plot codes as in Table 9.1.

a high-resolution vegetation map that documents the distribution of humid and deciduous forest, as well as chacoan scrubland, cerrado savanna and Pantanal wetlands (Figure 9.2). Our goal is to document the floristic differences among the various forest types, in order to better characterize the Chiquitano dry forest and its biodiversity. In addition, we demonstrate that the geographical distribution varies among species and that species distributions are linked to both climate and edaphic factors. Information on the distribution of individual species is essential for understanding which species will be impacted by climate change and which species might be able to adapt to changing climatic conditions. Finally, we show how the consolidation of databases from different projects and institutions can be used to enhance floristic analysis and conservation planning.

9.2 METHODS

The permanent plots incorporated in the study were established over 15 years by different institutions and individuals (Table 9.1 and Figure 9.1). Most of the plots are located within what is considered to be the Chiquitano dry forest, a seasonal forest that represents the transition between the humid Amazon and semi-arid Gran Chaco. However, the study also includes a few plots from adjacent regions including both lowland and montane localities, in order to document the transitional nature of this vegetation type and the geographical distribution of the species which characterize it.

TABLE 9.1
Characteristics of the Plot Localities Included in this Study; See Figures 9.1 and 9.2

Locality	Acronym	Number of Plots	Forest Type	Phenology	Ecoregion	Number of Species
Afamosam[1]	afamo	1	Upland	Deciduous	Chiquitano	31
Amboró National Park[1,2]	ambcb ambra ambsr ambrs	4	Cloud Upland	Evergreen	Yungas Preandean Amazon	39–70
Bosque Chimanes[3,4]	chimc chimj	2	Upland Inundated	Evergreen	Preandean Amazon	47, 87
Chuquisaca[1]	peña herba	2	Upland Cloud	Evergreen Deciduous	Tucumano- Boliviano	40, 44
CIMAL Sandoval[5]	Sand	3	Upland	Deciduous	Chiquitano	25–29
Estación Tucavaca[1,6]	tucav	1	Upland	Deciduous	Gran Chaco	29
CIMAL ex-Marabol[5]	mara	4	Upland	Deciduosu	Chiquitano	24–29
GTB Pipeline[1,7]	gtb 1 to 9	12	Upland	Deciduous	Gran Chaco	2–24
INPA[5]	Inpa 1 to 8	8	Upland	Deciduous	Chiquitano	29–42
La Chonta[5]	cho 1 to 12	12	Upland Inundated	Semi-Evergreen	Beni-Santa Cruz Amazon	46–64
Las Trancas[5]	ltr 1 to 18	18	Upland	Deciduous	Chiquitano	36–48
Noel Kempff Mercado National Park[1,4]	nk + misc codes	28	Upland Inundated	Evergreen Deciduous	Beni/Santa Cruz Amazon Cerrado	46–123
Río Colorado/Pilon Lajas[3,4]	Plrc	1	Upland	Evergreen	Cloud Preandean Amazon	78
CIMAL San Miguel [5]	Smig	12	Upland	Deciduous	Chiquitano	16–32
Santa Cruz Botanical Garden[1,2]	Sczjb	1	Upland	Deciduous	Gran Chaco	31
Suto[5]	Suto	8	Upland	Deciduous	Chiquitano	18–32

[1] Museo Noel Kempff Mercado
[2] New York Botanical Garden
[3] Herbario Nacional de Bolivia
[4] Missouri Botanical Garden
[5] Instituto Boliviano de Investigaciones Forestales (IBIF)
[6] Conservation International
[7] Capitanía de Alto y Bajo Izozog

The plots vary in design, typically being square (100 × 100 m) or rectangular (20 × 500 m); all used the same minimum diameter at breast height (dbh) of 10 cm following other protocols described by Adler and Synott (1992). All sample plots adopted standard botanical practices to ensure the proper taxonomic identification of each species and the collection of a voucher specimen that was later identified by an experienced botanist with access to a modern herbarium.

The standardization of scientific names among plots and study sites was accomplished using the TAXONSCRUBBER software application, which provides a semi-automated procedure for comparing and standardizing the orthographic variability that is common when combining data from different sources (Boyle, 2004). The application splits concatenated information into separate fields in order to facilitate comparisons among databases with different field structures and provides a standard format for the analysis. It then checks the spelling of each name as well as the scientific authority with a standard list of validly published names and authorities based on the International Plant Names Index (http://www.ipni.org/index.html) and the TROPICOS system at the Missouri

Botanical Garden (http://mobot.mobot.org/W3T/Search/vast.html). TAXONSCRUBBER flags all names that match the standard list and provides pull-down menus to review all unmatched names; orthographic variants were corrected manually and all unmatched names are treated as unidentified morphospecies.

The similarity among plots was evaluated using both a detrended correspondence analysis (DCA) and a canonical correspondence analysis (CCA). A DCA partitions the variation among ordinate axes based entirely on florisitic similarity, while a CCA provides a gradient analysis where the ordinate axes are weighted by taxa whose distribution are correlated with a set of defined environmental parameters (Hill and Gauch, 1980; Ter Braak, 1994; McCune and Mefford, 1999). The CCA analysis was applied separately using four different sets of environmental parameters: climate, geographical location, major geological unit and rock type. The climatic data were obtained from the interpolated values provided by the WORLDCLIM data set (Hijmans et al., 2004) and included mean annual precipitation (MAP), the mean precipitation of the driest quarter (MPDQ), the mean annual temperature (MAT), the mean temperature of the coldest quarter (MTCQ) and a measure of seasonal precipitation (coefficient of variation of precipitation [CVP]). The major geological regions were defined as the Andean piedmont, Brazilian (Precambrian) Shield and the Beni-Chaco alluvial plain. The geographical location included latitude and longitude as obtained by GPS instruments, while the rock types were obtained from the geological maps of Bolivia (Litherland, 1984; Suarez-Soruco, 2000).

The ordination of floristic similarity at the family and genus level used a matrix where the number of species for taxa within each plot was the data, while the analyses conducted at the species level were based on abundance values for individual species. In the latter case, all unidentified species were excluded from the analysis with the exception of generic determinations unique to a single plot or group of plots which were then treated as a species. The number of undetermined or partially determined species varied among plots, being greatest in those regions with high levels of species richness. The number of determined species ranged from a low of 50% in one plot established in the dry season in a poorly studied deciduous forest type Noel Kempff National Park (*nklt1*) and 100% in a non-diverse plot in the relatively well studied region of the Gran Chaco (*gtb-series*).

The ordination analyses were done in an iterative fashion, starting at the family level for all plots, with subsequent analyses based on genera and species conducted on sequentially circumscribed geographic areas. Plots with extreme differences in floristic composition that distorted the analyses by heavily weighting one or more ordinate axes were eliminated from the analysis as they were identified. The DCAs were conducted prior to the CCAs so as to identify the major floristic groupings within the region. The CCAs were then conducted as a direct gradient analysis as a test of generally recognized hypotheses in biogeography and of the ecoregional classifications that have been proposed for the region.

The vegetation map was made using orthorectified Landsat (TM and ETM) images obtained from the Global Land Cover Facility at the University of Maryland (http://glcf.umiacs. umd.edu/index. shtml). The map (Figure 9.2) is a mosaic of different studies conducted over several years and is the result of a continuous effort to map and characterize vegetation and land cover in the Bolivian lowlands. A variety of methodologies were used including both supervised classifications and unsupervised classifications. An electronic version of this map can be obtained from the Museo Noel Kempff web site (http://www.museonoelkempff.org).

9.3 RESULTS

The DCA of plots based on species richness at the family level demonstrated the large floristic differences between the montane and lowland forests. All of the lowland plots were tightly grouped together with the first axis differentiating the Andean plots due to the importance of several families that were absent from lowland plots: Actinidiaceae, Aquifoliaceae, Asteraceae, Betulaceae, Caprifoliaceae, Clethraceae, Cunoniaceae, Myrtaceae, Podocarpaceae, Cyatheaceae, Dicksoniaceae and

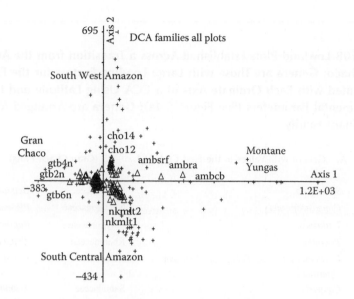

FIGURE 9.3 DCA of 118 plots based on species richness of 99 families. The montane Yungas forest plots (*amb series*) are separated on the first ordinate axis; triangles are plots and plus signs are families. Families weighting the first axis are provided in the text; plot codes as in Table 9.1.

Thymeleaceae (Figure 9.3). The second ordinate axis identified a gradient with western Amazonian plots weighted negatively and central Amazonian plots positively; plots from Chiquitania and the Gran Chaco clustered around the intercept of the two axes.

A DCA conducted on the generic matrix after eliminating the montane Yungas plots (three of the four plots situated in Amboró National Park; *amb-series*) identified three additional plots that were radically different from the rest. Two of these plots were situated to the south of Santa Cruz (*pena, yerba*) and were separated due to the presence of *Vassobia* (Solanaceae), *Sambucus* (Caprifoliaceae), *Xylosma* (Flacourtiaceae), *Pfaffia* (Amaranthaceae), *Juglans* (Juglandaceae), *Diatenopteryx* (Sapindaceae), *Citronella* (Icacinaceae), *Aralia* (Araliaceae) and *Carica* (Caricaceae). The third plot situated in the Gran Chaco (*gt1s*) was identified as unique due to the dominance of two species: *Copernicia alba* Morong and *Geoffroea* sp. The results of these two DCA analyses (Figure 9.3 and a second analysis which is not shown here) correspond to the ecoregional classification that recognizes the Yungas montane forests and the Tucumano-Boliviano forests.

After eliminating the five montane plots and the idiosyncratic plot from the Gran Chaco from the data matrix, a DCA based on species richness at the generic level demonstrated floristic groupings that correspond to the a priori classification of the plots into ecoregions (Amazon, Chiquitania and Gran Chaco) and identified several plots that were intermediate in floristic composition. Examination of the eigenvectors associated with each axis were used to identify the genera that weighted those axes and are considered to be characteristic of those ecoregions (Table 9.2). The DCA based on generic richness revealed little heterogeneity among plots within the Chiquitano dry forest, while showing a moderate amount of heterogeneity among plots from the Gran Chaco and a great deal of heterogeneity among Amazonian plots. This DCA analysis is not shown, because CCA analysis provides a better discrimination of plots and genera in regard to environmental variables (Figure 9.4a and 9.4b).

The CCA analysis of the generic data-set showed that genera are distributed across gradients correlated with both latitude and longitude (Figure 9.4a), as well as with mean annual precipitation and seasonality (Figure 9.4b). The trends first identified by the DCA conducted at the family level were more clearly shown by the CCA conducted at the generic level. There was discrimination among plots situated in the south-west Amazon on the alluvial plains and piedmont adjacent to the

TABLE 9.2
Genera from 108 Lowland Plots Established Across a Transition from the Amazon
to the Gran Chaco; Genera are Those with Large Loading Factors for the Eigens
Vectors Associated with Each Ordinate Axis of a CCA Using Latitude and Longitude
as the Environmental Parameters (See Figure 9.4a); Genera are Arranged Alphabetically
According to Plant Family

A. Genera with Values in the First CCA AX1 < −1.5 (Gran Chaco Plots)

Anacardiaceae	*Schinopsis*	Myrsinaceae	*Geissanthus*
Annonaceae	*Cremastosperma*	Nyctaginaceae	*Pisonia*
Palmae	*Trithrinax*	Polygonaceae	*Ruprechtia*
Bignoniaceae	*Tynanthus*	Rhamnaceae	*Ziziphus*
Cactaceae	*Browningia, Cereus, Echinopsis, Pereskia, Stetsonia*	Rutaceae	*Esenbeckia*
Capparidaceae	*Capparis*	Santalaceae	*Acanthosyris*
Euphorbiaceae	*Phyllanthus*	Sapindaceae	*Athyana*
Leguminosae	*Geoffroea, Mimozyganthus, Prosopis, Pterogyne*	Sapotaceae	*Sideroxylon*
Meliaceae	*Cabralea*	Zygophyllaceae	*Bulnesia*

B. Genera with Values in the First CCA > 1 (South Central Amazon Plots)

Anacardiaceae	*Tapirira*	Malpighiaceae	*Banisteriopsis, Byrsonima, Mascagnia*
Annonaceae	*Bocageopsis, Cardiopetalum, Ephedranthus, Onychopetalum*	Marcgraviaceae	*Norantea*
		Melastomataceae	*Mouriri*
Palmae	*Bactris, Mauritia, Maximiliana*	Meliaceae	*Ruagea*
Asteraceae	*Mikania*	Menispermaceae	*Anomospermum*
Bignoniaceae	*Callichlamys, Cydista, Jacaranda, Lundia, Mussatia, Paragonia*	Monimiaceae	*Siparuna*
Burseraceae	*Huberodendron, Crepidospermum, Dacryodes*	Moraceae	*Cecropia, Coussapoa, Maquira*
Celastraceae	*Elaeodendron*	Myristicaceae	*Osteophloeum*
Chrysobalanaceae	*Couepia, Licania, Parinari*	Myrsinaceae	*Myrsine*
Clusiaceae	*Kielmeyera, Vismia*	Myrtaceae	*Marcia*
Combretaceae	*Buchenavia*	Olacaceae	*Chaunochiton, Dulacia, Schoepfia*
Connaraceae	*Connarus*	Polygalaceae	*Bredemeyera, Moutabea*
Convolvulaceae	*Dicranostyles*	Polygonaceae	*Symmeria*
Dichapetalaceae	*Tapura*	Proteaceae	*Panopsis, Roupala*
Dilleniaceae	*Curatella, Doliocarpus*	Rhamnaceae	*Colubrina*
Euphorbiaceae	*Alchornea, Alchorneopsis, Aparisthmium, Chaetocarpus, Conceveiba, Hyeronima, Maprounea, Sebastiana, Pera*	Rosaceae	*Prunus*
Leguminosae	*Ateleia, Derris, Diptychandra, Hymenolobium, Macrolobium, Mimosa, Pterodon, Stryphnodendron, Vataireopsis*	Rubiaceae	*Amaioua, Chomelia, Coussarea, Coutarea, Faramea, Ferdinandusa, Pagamea, Randia, Uncaria*

TABLE 9.2
Genera from 108 Lowland Plots Established Across a Transition from the Amazon to the Gran Chaco; Genera are Those with Large Loading Factors for the Eigens Vectors Associated with Each Ordinate Axis of a CCA Using Latitude and Longitude as the Environmental Parameters (See Figure 9.4a); Genera are Arranged Alphabetically According to Plant Family (Continued)

B. Genera with values in the first CCA > 1 (South Central Amazon Plots)

Flacourtiaceae	*Banara*	Rutaceae	*Dictyoloma, Metrodorea*
Hippocrateaceae	*Anthodon, Cheiloclinium, Hippocratea, Prionostemma*	Sabiaceae	*Meliosma*
Humiriaceae	*Humiria, Sacoglottis*	Sapindaceae	*Matayba*
Icacinaceae	*Citronella*	Sapotaceae	*Ecclinusa*
Lacistemataceae	*Lacistema*	Simaroubaceae	*Simarouba*
Lauraceae	*Aiouea, Caryodaphnopsis, Endlicheria*	Tiliaceae	*Mollia*
Lecythidaceae	*Couratari, Eschweilera*	Vochysiaceae	*Erisma*
Loganiaceae	*Strychnos*		

C. Genera with values in the CCA AX2 < –1 (South West Amazon/Andean Piedmont Plots)

Acanthaceae	*Suessenguthia*	Meliaceae	*Cabralea, Swietenia, Trichilia*
Annonaceae	*Annona, Cremastosperma, Ruizodendron*	Moraceae	*Batocarpus, Clarisia, Coussapoa, Poulsenia*
Apocynaceae	*Peschiera*	Myristicaceae	*Otoba*
Palmae	*Iriartea, Scheelea*	Myrsinaceae	*Geissanthus*
Bignoniaceae	*Tynanthus*	Myrsinaceae	*Stylogyne*
Bombacaceae	*Cavanillesia*	Papilionoideae	*Machaerium*
Burseraceae	*Tetragastris*	Polygonaceae	*Triplaris*
Caricaceae	*Jacaratia*	Rubiaceae	*Genipa, Ixora*
Clusiaceae	*Calophyllum, Rheedia, Symphonia*	Sapindaceae	*Sapindus*
Euphorbiaceae	*Croton, Hura, Margaritaria, Richeria, Sapium*	Sterculiaceae	*Guazuma*
Leguminosae	*Albizia, Cyclolobium, Machaerium*	Tiliaceae	*Heliocarpus*
Flacourtiaceae	*Hasseltia*	Ulmaceae	*Ampelocera, Trema*
Lauraceae	*Licaria, Nectandra, Ocotea*	Verbenaceae	*Aegiphila*
		Violaceae	*Leonia*

Andes (*plrc, chimc, chimj, ambrs*) when compared to those situated in the southern central Amazon on the Brazilian Shield (*nk-series*), while those plots situated on the extreme western edge of the Brazilian shield (*cho-series*) were intermediate to these other two groups. The gradient identified based on geographical location was slightly different when the CCA was conducted using a precipitation data set (Figure 9.4b), reflecting the non-latitudinal nature of the precipitation gradient in the south-west Amazon. The floristic differences of the plots situated in the Gran Chaco were discriminated forming a gradient with the plots that were located in Chiquitania. The genera which weighted the first two ordinate axes of the CCA are shown in Table 9.2.

Next, the DCA and CCA were applied to a subset that included all of the plots from Chiquitania and specific plots from Amazon ecotone that had been identified as being the most similar to the Chiquitania plots (e.g. select plots from the *nkm* series) or which were geographically situated so as to indicate that they should be transitional in nature (e.g. *cho* series). Plots situated to the south in the Gran Chaco (*gto* series and *tucavaca*) were excluded so as to discriminate the gradients on the northern ecotone better (Figure 9.5). The DCA analyses confirmed the previously observed

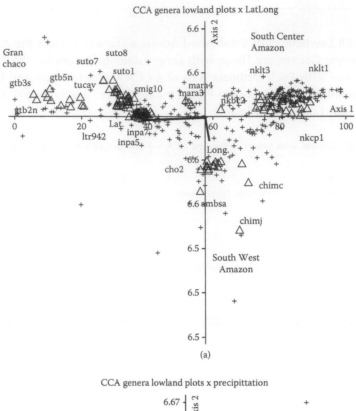

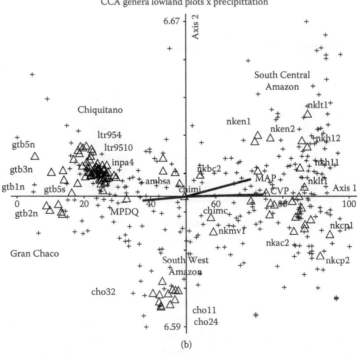

FIGURE 9.4 CCA of 114 plots based on species richness of 379 genera constrained by geographical position (a) and precipitation (b); triangles are plots and plus signs are genera. Table 9.2 provides the list of the genera that are weighted for Gran Chaco (*gtb series*), south-central Amazon (*nk series*) and the south-west Amazon (*chim, amb, plrc*); plot codes as in Table 9.1. MPDQ: mean precipitation of the driest quarter; CVP: coefficient of variation for precipitation; MAP: mean annual precipitation (from WORLDCLIM data set (Hijmans et al., 2004).

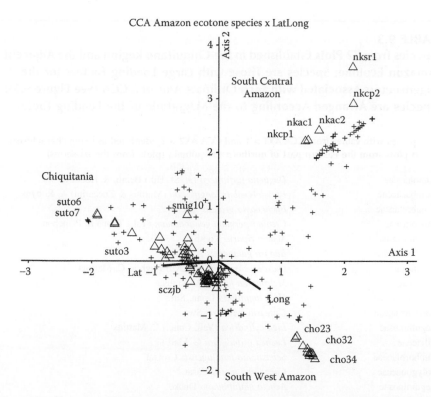

FIGURE 9.5 CCA of 72 plots and 277 species from the Chiquitano dry forest and the most similar transitional plots in adjacent Amazonian humid forest; triangles are plots and plus signs are species. Species with large axis loading factors are provided in Table 9.3; plot codes as in Table 9.1.

floristic differences between the south-central and south-west Amazon (*cho* series). The CCA analysis was used to identify the transitional species present along both the latitudinal and longitudinal gradients. Table 9.3 shows those species that weight the axes that separate the two subregions of the southern Amazon and which are present in the transitional plots; these are essentially Amazonian species at the extreme southern extent of their geographical distribution. There are also Chiquitano species present in these plots, and a comparison of the floristic composition of the plots from the south-central Amazon (*nkm* series) and all of the plots within the core area of Chiquitania revealed 79 species that were present in both regions. This CCA ordination also revealed that although most plots within Chiquitania are more or less tightly clustered around the intersection of the ordinate axes, a subset of plots from south-east Chiquitania (*suto* series) were offset from the rest on the second ordinate axis (Figure 9.5).

The next iteration was performed using the same plots situated in the center of the Chiquitano seasonal dry forest region, but excluded the transitional plots from the Amazon ecotone while including the plots situated in the Gran Chaco (*gtb* series). The DCA identified a transition from the Chiquitano to Gran Chaco on the first ordinate axis. However, unlike the northern ecotone where the floristically intermediate plots could be associated with a latitudinal gradient, the *gtb* plots in the south were all situated at the same latitude. Two sets of plots (*suto* and *sand*) were grouped separately from the majority of the plots from Chiquitania and the Gran Chaco. The CCA analyses that provided the best separation of the various groups was provided using rock type for the environmental parameters (Figure 9.6). The plots from the chaco (*gtb* series) were segregated on the first ordinate axis which was correlated with alluvial sediments ($r^2 = 0.912$, $p < 0.05$) and the *suto/sand* series were separated by the second ordinate axis which was highly correlated with

TABLE 9.3

Species from 72 Plots Established in the Chiquitano Region and the Adjacent Amazon Ecotone; Species are Those with Large Loading Factors for the Eigenvectors Associated with Each Ordinate Axis of a CCA (See Figure 9.5); Species are Arranged According to the Magnitude of the Loading Factor

A. Species with values from CCA AX1 > 1 and CCA AX2 > 2, identified as being characteristic in plots from the eastern part of northern Chiquitania (plots from the *nk*-series)

Annonaceae	*Duguetia furfuracea* (A. St. Hil.) Benth. & Hook. f.
Bombacaceae	*Pseudobombax longiflorum* (Martius & Zuccarini) A. Robyns
Combretaceae	*Buchenavia grandis* Ducke
Burseraceae	*Crepidospermum goudotianum* (Tul.) Triana & Planchon
Leguminosae	*Acacia riparia* Kunth
Flacourtiaceae	*Banara guianensis* Aublet
Ochnaceae	*Ouratea castaneifolia* (DC.) Engl. in C. Martius
Rutaceae	*Metrodorea flavida* K. Krause
Nyctaginaceae	*Neea amplifolia* Donn. Sm.
Flacourtiaceae	*Casearia sylvestris* Sw.
Leguminosae	*Inga cylindrica* (Vell. Conc.) C. Martius
Tiliaceae	*Luehea grandiflora* C. Martius
Euphorbiaceae	*Sebastiania huallagensis* Croizat
Polygonaceae	*Coccoloba mollis* Casar.
Leguminosae	*Acacia multipinnata* Ducke
Euphorbiaceae	*Sapium laurifolium* (A. Rich.) Griseb.
Annonaceae	*Duguetia marcgraviana* C. Martius
Celastraceae	*Maytenus floribunda* Pittier
Dichapetalaceae	*Tapura amazonica* Poeppig in Poeppig & Endl.
Euphorbiaceae	*Chaetocarpus echinocarpus* (Baillon) Ducke
Leguminosae	*Inga thibaudiana* DC.
Leguminosae	*Ormosia coarctata* Jackson
Proteaceae	*Roupala montana* Aublet
Vochysiaceae	*Qualea cordata* (C. Martius) Spreng.
Leguminosae	*Machaerium jacarandifolium* Rusby
Leguminosae	*Inga cayennensis* Sagot ex Benth.
Humiriaceae	*Sacoglottis mattogrossensis* Malme
Bombacaceae	*Chorisia integrifolia* Ulbr.
Euphorbiaceae	*Margaritaria nobilis* L. f.
Bignoniaceae	*Macfadyena unguis-cati* (L.) A. Gentry
Araliaceae	*Didymopanax morototoni* (Aublet) Decne & Planchon
Leguminosae	*Bauhinia rufa* (Bong.) Steudel
Palmae	*Attalea phalerata* Mart. ex Spreng.
Flacourtiaceae	*Casearia arborea* (Rich.) Urban
Rhamnaceae	*Rhamnidium elaeocarpum* Reissek

B. Species with values from CCA AX1 > 1 and CCA AX2 < 1, identified as being characteristic in plots from the western part of northern Chiquitania (plots from the *cho-series*)

Myrtaceae	*Eugenia florida* DC.
Moraceae	*Batocarpus amazonicus* (Ducke) Fosb.
Acanthaceae	*Suessenguthia multisetosa* (Rugby) Wassh.
Caricaceae	*Jacaratia spinosa* (Aublet) A. DC.
Euphorbiaceae	*Sapium glandulosum* (L.) Moroni

TABLE 9.3
Species from 72 Plots Established in the Chiquitano Region and the Adjacent Amazon Ecotone; Species are Those with Large Loading Factors for the Eigenvectors Associated with Each Ordinate Axis of a CCA (See Figure 9.5); Species are Arranged According to the Magnitude of the Loading Factor (Continued)

A. Species with values from CCA AX1 > 1 and CCA AX2 > 2, identified as being characteristic in plots from the eastern part of northern Chiquitania (plots from the *nk*-series)

Sapindaceae	*Cupania cinerea* Poeppig
Clusiaceae	*Rheedia brasiliensis* (C. Martius) Planchon & Triana
Palmae	*Syagrus sancona* Karsten
Apocynaceae	*Peschiera australis* (Müll. Arg. In C. Martius) Miers
Moraceae	*Pseudolmedia laevis* (Ruiz Lopez & Pavon) J.F. Macbr.
Rutaceae	*Zanthoxylum sprucei* Engl.
Leguminosae	*Inga edulis* C. Martius
Bombacaceae	*Cavanillesia hylogeiton* Ulbr.
Palmae	*Copernicia alba* Morong ex Morong & Britton
Nyctaginaceae	*Neea hermaphrodita* S. Moore
Moraceae	*Pseudolmedia laevigata* Trecul
Bombacaceae	*Ceiba pentandra* (L.) Gaertner
Meliaceae	*Guarea guidonia* (L.) Sleumer
Moraceae	*Pourouma cecropiifolia* C. Martius
Leguminosae	*Ormosia nobilis* Tul.
Elaeocarpaceae	*Sloanea terniflora* (Sesse & Mocino ex DC.) Standley
Clusiaceae	*Calophyllum brasiliense* Cambess.
Leguminosae	*Inga marginata* Willd.
Lauraceae	*Ocotea guianensis* Aublet
Sapindaceae	*Sapindus saponaria* L.
Leguminosae	*Acacia bonariensis* Gillies ex Hook. & Arn.

calcarious rocks ($r^2 = 0.83$, $p < 0.01$). The species identified as being characteristic of the plots from calcareous rocks are provided in Table 9.4.

The next iteration was made on a data-set that excluded the plots from both Amazon (*nkp-*, *cho-*) and the Gran Chaco (*gtb-*, *tucavaca*), as well as the plots from calcareous rocks identified in the previous analyses (*suto-m sand*). The DCA performed on this set of plots from what is considered to be the heart of the Chiquitano forest showed little heterogeneity and all of the plots were tightly clustered around the center of the first, second and third ordinate axes. The subsequent CCA analyses all showed some grouping which was best explained using latitude and longitude as the environmental matrix (Figure 9.7, Table 9.5). The taxa that are at the center of this last ordination procedure and which were not weighted with any ordinate axis, can be considered to be the typical or characteristic species of the Chiquitano dry forest (Table 9.5e).

9.4 DISCUSSION

9.4.1 BIOGEOGRAPHY AND ENDEMISM

The results confirm the hypothesis that the Chiquitano dry forest is a transition zone between the humid tropics and semi-arid subtropics and the distribution of species is related to climatic variables which are highly correlated with latitude. The species found in the chiquitano dry forest do not

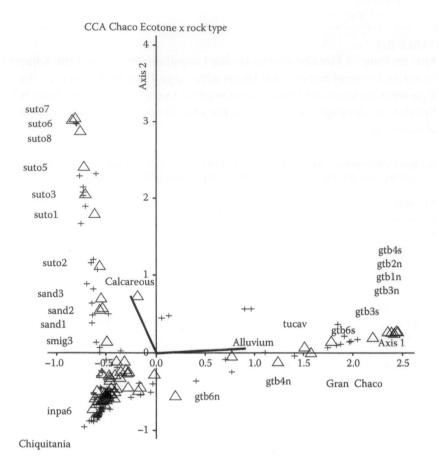

FIGURE 9.6 CCA of 167 species from 67 plots in Chiquitania and Gran Chaco. The species identified by this ordination as associated with alluvial soils (*gtb*) or calcareous rocks (*suto*) are provided in Table 9.4; plot codes as in Table 9.1.

represent a mixture of Amazonian and Gran Chaco species, but are a distinct assemblage of species when compared to the forests situated either north or south of the region (Table 9.6), thus also supporting the hypothesis that the Chiquitano dry forest merits recognition as a unique ecoregion (Olsen and Dinerstein, 1998; Ibisch et al., 2002).

The degree of endemism in the region is not well known, with only a few documented plant species, such as *Acosmium cardenasii* H.S. Irwin & Arroyo, *Cereus tacuaralensis* Cardenas and *Swartzia jorori* Harms (see Table 9.5e). In this study there were 11 undescribed taxa that were given morphospecies names by the botanists who identified the voucher specimens in the following genera *Phyllanthus, Acacia, Bauhinia, Inga, Lonchocarpus, Machaerium, Casearia, Myrcianthes, Myrciaria, Neea* and *Zanthoxylum*. All of these genera are large, taxonomically complex and their species are hard to identify, which lessens the probability that these morphospecies represent endemic species new to science.

The low level of endemism may simply be a taxonomic artefact that is the result of too few collections and too few taxonomists in Bolivia. However, a more likely explanation may lie in its recent geological past. Almost all of the species identified as characteristic of the core area of the Chiquitano dry forest (Table 9.5) are also part of the seasonal dry forest complex described by Prado and Gibbs (1993), Prado (2000) and Pennington et al. (2004). Mayle et al. (2004) hypothesized that the flora of this region is essentially young and the constituent species of this seasonal

TABLE 9.4

Species from 67 Plots Established in the Chiquitano Region and the Adjacent Gran Chaco Ecotone; Species are Those with Large Loading Factors for the Eigenvectors Associated with Each Ordinate Axis of a CCA (See Figure 9.6); Species are Arranged According to the Magnitude of the Loading Factor

A. Species with values for the CCA AX1 >1, identified as being important in plots from the northern sector of the Gran Chaco (*gtb-series*)

Leguminosae	*Prosopis nigra* (Griseb.) Hieron.
Leguminosae	*Mimozyganthus carinatus* (Griseb.) Burkart
Cactaceae	*Browningia caineana* (Cardenas) Cardenas
Leguminosae	*Prosopis nuda* Schinini
Capparidaceae	*Capparis salicifolia* Griseb.
Nyctaginaceae	*Bougainvillea praecox* Griseb.
Santalaceae	*Acanthosyris falcata* Griseb.
Polygonaceae	*Ruprechtia triflora* Griseb.
Leguminosae	*Lonchocarpus nudiflorens* Burkart
Leguminosae	*Acacia praecox* Griseb.
Cactaceae	*Cereus dayamii* Speg.
Cactaceae	*Pereskia sacharosa* Griseb.
Leguminosae	*Geoffroea spinosa* Jacq.
Bignoniaceae	*Tabebuia nodosa* (Griseb.) Griseb.
Bombacaceae	*Pseudobombax heteromorphum* (Kuntze) Kuntze
Capparidaceae	*Capparis tweediana* Eichler in C. Martius
Nyctaginaceae	*Neea hermaphrodita* S. Moore
Capparidaceae	*Capparis speciosa* Griseb.
Leguminosae	*Caesalpinia paraguariensis* (Parodi) Burkart
Bombacaceae	*Ceiba insignis* (Kunth) P. Gibbs & Semir
Opiliaceae	*Agonandra excelsa* Griseb.
Olacaceae	*Ximenia americana* L.
Ulmaceae	*Trema micrantha* (L.) Blume
Capparidaceae	*Capparis retusa* Griseb.

B. Species with values for CCA AX2 > 0.5, identified as being characteristic in plots associated with calcareous rock (suto, sand)

Sapotaceae	*Pouteria nemorosa* Baehni
Rutaceae	*Esenbeckia almawillia* Kaastra
Rubiaceae	*Pogonopus tubulosus* (A. Rich.) Schumann
Flacourtiaceae	*Lunania parviflora* Spruce ex Benth.
Sterculiaceae	*Guazuma ulmifolia* Lam.
Sapotaceae	*Chrysophyllum gonocarpum* (C. Martius & Eichler) Engl.
Rutaceae	*Zanthoxylum coco* Gillies ex Hook. f. & Arn.
Vochysiaceae	*Vochysia haenkeana* C. Martius
Leguminosae	*Sweetia fruticosa* Spreng.
Combretaceae	*Terminalia oblonga* (Ruiz Lopez & Pavon) Steudel
Leguminosae	*Machaerium scleroxylon* Tul.
Capparidaceae	*Capparis retusa* Griseb.
Anacardiaceae	*Spondias mombin* L.
Tiliaceae	*Lueheopsis duckeana* Burret

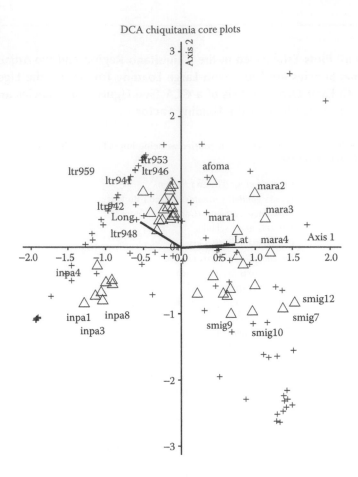

FIGURE 9.7 CCA of core plots from Chiquitania showing a spatial heterogeneity associated with proximity.

dry forest flora are relatively recent arrivals that migrated into the region from more northerly latitudes after the last glacial maximum. Regardless of its past, the chiquitano dry forest is situated at the crossroads of any potential scenario that describes how the different disjunct regions with taxa characteristic of this newly defined biome might have once been in contact. This includes the caatinga, the southern Andean piedmont, eastern Paraguay and the dry valleys of the central Andes.

9.4.2 ALPHA AND BETA DIVERSITY

Gentry (1995), using limited data from a few localities, suggested that the dry forests in this region were some of the most floristically diverse in the New World, a conclusion that was supported by a study that identified a total of 107 tree species as part of a single site inventory (Killeen et al., 1998) and a compendium of 237 trees known from herbaria specimens (Jardim et al., 2003). However, this study of 56 1-ha plots identified only 155 tree species, which suggests that the tree flora of the region might only be moderately diverse. Nonetheless, since it is a relatively open forest type, the diversity of the herbaceous, shrub and liana life forms is large, with up to 40% of all the species registered for the region (Killeen et al., 1998; Ibisch et al., 2002).

Plots from the Chiquitano dry forest region were consistently grouped together based on tree composition when compared to both Gran Chaco and Amazon plots (see Figures 9.5 and 9.6), and spatial heterogeneity was only displayed when all transitional or edaphically unique areas were eliminated from the analysis. This apparent lack of β-diversity may be an artefact and the result of

TABLE 9.5
Species from 43 Plots from Chiquitania Excluding Ecotonal Plots and Those Situated Over Calcareous Rock (See Figure 9.7); Species are Arranged According to the Magnitude of the Loading Factor

A. Species with CCA AX1 < –0.75 and CCA AX2 < –0.75), identified as being important in the north-west sector of Chiquitania (*inpa-series*)

Bignoniaceae	*Tabebuia roseoalba* (Ridley) Sandw.
Leguminosae	*Copaifera reticulata* Ducke
Leguminosae	*Machaerium saraense* Rudd
Polygonaceae	*Ruprechtia laxiflora* Meisn. in C. Martius
Leguminosae	*Machaerium pilosum* Benth.
Celastraceae	*Maytenus robustoides* Loes.
Bignoniaceae	*Tabebuia serratifolia* (M. Vahl) Nicholson
Tiliaceae	*Luehea candicans* C. Martius
Leguminosae	*Dalbergia riparia* (Mart.) Benth.
Nyctaginaceae	*Neea steinbachiana* Heimerl

B. Species with CCA AX1 < 0.0 and CCA AX2 > 0.75, identified as being important in the north-central sector of Chiquitania (*ltr-series*)

Vochysiaceae	*Qualea grandiflora* C. Martius
Leguminosae	*Caesalpinia paraguariensis* (Parodi) Burkart
Leguminosae	*Machaerium hirtum* (Vell. Conc.) Vell. Conc.
Lythraceae	*Physocalymma scaberrimum* Pohl
Violaceae	*Leonia glycycarpa* Ruiz, Lopez & Pavon
Rutaceae	*Zanthoxylum hasslerianum* (Chodat) Pirani
Moraceae	*Cecropia concolor* Willd.
Bombacaceae	*Pseudobombax marginatum* (A. St. Hil., A. L. Juss. & Cambess.) A. Robyns
Sterculiaceae	*Guazuma ulmifolia* Lam.
Vochysiaceae	*Vochysia haekeana* C. Martius
Nyctaginaceae	*Bougainvillea modesta* Heimerl
Leguminosae	*Machaerium villosum* J. Vogel

C. Species with CCA AX1 > 0.5 and CCA AX2 > 0.5 identified as being important in the south-central sector of Chiquitania (*mara-series*)

Capparidaceae	*Capparis speciosa* Griseb.
Flacourtiaceae	*Casearia aculeata* Jacq.
Leguminosae	*Acacia loretensis* J.F. Macbr.
Leguminosae	*Bauhinia rufa* (Bong.) Steudel
Rutaceae	*Zanthoxylum monogynum* A. St. Hil.
Bombacaceae	*Ceiba insignis* (Kunth) P. Gibbs & Semir
Ulmaceae	*Trema micrantha* (L.) Blumc

D. Species with CCA AX1 > 0.5 and CCA AX2 < 0.5, identified as being important in the south-east sector of Chiquitania (*smig-series*)

Capparidaceae	*Capparis speciosa* Griseb.
Anacardiaceae	*Schinopsis brasiliensis* Engl. in C. Martius
Flacourtiaceae	*Casearia aculeata* Jacq.
Leguminosae	*Acacia loretensis* J.F. Macbr.
Leguminosae	*Bauhinia rufa* (Bong.) Steudel
Rutaceae	*Zanthoxylum monogynum* A. St. Hil.
Bombacaceae	*Ceiba insignis* (Kunth) P. Gibbs & Semir
Anacardiaceae	*Spondias mombin* L.
Bignoniaceae	*Tabebuia impetiginosa* (C. Martius ex DC. in A. DC.) Standley

(continued)

TABLE 9.5

Species from 43 Plots from Chiquitania Excluding Ecotonal Plots and Those Situated Over Calcareous Rock (See Figure 9.7); Species are Arranged According to the Magnitude of the Loading Factor (Continued)

Ulmaceae	*Trema micrantha* (L.) Blume
Leguminosae	*Acacia polyphylla* DC.
Leguminosae	*Caesalpinia pluviosa* DC.

E. Core species that were found in all sectors of the Chiquitano region; CCA AX1 > –0.5 and < 0.5 and CCA AX2 < 0.5 and > –0.5)

Achatocarpaceae	*Achatocarpus nigricans* Triana
Anacardiaceae	*Spondias mombin* L.
Bignoniaceae	*Tabebuia chrysantha* (Jacq.) Nicholson
Bignoniaceae	*Tabebuia impetiginosa* (C. Martius ex DC. in A. DC.) Standley
Bombacaceae	*Ceiba samauma* (C. Martius) Schumann
Bombacaceae	*Chorisia speciosa* A. St. Hil.
Boraginaceae	*Cordia alliodora* (Ruiz Lopez & Pavon) Cham.
Cactaceae	*Cereus tacuaralensis* Cardenas
Cochlospermaceae	*Cochlospermum orinocense* (Kunth) Steudel
Leguminosae	*Acacia polyphylla* DC.
Leguminosae	*Acosmium cardenasii* H.S. Irwin & Arroyo
Leguminosae	*Anadenanthera colubrina* (Vell. Conc.) Brenan
Leguminosae	*Caesalpinia pluviosa* DC.
Leguminosae	*Guibourtia chodatiana* (Hassler) Leonard
Leguminosae	*Lonchocarpus guillemineanus* (Tul.) Malme
Leguminosae	*Machaerium acutifolium* J. Vogel
Leguminosae	*Swartzia jorori* Harás
Leguminosae	*Sweetia fruticosa* Spreng.

TABLE 9.6

Floristic Similarity of Groups of Plots in Different Ecoregions

	Andes	Piedmont	Chonta	Noel Kempff	Noel Kempff Dry Forest Types	Chiquitano	Chaco	Serrano
Andes		3	0	2	0	3	0	5
Piedmont	0.02		60	86	11	36	0	1
Chonta	—	0.29		53	17	33	2	2
Noel Kempff	0.00	0.14	0.10			79	8	11
Noel Kempff Dry Forest Types	—	0.05	0.14			44	6	6
Chiquitano	0.02	0.16	0.25	0.15	0.30		18	10
Chaco	—	—	0.01	0.01	0.04	0.10		6
Serrano	0.06	0.01	0.02	0.02	0.06	0.09	0.05	
Total species	96	297	112	901	137	155	201	59
Number of plots	3	4	12	23	5	67	13	2

In the upper right quadrant are the number of shared species between sites and in the lower right quadrant are the Sørensen Indices of similarity based upon shared species between sites. Andes: cloud forest >1000 m (ambcb, ambra, ambfa); piedmont: pre-Andean Amazon <500 m (plrc, chmj, chms, ambrs); chonta: western edge of Brazilian Shield (cho1 – 12); Noel Kempff: subset of nk-plots that exclude transitional forest; Noel Kempff Dry Forest: subset of nk- that were identified by ordination as transitional; Chiquitano: all plots from smig, mara, sand, ltr94, ltr95, inpa, afomasam; chaco: all plots from gtb, tucav; serrano: southern Andes (pena, herbal).

the data sets that were used for analyses. The Noel Kempff plots were part of a botanical and ecological inventory (Killeen et al., 2001) where an effort was made to establish plots in contrasting vegetation types, which included inundated, dwarf and liana-dominated habitats, as well as both semi-deciduous and evergreen forests. The Gran Chaco plots were systematically installed at predetermined intervals and identify a range of different plant communities that are related to edaphic conditions (R. Guillén, CABI, Bolivia, pers. comm.). For instance, one of the most radically different plots in the entire data-set was located in a wetland dominated by the palm *Copernicia alba* L.. In contrast, all of the Chiquitano dry forest plots were established by foresters who were interested in forest inventory and management (Dauber et al., 2003), thus they may represent a more homogeneous, closed-canopy, forest type on well-drained soils with high economic potential for hardwood species. In spite of the high overall similarity of the plots, the Chiquitano dry forest region has considerable edaphic variability that has not been sampled by this study, including sandstone mesetas, metamorphic and ultramaphic rocks and extensive inundated forests (Litherland, 1984). Most of the plots from Chiquitania included in this study, with the notable exception of the calcareous series, were situated on soils derived from granite or the very similar gneiss. Further exploration of the edaphic variability within the region will probably provide a much greater level of forest heterogeneity.

9.4.3 CLIMATE CHANGE

Mayle et al. (2000, 2004) and Burbridge et al. (2004) hypothesized that the species found within these seasonal dry forests are relatively recent arrivals that migrated into the region from the north as a response to millennial-scale climate change. The identification of a floristic gradient that coincides with a latitudinal and climatic gradient supports that hypothesis. However, this study also shows that the gradient has two sub-axes over which migration into the region from the north might have occurred.

Currently, there is a great deal of interest in monitoring how climate change will impact biological diversity as species are forced to shift geographical ranges in response to global warming. Ecotonal and transitional environments are more likely to be impacted by climate change as many species are at the extreme limits of their ecological requirements or where competition from differently adapted species places their survival at a disadvantage. The eastern lowlands of Bolivia offer a unique and valuable laboratory to study the impact of climate change on biodiversity. Several groups of scientists are working in eastern Bolivia with the objective of documenting how climate change may be impacting different aspects of ecosystem function across this climatic and ecological transition. The permanent plot network used for this study has been established largely for the purposes of biological inventory and forest management. However, it is an invaluable asset and should be integrated as a core facility into linking studies focusing on climate change, geochemistry and ecosystem function with biodiversity and biodiversity conservation.

9.4.4 CONSERVATION

The Chiquitano dry forest is threatened by land-use change and forest degradation due to logging and fire. In a recent unpublished study conducted for the Vice Ministry for the Environment, the total original extent of the Chiquitano forest was estimated at 125,000 km². Of this original total, an estimated 18,600 km² or 15% had been deforested prior to 2001 and deforestation was documented at a mean annual rate of 108,000 ha for the decade spanning 1991 to 2001 (Killeen, pers. obs.). Timber extraction may also threaten the long-term conservation of the forest. Current logging rates are relatively low, but increased logging could become unsustainable due to the very slow growth of the species in this forest. Most logging management plans contemplate only a 25-year harvest cycle, while the current crop of trees being exploited are all much older than 100 years and the very largest trees probably approach 500 years of age. Overexploitation of timber resources will spur deforestation, because once forests have no valuable timber, they will be cleared for pasture and converted to cattle ranches.

Both land-use change and logging promote fire, and in the fall of 2004 an estimated 1.5 million hectares was impacted by forest fires, most of which originated in forest being cleared for pasture and which spread into adjacent intact forest. Fire is a well-known ecological phenomenon in the chiquitano forest, but anecdotes from long-time residents indicate that fire frequencies are increasing from decadal-scale phenomena to a recurrent event that is impacting the region every few years: large-scale fires occurred in Chiquitania in 2004, 2000, 1998, 1994 and 1987 (Killeen, pers. obs.).

Conservation efforts are being pursued by the Chiquitano Forest Conservation Foundation, a private non-profit organization funded by private industry in collaboration with Bolivian and international conservation organizations. Several protected areas have been established, but these tend to be zoned for multiple use and contain within them logging concessions, land claims and indigenous lands. Consequently, efforts by the foundation focus on working with the various stakeholders who have the legal right to exploit forest resources. Key to this effort is an emphasis on working with municipal governments and local communities who under Bolivian laws have considerable leeway in monitoring natural resource use. In addition, research conducted by the Bolivian Forest Research Institute is aimed at the definition of more sustainable management practices. Likewise, the government and the private forest sector are working to open the timber market to new species so as to reduce the pressure on the few species that are currently harvested.

ACKNOWLEDGEMENTS

The Instituto Boliviano de Investigación Forestal would like to acknowledge the BOLFOR II sustainable forestry project (a joint effort of USAID and the Bolivian Ministry of Sustainable Development that is implemented by The Nature Conservancy), the Fundación para la Conservación del Bosque Chiquitano, the CIMAL and Suto timber companies for providing funding for the establishment and monitoring of the plots included in this study, and all the other timber companies for logistical support. The Noel Kempff Mercado Natural History Museum would like to acknowledge the Missouri Botanical Garden for its 25 years of support for botanical research in Bolivia.

REFERENCES

Adler, D. and Synott, T.J., *Permanent sample plot techniques for mixed tropical forest*, Tropical Forestry Papers No. 25, Oxford Forestry Institute, Oxford, 1992, 124.

Beck, S.G., Killeen, T.J., and E. García, E., Vegetación de Bolivia, in *Guía de Arboles de Bolivia*, Killeen, T.J., Beck, S.G., and E. García, E., Eds, Herbario Nacional de Bolivia & Missouri Botanical Garden, La Paz, Bolivia, 1993, 6.

Boyle, B. *TaxonScrubber (version 1.2)*, The SALVIAS Project, http://www.salvias.net/pages/taxonscrubber.html, 2004.

Burbridge, R.E., Mayle, F.E., and Killeen, T.J., 50,000 year vegetation and climate history of Noel Kempff Mercado National Park, Bolivian Amazon, *Quatern. Res.*, 61, 215, 2004.

Dauber, E. et al., *Tasas de incremento diamétrico, mortalidad y reclutamiento con base en las parcelas permanentes instaladas en diferentes regiones de Bolivia.* Proyecto de Manejo Forestal Sostenible, BOLFOR, Santa Cruz, Bolivia, 2003.

Gentry, A., Diversity and floristic composition of neotropical dry forests, in *Seasonally Dry Tropical Forests*, Bullock, S.H., Mooney, H.A., and Medina, E., Eds, Cambridge University Press, Cambridge, 1995, 146.

Gott. R., *Land Without Evil: Utopian Journeys Across the South American Watershed*, Verso Books, 1993.

Hijmans, R.J., Cameron, S., and Parra, J, *WorldClim, a square kilometer resolution database of global terrestrial surface climate (version 1. 2.)*, http://biogeo.berkeley.edu/, 2004.

Hill, M.O. and Gauch, H.G., Detrended correspondence analysis: an improved ordination technique, *Vegetatio*, 42, 47, 1980.

Ibisch, P.L. and Merida, G., Eds, *Biodiversidad: La Riqueza de Bolivia*, Editorial FAN, Santa Cruz, Bolivia, 2003.

Ibisch, P. L., Columba, K., and Reichle, S., *Plan de Conservación y Desarrollo Sostenible para el Bosque Seco Chiquitano, Cerrado y Pantanal Boliviano*, Fundación para la Conservación del Bosque Chiquitano, Editoral FAN, Santa Cruz, Bolivia, 2002.

Jardim, A., Killeen, T.J., and Fuentes, A., *Guía de los Arboles y Arbustos del Bosque Seco Chiquitano, Bolivia*, Editoral FAN, Santa Cruz, Bolivia, 2003, 324.

Killeen, T.J., Southeastern Santa Cruz, in *Centres of Plant Diversity - A Guide and Strategy for their Conservation. Vol. 3. the Americas*, Davis, S.D. et al., Eds, The World Wide Fund for Nature (WWF) and IUCN – The World Conservation Union, Oxford, UK, 1997, 417.

Killeen, T.J. and Nee, M., Un catálogo de las plantas sabaneras de Concepción, Santa Cruz, Bolivia, *Ecología en Bolivia*, 17, 53, 1991.

Killeen, T.J. et al., Diversity, composition, and structure of a tropical semideciduous forest in the Chiquitanía region of Santa Cruz, Bolivia, *J. Trop. Ecol.*, 14, 803, 1998.

Killeen, T.J. et al., Habitat heterogeneity on a forest-savanna ecotone in Noel Kempff Mercado National Park (Santa Cruz, Bolivia); implications for the long-term conservation of biodiversity in a changing climate, in *How Landscapes Change: Human Disturbance and Ecosystem Disruptions in the Americas*, Bradshaw, G.A. and Marquet P., Eds, Ecological Studies Vol. 162, Springer-Verlag, Berlin, 2001, 285.

Litherland, M., *Mapa Geológico del Area el Proyecto Precambrico*, Geological Survey of Bolivia and British Geological Survey, 1984.

Marengo, J.A. et al., Climatology of the low-level jet east of the Andes as derived from the NCEP-NCAR reanalyses: characteristics and temporal variability, *J. Climate*, 17, 2261, 2004.

Mayle, F.E., Burbridge, R., and Killeen, T.J., Millenial-scale dynamics of southern Amazonian rain forests, *Science*, 290, 2291, 2000.

Mayle, F.E. et al., Responses of Amazonian ecosystems to climatic and atmospheric carbon dioxide changes since the last glacial maximum, *Phil. Trans. R. Soc. Lond.* B, 359, 499, 2004.

McCune, B. and Mefford, M.J., *PC-ORD, multivariate analysis of ecological data. (version 4)*, MjM Software Design, Gleneden Beach, 1999.

Mostacedo, C.B. and Killeen, T.J., Estructura y composición florística del Cerrado en el Parque Nacional Noel Kempff Mercado (Santa Cruz, Bolivia), *Bol. Soc. Bot. México*, 60, 25, 1997.

Navarro, G. and Maldonado, M., *Geografía Ecológica de Bolivia: Vegetación y Ambientes Acuáticos*. Centro de Ecología Fundación Simon I, Patiño, Cochabamba, Bolivia, 2002, 719.

Navarro, G. et al., *Tipificación y caracterización de los ecosistemas del Parque Nacional Kaa-Iya del Gran Chaco (Departamento de Santa Cruz, Bolivia)*, Informe Técnico No. 36, Plan de Manejo, Proyecto Kaa – Iya, CABI-WCS-USAID, 1998.

Olsen, D.M. and Dinerstein, E., The global 200: a representation approach to conserving the earth's most biologically valuable ecoregions, *Conserv. Biol.*, 12, 502, 1998.

Pacheco, P. and Mertens, B., Land use change and agriculture development in Santa Cruz, *Bois et Forêt des Tropiques*, 280, 29, 2004.

Pennington, R.T. et al., Historical climate change and speciation: neotropical seasonally dry forest plants show patterns of both Tertiary and Quaternary diversification, *Phil. Trans. R. Soc. Lond.* B, 359, 515, 2004.

Prado, D.E., Seasonally dry forests of tropical South America: from forgotten ecosystems to a new phytogeographic unit, *Edin. J. Bot.*, 57, 437, 2000.

Prado, D.E. and Gibbs, P.E., Patterns of species distributions in the dry seasonal forests of South America, *Ann. Missouri Bot. Gard.*, 80, 902, 1993.

Ratter, J.A. et al., Observations on forests of some mesotrophic soils in central Brazil, *Revista Bras. Bot.*, 1, 47, 1978.

Steininger, M.K. et al., Clearance and fragmentation of tropical deciduous forest in the Tierras Bajas, Santa Cruz, Bolivia, *Conserv. Biol.*, 15, 127, 2001a.

Steininger, M.K. et al., Tropical deforestation in the Bolivian Amazon, *Environ. Conserv.*, 28, 127, 2001b.

Suárez-Soruco, M., Compendio de geología de Bolivia, *Revista Técnica de YPFB*, 18, 1, 2000.

Ter Braak, C.J.F., Canonical community ordination. Part I. Basic theory and linear methods, *Ecoscience*, 1, 127, 1994.

10 Inter-Andean Dry Valleys of Bolivia – Floristic Affinities and Patterns of Endemism: Insights from Acanthaceae, Asclepiadaceae and Labiatae

John R.I. Wood

CONTENTS

ABSTRACT

The various dry inter-Andean valleys in Bolivia are identified and characterized in terms of topography, climate and vegetation. The distribution of the families Labiatae, Asclepiadaceae and Acanthaceae in Bolivia is analysed. All three families show high levels of endemism in Andean

Bolivia and, in particular, in the dry inter-Andean valleys. However, the floristic affinities of the different families vary considerably, making generalizations about floristic composition of the valleys difficult. In Labiatae, the species occurring in the dry valleys are principally from genera with an essentially Andean distribution. In the case of Asclepiadaceae and Acanthaceae, there are significant chaco and cerrado elements. Genera and families with an amphitropical distribution show stronger links with the Argentinean flora. Disjunct distributions give only weak support for the Pleistocenic arc thesis. High levels of endemism imply the need for more protected areas.

10.1 INTRODUCTION

The Andes reach their widest point around 17°S in Bolivia and occupy an extensive area in the southwest of the country. Much of this consists of a high plateau, or *altiplano*, lying at around 3700 m, most of which is extremely arid and consists of undulating plain separated by relatively low mountain ranges and covered in puna-type vegetation interspersed with lakes, salt flats and scattered areas of cultivation. By contrast, the eastern slopes of the Andes are still covered in extensive areas of moist forest of various types and enjoy lengthy cloud cover as well as relatively high levels of rainfall. Between these two areas lie the dry inter-Andean valleys with their distinctive flora.

It is possible to identify three distinct areas of inter-Andean dry valleys in Bolivia (Figure 10.1): the Yungas of La Paz, the Río Grande basin and the Río Pilcomayo valley system. These are also

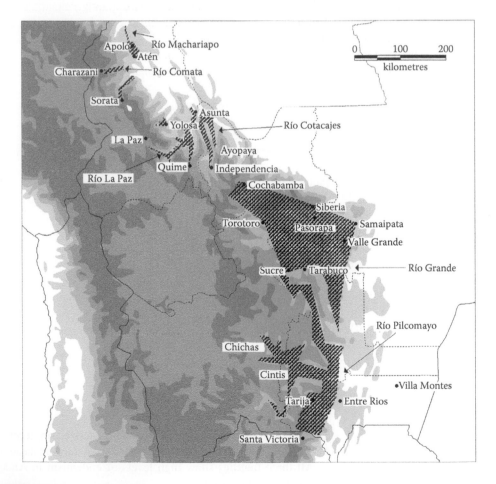

FIGURE 10.1 Inter-Andean dry valleys in Bolivia and principal places mentioned in the chapter. (Base map © Collins Bartholomew Ltd, 2004, reproduced by kind permission of Harper Collins Publishers.)

shown in the outline map of the distribution of seasonally dry vegetation given in Pennington et al. (2000), which conforms closely to the map in Killeen et al. (1993). All these areas have a prolonged dry season of about 8 months, and most rainfall is of short duration but often in the form of intense storms, with an annual average of around 500–600 mm (Killeen et al., 1993), or less in some regions (López, 2003b). Much of the area is covered in deciduous bushland, with cacti and leguminous trees and shrubs conspicuous. This is perhaps a degraded form of the true dry forest that still survives in restricted areas. There are also extensive areas largely devoid of trees and shrubs, including a few areas of ancient* grassland, particularly near Apolo in the north and along the north–south transition to the Tucuman-Bolivian forest in the extreme east.

10.2 INTER-ANDEAN DRY VALLEYS IN BOLIVIA

10.2.1 THE YUNGAS OF LA PAZ

These consist of mostly very steep-sided, often isolated, narrow valleys lying in areas of rain shadow. In general, the higher slopes receive significantly more cloud and rain, and are mostly covered in Andean cloud forest, except at the upper ends of the valleys, where vegetation of a puna type can be found. Vegetation is increasingly xerophytic towards the valley bottoms. Transition from one vegetation type to another often takes place over very short distances, both on the valley sides and on the ridges above, where one side may have a completely different vegetation from the corresponding slope on the other side. Five main dry valley systems can be identified in this region.

1. The northern valleys of Saavedra and Tamayo provinces. There are at least three distinct areas of dry valley vegetation here. The best known and best studied (Kessler and Helme, 1999) lies in the upper reaches of the Río Tuichi and its tributaries, the Río Ubito and the Río Machariapo, at around 1100–1400 m altitude near Apolo. This contains almost certainly 'the largest well-preserved example of Andean dry forests remaining in Bolivia and probably anywhere in the Andes' (Kessler and Helme, 1999). The second example is small and degraded but quite species rich, and lies in the valley of the Río Comata below Charazani, extending from around 1400 m to 3000 m. The third, most extensive but least typical, lies in the region between Apolo and Atén, and consists of a series of periodically burnt, cerrado-like, grassy plains and low hills with extensive areas of gallery forest at around 1350–1800 m in altitude.
2. The valleys in the Sorata region lying at altitudes from around 2700 m to as low as 600 m, mostly in Larecaja province. These are now very degraded, and there is much erosion in the region. At higher altitudes, there is little or no forest and relatively few shrub species, but some degraded forest and scrub survive at lower altitudes.
3. A small area of dry, seasonally burnt grassland merging into scrub and forest in the deep valley of Nor Yungas province near Yolosa, between about 850 m and 1500 m.
4. The Río La Paz and Río Bopi river systems mostly in Sud Yungas province. These dry valleys extend from well over 3000 m below La Paz to around 600 m near Asunta. There is much erosion in this area but also isolated patches of good deciduous forest in the Río Bopi valley, for example, and areas of rough, seasonally burnt grassland with some cerrado-like vegetation.

---

* The term *ancient* is used in this chapter to describe a habitat that is natural but which may have been altered or extended by human intervention in some way, such as burning or grazing. Ancient grassland is often very species-rich.

5. The Inquisivi valley and the neighbouring valleys of Ayopaya province in Cochabamba department descending into the Río Cotacajes. These valleys have a distinct flora and some excellent relict dry woodland, particularly in the Río Cotacajes itself. Dry valley vegetation of different kinds can be found from as low as 800 m to above 2800 m near Quime and Independencia, where, at higher altitudes in this zone, there is often a transition to Ceja de Monte Yungeña vegetation (Killeen et al., 1993). This is essentially a ridge-top vegetation, often on acid soil, with species of Ericaceae, *Clusia*, *Myrica* and *Weinmannia*, which is transitional between moist Andean cloud forest and more xerophytic vegetation types such as puna or dry valley vegetation.

10.2.2 THE RIO GRANDE BASIN

This is a large area drained mostly by the Río Grande and its principal tributary, the Río Mizque, and covers extensive areas of Cochabamba, Chuquisaca and Santa Cruz departments, with some areas of Potosí department. Important centres of population, including Cochabamba, Sucre and Vallegrande, are contained within this area. This whole region consists of a series of ridges and valleys bounded to the west by the altiplano and to the north-east by the cloud-covered and forested Andean escarpment. Flatter areas are often cultivated, particularly where irrigation is possible or on ridges where there is sufficient rainfall. The area contains some deep dry valleys, notably those of the Río Grande and Río Mizque themselves, but they are neither so steep-sided nor so isolated as in the Yungas; rainfall, and consequently the vegetation, is relatively uniform throughout much of the area.

Vegetation change is mostly gradual and related to altitude, altering slowly from a puna-type vegetation with only scattered shrubs and perennial herbs found on the higher slopes above 3000 m to the north and west of Cochabamba and to the west of Torotoro and Sucre, to that found at lower altitudes reaching to below 1000 m in the east. Much of the vegetation is what is best described as open dry bushland, but well-developed dry forest is found in some areas, notably near Pasorapa and in the more remote parts of the Río Grande valley. Along the north-east of this area, where the dry valleys meet the humid forests of the Amboró and Carrasco Parks at an altitude of 3000 m or less, there is a transitional zone, often of only a few hundred metres, with Ceja de Monte Yungeña vegetation. The eastern edge of this region is slightly to the west of the 65° longitude line and is marked by a broken line of moist Tucuman-Bolivian forest (Killeen et al., 1993), in which *Podocarpus parlatorei* Pilg., *Crinodendron tucumanum* Lillo and various species of Myrtaceae are common. Where this forest is present the demarcation is clear, but quite frequently there is an ill-defined transition to another dry vegetation type, the chaco-serrano forest (Killeen et al., 1993), which differs in little more than its denser tree cover and the presence of more climbers, especially various species of Bignoniaceae. Although this is the most extensive dry valley system in Bolivia, only parts have been studied intensively (Ibisch and Rojas, 1994; Antezana and Navarro, 2002), but a good general account of the vegetation is given in Navarro and Maldonado (2002).

10.2.3 THE RIO PILCOMAYO VALLEY SYSTEM

This area of dry valley vegetation is separated from the Río Grande basin by the ridge connecting Sucre and Tarabuco, and extending south-east along the Cordillera de los Sombreros. Much of this drainage system in Chuquisaca, Potosí and Tarija departments consists of deep, steep-sided, arid valleys sometimes forming narrow canyons. Vegetation is of the open bushland type and can be found from about 3000 m at its upper limits, where it merges into puna vegetation, to below 1000 m above Villa Montes, where it is clearly merging into the extensive dry scrub vegetation of the Chaco. However, in two areas the valleys open out. The first of these is in the Cintis of Chuquisaca and the neighbouring Chichas of Potosí department, where a relatively high-altitude dry valley vegetation, often called *prepuna*, has developed above about 2500 m. This has been carefully described (López, 2000; López and Beck, 2002) and is particularly rich in cacti but remarkable for its poverty in species from the three principal plant families under discussion in this chapter. The second is the basin in which the city of Tarija is situated at around 2000 m. Here are extensive areas of dry bushland and open woodland merging into puna-type vegetation on higher slopes. The eastern limits are marked by areas of Tucuman-Bolivian forest in the Entre Ríos region.

Exact definition of the Andean dry valleys is not easy. My interpretation appears to differ somewhat from that of López (2003b), who excludes the areas north of La Paz, as well as that of Navarro and Maldonado (2002), who distinguish between a biogeographical province of the Yungas, in which the northern valleys lie, and a Tucuman-Bolivian province in which the Río Grande and Pilcomayo valleys lie. However, the northern valleys are consistently included by others (Killeen et al., 1993; Kessler and Helme, 1999; Pennington et al., 2000). I also disagree with López about the lower limits of the dry valley vegetation. It is not easy to put an exact altitude on this, but it is certainly below 1000 m both in the Yungas of La Paz and in the eastern part of the Río Grande basin. Boundaries are not fixed either, and transitional zones are found in almost all areas where the dry valley vegetation changes to another type. Changes from one type to another are often relatively abrupt and may take place over only a few hundred metres. However, my own field observations tend to support the view that these transitional zones and other atypical habitats (moist forested gullies in otherwise dry areas, for example) are often precisely the places where very local endemic species occur. *Salvia amplifrons* Briq., *S. cardenasii* J.R.I. Wood, *Justicia chuquisacensis* Wassh. & J.R.I. Wood, *Lepechinia* sp. nov. (*Wood, Mendoza & Vidal* 19701), *Philibertia fontellae* Goyder, *Oxypetalum fuscum* Goyder & Fontella and *Aphelandra kolobantha* Lindau are all typical of these transitional habitats. It should also be noted that some characteristic endemic species of the dry valleys are also common in other vegetation types. A good example is the tree *Cardenasiodendron brachypterum* (Loes.) F.A. Barkley, which also occurs frequently in the chaco-serrano formation. These factors make it difficult to give precise figures about endemism and distribution.

10.3 FLORISTIC AFFINITIES AND ENDEMISM BY FAMILY

This chapter looks closely at the distribution of species of three plant families (Acanthaceae, Asclepiadaceae and Labiatae) that occur in the dry valleys of Bolivia. These families have been chosen not because they are particularly important in this zone but because they have been the subject of intense study by the author and a number of colleagues over many years. Papers on the Acanthaceae of Bolivia and neighbouring countries (Wasshausen, 1999; Wasshausen and Wood, 2001; Schmidt-Lebuhn, 2003; Wasshausen and Wood, 2003a,b, 2004a–c) and Asclepiadaceae (Goyder, 2003, 2004a,b; Goyder and Fontella 2005) have recently appeared, while others on Asclepiadaceae (D.J. Goyder, Royal Botanic Gardens, Kew, Richmond) and Labiatae (R.M. Harley, M. Mercado, A.N. Schmidt-Lebuhn and J.R.I. Wood) are in active preparation. The taxonomy and distribution of the species in these families is thus relatively well known, so insights on the floristic affinities of the Bolivian dry valleys taken from these families have a sound taxonomic foundation. This approach is different to that used by López, who drew his conclusions from a comprehensive list of species occurring in the Bolivian dry valleys, which he drew up from a variety of sources (López, 2003b).

Even with the three families under discussion, there are important uncertainties. At least two genera in Labiatae (*Minthostachys* and *Stachys*) are in urgent need of revision, and there is still considerable uncertainty over species delimitation in these genera in Bolivia and neighbouring countries. In Asclepiadaceae, the problem is more at the generic level. Whereas the species are now well delimited in most cases, this is not so at the generic level. In the case of the small-flowered genera such as *Cynanchum*, *Ditassa* and *Metastelma*, ongoing phylogenetic study has not yet resulted in clear generic definitions (D.J. Goyder, Royal Botanic Gardens, Kew, Richmond, pers. comm.). This makes comparing data from other countries in even quite recent publications difficult

(Brako and Zarucchi, 1993; Jorgensen and León-Yáñez,1999; Zuloaga and Morrone, 1999), as the generic concepts of the authors are not necessarily the same as those used here. One thing, however, is clear: the number of genera recognized in Asclepiadaceae will fall (Goyder, 2001, 2004a,b), and those hitherto regarded as endemic to Bolivia (*Corollonema*, *Dactylostelma*, *Fontellaea* and *Stelmatocodon*) will disappear into synonymy.

10.3.1 LABIATAE

Labiatae is a large, mainly temperate family, which is also well represented in tropical regions, particularly by the large genera, *Hyptis* in America and *Plectranthus* in the Old World. Table 10.1 shows that in Bolivia there are 14 native genera with 115 species. More than half the genera and nearly half the species (53) are essentially tropical in distribution. Of the remaining genera, one, *Minthostachys*, is entirely Andean; two, *Lepechinia* and *Hedeoma*, are found on the mountains of both North and South America (*Lepechinia* also in Hawaii); and the remaining three, *Clinopodium*, *Salvia* and *Stachys*, are widely distributed in both hemispheres in temperate regions and on mountains in the tropics. Nearly all species of Labiatae occurring in the dry valleys of Bolivia (48 out of a total of 56) belong to these essentially non-tropical genera.

In Bolivia, 27 Labiatae species, mostly from non-tropical genera, representing almost a quarter of the total of 115 species, are endemic to the country. All but three of these are Andean in distribution. Of Labiatae endemic to Bolivia, an impressive 74% (20 species) are endemic to the dry valleys or, in a few cases, to the transitional zones at their edge.

Table 10.2 shows the diversity of species in five of the seven non-tropical Labiatae genera. *Hedeoma* has been excluded because there is only a single Andean species, while *Stachys* has been excluded because of uncertain species delimitation, although it appears to conform to the pattern shown in the table. This shows very clearly that all five genera are more diverse to the north of Bolivia, and that the number of species steadily declines southwards from the greatest centre of diversity in Peru. This kind of distribution suggests that these genera have colonized southern South America along the Andean chain from the north, although no phylogenetic studies have been published to evaluate this.

TABLE 10.1
Labiatae Genera and Species in Bolivia

Genus	Native Species	Endemics	Andean Endemics	Species in Dry Valleys	Dry Valley Endemics
Clinopodium	7	2	2	6	2
Eriope	2	0	0	0	0
Hedeoma	1	0	0	0	0
Hypenia	2	0	0	0	0
Hyptidendron	3	0	0	1	0
Hyptis	41	3	0	6	0
Lepechinia	8	3	3	5	2
Marsypianthes	1	0	0	0	0
Minthostachys	6	3	3	?5	3
Ocimum	2	0	0	1	0
Peltodon	2	0	0	0	0
Salvia	30	14	14	26	12
Scutellaria	5	0	0	4	0
Stachys	5	2	2	2	1
Total	115	27	24	56	20

TABLE 10.2
Distribution of Five Genera of Andean Labiatae by Country

Country	Clinopodium	Lepechinia	Salvia	Scutellaria	Minthostachys
Ecuador	10	9	44	14	?3
Peru	26	11	74	15	?8
Bolivia	6	8	31	5	?6
Argentina and Chile	5	4	19	6	1

When we look at the phytogeographical elements represented by the 55 individual species occurring in the dry valleys, and using the same categories as those of López (2003b), we get the following results.

Category 1 (species endemic to Bolivia)	20
Category 2 (Andean species occurring from Bolivia southwards)	10
Category 3 (Andean species occurring in Peru, Bolivia and Argentina)	7
Category 4 (species occurring in the South American lowlands)	5
Category 5 (Andean species occurring from Bolivia northwards)	12
Category 6 (species widespread in South American lowlands)	0
Category 7 (amphitropical* species)	0
Category 8 (widespread neotropical)	2
Category 9 (pantropical or cosmopolitan species)	0

As well as showing the high level of endemism in Labiatae in the dry valleys of Bolivia, these figures also confirm that the family is almost entirely represented by Andean species in these valleys. These figures do not, however, confirm López's results that Andean species occurring from Bolivia southwards (category 2) are predominant, suggesting that there are considerable differences between individual families and possibly between different regions of the Bolivian dry valleys.

10.3.2 ASCLEPIADACEAE

Asclepiadaceae is a large tropical family that is here treated in the traditional sense as separate from Apocynaceae. Unlike most tropical families such as Acanthaceae, it shows a strong amphi-tropical tendency, particularly in the New World. It is most diverse around the tropics of Cancer and Capricorn, where it finds the drier habitats it tends to favour, but it is noticeably less diverse, for example, in Ecuador and Amazonian Brazil. There are 156 species known from Bolivia, but the number of genera is uncertain until the results of ongoing phylogenetic research become available. In Bolivia 45 species, representing more than a quarter of the total present, are endemic. This total is, however, artificially low because of a freak of history that gave the Los Toldos region to Argentina after a boundary realignment in the 1930s following the Chaco War. If the zone around Santa Victoria almost within sight of the Bolivian border had remained part of Bolivia, there would have been five additional endemic Asclepiadaceae, raising the number of endemic species to over 30% of the total. Of Asclepiadaceae endemic to Bolivia, 33 (or about 73%) are endemic to the Andean region and 23 (or just over 50%) are found in the dry valleys, although sometimes in transitional zones, principally with the Tucuman-Bolivian forest region.

Table 10.3 shows the distribution of Asclepiadaceae by genus in Bolivia. The generic concepts require some explanation, as they are here used in an aggregate sense. With *Matelea* is included

---

* Amphitropical species are those that are distributed around 20–25°N and 20–25°S but are absent from Equatorial regions. An alternative term is *amphi-Equatorial* (Daniel, 2005).

TABLE 10.3
Genera and Species of Asclepiadaceae in Bolivia

Genus	Native Species	Endemics	Andean Endemics	Species in Dry Valleys	Dry Valley Endemics
Asclepias	6	0	0	4	0
Barjonia	1	0	0	1	0
Blepharodon	6	2	0	1	0
Fischeria	2	0	0	0	0
Funastrum	3	1	1	2	1
Hemipogon	3	1	1	2	1
Macrocepis	1	0	0	0	0
Marsdenia	9	2	0	1	0
Matelea s.l.	25	9	7	7	3
Morrenia s.l.	7	0	0	3	0
Nephradenia	2	0	0	0	0
Oxypetalum	16	4	3	4	2
Philibertia	25	8	8	25	8
Schistogyne	3	1	1	3	1
Schubertia	3	1	0	1	0
Tweedia	1	0	0	0	0
Widgrenia	1	0	0	0	0
Miscellaneous small-flowered spp.	c.42	16	12	20	7
Total	156	45 + 5[a]	33 + 5[a]	74	23 + 5[a]

[a] The additional five in the total for endemic species indicates the additional five species that would be added to the total if the Santa Victoria area of Los Toldos, Argentina, had remained part of Bolivia. *Salvia atrocyanea* and *Dyschoriste axillaris* Wassh. & J.R.I. Wood would also be additional to the total in Labiatae and Acanthaceae, respectively.

Gonolobus as well as *Pseudibatia* and *Chthamalia*; with *Morrenia* is included *Araujia*; with *Oxypetalum*, *Metoxypetalum* and *Dactylostelma* (Goyder 2004a); and within *Philibertia* the range of genera included by Goyder (2004b), that is *Ambystigma*, *Fontellaea*, *Melinia*, *Mitostigma*, *Podandra* and *Stelmatocodon*. The small-flowered species is a blanket term to cover a variety of genera including *Amphistelma*, *Cynanchum*, *Ditassa*, *Metastelma*, *Petalostelma*, *Stenomeria* and *Tassadia*, the boundaries of which remain unclear until the results of ongoing research are published.

The only genus in this family in Bolivia that has a clearly Andean distribution is *Philibertia*. This is wholly South American and centred almost exactly on the tropic of Capricorn. Its distribution is shown in Table 10.4. It conforms closely with López's results (2003b), with eight Bolivian endemics and 15 species occurring in Bolivia and Argentina (category 2) but only two from Bolivia

TABLE 10.4
Distribution of *Philibertia* by Country

Country	Number of Spp.	Number of Endemics
Ecuador	1	0
Peru	6	2
Bolivia	25	8
Argentina (and Chile)	28	12
Paraguay and Brazil	0	0

TABLE 10.5
Distribution of *Morrenia* s.l. (Including *Araujia*) by Country

Country	Number of Spp.	Number of Endemics
Peru	0	0
Bolivia	7	0
Argentina	11	3
Paraguay	7	1
Brazil	6	0

northwards (category 5). When we look at the figures for all Asclepiadaceae species occurring in the dry valleys, a slightly different picture emerges.

Category 1 (species endemic to Bolivia)	23
Category 2 (Andean species occurring from Bolivia southwards)	30
Category 3 (Andean species occurring in Peru, Bolivia and Argentina)	2
Category 4 (species occurring in the South American lowlands)	14
Category 5 (Andean species occurring from Bolivia northwards)	4
Category 8 (widespread neotropical)	1

While these figures still support López's results, they also indicate the presence of a significant element from the lowlands of South America in the dry valleys of Bolivia. This is not at all surprising given the distribution of the generic groupings listed in the table. *Marsdenia* is pantropical, while *Asclepias*, the *Matelea* group and *Oxypetalum* are widespread in tropical America. However, two more specific phytogeographical patterns are hidden within these figures.

The first of these is a chaco element exemplified by the genus *Morrenia* and its close ally *Araujia*, which is shown in Table 10.5. This generic grouping is clearly centred on the Chaco region as defined by Prado (1993), and three species enter the dry valleys, one of which, *Morrenia odorata* (Hook. & Arn.) Lindl., is one of the most conspicuous species of the area extending north to the La Paz area (Figure 10.2). Another, *Araujia plumosa* Schltr., is remarkable for its occurrence in

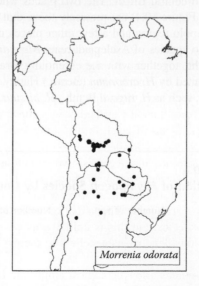

FIGURE 10.2 Distribution of *Morrenia odorata*, a species of the Chaco extending into the inter-Andean dry valleys.

FIGURE 10.3 Distribution of *Funastrum gracile*, a species of the Chaco extending deep into the inter-Andean dry valleys.

isolated populations in the most northerly of the dry valleys (that of the Río Machariapo north of Apolo), far from its centre of distribution in the Chaco and indicating strongly both the links between the dry valleys as defined in this chapter and the importance of the chaco element in their flora. Nor is the chaco element limited to *Morrenia* and its allies. The genus *Schubertia* shows a similar distribution, as does *Funastrum gracile* (Decne.) Schltr., which is widespread and typical of the Chaco lowlands but extends deep into the Andean dry valleys of Bolivia (Figure 10.3).

The second of these patterns is more surprising. This is a cerrado element shown in the distribution of the genus *Hemipogon*, which is virtually restricted to the cerrados of Brazil (Table 10.6). One species, *Hemipogon sprucei* E. Fourn., has a strikingly disjunct distribution, occurring in dry Andean valleys of Peru and Bolivia far from its main centre of distribution in the cerrados of eastern Bolivia and central Brazil. The two places where it is found in the Bolivian Andes (Apolo and Samaipata; Figure 10.4) are precisely those that show phytogeographical links with the cerrado. It is in the Apolo region and a few other places in the Yungas of La Paz where occur two other typical cerrado species of Asclepiadaceae, *Blepharodon lineare* (Decne.) Decne. and *Oxypetalum crispum* Wight, together with the essentially Brazilian cerrado Labiatae genus *Hyptidendron*, which is represented by *H. arboreum* (Benth.) Harley. Here too occur several species of *Hyptis* typical of the cerrado, such as *H. rugosa* Benth., *H. hirsuta* Kunth and *H. lantanifolia* Poit.

TABLE 10.6
Distribution of *Hemipogon* Species by Country

Country	Number of Spp.	Number of Endemics
Peru	1	0
Bolivia	3	1
Argentina	0	0
Paraguay	1	1
Brazil	12	10

FIGURE 10.4 Disjunct distribution of *Hemipogon sprucei*.

At Samaipata, *Hemipogon sprucei* is accompanied by another characteristic cerrado Asclepiadaceae, *Barjonia erecta* K. Schum., and yet another, *Oxypetalum capitatum* Mart., grows nearby. More dramatic confirmation of these links is found in the presence of two endemic species. One of these, *Hemipogon andinus* Rusby, is restricted to the grasslands of Apolo and a few other similar places in the Yungas of La Paz, and its existence provides evidence not only for the phytogeographical links of this region with the cerrados but also that the grasslands of this region are both ancient and natural. The endemic Acanthaceae, *Ruellia antiquorum* Wassh. & J.R.I. Wood, probably a fire climax species, similarly indicates that the grassy slopes of the Samaipata region are natural and suggests that they too have strong phytogeographical links with the cerrados.

10.3.3 ACANTHACEAE

Acanthaceae is another large, almost entirely tropical family. Table 10.7 shows that in Bolivia there are 166 native species in 29 genera. Over half the species (95) are contained in the two large pantropical

TABLE 10.7
Genera and Species of Acanthaceae in Bolivia

Genus	Native Species	Endemics	Andean Endemics	Species in Dry Valleys	Dry Valley Endemics
Aphelandra	11	5	5	1	1
Dicliptera	7	2	2	4	1
Dyschoriste	5	2	2	4	2
Justicia	67 (+ 5 subspp.)	19 (+ 5 subspp.)	17 (+ 2 subspp.)	11 (+ 2 subspp.)	5 (+ 2 subspp.)
Lepidagathis	3	0	0	0	0
Mendoncia	7	0	0	0	0
Ruellia	28	5	4	2	1
Stenandrium	3	0	0	2	0
Stenostephanus	9	7	7	0	0
Suessenguthia	5	4	3	0	0
19 genera with < 3 spp.	21	0	0	3	0
Total	166	44	40	27	10

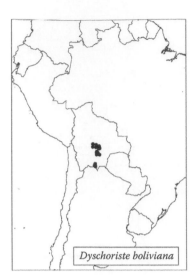

FIGURE 10.5 Distribution of *Dyschoriste boliviana*, which marks the transitional line between the Tucuman-Bolivian forest belt and the eastern edge of the dry valleys.

genera, *Justicia* and *Ruellia*. In Bolivia 44 Acanthaceae species, representing more than a quarter of the total of 166, are endemic, and all but four of these are Andean in distribution. Endemism in Bolivia is highest in the three entirely American genera characteristic of moist Andean hill forest: *Aphelandra*, *Stenostephanus* and *Suessenguthia*. These genera do occur sporadically in the lowlands but hardly enter the dry valleys. Of Acanthaceae endemic to Bolivia, only about 23% (10 species) are endemic to the dry valleys, and three of these are restricted to transitional zones (*Aphelandra kolobantha*, *Justicia chuquisacensis* and *Ruellia antiquorum*), while one (*Dyschoriste boliviana* Wassh. & J.R.I. Wood; Figure 10.5) is perhaps more common in the Tucuman-Bolivian forest zone.

Apart from *Aphelandra*, one species of which occurs in specialized habitats in the dry valleys, no genus of Acanthaceae with an Andean distribution is recorded from the dry valleys. All other genera represented are pantropical, with the two exceptions of *Anisacanthus* and *Tetramerium*, which are wholly American, with their centres of diversity in the northern hemisphere, and represented in Bolivia by a single species each. Again using the same categories as López (2003b) to look at the 31 species occurring in the dry valleys, we get the following results.

Category 1 (species endemic to Bolivia)	10
Category 2 (Andean species occurring from Bolivia southwards)	4
Category 3 (Andean species occurring in Peru, Bolivia and Argentina)	1
Category 4 (species occurring in the South American lowlands)	11
Category 5 (Andean species occurring from Bolivia northwards)	3
Category 8 (widespread neotropical)	2

Once again, these figures confirm the high level of endemism in the dry valleys of Bolivia (although the figure for the Andean region as a whole would be even higher in this case). In Acanthaceae, it is not always easy to distinguish between Andean and lowland species from southern South America (*Dicliptera jujuyensis* Lindau is a case in point), but those with a lowland distribution predominate, and this is not a surprising result for an essentially tropical family. They do not support López's findings that the largest group is composed of Andean species occurring from Bolivia southwards.

Within category 4 can be seen the same two phytogeographical elements identified under Asclepiadaceae, but the chaco element is predominant. Several species of *Justicia*, including

FIGURE 10.6 Distribution of *Dyschoriste venturii*, which extends from the Chaco into the dry inter-Andean valleys of Bolivia.

J. goudotii V.A.W. Graham, *J. saltensis* de Marco & Ruiz, *J. squarrosa* Griseb. and *J. thunbergioides* (Lindau) Leonard, as well as *Anisacanthus boliviensis* (Nees) Wassh. and *Dyshoriste venturii* Leonard (Figure 10.6), both representatives of essentially amphitropical genera, are good examples of species that are widely distributed in the Chaco lowlands. These species are all characteristic of the eastern, lower parts of the inter-Andean dry valleys of Bolivia, but all reach their highest altitude in this region.

The cerrado element is much less obvious, but its presence is shown by the occurrence of *Ruellia geminiflora* Kunth in the Apolo region and elsewhere in the Yungas of La Paz, and the abundance of *Justicia tocantina* (Nees) V.A.W. Graham centred on the Samaipata area (Figure 10.7), precisely the two areas where cerrado elements were found in Asclepiadaceae.

FIGURE 10.7 Disjunct distribution of *Justicia tocantina*, an essentially cerrado species occurring as a distinct species at the eastern end of the Bolivian dry valley system.

10.4 DISJUNCT DISTRIBUTIONS

Ever since a seminal paper by Prado and Gibbs (1993), there has been considerable interest in the disjunct distribution of certain dry forest species in tropical America (e.g. Pennington et al., 2000). It has been argued that these represent relict populations of widespread dry forest in the Neotropics during the late Pleistocene. Some support for this thesis is provided by the phytogeography of species in the three families under discussion here.

10.4.1 TRANS-ANDEAN LINKS

Although this chapter has shown that in one family, Labiatae, there are strong links at generic and species level with the vegetation of Peru, there are very few examples of trans-Andean links in this or the other families, and only one between the seasonally dry forests of northern Peru and the inter-Andean dry valleys of Bolivia. This is *Tetramerium wasshausenii* T.F. Daniel, the only South American representative of a mainly Mexican genus, which is known from north-west Peru and a few scattered localities in dry forest areas of Bolivia (Daniel, 1986), in the Yungas of La Paz, in the lower part of the Río Grande basin and in the Chaco (Figure 10.8). However, there are also isolated dry valleys on the western slopes of the Andes in central Peru, shown on the map in Bridgewater et al. (2003). There is a striking link between these dry Andean valleys and the main area of distribution of one of the most common and conspicuous species of the Bolivian inter-Andean dry valleys, *Salvia haenkei* Benth. (Figure 10.9). Nor is this link unique. If the three closely related and scarcely distinguishable species of *Salvia* sect. *Tomentellae* (*S. cuspidata* Ruiz & Pav. of Peru, *S. bangii* Rusby of the Río Grande basin in Bolivia and *S. gilliesii* Benth. of southern Bolivia, Argentina and Chile) are put together, exactly the same pattern of distribution is shown but with an extension south into Argentina and Chile. It seems probable that long isolation of the possible common ancestors of these species has resulted in a degree of separate evolution.

10.4.2 LINKS AROUND THE PLEISTOCENIC ARC

The distribution of *Hemipogon sprucei* (Figure 10.4) has already been commented on. This widespread and common species of cerrados in eastern Bolivia and Brazil occurs in isolated pockets in

FIGURE 10.8 Disjunct trans-Andean distribution of *Tetramerium wasshausenii* with isolated populations in north-west Peru and Bolivia.

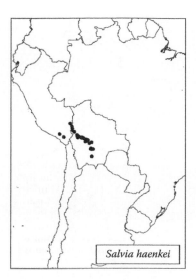

FIGURE 10.9 Disjunct trans-Andean distribution of *Salvia haenkei*.

the dry inter-Andean valleys of Peru (Tarapoto) and Bolivia (Apolo and Samaipata). This distribution conforms well to that described by Prado and Gibbs as the 'Pleistocenic arc type distribution' (Prado and Gibbs, 1993), even though *H. sprucei* is a cerrado rather than a true seasonally dry forest species. Another Asclepiadaceae, *Marsdenia macrophylla* E. Fourn., might also fit this pattern, being found along the Andes as well as in the Guianas, Brazil and Paraguay, but is hardly typical of the inter-Andean dry valleys under discussion here, where it is known from only two isolated localities.

The distribution of a number of species entering the dry valleys of Bolivia from the lowlands of South America (type 4 of López, 2003b) conforms loosely to the Pleistocenic arc distribution. At least three species of *Hyptis* occurring in the Apolo area of northern Bolivia, *H. hirsuta*, *H. lantanifolia* and *H. rugosa*, conform to this pattern, as do *Justicia corumbensis* (Lindau) Wassh. & C. Ezcurra and *J. glutinosa* (Bremek.) V.A.W. Graham, although these last two are not really typical of the inter-Andean dry forest valleys of Bolivia. More striking, however, are the distributions of two other species: *Justicia tocantina* and *J. aequilabris* (Nees) Lindau.

Justicia tocantina has a disjunct distribution in three separate areas. Its main centre is the cerrado area of central Brazil in Tocantins and Goiás states. There is an isolated population on Serro León in Paraguay and a much larger one in Bolivia, in the chaco-serrano and dry inter-Andean valleys from Samaipata south to Monteagudo (Figure 10.7). This distribution conforms to the Pleistocenic arc, but the populations have clearly been isolated for a long period, as those in Bolivia are distinguished by a series of minor morphological characters resulting in recognition at infraspecific rank as subspecies *andina* Wassh. & J.R.I. Wood (Wasshausen and Wood, 2004a).

Justicia aequilabris shows a similar kind of distribution, extending from Bahia state in Brazil to deep inside the Río Grande basin in Andean Bolivia (Figure 10.10). Once again, relatively minor but significant morphological differences point to the long isolation of these populations and allow recognition of subspecies. Subspecies *riograndina* Wassh. & J.R.I. Wood is common in the Río Grande basin in Bolivia, endemic to it, and readily distinguished from other subspecies by its acuminate bracts.

It is interesting that most species in these families that show a disjunct distribution, conforming loosely to the Pleistocenic arc, are essentially cerrado species rather than species of seasonally dry tropical forests. It is also in species such as *Justicia aequilabris* or genera such as *Hemipogon* where the best evidence for ongoing evolution of species can be discerned. However, until phylogenetic

FIGURE 10.10 Disjunct distribution of *Justicia aequilabris*, a dry forest species of central Brazil occurring in the Río Grande basin as an endemic subspecies, *riograndina*.

studies of the kind described by Pennington et al. (2004) are published for the three families under discussion here, this cannot be demonstrated conclusively. Finally, it is perhaps worth noting that although the two habitats are well defined, they occur in close geographical proximity in the Apolo area, in the Yungas of La Paz and in the Samaipata area, and climate change will have affected both.

10.4.3 AMPHITROPICAL DISTRIBUTIONS

Ruellia erythropus (Nees) Lindau is a common species of the Chaco region in South America extending into the inter-Andean dry valleys of Bolivia. However, it has a striking distribution, occurring in North America in dry forest in two separate parts of Mexico (Figure 10.11). A somewhat similar

FIGURE 10.11 Disjunct amphitropical distribution of *Ruellia erythropus* (data partly from Ezcurra, 1993).

distribution has been recently demonstrated for the essentially chaco species *Justicia ramulosa* (Morong) C. Ezcurra, which is found not only in Peru but in Honduras as well (Daniel, 2005). These are the only two species with an amphitropical distribution within the taxa studied here, but it may be of some significance. This kind of distribution has already been noted in the case of Asclepiadaceae and of the two genera *Dyschoriste* and *Anisacanthus* in Acanthaceae. It certainly occurs in the genus *Brongniartia* (Leguminosae) and the monospecific genus *Koeberlinia*, which is often separated into its own family, Koeberlinaceae.

10.5 PATTERNS OF ENDEMISM

Endemism in the dry inter-Andean valleys of Bolivia is not random. Although it is difficult, as yet, to pinpoint concentrations of endemic species within these valleys, it is possible to pick out distinct patterns of distribution based on our knowledge of the three families.

10.5.1 ENDEMISM IN ISOLATED VALLEYS

It is clear from an examination of the distribution of different species in *Justicia* that endemism is a feature of individual valleys (Figure 10.12) in the Yungas. *Justicia kessleri* Wassh. & J.R.I. Wood is unique to the Río Machariapo and Río Ubito north of Apolo. An undescribed species is found in the same area and also in the Río Comata below Charazani. In the dry valleys of Nor and Sud Yungas provinces, *J. rusbyana* Lindau is a common species but occurs nowhere else. In the valleys of Inquisivi and Ayopaya, it is replaced by *J. pluriformis* Wassh. & J.R.I. Wood. All these species are locally frequent but found nowhere outside a very restricted area in one or more valleys, and the pattern would be even more impressive if endemic *Justicia* species from moist forest in these valleys were also included. Other species from different genera and families show similarly restricted distributions to one or more of these valleys: *Aphelandra kolobantha* (Sorata, Quime and Ayopaya), *Philibertia fontellae* (Ayopaya and Quime), *P. peduncularis* (Benth.) Goyder (Charazani and Sorata) and *Minthostachys andina* (Britton ex Rusby) Epling (Sorata).

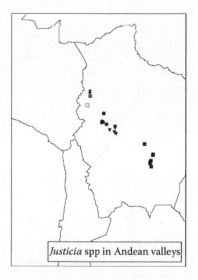

FIGURE 10.12 Localized distribution of endemic *Justicia* species in the dry valleys (★, *J. kessleri*; □, *Justicia* sp. nov. (Wood et al. 19826); ●, *J. rusbyana*; ▼, *J. pluriformis*; ■, *J. chuquisacensis;* the star in square represents two species in the same locality).

10.5.2 Endemism along the Transition to the Tucuman-Bolivian Forest

The distribution of *Dyschoriste boliviana* (Figure 10.5) illustrates clearly the distribution of a species that extends along a north–south axis from the Samaipata area southwards. Some species extend along this axis for only a short distance (*Justicia chuquisacensis*, Figure 10.12; *Philibertia tactila* Goyder), whereas a few, such as *Dyschoriste boliviana* or *Oxypetalum fuscum*, are frequent from Comarapa and Samaipata south to the border with Argentina. Others — such as *Salvia atrocyanea* Epling, *Philibertia multiflora* (T. Mey.) Goyder, *P. boliviana* (Baill.) Goyder and *P. boliviensis* (Schltr.) Goyder — are not strictly endemic, as they extend into Argentina for a short distance, but they conform closely to this pattern. This kind of distribution also suggests the importance of transitional zones in the distribution of endemics. Although difficult to demonstrate convincingly, my own field observations suggest that endemic species often occur at a point of change in the vegetation.

10.5.3 Endemics of the Río Grande Basin

The Río Grande basin is remarkable for a series of endemic species that are widespread and often frequent within a large part of this area, but unknown elsewhere. Figure 10.13 shows the distribution of one species with this kind of distribution, *Salvia praeclara* Epling. What is surprising is how very similar distributions are repeated in different genera in all three families. In Labiatae, *Salvia bangii*, *S. orbignaei* Benth., *S. retinervia* Briq., *S. sophrona* Briq., *Lepechinia bella* Epling and *Clinopodium axillare* (Rusby) Harley all conform to this pattern. In Asclepiadaceae, *Philibertia globiflora* Goyder, *P. velutina* Goyder, *Funastrum* sp. (*Wood* 9360) and *Metastelma* sp. (*Wood* 7655) are similar in distribution, while in Acanthaceae, *Justicia consanguinea* (Lindau) Wassh. & C. Ezcurra, *J. aequilabris* subsp. *riograndina*, *Dyschoriste prostrata* Wassh. & J.R.I. Wood and *Dicliptera cochabambensis* Lindau repeat this pattern. This last is the only anomalous species in this list, as it also occurs uniquely in the Tarija valley. The rarity of floristic links of this type between the Tarija area and the Río Grande basin is, in fact, noteworthy.

Although the endemic species mentioned in the previous paragraph are quite widespread in the Río Grande basin, there are also a good number with a much more restricted distribution.

FIGURE 10.13 Distribution of *Salvia praeclara*, one of many common endemic species in the Río Grande basin.

These vary from a new species of *Lepechinia* known only from a few plants near Siberia and *Philibertia religiosa* Goyder, known from two hills in the city of Cochabamba, to plants such as *Salvia graciliramulosa* Epling & Játiva, which is locally abundant on red sandstone outcrops in the Río Chico valley of Chuquisaca.

Very local endemic species from these families do occur in the Yungas of La Paz but are almost totally absent from the Río Pilcomayo drainage system. With 10 species of *Salvia* endemic to the Río Grande basin and these from different subsections of the genus (J.R.I. Wood, in prep.), and numerous endemics from different genera in the three families under discussion here, it is clear that this basin has been an important centre of speciation and is rich in endemics. What is not certainly known is how far this pattern is repeated over other families, and further research will be necessary to demonstrate this clearly.

10.6 CONCLUSIONS

10.6.1 ENDEMISM

Although endemism is significant in all three families discussed here, considerable variation can be observed between families in the number of endemic species, in their geographical distribution and in their frequency across different genera within a particular family. The same patterns will not necessarily be repeated in other families. However, most large and medium-sized genera in these families have at least one endemic species occurring in Bolivia, although endemism is much more marked in some genera than in others.

This chapter shows clearly that endemism in Acanthaceae, Asclepiadaceae and Labiatae in Bolivia is a significant phenomenon, with 116 out of a total of 437 accepted species (nearly 27%) in these three families endemic to the country. Endemism is concentrated in the Andean region, where 97 of the 116 endemic species are found. Within the Andean region, the dry valleys are especially rich in endemism, with 53 of the 97 Andean endemics, or almost 55%, found in the dry valleys. This last figure should be treated with some caution because of the inclusion of plants from transitional zones. However, it can safely be asserted that the dry Andean valleys are at least as important as centres of endemism as the moister valleys.

Within the dry valleys themselves, endemic species are an important proportion of the total flora. Fifty-three out of 157 species in these three families, or nearly a third of the total number occurring in the dry valleys, are endemic to them. This is higher than the total of 18.1% recorded by López (2003b) but significantly lower than the figure of 62.7% he noted for Cactaceae. This reinforces the comment made earlier about variation in the level of endemism between different families. However, the Cactaceae are likely to be atypical. Succulent families are notorious for their high level of speciation and for the narrow species concept employed by many students of these groups. Endemism is more marked in genera with an Andean distribution within South America, such as *Philibertia* or *Salvia*, than in essentially lowland genera.

Within the dry valleys themselves, endemism in the three families is more marked in areas of transition to another vegetation type, such as along the line of the Tucuman-Bolivian forest belt, in the isolated valleys of the La Paz Yungas and in the Río Grande basin, than in other areas. The Río Grande basin in particular seems to be an important centre of endemism.

10.6.2 FLORISTIC AFFINITIES

The flora of the inter-Andean dry valleys of Bolivia is a composite of various elements, of which the principal are Andean, chaco, cerrado and a widespread lowland element. Although this analysis does not give much support to the findings of various researchers, such as López (2003a,b) or López and Beck (2002), which suggest that the main floristic element in the dry valleys consists of Andean species distributed from Bolivia southwards, it does not provide very strong evidence against it either, as our sample is too small. There is, however, a strong suggestion that the composition of the flora in some parts of the dry valleys is influenced more by one element than is the case elsewhere. The more northern valleys of the Yungas appear to have a more important lowland element, as exemplified by the abundance and diversity of *Justicia* in this area. The flora of the Yungas, in particular that of the Apolo-Atén plains, and that of the Samaipata region has significant cerrado elements. That of the Río Grande basin below about 1500 m has an important chaco element. Thus it seems very probable that the southern Andean element would be of increasing importance in the southern inter-Andean valleys of the Río Pilcomayo drainage system, where López carried out most of his studies.

The distribution of individual families in the inter-Andean dry valley system is influenced, at least in part, by the general distribution of the family or genus. Thus the essentially tropical family, Acanthaceae, is better represented in the northern dry inter-Andean valleys of the Yungas than further south. Species numbers fall steadily the further away from the Equator one moves. In the case of Labiatae, species numbers also fall away from the Equator, but for other reasons. The Andean genera appear to have local centres of diversity in Peru, and their frequency falls southwards. The reverse is seen in the amphitropical family Asclepidaceae, where greater diversity is present in the southern valleys than those of the Yungas. The same pattern is seen in the amphitropical Acanthaceae genus, *Dyschoriste*.

Although a few trans-Andean links with northern and central Peru were noted, these families show few links at species level with the dry valleys further north in the Andes. This accords with the findings of Sarmiento (1975) and Bridgewater et al. (2003), who suggest that the Bolivian dry valleys constitute a different phytogeographical unit from those of northern Peru. Disjunct distributions of some lowland species along the Pleistocenic arc as described by Prado and Gibbs (1993) were noted particularly in Acanthaceae, where in two species, *Justicia tocantina* and *J. aequilabris*, long isolation of Andean dry valley populations has resulted in significant morphological divergence from populations in Brazil and elsewhere, allowing the recognition of subspecies. The infrequency of these links is noteworthy. It is possible that the herbs and undershrubs typical of the families under discussion in this chapter speciate more rapidly than the tree species discussed by Prado and Gibbs (1993), thus explaining both the development of endemic species and subspecies in the more isolated Andean valleys and the apparent rarity of links along the Pleistocenic arc.

10.6.3 Conservation Priorities

The Bolivian flora is still relatively poorly known. A significant number of the species mentioned in this chapter are newly described or still await publication. A recent paper by Wasshausen and Wood (2004a) described 21 new species of *Justicia* out of a total of 67 species occurring in Bolivia. This figure may be exceptionally high, but papers by Goyder (2004a) and Wood (in prep.) on *Philibertia* and *Salvia*, respectively, propose four new species out of a total of 27 for the former and seven out of a total of 30 for the latter. Nearly all the new species in these genera are from the Andean dry valleys, and nearly all have been collected in Bolivia for the first time within the past 20 years. This suggests that the endemic flora of the Bolivian dry valleys is still incompletely known, and many endemic species remain to be discovered. This opinion is confirmed by work on the genus *Eryngium* (Umbelliferae) (Mendoza and Watson, in prep.), which suggests that four out of a total of about 18 are new species.

The high level of endemism in the Andean dry valleys suggests that conservation efforts need to be directed to this area as well as to the humid forests of the Andean escarpment and other areas, where most of the protected areas in Bolivia are situated. At present, four areas containing Andean dry valleys are legally protected. The upper reaches of the Río Tuichi and its tributaries, the Río Ubito and the Río Machariapo, fall within the Madidi National Park. The Amboró National Park contains some dry valley areas along its southern fringe from Comarapa east to Samaipata, while two small, protected areas, those of El Palmar and Torotoro, lie almost entirely within the inter-Andean dry valley system. However, the latter was set up to conserve its caves and fossils rather than its vegetation.

The inter-Andean dry valleys contain significant centres of human population, and much of the original vegetation has been destroyed or modified as a result of agriculture and animal grazing. Few individual endemic species are under immediate threat, although many should be regarded as vulnerable because of their very restricted range or proximity to areas of rapid urbanization. There are still considerable areas of dry forest in good condition, especially in the Río Grande valley, and these merit conservation before they are seriously degraded. Efforts at conservation should perhaps be concentrated on these areas while the conservation status of individual species is evaluated. Because the area has long-established human settlements, conservation initiatives will require the active collaboration of local communities whose support will be necessary for any successful outcome. This underlines the important role that education, in the broadest sense, needs to play in order to raise public awareness and participation in conservation initiatives. A Darwin Initiative Project, in which Oxford University collaborates with four Bolivian institutions, will make recommendations for the conservation of certain areas in the Río Grande basin along the lines outlined in this paragraph.

ACKNOWLEDGEMENTS

I owe a special debt to David Goyder (Kew), Ray Harley (Kew), Dieter Wasshausen (Smithsonian Institution, Washington) and Alexander Schmidt-Lebuhn (Göttingen), who have discussed the taxonomy and distribution of the species in these families with me over many years and in some cases have shown me their unpublished results. At Oxford, I have benefited from the advice and help of my colleagues in the Department of Plant Sciences, in particular Robert Scotland, Colin Hughes, Tim Waters, Ruth Eastwood and Alex Wortley, and in the Bodleian Library that of Nigel James. In recent years my fieldwork, on which much of this chapter is based, has been financed through the Darwin Initiative.

REFERENCES

Antezana, C. and Navarro, G., Contribución al análisis biogeográfico y catálogo preliminar de la flora de los valles secos interandinos del centro de Bolivia, *Rev. Boliv. Ecol. Conserv. Ambient.*, 8, 25, 2002.

Brako, L. and Zarucchi, J.L., Eds, Catalogue of the flowering plants and gymnosperms of Peru, *Monogr. Syst. Bot. Missouri Bot. Gard.*, 45, 1, 1993.

Bridgewater, S. et al., A preliminary floristic and phytogeographic analysis of the woody flora of seasonally dry forests in northern Peru, *Candollea*, 58, 1, 2003.

Daniel, T.F., Systematics of *Tetramerium*, *Syst. Bot. Monogr.*, 12, 1, 1986.

Daniel, T.F., Catalog of Honduran Acanthaceae with taxonomic and phytogeographic notes, *Contr. Univ. Michigan Herb.*, 24, 51, 2005.

Ezcurra, C., Systematics of *Ruellia* (Acanthaceae) from South America, *Ann. Missouri Bot. Gard.*, 80, 747, 1993.

Goyder, D.J., The identity of *Hickenia* Lillo (Apocynaceae subfam. Asclepiadoideae), *Kew Bull.*, 56, 162, 2001.

Goyder, D.J., A synopsis of *Morrenia* Lindl. (Apocynaceae subfam. *Asclepiadoideae*), *Kew Bull.*, 58, 713, 2003.

Goyder, D.J., The identities of *Corollonema* Schltr., *Dactylostelma* Schltr. and *Metoxypetalum* Morillo (Apocynaceae: Asclepiadoideae), *Kew Bull.*, 59, 301, 2004a.

Goyder, D.J., An amplified concept of *Philibertia* Kunth (Apocynaceae: Asclepiadoideae) with a synopsis of the genus, *Kew Bull.*, 59, 415, 2004b.

Goyder, D.J. and Fontella, J., Notes on *Oxypetalum* R.Br. (Apocynaceae: Asclepiadoideae) in Bolivia and
    Peru, *Kew Bull.*, 60, XX, 2005.
Ibisch, P. and Rojas, P., Flora y vegetación de la provincia Arque, Departmento Cochabamba, Bolivia, *Ecol.
    Bolivia*, 22, 1, 1994.
Jorgensen, P.M. and León Yáñez, S., Eds, Catalogue of the vascular plants of Ecuador, *Monogr. Syst. Bot.
    Missouri Bot. Gard.*, 75, 1, 1999.
Kessler, M. and Helme, N., Floristic diversity and phytogeography of the central Tuichi Valley, an isolated
    dry forest locality in the Bolivian Andes, *Candollea*, 54, 341, 1999.
Killeen, T.J., García, E., and Beck, S.G., *Guía de Árboles de Bolivia*, Herbario Nacional de Bolivia–Missouri
    Botanical Garden, La Paz, 1993.
López, R.P., La Prepuna boliviana, *Ecol. Bolivia*, 34, 45, 2000.
López, R.P., Diversidad florística y endemismo de los valles secos bolivianos, *Ecol. Bolivia*, 38, 28, 2003a.
López, R.P., Phytogeographical relations of the Andean dry valleys of Bolivia, *J. Biogeogr.*, 30, 1659, 2003b.
López, R.P. and Beck, S.G., Phytogeographical affinities and life form composition of the Bolivian Prepuna,
    Candollea, 57, 77, 2002.
Navarro, G. and Maldonado, M., *Geografía Ecológica de Bolivia*, Centro de Ecología Simon Patiño, Cocha-
    bamba, 2002.
Pennington, R.T., Prado, D.E., and Pendry, C.A., Neotropical seasonally dry forests and Quaternary vegetation
    changes, *J. Biogeogr.*, 27, 261, 2000.
Pennington, R.T. et al., Historical climate change and speciation: neotropical seasonally dry forest plants show
    patterns of both Tertiary and Quaternary diversification, *Philos. Trans., Ser. B*, 359, 515, 2004.
Prado, D.E., What is the Gran Chaco vegetation in South America? 1. A review. Contribution to the study of
    the flora and vegetation of the Chaco, *Candollea*, 48, 152, 1993.
Prado, D.E. and Gibbs, P.E., Patterns of species distribution in the dry seasonal forests of South America,
    Ann. Missouri Bot. Gard., 80, 902, 1993.
Sarmiento, G., The dry plant formations of South America and their floristic connections, *J. Biogeogr.*, 2, 233,
    1975.
Schmidt-Lebuhn, A.N., A taxonomic revision of the genus *Suessenguthia* Merxm. (Acanthaceae), *Candollea*,
    58, 101, 2003.
Wasshausen, D.C., The genus *Stenostephanus* (Acanthaceae) in Bolivia, *Harvard Pap. Bot.*, 4, 279, 1999.
Wasshausen, D.C. and Wood, J.R.I., Further discoveries in the genus *Stenostephanus* (Acanthaceae) in Bolivia,
    Harvard Pap. Bot., 6, 449, 2001.
Wasshausen, D.C. and Wood, J.R.I., The genus *Dyschoriste* (Acanthaceae) in Bolivia and Argentina, *Brittonia*,
    55, 10, 2003a.
Wasshausen, D.C. and Wood, J.R.I., Notes on the genus *Ruellia* (Acanthaceae) in Bolivia, Peru and Brazil,
    Proc. Biol. Soc. Wash., 116, 263, 2003b.
Wasshausen, D.C. and Wood, J.R.I., Notes on the genus *Justicia* in Bolivia, *Kew Bull.*, 58, 769, 2004a.
Wasshausen, D.C. and Wood, J.R.I., An annotated checklist of the Acanthaceae of Bolivia, *Contr. US Natl
    Herb.*, 49, 1, 2004b.
Wasshausen, D.C. and Wood, J.R.I., Notes on the genus *Dicliptera* (Acanthaceae) in Bolivia and Peru, *Proc.
    Biol. Soc. Wash.*, 117, 117, 2004c.
Zuloaga, O. and Morrone, O., Eds, Catálogo de las plantas vasculares de la República Argentina. 11. Angiosper-
    mae (Dicotyledoneae), *Monogr. Syst. Bot. Missouri Bot. Gard.*, 74, 1, 1999.

11 Phytogeography and Floristics of Seasonally Dry Tropical Forests in Peru

Reynaldo Linares-Palomino

CONTENTS

ABSTRACT

This chapter presents an up-to-date view of the current biological knowledge and conservation status of seasonally dry tropical forests in Peru. The Peruvian seasonally dry tropical forests are distributed in three geographically distinct regions: 1) the north-western coast and western lower Andean foothills, 2) the dry inter-Andean valleys in northern, central and southern Peru, and 3) the Tarapoto and Huallaga valley region east of the Andes. Despite little biological information being available, each of these regions can be characterized by its own flora and environmental variables. All three regions show impressive figures of endemic species and floristic richness, but are highly threatened by human disturbance and little effort is being made to protect and conserve them. The floristic relationships of these forests with the seasonally dry tropical forests in adjacent Ecuador, with which they form a single phytogeographic unit, and with those in Colombia-Venezuela and Bolivia are discussed.

11.1 INTRODUCTION

Seasonally dry tropical forests in the Neotropics are highly threatened and very reduced ecosystems (e.g. Costa Rica, Quesada and Stoner, 2004; Colombia, Instituto Alexander von Humboldt (IavH), 1998; Ecuador, Dodson and Gentry, 1991; Bolivia, Parker et al., 1993). Their ecological and floristic value has been underestimated and neglected and little attention has been given to their conservation (Prado, 2003). The overall picture is no different for Peru. There, seasonally dry tropical forests were thought to be a very poor and homogeneous ecosystem. This, coupled with the fact that seasonally dry tropical forests are easily cleared through deforestation and fire and have a suitable seasonal climate, makes them attractive for cattle raising and agriculture. Consequently, much of the original forests remain today as fragments and remnants, especially in the dry inter-Andean valleys and the Tarapoto area.

Previous accounts on general vegetation in Peru almost never recognize all types of seasonally dry tropical forests. They either fail to include all the inter-Andean seasonally dry tropical forest variations, or the unique seasonally dry tropical forest formations in eastern Peru. This chapter presents a synthesis of Peruvian seasonally dry tropical forests from a floristic and botanical point of view, discussing the affinities between them as well as with other adjacent seasonally dry tropical forest formations.

11.2 SEASONALLY DRY FORESTS IN THE NEOTROPICAL REGION

11.2.1 Definition

The definition of seasonally dry tropical forests as used in this chapter will follow Pennington et al. (2000) with minor modifications (Linares-Palomino, 2004a). Thus, seasonally dry tropical forests receive less than 1600 mm of annual rainfall. However, the amount of rainfall is not evenly distributed throughout the year. A very marked dry season of more than five months is usually present, in which total rainfall is below 100 mm. With a few exceptions, these forests are of lower stature and basal area when compared to rainforests (Linares-Palomino and Ponce Alvarez, 2005). They are best represented at elevations below 1000 masl, but can occur as high as 2500–2800 masl (Linares-Palomino, 2004a). The chaco and savanna formations are not included within this definition as has been discussed previously (Linares-Palomino et al., 2003). It is necessary to point out that this is a very broad definition, including a complex mosaic of seasonally dry tropical forest habitats ranging from highly deciduous to semi-deciduous forests, riparian forest strips within the former, dense woody thickets and dry scrub vegetation, as well as the habitats that are produced when these inter-grade.

11.2.2 Distribution and Biogeography

The distribution of seasonally dry tropical forests in the neotropical region has been covered previously in general vegetation mapping studies (e.g. Smith and Johnston, 1945; and more recently Eva et al., 2002). However, the first specific study of seasonally dry tropical forests appears to be the work by Hueck (1959), who attempted to summarize the available information on seasonally dry tropical forest formations in South America (including the chaco and savannas). More recent accounts of seasonally dry tropical forests for the Neotropics are the works of Sarmiento (1975), Gentry (1995), Prado (2000), Pennington et al. (2000) and Linares-Palomino et al. (2003). According to the latter, based on floristic information, the seasonally dry tropical forests of the neotropical region can be subdivided into at least three different phytogeographic groups: 1) a Mesoamerican-Caribbean Group (including Mexico, Central America and northern South America), 2) an Ecuadorean-Peruvian Group and 3) a Bolivian-Argentinean Group (Figure 11.1). A fourth group including the extensive caatingas formations in north-eastern Brazil should be added, since they

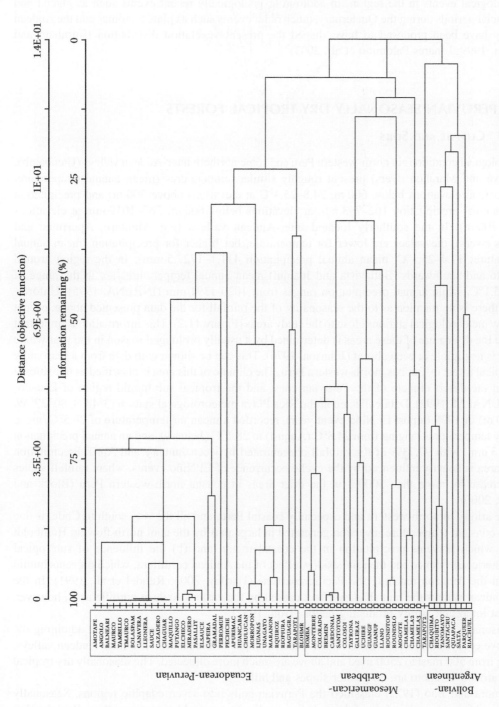

FIGURE 11.1 UPGMA cluster dendrogram for seasonally dry forests in the neotropical region (from Linares-Palomino et al., 2003).

have been included in the Pleistocenic Arc (Prado, 1991, 2003), a new phytogeographical unit in South America including seasonally dry tropical forests (Prado, 2000).

The origin of each of these groups has been attributed to complex historical, geological and climatological events in the region. In addition to geologically recent events such as glacial and interglacial periods during the Quaternary, much older events such as plate tectonics and the Andean orogeny have been proposed to have shaped the present vegetation distribution (Burnham and Graham, 1999; Linares-Palomino et al., 2003).

11.3 PERUVIAN SEASONALLY DRY TROPICAL FORESTS

11.3.1 CLIMATE AND SOILS

Climatological conditions in north-western Peru and some northern inter-Andean valleys (Utcubamba, Chamaya and Marañon rivers) present roughly similar temperatures (mean annual temperature: 23.4–25°C at elevations below 600 m, 24.8–25.4°C at elevations above 700 m) and precipitation (mean annual precipitation 162–793 mm at elevations below 600 m, 567–1019 mm at elevations above 700 m). In the southerly located inter-Andean valleys (e.g. Mantaro, Apurimac and Pampas rivers), the values are lower for temperature, but higher for precipitation (mean annual temperature: 17.4–25.1°C; mean annual precipitation 411.1–1727.5 mm). In the region around Tarapoto and southwards (Bellavista and Juanjuí) mean annual temperatures are in the range of 23.9–25.1°C, while annual precipitation ranges from 1020-1391 mm (INRENA, 1975). Unfortunately, there is no mention as to the seasonality of the rainfall for the data presented above, except for a few meteorological stations close to the study areas (Figure 11.2). This information is important because the dry nature of these areas is determined by a usually prolonged season in the year where rainfall is negligible or even absent (Johnson, 1976). This can be shown with data from a seasonally dry tropical forest in Tumbes, north-western Peru. The climate of this area is classified as transitional between the desert climate of the Peruvian coast and the tropical sub-humid region of Ecuador (CDC-UNALM, 1992). Data collected at the Rica Playa meteorological station (3°48′S, 80°27′ W, alt. 120 m) in 1999 (a post El Niño event year), recorded a mean air temperature of 26.5°C, mean monthly temperatures ranging from 24.8°C (August) to 28.1°C (January), and an annual precipitation of 582.3 mm, with 97.73% of the rainfall concentrated between January and April. Precipitation in this area is heavily influenced by the cyclic occurrence of El Niño events, where rainfall values can increase by more than 5000% in the drier areas in coastal north-western Peru (Block and Richter, 2000).

The aridity in north-western and especially coastal Peru, and all the way south to Chile, is due to (a) a constant temperature inversion generated in large part by the cool north-flowing Humboldt current which prevents precipitation in the coastal region and (b) the influence of subtropical atmospheric subsidence and the rain-shadow effect of the Andean cordillera, which prevents humid air from the Amazon reaching the Pacific coastline (Hartley, 2003; Rundel et al., 1991). In the inter-Andean valleys the rain-shadow effect caused by the eastern Andean cordilleras is, however, the most important aridity-producing factor (Troll, 1952).

Seasonally dry tropical forests are found in north-western Peru from sea level, adjoining the mangrove ecosystem, to 1800 masl on the western slopes of the Andes. The inter-Andean valleys, ranging from 500 masl to 2500 masl and above are much more dissected. The seasonally dry tropical forests around Tarapoto are on gentler slopes and hills.

Zamora and Bao (1972) classified the Peruvian soils into seven edaphic regions. Seasonally dry tropical forests occur in four of them. 1) Seasonally dry tropical forests in northern Peru develop on soils of the Yermosolic region, which are soils of the dry coastal lowlands up to 1000 masl. The region includes wide sedimentary plains, low hills, elevated marine terraces and many east–west

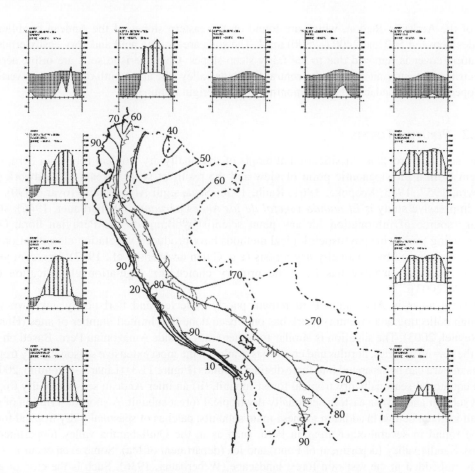

FIGURE 11.2 Percentage of annual rainfall between November and April in Ecuador, Peru and Bolivia. Dotted line is 2000 masl (modified from Johnson, 1976). Climate diagrams (CDC-UNALM, 1992; Lieth et al., 1999) show temperature and precipitation from July to June. The five upper diagrams are from stations in the equatorial seasonally dry tropical forest area, the three on the left are from inter-Andean valleys and the three on the right from the Huallaga valley area.

flowing alluvial valleys. It is characterized by yermosols, regosols, litosols, fluvisols, solonchaks, as well as primary andosols. Vertisols are, however, the soils which support most of the seasonally dry tropical forests in this area (CDC-UNALM, 1992). Some of the seasonally dry tropical forests in this area growing over 1000 masl reach the lowest part of the adjacent Litic edaphic region on the western slopes of the Andes, containing usually calcic yermosols. 2) Seasonally dry tropical forests in the inter-Andean river systems of the Marañon, Apurimac and Mantaro develop on soils of the Kastanosolic and Lito-Cambisolic region. The Kastanosolic region is predominant at altitudes between 2200 and 4000 masl in the inter-Andean valleys parallel to the Andean cordillera. The soils are the product of the accumulation of sandstone and calcareous rock sediments, the latter determining the usually fertile characteristics of these areas. This is why they have been commonly used for agricultural purposes since the pre-colonial era. The Lito-Cambisolic region is predominant at altitudes between 2200 and 3600 masl on very steep slopes, with rougher physiography and higher rainfall than the former. This region is dominated by litosols commonly associated with eutric and dystric cambisols. 3) Finally, seasonally dry tropical forests around Tarapoto develop on soils of the Acrisolic Region, which are found on the eastern slopes of the Andean cordillera at altitudes between 500 and 2200 (–2800) masl. The soils are usually acid and very easily eroded if vegetation cover is removed due to the fairly steep slopes. Characteristic soils are orthic acrisols and eutric and dystric nitosols. In the central Huallaga valley (the area south of Tarapoto), vertisols developed from expandable clays of montmorillonite origin dominate.

11.3.2 PERUVIAN FORESTS

There are a few previous classification attempts of seasonally dry tropical forests in Peru, most of them from a physiognomic point of view and the result of several years of fieldwork (e.g. Ferreyra, 1957, 1983; Koepcke, 1961; Rauh, 1958; Sagástegui Alva, 1989; Tovar, 1990). The most impressive study is *El mundo vegetal de los Andes peruanos* (Weberbauer, 1945), still a primary source of information for any plant scientist working on the Peruvian flora. Other studies using traditional phytosociological methods have produced vegetation classifications for some areas covered by seasonally dry forests (e.g. Galán de Mera et al., 1997). In recent years, however, satellite imagery has been the primary choice for vegetation classification (e.g. INRENA, 2003).

Several areas in Amazonian Peru remain poorly collected and floristically unknown since thorough collecting and inventory work has only been done in a limited number of sites (Honorio and Reynel, 2003). The situation is similar for other areas outside Amazonian Peru. Based on data from the few inventories, florulas and checklists available, the most extensive seasonally dry tropical forests in Peru can be separated into three distinct subunits (Figure 11.3) (Linares-Palomino, 2004b): (i) an equatorial seasonally dry tropical forest subunit, (ii) an inter-Andean seasonally dry tropical forest subunit and (iii) an eastern seasonally dry tropical forest subunit. A short description of each subunit is given below. In addition to these main subunits, patches of seasonally dry tropical forests can be found in several other areas in Peru, such as in the Quillabamba valley (department of Cusco), Sandia valley (department of Puno) and Ica (department of Ica). Some even occur as small islands embedded in the vast rain forest landscape (Weberbauer, 1936). Such is the case of areas in the Gran Pajonal (department of Ucayali), the Ene-Perené river confluence (department of Junin) and the Chanchamayo valley (department of Junin) (Figure 11.3F). These patches are of variable extension (some are just a few tens of hectares in size, as in Sandia and Ica) and, as might be expected, are little known. The exact extent of most of them is unknown and some areas might, in fact, constitute a vegetation type closer to savanna than to seasonally dry tropical forest (Pennington et al., Chapter 1).

Data from a recently compiled checklist of woody plants in Peruvian seasonally dry tropical forests (http://rbg-web2.rbge.org.uk/dryforest/database.htm for an online version of the checklist) show that the 10 most speciose families are Leguminosae with 108 species, Cactaceae with 35, followed by Bignoniaceae (19 spp.), Asteraceae (17 spp.), Euphorbiaceae (15 spp.), Polygonaceae (15 spp.), Capparidaceae (14 spp.), Rutaceae (14 spp.), Moraceae (12 spp.) and Rubiaceae (12 spp.) (Table 11.1). The overall contribution of Leguminosae is almost a quarter of the total number of species recorded in the checklist (total = 466 species). At the other extreme, 19 families are represented by a single species. This differs somewhat from Gentry's observations (1995) that listed, in decreasing order of prevalence based on number of species and genera, the most important neotropical seasonally dry forest families as: Leguminosae, Bignoniaceae, Rubiaceae, Sapindaceae, Euphorbiaceae, Flacourtiaceae and Capparidaceae. At the generic level, the checklist reveals that the most speciose genera are *Capparis* with 11 species, *Senna* with 10 and *Espostoa*, *Erythroxylum* and *Mimosa*, each with eight species (Table 11.2). It is noteworthy that as many as 146 genera are represented by only one species. The list highlights also the impressive degree of endemism of these forests: 112 species are endemic to Peru and 42 restricted to Ecuador and Peru (Table 11.3).

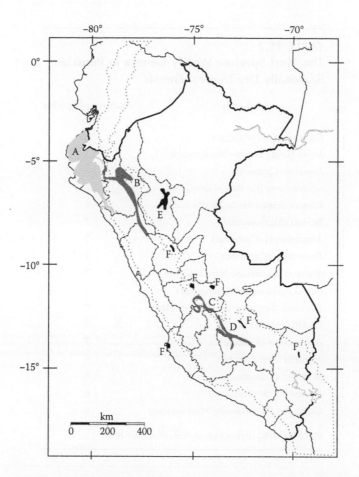

FIGURE 11.3 Distribution of seasonally dry tropical forests in Peru. A = Equatorial seasonally dry tropical forests; B = inter-Andean seasonally dry tropical forests of the Marañon system; C = inter-Andean seasonally dry tropical forests of the Mantaro system; D = inter-Andean seasonally dry tropical forests of the Apurimac system; E = eastern seasonally dry tropical forests; F = small isolated remnants. Dotted line is 2000 masl (modified from Linares-Palomino, 2004b).

TABLE 11.1
The 10 Most Speciose Families in Peruvian Seasonally Dry Tropical Forests

Family	Number of Species
Leguminosae	108
Cactaceae	35
Bignoniaceae	19
Asteraceae	17
Euphorbiaceae	15
Polygonaceae	15
Capparidaceae	14
Rutaceae	14
Moraceae	12
Rubiaceae	12

http://rbg-web2.rbg.org.uk/dryforest/database.htm

TABLE 11.2
The Most Speciose Woody Genera in Peruvian
Seasonally Dry Tropical Forests

Genus	Number of Species
Capparis (Capparidaceae)	11
Senna (Leguminosae-Caesalpinioideae)	10
Espostoa (Cactaceae)	8
Erythroxylum (Erythroxylaceae)	8
Mimosa (Leguminosae-Mimosoideae)	8
Tecoma (Bignoniaceae)	7
Armatocereus (Cactaceae)	7
Browningia (Cactaceae)	7
Acacia (Leguminosae-Mimosoideae)	7
Inga (Leguminosae-Mimosoideae)	7
Coursetia (Leguminosae-Papilionoideae)	7
Trichilia (Meliaceae)	7
Zanthoxylum (Rutaceae)	7
Tabebuia (Bignoniaceae)	6
Maytenus (Celastraceae)	6
Ipomoea (Convolvulaceae)	6
Calliandra (Leguminosae-Mimosoideae)	6

http://rbg-web2.rbge.org.uk/dryforest/database.htm

11.3.2.1 The Equatorial Seasonally Dry Tropical Forests

In Peru, seasonally dry tropical forests are distributed mainly adjacent to the Ecuadorean ones in the departments of Tumbes, Piura, Lambayeque and Cajamarca, with some small remnants in La Libertad (Figure 11.3A), forming a continuous unit. They constitute the most extensive seasonally dry tropical forest vegetation in the country, and probably also the least fragmented due to the presence of several protected areas (Figure 11.4). Based on preliminary floristic and structural

TABLE 11.3
Number of Woody Plant Species in Six Seasonally Dry Tropical Forest (SDTF) Regions in Peru

	Equatorial SDTF– Montane	Equatorial SDTF– Lowland	Inter-Andean SDTF– Marañon	Inter-Andean SDTF– Mantaro	Inter-Andean SDTF– Apurimac	Eastern SDTF– Tarapoto	SDTF Peru– Total
Endemic to Peru	24	12	69	8	11	12	112
Endemic to Ecuador–Peru	36	19	10	4	3	0	42
Total	193	103	184	50	65	108	466

http://rbg-web2.rbge.org.uk/dryforest/database.htm

FIGURE 11.4 Characteristic *Ceiba trichistandra* (A. Gray) Bakh. tree in the Equatorial seasonally dry tropical forest in Tumbes, Peru (photo R. Linares-Palomino).

studies (Linares-Palomino, 2004b) two main vegetation types can be tentatively distinguished. But a finer subdivision of these forests will not be attempted here since insufficient data are available. There have been, however, several authors who have attempted to classify the vegetation of some areas of the region (for a review of these see Linares-Palomino, 2004a).

The lowland seasonally dry tropical forests are located mainly close to the coast and at altitudes below 600 masl. The extreme formations of this type are areas where a few species dominate, as in the *algarrobales,* made up entirely of trees of the *Prosopis juliflora* (Sw.) DC./*Prosopis pallida* (Humb. & Bonpl. ex Willd.) Kunth complex. Elsewhere these forests have 1) low values of species richness, stem densities and canopy height when compared to the montane seasonally dry tropical forests (Linares-Palomino and Ponce Alvarez, 2005) and 2) are characterized by xeric vegetation species such as *Parkinsonia praecox* (Ruiz & Pav.) Hawkins, *Prosopis juliflora/Prosopis pallida*, *Caesalpinia glabrata* Kunth and several Cactaceae, such as the very characteristic *Armatocereus cartwrightianus* (Britton & Rose) Backeb. ex A.W. Hill (Table 11.4). Their woody flora, including trees and large shrubs, contains 103 species, of which 19 are endemic to the Ecuadorean-Peruvian seasonally dry tropical forest area, and 12 endemic to Peru (Table 11.3). The other vegetation type is the montane seasonally dry tropical forest, located mainly on the west-facing slopes of the western cordillera above 700 masl, reaching up to 1800 masl, as for example in the Cerros de Amotape mountains in Tumbes and Piura. These montane seasonally dry tropical forests have 1) up to twice

TABLE 11.4
Characteristic Woody Species in Six Seasonally Dry Tropical Forest Areas in Peru

Equatorial Seasonally Dry Tropical Forest — Lowland

Acacia macracantha Humb. & Bonpl. ex Willd. (Leguminosae-Mimosoideae)
Armatocereus cartwrightianus (Britton & Rose) Backeb. ex A.W. Hill (Cactaceae)
Bursera graveolens (Kunth) Triana & Planch. (Burseraceae)
Caesalpinia glabrata Kunth (Leguminosae-Caesalpinioideae)
Capparis mollis HBK (Capparidaceae)
Cordia lutea Lam. (Boraginaceae)
Ipomoea carnea Jacq. (Convolvulaceae)
Parkinsonia aculeata L. (Leguminosae-Caesalpinioideae)
Parkinsonia praecox (Ruiz & Pav.) Hawkins (Leguminosae-Caesalpinioideae)
Prosopis juliflora (Sw.) DC. (Leguminosae-Mimosoideae)
Prosopis pallida (Humb. & Bonpl. ex Willd.) Kunth (Leguminosae-Mimosoideae)
Scutia spicata (Humb. & Bonpl. ex Willd.) Weberb. (Rhamnaceae)

Equatorial Seasonally Dry Tropical Forest — Montane

Albizia multiflora (Kunth) Barneby & J.W. Grimes (Leguminosae-Mimosoideae)
Bursera graveolens (Kunth) Triana & Planch. (Burseraceae)
Ceiba trichistandra (A. Gray) Bakh. (Malvaceae-Bombacoideae)
Coccoloba ruiziana Lindau (Polygonaceae)
Cochlospermum vitifolium (Willd.) Spreng. (Cochlospermaceae)
Eriotheca ruizii (K. Schum.) A. Robyns (Malvaceae-Bombacoideae)
Erythrina smithiana Krukoff (Leguminosae-Papilionoideae)
Geoffroea spinosa Jacq. (Leguminosae-Papilionoideae)
Lafoensia acuminata (Ruiz & Pav.) DC. (Lythraceae)
Loxopterygium huasango Spruce ex Engl. (Anacardiaceae)
Pisonia macranthocarpa J. D. Smith (Nyctaginaceae)
Schmardaea microphylla (Hook.) Karsten ex Mueller (Meliaceae)
Tabebuia billbergii (Bureau & K. Schum.) Standl. ssp. *ampla* A. Gentry
  (Bignoniaceae)
Tabebuia chrysantha (Jacq.) G. Nicholson ssp. *chrysantha* (Bignoniaceae)
Terminalia valverdeae A.H. Gentry (Combretaceae)

Inter-Andean Seasonally Dry Tropical Forest — Marañon

Acacia macracantha Humb. & Bonpl. ex Willd. (Leguminosae-Mimosoideae)
Athyana weinmanniifolia (Griseb.) Radlk. (Sapindaceae)
Ceiba insignis (Kunth) P. E. Gibbs & Semir (Malvaceae-Bombacoideae)
Cordia iguaguana Melch. ex I.M. Johnst. (Boraginaceae)
Cyathostegia mathewsii (Bcnth.) Schery (Leguminosae-Papilionoideae)
Eriotheca discolor (Kunth) A. Robyns (Malvaceae-Bombacoideae)
Eriotheca peruviana A. Robyns (Malvaceae-Bombacoideae)
Geoffroea spinosa Jacq. (Leguminosae-Papilionoideae)
Hura crepitans L. (Euphorbiaceae)
Krameria lappacea (Domb.) Burdet & Simpson (Krameriaceae)
Llagunoa nitida Ruiz & Pav. (Sapindaceae)
Parkinsonia praecox (Ruiz & Pav.) Hawkins (Leguminosae-Caesalpinioideae)
Praecereus euchlorus (Weber ex K. Schum.) Backeb. ssp. *jaenensis*
  (Rauh ex Backeb.) Ostolaza (Cactaceae)
Rauhocereus riosaniensis Backeb. (Cactaceae)

(continued)

TABLE 11.4
Characteristic Woody Species in Six Seasonally Dry Tropical Forest Areas in Peru (Continued)

Saccellium lanceolatum Bonpl. (Boraginaceae)
Tecoma rosifolia HBK (Bignoniaceae)

Inter-Andean Seasonally Dry Tropical Forest — Mantaro

Acacia macracantha Humb. & Bonpl. ex Willd. (Leguminosae-Mimosoideae)
Anadenanthera colubrina (Vell.) Brenan var. *cebil* (Griseb.) Altschul
  (Leguminosae-Mimosoideae)
Bursera graveolens (Kunth) Triana & Planch. (Burseraceae)
Cedrela weberbaueri Harms (Meliaceae)
Eriotheca ruizii (K. Schum.) A. Robyns (Malvaceae-Bombacoideae)
Cyathostegia mathewsii (Benth) Schery (Leguminosae-Papilionoideae)
Ipomoea pauciflora M. Marten & Galetti ssp. *vargasiana* (O'Donell) McPherson
  (Convolvulaceae)
Caesalpinia glabrata Kunth (Leguminosae-Caesalpinioideae)
Parkinsonia praecox (Ruiz & Pav.) Hawkins (Leguminosae-Caesalpinioideae)
Prosopis pallida (Humb. & Bonpl. ex Willd.) Kunth (Leguminosae-Mimosoideae)

Inter-Andean Seasonally Dry Tropical Forest — Apurimac

Acacia aroma Hook. & Arn. (Leguminosae-Mimosoideae)
Acacia macracantha Humb. & Bonpl. ex Willd. (Leguminosae-Mimosoideae)
Anadenanthera colubrina (Vell.) Brenan var. *cebil* (Griseb.) Altschul
  (Leguminosae-Mimosoideae)
Aralia soratensis Marchal (Araliaceae)
Caesalpinia spinosa (Molina) Kuntze (Leguminosae-Caesalpinioideae)
Eriotheca ruizii (K. Schum.) A. Robyns (Malvaceae-Bombacoideae)
Erythrina falcata Benth. (Leguminosae-Papilionoideae)
Kageneckia lanceolata Ruiz & Pav. (Rosaceae)
Poissonia orbicularis (Benth.) Hauman (Leguminosae-Papilionoideae)
Prosopis juliflora (Sw.) DC. (Leguminosae-Mimosoideae)
Zapoteca caracasana (Jacq.) H. Hern. ssp. *weberbaueri* (Harms) H. Hern.
  (Leguminosae-Mimosoideae)
Ziziphus mistol Griseb. (Rhamnaceae)

Eastern Seasonally Dry Tropical Forest — Tarapoto

Albizia cf. *niopoides* (Benth.) Burkart (Leguminosae-Mimosoideae)
Apuleia leiocarpa (Vog.) Macbr. (Leguminosae-Caesalpinioideae)
Byrsonima crassifolia (L.) Kunth (Malpighiaceae)
Byrsonima spicata (Cav.) DC. (Malpighiaceae)
Cybistax antisyphilitica (Mart.) Mart. (Bignoniaceae)
Inga tenuicalyx T. D. Penn. (Leguminosae-Mimosoideae)
Jacaranda glabra (DC.) Bureau & K. Schum. (Bignoniaceae)
Manilkara bidentata (A.DC) Chev. ssp. *surinamensis* (Miq.) T.D. Penn. (Sapotaceae)
Platymiscium gracile Benth. (Leguminosae-Papilionoideae)
Rheedia spruceana Engl. (Clusiaceae)
Schinopsis peruviana Engl. (Anacardiaceae)
Steriphoma peruvianum Spruce ex Eichl. (Capparidaceae)
Tabebuia aurea (Manso) Benth. & Hook. ex S. Moore (Bignoniaceae)
Trichilia ulei C. DC. (Meliaceae)

the species richness and five times the stem density of the lowland seasonally dry tropical forests, 2) a dense canopy and forest of taller stature; 3) characteristic species *Ceiba trichistandra* (A. Gray) Bakh., *Eriotheca ruizii* (K. Schum.) A. Robyns, *Eriotheca discolor* (Kunth) A. Robyns and *Terminalia valverdeae* A. H. Gentry (Table 11.4). They thrive in much more humid conditions produced by fog, arising from the nearby Pacific Ocean, which generates a heavy epiphytic cover (usually of *Tillandsia* spp.). Their woody flora contains 193 species, of which 36 are endemic to the greater Ecuadorean-Peruvian seasonally dry tropical forest area, and 24 endemic to Peru (Table 11.3). In total, the equatorial seasonally dry tropical forests contain 227 woody species, of which 69 are shared between the lowland and montane variants.

From the information available, the overall characteristics of equatorial seasonally dry tropical forests are that the most conspicuous species are *Ceiba trichistandra*, *Eriotheca ruizii* (Bombacaceae), *Loxopterygium huasango* Spruce ex Engl. (Anacardiaceae), *Caesalpinia glabrata*, the *Prosopis juliflora/Prosopis pallida* complex, *Acacia macracantha* Humb. & Bonpl. ex Willd. (Leguminosae) and several Capparidaceae, especially *Capparis scabrida* Kunth and *Capparis flexuosa* (L.) L (Table 11.4). The importance of the Bombacaceae in these forests was first recognized by Gentry (1993) when he stated 'Bombacaceae, though less speciose, are represented by five different genera of large trees and are probably more dominant here than anywhere else on Earth'. Based on satellite image analysis, it has been estimated that these forests cover around 32,303.63 km^2, which corresponds to 58% of the total area in the departments of Tumbes, Piura and Lambayeque (Proyecto Algarrobo, 2003). This area, however, has been estimated without including information from the scattered and smaller seasonally dry tropical forest remnants in La Libertad and Cajamarca.

11.3.2.2 The Inter-Andean Seasonally Dry Tropical Forests

These are mostly fragmented remnants in the xeric valleys of the Huancabamba, Marañon, Apurímac and Mantaro rivers, and their respective tributaries (Figures 11.3B–D and Figure 11.5). Their total area is estimated at 3,106 km^2 (INRENA, 1995). The most interesting fragments can be found in the upper Marañon valley and its tributaries (e.g. Chamaya and Utcubamba rivers). This area is home to 184 woody plant species, of which 69 are endemic to Peru, many of which are rare, narrow endemics, restricted to small isolated areas (Table 11.3). Notably, these forests only have 10 species endemic to Ecuador and Peru. Some of the endemic species are *Browningia riosaniensis* (Backeb.) G. D. Rowley and *Praecereus euchlorus* (Weber ex K. Schum.) Backeb. subsp. *jaenensis* (Rauh & Backeb.) Ostolaza (Bridgewater et al., 2003). Recently, new species (*Parkinsonia peruviana* C. E. Hughes, Daza & Hawkins, *Ruprechtia aperta* Pendry and *Ruprechtia albida* Pendry) and even a new genus (*Maraniona* C. E. Hughes, G. P. Lewis, Daza and Reynel) have been described from the Marañon and adjacent valleys in seasonally dry tropical forest and scrub vegetation (Hughes et al., 2003, 2004; Pendry, 2003). These new taxa, together with other species restricted to these areas (e.g. *Coursetia cajamarcana* Lavin and *Coursetia maraniona* Lavin; Lavin, 1988) have begun to reveal the unique characteristics of these valleys, which are botanically very little explored and, unfortunately, highly threatened by deforestation (Linares-Palomino, 2004b). The inter-Andean seasonally dry tropical forests in the Mantaro and Apurimac valleys are, in comparison, less species rich (Table 11.3). Fifty species are reported for the Mantaro (8 endemic to Peru and 4 endemic to Ecuador-Peru), while the Apurimac is slightly richer with 65 species (11 endemic to Peru and 3 endemic to Ecuador-Peru). The flora of the subunit totals 243 woody species, of which only 13 are shared between the three valley systems (i.e. including their respective tributaries). Twenty-four species are shared between the Mantaro and Apurimac systems, 27 are shared between the Marañon and Apurimac and 16 between the Marañon and the Mantaro. Characteristic species for each of these areas are shown in Table 11.4.

FIGURE 11.5 Inter-Andean seasonally dry tropical forests in the Marañon valley, Cajamarca, Peru. The big trees with pale trunk are *Eriotheca peruviana* A. Robyns. Most of the shrubs in the foreground are *Croton* sp. (photo. R. Linares-Palomino).

11.3.2.3 The Eastern Seasonally Dry Tropical Forests

The seasonally dry tropical forest formations around the Tarapoto area, on the eastern slopes of the Andes, are biogeographically and floristically unique. They are mostly composed of fragments, apart from a few areas where local conservation efforts are maintaining more extensive patches (Figure 11.3E and Figure 11.6). Although their area has been estimated to be 5,394 km^2 (INRENA, 1975), this is probably an overestimation since much of these forests has been replaced by plantations, roads and cattle-grazing farms. Indeed, recent studies by the Instituto de Investigaciones de la Amazonia Peruana show that only a tenth of that area (528 km^2) is actually covered by seasonally dry tropical forests (Reategui, 2003). These forests have been botanically neglected for decades. However, recent rapid surveys have shown their unique value by highlighting several endemics. Bridgewater et al. (2003) reported the presence of three endemic species in a 0.2-ha plot (*Schinopsis peruviana* Engl., *Trichilia ulei* C. DC. and *Triplaris peruviana* Fisch. & Mey. ex C. A. Meyer). Outside the plot *Platymiscium gracile* Benth., *Lecointea* cf. *peruviana* J. F. Macbr. and *Inga tenuicalyx* T. D. Penn. were also reported as endemics (for a list of characteristic species see Table 11.4). Their woody flora has been reported to have 108 species, of which 12 are endemic to Peru (Table 11.3).

FIGURE 11.6 El Quinillal Municipal Conservation Area in the eastern seasonally dry tropical forests of the Huallaga valley, Juanjui, San Martin, Peru (photo M. León).

11.4 BIOGEOGRAPHICAL RELATIONSHIPS OF PERUVIAN SEASONALLY DRY TROPICAL FORESTS

The phytogeographical relationships among these forests are complex, not least due to the poor data available. Even Weberbauer, after three decades of travelling and collecting throughout Peru, could not entirely grasp the whole picture. He suggested, for example, that the xerophytic vegetation of the Marañon valley was closely related to the Brazilian caatingas (Weberbauer, 1914), and that the dry vegetation of north-western Peru had affinities not only with southern Ecuador, but also with the dry vegetation of Colombia and Central America. Previous research has explored the relationships between seasonally dry tropical forests in Peru using multivariate analyses (Linares-Palomino, 2004b). However, since that paper was published, new data have been gathered and added to the database, which now comprises 154 species from 44 inventoried sites. Additionally, a checklist of woody plant species in Peruvian seasonally dry tropical forests has been compiled (http://rbg-web2.rbge.org.uk/dryforest/database.htm), containing information about 466 tree and large shrub species (of more than 2 m height at maturity) for six seasonally dry tropical forest areas: Lowland Equatorial, Montane Equatorial, inter-Andean Marañon, inter-Andean Mantaro, inter-Andean Apurimac and Eastern (Table 11.3). These new data have been subjected to floristic and multivariate analyses (for detailed methods and explanation of multivariate analyses employed, see Linares-Palomino et al., 2004b). The number of species and inventories will undoubtedly increase in the next few years since several recent projects and plant collecting expeditions have collected relevant data from previously unworked areas. Much of these data from Peru are still being processed and identified (M. A. La Torre-Cuadros, A. Sabogal and A. Zegarra, pers. comm.).

From the inventory database, the most commonly reported species are *Capparis scabrida* with 31 occurrences (i.e. it was recorded at 31 of the 44 sites), *Bursera graveolens* (Kunth) Triana & Planch. with 25 occurrences, *Loxopterygium huasango* with 23 and *Eriotheca ruizii* and *Caesalpinia glabrata*, each with 22. These are the only species which occur in at least half of all the recorded sites. Although these numbers reflect the fact that there are many more inventoried sites in the equatorial seasonally dry tropical forests, they also show that these species are very dominant in these forests. All five species are characteristic members of seasonally dry tropical forests. *Bursera* is a characteristic seasonally dry tropical forest genus in the Neotropics, especially in Mexico. *Capparis* is the largest and ecologically most important genus in Capparidaceae, one of the very few families that seems to be better represented in dry forests than rain forests (Gentry, 1995). *Eriotheca ruizii* and *Loxopterygium huasango* are endemic species restricted to the seasonally dry tropical forests in south-western Ecuador and north-western and inter-Andean Peru. The former is found mainly on the slopes of hills and mountains in the area (Robyns, 1963) and one specimen, collected in the Apurimac valley in Peru, was collected at 2500 masl (W3 Tropicos online database). Data from the checklist reveal that the most widespread species are *Acacia macracantha*, *Buddleja americana* L. and *Eriotheca ruizii* that occur in all but the eastern seasonally dry tropical forests. In marked contrast, data from the inventories database and the checklist show that 84 species have been recorded at only one of the 44 study sites and that 297 species are recorded from just one of the six seasonally dry tropical forest areas evaluated in this chapter. The multivariate analyses reveal that the eastern seasonally dry tropical forests are unique and differ markedly from the other areas and that the inter-Andean seasonally dry tropical forests, especially those from the Marañon system, are closely related to the equatorial seasonally dry tropical forests (Figure 11.7).

11.4.1 BIOGEOGRAPHIC RELATIONSHIPS AMONG PERUVIAN SEASONALLY DRY TROPICAL FORESTS

The strongest affinities are among the Equatorial seasonally dry tropical forests and the inter-Andean seasonally dry tropical forests. The checklist data reveal 81 shared species between both areas. This link has been reported previously by studies of range data for plants (Bridgewater et al., 2003; see

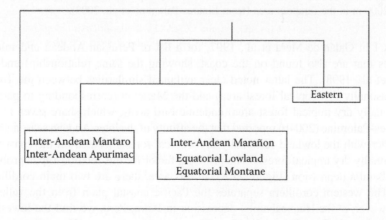

FIGURE 11.7 Results of the TWINSPAN analysis for Peruvian seasonally dry tropical forests using checklist data.

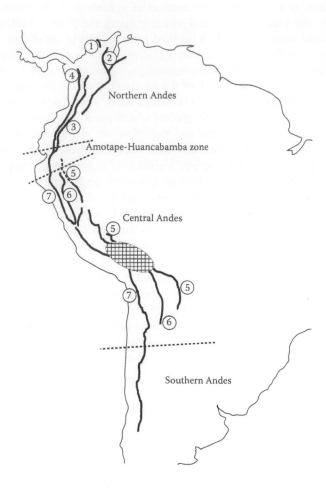

FIGURE 11.8 Simplified geographical division of the Andes. In the Northern Andes: 1. Cordillera (C) of Santa Marta. 2. Eastern C. 3. Central C. 4. Western C. In the Central Andes: 5. Eastern C. 6. Central C. 7. Western C. Hatched area represents the Peruvian-Bolivian altiplano region. Stippled line shows mountain ranges below 2000 m in the Tarapoto area (from Linares-Palomino et al., 2003).

also appendix I in Galán de Mera et al., 1997, for a list of Peruvian Andean endemics other than woody species that are also found on the coast, showing the same relationship) and bird species (Stattersfield et al., 1998). The latter noted clear avifaunal similarities between the Tumbesian (the Equatorial seasonally dry tropical forest area) and the Marañon (corresponding to part of the inter-Andean seasonally dry tropical forest area) endemic bird areas, which share seven restricted-range species. Linares-Palomino (2004b) reported that the affinity of the inter-Andean seasonally dry tropical forests is higher with the lowland seasonally dry tropical forests located on the coast than with the montane seasonally dry tropical forests. The principal factor for this relationship is probably the low-lying Huancabamba depression (Figure 11.8). In this area, there are two main cordilleras: western and central. The western cordillera separates the Pacific coastal plain from the valley of the Rio Marañon. However, at the Huancabamba depression in Piura, Peru, we find the lowest point in the entire Andes (Abra de Porculla) at 2,145 m. This area, the Amotape-Huancabamba Zone (*sensu* Weigend, 2004) has been repeatedly suggested as a north-to-south or south-to-north dispersal barrier for montane organisms (e.g. *Fuchsia*, Berry, 1982; Loasaceae, Weigend, 2002). However, it might have also provided a west-to-east dispersal opportunity for some lowland seasonally dry tropical forest species, moving from the coast to the Marañon valley or vice versa (Bridgewater et al., 2003; Linares-Palomino et al., 2003). Although termed lowland vegetation, some seasonally dry tropical forest species are found at high altitudes. Weberbauer (1936) reported typical seasonally dry tropical forest species growing as high as 2,800 masl in the Mantaro valley in central Peru. This suggests that seasonally dry tropical forest species may be able to cross the Porculla pass. Furthermore, seasonally dry tropical forest vegetation grows within a few kilometres of the summit of the pass (Bridgewater et al., 2003). The dispersal of seasonally dry tropical forest plants across the western cordillera in this area may have occurred recently, and may still be happening for some organisms.

The uniqueness of the eastern seasonally dry tropical forests has made floristic comparisons with the other seasonally dry tropical forests in Peru and adjacent countries difficult, revealing little relationship (Bridgewater et al., 2003; Linares-Palomino et al., 2003). Only 20 and 24 species are shared with the Equatorial and inter-Andean seasonally dry tropical forests, respectively. This condition is accentuated by a mosaic of vegetation formations in the area, which has resulted in characteristic wet forest species intergrading with seasonally dry tropical forest species to form a distinctive seasonally dry tropical forest formation. Additionally, this region has been under human disturbance for several decades, and natural or original vegetation is hard to find.

11.4.2 BIOGEOGRAPHIC RELATIONSHIPS OF PERUVIAN SEASONALLY DRY TROPICAL FORESTS

The seasonally dry tropical forests in south-western Ecuador and north-western Peru have been identified as a phytogeographically distinct area, in which the strongest floristic link is between the Equatorial seasonally dry tropical forest and the adjacent Ecuadorean forests (Gentry, 1992; Linares-Palomino et al., 2003). It has also been suggested that these forests have affinities with the xeric vegetation of the Galapagos Islands (Weberbauer, 1945; Svenson, 1945). The inter-Andean seasonally dry tropical forests do not show strong affinities with similar vegetation types outside Peru. Some data point to a weak relationship with Bolivian seasonally dry tropical forests, although this is very difficult to discuss due to scant information from inter-Andean valleys in both countries (Linares-Palomino et al., 2003). Nevertheless, some authors have suggested close relationships between the inter-Andean dry floras of both countries (e.g. Weberbauer, 1945).

Using the inventory database with data from Ecuador (161 species, 27 Sites), Colombia-Venezuela (214 species, 27 sites) and Bolivia (133 species, 10 sites), plus 44 Peruvian sites, reveals that of the 154 woody species from Peru, 50 are shared with neighbouring Ecuador (seasonally dry tropical forests in Azuay, El Oro, Guayas, Loja and Manabí). Only 29 species are shared with the seasonally dry tropical forests in Colombia and Venezuela and even fewer, 22 species, with the data subset from Bolivia. As might be expected, most of the shared species are widespread neotropical elements such as *Capparis flexuosa* (Capparidaceae), widespread from south-eastern USA and western Mexico through Mesoamerica and the Caribbean to coastal Colombia and Venezuela, continuing to Ecuador, Peru, southern Brazil, Bolivia and Argentina (Iltis and Ruiz-Zapata, 1998), Figure 11.9; *Acacia macracantha* (Leguminosae: Mimosoideae) present in the subtropical and tropical regions of North and Central America and the Caribbean islands and throughout northern and western South America from Guayana and Venezuela to Peru and Chile (Ebinger et al., 2000); *Geoffroea spinosa* Jacq. (Leguminosae: Papilionoideae) occurring in five disjunct areas of seasonally dry tropical forests in north-eastern Brazil, north-eastern Argentina, Paraguay, Bolivia, Ecuador and northern Peru, Galapagos and Colombia, Venezuela and the Antilles (Ireland and Pennington, 1999); *Pisonia aculeata* L. (Nyctaginaceae) found from Florida and Texas in the USA throughout the New World tropics and considered introduced in Africa, Asia and the Philippines (Steyermark and Aymard, 2003); and *Sapindus saponaria* L. (Sapindaceae) widespread from Mexico through Central America to most South America (Cordero and Boshier, 2003).

FIGURE 11.9 Distribution of *Capparis flexuosa* (L.) L. in the Neotropics, based on geographical coordinates from herbarium specimen labels.

11.4.3 Origin and Evolution of the Seasonally Dry Tropical Forests in Peru

Linares-Palomino et al. (2003) speculated that Andean uplift events and Quaternary climatic changes may have been important factors causing the weak floristic affinities between the Peruvian and the Bolivian-Argentinean seasonally dry tropical forests. Orographic factors could have caused isolation of the seasonally dry tropical forests of the dry inter-Andean valleys of southern Peru from those located in more southern regions, especially the ones in the Bolivian Andes. Even though as recently as 10 million years ago the elevation of the Bolivian altiplano was only half of its present height of 4000 m, organisms still had to disperse across at least two (eastern and central), and possibly three (western) of the cordilleras (see Figure 11.8) to cross between northern Peru and Bolivia. This suggests that the high peaks of the Andes have played a role in isolating these different seasonally dry tropical forest areas for over 10 million years.

Several authors have proposed a dynamic relationship between moist/wet forests and dry vegetation formations (savannas and seasonally dry tropical forests) in the neotropical region during glacial and inter-glacial cycles in the Quaternary (e.g. Haffer, 1982; Prado and Gibbs, 1993). The moist/wet forests have been proposed as expanding their geographical range during interglacial cycles (such as that of the present day), in which more humid and warm climates are predominant. As a consequence, the dry vegetation formations, once more widespread in distribution, have receded to areas of low rainfall, forming seasonally dry tropical forest and savanna refugia. During glacial maxima, evidence suggests that the climate was drier and cooler by around 2–8°C (Burnham and Graham, 1999) and the sea level lower by around 100–200 m (Gregory-Wodzicki, 2000). These events would have been important factors in promoting the retreat of moist/wet forests into refugia and the expansion of drier vegetation formations around the region. This model of Pleistocene vegetation shifts may be a more recent factor reinforcing the separation between coastal Pacific and Andean seasonally dry tropical forests in Ecuador and Peru, and those in Colombia and Venezuela. Gentry (1982) postulated that the super-humid Chocó region in western Colombia may have always kept these areas separate. The Chocó presently has a high record of precipitation (around 11.6 m/year) and during drier glacial maxima it may have remained a moist barrier to the expansion of the drier Pacific vegetational formations from South America and Caribbean Colombia and Venezuela (Gentry, 1982).

11.5 CONSERVATION AND MANAGEMENT ISSUES

The seasonally dry tropical forests in south-western Ecuador and adjacent Peru (especially the extreme north-western region) have been repeatedly recognized as biologically outstanding. They are part of the Tumbesian and inter-Andean valley seasonally dry tropical forest ecoregion (Figure 11.10), which has been considered as critically endangered and consequently included as one of the Global 200 priority ecoregions for global conservation (Olson and Dinerstein, 2002). Additionally it has been recognized as a centre of endemism (Cracraft, 1985) and a critical endemic bird area (Stattersfield, 1998). More recently, it has been included in the hotspot list of the world, together with the Colombian Chocó forests forming the Tumbes-Chocó-Magdalena Hotspot (Mittermeier et al., 2005). All these recognitions demonstrate the immense biological value of this area. However, it is also severely threatened, and as in most tropical regions, habitat loss and fragmentation by human activities are probably the principal threat (Table 11.5). Nonetheless, despite this recognition of the Tumbes area, good biological information on seasonally dry tropical forests in Peru as a whole is scant. The best we have comes from the lowland Pacific equatorial region, where the Dutch sponsored Proyecto Algarrobo (algarrobo is the vernacular name for the *Prosopis juliflora/Prosopis pallida* complex, one of the most important forest resources in this dry ecosystem) worked to strengthen development and conservation over the past 10 years. However, even they neglected the much more diverse and biologically interesting montane seasonally dry tropical forest in the region. As a consequence, and owing to the fact that Peruvian people tend to associate seasonally dry tropical forest only with this region, they assume that everything has already been done by the Proyecto Algarrobo. This assumption is certainly not true. As has been shown by recent descriptions of new species of birds (Apurimac valley, M. Kessler pers. comm.) and plants (Marañon valley,

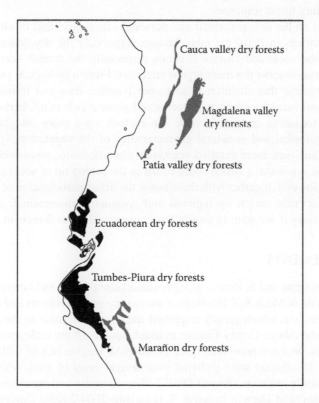

FIGURE 11.10 The Tumbes seasonally dry tropical forests ecoregion, modified from Olson and Dinerstein (2002).

TABLE 11.5
Main Threats to Seasonally Dry Tropical Forests in the Neotropical Region

Threat	Costa Rica (Quesada and Stoner, 2004)	Puná Island (Madsen et al., 2001)	North-West Peru (Linares-Palomino, pers. obs.)
Cattle	Deforestation to create space for cattle ranching and subsequent cyclic burning of pastures.	Introduction of goats and cattle.	Nomadic cattle and goat raising and cyclic burning of pastures.
Logging	Selective wood extraction.	Selective wood and firewood logging.	Selective wood and firewood logging.
Other	Recent uncontrolled development of tourism.	Introduction of exotic species.	

see section on the inter-Andean seasonally dry tropical forests), the inter-Andean valleys have still much to reveal. However, time is running out. Land use conversion from forest into agriculture is the major threat. Few, if any, intact forest patches remain in these valleys. There are no protected areas in any of these inter-Andean seasonally dry tropical forest areas, or in the eastern forests. Recently, the Area de Conservación Municipal El Quinillal, a very well preserved seasonally dry tropical forest south of Tarapoto in the eastern seasonally dry tropical forests, was created. Unfortunately, the existence of this conservation area is threatened by the heavy logging activities surrounding it and, since it is not managed by the state but by the financially poor local council, it is a conservation area on paper, but not on the ground. Conservation in these areas is a profound socio-economic issue. Most people living in these forgotten and isolated areas do not have alternatives for survival except the surrounding forest resources.

It is clear, based on the data presented and personal experience, that of all the seasonally dry tropical forest formations in Peru, the inter-Andean (especially the dry Marañon and Apurimac valleys and their tributaries) and eastern subunits (especially the forests near to El Quinillal in Juanjuí and Bellavista) deserve the most urgent attention. From a biological point of view, one of the first things to reverse this situation is to gather baseline data and initially produce simple checklists of flora and fauna based on inventories and general collecting, herbarium and museum work. Once this information is available, we can embark on a more integrated study of these ecosystems (e.g. ecological and structural characteristics of the vegetation, spatial and temporal dynamics, gradient analyses, more detailed vegetation classifications, assessments of the degree of endemism of various organisms), which could result in the setting up of some of these ecosystems as protected areas. However, together with these tasks, the direct participation of local communities, and increased conservation action by regional and national environmental, social and political bodies will be necessary if we want to ensure the existence of these forests in the next decades.

ACKNOWLEDGEMENTS

Thanks to R.T. Pennington and S. Ponce, who provided helpful ideas and comments on early drafts of the chapter. A. Galán de Mera, R.T. Pennington, two anonymous reviewers and the editors provided comments and suggestions which greatly improved the chapter. Thanks to the Missouri Botanical Garden for making the Alwyn Gentry Dataset available and to all my colleagues and assistants for help during fieldwork. M. Leon provided Figure 11.6 and M. Hughes helped with Figure 11.9. Much of the data used in this chapter were gathered over several years of work which were financially supported by the Darwin Initiative (Project 09/017 'Tree diversity and agroforestry development in the Peruvian Amazon' and Darwin Initative Scholarship 2004/2005), University of Edinburgh, Royal Botanic Garden Edinburgh, Friends of the Royal Botanic Garden Edinburgh in the UK and the Instituto Nacional de Recursos Naturales in Lima, Peru.

REFERENCES

Berry, P.E., The systematics and evolution of *Fuchsia*, section *Fuchsia* (Onagraceae), *Ann. Missouri Bot. Gard.*, 69, 1, 1982.

Block, M. and Richter, M., Impacts of heavy rainfall in El Niño 1997/1998 on the vegetation of Sechura desert in northern Peru, *Phytocoenologia*, 30, 491, 2000.

Bridgewater, S. et al., A preliminary floristic and phytogeographic analysis of the woody flora of seasonally dry forests in northern Peru, *Candollea*, 58, 129, 2003.

Burnham, R.J. and Graham, A., The history of neotropical vegetation: new developments and status, *Ann. Missouri Bot. Gard.*, 86, 546, 1999.

CDC-UNALM, *Estado de la Conservación de la Diversidad Natural de la Región Noroeste del Perú*, Centro de Datos para la Conservación - Universidad Nacional Agraria La Molina, Lima, 1992.

Cordero, J. and Boshier, D.H., *Árboles de Centroamerica*, Oxford Forestry Institute, Oxford, UK, 2003, 1079.

Cracraft, J., Historical biogeography and patterns of differentiation within the South American avifauna: areas of endemism, *Ornith. Monogr.*, 36, 49, 1985.

Dodson, C.H. and Gentry, A.H., Biological extinction in western Ecuador, *Ann. Missouri Bot. Gard.*, 78, 273, 1991.

Ebinger, J.E., Seigler, D.S., and Clarke, H.D., Taxonomic revision of South American species of the genus *Acacia* subgenus *Acacia* (Fabaceae: Mimosoideae), *Syst. Bot.*, 25, 588, 2000.

Eva, H.D. et al., *A Vegetation Map of South America*, European Commission Joint Research Centre, Luxembourg, 2002.

Ferreyra, R., Contribución al conocimiento de la flora costanera del norte Peruano (Departamento de Tumbes), *Bol. Soc. Argent. Bot.*, 6, 194, 1957.

Ferreyra, R., Los tipos de vegetación de la costa peruana, *Anales Jard. Bot. Madrid*, 40, 241, 1983.

Galán de Mera, A. et al., Phytogeographical sectoring of the Peruvian coast, *Glob. Ecol. Biogeogr. Lett.*, 6, 349, 1997.

Gentry, A.H., Phytogeographic patterns as evidence for a Chocó refuge, in *Biological Diversification in the Tropics*, Prance, G.T., Ed, Columbia University Press, New York, 1982, 112.

Gentry, A.H., Phytogeography, in *Status of Forest Remnants in the Cordillera de la Costa and Adjacent Areas of Southwestern Ecuador*, Parker, T.A. III and Carr, J.L., Eds, Conservation International, RAP Working Papers 2, 1992, 56.

Gentry, A.H., Overview of the Peruvian flora, in *Catalogue of the Flowering Plants and Gymnosperms of Peru*, Brako, L. and Zarucchi, J.L., Eds, Missouri Botanical Garden Press, Missouri, 1993, xxix.

Gentry, A.H., Diversity and floristic composition of neotropical dry forests, in *Seasonally Dry Tropical Forests*, Bullock, S.H., Mooney, H.A., and Medina, E., Eds, Cambridge University Press, Cambridge, UK, 1995, 146.

Gregory-Wodzicki, K.M., Uplift history of the Central and Northern Andes: a review, *GSA Bull.*, 112, 1091, 2000.

Haffer, J., General aspects of the refuge theory, in *Biological Diversification in the Tropics*, Prance, G.T., Ed, Columbia University Press, New York, 1982, 6.

Hartley, A.J., Andean uplift and climate change, *J. Geol. Soc. London*, 160, 7, 2003.

Honorio, E. and Reynel, C., *Vacíos de colección de la flora de los bosques húmedos del Perú*, Herbario de la Facultad de Ciencias Forestales de la Universidad Nacional Agraria La Molina, Lima, 2003, 87.

Hueck, K., Bosques secos de la zona tropical y subtropical de la América del Sur, *Bol. Instit. Forestal Lat. Am. Inv. Capacit.*, 4, 1, 1959.

Hughes, C.E., Daza Yomona, A., and Hawkins, J.A., A new Palo Verde (*Parkinsonia* – Leguminosae: Caesalpinioideae) from Peru, *Kew Bull.*, 58, 467, 2003.

Hughes, C.E. et al., *Maraniona*. A new dalbergioid legume genus (Leguminosae, Papilionoideae) from Peru, *Syst. Bot.*, 29, 366, 2004.

IavH (Instituto Alexander von Humboldt), *El bosque Seco Tropical (Bs-T) en Colombia*, Instituto Alexander von Humboldt, Colombia, 1998.

Iltis, H.H. and Ruiz-Zapata, T., *Capparis*, in *Flora of the Venezuelan Guayana, Vol. 4*, Berry, P.E., Holst, B.K., and Yatskievych, K., Eds, Missouri Botanical Garden Press, St. Louis, 1998, 134.

INRENA (Instituto Nacional de Recursos Naturales), *Mapa Ecológico del Peru Escala 1:1000000 con Guia Explicativa*, INRENA, Lima, 1975.

INRENA (Instituto Nacional de Recursos Naturales), *Mapa Forestal del Peru Escala 1:1000000 con Guia Explicativa*, INRENA, Lima, 1995.

INRENA (Instituto Nacional de Recursos Naturales), *Mapa de Bosques Secos del Departamento de Tumbes: Memoria Descriptiva*, Proyecto Algarrobo, Piura, 2003, 31.

Ireland, H. and Pennington, R.T., A revision of *Geoffroea* (Leguminosae - Papilionoideae), *Ed. J. Bot.*, 56, 329, 1999.

Johnson, A.M., The climate of Peru, Bolivia and Ecuador, in *Climates of Central and South America, World Survey of Climatology 12*, Schwerdtfeger, W., Ed, Elsevier Scientific, Amsterdam, 1976, 147.

Koepcke, H.W., Synökologische Studien and der Westseite der peruanischen Anden, *Bonner Geogr. Abh.*, 29, 1, 1961.

Lavin, M., Systematics of *Coursetia* (Leguminosae – Papilionoideae), *Syst. Bot. Monogr.*, 21, 1, 1988.

Lieth, H. et al., *Climate Diagram World Atlas*, CD-Series: Climate and Biosphere, Backhuys Publishers, Leiden, 1999.

Linares-Palomino, R., Los bosques tropicales estacionalmente secos: I. El concepto de los bosques secos en el Perú, *Arnaldoa*, 11, 85, 2004a.

Linares-Palomino, R., Los bosques tropicales estacionalmente secos: II. Fitogeografía y composición florística, *Arnaldoa*, 11, 103, 2004b.

Linares-Palomino, R. and Ponce-Alvarez, S.I., Tree community patterns in seasonally dry tropical forests in the Cerros de Amotape Cordillera, Tumbes, Peru, *For. Ecol. Manage.*, 209, 261, 2005.

Linares-Palomino, R., Pennington, R.T., and Bridgewater, S., The phytogeography of the seasonally dry tropical forests in Equatorial Pacific South America, *Candollea*, 58, 473, 2003.

Madsen, J.E., Mix, R.L., and Balslev, H., *Flora of Puna Island: Botanical Resources on a Neotropical Island*, Aarhus University Press, Aarhus, 2001.

Mittermeier, R.A. et al., *Hotspots Revisited: Earth's Biologically Richest and Most Threatened Terrestrial Ecoregions*, Conservation International, Washington, 392, 2005.

Olson, D.M. and Dinerstein, E., The global 200: Priority ecoregions for global conservation, *Ann. Missouri Bot. Gard.*, 89, 125, 2002.

Parker, T.A. III, et al., *The Lowland Dry Forests of Santa Cruz, Bolivia: A Global Conservation Priority*, Conservation International, RAP Working Papers 4, 1993.

Pendry, C., Nine new species of *Ruprechtia* (Polygonaceae) from Central and South America, *Edinburgh. J. Bot.*, 60, 19, 2003.

Pennington, R.T., Prado, D.E., and Pendry, C.A., Neotropical seasonally dry forests and Quaternary vegetation changes, *J. Biogeogr.*, 27, 261, 2000.

Prado, D.E., *A critical evaluation of the floristic links between chaco and caatingas vegetation in South America*, PhD thesis, University of St Andrews, St Andrews, UK, 1991.

Prado, D.E., Seasonally dry forests of tropical South America: from forgotten ecosystems to a new phytogeographic unit, *Edinburgh J. Bot.*, 57, 437, 2000.

Prado, D.E., As caatingas do América do Sul, in *Ecología e Conservação da Caatinga*, Leal, I.R., Tabarelli, M., and Silva, J.M.C., Eds, Editora Universitária da UFPE, Recife, 2003, 3.

Prado, D.E. and Gibbs, P.E., Patterns of species distribution in the dry seasonal forests of South America, *Ann. Missouri Bot. Gard.*, 80, 902, 1993.

Proyecto Algarrobo, *Resumen Ejecutivo 2003*, Proyecto Algarrobo, Chiclayo, Peru, 2003.

Quesada, M. and Stoner, K.E., Threats to the conservation of tropical dry forests in Costa Rica, in *Biodiversity Conservation in Costa Rica, Learning the Lessons in a Seasonal Dry Forest*, Frankie, G.W., Mata, A., and Vinson, S.B., Eds, University of California Press, Berkeley, Los Angeles, London, 2004, 266.

Rauh, W., Beitrag zur Kenntnis der peruanischen Kakteenvegetation, *Sitzungsber. Heidelb. Akad. Wiss. Math. Naturwiss. Kl.*, 1, 1, 1958.

Reategui, F., *Zonificación Ecológica Económica de la Región San Martín, Estudio Temático Preliminar – Forestal*, Instituto de Investigaciones de la Amazonia Peruana, Tarapoto, 2003, 41.

Robyns, A., Essai de monographie du genre *Bombax* s.l. (Bombacaceae), *Bull. Jard. Bot. Brux.*, 33, 145, 1963.

Rundel, P.W. et al., The phytogeography and ecology of the coastal Atacama and Peruvian deserts, *Aliso*, 13, 1, 1991.

Sagástegui Alva, A., *Vegetación y Flora de la Provincia de Contumazá*, CONCYTEC, Trujillo, 1989, 76.

Sarmiento, G., The dry plant formations of South America and their floristic connections, *J. Biogeogr.*, 2, 233, 1975.

Smith, A.C. and Johnston, I.M., A phytogeographic sketch of Latin America, in *Plants and Plant Science in Latin America, A New Series of Plant Science Books, Vol. 16*, Verdoorn, F., Ed, Chronica Botanica Co., Waltham, Massachusetts, 1945, 11.

Stattersfield, A.J. et al., *Endemic bird areas of the world, Priorities for biodiversity conservation*, BirdLife Conservation Series No. 7, BirdLife International, Cambridge, 1998.

Steyermark, J.A. and Aymard, G.A., Nyctaginaceae, in *Flora of the Venezuelan Guayana, Vol. 7*, Berry, P.E., Yatskievych, K., and Holst, B.K., Eds, Missouri Botanical Garden Press, St Louis, 2003, 101.

Svenson, H.K., Vegetation of the coast of Ecuador and Peru and its relation to the Galapagos Islands, I, Geographical relations of the flora, *Am. J. Bot.*, 33, 394, 1945.

Tovar, O., *Tipos de Vegetación, Diversidad Florística y Estado de Conservación de la Cuenca del Mantaro*, Centro de Datos para la Conservación, Lima, 1990, 70.

Troll, C., Die Lokalwinde der Tropengebirge und ihr Einfluß auf Niederschlag und Vegetation, *Bonner Geogr. Abh.*, 9, 124, 1952.

Weberbauer, A., Die Vegetationsgliederung des nördlichen Peru um 5° südl. Br. (Departmento Piura und Provincia Jaen des Departamento Cajamarca), *Bot. Jahrb. Syst.*, 50, 72, 1914.

Weberbauer, A., Phytogeography of the Peruvian Andes, *Field Mus. Nat. Hist., Bot. Ser.*, 13, 13, 1936.

Weberbauer, A., *El Mundo Vegetal de los Andes Peruanos*, Estac. Exper. Agric. La Molina, Editorial Lume, Lima, 1945.

Weigend, M., Observations on the biogeography of the Amotape-Huancabamba zone in northern Peru, *Bot. Rev.*, 68, 38, 2002.

Weigend, M., Additional observations on the biogeography of the Amotape-Huancabamba zone in northern Peru: defining the south-eastern limits, *Rev. Peru. Biol.*, 11, 127, 2004.

Zamora, C. and Bao, R., *Regiones Edáficas del Perú*, Oficina Nacional de Evaluación de Recursos Naturales (ONERN), Lima, 1972, 21.

12 Seasonally Dry Forests of Southern Ecuador in a Continental Context: Insights from Legumes

Gwilym P. Lewis, Bente B. Klitgaard and Brian D. Schrire

CONTENTS

ABSTRACT

Seasonally dry tropical forests (SDTF) of Ecuador have been variously referred to as bosque seco, bosque espinoso, matorral seco, deciduous or semideciduous forest, and dry cactus scrub, amongst others. Isolated areas of this vegetation type occur in the dry inter-Andean valleys of southern Ecuador and in coastal Ecuador to the west of the Andes. Together with similar SDTF pockets in northern Peru they form part of the Tumbesian Centre of Endemism, a region of fragmented diverse forest types. Recent studies by Pennington et al. (2000, 2004) have shown that SDTF are scattered throughout the Neotropics, and these have been interpreted as isolated fragments of earlier more widespread vegetation. The floristic component of these seasonally dry forests is quite distinct from the savannas, cerrado, chaco and llanos vegetation types of South America. The woody Leguminosae of southern Ecuador have been studied in detail (Lewis and Klitgaard, 2002) permitting a comparison of the local SDTF legume flora with other neotropical seasonally dry forests. While a small number of woody legumes are narrowly restricted endemics in southern Ecuadorean SDTF, many more are found also in similar SDTF in northern Peru. Still others are to be found either widely dispersed amongst, or disjunct between, pockets of this vegetation type throughout the Neotropics. Examples of these different distribution patterns are presented and discussed, and the neotropical SDTF are considered in the context of a recently defined global legume 'succulent biome'.

12.1 INTRODUCTION

Ecuador (Figure 12.1), including the Galápagos Islands, is the smallest of the Andean countries with a land area of approximately 283,000 km^2, about the size of the state of Colorado, USA, or somewhat larger than Great Britain (Neill, 1999a). The country straddles the Equator from about 1°30′N to 5°S latitude. Mainland Ecuador extends from about 75°20′W to approximately 81°W longitude. The country is divided into 21 political provinces, with each province being largely associated with one of four main geographical regions: (1) the Pacific coastal region (the costa) which includes the lower western slopes of the Andes below 1000 m elevation, (2) the Andes mountains above 1000 m (the Sierra), (3) the Amazon lowlands to the east of the Andes (the Oriente); this region includes the eastern slopes of the Andes up to 1000 m, and (4) the Galápagos Islands, a volcanic archipelago in the Pacific Ocean, approximately 1000 km west of mainland Ecuador. The Andes of southern Ecuador form a complex pattern of ridges, some of which trend east–west and some north–south. The Quaternary volcanoes of the central and northern Ecuadorean cordilleras are missing from southern Ecuador where the highest peaks and ridges barely surpass 4000 m (Neill, 1999a). This chapter focuses mainly on the SDTF of two southern provinces: Loja (largely in the Andean region, but with lowland forests in the south-west which are best considered as part of the coastal region) and El Oro (which extends from the coastal region to the western slopes of the Andes), and concentrates on the species distribution patterns of woody plants in family Leguminosae.

The Leguminosae, or bean and pea family, is divided into three subfamilies: Caesalpinioideae, Mimosoideae and Papilionoideae, and currently comprises 727 genera and 19,325 species worldwide (Lewis et al., 2005). Legumes vary in habit from ephemeral herbs to shrubs, vines, woody climbers and giant emergent forest trees; a few are aquatics. They are to be found as major components of most of the world's vegetation types and many have the ability to colonize marginal or barren lands because of their capacity to fix atmospheric nitrogen through root nodules (Sprent, 2001). The seeds, young green pods, leaves, roots and flowers of many legumes provide a protein-rich food source for humans and animals and the large-scale cultivation of some species produces very valuable cash crops. Many legume species in Ecuador have been introduced from elsewhere as garden ornamentals and showy street trees. Some of these have escaped cultivation and are now naturalized, adding to an already rich and diverse Ecuadorean legume flora.

Lewis and Klitgaard (2002) recognized 605 species in 126 genera of Leguminosae for Ecuador. Only one legume genus, the recently described *Ecuadendron* D. A. Neill (1998), is endemic to the country,

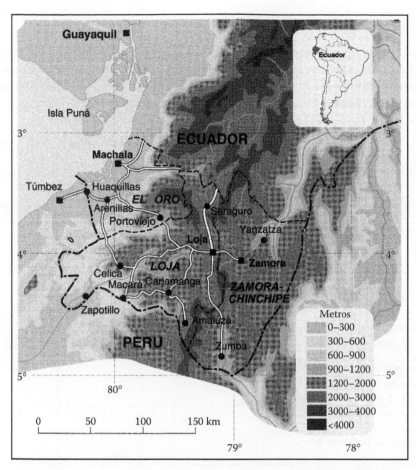

FIGURE 12.1 (See colour insert following page 208) The three southern provinces of Ecuador, El Oro, Loja and Zamora Chinchipe and the location of Ecuador in South America. Reproduced from Botánica Austro-ecuatoriana with permission from H. Balslev.

and this is only known in the provinces of Azuay, Esmeraldas and Guayas. The level of endemism among legume species in Ecuador is between 10 and 11% (Neill, 2000) with approximately 61 species considered to be endemic to the country. The SDTF of southern Ecuador together comprise a total of 61 species (63 taxa) (Table 12.1) of woody legume (lianas and perennial herbs are excluded in this survey, as are species that occur mainly in humid vegetation, but are also recorded from seasonally dry forests, e.g. four species of the mimosoid genus *Inga*). The 61 species are accommodated in 33 genera (Table 12.2), with 19 genera (two of which are not found outside South America) restricted to the Neotropics, three extending to Africa and 11 with a pantropical distribution.

12.1.1 Affinities of Southern Ecuadorean Seasonally Dry Tropical Forests: A Global Legume Succulent Biome

Schrire et al. (2005a, b) recognized a series of four major global biomes into which all legume genera can be accommodated based on the predilection of the majority of the range-restricted species in each genus. One novel legume biome resulting from their work is the aptly named Succulent biome, which is a fire-intolerant, succulent-rich and grass-poor, dry tropical forest, thicket and bushland biome prone to bimodal or erratic rainfall patterns. The SDTF of Pennington et al. (2000; 2004) and this chapter largely equate to the neotropical element of this highly fragmented global

TABLE 12.1

Woody Legume Species and Infraspecific Taxa (Lianas and Woody-Based Perennial Herbs Excluded) Occurring in the Seasonally Dry Tropical Forests of Southern Ecuador (the Tumbesian Zone and Inter-Andean Valleys) and Their Distribution Patterns, from Narrow Endemics to Widespread Neotropical (Taxa Arranged Alphabetically by Genus and Species within Each Legume Subfamily)

Taxon	SENE	EE	EPE	EPSP	ECD	EPCD	EPCVD	EPBD	SAD	ND
Caesalpinioideae										
Bauhinia aculeata L. ssp. *grandiflora* (A. Juss.) Wunderlin								+		
Bauhinia augusti Harms			+							
Bauhinia weberbaueri Harms			+							
Caesalpinia ancashiana Ulibarri			+							
Caesalpinia cassioides Willd.						+				
Caesalpinia glabrata Kunth			+							
Caesalpinia spinosa (Molina) Kuntze									+	
Chamaecrista glandulosa (L.) Greene var. *andicola* H.S. Irwin & Barneby								+		
Cynometra crassifolia Benth.	+									
Hoffmannseggia viscosa (Ruiz & Pav.) Hook. & Arn.								+		
Parkinsonia praecox (Ruiz & Pav.) Hawkins										+
Senna cajamarcae H.S. Irwin & Barneby										+
Senna hirsuta (L.) H.S. Irwin & Barneby var. *hirta* H.S. Irwin & Barneby										+
Senna huancabambae (Harms) H.S. Irwin & Barneby			+							
Senna incarnata (Pav. ex Benth.) H.S. Irwin & Barneby			+							
Senna macranthera (DC. ex Collad.) H.S. Irwin & Barneby var. *andina* H.S. Irwin & Barneby			+							
Senna mollissima (Humb. & Bonpl. ex Willd.) H.S. Irwin & Barneby var. *mollissima*			+							
Senna oxyphylla (Kunth) H.S. Irwin & Barneby var. *hartwegii* (Benth.) H.S. Irwin & Barneby		+								

SENE = southern Ecuador narrow endemic; EE = Ecuador endemic; EPE = Ecuador-Peru (Tumbesian) endemic; EPSP = Ecuador-Peru, with disjunction in southern Peru; ECD = Ecuador-Colombia disjunct; EPCD = Ecuador-Peru-Colombia disjunct; EPCVD = Ecuador-Peru-Colombia-Venezuela disjunct; EPBD = Ecuador-[± Peru]-Bolivia disjunct (not all the taxa in this column are recorded from Peru); SAD = South American disjunct, in some taxa the range extending to the caatingas of north-eastern Brazil; ND = neotropical disjunct. *Senna pistaciifolia* var. *picta* also occurs on the Galapagos Islands; *Senna cajamarcae* is known only from Ecuador, Peru and Panama; *Albizia multiflora* var. *multiflora* occurs in Ecuador, Peru, Colombia and Panama.

TABLE 12.1
Woody Legume Species and Infraspecific Taxa (Lianas and Woody-Based Perennial Herbs Excluded) Occurring in the Seasonally Dry Tropical Forests of Southern Ecuador (the Tumbesian Zone and Inter-Andean Valleys) and Their Distribution Patterns, from Narrow Endemics to Widespread Neotropical (Taxa Arranged Alphabetically by Genus and Species within Each Legume Subfamily) (Continued)

Taxon	SENE	EE	EPE	EPSP	ECD	EPCD	EPCVD	EPBD	SAD	ND
Senna pilifera (Vogel) H.S. Irwin & Barneby var. *subglabra* (S. Moore) H.S. Irwin & Barneby							+			
Senna pistaciifolia (Kunth) H.S. Irwin & Barneby var. *glabra* (Benth.) H.S. Irwin & Barneby								+		
Senna pistaciifolia (Kunth) H.S. Irwin & Barneby var. *picta* (G.Don) H.S. Irwin & Barneby			+							
Senna robiniifolia (Benth.) H.S. Irwin & Barneby										+
Senna spectabilis (DC.) H.S. Irwin & Barneby var. *spectabilis*									+	
Mimosoideae										
Acacia weberbaueri Harms			+							
Acacia macracantha Humb. & Bonpl. ex Willd.										+
Albizia multiflora (Kunth) Barneby & J.W. Grimes var. *multiflora*										+
Anadenanthera colubrina (Vell.) Brenan var. *cebil* (Griseb.) Altschul									+	
Calliandra taxifolia (Kunth) Benth.			+							
Calliandra tumbeziana J. F. Macbr.			+							
Chloroleucon mangense (Jacq.) Britton & Rose var. *mangense*										+
Leucaena trichodes (Jacq.) Benth.							+			
Mimosa acantholoba (Humb. & Bonpl. ex Willd.) Poir. var. *acantholoba*										+
Mimosa albida Humb. & Bonpl. ex Willd. var. *willdenowii* (Poir.)Rudd							+			
Mimosa caduca (Humb. & Bonpl. ex Willd.) Poir.			+							
Mimosa debilis Humb. & Bonpl. ex Willd. var. *aequatoriana* (Rudd) Barneby		+								
Mimosa loxensis Barneby	+									
Mimosa nothacacia Barneby			+							
Mimosa quitensis Benth.					+					
Mimosa townsendii Barneby	+									
Piptadenia flava (Spreng. ex DC.) Benth.										+
Pithecellobium excelsum (Kunth) Mart. ex Benth.			+							

(continued)

TABLE 12.1
Woody Legume Species and Infraspecific Taxa (Lianas and Woody-Based Perennial Herbs Excluded) Occurring in the Seasonally Dry Tropical Forests of Southern Ecuador (the Tumbesian Zone and Inter-Andean Valleys) and Their Distribution Patterns, from Narrow Endemics to Widespread Neotropical (Taxa Arranged Alphabetically by Genus and Species within Each Legume Subfamily) (Continued)

Taxon	SENE	EE	EPE	EPSP	ECD	EPCD	EPCVD	EPBD	SAD	ND
Pseudosamanea guachapele (Kunth) Harms										+
Zapoteca andina H.M. Hern.								+		
Zapoteca caracasana (Jacq.) H.M. Hern. spp. *weberbaueri* (Harms) H.M. Hern.						+				
Papilionoideae										
Aeschynomene scoparia Kunth			+							
Aeschynomene tumbezensis J.F. Macbr.			+							
Amicia glandulosa Kunth			+							
Centrolobium ochroxylum Rose ex Rudd			+							
Clitoria brachystegia Benth.	+									
Coursetia caribaea (Jacq.) Lavin var. *caribaea*										+
Coursetia caribaea (Jacq.) Lavin var. *ochroleuca* (Jacq.) Lavin								+		
Cousetia grandiflora Benth.			+							
Cyathostegia mathewsii (Benth.) Schery				+						
Dalea carthagenensis (Jacq.) J.F. Macbr. var. *brevis* (J. F.Macbr.) Barneby				+						
Erythrina poeppigiana (Walp.) O.F. Cook										+
Erythrina smithiana Krukoff		+								
Erythrina velutina Willd.							+			
Geoffroea spinosa Jacq.									+	
Gliricidia brenningii (Harms) Lavin			+							
Lonchocarpus atropurpureus Benth.							+			
Machaerium millei Standl.			+							
Myroxylon balsamum (L.) Harms										+
Piscidia carthagenensis Jacq.										+

succulent biome, although exact comparisons are problematic given the different approaches to defining these entities. The biomes of Schrire et al. (2005a,b) are floristic units derived by analysis of legume species distributions while SDTF have been delimited on the basis of a group of vegetation types underpinned by a distinctive set of environmental parameters.

The present-day Succulent biome (Figure 12.2) encompasses regions in both the Neotropics and the Paleotropics. The neotropical Succulent biome areas include semi-arid tropical to subtropical Mexico, Central America and the Caribbean (particularly the Greater Antilles: Wolfe, 1975; Lavin et al., 2001), linked frequently to circum-Amazonian Pleistocenic Arc dry forest elements

TABLE 12.2
Southern Ecuadorean Woody Legume SDTF Genera and Their Global Distribution (Genera Arranged Alphabetically within Each of the Three Subfamilies)

Genus	Distribution
Caesalpinioideae	
Bauhinia sensu stricto	Pantropical
Caesalpinia sensu stricto	Pantropical (succulent biome)
Caesalpinia (species other than s.s.)	Neotropical
Chamaecrista	Pantropical (+subtropical and temperate)
Cynometra	Pantropical
Hoffmannseggia	Neotropical
Parkinsonia	Neotropical + African (succulent biome)
Senna	Pantropical (+temperate)
Mimosoideae	
Acacia sensu lato	Pantropical
Albizia sensu lato	Pantropical
Anadenanthera	South American
Calliandra	Neotropical
Chloroleucon	Neotropical
Leucaena	Neotropical
Mimosa	Neotropical (most species) + Palaeotropical (esp. Madagascar)
Piptadenia	Neotropical
Pithecellobium	Neotropical
Pseudosamanea	Neotropical
Zapoteca	Neotropical
Papilionoideae	
Aeschynomene	Pantropical (+ subtropical)
Amicia	Neotropical
Centrolobium	Neotropical
Clitoria	Pantropical
Coursetia	Neotropical
Cyathostegia	South American
Dalea	Neotropical
Erythrina	Pantropical
Geoffroea	Neotropical (succulent biome)
Gliricidia	Neotropical
Lonchocarpus	Neotropical (+ 1 sp. ampi-Atlantic)
Machaerium	Neotropical (+ 1 sp. ampi-Atlantic)
Myroxylon	Neotropical
Piscidia	Neotropical (succulent biome)
TOTAL	DISTRIBUTION
2 genera	restricted to South America
17 genera	restricted to the Neotropics
3 genera	neotropical with extensions to Africa
11 genera	pantropical

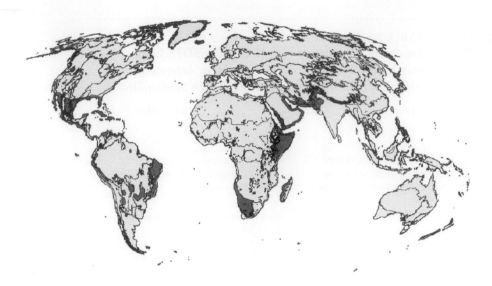

FIGURE 12.2 (See colour insert) Global distribution of the legume Succulent biome.

in South America, including the inter-Andean valleys of Ecuador and Peru, the Piedmont area of north-western Argentina and central Bolivia, the Misiones region of north-eastern Argentina and adjacent Paraguay, and the caatinga region of eastern Brazil (Prado and Gibbs, 1993; Pennington et al, 2000, 2004, Schrire et al., 2005a, b). Neotropical Succulent biome centres are linked with intervening fossil evidence from Tertiary tropical North America and Europe (Herendeen et al., 1992) across to the Old World, including: the Somalia-Masai regional centre of endemism (White, 1983) of the Horn of Africa, with dry forest and thicket arid corridor disjunctions and extensions through to the Nama-Karoo, succulent Karoo, desert and thicket regions (Cowling et al., 2005) of southern and south-western Africa, to western Madagascar, and Arabia to West Asia and north-west India (Schrire et al. 2005a, b and several references therein).

The genus *Caesalpinia* sensu stricto (as currently circumscribed, e.g. in Lewis et al., 2005) aptly demonstrates the Succulent biome distribution pattern (Figure 12.3, Table 12.3) with 12 species in the seasonally dry forests of the Caribbean, three in Mexican and Central American SDTF, two in South American SDTF (including *C. cassioides* Willd. in the SDTF of Colombia, southern Ecuador and northern Peru), four in north-east African and Arabian bushland and thicket, five in southern African succulent-rich bushland and thicket, one endemic to the dry forests of Madagascar, and one (for which precise habitat data are not known) only found in South-Central Africa.

Various methods of cladistic vicariance analyses (Schrire et al., 2005a,b) have all suggested that legume lineages confined to the succulent biome have given rise to sublineages occupying the other three major legume biomes: the Grass, Rain forest and Temperate biomes. Evidence also suggests that Rain forest biome clades may be the most recently derived in legumes, supporting the findings of Richardson et al. (2001) for the legume genus *Inga*, and Pendry (2004) and Pennington et al. (2004) for the Polygonaceae genus *Ruprechtia*. The results of Schrire et al. (2005a,b) and Lavin et al. (2005) are in agreement with the fossil record; suggesting a rapid diversification of nearly all major legume clades throughout much of the world during the Early Tertiary, with an origin in dry forests around the margins of the Tethys seaway. Fossil sites rich in legumes support a long association between legumes and seasonally dry areas (Herendeen et al., 1992; Herendeen, 2001). Many of the adaptations which characterize extant legume taxa are to a seasonally dry warm climate. These include: compound and deciduous leaves, high nitrogen metabolism, diverse and mobile chemical defences, hard seed testas, long seed dormancy and long

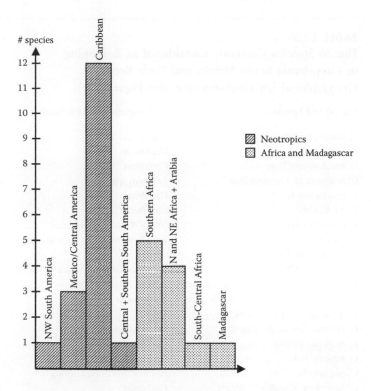

FIGURE 12.3 Distribution of *Caesalpinia* sensu stricto.

viability, the ability of seeds to store nitrogen, and various associations with ants (Schrire et al., 2005a and references therein). At the time of initial legume diversification there is also evidence of a seasonally dry tropical climate (Scotese, 2001) and the presence of deciduous forest in and around the margins of the Tethys seaway, further supporting a link between legume diversification and a seasonally dry environment. Remnants of Tethyan distributions are likely to be most evident in the Succulent biome because of the persistence of the drought-tolerant legume floras in this now fragmented global biome (Schrire et al., 2005a). Put another way, tropical dry areas have probably been occupied by legumes since their inception.

12.1.2 Description of the Seasonally Dry Tropical Forests of Southern Ecuador

This chapter adopts a broad definition of seasonally dry tropical forests (SDTF) following Murphy and Lugo (1995), Pennington et al. (2000, 2004), Bridgewater et al. (2003), and Linares-Palomino (2004a, b). These tree-dominated ecosystems are mostly deciduous during the strongly defined dry season which, in South American SDTF, is a period of at least five to six months receiving a total of less than 100 mm of rainfall. The floristic composition of the SDTF is quite distinct from that of the cerrado, chaco and llanos vegetation types in South America, even though isolated pockets of dry forest do occur in these other major vegetations and act as habitat stepping stones between major SDTF areas. In South America the SDTF occur on relatively fertile soils where total rainfall is less than 1600 mm per year (Gentry, 1995; Graham and Dilcher, 1995; Pennington et al. 2000; Bridgewater et al., 2003). Local precipitation can increase dramatically in the Pacific coastal dry forests during so-called El Niño years which occur on a three- to eight-year cyclical basis (Block and Richter, 2000; Jørgensen, pers. comm., 2005). Thorny species are common, grasses are scarce, and in the driest SDTF there is a marked increase in evergreen and succulent species, especially

---

TABLE 12.3
The 28 Species Currently Considered as Belonging
to *Caesalpinia* Sensu Stricto and Their Broad
Geographical Distributions (see also Figure 12.3)

Caesalpinia Species	Geographical Distribution
C. anacantha Urb.	Caribbean
C. bahamensis Lam.	Caribbean
C. barahonensis Urb.	Caribbean
C. bracteata G. Germishuizen	Southern Africa
C. brasiliensis L.	Caribbean
C. buchii Urb.	Caribbean
C. cassioides Willd.	North-West South America
C. dauensis Thulin	East & North-East Africa to Arabia
C. domingensis Urb.	Caribbean
C. erianthera Chiov.	East & North-East Africa to Arabia
C. glandulosopedicellata R. Wilczek	South-Central Africa
C. madagascariensis (R. Vig.) S. Senesse	Madagascar
C. melanadenia (Rose) Standl.	Mexico
C. merxmuellerana A. Schreiber	South Africa
C. monensis Britton	Caribbean
C. nipensis Urb.	Caribbean
C. oligophylla	North & North-East Africa
C. pauciflora (Griseb.) C. Wright	Caribbean
C. pearsonii L. Bolus	South Africa
C. pulcherrima (L.) Sw.	Mexico & Central America
C. reticulata Britton	Caribbean
C. rosei Urb.	Caribbean
C. rostrata N. E. Br.	Southern Africa
C. rubra (Engl.) Brenan	Southern Africa
C. secundiflora Urb.	Caribbean
C. sessiliflora S. Watson	Mexico & Central America
C. stuckertii Hassl.	Central & Southern South America
C. trothae Harms	North & North-East Africa

---

of Cactaceae (Gentry, 1995, Mooney et al., 1995). The two plant families Leguminosae and Bignoniaceae generally dominate the woody floras of the SDTF throughout their range, but there are exceptions where these are replaced by other families (e.g., see Bridgewater et al., 2003). The main areas of SDTF in the Neotropics are discussed in the introduction to this book and illustrated in Figure 12.4. In Ecuador the SDTF occur in the coastal region from Esmeraldas, southwards through Manabí and Guayas provinces to western El Oro and south-western Loja provinces (Figure 12.5), and also intermittently in the inter-Andean valleys along the mountain range.

The SDTF of southern Ecuador are referred to in the plentiful literature under several vegetation terminologies leading to confusion as to their full extent, their classification and nomenclature. This situation is no different in other parts of the Neotropics where SDTF are known by several different terms (e.g., see Trejo, 1996; Trejo and Dirzo, 2000; for Mexico). Broadly speaking, and in this chapter, SDTF in southern Ecuador encompass bosque espinoso, bosque caducifolio, bosque seco tropical, bosque seco montano, deciduous forest (although in southern Ecuador hardly any vegetation is completely deciduous as species of *Acacia* and *Ceiba* often retain leaves right through the dry season), matorral seco espinoso, matorral seco montano, bosque muy seco occidental, bosque seco oriental, semideciduous forest, mainly deciduous tropical thorn forest, dry cactus scrub, *Ceiba trichistandra* forest, *Ceiba pentandra* forest, Tumbesian dry forest and inter-Andean (or

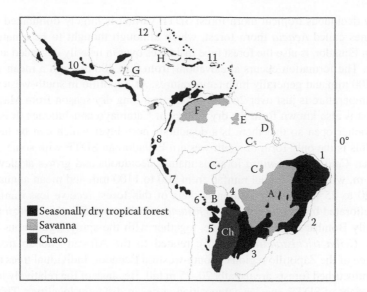

FIGURE 12.4 Schematic distribution of seasonally dry forests and savannas in the Neotropics. Seasonally dry forest: 1. caatingas; 2. south-east Brazilian seasonal forests; 3. Misiones nucleus; 4. Chiquitano; 5. Piedmont nucleus; 6. Bolivian inter-Andean valleys; 7. Peruvian and Ecuadorean inter-Andean valleys; 8. Pacific coastal Peru and Ecuador; 9. Caribbean coast of Colombia and Venezuela; 10. Mexico and Central America; 11. Caribbean Islands (small islands coloured black are not necessarily entirely covered by seasonally dry forests); 12. Florida. Savannas: (A) cerrado; (B) Bolivian; (C) Amazonian (smaller areas not represented); (D) coastal (Amapá, Brazil to Guyana); (E) Rio Branco-Rupununi; (F) Llanos; (G) Mexico and Central America; (H) Cuba. Ch: Chaco. Modified after Pennington et al. (2000) and Huber et al. (1987).

intermontane) dry forest and scrub, desert and semi-desert (all of these are discussed in varying amounts of detail in: Kessler, 1992; Parker and Carr, 1992; Best and Kessler, 1995; Neill, 1999b; Sierra, 1999; Madsen et al., 2001; Balslev and Øllgaard, 2002; and Lozano, 2002; also refer to Chapter 11). While all of these vegetation terminologies fit within the definition of SDTF used in this chapter, each variant tends to have its unique characteristics, including climate, altitude, soil type and precise species composition. The so-called Tumbesian centre of endemism (Best and Kessler, 1995) in which the SDTF of southern Ecuador (and northern Peru) occur is a highly complex environment of fragmented forest patches which includes some humid forests in addition to the many variants of dry forest.

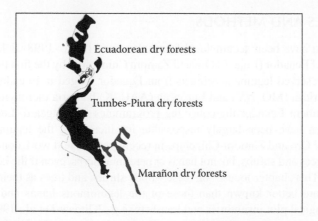

FIGURE 12.5 Distribution of SDTF in southern Ecuador and northern Peru (part of the Tumbes SDTF Ecoregion). Modified from Olson and Dinerstein (2002) by Linares-Palomino.

The mainly deciduous tropical thorn forest SDTF variant is largely dominated by *Acacia* and is thus sometimes called *Acacia* thorn forest, which although thought to be a natural vegetation type in southern Ecuador, is also the forest type that takes over in heavily degraded areas, especially of *Ceiba* forest. The formation occurs at elevations from 0 to 400 m. with a mean annual precipitation of 100–200 mm and generally high temperatures. At Zapotillo in south-western Ecuador the mean annual temperature is just over 26°C and there is a long dry season from May to December. This thorn forest is also known from the dry valleys of Catamayo and Jubones. It is of low stature (5–10 m), but rather open so that there is a deciduous herb layer which can be lush in the short rainy season. This is the only form of south-western Ecuadorean SDTF with some grasses present in the herb layer. *Ceiba trichistandra* forest is mainly deciduous and grows at elevations ranging from 0 to 1400 m, with mean annual rainfall from 200 to 1100 mm and mean annual temperatures ranging from 20 to 25°C. Lower elevation pockets of this forest receive less rainfall, but have a dry season ameliorated by coastal fog. *C. trichistandra* Bakh. and *Eriotheca ruizii* (K. Schum) A. Robyns in family Bombacaceae predominate, together with the spiny leguminous tree *Erythrina velutina* Willd. *Ceiba trichistandra* ('ceibo'), related to the African baobab tree, is the most characteristic tree of the Zapotillo region of south-western Ecuador. Individual trees can attain 35 m in height and undisturbed forests are usually 20–25 m tall. Because of the relatively large altitudinal range of this variant of SDTF, species composition varies at different localities. This forest type is encountered in Puyango on the El Oro-Loja province boundary and in the higher Catamayo valley of Loja (Best and Kessler, 1995). In the ground layer Cactaceae are common, but grasses are rare.

Ceiba pentandra forest is distinguished from *C. trichistandra* forest in having several of its canopy trees evergreen, and in having only a single tree, *C. pentandra* (L.) Gaertn. (the kapok), prominent, with all other trees being superficially similar to each other. This forest variant is less dry and less seasonal than *C. trichistandra* forest; herbs are rare but epiphytes are fairly common. Intermontane scrub and thorn forest occur in the dry valleys from 500 to above 2000 m altitude where mean annual precipitation is from 150 to 800 mm. The dry season can be eight months long, the environment arid, and cacti are common. Scrub vegetation in southern-most Ecuador (in the intermontane Chinchipe valley of Zamora Chinchipe and Loja provinces) is dominated by a thorny, woody legume flora which includes *Parkinsonia praecox* (Ruiz & Pav.) Hawkins, *Acacia macracantha* Humb. & Bonpl. ex Willd. and *Anadenanthera colubrina* (Vell.) Brenan var. *cebil* (Griseb.) Altschul, together with several members of the Cactaceae. Similar vegetation is encountered in the inter-Andean arid valleys of central eastern Loja. Here the leguminous tree *Caesalpinia spinosa* (Molina) O. Kuntze is frequent. Good soils, a paucity of weedy herbs and a reduction in pests has led to human settlement and agriculture in these valleys resulting in much of the original vegetation now being highly degraded or totally absent.

12.2 MATERIALS AND METHODS

Legume species data have been accumulated from two years' (1996–1998) field work in the three southern provinces of Ecuador (Loja, El Oro and Zamora Chinchipe) by the first two authors, together with the study of preserved legume specimens from Ecuador lodged in Ecuadorean (LOJA, QCA, QCNE), North American (MO, NY) and European (AAU, K) herbaria (acronyms follow Holmgren et al., 1990). In southern Ecuador the collecting programmes of Klitgaard, Lewis, Lozano, Neill, and Van den Eynden have been largely responsible for increasing the legumes recorded in the provinces of Loja, El Oro and Zamora-Chinchipe in recent years. Field work concentrated on woody species of legume (trees and shrubs, but not lianas or perennial herbs, even if the latter were somewhat woody at the base). This chapter focuses on these woody shrubs and trees as their distribution within the SDTF is therefore better known than those of the leguminous lianas and herbs. Field work included some ecological plot inventories and transects (e.g., Klitgaard et al., 1999). When mapping species distributions it is important that the taxonomic status of these taxa has been resolved within the group or genus in which they occur, so that plant names are consistent across the full geographical range of a taxon. Of particular value when cross checking species names and tracing provincial records was the *Catalogue of the Vascular Plants of Ecuador* (Jørgensen and León-Yánez, 1999). It is apparent that mapping the distribution of a single taxon in isolation is not as biogeographically revealing as mapping it together with its sister or other closely related taxa. For patterns of distribution to be meaningful, species with described infraspecific taxa need to be analysed in their entirety. In addition, species with apparently closely related sister species should be mapped together because the literature contains examples of plants that have been named more than once because of widely disjunct populations of the same species (see the example of *Pithecellobium excelsum* (Kunth) Mart. ex Benth. discussed by Pennington et al., 2000, and later in this chapter). The regional herbarium (Herbario LOJA Reinaldo Espinosa) in Loja, southern Ecuador, was especially important in terms of historical legume collections from the region. The legumes in the LOJA herbarium were named (and/or nomenclaturally updated) using the most recently published monographs and revisions of genera by recognized legume specialists, and by comparison of the specimens with authoritatively named material in the Herbarium of the Royal Botanic Gardens, Kew.

Generic concepts of legumes largely follow those presented in Lewis et al. (2005), while species and infraspecific taxa correspond to those used in the most recently published monographs and revisions.

12.3 FLORISTIC AFFINITIES OF SOUTHERN ECUADOREAN SDTF WOODY LEGUME SPECIES: EXAMPLES OF DISTRIBUTION PATTERNS

12.3.1 PATTERNS OF ENDEMISM

12.3.1.1 Southern Ecuador SDTF Narrowly Restricted Endemics

Woody legumes that are currently known to be restricted to southern Ecuador SDTF occur in two distinct areas: the inter-Andean valleys of Loja province and the 'bosque petrificado' of Puyango on the Loja-El Oro province boundary. In Puyango there are two narrowly restricted woody legume endemics: *Cynometra crassifolia* Benth. in subfamily Caesalpinioideae, and *Clitoria brachystegia* Benth. (Figure 12.6) in subfamily Papilionoideae. In contrast, the two legume endemics in the inter-Andean valleys: *Mimosa townsendii* Barneby and *M. loxensis* Barneby are members of subfamily Mimosoideae. When first described by Barneby (1991), *M. townsendii* was only known from the type specimen and unknown in fruit. Since then the species has been gathered several times (in flower and fruit) in semi-deciduous thorn scrub vegetation at a number of locations ranging from 1600 m to 2150 m altitude in the dry valleys of Loja. *Mimosa loxensis* was not scientifically described until two years after publication of Barneby's (1991) monograph (Barneby, 1993), although the type specimen of the species name was originally collected by Reinaldo Espinosa in 1947. The species appears to be a narrowly restricted endemic with a predilection for calcareous soils on steep gully slopes in thorn and cactus scrub between 1600 m and 2100 m altitude. Morphological characteristics of the species suggest a weak affinity with *Mimosa polycarpa* Kunth var. *redundans* Barneby, a taxon found as yet no further north than Cajamarca in Peru.

12.3.1.2 Less Narrowly Restricted Ecuadorean SDTF Woody Legume Endemics

Mimosa debilis Humb. & Bonpl. ex Willd. var. *aequatoriana* (Rudd) Barneby is locally plentiful in Pacific lowland Ecuador in the provinces of Esmeraldas, Guayas and El Oro and has once been gathered on the Isla San Cristóbal in the Galapagos archipelago (Barneby, 1991). *Senna oxyphylla* (Kunth) H.S. Irwin & Barneby var. *hartwegii* (Benth.) H.S. Irwin & Barneby (subfamily Caesalpinioideae) is, similarly, locally common on the coastal plains and lower foothills of the Andes around the Gulf of Guayaquil in Los Ríos, Guayas and El Oro provinces and in Manabí (Irwin and Barneby, 1982).

FIGURE 12.6 (See colour insert) *Clitoria brachystegia* Benth. (Leguminosae: Papilionoideae): a narrowly restricted endemic from the bosque petrificado at Puyango on the Loja-El Oro boundary. (Photo: G.P. Lewis.)

Erythrina smithiana Krukoff (subfamily Papilionoideae) is also apparently restricted to the seasonally dry forests of Ecuador.

12.3.1.3   Tumbesian Endemics: SDTF Woody Legume Taxa Restricted to Southern Ecuador and Northern Peru (Here Combining Coastal and Inter-Andean SDTF)

Seven caesalpinioid taxa are currently known from southern Ecuador and northern Peru: two *Bauhinia* species, *B. augusti* Harms and *B. weberbaueri* Harms, two *Caesalpinia sensu lato* species, *C. glabrata* Kunth and *C. ancashiana* Ulibarri and three taxa in the genus *Senna*: *S. macranthera* (Collad.) H.S. Irwin & Barneby var. *andina* H.S. Irwin & Barneby, *S. mollissima* (Willd.) H.S. Irwin & Barneby var. *mollissima* with a northerly range extension into Guayas, and *S. huancabambae* (Harms) H.S. Irwin & Barneby which just encroaches into the dry valleys of Azuay province in Ecuador. A fourth *Senna* species, *S. incarnata* (Benth.) H.S. Irwin & Barneby, is locally abundant in widely separated pockets along the crest and Pacific slopes of the Andes, being found as far north as Chimborazo in Ecuador and as far south as Lima in Peru, a range distribution from 2°15′S to 12°S.

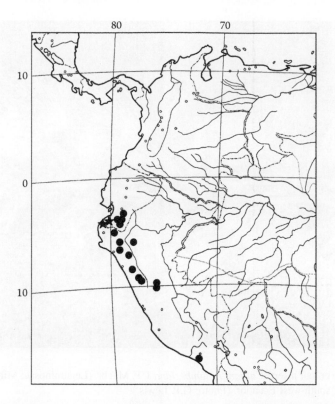

FIGURE 12.7 Distribution of *Calliandra tumbeziana* (★) and *C. taxifolia* (●). Redrawn from Barneby (1998).

Six mimosoid legumes are restricted in distribution to southern Ecuador and northern Peru: the little collected and poorly known *Acacia weberbaueri* Harms, two species of *Calliandra*, two of *Mimosa* and *Pithecellobium excelsum. Calliandra tumbeziana* J.F. Macbr. (Figure 12.7 and Figure 12.8) was, until recently (Barneby, 1998), known only from the Cerros de la Brea in the department of Tumbes, north-west Peru, but its subsequent discovery in the dry spiny forests of south-west Ecuador at 400–750 m altitude was not unexpected. *Calliandra taxifolia* (Kunth) Benth. (Figure 12.7) is recorded from the west-draining inter-Andean valleys of Azuay and Loja in Ecuador and from northern Peru in the departments of Amazonas and Huánuco across the Andean crest to the headwaters of the rivers Marañon and Huallaga (Barneby, 1998). It is also apparently disjunct in the department of Arequipa at 350–510 m altitude on the foothills of the Andes near Mollendo (16°30′S), although this population is recorded as having orange, orange-red or tangerine yellow (specimen: *Hughes* 2357) stamen filaments, rather than the simply red flowers of the Ecuadorean and northern Peruvian populations. The Mollendo population was originally described as *C. prostrata* Benth., a taxonomy followed by Macbride (1943). It was Barneby (1998) who considered the two species as being conspecific, but a detailed field study might overturn this view. *Mimosa caduca* (Willd.) Poir. occurs in the seasonally arid inter-Andean valleys of Loja province, Ecuador and in Cajamarca, Peru, and has an altitudinal range from 800 m to 2300 m. It is the only Andean species in *Mimosa* series *Bimucronatae* Barneby and its kinship is morphologically similar, but remotely allopatric, to Mexican, Brazilian and Antillean taxa in the series, rather than to other Andean taxa of similar growth habit and ecology (Barneby, 1991). Included in the same series of *Mimosa* is *M. hexandra* Micheli (see Barneby, 1991, p. 167, Map 16) which has a circum-Amazonian disjunct seasonally dry tropical forest distribution which includes Paraguay; Mato Grosso, Brazil; Formosa and Misiones, Bolivia and Argentina; the caatinga of north-east Brazil; the arid lowlands of north-east Colombia and north-west Venezuela and the Pacific lowlands of the Tehuantepec isthmus in

FIGURE 12.8 (See colour insert) *Calliandra tumbeziana* J. F. Macbr. (Leguminosae: Mimosoideae) from the dry spiny forests of south-west Ecuador. (Photo: G.P. Lewis.)

Oaxaca, Mexico. If *M. caduca* is correctly aligned systematically, and its postulated relationship to *M. hexandra* is confirmed, then the two together occupy an increased area of the disjunct pockets of seasonally dry forest around the periphery of the Amazon basin.

Mimosa series *Nothacaciae* Barneby contains two morphologically isolated species: *M. townsendii* (discussed above) endemic to southern Ecuador and *M. nothacacia* Barneby from Piura, province Huancabamba in northern Peru, and recently collected in Loja province, Ecuador, in both dry and humid montane forest and secondary scrub. *Pithecellobium excelsum* (Figure 12.9 and Figure 12.10), is also restricted in distribution to the seasonally dry forests and deciduous thorn scrub of Ecuador (Manabí, Guayas and El Oro) and northern Peru (Tumbes and Huancabamba, extending eastwards to the Marañon valley in Cajamarca). It is feebly distinct from *P. diversifolium* Benth. except in geography (Figure 12.9) (Barneby and Grimes, 1997), the latter apparently endemic to north-east Brazilian caatinga vegetation. As noted by Pennington et al. (2000), a wider appreciation of the former extent of SDTF could influence taxonomic decisions. Postulation of a once more extensive seasonally dry forest in the Neotropics (now restricted to disjunct fragments) would associate north-east Brazilian caatinga with the dry forests of Ecuador and Peru and in light of such a hypothesis there is less temptation to recognize these two *Pithecellobium* species as distinct. Furthermore, both *P. excelsum* and *P. diversifolium* closely resemble, and might be directly related to or independently derived from, *P. ungis-cati* (L.) Mart. ex Benth., a species of seasonally dry, deciduous or semideciduous woodland in Mexico, Central America, the Caribbean, and northern, coastal Venezuela and Colombia (Barneby and Grimes, 1997).

In Leguminosae subfamily Papilionoideae nine taxa are largely restricted in range to southern Ecuador and northern Peru SDTF: *Aeschynomene scoparia* Kunth, *A. tumbezensis* J.F. Macbr., *Amicia glandulosa* Kunth (a species that can also tolerate more humid areas), *Centrolobium ochroxylum* Rose ex Rudd, *Coursetia grandiflora* Benth., *Cyathostegia mathewsii* (Benth.) Schery (this also collected further south in Apurímac, Peru; pers. comm., Hughes, 2005), *Dalea carthagenensis* (Jacq.) J.F. Macbr. var. *brevis* (J.F. Macbr.) Barneby (also known from SDTF further south

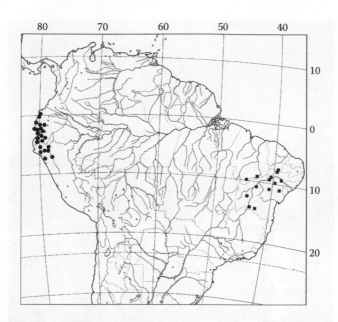

FIGURE 12.9 Distribution of *Pithecellobium excelsum* (●) and *P. diversifolium* (■). Redrawn from Barneby and Grimes (1997).

in Peru; pers. comm., Hughes, 2005), *Gliricidia brenningii* (Harms) Lavin, and *Machaerium millei* Standl. (Figure 12.11).

12.3.2 DISJUNCT SDTF WOODY LEGUME DISTRIBUTIONS

12.3.2.1 An Ecuador-Colombia SDTF Link

Of the 63 taxa of legume from seasonally dry tropical forest in southern Ecuador only one species, *Mimosa quitensis* Benth., is restricted to Ecuador and Colombia. The species is one of four in *Mimosa* series *Andinae* Barneby which as a group ranges in geography from Colombia through Ecuador to Peru (see the next legume distribution pattern for individual legume species which show this range). *Mimosa quitensis* grows from 2°30'N (Popayán in El Cauca, Colombia) to 3°35'S (in the province of Loja, Ecuador). The closely related *M. andina* Benth. is restricted to Ecuador (Azuay, Cañar and Chimborazo), but does not enter the southern provinces of Loja, El Oro and Zamora Chinchipe. The other two species in series Andinae, *M. weberbaueri* Harms and *M. montana* Kunth, are restricted to the dry inter-Andean valleys of central and northern Peru.

12.3.2.2 An Ecuador-Peru-Colombia SDTF Distribution Pattern

Caesalpinia cassioides Willd. (Figure 12.12) has long been known to occur in the inter-Andean valleys of southern Colombia and northern Peru, but was, until 1999, unknown in Ecuador. The species is now recorded from two different inter-Andean valley localities of southern Ecuador, between 1500 m and 1900 m altitude, on dry rocky cactus-rich slopes and in mainly deciduous forest. *Zapoteca caracasana* (Jacq.) H. Hern. subsp. *weberbaueri* (Harms) H. Hern. (subfamily Mimosoideae, Figure 12.13 and Figure 12.14), is recorded from scrub vegetation in SDTF in three disjunct areas of north-west South America: the dry valleys of central Colombia (departments of Antioquia, Cauca, Cundinamarca, Santander, Tolima and Valle); coastal and inter-Andean central and southern Ecuador (provinces of Chimborazo, Guayas and Loja) and northern Peru (Tumbes and Lambayeque); and central Peru (Apurímac, Cuzco and Huancavelica). The taxon occupies a wide altitudinal range, from 0 to 2500 m (Hernández, 1989). The related subspecies,

FIGURE 12.10 (See colour insert) *Pithecellobium excelsum* (Kunth) Mart. ex Benth. (Leguminosae: Mimosoideae): restricted to the SDTF of Ecuador and northern Peru. (Photo: G.P. Lewis.)

Zapoteca caracasana subsp. *caracasana*, has an allopatric distribution when compared with subsp. *weberbaueri*, and does not occur in Ecuador, although the two taxa come into close proximity in Colombia and Peru (Figure 12.13). Subspecies *caracasana* is known from Hispaniola and northern Venezuela (including the Isla Margarita) from where subsp. *weberbaueri* is absent. It is worth noting that the paper on the systematics of *Zapoteca* by Hernández (1989) relied heavily on collections housed in North American herbaria while South American herbaria, especially small regional ones, were mostly not surveyed. In addition, in South America, the genus is relatively poorly known and under-collected and in several herbaria it is likely that specimens are still to be found hidden away in the genus *Calliandra*, from which *Zapoteca* was segregated. As a result, it is possible that some of the apparent gaps in geographical range might be filled by an intensive study of regional herbaria in north-west South America. A second mimosoid taxon, *Albizia multiflora* (Kunth) Barneby & J.W. Grimes var. *multiflora*, has a similar distribution to *Zapoteca caracasana* subsp. *weberbaueri*, but is, in addition, recorded in a disjunct location in central Panama (province Colón). It is locally plentiful in south-west Ecuador (Guayas, El Oro, S Manabí and S Loja) from 0 to 700 m altitude and in the inter-Andean valleys of north-west Peru (Tumbes and Piura), as well as in the Marañon valley in Cajamarca and Jaén.

FIGURE 12.11 (See colour insert) *Machaerium millei* Standl. (Leguminosae: Papilionoideae) restricted to southern Ecuador and northern Peru. (Photo: G.P. Lewis.)

12.3.2.3 An Ecuador-Peru-Colombia-Venezuela SDTF Pattern

Leucaena trichodes (Jacq.) Benth. (Figure 12.15) is the only species in the genus which grows naturally south of the Equator and is the only indigenous species in South America where it occurs across the northern coastal regions of Colombia and Venezuela and in the coastal provinces of Ecuador. It also occurs in the dry forests of Loja province in Ecuador and is to be found as far as 13°S in the departments of Cuzco and Apurímac in Peru (Hughes, 1998). No discrete characters separate *L. trichodes* from *L. macrophylla* Benth.; the latter has two subspecies which occur only in Mexico. The two species together form a highly disjunct and somewhat variable alliance which shares pollen and anther characteristics. Hughes (1998) suggested that the treatment of both subspecies of *L. macrophylla* as subspecies of *L. trichodes* would be a reasonable taxonomic alternative. *L. macrophylla* subsp. *istmensis* C.E. Hughes is a taxon from south central lowland Mexico; *L. macrophylla* subsp. *macrophylla* is from the highlands and coastal foothills of west-central Mexico. Both subspecies occur in a range of habitats from dry deciduous forest to mid elevations in moist oak forest. Hughes et al. (2002) provided a phylogeny for *Leucaena* that shows *L. trichodes* to be sister to *L. macrophylla* with the two together forming a derived group nested within a larger Mexican group.

FIGURE 12.12 (See colour insert) *Caesalpinia cassioides* Willd. (Leguminosae: Caesalpinioideae) from the inter-Andean valleys of southern Colombia, southern Ecuador and northern Peru. (Photo: G.P. Lewis.)

12.3.2.4 An Ecuador-[Peru]-Bolivia SDTF Pattern

A number of legumes have a distribution which ranges from southern Ecuador intermittently through (or apparently totally by-passing) the dry inter-Andean valleys of Peru into the SDTF of Bolivia, but are not known to occur outside these three countries. Examples of such a pattern include, in subfamily Caesalpinioideae, *Bauhinia aculeata* L. subsp. *grandiflora* (A. Juss.) Wunderlin, *Chamaecrista glandulosa* (L.) Greene var. *andicola* H.S. Irwin & Barneby, *Hoffmannseggia viscosa* (Ruiz & Pav.) Hook. & Arn., *Senna pistaciifolia* (Kunth) H.S. Irwin & Barneby var. *picta* (G. Don) H.S. Irwin & Barneby; in subfamily Mimosoideae, *Zapoteca andina* H. Hern. (from Azuay and Loja in southern Ecuador, Cajamarca, La Libertad and Piura in northern Peru and widely disjunct in La Paz, Bolivia); in subfamily Papilionoideae, *Coursetia caribaea* (Jacq.) Lavin var. *ochroleuca* (Jacq.) Lavin. Pennington et al. (2004) also highlighted this disjunct pattern between northern Peru and Bolivia.

12.3.2.5 A Circum-Amazonian Disjunct SDTF Pattern

One legume taxon, *Anadenanthera colubrina* (Vell.) Brenan var. *cebil* (Griseb.) Altschul (subfamily Mimosoideae), has a geographical range which extends from the dry forests of the south-west

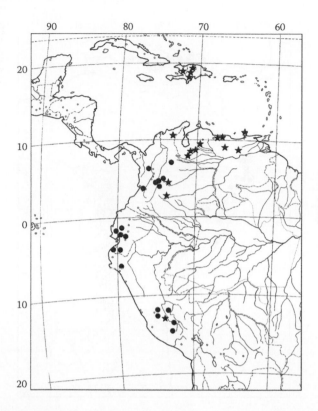

FIGURE 12.13 Distribution of *Zapoteca caracasana* subsp. *weberbaueri* (●) and subsp. *caracasana* (★). Redrawn from Hernández (1989).

FIGURE 12.14 (See colour insert) *Zapoteca caracasana* (Jacq.) H. Hern. subsp. *weberbaueri* (Harms) H. Hern. (Leguminosae: Mimosoideae): from SDTF in Colombia, Ecuador and Peru. (Photo: G.P. Lewis.)

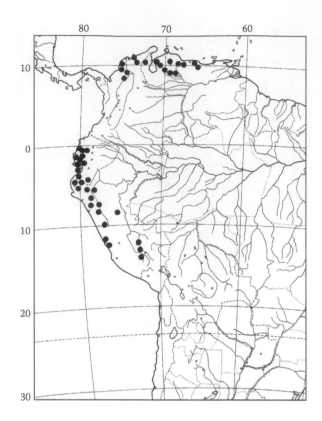

FIGURE 12.15 Distribution of *Leucaena trichodes*. Redrawn from Hughes (1998).

corner of Ecuador around the southern periphery of the Amazon basin in the disjunct SDTF areas of the so-called Pleistocenic Arc of Prado (1991) and Prado and Gibbs (1993) into the dry caatinga vegetation of north-eastern Brazil (Figure 12.16). This range encompasses three continuous distribution nuclei in South America: the major caatingas nucleus in north-east Brazil, and the Misiones and Piedmont nuclei (Figure 12.4) on either side of the chaco; the latter a region of regular frosts where *Anadenanthera* does not grow. From the main nuclei of SDTF radiate various discontinuous extensions where *Anadenanthera colubrina* var. *cebil* is frequently encountered; these include the dry forests of the Bolivian chiquitano, the inter-Andean valleys of Bolivia, and intermittently northwards through the dry valleys of Apurímac, Mantaro, Huállaga and Marañon in Peru into the dry valleys of southern Ecuador. Comparing Figures 12.16 and 12.4 it is clear that the distribution of *Anadenanthera colubrina* var. *cebil* in South America closely mirrors that of the SDTF. The distribution of the typical variety of *Anadenanthera*, *A. colubrina* var. *colubrina*, further extends the range of the species, occurring sympatrically with variety *cebil* in Misiones, Argentina and in southern Bahia, Brazil, but also allopatrically in Rio de Janeiro, São Paulo and Paraná states of southern Brazil. The typical variety, which apparently evolved from variety *cebil* (according to Altschul, 1964), occupies moister forests than those preferred by variety *cebil*, suggesting that a wet forest taxon has evolved from a dry forest one.

A similar, but apparently less continuous South American SDTF pattern is displayed by *Geoffroea spinosa* Jacq. (Ireland and Pennington, 1999; Figure 12.17). This species, in subfamily Papilionoideae, has five disjunct areas of distribution, all within the deciduous or semi-deciduous seasonally dry forests: the caatingas of north-east Brazil; north-east Argentina (east of the Gran Chaco region), Paraguay and Bolivia (west of the Chaco); north-western Peru together with coastal and south-western inter-Andean Ecuador; the Galapagos archipelago; and Colombia, Venezuela

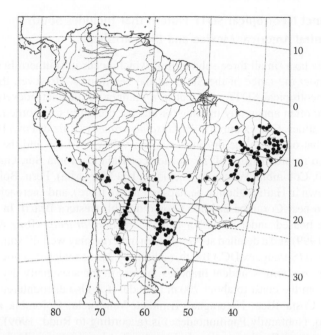

FIGURE 12.16 Distribution of *Anadenanthera colubrina* var. *cebil*. Redrawn from Prado (1991), with herbarium voucher data and literature sources combined.

and part of the Antilles. Ireland and Pennington (1999) demonstrated that geographical isolation does not appear to have caused any morphological diversification within the species; rather, the observed variation in morphology within *G. spinosa* is probably ecotypic, a phenomenon common in South American dry forest species. It is unknown whether populations of *G. spinosa* in the Galapagos Islands are the result of long distance (water) dispersal, or if the species was introduced to the archipelago by humans for its edible fruits.

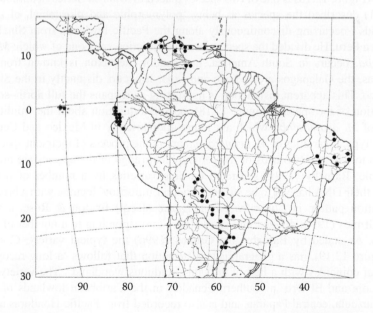

FIGURE 12.17 Distribution of *Geoffroea spinosa*. Redrawn from Ireland and Pennington (1999).

12.3.2.6 A Disjunct Neotropical SDTF Pattern that Includes South and Central America, Mexico and the Caribbean

A number of legume taxa (in all three subfamilies) which occur in southern Ecuador have a broad distribution throughout the range of the SDTF of the Neotropics. However, they do not occur in every present-day neotropical SDTF area, but in a number of disjunct pockets of seasonally dry vegetation. They are often apparently missing (or at least presently unrecorded) in intervening dry forest and scrub localities. One example is *Parkinsonia praecox* (Ruiz & Pav.) Hawkins (subfamily Caesalpinioideae) that occurs in the dry inter-Andean valleys of Loja and the dry coastal forests of Guayas. It is also recorded from SDTF in Mexico (Baja Califonia Sur, Guerrero, Michoacan, Oaxaca and Sonora), Colombia, Venezuela (including Isla Margarita), Peru, Bolivia and Argentina. In addition, it is known in Haiti (although perhaps not native there), and encroaches into the savanna chaquena vegetation near Corumbá in Mato Grosso do Sul, western Brazil. In many herbaria this species will still be housed under its earlier name *Cercidium praecox* (Ruiz & Pav.) Harms (but see Hawkins et al., 1999) and a detailed survey of such material may well fill some of the distribution gaps. *Piptadenia flava* (Spreng. ex DC.) Benth. (subfamily Mimosoideae), a prickly-stemmed shrub to small tree, sometimes with scandent branches, occurs in the seasonally dry forests of Guayas, El Oro and Loja, from the coast to about 700 m altitude. It is also encountered in the SDTF of El Salvador, Mexico, Costa Rica, Nicaragua, Trinidad, Venezuela, Colombia and Peru. *Piscidia carthagenensis* Jacq. (subfamily Papilionoideae) is (according to Rudd, 1969) the only species in the genus which occurs in South America. In Ecuador it is known from SDTF vegetation in Loja, El Oro, Guayas and the Galapagos Islands. The species is also recorded from dry forests in Mexico, Costa Rica, Belize, El Salvador, Guatemala, Honduras, Puerto Rico, Tobago, the Lesser Antilles, Panama, Venezuela, Colombia and northern Peru. The six other species of *Piscidia* are together confined to the Caribbean, Mexico and parts of Central America (Rudd, 1969) and all have a predilection for seasonally dry vegetation, including coastal limestone scrub. Four of the six are narrowly restricted: two (*P. cubensis* Urb. and *P. havanensis* (Britton & Wilson) Urb. & Ekman) in Cuba, one (*P. ekmanii* Rudd) in Haiti, and one (*P. mollis* Rose) in Sonora and Sinaloa, Mexico. The genus is in need of a taxonomic update and it is quite possible that the geographical range of some species will be extended by recent collections. *Mimosa acantholoba* (Willd.) Poir. (subfamily Mimosoideae) (Figure 12.18) is one of two species placed in *Mimosa* Series *Acantholobae* Barneby. Barneby (1991) described the species as being 'polymorphic and pluriracial, of seasonally dry scrub-woodlands', occurring discontinuously along the Pacific lowlands from Sinaloa in Mexico to north-western Peru. He divided the species into five varieties, only one of which: *M. acantholoba* var. *acantholoba*, occurs in South America. In Ecuador the taxon is known from Esmeraldas, Manabi, Guayas, the Galapagos archipelago and Loja. It occurs disjunctly in the Sierra Madre in Sinaloa, Mexico. This apparent disjunction is interesting as it spans the full north–south extension of the distribution of the species (Figure 12.18), posing a question about the validity of the other four varieties of *M. acantholoba* which are found allopatrically in Mexico and Central America. If the four non-typical varieties of *M. acantholoba* are in the process of incipient speciation it poses the question as to why the two ends of the geographical range of the species are morphologically indistinguishable. Barneby (1991) separated the five varieties by a number of flower and fruit characters, but their DNA has yet to be studied. Another mimosoid legume with a broad neotropical SDTF distribution pattern is *Chloroleucon mangense* (Jacq.) Britton & Rose, a species which Barneby and Grimes (1996) divided into six varieties corresponding to a number of locally differentiated forms. As stated by Barneby and Grimes (1996) the typical variety, *C. mangense* var. *mangense* (Figure 12.19), has a geographical distribution that follows 'a long-recognized pattern of dispersal that coincides with a belt of drought-deciduous woodland'. The variety occurs in the provinces of Loja and El Oro in southern Ecuador, in the Caribbean lowlands of Colombia, in north-west Venezuela, central Panama, and is also recorded from Pacific Honduras and Chiapas in Mexico. The related taxon *C. mangense* var. *mathewsii* (Benth.) Barneby & J.W. Grimes

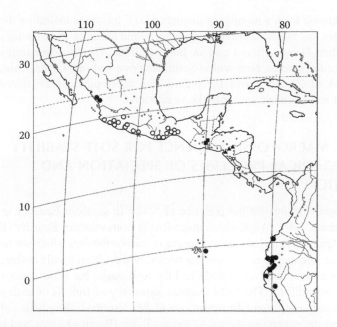

FIGURE 12.18 Distribution of *Mimosa acantholoba* var. *acantholoba* (●), var. *eurycarpa* (○), var. *liesneri* (▽), var. *molinarum* (▲) and var. *platycarpa* (■). Redrawn from Barneby (1991).

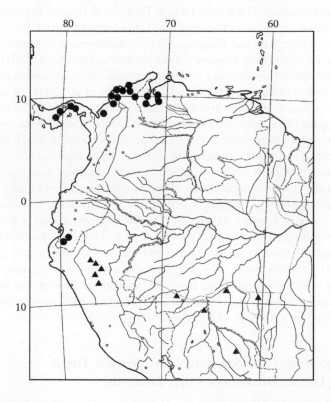

FIGURE 12.19 Distribution of *Chloroleucon mangense* var. *mangense* (●) and var. *mathewsii* (▲). Redrawn from Barneby and Grimes (1996).

(Figure 12.19) is known from a number of disjunct SDTF locations including the middle Huallaga valley in department San Martin, Peru, the upper Purús and Madeira-Beni basins in Acre, Rondônia and north-west Mato Grosso, Brazil and in Beni, Bolivia. The four remaining varieties further extend the range of the species, both within South America and into Central America, Mexico and the Caribbean and all closely mirror the present-day neotropical SDTF distribution, with only occasional extensions into more savanna-like vegetation.

12.4 LEGUME MACROFOSSIL EVIDENCE FOR SDTF STABILITY AND HISTORICAL PROCESSES OF SPECIATION AND EXTINCTION

There is strong fossil evidence for the presence of SDTF in southern Ecuador in the mid-Miocene (Burnham and Graham, 1999). Angiosperm macrofossils from southern Ecuador (Burnham 1995a, b; Burnham and Graham, 1999) include several genera in various families which are well-known present-day SDTF taxa. The legume genus *Tipuana* has been identified from fossils in three areas of southern Ecuador: the Cuenca basin, the Nabon basin and the Loja basin, but is not known from present-day Ecuadorean or Peruvian SDTF. The fossil material, especially of fruit, is of such good quality that a new species, *Tipuana ecuatoriana* Burnham, could be described (Burnham, 1995a). Present-day *Tipuana* is monospecific, containing the single species *T. tipu* (Benth.) Kuntze, and is only known from the Piedmont nucleus of SDTF in southern Bolivia and northern Argentina. Fruits of the extant species are, however, remarkably similar to those of the fossil *Tipuana*. A new, monospecific genus, *Maraniona*, also considered to be closely related to *Tipuana*, has recently been described from the SDTF of the Marañon valley of northern Peru (Hughes et al., 2004). The apparent close relationship between *Maraniona lavinii* C.E. Hughes, G.P. Lewis, Daza & Reynel and *Tipuana* suggests that the habitat in which the fossil *T. equatoriana* once grew might well have been similar to the present-day SDTF in the same general area (Burnham and Carranco, 2004). The closeness of match between the extant Andean species and the fossils from Miocene deposits in nearby inter-Andean SDTF suggest a period of at least 10 million years of evolutionary and morphological stasis (Burnham and Carranco, 2004; Hughes, 2005). A cladistic biogeographic study by Pennington et al. (2004) of several genera, each having high levels of species endemism in different areas of neotropical SDTF, revealed a strong sister relationship between the Ecuadorean and Peruvian SDTF and the Andean Piedmont SDTF nucleus of northern Argentina and southern Bolivia. This Ecuador-Peru-Piedmont SDTF distribution pattern, together with the extant and fossil distribution of *Tipuana* are perhaps suggestive of a more widely distributed SDTF in the Miocene, which was subsequently disrupted and fragmented by Andean orogeny and associated change in local climate. Furthermore, the close similarity of the fruits of the Anacardiaceae fossil *Loxopterygium laplayense* Burnham and Carranco, from the Loja and Cuenca Basins of southern Ecuador, to extant Andean SDTF *Loxopterygium* species is also indicative of long dry forest stasis (Burnham and Carranco, 2004). Long-term stability of the SDTF vegetation in the region would presumably result in an accretion of endemic species through the natural processes of speciation triggered by mountain uplift, forest fragmentation and isolation. The long, narrow, seasonally dry Marañon valley of northern Peru, which runs north to south for 250 km, has accumulated clusters of closely related, narrowly restricted endemic species (Hughes, 2005) supporting the hypothesis of local allopatric speciation in fragmented pockets of SDTF.

12.4.1 DISCUSSION: HOW ARE SEASONALLY DRY TROPICAL FOREST WOODY LEGUME DISTRIBUTIONS BEST EXPLAINED?

Endemism can be defined as 'the condition of being restricted to a particular area with a prescribed extent' (Williams et al., 1996). The distribution of endemic plant species (those with restricted ranges)

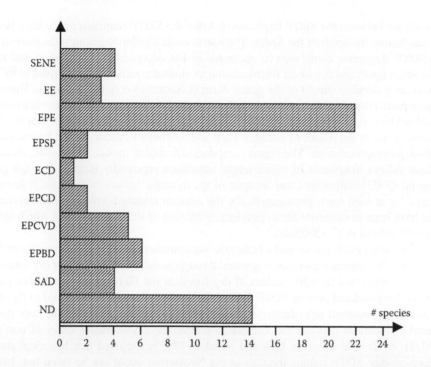

FIGURE 12.20 Neotropical distribution patterns of the 63 southern Ecuadorean woody legume seasonally dry tropical forest taxa (none occur outside the Neotropics). SENE = southern Ecuador narrow endemic; EE = Ecuador endemic; EPE = Ecuador-Peru (Tumbesian) endemic; EPSP = Ecuador-Peru, with disjunction into southern Peru; ECD = Ecuador-Colombia disjunct; EPCD = Ecuador-Peru-Colombia disjunct; EPCVD = Ecuador-Peru-Colombia-Venezuela disjunct; EPBD = Ecuador-(+/- Peru)-Bolivia disjunct; SAD = South American disjunct, in some taxa the range extending to the caatingas of north eastern Brazil; ND = neotropical disjunct.

is non-random (Kessler, 2002), but, especially for tropical species, the factors determining the range sizes of the species and the distribution of endemism are still poorly understood for many plant families and geographical regions. Historical events over broad time scales, particularly when and where speciation took place, have significant impact on plant range sizes (Young et al., 2002). Further complicating the study of endemism is that so-called centres of endemism often correspond to areas of intensive botanical collecting rather than to natural patterns (Nelson et al., 1990; Kessler, 2002).

Only four species of woody legume in the southern Ecuadorean SDTF are currently considered to be narrowly restricted Ecuadorean endemics (Table 12.1): two in the SDTF of Puyango on the Loja-El Oro border and two in the inter-Andean valleys of Loja. Considering all 63 Ecuadorean seasonally dry tropical forest (all variants included) woody legume taxa (Table 12.1), two main distribution patterns emerge (Figure 12.20, Table 12.1): 22 are restricted to the dry vegetation of southern Ecuador and northern Peru, part of the so-called Tumbesian zone of endemism (here including the outlying higher altitude fragments of SDTF in the inter-Andean valleys of both countries), while 14 are widely, but disjunctly, distributed throughout the neotropical SDTF.

The restriction of so many taxa to the Tumbesian zone highlights the strong geological and ecological links between southern Ecuador and northern Peru (Kessler, 1992; Best and Kessler 1995). Bridgewater et al. (2003) compared northern Peru SDTF plot data with our data from southern Ecuador and concluded that the northern Peruvian Tumbes and Marañon dry forests and those of southern Ecuador may constitute a distinct phytogeographical unit. They noted that the eastern and western Andean massifs and the mesic forests that cover them are obvious present-day barriers preventing species migration between the SDTF fragments, but that the SDTF remnants might have been joined up in the past before the uplift of the Andes. The low floristic similarity amongst the present northern Peruvian SDTF fragments could then be attributed to the occurrence of speciation and extinction since the Andean uplift and dry forest fragmentation. A similar conclusion was arrived at by Weigend (2002), who, in a detailed survey of the genus *Nasa* (Loasaceae) concluded that the Huancabamba depression (a partial interruption of the Andean chain by the Río Chamaya drainage system) in northern Peru should not be considered a biogeographical barrier to montane taxa, but rather that the Amotape-Huancabamba zone in the Andes of northern Peru and southern Ecuador should be considered as a distinct phytogeographical zone. The region comprises a complex mosaic of habitats, including arid inter-Andean valleys, fragments of which might have been repeatedly isolated through geological time. Weigend (2002) commented that 'in spite of the dynamic history of the area, it has also been a refuge area for at least some phylogenetically (by Andean standards) old taxa, and certain habitat types must have been in existence for a considerable amount of time', a statement which agrees with the findings of Schrire et al. (2005a,b).

Best and Kessler (1995) questioned whether the high number of endemic bird species in the SDTF vegetation of the Tumbesian region has originated through isolation of this area of dry forest that was presumed to be connected to other pockets of dry forest in the Pleistocene, or whether the species colonized this long-isolated area of SDTF from surrounding habitats, then adapted to the new environment and finally reached reproductive isolation. Others (e.g., Marchant, 1958) have pointed out the avifaunal connections between the Tumbesian region, the dry Marañon valley of northern Peru and the SDTF of Central America. Best and Kessler (1995) concluded that historical connections between present-day SDTF refugia throughout the Neotropics could not be ruled out, but that the present assemblage of bird species in the Tumbesian region originated from a wide variety of sources over a long time period. Distinctive species in the area suggest a long evolutionary history, but disjunct populations of other, little-differentiated species throughout the dry regions of the Neotropics indicate more recent vicariance events.

Prado and Gibbs (1993) offered a vicariance explanation for the disjunct distribution of numerous species in many different plant families in the SDTF of South America. They proposed that the repeated, and often widespread, distribution patterns are the result of the separation of a once wider common historical distribution rather than the result of dispersal from different areas. They concluded that 'these fragmentary and mostly disjunct distributional patterns are vestiges of a once extensive and largely contiguous seasonal woodland formation'. It has also been suggested that the patchy distribution of dry forest species supports Gentry's (1979, 1982) theory of reverse refuges which similarly assumes that present-day discontinuities in the distribution of dry forests, together with the occurrence of morphologically differentiated and geographically restricted dry forest populations, result from a shift towards more mesic conditions after climatic change in the Quaternary (Hernández, 1989).

In contrast to the Tumbesian endemics, 14 species of woody legume which occur in the SDTF of southern Ecuador are disjunctly distributed throughout the Neotropics, but their widespread distribution patterns are not identical, i.e. each of the 14 species does not occupy exactly the same set of SDTF nuclei. If all 14 distribution patterns were the result of vicariance with a common cause we might expect more congruence amongst them. It seems more parsimonious to suggest that these widespread, but not completely similar, distributions have been generated by at least some dispersal and not solely by fragmentation of a once more continuous belt of seasonally dry tropical forest.

At the global level, Schrire et al. (2005a, b) noted that the present-day pattern of legume distribution (i.e. at the level of genus and above) appears to be the result of global dispersal within and between neighbouring biomes. Dispersal has enormous consequences over large spatial and temporal scales (Hubbell, 2001) and the four legume biomes (Succulent, Grass, Rain forest and Temperate) identified by Schrire et al. (2005a, b) could be the result of dispersal assembly, where taxa with similar ecological preferences have, over a given timeframe, dispersed to similar ecological settings worldwide. At the biome (metacommunity) -level the processes of ecological drift, immigration, extinction, and resident speciation have played out over enormous spatial and temporal scales (Hubbell, 2001; Lavin et al., 2004, 2005).

Schrire et al. (2005a, b) also concluded, however, that the abundance of legume genera confined to similar regions within continents, and the lack of intercontinentally distributed legume species among taxa centred within the Succulent biome regions (which largely include the neotropical SDTF) of the world are probably the result of a single process: the contraction in size of the Succulent biome during recent geological times. The Succulent biome has apparently been restricted in size as a target area for sufficient time to reduce successful immigration into local regions of this metacommunity. The drought tolerance of the biome is probably the most important factor limiting immigration, especially from other biome elements that comprise taxa with a higher water requirement. Due to restricted dispersal (immigration) into local regions of the Succulent biome there has likely been less extinction of resident taxa so that standing diversity is generated mostly by endemic speciation (Lavin et al., 2004, 2005; Chapter 19). At the species level, local areas of the Succulent biome (including neotropical SDTF) thus house suites of distinct endemic taxa.

Most likely, the present-day woody legume flora in the SDTF of southern Ecuador results from a combination of dispersal and vicariance acted out over a long time frame.

12.5 CONSERVATION

The tropical Andes is number one out of a total of 25 global 'biodiversity hotspots' having a total of 20,000 endemic plant species which together comprise 6.7% of all plant species worldwide (Myers et al., 2000). The tropical Andes are also top of the list for endemic vertebrates and Myers et al. (2000) consider the region as a 'hyper-hot' candidate for conservation support. Mittermeier et al. (1999) included dry forests in a western Ecuador biodiversity hotspot and extended it into northwestern coastal Peru. Mittermeier et al. (2005) included the seasonally dry forests of southern Ecuador in a Tumbes-Chocó-Magdalena hotspot. According to Stattersfield et al. (1998) the forests of the Tumbesian region represent one of the richest and most threatened biotic sites on earth. At least 282 Ecuadorean endemic plant species qualify as being critically endangered using the IUCN threat category (Pitman et al., 2001). Of these, 60% are concentrated in the Andes and 35% in the coastal region. If a plant species is restricted to a single country its protection is largely in the hands of the policy makers of that country, so that in situ conservation becomes essentially a local activity (Stein, 2002). The conservation status of the entire endemic flora of Ecuador was evaluated by Valencia et al. (2000) in the *Libro Rojo de las Plantas Endèmicas del Ecuador*. Neill (2000) listed the endemic Leguminosae (Fabaceae) which included the following species that occur in the SDTF of the southern provinces: *Mimosa townsendii, Centrolobium ochroxylum* (this species has recently been collected in northern Peru and is thus no longer endemic to Ecuador), *Clitoria brachystegia, Erythrina smithiana* and *Gliricidia brenningii* (this is also known from northern Peru; Lavin and Sousa, 1995). Attention to these endemic species, that often are species for which we have limited knowledge, has generated improved distribution data and, as indicated here, removed them from the list of national endemics.

As accelerated habitat conversion by human activity places increasing pressure on natural systems it is endemic species with highly restricted ranges that are likely to be at greatest risk of extinction (Stein, 2002). This is especially so with species endemic to SDTF as the land on which such vegetation grows is highly prized for agriculture because of the relatively fertile soils. In South America the SDTF ecosystem is highly threatened (Janzen, 1988), but has received far less attention from ecologists and conservationists relative to that given to rain forests (Janzen, 1988; Mooney et al., 1995).

If in situ conservation is to protect the majority of the narrowly restricted endemic species this will require protection of different pockets of SDTF vegetation, a fact that conservation strategists are beginning to recognize. Bridgewater et al. (2003) found little floristic similarity between several fragments of SDTF in northern Peru and also concluded that each area would need protection if most rare species are to be protected. Even the north-west Peru Biosphere Reserve

(226,300 hectares) will only partly protect the huge diversity in the Tumbesian region of northern Peru and southern Ecuador. Trejo and Dirzo (2002) also concluded that if the floristic diversity (particularly the narrowly restricted local endemics) of the SDTF vegetation in Mexico is to be adequately protected then a network of numerous reserves will have to be set up throughout the country. If most of the narrowly restricted endemic species in SDTF are to be adequately protected then the network of conservation areas will need to be expanded to the whole of the Neotropics, and on a global scale to the whole of the Succulent biome across its broad geographical range.

In the SDTF of coastal Ecuador (Manabí, Guayas and El Oro) few remnants of the original dry forest vegetation remain, most having been highly disturbed by agriculture, timber extraction and grazing (Neill et al., 2000). The most important reserves are in Machalilla National Park in Manabí (where Josse and Balslev (1994) carried out quantitative forest inventories) and the Bosque Protector Cerro Blanco in Guayas province, located on the outskirts of Guayaquil, but in both these areas the forest has been severely disturbed by human activity (Neill et al., 2000). Best and Kessler (1995) concluded that for the Tumbesian region as a whole the main problem is just how little forest there is left, particularly in the provinces of Loja, El Oro and Azuay where there are 'few forest areas large enough to justify establishing a reserve'. Generally, however, every remaining forest patch plays an important ecological role as a remaining refuge for narrowly restricted, threatened endemics, as well as for more widespread species. Josse and Balslev (1994) also concluded that conservation of the dry forests of western Ecuador is worthwhile, even after a high degree of disturbance, because these forests regenerate relatively quickly after disturbance. However, to date, no viable plan has been found to effectively conserve, preserve, or at least secure that such areas are exploited in a sustainable way (Jørgensen, pers. comm.).

The SDTF of south-western Loja province have poorer soils (derived from weathered metamorphic rocks) than those of the more northerly coastal provinces and have thus been less exploited for agriculture and carry a much lower human population density (Cabrera et al., 2002). As noted by Neill et al. (2000), the areas of tropical dry forest in the Zapotillo region 'are not pristine wilderness, but they are the best-preserved examples of this habitat type in western Ecuador'. Neill et al. (2000) suggested that an intact area of the Zapotillo SDTF, ideally a reserve of several thousand hectares, should be given priority consideration as a conservation area and that surrounding lands should form part of a broader conservation strategy which includes sustainable use by local inhabitants. Best and Kessler (1995) had, likewise, earlier suggested that sustainably used marginal areas should be set up as buffer zones around any designated reserves in an attempt to reduce encroachment into the protected forest fragments. In Africa private game reserves abutting on to national reserves has already proved to be a useful model. There has been some recent progress towards Ecuadorean dry forest conservation with reserves having been set up by Nature and Culture International (also known as the Fundación San Francisco) in south-western Loja (Neill, pers. comm.).

It is probable that during the Pleistocene the SDTF of southern Ecuador and northern Peru were more stable than other areas, especially compared to climatic transition zones that were more susceptible to fluctuations in temperature and humidity (Best and Kessler, 1995). Essentially, the assumption is that very dry areas can become even drier and hotter, or slightly cooler and more humid without having a significant effect on the resident flora and fauna. As pointed out by Best and Kessler (1995), if this holds true under future climate change, then this has important implications for conservation action and priority. Designated protected areas in the SDTF are more likely to preserve their present-day biodiversity than areas more susceptible to climatic change where threatened plants and animals with a higher water requirement are more likely to go extinct.

12.6 CONCLUSIONS

The SDTF of southern Ecuador contain no endemic legume genera, and only four endemic species, divided between the lowland SDTF and the inter-Andean SDTF. The majority of the woody legume taxa (22 out of 63) are, however, restricted to southern Ecuador and northern Peru and constitute part of the Tumbesian centre of endemism, here including the fragmented higher altitude inter-Andean valleys. The number of woody legumes confined to southern Ecuador and northern Peru confirm the findings of others (e.g., Best and Kessler, 1995; Weigend, 2002; Bridgewater et al., 2003; Linares-Palomino, Chapter 11) in highlighting the region as a phytogeographical zone and a centre of endemism. Nevertheless, this area is a highly complex mosaic of inter-digitating vegetation types that include several variants of SDTF. Furthermore, much of the dry forests are now very fragmented as a result of agricultural systems which have led to highly reduced pockets of this vegetation type and apparent narrow endemism. Although SDTF vegetation is not naturally fire prone (due to a lack of grasses) it burns readily once cut down to make way for farming and cattle grazing. The legume floras of the lowland coastal seasonally dry forests and the inter-Andean dry forests are largely distinct, and the Andean valleys have been further fragmented by natural processes (mountain uplift and climate change) through geological time resulting in local speciation and endemism.

The Ecuadorean SDTF form one small part of the neotropical extent of this vegetation type, and this, in turn, is part of a global, but fragmented, Succulent biome (Lavin et al., 2004; Schrire et al., 2005a, b). High levels of geographical phylogenetic structure are being demonstrated in the endemic lineages found in this Succulent biome (Lavin, Chapter 19; Lavin et al., 2004; Pennington et al., 2004). As noted by Lavin et al. (2004), phylogenetic analysis combined with the neutral ecological theory of Hubbell (2001) can, together, help us to better understand the individual contributions of immigrant and resident lineages to the observed species diversity in the present-day Succulent biome forest fragments. Of particular interest are the small pockets of seasonally dry tropical forest to be found as islands in other major vegetation types, such as cerrado. These have probably acted as stepping stones for legume dispersal over long time periods and should be areas for targeted collecting and floristic analysis.

What are needed for legumes in Ecuador are more ecological plot inventories and dry forest transects, so that a better picture of local species composition and abundance is built up. In addition, the collection of fertile legume specimens from poorly known areas is necessary if gaps in the distribution data are to be filled. With more baseline data about what plants grow where, and in what numbers, we can then better explore the relationships between individual fragments of seasonally dry tropical forest. The phylogenetic and statistical tools are available to us, but without carefully targeted plant collecting and rigorous identification the baseline plant diversity data will continue to be sparse. Once the species distribution and abundance data are more robust there is then a need for more species-level phylogenies that include southern Ecuadorean legumes, similar to those published for *Coursetia* by Lavin et al. (2003) and for *Leucaena* by Hughes et al. (2002). These data and analyses will help to answer the question as to whether the narrow endemics might be confined to a limited area because of a close (and possibly long-standing) association with local climate and soil type or because they are neoendemics that have, as yet, failed to invade other SDTF sites.

ACKNOWLEDGEMENTS

The authors thank Peter Jørgensen, Colin Hughes and Toby Pennington for reading earlier drafts of this chapter and for helpful comments which have improved the final version. We also thank Henrik Balslev for permission to reproduce Figure 12.1 and Reynaldo Linares-Palomino for permission to use the map in Figure 12.5.

REFERENCES

Altschul, S. von Reis, A taxonomic study of the genus *Anadenanthera*, *Contribs. Gray Herb. Univ. Havard*, 193, 1, 1964.

Balslev, H. and Øllgaard, B., Mapa de vegetación del sur de Ecuador, in *Botánica Austroecuatoriana*: Aguirre M.Z. et al., Eds, Abya-Yala, Quito, 2002, 51.

Barneby, R.C., Sensitivae Censitae, a description of the genus *Mimosa* Linnaeus (Mimosaceae) in the New World, *Mem. New York Bot. Gard.*, 65, 1, 1991.

Barneby, R.C., Increments to the genus *Mimosa* (Mimosaceae) from South America, *Brittonia*, 45, 328, 1993.

Barneby, R.C., Silk tree, guanacaste, monkey's earring: a generic system for the synandrous Mimosaceae of the Americas, part III, *Calliandra, Mem. New York Bot. Gard.*, 74, 1, 1998.

Barneby, R.C. and Grimes, J.W., Silk tree, guanacaste, monkey's earring: a generic system for the synandrous Mimosaceae of the Americas, part I. *Abarema, Albizia*, and allies, *Mem. New York Bot. Gard.*, 74, 1, 1996.

Barneby, R.C. and Grimes, J.W., Silk tree, guanacaste, monkey's earring: a generic system for the synandrous Mimosaceae of the Americas, part II. *Pithecellobium, Cojoba*, and *Zygia, Mem. New York Bot. Gard.*, 74, 1, 1997.

Best, B.J. and Kessler, M., *Biodiversity and Conservation in Tumbesian Ecuador and Peru*, BirdLife International, Cambridge, 1995.

Block, M. and Richter, M., Impacts of heavy rainfall in El Niño 1997/1998 on the vegetation of Sechura desert in northern Peru, *Phytocoenologia*, 30, 491, 2000.

Bridgewater, S. et al., A preliminary floristic and phytogeographic analysis of the woody flora of seasonally dry forests in northern Peru, *Candollea*, 58, 129, 2003.

Burnham, R.J., A new species of winged fruit from the Miocene of Ecuador: *Tipuana ecuatoriana* (Leguminosae), *Amer. J. Bot.* 82, 1599, 1995a.

Burnham, R.J., Middle and Late Miocene plant fossils from central and southern Ecuador, *Amer. J. Bot.* 82, 84, 1995b.

Burnham, R.J. and Carranco, N.L., Miocene winged fruits of *Loxopterygium* (Anacardiaceae) from the Ecuadorian Andes, *Amer. J. Bot.*, 91, 1767, 2004.

Burnham, R.J. and Graham, A., The history of neotropical vegetation: new developments and status, *Ann. Missouri Bot. Gard.*, 86, 546, 1999.

Cabrera, O. et al., Estado actual y perspectivas de conservación de los bosques secos del sur-occidente ecuatoriano, in *Botánica Austroecuatoriana*: Aguirre M.Z. et al., Eds, Abya-Yala, Quito, 2002, 63.

Cowling, R.M., Proches, S., and Vlok, J.H.J., On the origin of southern African subtropical thicket, *S. African J. Bot.*, 71, 1, 2005.

Gentry, A., Distribution patterns of neotropical Bignoniaceae: some phytogeographic implications, in *Tropical Botany*, Larsen, K. and Holm-Nielsen, L., Eds, Academic Press, London, 1979, 339.

Gentry, A., Phytogeographic patterns as evidence for a chocó refuge, in *Biological Diversification in the Tropics*, Prance, G., Ed, Columbia University Press, New York, 1982, 112.

Gentry, A. H., Diversity and floristic composition of neotropical dry forests, in *Seasonally Dry Tropical Forests*, Bullock, S.H., Mooney, H.A., and Medina, E., Eds, Cambridge University Press, Cambridge, 1995, 146.

Graham, A. and Dilcher, D., The Cenozoic record of tropical dry forest in northern Latin America and the southern United States, in *Seasonally Dry Tropical Forests*, Bullock, S.H., Mooney, H.A., and Medina, E., Eds, Cambridge University Press, Cambridge, 1995, 124.

Hawkins, J.A. et al., Investigation and documentation of hybridization between *Parkinsonia aculeata* and *Cercidium praecox* (Leguminosae: Caesalpinioideae), *Pl. Syst. Evol.*, 216, 49, 1999.

Herendeen, P.S., Crepet, W.L., and Dilcher, D.L., The fossil history of the Leguminosae: phylogenetic and biogeographic implications, in *Advances in Legume Systematics, Part 4, the Fossil Record*, Herendeen, P.S. and Dilcher, D.L., Eds, Royal Botanic Gardens, Kew, 1992, 303.

Herendeen, P.S., The fossil record of the Leguminosae: recent advances, in *Legumes Down Under: The Fourth International Legume Conference, Abstracts*, Australian National University, Canberra, 2001, 34.

Hernández, H.M., Systematics of *Zapoteca* (Leguminosae), *Ann. Missouri Bot. Gard.*, 76, 781, 1989.

Holmgren, P.K., Holmgren, N.H., and Barnett, L.C., *Index Herbariorum*. 8th ed. New York, 1990.

Hubbell, S.P., *The Unified Neutral Theory of Biodiversity and Biogeography*, Princeton, University Press, 2001.

Hughes, C., Monograph of *Leucaena* (Leguminosae-Mimosoideae), *Syst. Bot. Monogr.*, 55, 1, 1998.

Hughes, C.E. et al., Divergent and reticulate species relationships in *Leucaena* (Fabaceae) inferred from multiple data sources: insights into polyploidy origins and nrDNA polymorphism, *Amer. J. Bot.*, 89, 1057, 2002.

Hughes, C.E. et al., *Maraniona*. A new dalbergioid legume genus (Leguminosae, Papilioinoideae) from Peru, *Syst. Bot.*, 29, 366, 2004.

Hughes, C.E., Four new legumes in forty-eight hours, *Oxford Plant Systematics*, 12, 6, 2005.

Ireland, H. and Pennington, R T., A revision of *Geoffroea* (Leguminosae- Papilionoideae), *Edinburgh J. Bot.* 56, 329, 1999.

Irwin, H.S. and Barneby, R.C., The American Cassiinae, a synoptical revision of Leguminosae tribe Cassieae subtribe Cassiinae in the New World, *Mem. New York Bot. Gard.*, 35, 1, 1982.

Janzen, D.H., Tropical dry forests, the most endangered major tropical ecosystem, in *Biodiversity*, Wilson, O.E., Ed, National Academic Press, Washington, DC, 1988, 130.

Jørgensen, P.M. and León-Yánez, S., Eds, *Catalogue of the Vascular Plants of Ecuador*, Monogr. Syst. Bot. Missouri Bot. Gard., 75, Missouri Botanical Garden Press, St Louis, 1999.

Josse, C. and Balslev, H., The composition and structure of a dry, semideciduous forest in western Ecuador, *Nordic J. Bot.*, 14, 425, 1994.

Kessler, M., The vegetation of south-west Ecuador, in *The Threatened Forests of South-West Ecuador*, Best, B.J., Ed, Biosphere Publications, Leeds, 1992, 79.

Kessler, M., Environmental patterns and ecological correlates of range size among bromeliad communities of Andean forests in Bolivia, *Bot. Rev.* 68, 100, 2002.

Klitgaard, B.B. et al., Composición florística y estructura del bosque petrificado de Puyango, *Herbario Loja* 3, 25, 1999.

Lavin, M. and Sousa S.M., Phylogenetic systematics and biogeography of the tribe Robinieae (Leguminosae), *Syst. Bot. Monogr.*, 45, 85, 1995.

Lavin, M., Herendeen, P.S. and Wojciechowski, M.F., Evolutionary rates analysis of Leguminosae implicates a rapid diversification of lineages during the Tertiary, *Syst. Biol.*, 54; 530, 2005.

Lavin, M. et al., Identifying Tertiary radiations of Fabaceae in the Greater Antilles: alternatives to cladistic vicariance analysis, *Int. J. Pl. Sci.*, 162, S53, 2001.

Lavin, M. et al., Phylogeny of robinioid legumes (Fabaceae) revisited: *Coursetia* and *Gliricidia* recircumscribed, and a biogeographical appraisal of the Caribbean endemics, *Syst. Bot.* 28, 387, 2003.

Lavin, M. et al., Metacommunity process rather than continental tectonic history better explains geographically structured phylogenies in legumes, *Philos. Trans., Ser. B* 359, 1509, 2004.

Lewis, G.P. and Klitgaard, B.B., Leguminosas del sur de Ecuador, in *Botánica Austroecuatoriana*: Aguirre M. Z. et al., Eds, Abya-Yala, Quito, 2002,185.

Lewis, G., Schrire, B., Mackinder, B., and Lock, M., Eds, *Legumes of the World*, Royal Botanic Gardens, Kew, London, 2005.

Linares-Palomino, R., Los bosques tropicales estacionalmente secos: I. El concepto de los bosques secos en el Perú, *Arnaldoa*, 11, 85, 2004a.

Linares-Palomino, R., Los bosques tropicales estacionalmente secos: II. Fitogeografía y composición florística, *Arnaldoa*, 11, 103, 2004b.

Lozano C., P.E., Los tipos de bosque en el sur de Ecuador, in *Botánica Austroecuatoriana*: Aguirre M. Z. et al., Eds, Abya-Yala, Quito, 2002, 29.

Macbride, J.F., Flora of Peru, family Leguminosae, *Botanical Series Field Mus. Nat. Hist., Bot. Ser.*,13, 73, 1943.

Madsen, J.E., Mix, R., and Balslev, H., *Flora of Puná Island, Plant Resources on a Neotropical Island*, Aarhus University Press, Aarhus, 2001.

Marchant, S., The birds of the Santa Elena Peninsula, S. W. Ecuador, *Ibis*, 100, 349, 1958.

Mittermeier, R.A. et al., *Hotspots, Earth's Biologically Richest and Most Endangered Terrestrial Ecosystems*, Conservation International, Mexico City, 1999, 123.

Mittermeier, R.A. et al., *Hotspots Revisited: Earth's Biologically Richest and Most Threatened Terrestrial Ecoregions*, Conservation International, Washington, 392, 2005.

Mooney, H.A., Bullock, S.H., and Medina, E., Introduction, in *Seasonally Dry Tropical Forests*, Bullock, S.H., Mooney, H.A., and Medina, E., Eds, Cambridge University Press, Cambridge, 1995, 1.

Murphy, P. and Lugo, A.E., Ecology of tropical dry forest, *Annual Rev. Ecol. Syst.*, 17, 67, 1995.

Myers, N. et al., Biodiversity hotspots for conservation priorities, *Nature*, 403, 853, 2000.

Neill, D.A., *Ecuadendron* (Fabaceae: Caesalpinioideae: Detarieae): A new arborescent genus from western Ecuador, *Novon*, 8, 45, 1998.

Neill, D.A. Geography, in *Catalogue of the Vascular Plants of Ecuador*, Monogr. Syst. Bot. Missouri Bot. Gard., 75, Jørgensen, P.M. and León-Yánez, S., Eds, Missouri Botanical Garden Press, St Louis, 1999a, 2.

Neill, D.A., Vegetation, in *Catalogue of the Vascular Plants of Ecuador*, Monogr. Syst. Bot. Missouri Bot. Gard., 75, Jørgensen, P.M. and León-Yánez, S., Eds, Missouri Botanical Garden Press, St Louis, 1999b, 13.

Neill, D.A., Fabaceae, in *Libro Rojo de las Plantas Endémicas del Ecuador 2000,* Valencia, R. et al., Eds, Quito, 2000, 196.

Neill, D.A. et al., Observations on the conservation status of tropical dry forest in the Zapotillo area, Loja province, Ecuador, 2000 (http://www.mobot.org.).

Nelson, B.W. et al., Endemism centers, refugia and botanical collection density in Brazilian Amazonia, *Nature,* 345, 714, 1990.

Olson, D.M. and Dinerstein, E., The global 200: priority ecoregions for global conservation, *Ann. Missouri Bot. Gard.,* 89, 125, 2002.

Parker, T.A. and Carr, J.L., Eds, Status of forest remnants in the Cordillera de la Costa and adjacent areas of southwestern Ecuador, *RAP Working Papers,* 2, 1, 1992.

Pendry, C.A., Monograph of *Ruprechtia* (Polygonaceae), *Syst. Bot. Monogr.,* 67, 1, 2004.

Pennington, R.T., Prado, D.E., and Pendry, C.A., Neotropical seasonally dry forests and Quaternary vegetation changes. *J. Biogeogr.,* 27, 261, 2000.

Pennington, R.T. et al., Historical climate change and speciation: neotropical seasonally dry forest plants show patterns of both Tertiary and Quaternary diversification. *Philos. Trans.,* Ser. B, 359, 515, 2004.

Pitman, N.C.A. et al., Extinction-rate estimates for a modern Neotropical flora, *Conservation Biology,* 16, 1427, 2001.

Prado, D.E., *A critical evaluation of the floristic links between chaco and caatingas vegetation in South America*, PhD thesis, University of St Andrews, 1991, 1.

Prado, D.E. and Gibbs, P.E., Patterns of species distributions in the dry seasonal forests of South America, *Ann. Missouri Bot. Gard.,* 80, 902, 1993.

Richardson, J.A. et al., Recent and rapid diversification of a species-rich genus of neotropical trees, *Science,* 293, 2242, 2001.

Rudd, V.E., A synopsis of the genus *Piscidia* (Leguminosae), *Phytologia,* 18, 473, 1969.

Schrire, B.D., Lavin, M., and Lewis, G.P., Global distribution patterns of the Leguminosae: insights from recent phylogenies, *Biol. Skr.,* 55, 375, 2005a.

Schrire, B.D., Lewis, G.P., and Lavin, M., Biogeography of the Leguminosae, in *Legumes of the World,* Lewis, G. et al., Eds, Royal Botanic Gardens, Kew, 2005b, 21.

Scotese, C.R., *Atlas of Earth History, vol. 1, Paleogeography,* PALEOMAP Project, Arlington, Texas (cited from PALEOMAP website, http://www.scotese.com), 2001.

Sierra, R., Ed, *Propuesta Preliminaria de un Sistema de Clasificación de Vegetación para el Ecuador Continental*, Proyecto INEFAN/GEF-BIRG & EcoCiencia, Quito, 1999.

Sprent, J.I., *Nodulation in Legumes,* Royal Botanic Gardens, Kew, 2001.

Stattersfield, A.J. et al., *Endemic Bird Areas of the World, Priorities for Biodiversity Conservation,* BirdLife Conservation Series No. 7, BirdLife International, Cambridge, UK, 1998, 210.

Stein, B.A., Foreword, *Bot. Rev.,* 68, 1, 2002.

Trejo, I., Características del medio físico de la selva baja caducifolia en México, *Investigaciones Geográficas Boletín Instituto de Geográfia* (special issue), 4, 95, 1996.

Trejo, I. and Dirzo, R., Deforestation of seasonally dry tropical forest: a national and local analysis in Mexico, *Biol. Conservation,* 94, 133, 2000.

Trejo, I. and Dirzo, R., Floristic diversity of Mexican seasonally dry tropical forests, *Biodiver. Conservation,* 11, 2063, 2002.

Valencia, R. et al., Eds, *Libro Rojo de Las Plantas Endémicas del Ecuador 2000,* Herbario QCA, Pontifica Universidad Católica del Ecuador, Quito, 2000.

Weigend, M., Observations on the biogeography of the Amotape-Huancabamba zone in northern Peru, *Bot. Rev.,* 68, 38, 2002.

White, F., *The Vegetation of Africa, a Descriptive Memoir to Accompany the UNESCO/AETFAT/UNSO Vegetation Map of Africa,* UNESCO, Paris, 1983.

Williams, P. et al., A comparison of richness hotspots, rarity hotspots and complementary areas for conserving diversity using British birds, *Conservation Biology,* 10, 155, 1996.

Wolfe, J.A., Some aspects of plant geography of the Northern hemisphere during the late Cretaceous and Tertiary, *Ann Missouri Bot. Gard.* 86, 264, 1975.

Young, K.R. et al., Plant evolution and endemism in Andean South America: an introduction, *Bot. Rev.,* 68, 4, 2002.

13 Mexican and Central American Seasonally Dry Tropical Forests: Chamela-Cuixmala, Jalisco, as a Focal Point for Comparison

Emily J. Lott and Thomas H. Atkinson

CONTENTS

ABSTRACT

In this chapter we discuss the distribution of seasonally dry tropical forests (SDTFs) in Mexico and Central America, emphasize their distinctive characteristics, and provide a summary of floristic and diversity studies of several sites. As a particular case, we analyze floristic affinities and diversity of the Chamela-Cuixmala Biosphere Reserve and adjoining area in Jalisco, Mexico, with respect to different taxonomic levels and life forms. The analysis is based on 1109 species (1064 native), 570 genera (544 native) and 115 families from all the biotic communities in the region from sea level to c.500 m. The native flora analyzed consists of 229 species of trees, 227 species of shrubs, 40 epiphytes, 371 herbs and 197 vines (86 herbaceous and 111 woody). At Chamela, about 41.5% of the species are distributed on the Pacific slope from Sonora and Baja California Sur to Central America, and it appears that this coastal flora is a relatively homogeneous and continuous floristic province. There is an (as yet) imprecisely defined zone of endemism which may encompass the area from Jalisco to the Isthmus of Tehuantepec in Oaxaca. Comparisons are made with diversity and floristic affinities of other Mexican and Central America SDTF sites. Based on the combination of species richness, endemism levels and other factors, we suggest that further conservation efforts might best be focused on the Pacific slope states of Michoacán, Guerrero and Oaxaca.

13.1 INTRODUCTION

This chapter has four principal aims. First, we provide an overview of the main areas of seasonally dry tropical forest (SDTF) in Mexico and Central America. Second, we summarize studies of floral diversity in these areas. Third, as an exemplar, we examine in detail the flora of one of the best studied areas of Mexican SDTF, the Chamela-Cuixmala Biosphere Reserve. Finally, we examine the distributions of the species comprising the Chamela flora, which to some extent allows us to make biogeographical generalisations about Mexican SDTF.

13.2 SUBTYPES WITHIN SEASONALLY DRY TROPICAL FOREST

This book uses a broad definition of SDTF which covers a gamut of vegetation types (Pennington et al., Chapter 1). In this chapter we do distinguish at times between tropical deciduous forest (*selva baja caducifolia* of Miranda and Hernández X., 1963; *bosque tropical caducifolio* of Rzedowski, 1978), tropical semideciduous forest (*selva mediana subcaducifolia* or *subperennifolia*, in part; SM hereafter), and thorn forest (*bosque espinoso*), as have Gordon et al. (2004) and others. The SM typically occurs on deeper soils along drainages with more available moisture than on slopes (Rzedowski, 1978).

13.3 DISTRIBUTION OF SDTF IN MEXICO AND CENTRAL AMERICA

SDTFs on the east coast of Mexico are isolated from each other; they are spottily distributed in Tamaulipas, San Luis Potosí and Veracruz, and are extensive in areas of the Yucatán Peninsula, particularly the north-western part (Figure 13.1). On the Pacific slope they are found in a nearly continuous band from northern Sonora and western Chihuahua to Oaxaca, and also in the southern part of the Baja California Peninsula. In the interior of the country, they are mainly found in the Tehuacán-Cuicatlán and Río Balsas valleys, and the Central Depression of Chiapas (Rzedowski, 1978). In Central America, there are remnant patches of SDTF on the Pacific slope in southern Guatemala, and from El Salvador and Honduras to Costa Rica.

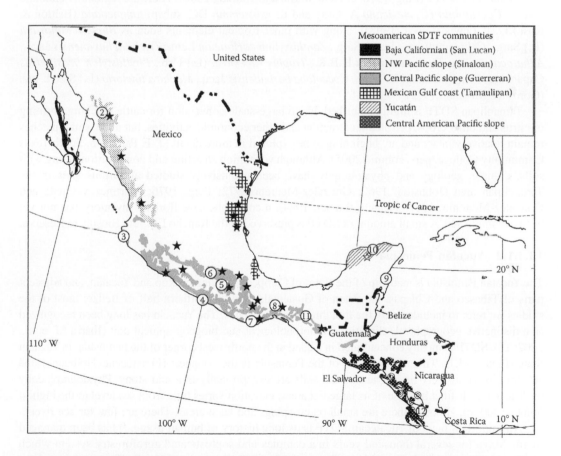

FIGURE 13.1 Distribution of SDTF in Mexico and Central America (modified from Reichenbacher et al., 1998). Numbered locations: 1: Cape Region, Baja California Sur; 2: Río Cuchujaqui (near Alamos), Sonora; 3: Chamela Bay Region, Jalisco; 4: Costa Grande Region, Guerrero; 5: Cañón del Zopilote, Guerrero; 6: Sierra de Nanchititla, Mexico; 7: Zimatán, Oaxaca; 8: Nizanda, Oaxaca; 9: Sian Ka'an, Quintana Roo; 10: area of SDTF, Yucatán; 11: Central Depression, Chiapas; 12: Guanacaste Province, Costa Rica. Stars represent locations from Trejo and Dirzo (2002). Circles represent locations from Gillespie et al. (2000).

There is a profusion of terms referring to these floristic areas in biogeographical and conservation literature; in this chapter we refer to Sonoran, Tamaulipan, Baja Californian, Central Pacific coast (Jalisco, Colima, Michoacán, Guerrero and Oaxaca), Balsas, Central Depression of Chiapas, Yucatecan and Central American SDTFs (Figure 13.1).

13.3.1 Mexico's East Coast

On the Atlantic slope, we have chosen to discuss only two areas, Tamaulipas and Yucatán, omitting sites in San Luís Potosí and Veracruz with which we are less familiar.

13.3.1.1 Tamaulipan SDTF

In north-eastern Mexico, at the northernmost limit of SDTF on the Gulf coast where it crosses the Tropic of Cancer, Tamaulipan SDTF is of interest because of its transitional nature and the complex mosaic it forms with Tamaulipan Scrub. There are fewer columnar cacti, no arborescent *Ipomoea* or *Jatropha*, no *Forchhammeria, Fouquieria, Alvaradoa, Haematoxylum, Manihot* or *Vallesia*. Some tropical tree genera which occur here but are absent or scarce in Sonoran SDTF are *Amyris, Beaucarnea, Brosimum* and *Phoebe* (Puig, 1976, cited in Martin and Yetman, 2000). There are regionally endemic taxa of *Caesalpinia* (*C. mexicana* A. Gray and *C. exostemma* DC. subsp. *tampicoana* (Britton & Rose) G. P. Lewis (Lewis, 1998)), along with other tropical elements such as *Bursera simaruba* (L.) Sarg., *Celtis iguanaea* (Jacq.) Sarg., *Zanthoxylum caribaeum* Lam., *Trichilia havanensis* Jacq., *Achatocarpus* sp., *Ficus cotinifolia* H.B.K., *Trophis racemosa* (L.) Urb., *Phyllostylon brasiliensis* Capanema, *Guazuma ulmifolia* Lam., *Coccoloba barbadensis* Jacq., *Maclura tinctoria* (L.) Steud., etc. (González-Medrano,1972).

Tamaulipan SDTF is highly disturbed due to large-scale conversion for cattle ranching, farming of citrus and other crops, petroleum extraction and other economic activities, but a few small patches remain relatively intact and are beginning to be explored in more detail (T. F. Patterson, South Texas Community College, pers. comm., 2005). Although vegetation structure and composition, along with soils, climate, geology and physiography, have been extensively studied at various sites in the Tamaulipan area (Johnston, 1960; González-Medrano,1972; Puig, 1976; Martínez y Ojeda and González-Medrano, 1977; Valiente Banuet, 1984), a comprehensive floristic inventory has not yet been accomplished. A small amount of SDTF is preserved in the Rancho El Cielo Biosphere Reserve.

13.3.1.2 Yucatán Peninsula

The Yucatán Peninsula is made up of the states of Campeche, Quintana Roo and Yucatán, and adjacent parts of Tabasco and Chiapas, the Peten of Guatemala, and the northern half of Belize; most of the studies we refer to included only the first three Mexican states. The Yucatán has long been recognized as a distinctive geographical, climatic, geomorphological and biogeographical unit (Ibarra M. et al., 2002: 18). SDTF is mostly concentrated in a band in the north-west corner of the peninsula, in Yucatán State (Figure 13.1). This northern end of the Peninsula is the youngest (Pleistocene–Holocene); the substrate is exclusively limestone and the soils are exceptionally thin and stony. Physiographically the Yucatán is distinct because of its large flat areas: elevation varies little from sea level to the highest point (*c.*400 m), although there are small outcroppings and karst areas. There are few surface rivers.

An important facet of Yucatecan SDTF is its long history of human usage. It has been managed by the Maya for several thousand years in a complex and sophisticated agroforestry system which supported a dense population (Gómez-Pompa et al., 1987; Primack, 1998). Studies on the comparison of floristic composition of archeological sites (forest growing on ruins, mounds, pyramids, and in small patches of strikingly different species makeup) with the flora of surrounding areas (Thien et al., 1982; White and Darwin, 1995; White and Hood, 2004), and investigations into how forest management and agricultural traditions of the Maya have affected the forests of today (Gómez-Pompa et al., 1987; Rico-Gray et al., 1985; and others) are numerous and ongoing.

Some projects which have contributed to floristic knowledge of the peninsula are the *Etnoflora Yucatanense* (Sosa et al., 1985), an illustrated flora, and a recent checklist of the Peninsula (Durán et al., 2000) which lists 2477 species in 992 genera of 182 families. This is an area which boasts a wealth of floristic and ethnobiological information, and also some of the largest forest reserves in Mesoamerica, although the proportion of SDTF in reserves is relatively small.

13.3.2 PACIFIC SLOPE

We will first identify some of the local areas within this SDTF belt while recognizing that the Pacific slope is home to a distinctive, fairly homogeneous and nearly continuous floristic province which stretches from Sonora and Baja California to Costa Rica.

13.3.2.1 Sonoran SDTF

In the north-west of Mexico SDTF occurs in southern Sonora, western Chihuahua and north-western Sinaloa (Figure 13.1). SDTF here was previously referred to by H. S. Gentry as short-tree forest (1942) and later as Sinaloan deciduous forest (1982). The SDTF is bordered on the north by Sonoran Desert, to the east by the Sierra Madre Occidental and to the west by the coastal plain of the Sea of Cortéz.

In addition to numerous studies of the ecology, vegetation and ethnobotany of this zone, there are several excellent local floristic treatments (Gentry, 1942, Martin et al., 1998; Van Devender et al., 2000; Felger et al., 2001) and popular natural history books (e.g., Bowden, 1993). Some tropical genera and species are at the northernmost part of their range here, and characteristic SDTF trees such as *Ceiba, Tabebuia, Haematoxylum, Lysiloma, Pseudobombax* and *Guaiacum coulteri* A. Gray are either noticeably smaller in stature or shrubbier than they are in wetter areas further to the south (Martin et al., 1998). Within this zone is the Río Cuchujaqui drainage, a small area in southern Sonora and northern Sinaloa studied by Van Devender et al. (2000) (Table 13.1). Since the early 1970s Sonoran SDTFs have suffered large-scale conversion to Buffelgrass (*Cenchrus ciliaris* L.) pasture (Martin and Yetman, 2000), which in part prompted establishment of the Alamos-Río Cuchujaqui Biosphere Reserve in 1995.

13.3.2.2 Baja California Sur

The Cape region of Baja California is near the northern limits of the SDTF and is isolated from the continent by the Gulf of California (or Sea of Cortéz) to the east and from other areas of similar vegetation by the deserts of the Peninsula and by the coastal plain of Sonora. SDTF is found on the lower to mid-slopes and steep canyon walls of the Sierra de la Laguna and the Sierra Giganta which are topped by oak-pine forest, in a variety of soil types (all igneous-derived) and situations (León de la Luz et al., 1999). This is one the driest Pacific SDTF areas in Mexico (< 500 mmaap) according to Rzedowski (1978); it is fairly well preserved and is included in a 112,437 ha biosphere reserve.

13.3.2.3 Central Pacific Slope

The history of botanical exploration on the Pacific slope of Mexico dates from 1790 (McVaugh, 1972). Since the synopsis of Central American botany by Hemsley (1879–1888) it has been recognized that the Pacific coast of Mexico boasts a rich and distinctive flora. This is the largest extension of Mexican SDTF, and parts of it are still relatively intact. In recent years there has been a significant advance in the knowledge of the flora owing to the greater access which began with the paving of a coastal highway in 1974, and to the consequent increase in field work. A large-scale floristic work, *Flora Novo-galiciana*, covers Jalisco, Colima, Aguascalientes, and parts of Nayarit, Durango, Zacatecas, Guanajuato and Michoacán (McVaugh, 1983–present); a flora of Guerrero is in progress, and a number of important local checklists have been published (Table 13.1).

TABLE 13.1
Summary of Floristic Checklists

Site	Area (km²)	Number of Families	Number of Genera	Number of Species	Elevation (m)	Vegetation Types	Source
Río Cuchujaqui, Son.	46	115	429	736	220 – 400	SDTF, thorn scrub, oak forest, riparian, aquatic	Van Devender et al., 2000
Sa. de Nanchititla, Méx.	13.2	87*	208	288	600 – 1400	SDTF, gallery forest, 80% secondary	Zepeda and Velázquez, 1999
Dep. Cent., Chis.	9000	103	489	998	200 – 1500	SDTF, thorn scrub	Reyes-García and Sousa, 1997
Dist. Tehuantepec, Oax.	6600	154	776	1720	0 – 1800	SDTF, thorn scrub, conifer forest, gallery forest, oak forest, palm forest, grassland, aquatic, halophytic, coastal	Torres-Colin et al., 1997
Venta Vieja, Cañón del Zopilote, Gro.	38	77*	222	307	700 – 1100	SDTF	Gual-Díaz, 1995
Costa Grande, Gro.	2,500	135	527	1047	0 – 300	SDTF, thorn scrub, moist forest, halophytic, aquatic, coastal	Peralta-Gómez et al., 2000
Zimatán, Oax.	713	144	668	1384	0 – 2580	SDTF, thorn scrub, subhumid and humid forest, moist montane forest, oak-pine forest	Salas-Morales et al., 2003
Nizanda, Oax.	85	117*	458	746 (380)	90 – 500	SDTF, gallery forest, savannas, aquatic, agricultural	Pérez-García and Meave, 2001
Cabo Baja Calif. Sur	8500	130 (92)	522 (312)	943 (454)	0 – 2100	SDTF, oak forest, oak-pine forest	León de la Luz et al., 1999
Chamela-Cuixmala, Jal.	350	125	572	1149 (739)	0 – 500	SDTF, thorn scrub, riparian, aquatic, halophytic, coastal	Lott, 2002
Río Balsas Basin	11,2320	202	1246	4442	0 – 2800	SDTF, thorn scrub, oak, pine, moist montane forest, aquatic, coastal	Fernández et al., 1998
Sian Ka'an, Q. Roo	5280	112	470	850	0 –	SDTF, seasonally inundated, savannas, swamps, coastal	Durán and Olmsted, 1987
Lowland Guanacaste Prov., Costa Rica		121*	642	1156		SDTF and associated riparian vegetation	Janzen and Liesner, 1980

*The number of families reflects consolidation of Leguminosae
SDTF here includes SM
Numbers in parentheses for families, genera, and species are for SDTF only (not stated in most cases)

13.3.2.4 Chamela

A focal point of research in SDTF of Mexico's Pacific slope, the Estación de Biología Chamela (EBCH) was established in 1971 by the Instituto de Biología, Universidad Nacional Autónoma de México (IBUNAM), upon the donation of 1600 ha near the settlement of Chamela, Jalisco, on the Pacific Coast between Puerto Vallarta and Barra de Navidad. The station was established for several purposes: to conserve an area of well-preserved SDTF in the vicinity, to learn about the structure and function of the ecosystems in the protected area, to support research, teaching and extension services, and to understand the social and economic problems of the surrounding region and seek solutions to them (Noguera et al., 2002). In 1993, another parcel of 1700 ha was added on the north-west side of the original rectangle. In this same year, the EBCH and Ecological Foundation of Cuixmala jointly formed the Chamela-Cuixmala Biosphere Reserve, as established by presidential degree; the combined area of the EBCH and the Biosphere Reserve now comprises *c*.13,142 ha (Noguera et al., 2002), including two protected turtle-nesting beaches, and small areas of SM, riparian, dune, mangroves, salt marsh and thorn scrub areas, but vegetation is principally the SDTF sub-type known in Mexico as *selva baja caducifolia* (Miranda and Hernández X., 1963) or *bosque tropical caducifolia* (Rzedowski, 1978). As of 2000, research in the Chamela area had resulted in 359 scientific articles and 124 theses. A list of these is available at the EBCH website, http://www.ibiologia.unam.mx/ebchamela/. Longterm ecological transect studies have been established on the EBCH, one of them since the mid-1970s. The flora (and indeed the overall ecology) of the Chamela-Cuixmala region is probably the best known of the Mexican Pacific Slope (Pérez Jiménez et al., 1981; Lott, 1985, 1993; Martínez-Yrízar et al., 2000:21; Van Devender et al., 2000; Lott, 2002; Noguera et al., 2002; and others).

13.3.2.5 Michoacán, Guerrero and Oaxaca

Botanical exploration of this part of the central coast, especially Oaxaca (Pérez-García et al., 2001; García-Mendoza, 2004) and Guerrero, is in an extremely active phase; the large area, highly complex and rugged topography, outstanding diversity and the sheer size of the flora mean that floristic inventory is still in progress. The Isthmus of Tehuantepec in Oaxaca is the southernmost limit of many of the SDTF species.

13.3.3 Río Balsas Basin

The Río Balsas Basin is an east–west-trending, distinct, large area of western Mexican SDTF which is joined to the Pacific coast only by the mouth of the Balsas River in Michoacán (border with Colima). The main Pacific slope SDTF and the Balsas Depression are separated by the coastal mountain ranges which run parallel to the coast at this point. There are many species in common with the Pacific SDTF, but there is also an exceptionally high level of endemism, 30–45% in some sites (see below).

13.3.4 Central America

The destruction of Central America's SDTFs was called to world attention by Janzen (1988), who estimated that the original Pacific SDTFs have been reduced to only 0.1% of their former range. According to Stevens et al. (2001), SDTF in Nicaragua has undergone several waves of alteration; the latest has occurred in the last 50 years, when the forests were cut again to make way for crops. They estimate that less than 1% of SDTF remains there, and of this little, if any, is in its natural state (Stevens et al., 2001). Gillespie et al. (2000) in Nicaragua and Costa Rica, and Gordon et al. (2004) in Honduras, among others, have also concluded that there is little remaining of mature SDTF in Central America. Thus it is too late to know more of the original extent and composition of these SDTFs, but there is much of vital importance to be learned of their restoration and regeneration.

There are large-scale floristic works, such as a flora of Guatemala, a partial flora of Costa Rica and the *Manual of the Flora of Costa Rica* (Hammel et al., 2003), *Flora de Nicaragua*

(Stevens et al., 2001 — present), and a series of guides to trees of Costa Rica from which information on SDTF could be gleaned with time and effort, but for rapid comparison of floristic composition of Central American and Mexican SDTFs, aside from species lists from vegetation and diversity studies, many studies have used Janzen and Liesner's (1980) widely available, informative and reasonably complete checklist of lowland Guanacaste Province (Figure 13.1).

13.4 LOCAL DIVERSITY: CHECKLISTS AND TRANSECTS

In Table 13.1 we show a summary of floristic checklists with the sites arranged from north to south, beginning with the Río Cuchujaqui in Sonora. The size of the sites ranges from 13.2 to 9000 square kilometers; the elevational range is from sea level to 2800 m; other vegetation types included are coastal, riparian, and aquatic to oak-pine, conifer and cloud forest. Average annual precipitation ranges from 684 mm in the Cañón de Zopilote, Guerrero, to 1100 mm in the Sierra de Nanchititla site, Mexico, but many of the study areas are far from established climatological stations and so precipitation data are of necessity extrapolated from those of the nearest stations, from maps, or else are not reported. Although each site includes a component of SDTF, the disparity in size of area and the variety of other vegetation types surveyed makes it difficult to compare the SDTFs of these areas quantitatively.

For this reason, we turn to a comparison of floristic records, which, though they will not reflect all species present in an area, should give a roughly comparable representation from one site to another when the same sampling protocol is followed. Irrespective of the total number of species in the region, it is possible to examine the floristic diversity of smaller areas and compare them with that of other neotropical forests. Information from these floristic sources was used in our biogeographical analysis of the Chamela flora (see below).

13.4.1 MEXICO IN A NEOTROPICAL CONTEXT

A.H. Gentry (1982, 1988, 1995) attempted to explain the patterns of diversity and of floristic composition in neotropical forests based on climatic factors. He used 28 sites of 0.1 ha (with 10 transects of 2×50 m each) to measure the diversity of woody individuals with a dbh of ≥ 2.5 cm. Based on his results, in 1982 and 1988 Gentry postulated that the local species diversity for humid neotropical forests was positively correlated with the quantity of total annual precipitation. Nevertheless, the *alpha*-diversity on the Chamela station (92 and 83 spp./0.1 ha; Lott et al., 1987) was much higher than that predicted by the Gentry model based on total annual rainfall. Gentry later suggested that the unexpected richness of the SDTF is a phenomenon of the subtropics and not 'a biogeographic peculiarity' (Gentry, 1995:155; Trejo and Dirzo, 2002) of Mexico, based on the results of a sample in Quiapaca, Bolivia, whose local diversity ($c.86$ spp./0.1 ha) was comparable to that of Chamela (but see Killeen et al., Chapter 9).

13.4.1.1 Mexico-Wide Studies

Trejo (1998) and Trejo and Dirzo (2002), in a landmark study of 20 SDTF sites in Mexico following the basic methodology of Gentry, found that climatic factors alone were not sufficient to explain the differences in diversity between sites. In contrast to Gentry (1982), Trejo and Dirzo's study (2002) included sites from the same biogeographical region, so that climatic factors could be separated from historic and biogeographical ones; they chose sites which included the whole range of environmental variation (precipitation, soils, temperature, rainfall, etc.) in which SDTF is found in Mexico. An important difference in Trejo and Dirzo's sampling protocol was their inclusion of all trees, shrubs and lianas with a dbh ≥ 1 cm, instead of dbh ≥ 2.5 cm. This allowed a more complete picture of the composition and floristic diversity of this forest type, where smaller-stemmed individuals of trees, shrubs and small lianas are an important component (Trejo, 1998; Trejo and Dirzo, 2002), and still yielded a data set that could be compared with Gentry's. For individuals with stems ≥ 2.5 cm dbh,

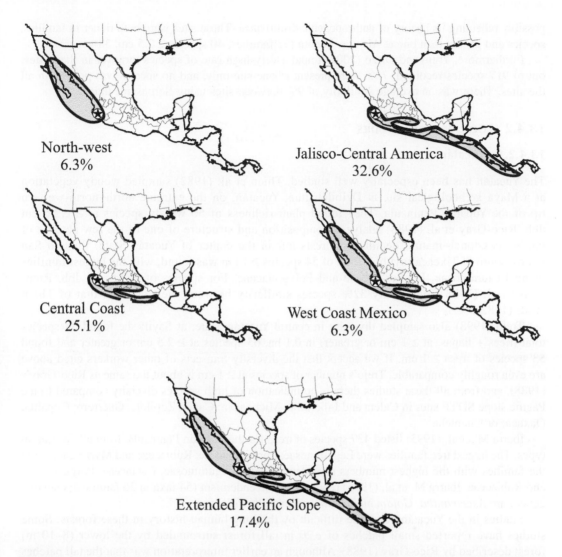

North-west
6.3%

Jalisco-Central America
32.6%

Central Coast
25.1%

West Coast Mexico
6.3%

Extended Pacific Slope
17.4%

FIGURE 13.2 Definition of subregions of distribution patterns restricted to the Pacific slope of Mexico (modified from Lott and Atkinson, 2002). More precise definitions of the subregions are found in Table 13.2. The percentages are of the Pacific slope flora (Table 13.3, subtotal). Location of Chamela-Cuixmala is indicated by a black star in a white circle.

Trejo and Dirzo (2002) found an average of 58 spp./0.1 ha (a range of 22 to 97) in the 20 sites. The forests of Caleta, Michoacán (97 spp./0.1 ha), and Copalita, Oaxaca (86 spp./0.1 ha), both located on the Pacific Coast (Figure 13.1), had the greatest species diversity and these values were similar to those previously reported from Chamela (Upland 1, 92; Upland 2, 83; Lott et al., 1987). The site at Infiernillo, Michoacán, along the Río Balsas a short distance from the coast, surpassed the number of species expected according to the rainfall model proposed by Gentry by 150%, and this tendency to present more species than those predicted by the model was also found in other sites, where nearly double the predicted number was found.

Two Tamualipan sites, at Las Flores and El Pensil, had only half the number of species predicted by Gentry's model, and they are the two sites with highest annual precipitation, 1370 and 1350 mm respectively. Las Flores (31 families, 55 genera, 61species ≥ 1 cm, 48 species ≥ 2.5 cm), had the third highest total number of species ≥ 30 cm dbh (10), and El Pensil (33 families, 52 genera, 57 species ≥ 1 cm, 41 species ≥ 2.5 cm) was the third highest in number of lianas ≥ 1 cm dbh or greater, possibly reflecting its history of anthropogenic disturbance. These sites were both richer in families, species and genera than one at Alamos, Sonora (22 families, 40 species ≥ 2.5 cm, 38 genera).

Furthermore, Trejo and Dirzo (2002) found a very high rate of species turnover in their sites: out of 917 species recorded, 72% were present at one site only, and no species was present in all the sites. There was an average similarity of 9% between sites using Sørensen's index.

13.4.2 LOCAL MEXICAN STUDIES

13.4.2.1 Yucatán

The Yucatán has been especially well studied. Thien et al. (1982) sampled woody vegetation at a Maya archeological site at Dzibilchaltun, Yucatán, on the extreme north–north-western tip of the Yucatán Peninsula. They found plant richness of 50 woody species at ≥ c.2.5 cm dbh. Rico-Gray et al. (1988) analyzed composition and structure of one of the few patches of late successional–mature deciduous forests left in the center of Yucatán State, Rancho San Pedro, south of Tixcacaltuyub. A total of 54 species ≥ 1 cm was found, with the largest families being Leguminosae, Euphorbiaceae and Polygonaceae. For species of ≥ 7.5 cm dbh, Rico-Gray et al. (1988) found only 32% species similarity between their site and that of Thien et al. (1982).

Trejo (1998) also sampled diversity in central Yucatán State, at Sayil: she found 83 species total (trees + lianas, at ≥ 1 cm or greater) in 0.1 ha, 65 species at ≥ 2.5 cm or greater and found 53 species of trees ≥ 1 cm. If we accept that the diversity transects of other workers cited above are even roughly comparable, Trejo's number of species at ≥ 1 cm is about the same as Rico-Gray's (1988), and from all these studies there is no indication of high species diversity compared to the Pacific slope SDTF sites in Caleta and Infiernillo, Michoacán; Cañón Zopilote, Guerrero; Copalita, Oaxaca; or Chamela.

Ibarra M. et al. (1995) listed 437 species of trees for the Yucatán Peninsula, from all vegetation types. The largest tree families were Leguminosae, Euphorbiaceae, Rubiaceae and Myrtaceae, while the families with the highest numbers of endemics were Leguminosae, Cactaceae, Polygonaceae and Rubiaceae. Ibarra M. et al. (1995) reported 12.3% endemism (54 taxa in 26 families); endemic genera are *Asemnantha, Goldmanella, Harleya* and *Plagiolophus*.

Studies in the Yucatán are made difficult by the long human history in these forests. Some studies have reported small patches of c.20 m tall forest surrounded by the lower (8–10 m) forest described by Rico-Gray (1988). Although an earlier interpretation was that the tall patches were remnants of primary forest, further study showed them to be dominated by *Brosimum, Manilkara, Sabal* and others known to have been used by the Maya, and still used in their gardens. Some of these patches are ringed by low stone walls (*pet kot*) (Gómez-Pompa et al., 1987).

Phytogeographical relationships of trees from all forest types of Yucatán were studied by Ibarra M. et al. (2002). Of the endemic tree species, about 80% are found in more than one of the five Mexican Yucatán reserves they analyzed, and 19 species (29.7%) in just one. However, 12 species (18.8%) were not found in any of the five reserves.

13.4.3 CENTRAL AMERICAN STUDIES

Gillespie et al. (2000) sampled seven sites in fragments of SDTF in Costa Rica (Santa Rosa National Park in Guanacaste Province, and Palo Verde) and in Nicaragua (Cosiguina, Masaya, Chacocente and Ometepe), all in conservation areas. They also noted the overall size of the forested area, environmental variables, annual precipitation and degree of human disturbance. Using Gentry's (1982) method, in these seven sites they found a total of 204 species and 1484 individuals ≥ 2.5 cm, with an average of c.56 species for the seven sites (ranging from 44 to 75 species), remarkably close to Trejo and Dirzo's (2002) average of 58 species for 20 Mexican sites. Santa Rosa National Park was the richest site, with 33 families, 60 genera and 75 species. Three other sites, Cosiguina, Masaya, and Ometepe (all in Nicaragua), were significantly less rich than Santa Rosa, and species richness was not significantly different between Santa Rosa and Palo Verde, La Flor, and Chacocente, although there were some significant differences in numbers of tree, shrub or liana species.

Gillespie et al. (2000) then compared their results to 21 other data sets from the West Indies, Mexico (Chamela), northern South America, the Pacific coast of South America, and the 'southern subtropics' of Argentina, Bolivia and Paraguay. They found no significant difference between the diversity of their Central American sites and that of many other neotropical forests, and they stated that the species richness of their sites justifies conservation status. Except for Santa Rosa, Costa Rica, which ranked eighth in total species, tree and shrub species diversity, and Palo Verde, with 65 spp./0.1 ha, the Central American sites were in the bottom half of all the neotropical sites in species richness. No SDTF transect site in Trejo and Dirzo (2002) or Lott et al. (1987) had Bignoniaceae or Sapindaceae in the top two families (although three out of the 20 Trejo and Dirzo sites did have Bignoniaceae in third place). In Gillespie et al.'s (2000) transects, 20% of all stems belonged to Bignoniaceae and Sapindaceae. The level of anthropogenic disturbance may have greatly influenced family composition and vine species (Gillespie et al., 2000; T.W. Gillespie, UCLA, pers. comm., 2005), perhaps because, as Gillespie et al. point out, some families of vines are more resistant to fire than others. The seven Central American sites of Gillespie et al. (2000) were all highly disturbed and had significantly lower density per area than the other neotropical sites they used for comparison. They found a significant correlation between anthropogenic disturbance and species richness (see studies cited in Gillespie et al., 2000).

Gordon et al. (2004) used a plotless rapid species inventory method in mature Oaxacan and highly disturbed, fragmentary Honduran sites. Forest types (hillside, riparian, beach, fallow fields, etc.) were distinguished and surveyed separately. Each site was searched for 4.5 person-hours; woody plants with stem diameters $\geq$ 2 cm at ground level, including woody stumps but excluding lianas which do not also appear as shrubs, were counted. In 43 sample sites, they found 375 species in 70 families and 210 genera; the Oaxacan sites were more diverse than the Honduran ones at all levels. The findings of Gordon et al. (2004) at the family level were very different from those of Gillespie et al. (2000). At the Honduran sites, the largest families were Leguminosae (37 spp.), Euphorbiaceae and Rubiaceae, with the largest genera *Ficus* (5), *Cordia* (5), *Annona*, *Senna*, *Solanum*, *Trichilia* (4 spp. each). In Oaxaca, the largest families were Leguminosae (64 spp.), Euphorbiaceae (19), Rubiaceae (10); the largest genera were *Bursera* (8), *Caesalpinia* (8), 7 species each in *Acacia*, *Croton* and *Lonchocarpus*, *Ficus* (6), *Cordia* (5), *Capparis* and *Jatropha* (4 each). The difference in the most speciose genera at each of these sites may reveal something of their degrees of disturbance. In our experience it is rare to find *Bursera*, *Lonchocarpus* and *Jatropha* (three of the largest genera in the Oaxacan sites) in highly disturbed sites except as remnants, whereas some species of *Cordia* and *Trichilia*, and many *Senna* and *Solanum* species are commonly associated with disturbed or secondary vegetation.

Based on analysis of their results, Gordon et al. (2004) suggested that anthropogenic disturbance and successional status partly explain the latitudinal gradient in species diversity discussed by Gentry (1982, 1988, 1995). The more species rich northern Chamela data sets he used for comparison were taken from relatively more mature undisturbed forest than the highly disturbed southern Guanacaste forests, which were forest remnants on cattle pastures maintained by fire.

13.5 COMPOSITION AND DIVERSITY OF THE CHAMELA FLORA

Direct comparisons between Chamela and other sites are difficult and somewhat tentative partly because while the floristic inventory process at Chamela (at least on the EBCH) and in Sonora is fairly advanced, many other newer reserves and study sites in Mexico are in an accelerating intensive inventory phase, with many additional species being found and new ones being studied and described.

The difficulty with comparisons of diversity studies of Central American sites is the fragmented and highly disturbed condition of what remains there.

13.5.1 LOCAL DIVERSITY

13.5.1.1 Families

To date 125 families of vascular plants have been reported from Chamela-Cuixmala (see floristic checklist, Lott, 2002). The most diverse native families in Chamela are in general those most common in the continental dry forests of the Neotropics (Gentry, 1995) and of Mexico (Trejo and Dirzo, 2002). The family with the greatest number of species (160 species; 14.0% of the flora) is Leguminosae (Caesalpiniodeae: 33 species; Mimosoideae: 47 species; Papilionoideae: 80 species) followed by Euphorbiaceae, with 94 species. These two families make up almost a fourth (22.2%) of the total flora. They are followed in importance by: Compositae (62 spp.), Gramineae (57 spp.), Convolvulaceae (40 spp.), Malvaceae (39 spp.), Solanaceae (29 spp.), Rubiaceae (29 spp.), Acanthaceae (27 spp.), Bromeliaceae (26 spp.), Cucurbitaceae (23 spp.), Verbenaceae (23 spp.) and Boraginaceae (22 spp.). All these, except for Bromeliaceae and Cucurbitaceae, are among the 20 most diverse families in the SDTF sites of Gentry (1995). While Gentry (1995) found that continental dry forests are 'dominated' by Leguminosae and Bignoniaceae and that 'Bignoniaceae are the indisputed number two family of woody plants of neotropical dry forests, averaging ...twice as many as third place Rubiaceae' (1995: 170–179), the Bignoniaceae and Rubiaceae do not seem to play as large a part in Mexican SDTFs, at least as represented in the transects of Trejo (1998).

In almost all neotropical forests the legume family has been found to be the richest in species (Gentry, 1988; Villaseñor and Ibarra M., 1998) and in all the SDTF sites discussed by Gentry (1995), Euphorbiaceae and Gramineae were among the four most diverse families, and Convolvulaceae and Compositae among the first seven. It is less common that Euphorbiaceae occupies second place (Gentry, 1995), but Chamela shares this characteristic with forests of the Antilles and with other sites on the Pacific slope of México (Trejo and Dirzo, 2002), as well as the Oaxacan and Honduran sites of Gordon et al. (2004).

The family Myrtaceae, of great importance in Antillean SDTF (Gentry, 1995), in our area is represented by only two genera, *Eugenia* and *Psidium*, and four species. Of the three species of *Eugenia* reported in Chamela two are endemic to Jalisco, and *Psidium sartorianum* (Berg) Ndzu. is the only species of the family which Chamela shares with the Antilles. According to J. Ratter (Royal Botanic Garden Edinburgh, pers. comm., 2005), 'In general, Myrtaceae seem to be calcifuges but *P. sartorianum* is a calcicole exception'. Since the Chamela area soils are igneous-derived, not limestone like Yucatán and Cuba, perhaps its presence at Chamela indicates that *P. sartorianum* has wider edaphic tolerances than many other Myrtaceae.

There is not a complete correspondence between species richness in the flora and abundance of individuals in the forest (Lott et al., 1987). There are families which may be considered typical of arid or semiarid conditions such as Burseraceae, Cactaceae, Capparaceae and Zygophyllaceae (Rzedowski, 1978); these are not found among the most important families in number of species, but certain of their species are very abundant in the SDTF.

Few species belonging to the families Aizoaceae, Basellaceae, Campanulaceae, Cannaceae, Caryophyllaceae, Cruciferae, Hydrophyllaceae, Martyniaceae, Onagraceae (except for *Hauya*), Papaveraceae and Umbelliferae occur in the Chamela region, and these are represented only by species of ephemeral habitats such as riparian (riverbeds or sandbars which are subject to violent changes by seasonal movement of water), and areas with a high level of anthropogenic disturbance. These families are primarily of more boreal affinities (Good, 1974).

The pteridophytes of Chamela have their affinities to the south. All of them (except *Cheilanthes lozanii* (Maxon) R. & A. Tryon), are of wide distribution in Mexico, Central America and South America (Lott, 2002). The richness of this group in other areas of SDTF is also somewhat distinct compared to Chamela. In the state of Sonora, near the northernmost limit of SDTF, Van Devender et al. (2000) reported 18 species of pteridophytes for the flora of the Río Cuchujaqui, a region of less precipitation, but with appreciable winter rains from January to March, and Martin et al. (1998) reported 48 species of pteridophytyes in the flora of the Río Mayo (the larger region of which the Cuchijaqui drainage is a part), of which only five are also found in Chamela. These differences from Chamela are possibly due in part to the higher altitude and more sheltered canyon environment of the Sonoran sites. The narrow canyons provide protection from cold northerly winds and from long exposure to the sun (Martin and Yetman, 2000). At the other geographical extreme, Janzen and Liesner (1980) reported 71 species of pteridophytes for the flora of Guanacaste Province, Costa Rica, of which only six species are shared with Chamela. Gentry (1995) considered that this reflects an ecological, not a phytogeographical difference, owing to the fact that in Guanacaste there is a gallery forest which provides a microhabitat similar to more humid forests, and thus more favorable for ferns.

13.5.1.2   Genera

At the generic level the flora of Chamela-Cuixmala is diverse for an area its size, with 555 genera. The most speciose genera are *Ipomoea* (26), *Tillandsia* (17), *Croton* (16), *Mimosa* (13), *Cyperus* (13), *Acalypha* (13), *Solanum* (12), *Lonchocarpus* (13), *Phyllanthus* (11), *Euphorbia* (11), *Cordia* (11), *Acacia* (11) and *Senna* (11). A notable characteristic is that four of the largest thirteen genera, *Croton*, *Acalypha*, *Phyllanthus*, and *Euphorbia* (51 species in total), belong to the family Euphorbiaceae. Among the most diverse genera (> 10 spp.), *Lonchocarpus* is the only entirely arborescent one, and almost all its species in Chamela are endemic to the Pacific slope, besides being very abundant (Lott et al., 1987). *Bursera* (9 spp.), *Caesalpinia* (9 spp.) and *Jatropha* (7 spp.), are genera which play an important role due to their abundance and number of species in the SDTF. The other most diverse genera are represented by herbaceous species (*Cyperus,* and *Solanum* in part), epiphytes (*Tillandsia*), shrubs (*Croton, Acalypha, Mimosa,* and *Acacia* in part), or herbaceous climbers (*Ipomoea* in part).

Some of these genera are associated with specific communities or habitats. *Cyperus*, *Solanum* and *Phyllanthus*, which also are important in humid forests, are typically found here in riparian sites or in the understory of more humid microhabitats. The genera present in Chamela which are important in arid zones are principally of cosmopolitan or pantropical distribution such as *Acacia*, *Aristida*, *Caesalpinia*, *Cassia*, *Jatropha*, *Mimosa*, *Randia*, *Salvia* and *Solanum* (Rzedowski, 1973). Several genera which are very diverse in other Mexican SDTF sites, such as *Bursera*, *Jatropha*, *Lonchocarpus*, *Croton* and *Ipomoea* (Trejo and Dirzo, 2002), contribute a relatively high proportion of species endemic to our region. However, it should be noted that *Lonchocarpus* is also very diverse in lowland tropical humid forests of Mexico, more so than in similar forests of South America (Wendt, 1993).

In the Chamela-Cuixmala region the diversity of species of *Ipomoea* is especially noteworthy. Of the 26 species of the genus in our area, only two or three are exotic. Also, some of the species of *Ipomoea* are found in disturbed sites, where they play an important role in secondary succession. A group of *c*.14 arborescent, shrubby and stout liana species of *Ipomoea* (Carranza et al., 1998; Carranza and McDonald, 2004; McDonald, 1991), represented in Chamela by *I. wolcottiana* Rose, is practically restricted in distribution to Mexico and Central America in SDTFs and in arid scrub (*matorral xerófilo*). In Mexico there are approximately 146 species of the genus *Ipomoea*, 77 of which are found in the south-west of the country, from Jalisco to Oaxaca. Of these, 25 are endemic, almost all in SDTF.

13.5.2   Diversity of Life Forms

13.5.2.1   Trees (229 Species)

Trees here are classified as woody individuals 4 m or more tall, single stemmed, or branched above 1.3 m breast height. Not surprisingly, the family with the greatest number of trees is Leguminosae

(57 species), with 23.4% of the trees of the total flora. Representative genera include *Albizia*, *Caesalpinia*, *Lonchocarpus*, *Lysiloma*, *Mimosa*, *Platymiscium* and *Poeppigia*. Euphorbiaceae (26 spp.) is the second largest arborescent family. Families of trees which are common in Chamela but absent in the Sonoran SDTF (Río Cuchujaqui, Van Devender et al., 2000; Felger, 2001) include Anacardiaceae, Annonaceae, Nyctaginaceae, Polygonaceae and Solanaceae. On the other hand, the most important families of trees in the Río Cuchujaqui SDTF are Cactaceae, Convolvulaceae (arborescent species of *Ipomoea*), Moraceae (*Ficus*), Fouquieriaceae and Zygophyllaceae; all these are present in Chamela-Cuixmala except for Fouquieriaceae, a family which is associated with greater aridity. Felger et al. (2001) reported 141 species of SDTF trees (51%) of the total in *Trees of Sonora*, whereas 229 tree species are reported from Chamela SDTF alone.

13.5.2.2 Shrubs (227 Species)

The differentiation between shrubs and trees is arbitrary. For purposes of this study we interpreted shrubs as woody individuals (usually up to 4–6 m tall), with various branches or trunks from the base or near the base, or shrubs of modest stature and single stems ('*vara*' or treelet type). The families with the greatest number of shrub species are Leguminosae (19%), Euphorbiaceae (13.9%), Malvaceae (6.3%), Rubiaceae (5.5%), Cactaceae (4.6%), Compositae (4.6%), Acanthaceae (4.2%) and Solanaceae (4.2%). Some very conspicuous shrub genera in Chamela-Cuixmala are *Acacia*, *Acalypha*, *Bauhinia*, *Caesalpinia*, *Capparis*, *Casearia*, *Croton*, *Erythroxylum*, *Indigofera*, *Mimosa*, *Piper*, *Psychotria* and *Senna*.

13.5.2.3 Lianas and Vines (197 Species)

There are many species of vines and we recognize two types of scandent plants: herbaceous climbers (86 species; stems strictly herbaceous), and woody climbers (111 species; base and some stems woody). The mechanisms by which these species climb are diverse: tendrils, twining, hooks and spines, scrambling, and having rough-textured leaves which prevent slippage (see Burnham and Lott, 1999).

Of the 25 climbing species of *Ipomoea*, two are lianas, *I. bombycina* (Choisy) Benth. & Hook. F. and *I. bracteata* Cav. The rest are herbaceous, slender-stemmed (< 1 cm) climbers with fleshy roots. These climbing heliophytes take advantage of the insolation of the open canopy and of clearings to grow rapidly at the beginning of the rainy season. Their growth is correlated with the seasonality to which the SDTF is subject.

Other important genera of lianas and woody climbers are *Arrabidaea*, *Clytostoma*, *Cydista* (Bignoniaceae); *Heteropterys*, *Hiraea*, *Tetrapterys* (Malpighiaceae); *Forsteronia*, *Marsdenia*, *Prestonia* (Asclepiadaceae); *Paullinia*, *Serjania* (Sapindaceae); *Combretum* (Combretaceae); *Liabum*, *Otopappus* (Compositae); *Entada* (Leguminosae); and *Hemiangium*, *Hippocratea*, and *Pristimera* (Hippocrateaceae). The most speciose families of herbaceous climbers in Chamela-Cuixmala are Cucurbitaceae, Leguminosae, Convolvulaceae and Dioscoreaceae.

As for the distribution patterns of the climbers within the forest, Lott et al. (1987) found that lianas are three times more numerous in the arroyo (a gully or seasonally dry streambed) forest than on the hillsides. This is possibly explained by the greater abundance of clearings in the arroyos and also because there is greater availability of moisture (G. Cabbalé, Institute de Botanique, Univ. Montpellier, pers. comm.).

13.5.2.4 Epiphytes (40 Species)

Epiphytes compose 3.9% of the total flora. Here we include parasitic shrubby hemiepiphytes of the families Loranthaceae and Viscaceae as well as more 'conventional' epiphytes. All species of *Ficus* were classified as trees, although some of them begin life as hemiepiphytes.

Among the families which make up the group of epiphytes are Bromeliaceae, Loranthaceae/Viscaceae, Orchidaceae and Cactaceae. Bromeliaceae is the most important family in terms of number of species present, and probably in number of individuals as well. In this family *Tillandsia* (17 species) is the predominant genus, with some species abundant and showy (for example, *T. fasciculata* Sw. var. *venosispica* Mez ex DC., perched high in the canopy with its conspicuous and attractive red and yellow inflorescence, or *T. paucifolia* Baker which appears as dense twisted gray chains of individuals). In this group, slope and insolation noticeably influence local abundance (Medina, 1995; S. Bullock, Centro de Investigación Científica y de Educación Superior de Ensenada, Mexico, pers. comm.).

There are no reports of epiphytic ferns in Chamela and the same scarcity of epiphytic ferns was found in all the SDTF sites sampled by Gentry (1995).

13.5.2.5 Herbs (366 Species)

Herbs are the predominant life form in the flora, with 371 species (34.9% of the total of native species). The largest families are Gramineae, Compositae, Euphorbiaceae, Leguminosae, Malvaceae, Cyperaceae, Acanthaceae, Solanaceae, Labiatae, Nyctaginaceae, Polypodiaceae and Amaranthaceae. The abundance of herbs is most noted in the rainy season, their period of vegetative growth. There is a tendency to consider the herbaceous plants as annual. Nevertheless, and although we have no exact data, the herbaceous natives apparently are predominantly perennial and survive the annual dry season by means of succulent or woody roots. In contrast to communities where herbaceous plants are principally found in disturbed areas, in the SDTF, many of these species form an important part of the understory.

13.6 PHYTOGEOGRAPHIC ANALYSIS OF DISTRIBUTION PATTERNS AND ENDEMISM

Although knowledge of the flora of Chamela-Cuixmala is sufficiently advanced to permit an analysis of the patterns of geographical distribution, the full neotropical distributions of many of the species are still uncertain and therefore any analysis of the biogeographical patterns must be considered tentative. Nevertheless, we are reasonably confident that the phytogeographical affinities reported here will be sufficiently reliable not to be changed by future discoveries.

The biogeographical patterns we have recognized are presented in Table 13.2. The geographical distributions of the Chamela flora used here were based on critical review of herbarium records and on the recent literature. Only the 1064 species considered native were included in the analysis. We excluded species with unresolved taxonomic problems, species probably new to science but not yet adequately studied, species not determined because of lack of sufficient material for identification (*Daphnopsis* sp., *Schoepfia* sp.), or for lack of other kinds of reliable information. This assured a conservative estimate of the number of endemic species.

The floristic assemblages forming the biogeographical units in Table 13.2 are based on overall distributions with a given species assigned only to the unit where it is regarded as most characteristic. However, this does not mean that species are restricted to a single unit; they may be found in others in which they are less characteristic. In the following pages we discuss these patterns in detail.

13.6.1 PACIFIC SLOPE

A high percentage of the flora of the region is limited, in different degrees, or is endemic to the Pacific slope of Mexico and Central America (40.7%, Table 13.3; Figure 13.1). The most important subdivisions include the *strictly local* element (Jalisco: 2.6%), Central Coast (Jalisco-Oaxaca: 9.7%), and Pacific slope of Mexico (Baja California to Oaxaca: 8.4 %). *Leucaena lanceolata* S. Wats. is a good example of a wide-ranging Pacific slope species; it ranges from Sonora and Baja California to Chiapas, with a small outpost on the coast of Veracruz. It also has a variety, *L. lanceolata* var. *sousae* (S. Zárate) C.E. Hughes, which is restricted to coastal Michoacán, Guerrero and Oaxaca (Hughes, 1998).

TABLE 13.2
Definition of Biogeographic Patterns

Principal Pattern	Secondary Pattern	Definition of Secondary Pattern
Pacific coast[a]	North-west	Jalisco to Sonora and Baja California Sur. Includes the Islas Tres Marías
	Jalisco (only)	Exclusively Jalisco. In many cases these are relatively recently described species known only from the type locality.
	Central coast	Jalisco to Oaxaca (effectively to the Isthmus of Tehuantepec)
	Mexico	Occurs in the northwest and central coast
	Chiapas-Guatemala	Jalisco to Guatemala
	Central America	Jalisco to Panama along the Pacific Coast
	Widespread dist.	Occurs in the northwest and central coast of Mexico, and also from Chiapas to Panama
Mexico[b]	Bicoastal	Occurs on the Pacific and Atlantic coasts
	Interior	Occurs on the Pacific and Atlantic coasts but also found in the interior of Mexico
Widespread neotropical[c]	To Central America	Occurs on the Pacific and Atlantic coasts. May include areas bordering the Gulf of Mexico and the Atlantic coast of the U.S.A.
	To South America	Occurs on the Pacific and Atlantic coasts but extending to South America

[a]The species included here are not found outside the Pacific slope of Mexico and Central America.
[b]Limited to Mexico in the broad sense, including US border states.
[c]Present on both coasts, may also occur in the Caribbean.

In general the endemism is at the level of species, not of genus or family. McVaugh (1961), in a study of some Euphorbiaceae of Nueva Galicia, commented that the western part of the region includes an important endemic element, and that perhaps half of the endemic or nearly endemic species were found in the area of very strong dry periods. McVaugh pointed out that the closest relationships of the SDTF in the coast of Nueva Galicia are with Michoacán and the Sierra Madre del Sur. Many species have distributions ranging up to central Sinaloa, and others also to the extreme southern end of Baja California, but there is little relationship with the flora of the coastal plain of Sonora. Later McVaugh (1983: 1) reiterated this conclusion: 'Relatively few species are strictly confined within the borders of Nueva Galicia, but approximately half of all the known species have their primary areas of distribution on the Pacific side of Mexico, north to central Sinaloa (a few to Sonora, southwestern Chihuahua or southern Arizona), and southeastward to Guerrero or Oaxaca'.

According to Rzedowski (1978) and Gentry (1995), there are numerous genera endemic to the Pacific slope, and Trejo (1998) found 20 genera endemic to Mexican SDTF out of 368 in her transects. Of the four endemic genera of the Pacific slope reported by Lott (1993) for Chamela, *Celaenodendron, Chalema, Dieterlea* and *Mexacanthus,* three are still considered valid today because *Celaenodendron* was synonymized with *Piranhea* (Radcliffe-Smith and Ratter, 1996). *Dieterlea* includes two species (McVaugh, 2001), both native to the Pacific slope of Mexico. *Holographis, Lasiocarpus, Cheileranthemum* (3 species), *Lagascea, Conzattia* (2 species) and *Willardia* also have been cited as genera principally endemic to Mexico, though *Willardia* is now accepted as a section (with six species) of the genus *Lonchocarpus* (Sousa S., 1992). None of the three genera endemic to the Río Balsas Basin (*Backebergia* (Cactaceae), *Haplocalymma* (Asteraceae), and *Pseudolopezia* (Onagraceae)), are yet known to occur in Chamela-Cuixmala.

Myrospermum was cited by Gentry (1995: 188) as the only SDTF-restricted genus found in Costa Rican diversity samples which does not also occur in western Mexico (the genus has since been

TABLE 13.3
Tabulation of Life Forms and Geographical Distributions

Principal Pattern	Secondary Pattern	Tree	Shrub	Epiphyte	Herb	Woody Vine	Herbaceous Vine	Total
Pacific slope	Northwest (BCS-Jal)	4.4 (10)	3.5 (8)	0 (0)	1.3 (5)	1.8 (2)	3.5 (3)	2.6 (28)
	Jalisco only	3.5 (8)	11.5 (26)	0 (0)	3 (11)	4.5 (5)	4.7 (4)	5.1 (54)
	Central Coast (Jal-Oax)	16.2 (37)	12.3 (28)	20 (8)	4.9 (18)	10.8 (12)	9.3 (8)	10.4 (111)
	Widespread Mexican coast (BCS-Oax)	4.4 (10)	2.2 (5)	0 (0)	2.7 (10)	0.9 (1)	2.3 (2)	2.6 (28)
	Jalisco-Chiapas-Guatemala	9.6 (22)	18.1 (41)	10 (4)	10.5 (39)	9 (10)	7 (6)	11.5 (122)
	Jalisco-Central America	2.6 (6)	0.4 (1)	2.5 (1)	2.2 (8)	1.8 (2)	4.7 (4)	2.1 (22)
	Widespread BCS-Central America	8.7 (20)	5.3 (12)	10 (4)	5.9 (22)	11.7 (13)	7 (6)	7.2 (77)
	Subtotal	*49.3 (113)*	*53.3 (121)*	*42.5 (17)*	*30.5 (113)*	*40.5 (45)*	*38.4 (33)*	*41.5 (442)*
México	Bicoastal	2.2 (5)	1.3 (3)	2.5 (1)	2.4 (9)	2.7 (3)	1.2 (1)	2.1 (22)
	Interior	2.2 (5)	3.1 (7)	2.5 (1)	2.2 (8)	1.8 (2)	2.3 (2)	2.3 (25)
	Subtotal	*4.4 (10)*	*4.4 (10)*	*5 (2)*	*4.6 (17)*	*4.5 (5)*	*3.5 (3)*	*4.4 (47)*
Widespread neotropical	To Central America	16.2 (37)	15.4 (35)	17.5 (7)	10 (37)	19.8 (22)	8.1 (7)	13.6 (145)
	To South America	30.1 (69)	26.9 (61)	35 (14)	55 (204)	35.1 (39)	50 (43)	40.4 (430)
	Subtotal	*46.3 (106)*	*42.3 (96)*	*52.5 (21)*	*65 (241)*	*55 (61)*	*58.1 (50)*	*54 (575)*
	Grand total	*21.5 (229)*	*21.3 (227)*	*3.8 (40)*	*34.9 (371)*	*10.4 (111)*	*8.1 (86)*	*100 (1064)*

Note: In each cell, the first number is the percentage for that particular life form, and the second (in parentheses) is the number of species. Both the geographical patterns defined here and the life forms represent mutually exclusive categories reported from the Río Balsas Basin (Fernández et al., 1998)), and he listed ten 'western Mexican dry-forest-restricted genera which do not reach Guanacaste: *Amphipterygium, Apoplanesia, Comocladia, Elaeodendron, Hintonia, Lagrezia, Pachycereus, Peniocereus, Recchia* and *Stenocereus*'. Trejo (1998) found 20 genera endemic to Mexico out of 368 found in her transects. So from the generic point of view, Mexico's SDTF appears to be much higher in endemism than that of southern Central America.

13.6.1.1 North-West (Baja California [BCS] - Jalisco)

There is a low percentage (Table 13.3: 28 species, 2.6%) of species restricted to the distribution pattern of Jalisco to Sonora (or further north) and Baja California Sur. In fact, when we examine this group of species we find that only species such as *Amaranthus palmeri* S. Wats., *Diospyros* aff. *rosei* Standl., *Gomphrena sonorae* Torr., *Randia mollifolia* Standl. and *Sida alamosana* S. Wats. are distributed from Jalisco to Sinaloa, Sonora or Baja California Sur, and that the greater proportion of the species in this pattern are restricted to Jalisco and Nayarit. This suggests that the principal zone of endemism in reality is the central coast of the Pacific (Jalisco to Oaxaca) and that Chamela is in the extreme north, not in the center.

There is the strong likelihood of a real relationship (not an artifact of collection) between the Islas Tres Marías and the coast of Nayarit and Jalisco. Some examples of species with this distribution pattern are *Astrocasia peltata* Standl., *Cephalocereus purpusii* Britton & Rose, *Esenbeckia nesiotica* Standl., *Jatropha malacophylla* Standl., *Matayba spondioides* Standl., *Peniocereus rosei* González-Ortega, *Trixis pterocaulis* B. L. Rob. & Greenm. and *Piranhea* (*Celaenodendron*) *mexicana* (also reported from Colima). This relationship may reflect a common geological history before the separation of the Baja California Peninsula (T.R. Van Devender, Arizona-Sonora Desert Museum, pers. comm.).

Another interpretation that can be drawn from these data is that there are practically no species in Chamela-Cuixmala whose principal distribution is the arid zones of northern Mexico (see also the section on Mexican Distribution, below). Based on the main part of their distributions, apparently the species of the coastal region extend their area of distribution toward the desert and not the reverse. Studies in Sonora by Martin et al. (1998) and by Van Devender et al. (2000) are clarifying the floristic relationship between Chamela-Cuixmala and the northern limit of SDTF in Sonora and Baja California Sur.

Van Devender et al. (2000) reported 734 species in 429 genera and 115 families in the flora of the Río Cuchujaqui in the south of Sonora. These authors found that 22.4% (251 species) of the species of Chamela-Cuixmala (taken from Lott, 1993, not 2002) also occur in that region, although this figure includes all the species of wide distribution (principally herbs). On the other hand, the flora of Río Cuchujaqui shares only 10% of the species of trees and 5.5% of the woody vines, confirming that although physiognomically the two forests are similar, their floristic composition is very different. Some endemic taxa in Sonoran SDTF (including parts of neighboring Chihuahua or Sinaloa) are *Ageratina sandersii* B.L. Turner, *Anisacanthus thurberi* (Torr.) A. Gray, *Bouteloua quiriegoensis* Beetle, *Brongniartia alamosana* Rydb., *Bunchosia sonorensis* Rose, *Caesalpinia palmeri* S. Wats., *C. caladenia* Standl., *Cardiospermum cuchujaquense* M.S. Ferrucci & Acev.-Rodr., *Coccoloba goldmanii* Standl., *Echinocereus stoloniferus* W.T. Marshall var. *stoloniferus, Euphorbia alatocaulis* V. W. Steinm. & Felger, *E. dioscoreoides* subsp. *attenuata* V.W. Steinm. (better regarded as a species, V. Steinmann, Rancho Santa Ana Botanical Garden, California, pers. comm., 2005), *Galactia* sp. nov., *Haematoxylum* sp. nov., *Havardia mexicana* (Rose) Britton & Rose, *Holographis pallida* Leonard & Gentry, *Lysiloma watsonii* Rose, *Mammillaria standleyi* (Britton & Rose) Orcutt, *Opuntia thurberi* Engelm. and *Tetramerium yaquianum* T.F. Daniel.

13.6.1.2 Jalisco Only

Due to the low number of species (27 species, 2.6% of the total) present in Chamela which apparently are known only from the state of Jalisco, (even if we consider the 44 species, 6.8% of the total, endemic to Jalisco and one other neighboring state (E.J. Lott, unpublished database, and L. Hernández López,

Centro Universitario de Ciencias Biológicas y Agropecuarias, Jalisco, Mexico, pers. comm.), it is doubtful that Chamela-Cuixmala in itself represents a center of very restricted and isolated endemism, but rather it forms part of a related coastal belt. These data suggest that Chamela and the two sites studied by Trejo and Dirzo (2002) in our 'Pacific slope: Central Coast' region are among the most diverse SDTFs of Mexico, but more in-depth work is needed to determine the similarity of floristic composition between them. Many of these restricted species are recently described (or yet to be described) (Lott, 1993), and at least some are probably actually wider-ranging. In the case of some supposed local endemics, known only from the Chamela region when first described (such as *Chalema synanthera* Dieterle, *Dieterlea fusiformis* E.J. Lott, *Manfreda chamelensis* E.J. Lott & Verhoek, *Croton chamelensis* E.J. Lott, *Jatropha bullockii* E.J. Lott, *J. chamelensis* Pérez-Jimenez and *Verbesina lottiana* B.L. Turner & Olsen), within a few years of their publication they were found in sites further away on the Pacific Coast. Some of the endemic species of the region will indeed be found to be of very limited distribution or to be actually rare plants. Hernández López (1995:14) calculated that, in the endemic flora of Jalisco, almost 50% of the endemic species have been reported from only one or two populations, followed by species with an intermediate number of populations, and in third place, those of very wide distribution in Mexico. This proportion is the opposite of that calculated for the flora of Mexico as a whole by Rzedowski (1991a), who proposed that the majority of the total endemic flora would consist of species of relatively wide distribution, followed in number by those of intermediate distribution.

Some species of Chamela-Cuixmala which have not yet been reported from other localities are: *Bourreria rubra* E.J. Lott & J.S. Miller, *Schaefferia lottiae* Lundell, *Bonamia mexicana* McDonald, *Acacia chamelensis* L. Rico, *Acalypha gigantesca* McVaugh, *Lonchocarpus minor* Sousa, *Styphnolobium protantherum* Sousa & Rudd, *Malpighia emiliae* W.R. Anderson, *Mirabilis russellii* Le Duc, *Lantana jaliscana* Moldenke and *Rhynchosia delicatula* O. Te'llez and M. Sousa.

Recently, several floristic projects in Colima and Jalisco have sought a better understanding of patterns of local endemism. Based on a census of herbarium collections and a critical revision of the literature, Hernández López (1995) made a preliminary inventory of the endemic flora of the state of Jalisco; for the state she found that 9% of the estimated total flora (*c.*5500 species) was endemic to Jalisco and one neighboring state ('near-endemics'), and another 5.4% of species are strictly limited to the state ('strict endemics'), giving a total of approximately 14.4% endemic to Jalisco and to one or another neighboring state. This same author indicated that Chamela was one of the four important centers of endemism in Jalisco-Colima, but cautioned that possibly this was due to intensive collecting in the general area of the EBCH relative to other parts of the state, and we think that this is most likely the case. A program of collecting in targeted areas of Jalisco which were poorly represented in herbaria has resulted from Hernández López's study.

There is little evidence of a tight relationship between the coast of Jalisco and the Cape region of Baja California Sur. The latter region has a very interesting endemic component: in the SDTF, endemic genera are *Carterella* (Rubiaceae), *Clevelandia* (Scrophulariaceae) and *Faxonia* (Asteraceae; possibly extinct). Aside from its own endemic species (37 (12.3%) out of 454 species reported by León de la Luz et al., 1999), the principal component of the SDTF flora of the Cape region consists of a subset of the flora that we call endemic to the Pacific slope, some of which is shared with Chamela-Cuixmala. Other shared species are of Sonoran distribution or of wide neotropical distribution.

13.6.1.3   Central Coast (Jalisco-Oaxaca)

The group of species whose distribution is limited to the coast from Jalisco to Oaxaca (Table 13.3: 10.4%) is the largest component of the group restricted to the Pacific slope. Lott et al. (1987) estimated that 16% of the species of the SDTF (excluding SM) of the EBCH were endemic to the coast of Sinaloa-Guerrero. In the present work we did not calculate that area separately, but we found that 14.9% (Table 13.4) of the species belonging to the SDTF in Chamela are restricted to Jalisco-Oaxaca, a figure very similar to the former one. If we expand our concept of the coast to

TABLE 13.4
Tabulation of Patterns of Geographical and Plant Communities

Principal Pattern	Secondary Pattern	SDTF	Semi-Evergreen Forest (SM)	Halophytes	Beach	Riparian	Aquatic	Disturbed Areas
Pacific slope	Northwest (BCS-Jal)	3.9 (22)	2.1 (4)	0 (0)	0 (0)	1 (1)	0 (0)	1.3 (3)
	Jalisco only	6.4 (36)	4.2 (8)	2.6 (1)	5.7 (2)	2.9 (3)	0 (0)	0.9 (2)
	Central Coast (Jal-Oax)	14.9 (84)	8.4 (16)	0 (0)	8.6 (3)	3.9 (4)	0 (0)	4.5 (10)
	Widespread Mexican coast (BCS-Oax)	14.4 (81)	13.2 (25)	10.3 (4)	11.4 (4)	9.7 (10)	7.7 (1)	4.9 (11)
	Jalisco-Chiapas-Guatemala	2.8 (16)	2.1 (4)	0 (0)	2.9 (1)	2.9 (3)	0 (0)	2.7 (6)
	Jalisco-Central America	2.5 (14)	3.2 (6)	0 (0)	2.9 (1)	1 (1)	0 (0)	1.3 (3)
	Widespread BCS-Central America	7.3 (41)	8.4 (16)	2.6 (1)	14.3 (5)	6.8 (7)	0 (0)	3.6 (8)
	Subtotal	*52.3 (294)*	*41.6 (79)*	*15.4 (6)*	*45.7 (16)*	*28.2 (29)*	*7.7 (1)*	*19.3 (43)*
México	Bicoastal	2.5 (14)	1.6 (3)	0 (0)	2.9 (1)	1 (1)	0 (0)	0.9 (2)
	Interior	2.8 (16)	2.6 (5)	2.6 (1)	2.9 (1)	3.9 (4)	0 (0)	1.3 (3)
	Subtotal	*5.3 (30)*	*4.2 (8)*	*2.6 (1)*	*5.7 (2)*	*4.9 (5)*	*0 (0)*	*2.2 (5)*
Widespread neotropical	To Central America	12.6 (71)	18.4 (35)	5.1 (2)	5.7 (2)	12.6 (13)	7.7 (1)	12.6 (28)
	To South America	29.7 (167)	35.8 (68)	76.9 (30)	42.9 (15)	54.4 (56)	84.6 (11)	65.9 (147)
	Subtotal	*42.3 (238)*	*54.2 (103)*	*82.1 (32)*	*48.6 (17)*	*67 (69)*	*92.3 (12)*	*78.5 (175)*
Grand Total		*562*	*190*	*39*	*35*	*103*	*13*	*223*

Note: In each cell, the first number is the percentage for that particular community, and the second is the number of species (in parentheses). The plant communities do not represent mutually exclusive categories (a species may exist in more than one community).

include those species restricted to some part of the region from Sonora and the Cape of Baja California Sur to Oaxaca, this number increases to 25.2%, (Table 13.4: total North-west, Jalisco only and Central Coast categories).

13.6.1.4 Río Balsas Basin

There is a very large number of species which are endemic to the Río Balsas Basin (Fernández et al., 1998). Trejo and Dirzo (2002) estimated that about 45% of the species at their site in Cañón del Zopilote, Guerrero, and 30% of the species of the site in Infiernillo, Michoacán, are restricted to the Balsas basin. Several workers (Kohlman and Sánchez C., 1984; Rzedowski, 1991a) consider that the Balsas Basin represents a distinct area of endemism. There are various indications that species-level endemism in Río Balsas Basin SDTF is high: of 45 species of *Bursera* reported in Fernández et al. (1998), around half are endemic. Every zone we have looked at has at least one endemic species of the *Caesalpinia Poincianella-Erythrostemon* group: the Balsas basin has five endemic species (Lewis, 1998); *Brongniartia* and *Desmodium* are also very speciose here.

13.6.1.5 Jalisco-Chiapas-Guatemala

The number of species whose distribution extends along the southern Pacific slope from Jalisco to Chiapas-Guatemala (Table 13.3: 11.5%) or to Central America (Table 13.3: 2.1%) totals 13.6%. This figure is significantly greater than the contribution of species which are found from the Northwest to Jalisco (2.6%), and also supports the idea that the center of the Pacific Coast distribution group is to the south of Chamela.

13.6.1.6 Conclusions on the Endemic Element of the Pacific Slope

We have demonstrated that in the Chamela flora there is an important element of species restricted to different subdivisions of the Pacific slope of Mexico and of Central America, which together form a distinctive flora. Apparently Chamela-Cuixmala and the coast of Jalisco are not an important center of endemism of the region as had been suggested earlier (Gentry, 1995; Hernández López; 1995) but rather the center of endemism extends from the coastal region of Jalisco to Oaxaca. Adding species which are restricted to the coast of Jalisco and the North-west or of Jalisco to Central America has a relatively unimportant effect, which supports the idea that Jalisco-Oaxaca is the center of distribution of this element. Once we have enough information to analyze the SDTFs of Oaxaca in the same manner, it may be that the center will be further to the south. The principle of the mid-range effect (Colwell and Lees, 2000) predicts that highest species diversity should be in geographically central areas. This would imply that the centre of SDTF species diversity on the Pacific slope should be in Guerrero, midway between Sonora in the north and Costa Rica in the south. Our data do not contradict this.

The Central Coast of Mexico (Jalisco-Oaxaca) is delimited to the North-west by arid zones. SDTF in the Sonoran zone is rather restricted to hillsides and canyons at intermediate elevations (200–500 m), not to the coastal plain (Martin et al., 1998, Van Devender et al., 2000). Towards the south-east, the Isthmus of Tehuantepec marks a transition to regions of greater annual precipitation distributed over a longer annual period. What is most characteristic of the present-day environment of the endemic element of the Pacific slope is the marked dry season, and to a lesser degree its low elevation.

Although we have not done an analysis of endemism of the Balsas River Basin based on the recent checklist (Fernández et al., 1998) and other floristic studies, every indication is that the endemic element is substantial. Relative to the Coast, it has a greater topographic complexity and in many parts has a more extreme climate as far as precipitation is concerned (Trejo, 1999; Trejo and Dirzo, 2002) which might help to explain its level of endemicity.

13.6.2 Distribution in Mexico

The species known from the region of Chamela-Cuixmala whose distribution is restricted to Mexico (in the sense of 'Megaméxico III' of Rzedowski, 1991a), but which include the Gulf of Mexico or the Caribbean are relatively few (Table 13.3: 4.4%). Although there are many species in common between the coast of Jalisco and the coast of the Gulf of Mexico, only 2.1% of them are restricted to Mexico. The rest are of broad neotropical distribution.

13.6.3 Broad Neotropical Distribution

The other large group of species of our region, almost the same size as that of the Pacific slope flora, is that of broad neotropical distribution (54%, Table 13.3). Within this group the greater part are distributed to southern South America (40.4% of the total, 80% of the subcategory). Among those species of continuous distribution from Mexico to South America, some are characteristic of secondary vegetation and disturbed sites, such as *Byttneria aculeata* Jacq., *Celtis iguanaea* (Jacq.) Sarg., *Cochlospermum vitifolium* (Willd.) Spreng., *Cordia alliodora* (Ruíz & Pav.) Oken, *Guazuma ulmifolia* Lam., *Lasiacis ruscifolia* (H.B.K.) Hitchc., *Sapindus saponaria* L., *Tabebuia rosea* (Bertol.) DC., *Trema micrantha* (L.) Blume, and *Trichilia havanensis* Jacq. (Rzedowski,1978).

In their physiognomy the SDTFs of the Pacific slope and those of the Caribbean and Yucatán appear to be similar at first sight (Trejo, 1998; C. Chiappy J., Instituto de Ecología, Veracruz, pers. comm.), but the species shared between the Pacific and the Caribbean are generally species of wider distribution. In the floristic regions delimited by Rzedowski (1978), Chamela is included in the Caribbean region, which comprises the Pacific Coast, the Atlantic Coast to Yucatán, and the Antilles. This classification is based on species which are common to the whole region, and not on species *restricted* to it. That is to say, many of the species which are common to both coasts of Mexico are of even wider distribution, and extend to South America in many cases. In spite of the fact that there are various mentions in the literature (for example, Standley, 1936, cited in Rzedowski, 1978) of possible phytogeographical relationships between the Pacific slope flora and that of the Caribbean or of Yucatán, we find no evidence for those suggestions based on species restricted to those regions. *Enriquebeltrania crenatifolia* (Miranda) Rzedowski (Euphorbiaceae) was one of the few species which was thought to occur only in Yucatán and on the Pacific slope, at Chamela, but the Chamela population is now thought to represent a new species (V. Steinmann, pers. comm., 2005). A related issue concerns the degree of relationship of the Yucatecan to the Antillean flora (Standley 1936; Rzedowski 1978). Miranda (1958) and Estrada-Loera (1991) argued that the closest relationship was with Central America, but in a study of 434 tree species from all vegetation types, Ibarra M. et al. (2002) found that the strongest floristic affinities were with Mesoamerica, and affinities with the Antilles weaker.

13.6.3.1 Distribution and Life Forms

The relationship between life form and distribution pattern is shown in Table 13.3. It is interesting that the percentage of trees (49.3%) and shrubs (53.3%) whose ranges are restricted to the Pacific slope is higher than the proportion of the Pacific slope group in total (41.5%) and the proportion of herbaceous plants (herbs 30.5%, herbaceous vines 38.4%) is less. Compared with the species of broad neotropical distribution, those of restricted distribution ('endemics') are disproportionately of the woody element. Many of the widely distributed species of herbaceous plants (65% of the herbs) are ephemeral weeds of disturbed areas or ruderal situations. In a pioneering series of studies, Rzedowski (1991a) made an estimation of the distribution of the endemic elements in the flora of Mexico according to life form: in the categories of herbaceous plants (excluding epiphytes and aquatic and subaquatic plants) and of shrubs, approximately 60% of the species are endemic to Mexico (Rzedowski, 1991a, Table 5). Herbaceous plants have been relatively little studied in Chamela-Cuixmala and in Mexican and Central American SDTF in general, and the significance of this apparent contrast is not clear.

13.6.4 Distribution of Species of Different Communities

We have already mentioned the two principal biotic communities in our area of interest, the SDTF and the SM. There also exists a variety of aquatic, riparian, halophytic and coastal communities, although these are much more limited in their original extent and have suffered more destruction due to human activity. The *Orbignya* (*Attalea*) forests of the area are almost completely replaced by coconut and banana plantations. The phytogeographical affinities of the species which make up the plant communities of Chamela-Cuixmala are summarized in Table 13.4. It is important to note that our *phytogeographical classifications are mutually exclusive* (each species is only classified in a single group), *but the communities are not* (a species can be found in more than one community).

In general the aquatic plants (*Eichhornia, Lemna, Marsilea, Wolffia, Pistia, Thalia, Typha*), plants of marshy places (*pantanos*) (Cyperaceae such as *Cyperus articulatus* L., *C. ligularis* L., *C. regiomontanus* Britton), the halophytes (the mangroves *Rhizophora mangle* L., *Avicennia germinans* (L.) L., *Laguncularia racemosa* (L.) Gaertner f., *Conocarpus erecta* L.), the plants of the littoral (*Ipomoea pes-caprae* (L.) Sweet) and those of salt flats (*salinas*) and the surrounding marshes (*Batis maritima* L., *Hippomane mancinella* L., *Phyllanthus elsiae* Urb.), are widely distributed (Good, 1974; Rzedowski, 1978; Castillo and Moreno-Casasola, 1998). That is, these habitats do not include species which are endemic to this area. An exception to this generalization is *Abronia maritima* Nutt. ex Wats., which is distributed from the central coast of California and Sonora to Baja California, Sinaloa and Jalisco, on coastal dunes. Those communities associated with water as a group contain few species in comparison with the SDTF (Rzedowski, 1991b).

Those species associated with disturbed areas are disproportionately of wide neotropical distribution (171 out of 224), although there is a large number of them which are of exotic origin (42). In general, there is little penetration of exotic species in natural (primary) communities of any kind in Chamela-Cuixmala.

The SDTF is the community with the greatest number of species in the region (562 natives, Table 13.4). In this community the species which are restricted to the Pacific slope are much more important (52.3%) than the species of wide distribution (29.7%), very different from their proportions in the total flora (42.3% and 54.2%, respectively).

Probably owing to its lesser local extension, the SM has fewer species than the SDTF (190). The respective contributions of the Pacific elements and of the wide neotropical ones are similar to their representations in the total flora (39.2% vs. 40.7% and 52.2% vs. 51.8% respectively; Tables 13.3 and 13.4). The importance of the element with restricted distribution (Table 13.4: 39.2%) in this community was surprising to us, probably because of the visual impression made by such conspicuous species of widespread distribution as *Astronium graveolens* Jacq., *Tabebuia impetiginosa* (Mart.) Standley, *Ceiba pentandra* (L.) Gaertn., *Brosimum alicastrum* Sw., and *Ficus insipida* Willd.

13.6.5 Biogeographical Conclusions

In summary, the flora of the Chamela region is composed of a group of species restricted to the Pacific slope of Mexico, and of another group of wide neotropical distribution. These two groups account for 92.5% of the total flora. Our tabulations do not support suggestions of close phytogeographical affinities between our region and the Yucatán Peninsula or the Antilles. Their similarities are based solely on shared species of broad neotropical distribution and a certain physiognomic similarity, not on a common history of endemism. On the other hand, neither is there evidence of a strong relationship between the coast of Jalisco and the Cape Region of Baja California Sur. The Cape Region is near the northernmost limits of the SDTF, isolated from the continent by the Sea of Cortéz, and it is in contact with zones of greater aridity, therefore it is not surprising that is has its own, evidently higher, level of endemism. Its flora also is composed of Pacific Coast elements, some of them shared with Chamela-Cuixmala. Other shared species are basically a few of Sonoran distribution.

We have seen that, compared to other sites, Chamela-Cuixmala is not an important center of endemism, but it is a fairly rich center of diversity. In terms of plants narrowly restricted to the coast of Jalisco, the majority are associated with SDTF, the most extensive plant community in the region.

13.7 CONSEQUENCES FOR CONSERVATION

Current rates of deforestation of SDTFs in Mexico are extremely high (Trejo and Dirzo 2000 and others), and only about a quarter of the area covered by SDTF is estimated to retain a fair degree of integrity. In light of the diversity and high turnover of species from one site to another in Pacific slope flora, and given the degree of threat which SDTFs are facing today, the best strategy for conservation is the protection of a variety of reserves spread throughout the range of SDTFs to include all the variations in environmental conditions such as microclimate, soils, and populations of the flora.

Gentry (1995:184) stated that: 'For conservation, areas of endemism are more important than areas of high diversity. Mexico's dry forests have higher endemism than other communities and deserve the attention they are receiving'. However, not all SDTF sites are 'deserving of attention' for the same reasons. As Gillespie et al. (2000) and Gordon et al. (2004) have shown, highly disturbed and secondary SDTFs and fallows can act as refuges of species for later re-colonization of recovering areas.

Dinerstein et al. (1995) assigned Endangered/Regionally Outstanding category ('Level I, highest priority at regional scale for biodiversity conservation') to two Mexican SDTF areas, Jalisco and Balsas, and Endangered/Bioregionally Outstanding to Oaxacan SDTFs. Sinaloan SDTF, also Bioregionally Outstanding, is listed as Vulnerable, and other sites mentioned in this chapter are in the Locally Important category of Biological Distinctiveness. In light of the evident diversity and endemism of Michoacán, Guerreran and Oaxacan SDTFs, it would be well to include them in the first group with the Balsas, perhaps removing Jalisco to a lower level as well. Fortunately, Jalisco has several stable biosphere reserves.

In the Chamela region, riparian, aquatic, coastal communities and palm forest (now almost entirely extirpated in the area, before their associates were well known) are of very limited extension and contain many plant species limited to them; nevertheless, these species are almost all of wide distribution. In this sense, their conservation is important to ensure the survival of the fauna which directly depends upon them and uses them for corridors, and for the hydrology of the region, not for the protection of endemic plant species. In addition, from a local viewpoint, it is important to conserve them because of their rarity in the region.

Endemic genera of Cactaceae need to be studied to ensure their conservation. Every zone of Mexican SDTF has endemic genera or at least species of cacti. They are important structural elements (as witnessed by the fact that they are the third most diverse family in the transects of Trejo (1998)) and the need to understand their diversity and distribution is critical.

Continued floristic inventories, fundamental to understanding a site, are to be encouraged and supported in SDTF, as are the resulting checklists which indicate for each species the vegetation types or habitats in which it occurs, as has been done by León de la Luz et al. (1999), Pérez-García et al. (2001) and others.

Mexican protected areas which include some SDTF are too numerous to list here. The degree to which each site is actively protected is not known, but a list which includes National Parks, Areas of Faunistic and Floristic Protection, Biosphere Reserves and Areas of Natural Resources Protection is available at http://www.vivanatura.org/BiodiversityConservationANP.html

ACKNOWLEDGEMENTS

We thank Patricia Balvanera, Steve Bullock, Fernando Chiang, Tom Daniel, Patricia Dávila, Elvira Durán, Richard Felger, Thomas Gillespie, Jamie Gordon, Leticia Hernández López, Mitchell Provance, Martin Quigley, Lourdes Rico, Andy Sanders, Mario Sousa, Victor Steinmann, Irma Trejo,

Tom Van Devender, Tom Wendt and David White for sharing unpublished data, literature, maps, and for helpful discussions. Irma Trejo and Steve Bullock graciously reviewed an earlier version of the manuscript and offered valuable suggestions for its improvement. Tom Wendt, Colin E. Hughes, Jim Ratter, and an anonymous reviewer read the current version: their thoughtful comments have greatly improved this paper. Paul Fryxell, Rogers McVaugh and Grady Webster were especially generous in allowing us access to their unpublished manuscripts for *Flora Novo-Galiciana*. We are grateful to our editors who have patiently contributed many helpful criticisms. Special thanks are due to curators Andy Sanders (UCR), Mario Sousa and Gerardo Salazar (MEXU) and Tom Wendt (TEX-LL) for herbarium and other support, and again, to all the collectors (especially Ma. Guadalaupe Ayala, Steve Bullock, Alfredo Pérez Jiménez and Arturo Solís Magallanes) and taxonomic specialists, our collaborators, without whose contributions this perspective on the phytogeographical patterns of the Chamela-Cuixmala flora would not be possible (see list in Lott, 2002).The senior author was employed by the Herbario Nacional, IBUNAM (1980–1986), and by the Herbarium, University of California at Riverside (1990–1992), during part of the fieldwork which inspired this paper. The Plant Resources Center of the University of Texas at Austin, where the senior author is a Research Associate, is gratefully acknowledged.

REFERENCES

Bowden, C., *The Secret Forest*, University of New Mexico Press, Albuquerque, 1993.

Burnham, R. and Lott, E.J., *Guia práctica para las plantas trepadoras de la Bahía de Chamela, Jalisco, México*, Field Museum, Chicago (iii, 207 pp. Single-sided, xeroxed), 1999.

Carranza, E. and McDonald, J.A., *Ipomoea curprinacoma* (Convolvulaceae); a new morning glory from southwestern Mexico, *Lundellia*, 7, 1, 2004.

Castillo, S. and Moreno-Casasola, P., Análisis de la flora de dunas costeras del litoral atlántico de México, *Acta Bot. Mex.*, 45, 55, 1998.

Colwell, R.K. and Lees, D.C., The mid-domain effect; geometric constraints on the geography of species richness, *Trends Ecol. Evol.*, 15, 70, 2000.

Dinerstein, E. et al., *A conservation assessment of the terrestrial ecoregions of Latin America and the Caribbean*, World Bank, Washington, D.C., 1995.

Durán, R. and Olmsted, I., *Listado florístico de la Reserva de Sian Ka'an*, Amigos de Sian Ka'an, Puerto Morelos, Q. Roo, 1987.

Estrada-Loera, E., Phytogeographic relationships of the Yucatán Peninsula, *J. Biogeogr.*, 18, 687, 1991.

Felger, R., Johnson, M.B., and Wilson, M.F., *The Trees of Sonora, Mexico*, Oxford University Press, Oxford, 2001, 391.

Fernández, R. et al., Listado florístico de la cuenca del Río Balsas, México, *Polibotánica*, 9, 1, 1998.

García-Mendoza, A.J., Ordóñez, M. de J., and Briones-Salas, M., Eds, *Biodiversidad de Oaxaca*, Instituto de Biología, UNAM, México, 2004.

Gentry, A.H., Patterns of neotropical plant species diversity, *Evol. Biol.*, 15, 1, 1982.

Gentry, A.H., Changes in plant community diversity and floristic composition on environmental and geographical gradients, *Ann. Missouri Bot. Gard.*, 75, 1, 1988.

Gentry, A.H., Diversity and floristic composition of neotropical dry forests, in *Seasonally Dry Tropical Forests*, Bullock, S.H., Mooney, H.A., and Medina, E., Eds, Cambridge University Press, Cambridge, 1995, 146.

Gentry, H.S., *Río Mayo plants; a study of the flora and vegetation of the valley of the Río Mayo, Sonora*, Carnegie Institution of Washington Publication no. 527, 1942.

Gentry, H.S., Sinaloan deciduous forest, *Desert Plants*, 4, 73, 1982.

Gillespie, T.W., Grijalva, A., and Farris, C.N., Diversity, composition, and structure of tropical dry forests in Central America, *Plant Ecology*, 147, 37, 2000.

Gómez-Pompa, A., Flores, J.S., and Sosa, V., The 'Pel Kot': a man-made tropical forest of the maya, *Interciencia*, 12, 10, 1987.

González-Medrano, F., La vegetación del nordeste de Tamaulipas, *Anales Inst. Biol. Univ. México, Bot.*, 43, 11, 1972.

González, S. E., Dirzo, R., and Vogt, R.C., Eds, *Historia Natural de Los Tuxtlas*, Instituto de Biología, UNAM, México, 1997.

Good, R., *The Geography of the Flowering Plants*, Longman, London, 1974, 557.

Gordon, J. et al., Assessing landscapes; a case study of tree and shrub diversity in the seasonally dry tropical forests of Oaxaca, Mexico and southern Honduras, *Biol. Cons.*, 117, 429, 2004.

Gual Díaz, M., Cañón del Zopilote (Area Venta Vieja), *Estudios Florísticos de Guerrero* 6, Facultad de Ciencias, UNAM, México, 1995.

Hammel, B., et al., *Manual de plantas de Costa Rica; Gimnospermas y monocotiledoneas (Agavaceae – Musaceae)*, Monogr. Syst. Bot. 92, Missouri Botanical Garden, St Louis, 2, 2003.

Hemsley, W. B., Botany, in *Biología Centrali-Americana*, Godwin, F.D. and Salvin, O., Eds, R.H. Porter and Dulan, London. 1879–1888, 5 vols.

Hernández López, L., *The endemic flora of Jalisco, Mexico; centers of endemism and implications for conservation*, MS thesis, Institute of Environmental Studies, University of Wisconsin, Madison, 1995, 76.

Hughes, C.E., Monograph of *Leucaena* (Leguminosae: Mimosoideae), *Syst. Bot. Monogr.*, 55, 1998.

Ibarra M., G., Villaseñor, J.L., and Durán, R., Riqueza de especies y endemismo del componente arbóreo de la Peninsula de Yucatán, México, *Bol. Soc. Bot. Mex.*, 57, 49, 1995.

Ibarra M., G. et al., Biogeographical analysis of the tree flora of the Yucatán Peninsula, *J. Biogeogr.*, 29, 17, 2002.

Janzen, D.A., Tropical dry forests; the most endangered major tropical ecosystem, in *Biodiversity*, Wilson, E.O., Ed, National Academic Press, Washington, D.C., 1988, 130.

Janzen, D.H. and Liesner, R., Annotated check-list of plants of lowland Guanacaste Province, Costa Rica, exclusive of grasses and non-vascular cryptogams, *Brenesia*, 18, 15, 1980.

Johnston, M.C., Investigaciones sobre la vegetación y flora de la provincia florística tamaulipeca, in *Simposio sobre el tema Vegetación de México, Resumenes de los trabajos presentados*, Primer Congreso Mexicano de Botánica, 24 a 26 de octubre de 1960, Mexico, 1960.

Kohlmann, B. and Sánchez-Colon, S., Estudio aerográfico del género *Bursera* Jacq. ex L. (Burseraceae) en México; una síntesis de métodos, in *Métodos cuantitativos en la biogeografía*, Ezcurra, E. et al., Eds, Instituto de Ecología, México, 1984, 41.

León de la Luz, J. L. et al., Listados florísticos de México 18, *Flora de la región del Cabo de Baja California Sur*, Instituto de Biología, Universidad Nacional Autónoma de México, México, 1999, 39.

Lewis, G. P., *Caesalpinia, A Revision of the Poincianella-Erythrostemon Group*, Royal Botanic Gardens, Kew, 1998, 233.

Lott, E.J., Listados florísticos de México 3, *La Estación de Biología Chamela, Jalisco*, Instituto de Biología, Universidad Nacional Autónoma de México, México, 1985 (1986).

Lott, E.J., Annotated checklist of the vascular flora of the Chamela Bay region, *Occas. Pap. California Acad. Sci.*, 148, 1, 1993.

Lott, E.J., Listado anotado de las plantas vasculares de Chamela-Cuixmala, in *Historia Natural de Chamela*, Noguera, F.A. et al., Eds, Instituto de Biología, UNAM, México, 2002, 99.

Lott, E.J., Bullock, S.H., and Solís Magallanes, J.A., Floristic diversity and structure of upland and arroyo forests of coastal Jalisco, *Biotropica*, 19, 228, 1987.

Martin, P.S. et al., Eds, *Gentry's Río Mayo Plants; the Tropical Deciduous Forest and Environs of Northwest Mexico*, University of Arizona Press, Tucson, 1998.

Martin, P.S. and Yetman, D.A., Introduction and prospect, secrets of a tropical deciduous forest, in *The Tropical Deciduous Forest of Alamos; Biodiversity of a Threatened Ecosystem in Mexico*, Robichaux, R.H. and Yetman, D.A., Eds, University of Arizona Press, Tucson, 2000, 1.

Martinez y Ojeda, E. and González Medrano, F., Vegetación del sudeste de Tamaulipas, México, *Biotica*, 2, 1, 1977.

Martínez-Yrízar, A., Búrquez, A., and Maass, M., Structure and functioning of tropical deciduous forest in western Mexico, in *The Tropical Deciduous Forest of Alamos; Biodiversity of a Threatened Ecosystem in Mexico*, Robichaux, R.H. and Yetman, D.A., Eds, Univ. of Arizona Press, Tucson, 2000, 2.

McDonald, A., Origin and diversity of Mexican Convolvulaceae, *Anales Inst. Biol. Univ. Nac. México, Bot.* 62, 65, 1991.

McVaugh, R., Euphorbiaceae novae Novo-Galicianae, *Brittonia*, 13, 145, 1961.

McVaugh, R., Botanical exploration in Nueva Galicia, Mexico, from 1790 to the present time, *Contr. Univ. Michigan Herb.*, 9, 205, 1972.

McVaugh, R., Gramineae, *Fl. Novo-Galiciana; a descriptive account of the vascular plants of western Mexico*, University of Michigan Press, Ann Arbor, 14, 1983.

McVaugh, R., Limnocharitaceae to Typhaceae, *Fl. Novo-Galiciana; a descriptive account of the vascular plants of western Mexico*, University of Michigan Press, Ann Arbor, 13, 1993.

McVaugh, R., Ochnaceae to Loasaceae, *Fl. Novo-Galiciana; a descriptive account of the vascular plants of western Mexico*, University of Michigan Herbarium, Ann Arbor, 3, 2001.

Medina, E., Diversity of life forms of higher plants in neotropical dry forests, in *Seasonally Dry Tropical Forests*, Bullock, S.H., Mooney, H.A., and Medina, E., Eds, Cambridge University Press, Cambridge, 1995, 221.

Miranda, F., Estudios sobre la vegetación, in *Los recursos naturales del sureste y su aprovechamiento*, Beltran, E., Ed, Instituto Mexicano de Recursos Naturales Renovables, México, D.F., 2, 1958, 215.

Miranda, F. and Hernández, X. E., Los tipos de vegetación de México y su clasificación, *Bol. Soc. Bot. Mex.*, 28, 29, 1963.

Noguera, F.A., Vega, R. J.H., and Garcia Aldrete, A.F., Eds, Introducción, in *Historia Natural de Chamela*, Instituto de Biología, UNAM, Mexico, 2002, xv.

Peralta-Gómez, S., Diego-Pérez, N., and Gual-Díaz, M., Listados florísticos de México 19, *La Costa Grande de Guerrero*, Instituto de Biología, UNAM, México, 2000, 44.

Pérez-García, E.A., Meave, J., and Gallardo, C., Vegetación y flora de la región de Nizanda, Oaxaca, México, *Acta Bot. Mex.*, 56, 19, 2001.

Pérez Jiménez, L.A., Cervantes, S. L.M., and Solís Magallanes, J.A., *Lista preliminar de las fanerogamas de la región de Chamela, Jal.*, Instituto de Biología UNAM, México [mimeographed], 1981, 16.

Primack, R.B., *Timber, Tourists, and Temples; Conservation and Development in the Maya Forest of Belize, Guatemala, and Mexico*, Island Press, Washington, 1998.

Puig, H., *Vegetation de la Huasteca, Mexique*, Etudes Mesoamericaines 5, México, D.F., Mission Archeologique et Ethnologique Francaise au Mexique, 1976.

Radcliffe-Smith, A. and Ratter, J., A new *Piranhea* from Brazil, and the subsumption of the genus *Celaenodendron* (Euphorbiaceae–Oldfieldioideae), *Kew Bull.*, 51, 543, 1996.

Reichenbacher, F., Franson, S.E., and Brown, D.E., *The biotic communities of North America*, U.S. Dept. of the Interior, Environmental Protection Agency and University of Utah Press, Salt Lake City, 1998, 1: 10, 000, 000 scale map.

Reyes-García, A. and Sousa, S. M., Listados florísticos de México 12, *Depresión central de Chiapas; La Selva Baja Caducifolia*, Instituto de Biología, Universidad Nacional Autónoma de México, México, 1997, 41.

Rico-Gray, V., Gómez-Pompa, A., and Chan, C., Las selvas manejadas por los mayas de Yohaltún, Campeche, México, *Biotica*, 10, 321, 1985.

Rzedowski, J., Geographical relationships of the flora of Mexican dry regions, in *Vegetation and Vegetational History of Northern Latin America*, Graham, A., Ed, Elsevier, Amsterdam, 1973, 61.

Rzedowski, J., *Vegetación de México*, Limusa, México, 1978.

Rzedowski, J., El endemismo en la flora fanerogámica de México, *Acta Bot. Mex.*, 15, 47, 1991a.

Rzedowski, J., Diversidad y orígenes de la flora fanerogámica de México, *Acta Bot. Mex.*, 14, 3, 1991b.

Salas-Morales, S.H., Saynes-Vásquez, A., and Schibli, L., Flora de la costa de Oaxaca, México; lista florística de la región de Zimatán, *Bol. Soc. Bot. Méx.*, 72, 21, 2003.

Sosa, V. et al., Lista florística y sinonimia maya, *Etnoflora Yucatanense*, INIREB, Xalapa, Veracruz, México, 1985, 1.

Sousa, S. M., *Willardia*, una nueva sección del género *Lonchocarpus* (Leguminosae), *Anales Inst. Biol. Univ. Nac. México Bot.*, 63, 147, 1992.

Standley, P.C., Las relaciones geográficas de la flora mexicana, *Anales Inst. Biol. Univ. Nac. México*, 7, 9, 1936.

Stevens, W. D. et al., Eds, *Flora de Nicaragua, Introducción, Gimnospermas y Angiospermas (Acanthaceae – Euphorbiaceae)*, Mongr. Syst. Bot. 85, Missouri Botanical Garden, St Louis, 1, 2001.

Thien, L.B., Bradburn, A.S., and Welden, A.L., The woody vegetation of Dzibilchatun, a Maya archaeological site in northwest Yucatan, Mexico, with an annotated checklist of plants by Bradburn, A.S. and Darwin, S.P., *Tulane Univ. Middle American Research Institute Occas. Pap.*, 5, 1, 1982.

Torres-Colín, R. et al., Listados florísticos de México XVI, *Flora del Distrito de Tehuantepec, Oaxaca*, Instituto de Biología, Universidad Nacional Autónoma de México, México, 1997, 68.

Trejo, I.V., *Distribución y diversidad de selvas bajas de México; relaciones con el clima y el suelo*, Tesis Doctoral, Facultad de Ciencias, UNAM, México, 1998.

Trejo, I.V., El clima de la selva baja caducifolia en México, *Investigaciones Geográficas. Bol. Inst. Geogr. UNAM*, México, 39, 40, 1999.

Trejo, I.V. and Dirzo, R., Floristic diversity of Mexican seasonally dry tropical forests, *Biodiver. Cons.*, 11, 2063, 2002.

Valiente Banuet, A., *Análisis de la vegetación de la región de Gómez Farias, Tamaulipas*, Tesis de Licenciatura, Facultad de Ciencias, UNAM, México, D. F., 1984, 92.

Van Devender, T.R. et al., Vegetation, flora and seasons of the Río Cuchujaqui, a tropical deciduous forest near Alamos, Sonora, in *The Tropical Deciduous Forest of Alamos, Sonora; Biodiversity of a Threatened Ecosystem in Mexico*, Robichaux, R.H. and Yetman, D.A., Eds, University of Arizona Press, Tucson, 2000, 3.

Villaseñor, J.L. and Ibarra M., G., La riqueza arbórea de México, *Bol. Inst. Bot., IBUG*, Epoca III (Volumen en homenaje a la profesora Luz María Villarreal de Puga), 5, 95, 1998.

Wendt, T., Composition, floristic affinities, and origins of the canopy tree flora of the Atlantic slope rain forests, in *Biological Diversity of Mexico; Origins and Distribution*, Ramamoorthy, T.P. et al., Eds, Oxford University Press, New York, 1993, 595.

White, D.A., and Darwin, S., Woody vegetation of tropical lowland deciduous forests and Mayan ruins in the north-central Yucatan peninsula, Mexico, *Tulane Stud. Zool. Bot.*, 30, 1, 1995.

White, D.A. and Hood, C.S., Vegetation patterns and environmental gradients in tropical dry forests of the northern Yucatan Peninsula, *J. Veg. Sci.*, 15, 151, 2004.

Zepeda, C. and Velázquez, E., El bosque tropical caducifolio de la vertiente sur de la Sierra de Nanchititla, Estado de México; la composición y la afinidad geográfica de su flora, *Acta Bot. Mex.*, 46, 29, 1999.

14 What Determines Dry Forest Conservation in Mesoamerica? Opportunism and Pragmatism in Mexican and Nicaraguan Protected Areas

James E. Gordon, Evan Bowen-Jones and Marco Antonio
González

CONTENTS

ABSTRACT

The status and biogeography of Mesoamerican seasonally dry tropical forests (SDTF) are briefly described. Two areas with SDTF conservation initiatives, one in Oaxaca, Mexico, the other in southern Nicaragua, are described. The degree to which scientific practices have influenced their establishment is discussed. The reasons for the selection of these areas as sites for biodiversity conservation are shown to be largely determined without recourse to scientific procedures such as biodiversity assessment or systematic reserve selection. It is argued that such opportunistic reserve selection, especially where local concerns are taken into account, is inevitable and can contribute much to the conservation of Mesoamerican SDTF.

14.1 INTRODUCTION

Seasonally dry tropical forests (SDTF) are taken here to conform to the definition of dry deciduous woodland (*bosque tropical caducifolio*) of Rzedowski (1981), that is, strongly drought deciduous, closed-canopy broadleaved tropical forest. They are considered to be amongst the world's most endangered terrestrial ecosystems (Lerdau et al., 1991; Miles et al., in press). Those of Mesoamerica, a region defined broadly here to include the countries of Central America, Panama and Mexico, are of particular concern (Janzen, 1988; Mooney et al., 1995).

Seasonally dry tropical forest has been under-studied and is under-protected compared to tropical rain forests that have received much greater attention in scientific and popular discourses. Fortunately, attention has begun to be directed towards the forests and other land uses that make up the SDTF landscapes of Mesoamerica and their diversity is now becoming better, if still incompletely, understood (Bullock et al., 1995; Gillespie, 1999; Sorensen and Fedigan, 2000; Gillespie and Walter, 2001; Balvanera et al., 2002; Trejo and Dirzo, 2002; Gordon et al., 2003, 2004; Frankie et al., 2004).

Areas subject to seasonal drought are thought to have always been preferred by agriculturalists (Murphy and Lugo, 1986) as drought suppresses pests and diseases, and enables the use of fire as a land management tool. Hence there has been a long association between the location of SDTF and areas of human occupation, resulting in widespread forest loss and fragmentation (Maass, 1995). In Mexico recent estimates suggest that only 27% of the original area of SDTF remains intact, with a similar quantity remaining in a much altered state (Trejo and Dirzo, 2000). In Central America, where the majority of the human population lives on the seasonally dry Pacific side of the country, SDTF loss has been especially acute. Central American SDTF now covers less than 2% of what is assumed to have been its original distribution and is 'totally fragmented' (Janzen, 1983). Throughout Mesoamerica it continues to be threatened by conversion for agriculture and livestock, poorly managed forest product extraction and expansion of infrastructure. Fire, in particular, is problematic in the management of Mesoamerican SDTF, as despite its apparent susceptibility, few of the tree species of this forest type appear to be adapted to fire. For example, thin bark rather than thick fire-resistant bark is the norm. This has led Janzen (2004) to suggest that all fires in north-eastern Costa Rican SDTF are anthropogenic in origin (but see also Middleton et al. [1997].). The continued association between humans and SDTF makes the conservation of this forest type a complex goal, with opportunities for the preservation and strict protection of substantial areas of undisturbed forest being few. Seasonally dry tropical forest conservation therefore faces a particular set of challenges (Quesada and Stoner, 2004) that are likely to require multiple solutions.

Despite a long history of anthropogenic disturbance, the interactions between SDTF resources and humans are not always negative. Significant areas of SDTF have survived, usually in an altered state, in Mesoamerica where agriculture was established centuries before the European conquest. This suggests that, at least in some places, humans must have utilized these resources sustainably. A rich ethnobiology has begun to be uncovered for these forests (Bye, 1995; Robichaux and Yetman, 2000; Yetman et al., 2000) and the conservation of SDTF biodiversity based on positive human–forest synergies is now being explored (Boshier et al., 2004; Sánchez-Azofeifa et al., 2005). Given extensive fragmentation, the diversity of land uses and the high human population densities associated with today's SDTF in Mesoamerica, such exploration will be crucial to the conservation of the biodiversity of this forest type.

Whilst an increased interest on the part of conservation biologists has enhanced our understanding of the importance of the region's SDTF and the magnitude of the threats it faces, reviews of the conservation initiatives that have tried to mitigate its loss are lacking. Even a simple inventory of protected areas containing SDTF in the region is not currently available, making progress difficult to assess. The task is made more difficult by the large number of political institutions that have jurisdiction over the areas that this forest type occupies, and by the different legislative frameworks within which forest protection takes place across Mesoamerica. As a result many initiatives remain largely unknown outside, and even within, their respective countries, making the comparative analysis of Mesoamerican SDTF conservation challenging. Against this background, two SDTF conservation initiatives have come to prominence in the scientific literature: the Guanacaste Conservation Area in Costa Rica and the Chamela-Cuixmala Biosphere Reserve in Jalisco, Mexico. The aim of this chapter is to review briefly the biogeography of Mesoamerican SDTF and to compare and contrast two more recently established and lesser-known SDTF conservation areas in Mesoamerica. We do not propose that these are necessarily models to be followed elsewhere, but offer them as examples of the varied but rational routes to SDTF conservation. It will be argued that because of the complexity of social and historical factors that influence Mesoamerican SDTF, scientific interests cannot be expected to play the dominant role in setting local conservation agendas in SDTF landscapes.

14.2 BIOGEOGRAPHY

A useful point of departure for interpreting Mesoamerican plant biogeography is the floristic interchange resulting from the formation of the Central American isthmus that united the continents of North and South America during the late Pliocene, approximately three million years ago. It is often assumed that until this point the biotas of these land masses had separate evolutionary histories following the break-up of Pangaea at the end of the Jurassic period, approximately 150 million years previously (Coates, 1997). The non-random nature of the floristic interchange has resulted in an unequal contribution of neotropical (South American) and boreal (North American) taxa to the various vegetation types found in modern Mesoamerica. A relatively larger contribution is made to the flora of the temperate forests of the highlands by taxa with affinities to North America than is made to lowland, tropical vegetation types. Evidence for this is found in the predominance of boreal genera such *Quercus* and *Pinus* amongst the tree diversity of the highlands of much of Mesoamerica (Rzedowski, 1981; Galindo-Leal et al., 2000) which contrasts with their near complete absence from lowland forests. The tree floras of the lowland, tropical forests of the region, which include SDTF, have usually been considered to be of principally neotropical affinity and hence have been described as northern extensions of South American floral assemblages. In particular, Mexican moist forests are relatively species poor compared to their counterparts in Central and South America from which their floras are thought to be derived (Toledo, 1982).

However, this simplistic interpretation has come under revision based on reinterpretation of palaeontological evidence and it is now suggested that a significant proportion of lowland Mesoamerica's tree diversity may have a boreotropical origin that predates the closure of the isthmus of Central America. It is proposed that interchange may have been possible during a period of close geographical proximity between the North American and Eurasian tectonic plates in the late Tertiary (Lavin and Luckow, 1993; Wendt, 1993). Taxa of the Mesoamerican SDTF whose affinities may be boreotropical include one of its most indicative and speciose genera, *Bursera*, whose closest generic relatives are now found in Africa, with Eurasia being proposed as a land bridge for its migration to North America (Weeks et al., 2005).

Speciation after the joining of North and South America is also controversial. Pleistocene climatic change, during alternate wet and dry climatic periods caused by glaciations, has been proposed as a driver of speciation in North and South American SDTF (Pennington et al., 2000). Whilst this refuge theory has been criticized (Mayle, 2004), evidence suggests that it may still be relevant to the biogeography of Mexican SDTF (Lavin and Luckow, 1993).

Based on the classification of Ceballos (1995), modern Mesoamerican SDTF can be divided into three principal phytogeographic areas:

1. The Central American forests that extend along the Pacific coast from western Guatemala to north-west Costa Rica and to which should be added the isolated SDTF on the Pacific coast of Panama;
2. The forests of the north and north-west of the Yucatán peninsula of Mexico;
3. The western Mexican dry forests that extend north from the isthmus of Tehuantepec to Sonora (beyond what is usually considered Mesoamerica) and include the Balsas river valley of Mexico.

The forests of these three phytogeographic areas share broad similarities in the composition of their woody diversity (see Lott and Atkinson, Chapter 13), the Leguminosae (particularly *Cassia, Senna, Lonchocarpus* and *Acacia*) being the dominant family by a number of species. Important contributions are also made by Bignoniaceae (particularly amongst the lianas), Euphorbiaceae (especially *Croton*), Rubiaceae and Boraginaceae (especially *Cordia*). These and many other families found in SDTF are also well represented in the wet tropical forests of the region, leading Gentry (1995) to propose that, at the familial level, SDTF floras are essentially depauperate analogues of wet forests. However, he goes on to note three important exceptions to this: Cactaceae, Capparidaceae and Zygophyllaceae, all families that are present in Mesoamerican SDTF but largely absent from wet forests. Between these three phytogeographic areas, the strongest floristic affinities appear to be between the Yucatán peninsula and Central America (Estrada-Loera, 1991; Ibarra-Manríquez et al., 2002). A suggested explanation of this is that the swamps of Tabasco, the mountains of Chiapas and the isthmus of Tehuantepec have been a barrier to floral exchange with the western Mexican SDTF. Western Mexican SDTF appears to have a greater affinity with the SDTF of northern South America than with either Central America or the Yucatán peninsula (Gentry, 1995). Gentry (1995) offers evidence that a greater proportion of genera are uniquely shared by Chamela in Mexico and SDTF in Venezuela and Colombia than are shared by Chamela and the forests of Guanacaste, Costa Rica. Lott and Atkinson (Chapter 13) discuss the floristic affinities of western Mexican dry forests in more detail.

In contrast to their relatively depauperate floras compared to tropical wet forests, Mesoamerican SDTF is relatively rich in endemics and this is especially so in western Mexico (Gentry, 1995). Indeed endemism in Mexico appears to increase along a gradient from mesic to xeric habitat types (Challenger, 1998) with desert habitats being, proportionately, the richest of all. Allopatric speciation driven by fragmentation and isolation during the Pleistocene climatic changes has been proposed as a possible reason for high SDTF endemism (Gentry, 1982). This isolation may have been exacerbated in the forests of western Mexico that are contiguous with the floristically very distinct temperate pine and oak forests and hence isolated from the tropical wet forests from which their flora is thought to be derived (Ceballos, 1995). Whatever the reason this high endemism makes the conservation of these SDTF a high priority for international conservation.

It should be noted that the division of Mesoamerican SDTF into three blocks obscures smaller, isolated areas of SDTF on Mexico's gulf coast in Tamaulipas and Veracruz and in interior valleys of Mexico, Guatemala and Honduras. The latter have largely been overlooked (but see Trejo and Dirzo, 2002), perhaps because such valleys have been preferred sites of human settlement since pre-Colombian times and little forest remains. However, given their isolation, it might be predicted that high levels of endemism would be found in the remaining fragments (Gentry, 1992) and hence these forests might deserve particular conservation attention.

Dinerstein et al. (1995) used expert opinion to broadly define and assess habitat types, called ecoregions, across Latin America. Their assessment of conservation priorities categorized the Central American Pacific dry forest ecoregion as 'bioregionally outstanding' and 'critically endangered', whilst the SDTF of Yucatán was categorized as 'locally important and endangered'. The same assessment divided the western Mexican SDTF into the Jalisco dry forest and the Balsas dry forest and described both as 'regionally outstanding', 'endangered' and 'of highest priority at a regional scale'. There is therefore little doubt that the conservation of this forest type is a priority.

It can therefore be concluded that phytogeographers and conservation biologists have together provided a reasonable, although as yet incomplete, understanding of patterns of Mesoamerican SDTF floristic diversity and of associated conservation priorities between regions. Below we consider local SDTF conservation initiatives within Mesoamerica.

14.3 CONSERVATION

14.3.1 ESTABLISHED SDTF INITIATIVES

The Guanacaste Conservation Area and the Chamela-Cuixmala Biosphere Reserve are well established and make important contributions to SDTF conservation. There are, however, reasons to believe that these two established reserves cannot be taken to be models for SDTF conservation throughout Mesoamerica. Firstly, the removal of human populations that would be required in order to create similar large protected areas is not likely to prove practical elsewhere. Secondly, science and scientists have been, to an important degree, central to the generation and modification of the social capital needed to maintain these protected areas. This has been possible because of the influence scientists have, in these locations and elsewhere, over popular and political discourses (Takacs, 1996), over the creation and functioning of institutions (Raustiala, 1997) and their ability to attract funding. Given that the resources available to scientists are limited, it is unlikely that Mesoamerican SDTF conservation would advance much beyond these areas if other means of generating social capital could not be found. It is, therefore, a welcome development that other conservation initiatives for this forest type have arisen that have been less directly reliant on scientists and their need to find sites at which to practice biology and conservation.

Despite the influence of scientists on the establishment and management of the Guanacaste Conservation Area and the Chamela-Cuixmala Biosphere Reserve, neither can be said to have been chosen as a result of the comparative assessment of the biodiversity of these and other possible reserve locations. Amongst the initial reasons for the establishment of these protected areas were the protection of a historic monument in Guanacaste and the donation of land by a benefactor in Jalisco. Since their establishment, much interest has been stimulated in how reserve sites are selected in response to the widely held perception that reserves are declared with little regard to their biodiversity content, relative to other actual or potential reserve sites, leading to a sub-optimal allocation of the resources available for biodiversity conservation (Margules et al., 1988; Margules and Pressey, 2000). Such reserve selection is referred to as opportunistic or ad hoc, in contrast to a more scientifically driven systematic or rational selection based on the comparative assessment of forests and their constituent species (Pressey, 1994; Sutherland et al., 2004).

14.3.2 SDTF CONSERVATION IN OAXACA, MEXICO

The state of Oaxaca on the Pacific coast of southern Mexico is of global importance as a biodiversity hotspot. The state's mountainous topography and various climatic regimes have given rise to a remarkable diversity of habitats, ranging from high-altitude semi-desert matorral to lowland tropical rain forest. It includes an endemic bird area (ICBP, 1992) and is part of the Mesoamerican biodiversity hotspot (Myers et al., 2000). Because of its internationally recognized importance, the area has attracted significant national and international funding for biodiversity conservation as well as interest from both local and international NGOs. It contains parts of several priority

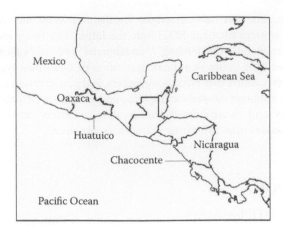

FIGURE 14.1 The locations of Huatulco and Chacocente.

ecoregions including the Balsas dry forest, which extends east to the coast of Oaxaca as far as the municipality of Santa María Huatulco and the tourist resort of Bahías de Huatulco. Together this municipality and tourist resort are here referred to as Huatulco (Figure 14.1). The SDTF of Huatulco is therefore part of the western Mexican SDTF which is described by Ceballos et al. (1998) as being amongst the most important for mammalian diversity conservation in Mexico. Restricted range tree and shrub species from these forests include *Achatocarpus oaxacanus* Standl., *Bursera instabilis* McVaugh & Rzed., *Caesalpinia coccinea* G.P.Lewis & J.L.Contr., *Gyrocarpus mocinnoi* Espejo, *Hibiscus kochii* Fryxell and *Jatropha alamani* Müll.Arg.

The varied landscape of the coastal lowlands of Huatulco is composed of both communally and privately held agricultural land. Some intensive agriculture is practiced but basic grain production interspersed with fallows that are left long enough to allow forest regeneration is the norm. Forest cover is therefore fragmented by agriculture and composed of a mixture of intervened forest and regenerating secondary forest of varying stages. The floristic diversity of the SDTF in the coastal region of Oaxaca has recently been surveyed by Salas-Morales et al. (2003) but no systematic faunal surveys have yet been published. In the SDTF around Huatulco two separate conservation initiatives relevant to SDTF have emerged, both of which are discussed here.

14.3.2.1 Huatulco National Park

Huatulco National Park was created by federal decree in 1998 after the federal trust for tourism development, FONATUR, had expropriated a stretch of coast to develop the Bahías de Huatulco tourist resort. The park was created in order to exert control over the use of the coastline in response to concerns raised by environmentalists about the rapid expansion of the tourist infrastructure. Six thousand hectares of the land expropriated by FONATUR and a similar area of contiguous marine habitat thus came to form Huatulco National Park which remains property of the trust.

The terrestrial area of the park is primarily SDTF, in various successional states, interspersed with occasional small areas of savanna and with dunes and mangroves surrounding the rocky bays. Prior to expropriation the land had been under control of local farmers, with much of the area held in communal tenure. Communities were compensated for loss of expropriated land and relocated further inland where they colonized other SDTF areas. One community refused to leave the newly designated park and continues to live partially within its official boundaries. The rest of the terrestrial area remains well protected, and extractive use is not permitted within the reserve.

As a national park, it is under the jurisdiction of the National Commission for Protected Natural Areas (CONANP). This is a federal institution that responds to national rather than local agendas, and which has contracted the majority of the park's senior staff in Huatulco from outside of the state of Oaxaca.

The influence of biological information on directing the establishment and management of this protected area has, to date, been limited. Awareness of the need for SDTF conservation in Mexico has undoubtedly grown in recent years, not least because of scientific interest (e.g. Murphy and Lugo, 1995; Ceballos et al., 1998; Trejo and Dirzo, 2002); however, the SDTF area of Huatulco National Park was not selected using the principles of comparative or systematic reserve selection (Pressey et al., 1993). Nor, with the limited resources available to the park's officials, has there been the opportunity for anything other than the most basic management. A systematic floristic inventory, which is perhaps a baseline requirement for monitoring, has only recently been initiated and is to be carried out by a local NGO, SERBO A.C. (A. Saynes, pers. comm.).

There is local representation in the park's steering committee as it is presided over by the municipality's commissary of communal resources. However, a sense of local involvement in park affairs is not prevalent. This is not surprising given that the establishment of the park was through land expropriation from local people, and that the legal and administrative frameworks of the park are set in Mexico City. This is compounded by the demands made on park staff by the management of tourism, which at times takes inevitable precedence over the needs of locals. Nonetheless, a stable and relatively well-funded institution has been created which continues to protect a small but significant area of SDTF.

14.3.2.2 Community Protected Areas in Huatulco

Within the forest-farm matrix that comprises Huatulco's landscape outside Huatulco National Park, 13 areas of forest in a variety of successional states have been set aside as natural reserves by the local assembly of Huatulco's agrarian communities. The 13 areas are not contiguous with the park and are institutionally separate. With the aid of a locally based NGO, *Grupo Autonomo para la Investigación Ambiental A.C.* (GAIA) and funding from various international institutions including UNDP and the World Wide Fund for Nature (UK), the reserves have been designated as the Communal System of Protected Areas (SCAP) to be maintained under forest cover indefinitely. The SCAP was initially recognized as protected by the communal and municipal authorities. Recently, and following intensive lobbying by the community and GAIA, it has received formal recognition as a communally protected area by CONANP. This gives a nationally recognized legal basis to the SCAP and consequently increases funding opportunities from Mexican institutions but without loss of legal or administrative control by the local community.

The majority of the SCAP's reserves are largely composed of SDTF but two that are at slightly higher altitudes in the north-west of the municipality contain predominantly semideciduous forest (*bosque tropical semicaducifolio*) (Rzedowski, 1981). The sites fall within an area of approximately 250 km^2 with altitudes varying between 20 m and 1020 m above sea level. In total they occupy approximately 10,000 ha, the largest containing 4780 ha, the smallest 15 ha.

The intention is for the SCAP to be managed by and for the local communities that have created it and own it. The implication of this is that various uses will be permitted whilst forest cover and diversity is maintained. The precise reasons for the designation of each area vary but include watershed protection, soil conservation and the potential for future sustainable extractive management of construction timber for local use. Non-use values, such as the aesthetic and even the ethical value of maintaining and conserving a diverse landscape were also important influences on local people's thinking in relation to establishment of the SCAP. Watershed protection, in particular, is currently of much interest in the area as the Bahías de Huatulco resort is highly water demanding and mechanisms for private-sector payment for this ecosystem service are a potential revenue source for local communities. The formal recognition of these reserves may also have served to strengthen communal and individual tenure over land held under forest cover in an area with a recent history of land expropriation. What the motivations for reserve designation have in common is that they reflect the concerns and needs of local rural populations rather than those of conservation biologists.

These two conservation initiatives, the Huatulco National Park and the SCAP, whilst geographically close and containing essentially similar SDTF, are very distinct in their histories, aims and approaches. Perhaps the most important distinction is that the SCAP aims to respond to a more local agenda, that of improving rural livelihoods through forest conservation, whilst Huatulco National Park seeks to address a global concern for the loss of SDTF biodiversity. What they have in common is their relative detachment from the tenets of conservation biology. Certainly initial interest in conservation in the region can be claimed to have been at least partially stimulated by broad-scale conservation assessments such as those of the WWF's Global 200 (Dinerstein et al., 1995) which have consequently attracted funding to the region. However, the precise location of the protected areas owes nothing to detailed biodiversity assessment or systematic reserve selection and everything to the prevailing socio-economic conditions and the opportunities that these have provided. These initiatives can therefore be considered ad hoc by the definition of Pressey (1994). Furthermore, and in contrast to the Guanacaste Conservation Area and Chamela-Cuixmala Biosphere Reserve, managers of these areas have felt little need, or had little opportunity, to call upon conservation biologists to aid with planning and management. The control of beach-orientated tourism in the Huatulco National Park and the promotion of rural development around the SCAP have proved to be far more pressing concerns. Despite this, it is likely that both initiatives will continue to contribute significantly to the conservation of globally important biodiversity in the region (Gordon et al., 2004).

14.3.3 NICARAGUAN DRY FOREST AS A CONSERVATION PRIORITY

Mesoamerica contains 7% of the world's biodiversity in 0.5% of the world's land area. Within Central America, Nicaragua has particular biological importance, being the largest Central American country, and after Belize, the nation with the lowest population density (UNEP, 1996). There is more forest left in Nicaragua than any other Mesoamerican country, after Mexico, and although much of this is evergreen rain forest on the Caribbean side of the country, it also has some of the largest remaining areas of SDTF in the region. Despite its size, Nicaragua has a slightly lower recorded biological diversity than neighbouring countries, perhaps because of its lower altitudinal variation.

As a response to the multiple threats facing Mesoamerican biodiversity, the Mesoamerican biological corridor (MBC) was proposed to link natural habitats in a contiguous landscape unit from southern Mexico to northern Colombia. The MBC is now the largest multilateral conservation programme in the world. In Nicaragua the government identified the Pacific SDTF as an element of this landscape conservation initiative, but because of limited resources the area was removed from the implementation phase of the programme. The Nicaraguan element is now concentrated on the tropical lowland forest of the Atlantic coast. Despite this setback, the Government of Nicaragua did view SDTF as a conservation priority and the Rio Escalante-Chacocente Wildlife Refuge, referred to here as Chacocente, was gazetted in 1983, and is now considered an integral part of Nicaragua's conservation strategy. The refuge contains 14,800 ha (4800 ha core and 10,000 ha buffer zone) of SDTF habitat, not all of which is forested, and 12 km of coastal habitat.

14.3.3.1 The Biological Importance of Chacocente

Chacocente was not established as a result of assessment of global or regional SDTF biodiversity priorities. It was established because it contains one of Mesoamerica's most important *arribada* (mass nesting) beaches for the olive Ridley turtle (*Lepidochelys olivacea* Eschscholtz), a species which is listed as endangered by IUCN. This is one of only seven such beaches known globally for this species. The critically endangered hawksbill turtle (*Eretmochelys imbricate* L.) and leatherback turtle (*Dermochelys coriacea* Vandelli) have also been known to nest on this beach.

Twenty years after its establishment, biodiversity baseline data are far from complete for Chacocente as this area has received relatively little research attention. However, Gillespie et al. (2000) carried out a comparative assessment of woody plant diversity in multiple SDTF sites in Nicaragua and Costa Rica and determined that Chacocente should be given the highest conservation priority amongst Nicaraguan SDTF sites. The reserve is known to contain several tree species of international conservation concern including: *Bombacopsis quinata* (Jacq.) Dugand, *Cedrela odorata* L., *Swietenia humilis* Zucc., *Guaiacum sanctum* L., *Lonchocarpus phlebophyllus* Standl. & Steyerm. and *Platymiscium pleiostachyum* Donn. Sm. (Otterstrom, 2003). Additionally, Gillespie and Walter (2001) compared bird species richness across SDTF reserves in Central America and found that Chacocente contained the highest recorded number of restricted range dry forest birds (18) amongst the seven SDTF sites in Nicaragua and Costa Rica that they sampled. Their assessment included the much larger and better-preserved Santa Rosa National Park in the Guanacaste Conservation Area.

14.3.3.2 The Human Dimensions of the Conservation of Chacocente Wildlife Refuge

Nationally 90% of the natural areas protected by public legislation in Nicaragua are on privately owned land. In Chacocente this rises to nearly 100% with many of the individual owners also living within the boundaries of the refuge's core. Over 60% of the reserve has now been acquired by a group of international investors, reputedly as a speculative tourism investment, whilst the remaining land is owned by local people who are dependent on small-scale agriculture and the extraction of forest and coastal resources (firewood, game and turtle eggs). The mean annual income of residents in the area is one of the lowest in Nicaragua at less than US$150 and ecosystem services are crucial to these communities in what is one of the poorest countries in the Americas.

Nicaragua's Sandinista government began a programme of agricultural reform in the 1980s that involved the formation of land cooperatives, many of which still exist in the environs of Chacocente. War veterans were settled in at least one community in the buffer zone although the majority of Chacocente's residents are not recent immigrants (Otterstrom, 2003) and were born in or around Chacocente. In general, the buffer zone communities have been established for several generations and therefore predate the formal establishment of the refuge. As well as the nearly 500 local people living in the refuge, there are approximately 4000 people living on its periphery in three village land cooperatives whose incomes depend on agriculture, forest and coastal resources and wages from seasonal labour in nearby Costa Rica. Thus, these 4500 people and the group of international investors will in one way or another determine the fate of the reserve.

14.3.3.3 Current Management of the Reserve

Given that the appropriate use of Chacocente's natural resources, and particularly the harvesting of turtle eggs, is crucial to the future of the refuge and its species, it is perhaps strange that the original management objectives of the reserve gave little consideration to human activities. Nicaragua's Ministry of Natural Resources (MARENA) whose legal mandate is to facilitate the incorporation of civil society into the management of protected areas, defines the management objectives of wildlife refuges as:

- to conserve the habitats, flora and fauna of national and/or international interest;
- to improve understanding of the main activities associated with sustainable resource use through scientific investigation and monitoring of the species in the area;

- to delimit areas for educational ends and permit the public to appreciate the characteristics of the habitats being protected and wildlife management activities;
- to conserve and manage habitats for the protection of resident or migratory species of national, regional or world interest.

However, none of these management objectives specifically mention involvement of the people that live within the refuge's boundaries, or within its buffer zone. Neither the private sector, represented by the international investors who control part of the park, nor the communities that have been there for generations have had significant involvement in its management to date. Indeed, there was little effective dissemination of information about the purpose of the reserve when it was gazetted, and there has been an ongoing lack of clarity over demarcation of the core and buffer zone areas that has led to problems of agricultural encroachment.

14.3.3.4 Ways Forward for Chacocente

MARENA is working with Fauna and Flora International and local NGOs under a recently secured UNDP global environment facility medium-sized project (MSP) on a different approach to co-management for Chacocente. The project, which is due to run for four years starting from 2005, will attempt to rectify the lack of models that encourage formalized private sector and community collaboration on management of Nicaraguan reserves. One of its primary objectives is to establish a multi-stakeholder management entity for the reserve, which up until now has only benefited from a skeleton staff of reserve guards. This will assist MARENA in implementing its strategy for the national protected areas system (SINAP) which promotes efficient decentralized management of protected areas, administered through co-management agreements with third parties and supervised by MARENA.

However, the MSP preparation process identified little awareness amongst community members of the purpose and approach of conservation in Chacocente and consequently they hold a 'neutral-to-negative' impression of the refuge. Not only is there a basic lack of awareness about the reserve itself, but crucially there is also a lack of understanding of the link between environmental services and the maintenance of incomes. This is as true of large landowners as it is of smallholders. A complex set of factors that contribute to problems of over-hunting, uncontrolled firewood collection, logging, turtle egg collection and encroachment was identified. The root causes were identified as poverty and the lack of biodiversity-friendly livelihood options that result from insufficient awareness of the important role biodiversity plays in natural resource management.

These issues must be addressed by including rather than excluding people from the decision-making processes if there is to be a reasonable chance that the conservation of this important area can be achieved. If management ignores the people who live within and around the reserve it will eventually be undermined by them. A first positive step has been made through co-management legislation that allows revenues generated in a protected area to be reinvested in conservation and social development in the same protected area (Barany et al., 2002).

14.4 DISCUSSION

In Huatulco and Chacocente forest conservation is evolving under very different circumstances. One of the most striking differences between the two areas is the degree to which local communities are dependent on the direct extraction of forest products from protected areas.

The landscape of Huatulco is one in which there remain large areas of forest, although much of it secondary in nature, outside both the national park and the SCAP. Thus the prevention of forest product extraction from the park and the sustainable management of the SCAP and other forested areas are made more easily achievable objectives. In the long term the successful conservation of SDTF in Huatulco will depend upon the effects of increasing tourism in the Bahías de

Huatulco. Outside of the immediate area claimed by FONATUR, the influence of tourism on forests is likely to be indirect, but not necessarily insignificant. The tourism industry provides employment that, for many, is an attractive alternative to agriculture. This may reduce the pressure of agriculture expansion on SDTF and allow forest fallow periods to be maintained and even enhanced.

However, outside the national park, traditional and communal management practices continue to contribute to the maintenance of forest cover. Should the importance of traditional land tenure be reduced by changing economic circumstances, a particular concern where long-term speculative investment in land for tourism development is concerned, new forms of management may take their place with unpredictable consequences for biodiversity. Already there has been a small expansion of intensive papaya production in Huatulco partially stimulated by the demands of the tourism industry. Alongside growing interest in charging for water management services as a way of capturing tourist revenues for local communities, there is also interest in the potential of ecotourism to contribute to rural incomes. If managed correctly there may be much positive benefit in these initiatives for the conservation of the remaining SDTF, provided revenue derived from forest conservation accrues to the individuals and communities that bear the cost of conservation and the traditional management practices that have maintained Huatulco's biodiversity are not undermined.

The very different institutional frameworks within which the SCAP and Huatulco National Park are embedded has the advantage of ensuring the continued functioning of one should the other be adversely affected by institutional change. However, it is also likely to be a hindrance to the integrated management of these forests. By international standards these protected forest areas are small and unlikely to be able to sustain all the species currently found at each site in the long term (Nunney and Campbell, 1993). More integrated management, with the identification of fragmented populations of threatened species across these various patches of forest, might provide a more certain future for many of these species. As already noted, the SCAP also contains semideciduous forest and it is possible that this may complement the deciduous forest of the park by providing habitat for drought-intolerant species that require more than one habitat type to complete their lifecycles. From a biological perspective there may be much to gain from more 'joined-up' management of the SCAP and the national park.

However, crucial to the long-term prospects for both protected areas in Huatulco will undoubtedly be the matrix of farm and forest which make up the wider landscape in which they are embedded. This landscape of small settlements, secondary forest successions and farmland with scattered trees blurs the distinction between forest and non-forest and probably provides suitable habitat for a great many of the species of the region (Boshier et al. 2004; Gordon et al., 2004). A recently implemented project funded by the Ford Foundation is attempting to promote and diversify sustainable land use management in Huatulco outside the protected areas. The extent to which this might directly aid biodiversity within the SCAP is yet to be determined.

Chacocente, by contrast, is under far more immediate pressure from local communities. Here, there is also considerable potential for tourism to provide a source of revenue, both through trekking in the forest, and expedition-style tourism based on turtle monitoring. The rational exploitation of turtle eggs must also be a given serious consideration. At Ostional in northern Guanacaste turtle eggs are harvested for the local communities under a sustainable management regime, and hatching rates have been maintained at what are considered optimal levels, avoiding the hatching declines seen in unharvested sites on the same coast (D. Chacon, ANAI, Costa Rica, pers. comm.). A mix of land uses is likely to be required if the refuge is to be successfully managed. The identification of areas that harbour key species and the prescription of suitable extraction limits for their management may be a key scientific contribution that can be made in Chacocente. This also implies a more sophisticated approach than the current dichotomy between core and periphery of the reserve.

The motivation for conserving SDTF in both Huatulco and Chacocente was never simply the maximization of conserved biodiversity, and consequently the maintenance of species diversity cannot be disassociated from any number of other environmental services which SDTF can perform.

In the case of Chacocente, the scientific rationale of maintaining connectivity between reserves (Bennett, 1999) that underpins the Mesoamerican biological corridor proved not to be politically convincing and instead the high value attached to charismatic species, turtles, was crucial to its establishment. The comparative assessment of its plant and bird diversity (Gillespie, 2000; Gillespie et al., 2000) was not attempted until after decisions concerning gazetting of the reserve had been made, less still were various possible sites subjected to any systematic reserve selection procedure.

The establishment of SDTF reserves in Huatulco and Chacocente within their broadly defined ecoregions has been opportunistic with respect to SDTF diversity. Whilst the beach, through its attractiveness to turtles, was crucial to the establishment of Chacocente, it was also important to SDTF conservation in Huatulco National Park because of its attractiveness to tourists. In both cases, the conservation of non-SDTF habitat provided opportunities for protection to be extended to SDTF. Huatulco's SCAP was given formal recognition not because the forest therein had been assessed as the most biodiverse in the area but because the individuals and communities with tenure over them decided these sites were amongst those that were least appropriate for conversion to agricultural use in the immediate future.

Despite this there is little doubt that these initiatives have great potential to contribute to the conservation of Mesoamerican SDTF diversity. Plant and bird assessments demonstrate this for Chacocente whilst for Huatulco the assessment of restricted range tree and shrub species of Gordon et al. (2004) showed that two forested areas protected by the SCAP were amongst the most important in coastal Oaxaca. However, these assessments took place after the establishment of the respective reserves; just as inventories of dry forest floras in Guanacaste and the Chamela biological station (the precursor of the Chamela-Cuixmala biosphere reserve) followed establishment of the first protected areas in these localities (Janzen and Liesner, 1980; Lott, 1993). It therefore appears that for this forest type, reserve location often determines where biological assessments should take place instead of vice versa. Rather than interpret this phenomenon as a flaw in regional SDTF conservation planning (Margules and Pressey, 2000; Pressey, 1994; Pressey et al., 1993) it is perhaps much better reformulated as an inevitable consequence of SDTF conservation being carried out within the broader reality of the needs and opportunities presented by social, political and economic processes.

Indeed, it is highly unlikely that prescriptive conservation planning based uniquely on biodiversity assessment and systematic reserve selection would be useful given competing demands on forest resources in Mesoamerican SDTF. More realistically it is suggested here that comparative assessments within ecoregions, or similar broad-scale assessments, be used to refine existing reserve networks and their management by identifying important elements of biological diversity, particularly those that cannot realistically be conserved in protected areas and thus require alternative solutions. Whilst priorities *between* ecoregions can be meaningfully set based primarily on biodiversity and threat assessment, the location of conservation initiatives *within* priority ecoregions in densely populated, underdeveloped areas will continue to be decided by predominantly non-biological criteria, as was the case for the sites discussed here. Identification of sites with the high levels of the social capital needed to sustain a protected area, whether generated by scientists or not, is no less important than the identification of the most outstanding sites of biodiversity. We contend that more is achieved by successfully conserving a site of relatively lower biodiversity importance than by failing to conserve one of relatively higher importance.

In these circumstances socio-economically determined reserve selection might better be described as pragmatic or realistic rather than ad hoc or opportunistic. A positive consequence of this reformulation might also be the reinterpretation of socio-economic conditions as opportunities rather than constraints. Assessing what *can* be conserved is at least as important as deciding what *should* be conserved. If this is done in a manner that builds on social and cultural linkages to biodiversity, it is likely to further the cause of conserving globally significant biodiversity in the human dominated landscapes of the Mesoamerica SDTF.

ACKNOWLEDGEMENTS

During field work that led to this work JEG was support by the UK government's Economic and Social Research Council and Natural Environment Research Council (R42200134147) and by the European Commission as part of the BIOCORES Project (PL ICA4-2000-10029). EBJ would like to thank colleagues at FFI, the communities, organizations and individuals who have been involved in the long planning process for the Chacocente GEF/UDNP MSP project. MAG was supported by the Ford Foundation. We thank, A. Entwistle, E. Lott, T. Pennington and an anonymous reviewer for useful comments on earlier drafts of the chapter.

REFERENCES

Balvanera, P. et al., Patterns of beta-diversity in a Mexican tropical dry forest, *J. Veg. Sci.*, 13, 145, 2002.

Barany, M.E. et al., Resource use and management of selected Nicaraguan protected areas: a case study from the Pacific region, *Nat. Area. J.*, 22, 61, 2002.

Bennett, A.F., *Linkages in the landscape: the role of corridors and connectivity in wildlife conservation*, IUCN, Gland, Switzerland and Cambridge, 1999.

Boshier, D.H., Gordon, J.E., and Barrance, A.J., Prospects for *circa situm* tree conservation in Mesoamerican dry forest agro-ecosystems, in *Biodiversity Conservation in Costa Rica, Learning the Lessons in a Seasonal Dry Forest,* Frankie, G.W., Mata, A., and Vinson, S.B., Eds, University of California Press, Berkeley, 2004, 210.

Bullock, S.H., Mooney, H.A., and Medina, E., Eds, *Seasonally Dry Tropical Forest*, Cambridge University Press, Cambridge, 1995.

Bye, R., Ethnobotany of the Mexican tropical dry forest, in *Seasonally Dry Tropical Forests,* Bullock, S.H., Mooney, H.A., and Medina, E., Eds, Cambridge University Press, Cambridge, 1995, 423.

Ceballos, G., Vertebrate diversity, ecology, and conservation in neotropical dry forests, in *Seasonally Dry Tropical Forests,* Bullock, S.H., Mooney, H.A., and Medina, E., Eds, Cambridge University Press, Cambridge, 1995, 195.

Ceballos, G., Rodríguez, P., and Medellín, R.A., Assessing conservation priorities in megadiverse Mexico: mammalian diversity, endemicity and endangerment, *Ecol. Appl.*, 8, 8, 1998.

Challenger, A., *Utilización y conservación de los ecosistemas terrestres de México: pasado, presente y futuro,* CONABIO, Mexico City, 1998.

Coates, A.G., Ed, *Central America: a Natural and Cultural History,* Yale University Press, New Haven and London, 1997.

Dinerstein, E. et al., *A conservation assessment of the terrestrial ecoregions of Latin America and the Caribbean,* World Bank/WWF, Washington DC, 1995.

Estrada-Loera, E., Phytogeographic relationships of the Yucatán Peninsula, *J. Biogeogr.*, 18, 687, 1991.

Frankie, G.W., Mata, A., and Vinson, S.B., *Biodiversity Conservation in Costa Rica, Learning the Lessons in a Seasonal Dry Forest,* University of California Press, Berkeley, 2004.

Galindo-Leal, C., Fay, J.P., and Sandler, B., Conservation priorities in the greater Calakmul region, Mexico: correcting the consequences of a congenital illness, *Nat. Area. J.*, 20, 376, 2000.

Gentry, A.H., Neotropical floristic diversity: phytogeographical connections between Central and South America, Pleistocene climatic fluctuations, or an accident of the Andean orogeny? *Ann. Missouri Bot. Gard.*, 69, 557, 1982.

Gentry, A.H., Tropical forest biodiversity: distributional patterns and their conservational significance, *Oikos*, 63, 19, 1992.

Gentry, A.H., Diversity and floristic composition of neotropical dry forests, in *Seasonally Dry Tropical Forests,* Bullock, S.H., Mooney, H.A., and Medina, E., Eds, Cambridge University Press, Cambridge, 1995, 146.

Gillespie, T.W., Life history characteristics and rarity of woody plants in tropical dry forest fragments of Central America, *J. Trop. Ecol.*, 15, 637, 1999.

Gillespie, T.W., Rarity and conservation of forest birds in the tropical dry forest region of Central America, *Biol. Conserv.*, 96, 161, 2000.

Gillespie, T.W., Grijalva, A., and Farris, C.N., Diversity, composition, and structure of tropical dry forests in Central America, *Plant Ecol.*, 147, 37, 2000.

Gillespie, T.W. and Walter, H., Distribution of bird species richness at a regional scale in tropical dry forest of Central America, *J. Biogeogr.*, 28, 651, 2001.

Gordon, J.E. et al., Trees and farming in the dry zone of southern Honduras II: the potential for tree diversity conservation, *Agroforest. Syst.*, 59, 107, 2003.

Gordon, J.E. et al., Assessing landscapes: a case study of tree and shrub diversity in the seasonally dry tropical forests of Oaxaca, Mexico and southern Honduras, *Biol. Conserv.*, 117, 429, 2004.

Ibarra-Manríquez, G. et al., Biogeographical analysis of the tree flora of the Yucatán Peninsula, *J. Biogeogr.*, 29, 17, 2002.

ICBP, *Putting Biodiversity on the Map: Priority Areas for Global Conservation,* International Council for Bird Preservation, Cambridge, 1992.

Janzen, D.H., *Costa Rican Natural History,* University of Chicago Press, Chicago, 1983.

Janzen, D.H., Tropical dry forests: the most endangered major tropical ecosystems, in *Biodiversity,* Wilson E.O., Ed, National Academy Press, Washington DC, 1988, 130.

Janzen, D.H., Ecology of dry-forest wildland insects in the Area de Conservación Guanacaste, in *Biodiversity Conservation in Costa Rica, Learning the Lessons in a Seasonal Dry Forest,* Frankie, G.W., Mata, A., and Vinson, S.B., Eds, University of California Press, Berkeley, 2004, 80.

Janzen, D.H. and Liesner, R., Annotated checklist of plants of lowland Guanacaste Province, Costa Rica, exclusive of grasses and non-vascular cryptograms, *Brenesia*, 18, 15, 1980.

Lavin, M. and Luckow, M., Origins and relationships of tropical North America in the context of the Boreotropics hypothesis, *Am. J. Bot.*, 80, 1, 1993.

Lerdau, M., Whitbeck, J., and Holbrook, N.M., Tropical deciduous forest: death of a biome, *Trends Ecol. Evol.*, 6, 201, 1991.

Lott, E.J., Annotated checklist of the vascular flora of the Chamela Bay region, Jalisco, Mexico, *Occas. Pap. Calif. Acad. Sci.*, 148, 1, 1993.

Maass, J.M., Conversion of tropical dry forest to pasture and agriculture, in *Seasonally Dry Tropical Forests,* Bullock, S.H., Mooney, H.A., and Medina, E., Eds, Cambridge University Press, Cambridge, 1995, 399.

Margules, C.R., Nicholls, A.O., and Pressey, R.L., Selecting networks of reserves to maximize biological diversity, *Biol. Conserv.*, 43, 63, 1988.

Margules, C.R. and Pressey, R.L., Systematic conservation planning, *Nature*, 405, 243, 2000.

Mayle, F.E., Assessment of the Neotropical dry forest refugia hypothesis in the light of palaeoecological data and vegetation model simulations, *J. Quaternary Sci.*, 19, 713, 2004.

Middleton, B.A. et al., Fire in a tropical dry forest of Central America: a natural part of the disturbance regime? *Biotropica*, 29, 515, 1997.

Miles, L. et al., A global overview of the conservation status of tropical dry forests, *J. Biogeogr.,* in press.

Mooney, H.A., Bullock, S.H., and Medina, E., Introduction, in *Seasonally Dry Tropical Forests,* Bullock, S.H., Mooney, H.A., and Medina, E., Eds, Cambridge University Press, Cambridge, 1995, 1.

Murphy, P.G. and Lugo, A.E., Ecology of tropical dry forest, *Annu. Rev. Ecol. Syst.*, 17, 67, 1986.

Murphy, P.G. and Lugo, A.E., Dry forests of Central America and the Caribbean, in *Seasonally Dry Tropical Forests,* Bullock, S.H., Mooney, H.A., and Medina, E., Eds, Cambridge University Press, Cambridge, 1995, 9.

Myers, N. et al., Biodiversity hotspots for conservation priorities, *Nature*, 403, 853, 2000.

Nunney, L. and Campbell, K.A., Assessing minimum viable population size: demography meets population genetics, *Trends Ecol. Evol.*, 8, 234, 1993.

Otterstrom, S., *Conservation of dry forest and coastal biodiversity of the Pacific Coast of the southern Nicaragua: building private and public partnerships*, UNDP/GEF/Fauna and Flora International/MARENA/DED/PSO/GTZ, 2003.

Pennington, R.T., Prado, D.A., and Pendry, C.A., Neotropical seasonally dry tropical forest and Pleistocene vegetation changes, *J. Biogeogr.*, 27, 261, 2000.

Pressey, R.L., *Ad hoc* reservations: forward or backward steps in developing representative reserve systems? *Conserv. Biol.*, 8, 662, 1994.

Pressey, R.L. et al., Beyond opportunism: key principles for systematic reserve selection, *Trends Ecol. Evol.*, 8, 124, 1993.

Quesada, M. and Stoner, K.E., Threats to the conservation of tropical dry forest in Costa Rica, in *Biodiversity Conservation in Costa Rica, Learning the Lessons in a Seasonal Dry Forest,* Frankie, G.W., Mata, A., and Vinson, S.B., Eds, University of California Press, Berkeley, 2004, 266.

Raustiala, K., Domestic institutions and international regulatory cooperation: comparative responses to the Convention on Biological Diversity, *World Polit.,* 49, 482, 1997.

Robichaux, R.H. and Yetman, D.A., Eds, *The Tropical Deciduous Forest of Los Alamos: Biodiversity of a Threatened Ecosystem in Mexico,* University of Arizona Press, Tucson, 2000.

Rzedowski, J., *Vegetación de México,* Editorial Limusa, México D.F., 1981.

Salas-Morales, S.H., Saynes-Váquez, A., and Schibli, L., Flora de la costa de Oaxaca: lista florística de la región de Zimatán, *Bol. Soc. Bot. México,* 72, 21, 2003.

Sánchez-Azofeifa, G.A. et al., Need for integrated research for a sustainable future in tropical dry forest, *Conserv. Biol.,* 19, 285, 2005.

Sorensen, T.C. and Fedigan, L.M., Distribution of three monkey species along a gradient of regenerating tropical dry forest, *Biol. Conserv.,* 92, 227, 2000.

Sutherland, W.J. et al., The need for evidence-based conservation, *Trends Ecol. Evol.,* 19, 305, 2004.

Takacs, D., *The Idea of Biodiversity: Philosophies of Paradise,* Johns Hopkins University Press, Baltimore and London, 1996.

Toledo, N.V., Pleistocene changes of vegetation in tropical Mexico, in *Biological Diversification in the Tropics,* Prance, G.T., Ed, Columbia University Press, New York, 1982, 93.

Trejo, I. and Dirzo, R., Deforestation of seasonally dry tropical forest: a national and local analysis in Mexico, *Biol. Conserv.,* 94, 133, 2000.

Trejo, I. and Dirzo, R., Floristic diversity of Mexican seasonally dry tropical forests, *Biodivers. Conserv.,* 11, 2063, 2002.

UNEP, *Status of protected area systems in the wider Caribbean region,* Technical Report No. 36, Caribbean Environment Programme, 1996.

Weeks, A., Daly, D.C., and Simpson, B.B., The phylogenetic history and biogeography of the frankincense and myrrh family (Burseraceae) based on nuclear and chloroplast sequence data, *Mol. Phylogenet. Evol.,* 35, 85, 2005.

Wendt, T., Composition, floristic affinities and origins of the canopy tree flora of the Mexican Atlantic slope rain forests, in *Biological Diversity of Mexico: Origins and Distribution,* Ramamoorthy, T.P., Bye, R., Lot, A., and Fa, J., Eds, Oxford University Press, Oxford, 1993, 595.

Yetman, D.A. et al., Monte Mojino: Mayo people and trees in southern Sonora, in *The Tropical Deciduous Forest of Alamos: Biodiversity of a Threatened Ecosystem,* Robichaux, R.H and Yetman, D.A., Eds, University of Arizona Press, Tucson, 2000, 142.

15 Botanical and Ecological Basis for the Resilience of Antillean Dry Forests

Ariel E. Lugo, Ernesto Medina, J. Carlos Trejo-Torres and Eileen Helmer

CONTENTS

ABSTRACT

Dry forest environments limit the number of species that can survive there. Antillean dry forests have low floristic diversity and stature, high density of small and medium-sized trees, and are among the least conserved of the tropical forests. Their canopies are smooth with no emergent trees and have high species dominance. Antillean dry forests occur mostly on limestone substrate, exposing them to more water stress and nutrient limitations than other dry forests. They also experience periodic hurricanes and anthropogenic disturbances. Many of the attributes that allow plants to survive in the stressful environment of the dry forest also provide resilience to disturbance. We attribute the high resilience of Antillean dry forests to the diversity of life forms, a high resistance to wind, a high proportion of root biomass, high soil carbon and nutrient accumulation belowground, the ability of most tree species to resprout and high nutrient-use efficiency. However, opening the canopy, eroding the soil and removing root biomass decreases forest resilience and allows alien species invasion.

15.1 INTRODUCTION

We may conclude that neotropical dry forests are intrinsically fascinating ecosystems, perhaps not so much for their diversity as for the coordinated way in which their relative low species diversity is organized.

Gentry, 1995: 189

Dry forests, seasonal and non-seasonal, are prominent features of the Caribbean landscape because they mostly occur on or near the coastal zone of the islands where they are visible to casual observers. Antillean dry forests occur on soils that can be shallow and rocky. They are not only exposed to low and variable rainfall, but also to winds that range from a breeze to hurricane-level intensity. Net primary productivity and biomass of these forests are low compared to tall moist and rain forests (Murphy and Lugo, 1995), but paradoxically Antillean dry forests are highly resilient to hurricanes, tree pruning, cutting and drought. A question of ecological interest, which builds on Gentry's quote (above), is how do the coordinated functions of Antillean dry forests species lead to high resilience?

Most of the ecosystems at tropical and subtropical latitudes are seasonally stressed by drought (Murphy and Lugo, 1986a). Research on population and ecosystem dynamics, and conservation efforts, however, rarely address the moisture gradient of forests from very wet to very dry. Indeed, the term tropical forest is usually understood as moist or rain forest. The extent of forests in the drier tropics and their characteristics are problematic subjects for those conducting global assessments using remote sensing techniques. In addition, dry forest degradation and conversion is far more advanced than that of wet forests, and yet they are less studied and less publicized (Lerdau et al., 1991; Sánchez-Azofeifa et al., 2005). Janzen (1988) noted that less than 2% of dry forest remains intact in Mesoamerica, and less than 0.1% is explicitly protected. This is unfortunate since tropical forests with prolonged annual drought occupy a greater area than forests in moist and wet regions, and have been of greater use to humans (Murphy and Lugo, 1986a).

What are tropical dry forests? We use the dry forest designation *sensu* Holdridge (1967) as described for the Neotropics by Murphy and Lugo (1995). In the simplest terms, dry tropical or subtropical forests occur in frost-free regions with a pronounced dry season and/or low annual rainfall. In the Holdridge life zone system, frost-free environments include both tropical and subtropical designations. However, because the Holdridge system does not account for seasonality, we expand the definition of dry forest to include forests receiving higher annual rainfall than the system specifies but behaving as dry forests because of an extended dry season. The dry season may be severe enough to select for drought-deciduous or even evergreen, drought-tolerant trees. Drought-deciduousness is, however, the principal adaptive mode found in this type of forest although at the dry extremes small evergreen trees may be important.

The vegetation of the Caribbean is classified into three phytogeographical units (Mexico and Central America, northern Venezuela/Colombia and the Antillean subregion), which in turn are subdivided into provinces according to vegetation affinities (Gentry, 1982; Samek, 1988). Borhidi (1996), however, lumped the two continental units into one. Our geographical focus is on the Antillean subregion, specifically on the dry forests. Lugo et al. (2000) reviewed the literature on the vegetation of the Antillean subregion while Bullock et al. (1995) reviewed the literature on seasonal dry tropical forests, including those of the Antillean subregion. Van der Maarel (1993) included reviews on coastal dry vegetation of the world, including two chapters dedicated to the Caribbean (Borhidi, 1993; Stoffers, 1993).

In this chapter we examine the floristic composition and botanical basis for the resilience of Antillean dry forests, and we explore their distribution and conservation status. We caution that the scarcity of research in the Antilles causes our presentation to be heavily influenced by Gentry (1995) and our research in the Guánica dry forest in southern Puerto Rico. We start with a description of the environmental setting of Antillean dry forests.

15.2 THE ENVIRONMENT OF ANTILLEAN DRY FORESTS

Antillean dry forests function in the context of three environmental conditions that together appear unique in the Neotropics. The first is common to all dry forests, namely, the low and seasonal availability of moisture. Drought stress has been characterized by climatic variables, mostly precipitation and temperature, as the balance between precipitation and evaporation and these are used in mapping the distribution of dry forests and correlating physiological and phenological patterns with climate (Murphy and Lugo, 1986a; Mateucci, 1987). The fundamental variable of soil moisture has been sparsely documented, and this hinders inter-site comparisons. The subject is complicated since soil physical characteristics as well as topography have great importance in dry forests as determinants of variation in water availability. Soils on limestone substrates are frequently very well drained, having a tendency to dry out quickly after rains, and they are therefore often drier than other soil types under the same rainfall regime.

Medina (1995) showed that the number of dry months, estimated through the Bailey Index (BI), was more indicative of water stress to plants than rainfall. The BI is calculated using basic evaporation laws with the formula:

$$BI = 0.18p/1.045t,$$

where p is the mean monthly precipitation in cm and t is the mean monthly temperature in °C.

The BI for which potential evapotranspiration equals rainfall during a given period is 6.37. This number allows separation of humid and dry climatic realms (Bailey, 1979). Dry forests not only experience more dry months than do moist or wet forests, but also more within-year variability of the Bailey Index (Medina, 1995). Soil moisture can also vary significantly during the wet season because of the sporadic occurrence of rains. The moisture limitations of Antillean forests are exacerbated by edaphic conditions. Gentry (1995) believed that many of the peculiarities of Antillean dry forests were due to edaphic factors, particularly the prevalence of limestone. Rocky limestone soils have low water holding capacity and nutritional limitations imposed by their calcareous composition. Growing on calcareous soils not only increases the potential water stress due to their comparatively low water retention capacity, but also restricts the availability of phosphorus (P).

The combination of low moisture availability, high number of dry months and dry days, variation in the availability of moisture and of occurrence of dry days in annual and decadal cycles makes the Antillean dry forest a highly stressful environment for plants. Wind exposure in coastal dry forests increases moisture requirements, putting further demand on water availability. The situation is often exacerbated by exposure to salt spray close to the sea (Stoffers, 1993). Salt spray causes death of exposed new leaves produced during the dry season, and this can be easily observed in wind exposed coastal forests. Flooding with seawater during tropical storms and hurricanes produces a more pronounced effect, causing the death of whole plants when the water percolates down to the root zone. Wright (1992) suggested that drought, and possibly nutrient stress, control floristic composition and relative density by limiting the number of species that can cope physiologically with such an environment. He also stated that seasonal water deficits contribute to mortality and limit diversity in epiphytes, terrestrial herbs and understory shrubs. These moisture limitations result in slow tree diameter growth rates (Weaver, 1979; Murphy and Lugo, 1986a).

The second environmental condition that helps delimit the unique environment of the Caribbean is the persistent exposure to hurricane conditions. The passage of hurricanes over the region is a large-scale, periodic and high-intensity disturbance to forests. A forest will experience anywhere from 10 to 70 hurricanes per century depending on its location in the Caribbean (Neuman et al., 1978). Hurricanes involve exposure to two environmental factors: water and wind. While the passage of hurricanes mitigates short-term water shortages for dry forests, their winds cause tree-fall and other structural changes and stimulate succession. Because of its geographical location, the Caribbean is also continuously exposed to trade winds and these moderate air temperatures, redistribute aerosols, dust, nutrients and organisms throughout the region, and influence the structure of forest canopies. Epiphytic plants benefit from the moisture and nutrient inputs of trade winds to Caribbean forests (Lugo and Scatena, 1992).

The third environmental factor that helps shape present-day Antillean dry forests is human activity. Humans have been active in the Caribbean for over four millennia (Rouse, 1992) and they have modified and converted dry forests through cutting, grazing and burning, resulting in species extinction and introduction of alien species. Because of human activity, it is unlikely that there is any primary Antillean dry forest left. Most stands have been modified in some way.

15.3 FLORISTICS

Gentry (1995) compared the floristics of 28 neotropical dry forests (using data from 0.1 ha per site) and found that Antillean dry forests were different from continental dry forests in terms of structural and floristic composition. Antillean dry forests have more sclerophyllous leaves and higher small and medium tree densities (2.5–20 cm diameter at breast height (dbh)) than their continental counterparts. While dry tropical forests have lower species richness than moist or wet tropical forests, those from the Antilles have even lower species richness than their continental counterparts. Gentry (1995) also reported less than a third as many species of lianas (22/0.1 ha), but a similar number of tree species (46/0.1 ha with dbh $\geq$ 2.5 cm), in Antillean dry forests compared to continental sites. Although Gentry did find a low epiphytic species occurrence in dry forests, he nevertheless considered their absence, or near absence, as one of the most distinctive structural features of this vegetation type. However, Murphy and Lugo (1986b) observed a higher biomass of epiphytes in the Guánica Forest in Puerto Rico than in the moist forests of the Venezuelan Amazon, and they and Medina (1995) attributed this difference to the effect of sea breezes with high atmospheric humidity and the incidence of dew.

Antillean dry forests have almost as many plant families (26) as their Equatorial mainland counterparts (27.9). Myrtaceae is the pre-eminent West Indian dry forest family, averaging four species per sample examined by Gentry (1995). Cactaceae, Capparaceae, Zygophyllaceae and Euphorbiaceae are common families closely associated with dry forests in general. Families like Cucurbitaceae, Asclepiadaceae and Passifloraceae as well as papilionoid legumes are disproportionately well represented in dry forests compared to moist and wet forests. The first three of these families are predominantly herbaceous climbers as are many papilinoids found in this habitat, and their abundance reflects the success of this growth-type in dry forests. Also common are shrubby members of families that are predominantly arboreal or scandent in other habitats.

Among the genera over-represented in the Antilles, Gentry listed *Coccoloba*, *Eugenia*, *Erythroxylum* and to a lesser extent *Drypetes* and *Casearia*. Gentry (1995) found that the level of generic endemism in Antillean dry forests was low compared to that of northern Venezuela and Colombia. Regarding the level of endemism in the dry forests of Cuba, Borhidi (1993) found a greater number of endemics in dry forests on serpentine (ultramaphic) soils on eastern Cuba, higher endemism in littoral thickets on limestone than in littoral forests on sandy beaches, a tendency for endemic species to increase towards the mature phases of vegetation succession, and that the endemic species are mostly local endemics.

Gentry (1995) observed that flowering and seed dispersal in dry forests contrasted with those of moist and wet forests. Two-thirds to three-quarters of dry forest woody taxa had conspicuous flowers pollinated by specialized pollinators such as large bees, hummingbirds or hawkmoths. Seed dispersal for woody plants in dry forests is by wind, occurring in about 80% of lianas and 5–33% of trees. In fact, of families with species showing both wind and animal dispersal, those dispersed by wind are predominant in dry forests. These charactersitics apply to most tropical dry forests, but have not been studied in Antillean dry forests (though see Gillespie, Chapter 16).

Knowledge about the number of plant species in the Antilles is incomplete. This is also true for the dry forests. An estimate for the number of native plant species in the Antillean Archipelago is 12,000 (Myers et al., 2000). Part of the problem of fully documenting the richness of Antillean dry forest plants is that most ecological studies record only plants above a certain size (dbh) and this causes many shrubs, herbs, vines and epiphyte species to be missed since they do not reach the minimum size required. However, the preponderance of herbs, lianas and shrub species in dry forests in comparison to those in moist and wet forests is well known (Gentry and Dodson, 1987).

Irrespective of the size of the Antillean flora, half of the species are considered as endemic to the region, many of them endemic to a single island or to a small region inside one island. Several examples from dry forests in Puerto Rico demonstrate that in spite of the modification, fragmentation and conversion of dry forest stands, new species and records of species continue to be documented by botanical investigations.

- In the dry forests of the south-west of Puerto Rico (Sierra Bermeja, Cabo Rojo, Guánica) six new species of shrubs or small trees were described up to 1990 (Little and Wadsworth, 1964; Little et al., 1974). Also, Acevedo Rodríguez (1999) named the rare vine *Marsdenia woodburyana* Acev.-Rodr., and Kuijt et al. (2005) catalogued the parasitic epiphyte *Dendrophtora bermejae* Kuijt, Carlo & Aukema.
- In Mona Island, the herbaceous species *Chamaesyce orbifolia* Alain, the orchid *Psychilis monensis* Sauleda and the shrub *Lobelia vivaldii* Lammers & Proctor were published by Liogier (1980), Sauleda (1988) and Lammers and Proctor (1994), respectively.
- In the dry forests of north-eastern Puerto Rican Bank, which includes the Virgin Islands, the new orchid *Psychilis macconnelliae* Sauleda was published by Sauleda (1988), and the shrubs *Eugenia earhartii* Acev.-Rodr., *Machaonia woodburyana* Acev.-Rodr. and *Malpighia woodburyana* Vivaldi (the latter is present in south-west Puerto Rico) were published by Acevedo Rodríguez (1993).

The early phytogeographical literature of the Antilles (mostly of the smaller islands) was reviewed by Stoffers (1993) who provided species lists for three communities on sandy beach substrates, two on sand spits and four on rock pavement. A more comprehensive analysis is that of Borhidi (1993) for Cuba. Borhidi segments the 5746 km coastline of Cuba into six main areas according to geography. The age of the substrates are about 30 million years, and most are on limestone substrate except for the eastern portion between Moa and Baracoa that also contains ultramaphic soils. Borhidi identifies six main vegetation zonation types for which he presents vegetation profiles and species lists. The main zonation types are sandy beaches, low limestone benches, high limestone benches, coastal semi-desert terrace areas without wind effects, semi-desert areas with strong wind effects and the serpentine coast where in one 20 m × 20 m plot he found 80 species of which 73 were endemic. In all, Borhidi provides information for 16 littoral communities in Cuba: six on beaches, seven on rocky semi-arid shores and three on rocky arid shores. The works of Borhidi (1993, 1996) and Bisse (1988) contain species lists and phytogeographical data for Cuban vegetation.

A comparative overview of the flora of Antillean dry forests has been made in two studies based on parsimony analyses of species assemblages. These analyses search for the most parsimonious arrangement of shared species among geographical locations with the purpose of revealing biogeographical affinities in a hierarchical pattern (Rosen and Smith, 1988; Rosen, 1992).

Trejo-Torres and Ackerman (2001) studied the Antillean archipelago based on its orchid flora. Islands dominated by dry forests (those flat, low-lying, calcareous islands such as the Bahamas archipelago and the Cayman islands), form regional island groups with cohesive floristic assemblages. Islands physically similar to these archipelagos but found in other areas such as Mona (between Puerto Rico and Hispaniola), Anegada (north-east Virgin Islands) and Isla de la Juventud (south-west of Cuba) affiliate to the Bahamas and Cayman islands instead of to their respective geographical neighbours. The conventional wisdom is that larger islands and regional archipelagoes are biogeographical units while small islands and small archipelagoes are subordinate to the other major neighbouring units (Borhidi, 1996; Samek, 1988). However, parsimony analyses for orchids indicate that floristic groups are concordant with physical factors of the islands (i.e. ecological settings) rather than with geological or geographical membership. This pattern is also understandable given the high vagility of dust seeds of orchids, and is hence expected for other vagile, passively dispersed organisms.

The study of limestone forests shows that dry forests can have closer floristic affinities with dry forests of other islands rather than with moist or wet forests on limestone substrate of their own island (Trejo-Torres and Ackerman, 2002), even when all of them share physiognomic and structural features (Dansereau, 1966; Chinea, 1980). This simple observation also undermines the traditional idea that islands are equivalent to biogeographical floristic units. Instead, it seems that specific biota in complex islands have close affinities to equivalent biota in other islands. Meanwhile, in the study of orchids, regional groups of islands form cohesive floristic units, but other distant non-neighbouring islands can also have close affinities to them, rather than to their neighbouring islands (Trejo-Torres and Ackerman, 2001).

The importance of environmental factors as determinants of floristic composition of Antillean dry forests connects the floristic component of these forests to their response to environmental disturbances and hence to resilience, the temporal response to disturbance. This is particularly true for those resilience attributes that are species specific such as the ability to resprout or the nutrient-use efficiency. Resilience in Antillean dry forests is our next topic.

15.4 RESILIENCE OF ANTILLEAN DRY FORESTS

Ewel (1971, 1977) studied the resilience to cutting and herbicides of a dry forest stand in Guánica Forest. He conducted short-term experiments (1 to 3 years) in dry, moist and wet forests in the Caribbean and Central America and found that while dry forests had a slower rate of recovery than wet forests, their resilience was higher because it had less structure to restore after the disturbance. As an example, in terms of the total amount of a variety of structural parameters, the recovery at Guánica forest was on average 39% that of the wet forest at Osa Peninsula in Costa Rica. However, recovery of Guánica forest was 24 times more complete than Osa when assessed in terms of the expected structural values at maturity at each site (Ewel, 1971).

Thirteen years after Ewel's experiments, Dunevitz (1985) assessed the Guánica site and found that the forest recovered more quickly when it was cut than when herbicides were applied. She also evaluated six parameters of forest structure in cut sites and found the following rates of recovery (shown in parentheses) expressed as a percentage of the same parameter measured in an adjacent control stand: number of mature forest species present (59%), basal area (7–40% depending on tree diameter, with a smaller percentage for larger trees), mean height (40%), aboveground biomass (43%), leaf area index (64%) and epiphyte density (11%). Notably, the cut site had 60% more tree species than the control sites due to the invasion of pioneer species. Regeneration was by 81% stem sprouts, 3% root sprouts and 16% seed germination.

Molina Colón (1998) assessed the resilience of the same dry forest vegetation at Guánica by measuring regrowth and forest recovery after abandonment of agriculture, dwellings and charcoal production. Her study examined the effects of long-term recovery 45 years after the abandonment of land uses that had lasted over a century. Molina Colón confirmed Ewel's observation that canopy parameters such as leaf area index and leaf biomass recovered faster than other forest attributes such as stem biomass or basal area. She also showed that forest resilience was higher after disturbances that removed only aboveground biomass relative to disturbances that removed both above- and belowground biomass. *Leucaena leucocephala* (Lam.) de Wit, a naturalized alien species, dominated sites previously used for agriculture, houses and a baseball park.

Van Bloem et al. (2003) found that the dry forest was resistant to a category three hurricane. After the hurricane, more trees showed no effect than those affected in some way by the hurricane. The forest experienced low tree mortality (2% of stems) and low structural changes such as defoliation, stem breakage and loss of basal area. Wunderle et al. (1992) reported similar results for dry limestone forests in Jamaica after the passage of hurricane Gilbert, and attributed to hurricanes some of the observed characteristics of Caribbean avifauna, including its distribution. Van Bloem et al. (2003) suggested that hurricane winds damage larger trees and cause basal sprouting, thus contributing to the maintenance of the low and dense structure of these forests.

Dry forests in Puerto Rico were also found to be resilient to forest fragmentation. Fragments as small as <1 ha retained species richness and composition of trees (including representation of mature forest species), reptiles and termites (Molina Colón, 1998; Genet, 1999; Genet et al., 2001; Ramjohn, 2004). Ramjohn (2004) found 12 rare or endangered species (out of 53 in the south-west of Puerto Rico) in at least one of 40 fragments he examined. However, wood decomposition in the small fragments was slower than in controls (Genet et al., 2001), suggesting that fragmentation might influence some ecosystem functions while not affecting species composition.

Dickinson et al. (2001) proposed a combination of factors associated with seasonal drought and low annual rainfall in dry limestone forests of the Yucatán peninsula. They suggested that low annual rainfall, seasonal drought, low storm frequency, low epiphyte loads, large diameter-to-height ratios and strong anchorage of trees led to low rates of canopy disturbance between disturbance events in dry forests. We modified their model (Figure 15.1) to make it more applicable to Antillean dry forests. In the Antilles, the biomass of epiphytes can be significant in some locations due to the favourable environment produced by the trade winds and night water condensation in the canopy. Also, storms are more frequent than in the Yucatán, and fire is not a natural factor in the Antillean dry forests. Nevertheless, the general notion of a wind-resistant

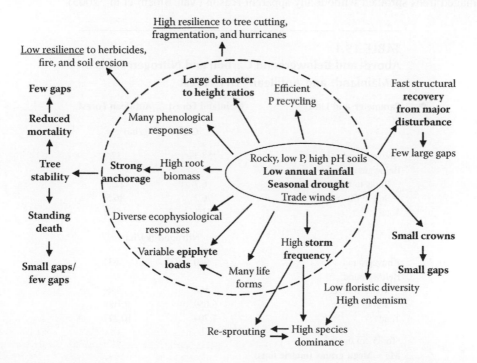

FIGURE 15.1 The combination of factors believed to contribute to high resilience of Antillean dry forests (modified from Dickinson et al., 2001). Bold entries represent information in Dickinson et al. and normal print text has been added.

canopy that has few and small gaps is also typical of Antillean dry forests. Tree stability through proper anchorage and standing tree mortality are also elements in common between Yucatán and Antillean dry forests. These common characteristics of dry forests lead us to a discussion of the botanical basis of their resilience.

15.5 BOTANICAL BASIS OF RESILIENCE

15.5.1 Community Attributes

Antillean dry forests are characterized by their lower height and higher tree density than mainland dry forests (Murphy and Lugo, 1986a, 1995; Gentry, 1995). The biomass distribution is also different (Table 5.1). Antillean dry forests have lower aboveground biomass but higher belowground biomass than continental dry forests. The total accumulation of nitrogen is higher in the Antillean forest, particularly in the belowground compartments but not aboveground. The weighted nitrogen concentration for the stand as a whole (including soil depth to 80 cm) is higher in the Antillean forest than continental forests (35 mg/g vs. 27 mg/g). This means that a gram of organic matter in Antillean dry forests contains on average 1.3 times more nitrogen than a gram of organic matter in continental dry forests.

A large number of tree species in Antillean dry forests sprout after cutting, and this accelerates the re-establishment of tree species after experimental (Ewel, 1971, 1977; Dunevitz, 1985; Murphy and Lugo 1986a) or subsistence cutting of stands (Molina Colón, 1998; McDonald and McLaren, 2003). Dunphy et al. (2000) reported that most tree species in the Guánica forest coppiced naturally as well as after cutting. They found 43% of all trees in the forest had multiple stems accounting for 58% of the total basal area. McDonald and McLaren (2003) found that only three out of 51 species tested in a Jamaican dry forest did not coppice. After a hurricane struck Guánica Forest, undamaged trees sprouted without any apparent reason (Van Bloem et al., 2003).

TABLE 15.1
Above- and Belowground Carbon and Nitrogen Pools in a Mainland[a] and Antillean[b] Dry Forest

Parameter and Unit	Mainland Forest	Antillean Forest
Carbon (Mg/ha)		
Aboveground	58.3	37.8
Belowground		
Roots to 60 cm	6.7	22.5[a]
Soil to 60 cm	76.2	86.5[a]
Total	141.1	146.8
Nitrogen (kg/ha)		
Aboveground	940	631
Belowground		
Roots to 60 cm	106	546[a]
Soil to 60 cm	6,659	9,100[a]
Total	7,704	10,277

[a]To 85 cm depth

Mg = Mega grams (metric tons)

Source: [a]Jaramillo et al., 2003
         [b]Lugo and Murphy, 1986; Murphy and Lugo, 1986a

Lugo (1991) showed that all dry forest stands in Puerto Rico and Mona Island have high species dominance. In a mature forest at Guánica Forest for example, Molina Colón (1998) reported that from a total of 36 species, five accounted for 66% of total basal area, and six accounted for 66% of total tree density (Figure 15.2). Part of this dominance is due to the low floristic diversity of Antillean dry forests, but Lugo also attributed high dominance as a response to hurricane and anthropogenic disturbances and drought. Given the harsh environmental conditions of Antillean dry forests, those few species that thrive under the stress are able to dominate sites. Miller and Kauffman (1998) studied the effect of anthropogenic disturbances on species importance and dominance in Mexican dry forests. When exposed to slashing, burning, cultivation and grazing, the experimental sites had a reduction of species number and an increase in the dominance of the surviving species. The experiments also showed the importance of high root turnover to forest resilience (Castellanos et al., 2001).

The community attributes discussed above all contribute to forest resilience, each attribute contributing to resilience to one or more type of disturbance. For example, the low stature of the forest offers a reduced physical barrier to wind, which in turn gives strong resistance to high winds. The low fraction of biomass aboveground means that the necessity to rebuild after wind disturbance is comparatively small. This recovery is accelerated by the resprouting ability of most dry forest species. Moreover, the high amount of root biomass supports the recovery of the above-ground portion by supplying water, nutrients and organic matter reserves. The high proportion of root biomass is also adaptive in capturing scarce soil water resources. Similarly, high soil organic matter stores available nutrients for vegetation, and the high proportion of below-ground nutrients allows for multiple turnovers of aerial vegetation (Lugo and Murphy, 1986). These factors provide resilience to pruning and fuel-wood harvesting.

15.5.2 Ecophysiology

15.5.2.1 Life Forms

Medina (1995) compiled the variety of life forms in dry forests (Table 15.2) and observed that diversity of life forms in dry forests appeared higher than in other types of tropical forests. He attributed this diversity to habitat (including surface) heterogeneity coupled with strong rainfall seasonality. This combination of factors leads to broad gradients of water availability so that each life form finds a sector of the gradient to which its particular physiological responses are competitively adaptive (Lugo et al., 1978; Medina, 1995). The diversity of life forms is accompanied by diversity in structure and physiology, for example, plant habit, leaf size, drought tolerance and growth seasonality (Medina, 1995). Sap concentration increases with aridity (Medina, 1995).

15.5.2.2 Phenology

Phenology of tropical deciduous forests is characterized by phenomena determined by pulses, such as the sudden increase in water availability at the beginning of the rainy season, moisture availability after sporadic rainfall events, and those triggered by slow reductions in resources availability, such as the decrease in water availability at the end of the rainy season (Table 15.3). These phenomena can be initiated, particularly near the outer fringes of the tropical regions, by photoperiod and temperature changes at the beginning and end of the rainy season.

Activation of the soil microbiota leads to rapid cascades of decomposition of below- and above-ground litter produced during the previous growing season, and active fluxes of gases derived from soil respiration (CO_2, NO, N_2O) (Matson and Vitousek, 1995). In these processes significant amounts of nutrients are released which can be leached or taken up by newly formed fine roots. Much research is still required to understand the synchronization of events related to water supply, particularly processes of decomposition and release of nutrients and trace gas fluxes.

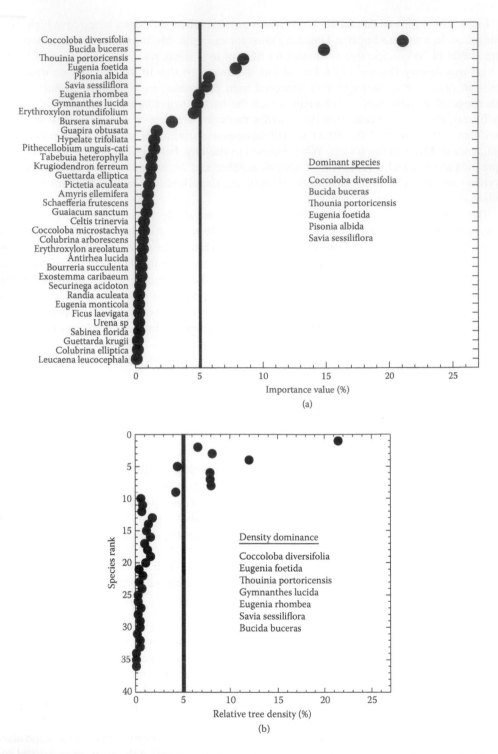

FIGURE 15.2 Importance values based on (A) relative density + relative basal area, (B) relative density and (C) relative basal area of the most common tree species in a mature dry forest in Guánica, Puerto Rico (from Molina Colón, 1998). Data are for trees with dbh ≥ 5 cm. The total number of species was 36 in a sampled area of 400 m². Absolute basal area was 19.1 m²/ha and absolute tree density was 5085/ha. The most important species representing >5% of the relative values are shown in each graph in decreasing order.

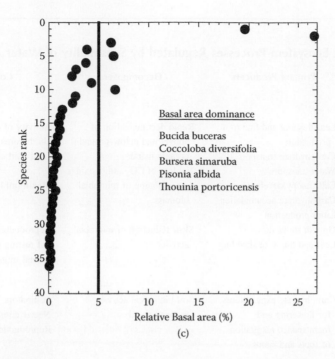

Basal area dominance

Bucida buceras
Coccoloba diversifolia
Bursera simaruba
Pisonia albida
Thouinia portoricensis

(c)

FIGURE 15.2 (Continued).

TABLE 15.2
Dominant Life Forms and Functional Attributes of Higher Plants from Tropical Dry Forests

Life Form and Morphological Features	Functional Attributes
Trees and Shrubs	
Mesophyllous	Deciduous
Sclerophyllous	Evergreen
Succulent	C_3 and CAM photosynthesis
Herbaceous	
Grasses	Annual, C_3 and C_4 photosynthesis
Dicotyledonous	Perennial
Succulents	C_3 and CAM photosynthesis
Vines	
Herbaceous	
Woody	
Mesophyllous	Deciduous
Sclerophyllous	Evergreen
Succulent	C_3 and CAM photosynthesis
Epiphytes	
Mesophyllous	Deciduous
Sclerophyllous	Evergreen
Succulent	C_3 and CAM photosynthesis
Parasites	Xylem-tapping mistletoes

Source: Medina, 1995

TABLE 15.3

Characteristic Ecosystem Processes Regulated by Seasonality of Water Availability

Phases	Primary Producers	Decomposers	Consumers
Rainy Season			
Beginning	Leaf flushing and fine root production Carbohydrate transport Water transport	Rapid decomposition of above- and below-ground organic matter Net loss of CO_2 and N oxides	Activity of detritivores, herbivores, and seed and fruit eaters
Middle	Full canopy development Carbohydrate accumulation Litter production	Accumulation of microbial biomass	Maximum herbivore activity
End	Growth slows down Leaf and fine root shedding	Slow reduction of microbial activity	Reproduction and beginning of resting phase Faunal migrations
Dry Season			
	Carbohydrate expenditures for flowering and maintenance respiration of roots and stems	Soil biological activity reduced or nil	Pollinators Nectar consumption Reproduction

Note: Ecosystem adaptive features are related to the duration and variability of the growing period (beginning to end of the rainy season, lasting 4–10 months)

Leaf flushing takes place at the beginning of the rainy season as a consequence of full hydration of meristems, while leaf shedding takes place at the end of the rainy season as a consequence of leaf dehydration and senescence (Frankie et al., 1974; Reich and Borchert, 1984). Similar processes take place by the production and sloughing of fine roots. Activity of insect herbivores (leaf and seed eaters) increases dramatically at the beginning of the rainy season, while activity of pollinators is high during the dry season, when numerous deciduous tree canopies are covered with dense flower crops (Janzen, 1981; Filip et al., 1995; Borchert et al., 2004). Essential determinants for the long-term dynamics of these forests are the duration and variability of the four phases of water availability (Table 15.3).

15.5.2.3 Nutrients

In the tropics, seasonal dry forests and savannas occur under the same climatic conditions. Separation of seasonal forests and savannas under undisturbed environmental conditions is possible on the basis of the fertility status of the soil. In the Neotropics, deciduous forests are found on soils of significantly higher fertility than savannas (Ratter et al., 1973; Furley et al., 1988; Oliveira-Filho and Ratter, 2002). Sarmiento (1992) developed a simple model based on the number of months with water deficit and on soil fertility, which allows the separation of evergreen forests, deciduous forests and seasonal savanna systems. The extension of deciduous forests has decreased because of deforestation for agriculture and pastures. Deforestation is associated in the tropics with an increase in the frequency of fires, leading to nutrient impoverishment and to the encroachment of fire-tolerant savanna communities. In Antillean forests, the soil nutritional conditions are not as favourable as on the continent due to the predominance of calcareous substrates. Low phosphorous availability probably accounts for the higher proportion of evergreen sclerophyllous trees in Antillean dry forests compared to dry forests in Central and South America (Lugo and Murphy, 1986).

Only a small percentage of the nitrogen (N) (10%), phosphorous (P) (2%) and potassium (K) (3%) in the Guánica forest stand in Puerto Rico was stored above ground (Lugo and Murphy, 1986).

TABLE 15.4
Percentage of Nutrient Demand Required for Leaf Growth (Measured by Either Area or Weight) that Could be Satisfied by Retranslocation in Eight Selected Tree Species of Tropical Dry Forests in India

	Nutrient Demand Potentially Satisfied by Retranslocation (%)	
Nutrient	Area	Weight
Nitrogen	50–100	46–80
Phosphorus	29–100	20–91
Potassium	29–100	20–57

Source: Lal et al., 2001

Trees retranslocated the N and P required for above-ground net primary production (30% and 65% respectively), while immobilizing P in dead roots. Lal et al. (2001) found that nutrient retranslocation by dry forest species in India could account for a significant portion of the nutrient demand for leaf and biomass expansion during the dry season (Table 15.4). These dry season leaves were capable of taking advantage of the changing moisture conditions at the onset of the rainy season. The result was that they maximized the length of time biomass accumulation could take place through net photosynthesis. Moreover, the ratio of organic matter to phosphorus in litterfall was 6057 to 1 in a mature stand in Guánica forest, one of the three highest P-use efficiencies reported in the literature (Lugo and Murphy, 1986). The storage and cycling of P in these dry forests contribute to resilience by facilitating organic matter production and use in spite of low P availability.

15.5.2.4 Limestone Substrates

In Caribbean islands, a thick layer of limestone deposits of marine origin up to several hundred meters deep may cover volcanic substrates. On top of these deposits a variety of plant communities develop according to rainfall, from open thorn thickets to predominantly deciduous forests with a similar tree flora throughout the lowlands of the Caribbean (Asprey and Robbins, 1953; Kelly et al., 1988; Tanner, 1977). Soil development on top of hard limestone leads to formation of terra rossa above soft rendzinas. Limestone also presents strong hardening processes. The dissolution process of the limestone leads to the formation of deep holes (dolines). In time, these holes can be connected and form wide flat areas surrounded by steep walls of limestone covered with hardened calcareous crusts. In the Greater Antilles, this process is well developed particularly in the moister areas and has lead to the Mogotes formation (Lötschert, 1958; Zanoni et al., 1990; Borhidi, 1996; Lugo et al., 2001).

Limestone is homogeneous in chemical composition, mainly containing calcium carbonate, with varying amounts of magnesium (dolomites) and clay (Marschner, 1995; Lugo et al., 2001). On limestone bedrock a variety of calcium-rich soils develop, particularly the rendzinas. Climatic differences determine variations in ion accumulations in the upper soil layers. In drier climates, rapid evaporation causes water-soluble salts to accumulate at the surface. In more humid regions, rainfall leaches out both nutrients and calcium carbonate leading to a 'secondary acidification'.

Water and nutrient relations differ markedly at the base, slope and tops of Mogotes, determining changes in density and species composition of the woody vegetation (Weaver, 1979; Chinea, 1980; Serrano et al., 1983; Álvarez Ruiz et al., 1997). The limestone areas in Puerto Rico have been divided into northern, southern and dispersed limestone areas (Lugo et al., 2001). Since little is known about the dispersed limestone areas, we will concentrate our attention on the northern and southern limestones. The northern limestone area includes the Karst belt, while the southern limestone area occupies some of the lower rainfall areas in Puerto Rico. On these limestone areas, three large geoclimatic units are differentiated according to their rainfall regime, supporting dry, moist and wet forest types. These forests grow under similar pedological conditions but intensity of carbonate dissolution varies in association with the amount of run-off and of percolating waters acidified by CO_2 from the air and root and microbial respiration. In these limestone forests, structural development and biomass production are tightly associated with rainfall regime (Lötschert, 1958; Lugo et al., 1978; Chinea, 1980; Medina and Cuevas, 1990).

15.6   VEGETATION ON CALCIUM CARBONATE-RICH SUBSTRATES

The occurrence or exclusion of plant species in karst environments has been studied in detail in temperate latitudes, and this has led to the ecological classification of plants as *calcicoles*, species frequently growing on calcium carbonate-rich soils, and *calcifuges*, species that avoid, or are not found, on those soils (Kinzel, 1983). This differentiation, however, is not necessarily related to physiological properties or modification of the competitive ability of those plant species under natural conditions.

Calcium carbonate-rich soils generally have lower availability P and iron (Fe) because of the formation of insoluble salts, and may create physiological stress in the plants due to the accumulation of large amounts of calcium (Ca) ions. Often these soils generate nutrient stress through the following:

- Inhibition of K uptake by high concentrations of Ca ions in the soil solution.
- Low P availability because the high pH and high concentration of Ca leads to the formation of insoluble calcium-phosphate. However, activity of roots and rhizospheric microorganisms may increase P availability through the excretion of organic acids.
- Reduction of Fe availability, which in calcifuge species cultivated on calcareous soils causes a yellowish discoloration of leaves (lime-induced chlorosis).

Calcium concentrations in the cytoplasm have to be below the micromolar level for Ca to act as a secondary messenger in the regulation of membrane permeability and root growth (Marschner, 1995). The excess of Ca availability in karst soils can be precipitated around roots or cells (as calcium carbonate or calcium oxalate), or is maintained in soluble conditions within the vacuoles, frequently as chelates of malate and citrate (Kinzel, 1989).

Higher plants have been classified into two groups according to their soluble K and Ca concentrations. Those with a soluble K/Ca molar ratio below one are termed *calciotrophs*, whereas *calciophobes* are species containing little or no soluble Ca, either because it is precipitated inside the cell as calcium oxalate, or outside the plant tissues as calcium carbonate or calcium oxalate (Kinzel, 1983, 1989). Most of the species investigated are of temperate climates and mainly herbaceous. Calciotrophic species are frequent within the Leguminosae, and calciophobes have been described within the Polygonaceae. These families are well represented in Caribbean dry forest on calcareous soils but have not been investigated yet. Other important families that deserve to be analyzed because of their adaptation to karst soils in the Caribbean are the Palmae, Piperaceae, Moraceae and Rhamnaceae.

In calcium-rich environments, both physiological types (or physiotypes *sensu* Kinzel, 1983) can coexist if water availability is high. In semi-arid environments, calciophobic plants cannot use Ca as an osmoticum, therefore they osmoregulate by accumulating large amounts of photosynthetically expensive organic compounds, particularly sugars. This physiological constraint may explain the calcifuge character of those species (Kinzel, 1989; Marschner, 1995). Mycorrhizal fungi play a role in adaptation of perennial plant species to calcareous soils through the release of siderophores that enhance iron acquisition.

The ecophysiological aspects of Antillean dry forests that we have just reviewed reveal complex adaptations to the diverse environmental conditions typical of these forests. The environmental complexity is revealed in a list of conditions that change spatially and temporally. For example: (1) variable gradients of soil moisture, (2) variable nutrient availability, (3) chemical challenges associated with high soil pH and calcium concentration, (4) pulses of resource availability associated with temperature and rainfall seasonality, (5) limited space for tree anchorage and roots due to rocky soils, (6) periodic hurricanes and others.

Ecophysiological responses that contribute to resilience within this environmental complexity include: (1) a diversity of plant life forms where different life forms function optimally on different sectors of the environmental gradients (Table 15.2), (2) metabolic specialization to deal with calcium-rich and high pH soils, (3) capacity to concentrate and use P at high levels of efficiency, (4) distribution of biomass and chemical resources in live vegetation and soils, (5) a variety of phenological clocks to deal with seasonality and climate variability, (6) stem and root sprouting and vegetative propagation capabilities, (7) a variety of gas exchange mechanisms in plants and others.

15.7 THE DISTRIBUTION, AREA AND CONSERVATION STATUS OF CARIBBEAN DRY FORESTS

Dry forests are widespread in the Antillean islands. A general distributional scheme for the Antilles includes:

- The Greater Antilles (Cuba, Hispaniola, Jamaica and Puerto Rico). In these islands, dry forests are of limited extension compared to other vegetation types, and are typically found on narrow strips of limestone on the northern and southern coastal areas. In Cuba, isolated areas of dry forests are also found along the centre of the island over ultramaphic soils, many times associated with limestone. Puerto Rico also contains dry forests on ultramaphic soils in the vicinity of dry forests over limestone. A peculiarity of Hispaniola (Dominican Republic and Haiti) is that dry forests are also present in interior valleys at the bases of the major mountain ranges.
- The islands of the Virgin Islands platform (British Virgin Islands, U.S. Virgin Islands, including St. Croix — which is outside the platform — and the Puerto Rican islands of Vieques and Culebra). These are volcanic, low mountainous islands with elevations up to 521 m. Dry forests and scrubs dominate except on islands with protected uplands, ravines and valleys, which sustain moist vegetation.
- In the Lesser Antilles two groups of islands are distinguished based on geomorphology: the volcanic arc and the limestone arc. The volcanic arc (from Saba to Grenada) is comprised of topographically abrupt islands up to 1467 m elevation, where dry forests are limited to reduced coastal areas, mainly on the leeward sides.The limestone arc of the Lesser Antilles (from Sombrero to Marie-Galante) as well as Barbados in the south, comprise islands up to 402 m elevation. They have similar conditions to the islands of the Virgin Islands platform in terms of dry forest coverage.
- Dry forests and shrubby vegetation dominate the flat and low-lying (up to 100 m elevation) limestone islands like the extensive Bahamas archipelago, the Cayman islands, Mona and Anegada (the last an atypical member of the otherwise volcanic Virgin Islands).

Brown and Lugo (1980) estimated that dry forests comprised 50%, 23%, 14% and 10% of the area of Bahamas, Dominican Republic, Puerto Rico and Trinidad, respectively. More recently, Helmer (2004) analyzed the areas of the various geoclimatic zones of Puerto Rico in relation to the level of development, forest cover and conservation as public lands (Table 15.5). Her analysis shows that 15% of Puerto Rico is in the dry forest life zone. A total of 24% of the island's development has occurred in this life zone, which means that the development burden is disproportionately high

TABLE 15.5
**Areas of Ecological Zones of Puerto Rico, Zonal Proportion of Island-Wide Land
Development, Proportion of Each Zone Under Protection, Upland Non-Cultivated
Woody Vegetation within Zone, and Areas and Percentage of Each Zone that is Protected
Upland Woody Vegetation**

Ecological Zone	% of Island-wide Development	Area of Ecological Zone (ha)	% Protected	Upland Woody Area (ha)	Protected Upland Woody Vegetation	
					Area (ha)	% of zone
Dry-alluvial	10.5	45,179	5.5	5.368	211	0.5
Dry-volcanic/ sedimentary/ limestone	11.8	82,379	4.6	30,441	3319	4.0
Dry-moist serpentine (ultramaphic)	1.9	6.411	25.1	3.690	1517	23.7
Total for dry forest zone	24.2	133,969	6.0	39,748	5047	3.8
Rest of the island	75.8	735,187	5.1	325,412	28,979	3.9
Total for Puerto Rico	100	869,156	5.2	365,160	34,026	3.9

Source: Helmer, 2004

in relation to the geographic representation of the zone. The dry forest life zone of Puerto Rico is 30% forested, has 6% of the land under conservation protection and 3.8% of its forest cover is protected. Within the dry forest zone, the alluvial portion has the lowest forest cover (12%) and lowest area under protection. This situation contrasts with the serpentine or ultramaphic geoclimatic zone, which is 58% forested, and has the highest proportion of land cover under conservation protection. Dry forest over ultramaphic soil is quite rare ecologically and floristically and this region of Puerto Rico, which also includes moist and wet forest, is well protected for its biodiversity value. This, together with its unsuitability for agriculture, is why the proportion of island development that has occurred on this zone is the lowest among those reported in Table 15.5.

In general, Helmer's analysis for Puerto Rico reflects the situation elsewhere of Antillean dry forests. These forests suffer a disproportionate effect of human activity because human populations tend to be high in this life zone (Murphy and Lugo, 1986a). In addition, humans convert Antillean dry forests to alternative uses particularly where soils are nutrient rich, as is the case of forests on alluvial soils. Moreover, local people use these dry forests as a source of fuel-wood and grazing for goats and cattle, which results in altered forest structure and species composition (McDonald and McLaren, 2003).

In some Antillean dry forests, particularly those recovering from disturbance, trees can have low stature or coppiced growth form, and succulents are common. Consequently, there are certain classification systems that would categorize such dry forests incorrectly as shrub-, scrub-, or woodlands. These terms, often regarded as derogatory, can lead to less overall regulatory protection from human exploitation. Several global classification systems, for example, use a height threshold to determine whether a woody vegetation formation is a shrubland or forest, which is inappropriate for Antillean dry forests. The Land Cover Classification System (LCCS) (Di Gregorio and Jansen, 2000) uses a minimum forest height of 3 m, which would mistakenly classify a large proportion of recovering Antillean dry forest as shrubland. The UN Food and Agriculture Organization (FAO) forest classification system (FAO 1996, 1998) uses a 5 m height threshold to distinguish forest from shrubland. A recent global map of terrestrial ecosystems (Olson et al., 2001) refers to some Antillean dry forest

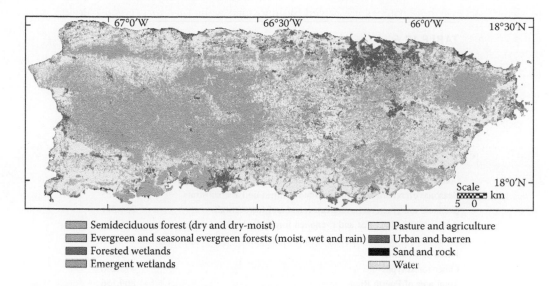

Semideciduous forest (dry and dry-moist)
Evergreen and seasonal evergreen forests (moist, wet and rain)
Forested wetlands
Emergent wetlands

Pasture and agriculture
Urban and barren
Sand and rock
Water

FIGURE 15.3 (See colour insert following page 208) Distribution of woody vegetation formations typical of dry and dry-to-moist forest formations of Puerto Rico (Helmer et al., 2002). Dry forest formations include woodlands, shrublands and forests that are drought deciduous or semi-deciduous. Unlike most other maps of Puerto Rican forest formations, this map recognizes that dry forest types dominate karst substrate along the north-western coast of the island, which receives less rainfall than inland sites. Based on field observation, Helmer et al. (2002) delineated this area with maps of total annual precipitation and geology, assuming that over karst substrate, where annual rainfall is less than 1500 mm, drier forest formations dominate on limestone hills.

formations as cactus scrub. Even Beard (1955), for example, refers to young secondary dry forests with names like cactus bush, logwood thicket, logwood-acacia bush, *Leucaena* thicket and thorn savanna. This discussion underscores another reason why tropical dry forests fail to receive the attention they deserve. It appears that even their classification and identification are still inconsistent and this hinders the development of a coherent conservation argument for these forests.

A map of forest types for Puerto Rico (Colour Figure 15.3) shows that 44% of the potential forest cover of Puerto Rico is either dry or seasonally dry forest (Table 15.6). This new perception of the forests of this well-studied island is based on more rainfall data than previously available and field verification of vegetation physiognomy (Helmer et al., 2002). The results of the analysis underscore the point already discussed above, that drought-adapted vegetation is more common in the tropics than previously believed.

15.8 FUTURE FORESTS: ALIEN AND ENDEMIC SPECIES COEXISTING?

We have discussed how community structure and ecophysiology of dry forests are important to their resilience to natural and anthropogenic disturbances. We have also reviewed the floristic composition of these forests and its relation to resilience through taxon specialization, endemism and life form diversity. However, given the high level of global anthropogenic change predicted for the future, the last question we address is whether the Antillean dry forests will sustain or lose their resilience in the future through irreversible changes to their biota and to below-ground accumulation of nutrients, biomass and soil organic carbon.

It is clear that the Antillean dry forests are vulnerable to human exploitation and degradation. Typically, wood is extracted for charcoal production and domestic animals graze heavily within

TABLE 15.6
Areas of Forest Types in Puerto Rico as Generalized from
Helmer et al. (2002) and Figure 15.3

Forest Type	Area (ha)
Mixed evergreen drought-deciduous shrubland with succulents	1,052
Mixed deciduous evergreen forest on serpentine[a] substrate	3,535
Mixed deciduous evergreen forest on serpentine[a] substrate	90
Drought deciduous and semi-deciduous forest and woodland/shrubland on karst and other substrates	46,509
Seasonal evergreen and semi-deciduous forest on karst substrate	52,283
Seasonal evergreen and evergreen forest and forest/shrub	256,371
Active sun/shade coffee and evergreen forest/shrub	26,879
Area of dry and seasonally dry forests	386,719
Forested wetlands	7,157
Other forest types	475,280
Total area of Puerto Rico	869,156

[a]ultramaphic

forest stands. Fire is frequently introduced and this results in destruction of trees and invasion by alien grasses. Degraded dry forests are characterized by a combination of the following: compacted soils, loss of soil organic matter, erosion, grazed or trampled understorey vegetation, species composition altered towards the dominance of a few species that are usually thorny, fewer trees, ragged canopies and invasion of alien tree species. Such degraded stands are common in the Caribbean and it is very difficult to reverse these changes because once a fire regime is in place and the soil loses its fertility, succession is arrested at a grassy or thorny scrubland stage. However, management actions in Guánica forest demonstrate that the rehabilitation of degraded dry forests is possible (Chinea, 1990; Wadsworth, 1990).

For rehabilitation to be successful in such areas it is necessary to remove grazing animals and fire. Tree planting and maintenance then follow these changes in disturbance regime until trees are re-established at the site. After that, natural succession can proceed to mature stages but not necessarily to the original species composition. In Puerto Rico, this formula has proven effective in the dry forest region of the island. A recent island-wide forest inventory documents the emerging new dry forests on the landscape (Lugo and Brandeis, 2005). The authors found that dry forests, although degraded, do not have an alien species that exerts dominance, as does *Spathodea campanulata* P.Beauv. elsewhere in Puerto Rico (Table 15.7). Apparently, the harsh natural conditions

TABLE 15.7
Percentage of Endemic and Alien Tree Species, Contribution of Aliens to the Importance Value (IV), and Number of Species (Alien and Native) by Size Class in Dry Forests in Puerto Rico (Data for 2002;.Based on 21 Random Plots and 1.47 ha Sample

Total Number of Species (%)		IV (%)	Number of Species Sampled			
Endemic Species	Alien Species	Alien Species	Seedlings	Saplings	Trees	Total
5.7	13.6	7.8	65	56	30	88

Source: Lugo and Brandeis (2005)

of dry forests prevent the dominance of alien species, but not their presence, as aliens contributed to 13.6% of the dry forest tree species vs. 5.7% for endemic species, with the remaining species native non-endemics. Alien species combined accounted for 7.8% of the importance value of all species in dry forests (in contrast with 40–65% in other forest types in the island.

Bursera simaruba (L.) Sarg. and *Andira inermis* (W. Wright) DC, both native, were the most important dry forest species, and both species show signs of regeneration. *Leucaena leucocephala*, a naturalized alien, however, dominates the sapling and seedling classes. Of the 88 species found by Lugo and Brandeis (2005) in dry forest, 75 were native. They reported endemic species to Puerto Rico such as *Psidium insulanum* Alain, *Thouinia striata* Radlk., *Rondeletia inermis* (Spreng.) Krug & Urb. and *Tabebuia haemantha* (Bertol. ex Spreng.) DC, of which the last three species are regenerating (Table 15.7). *Prosopis pallida* (Humb. & Bonpl. ex Willd.) Kunth., *Pithecellobium dulce* (Roxb.) Benth., *Albizia lebbeck* (L.) Benth., *Tamarindus indica* L., *Spathodea campanulata*, *Melicoccus bijugatus* Jacq., *Acacia farnesiana* (L.) Willd., *Persea americana* Mill., *Sterculia apetala* (Jacq.) H. Karst., and *Parkinsonia aculeata* L., in order of importance, were the alien species found in dry forest as upperstorey species. Most of these also showed sapling and seedling regeneration in undisturbed conditions.

From the above, it appears that alien species coexist with native and endemic species in dry forests. In fact these pioneer alien species appear to facilitate (sensu Callaway, 1995) the establishment of native species on degraded forestlands (Molina Colón, 1998). However, the alien species do not appear to dominate natural stands, although they do so in the colonization of sites where all vegetation has been removed. In this instance, the alien species form forest stands that slowly mature into communities with mixed species composition, including aliens and native species. It is possible that these new forests (sensu Lugo and Helmer, 2004) represent the forests that will prevail in human-dominated landscapes, assuming that humans learn to manage fire and grazing there. The level of resilience of these new forests will require additional research, particularly as they become the predominant vegetation of the Antillean dry forest.

ACKNOWLEDGEMENTS

This work was developed in cooperation with the University of Puerto Rico. We thank D.J. Lodge, S. Van Bloem, P.G. Murphy, F.H. Wadsworth, R.T. Pennington, J. Ratter and an anonymous reviewer for reviewing the manuscript. We also thank L. Sánchez and Mildred Alayón for contributing to the production of the manuscript.

REFERENCES

Acevedo Rodríguez, P., Additions to the flora of St. John, United States Virgin Islands, *Brittonia*, 45, 130, 1993.
Acevedo Rodríguez, P., West Indian novelties: a new species of *Marsdenia* (Asclepiadaceae) from Puerto Rico and a new name for a Jamaican species of *Calyptranthes* (Myrtaceae), *Brittonia*, 51, 166, 1999.
Álvarez Ruiz, M., Acevedo Rodríguez, P., and Vázquez, M., Quantitative description of the structure and diversity of the vegetation in the limestone forest of Río Abajo, Arecibo-Utuado, Puerto Rico, *Acta Científica*, 11, 21, 1997.
Asprey, G F. and Robbins, R.G., The vegetation of Jamaica, *Ecol. Monogr.*, 23, 89, 1953.
Bailey, H.P., Semi-arid climates: their definition and distribution, in *Agriculture in Semi-Arid Environments*, Hall, A.E., Cannell, G.H., and Lawton, H.W., Eds, Springer Verlag, Berlin, 1979, 73.
Beard, J., The classification of tropical American vegetation types, *Ecology*, 36, 89, 1955.
Bisse, J., *Árboles de Cuba*, Ministerio de Cultura, Editorial Científico-Técnica, Habana, Cuba, 1988.
Borchert, R. et al., Environmental control of flowering periodicity in Costa Rican and Mexican tropical dry forests, *Global Ecol. Biogeogr.*, 13, 409, 2004.
Borhidi, A., Dry coastal ecosystems of Cuba, in *Dry Coastal Ecosystems: Africa, America, Asia and Oceania*, Van der Maarel, E., Ed, Elsevier, Amsterdam, The Netherlands, 1993, 423.

Borhidi, A., *Phytogeography and Vegetation Ecology of Cuba,* 2nd ed., Akadémiai Kiadó, Budapest, Hungary, 1996.

Brown, S. and Lugo, A.E., Preliminary estimate of the storage of organic carbon on tropical forests ecosystems, in *The Role of Tropical Forests on the World Carbon Cycle,* Brown, S., Lugo, A.E., and Liegel, B., Eds, U.S. Department of Energy, CONF-800350 UC-11, Virginia, 1980, 65.

Bullock, S.H., Mooney, H.A., and Medina, E., Eds, *Seasonally Dry Tropical Forests,* Cambridge University Press, Cambridge, 1995.

Callaway, R.M., Positive interactions among plants, *Bot. Rev.,* 61, 306, 1995.

Castellanos, J. et al., Slash-and-burn effects on fine root biomass and productivity in a tropical dry forest ecosystem in México, *Forest Ecol. Manag.,* 148, 41, 2001.

Chinea, J.D., *The forest vegetation of the limestone hills of northern Puerto Rico,* Master's thesis, University of Cornell, New York, 1980.

Chinea, J.D., Árboles introducidos a la reserva de Guánica, Puerto Rico, *Acta Científica,* 4, 51, 1990.

Dansereau, P., Description and the integration of plant communities, in *Studies on the Vegetation of Puerto Rico, Special Publication 1,* Dansereau, P. and Buell, P.F., Eds, Institute of Caribbean Science, University of Puerto Rico, Puerto Rico, 1966, 3.

Dickinson, M.B., Hermann, S.M., and Whigham, D.F., Low rates of background canopy-gap disturbance in a seasonally dry forest in the Yucatan Peninsula with a history of fires and hurricanes, *J. Trop. Ecol.,* 17, 895, 2001.

Di Gregorio, A. and Jansen, L.J.M., *Land cover classification system (LCCS): classification concepts and user manual.* GCP/RAF/287/ITA Africover - East Africal Project and Soil Resources, Management and Conservation Service, Food and Agriculture Organization, Rome, 2000.

Dunevitz, V.L., *Regrowth of clearcut subtropical dry forest: mechanisms of recovery and quantification of resilience,* Master's thesis, Michigan State University, Department of Botany and Plant Pathology, East Lansing, 1985, 110.

Dunphy, B.K., Murphy, P.G., and Lugo, A.E., The tendency for trees to be multiple-stemmed in tropical and subtropical dry forests: studies of Guánica forest, Puerto Rico, *Trop. Ecol.,* 41,161, 2000.

Ewel, J.J., *Experiment in arresting succession with cutting and herbicides in five tropical environments,* Dissertation, University of North Carolina, Chapel Hill, 1971.

Ewel, J.J., Differences between wet and dry successional tropical ecosystems, *Geo-Eco-Trop,* 1, 103, 1977.

FAO, *Forest resources assessment 1990, survey of tropical forest cover and study of change processes,* FAO Forestry Paper 130, Rome, Italy, 1996, 1.

FAO, *FRA 2000: Guidelines for assessments in tropical and sub-tropical countries,* FAO working paper No. 2, Rome, Italy, 1998, 1.

Filip, V. et al., Within and between year variation in the levels of herbivory on the foliage of trees from a Mexican deciduous forest, *Biotropica,* 27, 78, 1995.

Frankie, G.W., Baker, H.H., and Opler, P.A., Comparative phenological studies of trees in tropical wet and dry forests in the lowlands of Costa Rica, *J. Ecol.,* 52, 881, 1974.

Furley, P.A., Ratter, J.A., and Gifford, D.R., Observations on the vegetation of eastern Mato Grosso. III. Woody vegetation and soil of the Morro de Fumaça, Torixoreu, Brasil, *Proc. R. Soc. Lond.* B, 203, 191, 1988.

Genet, K.S., *The resilience of lizard communities to habitat fragmentation in dry forests of southeastern Puerto Rico,* Master's thesis, Michigan State University, Department of Zoology, East Lansing, 1999, 132.

Genet, J.A. et al., Response of termite community and wood decomposition rates to habitat fragmentation in a subtropical dry forest, *Trop. Ecol.,* 42, 35, 2001.

Gentry, A.H., Neotropical floristics diversity: phytogeographical connections between Central and South America, Pleistocene climatic fluctuations, or an accident of the Andean orogeny? *Ann. Missouri Bot. Gard.,* 69, 557, 1982.

Gentry, A.H., Diversity and floristic composition of Neotropical dry forests, in *Seasonally Dry Tropical Forests,* Bullock, S. H., Mooney, H.A., and Medina, E., Eds, Cambridge University Press, Cambridge, 1995, 146.

Gentry, A.H. and Dodson, C.H., Diversity and biogeography of the neotropical vascular epiphytes, *Ann. Missouri Bot. Gard.,* 74, 205, 1987.

Helmer, E., Forest conservation and land development in Puerto Rico, *Landscape Ecol.,* 19, 29, 2004.

Helmer, E.H. et al., Mapping forest type and land cover of Puerto Rico, a component of the Caribbean biodiversity hotspot, *Caribb. J. Sci.,* 38, 165, 2002.

Holdridge, L.R., *Life Zone Ecology,* Tropical Science Center, San José, Costa Rica, 1967.

Janzen, D.H., Patterns of herbivory in a tropical deciduous forest, *Biotropica*, 13, 271, 1981.

Janzen, D.H., Tropical dry forests: the most endangered major tropical ecosystems, in *Biodiversity*, Wilson E.O., Ed, National Academy Press, Washington, DC, 1988, 130.

Jaramillo, V.J. et al., Biomass, carbon, and nitrogen pools in Mexican tropical dry forest landscapes, *Ecosystems*, 6, 609, 2003.

Kelly, D.L. et al., Jamaican limestone forests: floristics, structure and environment of three examples along a rainfall gradient, *J. Trop. Ecol.*, 4, 121, 1988.

Kinzel, H., Influence of limestone, silicates and soil pH on vegetation, in *Physiological Plant Ecology III. Responses to Chemical and Biological Environment*, Lange, O.L. et al., Eds, Springer Verlag, Berlin, 1983, 201.

Kinzel, H., Calcium in the vacuoles and cell walls of plant tissue: forms of deposition and their physiological and ecological significance, *Flora*, 182, 99, 1989.

Kuijt J., Carlo, T.A., and Aukema J.E., A new endemic species for Puerto Rico: *Dendrophthora bermejae*, *Contr. Univ. Michigan Herb.*, 24, 115, 2005.

Lal, C.B. et al., Foliar demand and resource economy of nutrients in dry tropical forest species, *J. Veg. Sci.*, 12, 5, 2001.

Lammers, T.G. and Proctor, G.R., *Lobelia vivaldii* (Campanulaceae: Lobelioideae) a remarkable new species of sect. Tylomium from Isla de Mona, Puerto Rico, *Brittonia*, 46, 273, 1994.

Lerdau, M., Whitbeck, J., and Holbrook, N.M., Tropical deciduous forests: death of a biome, *Trends Ecol. Evol.*, 6, 201, 1991.

Liogier, A.H., Novitates Antillanae, VIII, *Phytologia*, 47, 167, 1980.

Little, E.L. and Wadsworth, F.H., *Common trees of Puerto Rico and the Virgin Islands*, Agriculture Handbook 249, USDA Forest Service, Washington, DC, 1964.

Little, E.L., Woodbury, R.O., and Wadsworth, F.H., *Trees of Puerto Rico and the Virgin Islands*, volume 2, USDA Forest Service Agriculture Handbook 449, Washington, DC, 1974, 1024.

Lötschert, W., Die Übereinstimmung von geologischer Unterlage und Vegetation in der Sierra de los Organos (Westkuba), *Ber. Deutsch. Bot. Ges.*, 71, 55, 1958.

Lugo, A.E., Dominancia y diversidad de plantas en Isla de Mona, *Acta Científica*, 5, 65, 1991.

Lugo, A.E. and Brandeis, T., A new mix of alien and native species coexist in Puerto Rico's landscapes, in *Biotic Interactions in the Tropics*, Burslem, D., Pinard, M., and Hartley, S., Eds, Cambridge University Press, Cambridge, 2005, 484.

Lugo, A.E. and Helmer, E., Emerging forests on abandoned land: Puerto Rico's new forests, *Forest Ecol. Manag.*, 190, 145, 2004.

Lugo, A.E. and Murphy, P.G., Nutrient dynamics of a Puerto Rican subtropical dry forest, *J. Trop. Ecol.*, 2, 55, 1986.

Lugo, A.E. and Scatena, F.N., Epiphytes and climate change research in the Caribbean: a proposal, *Selbyana*, 13, 123, 1992.

Lugo, A.E., Colón, J.F., and Scatena, F.N., The Caribbean, in *North American Terrestrial Vegetation*, Barbour, M.G. and Billings, W.D., Eds, Cambridge University Press, Cambridge, 2000, 593.

Lugo, A.E. et al., Structure, productivity, and transpiration of a subtropical dry forest in Puerto Rico, *Biotropica*, 10, 278, 1978.

Lugo, A.E. et al., *Puerto Rican karst - a vital resource*, USDA Forest Service, General Technical Report WO-65, Washington, DC, 2001, 1.

Marschner, H., *Mineral Nutrition of Higher Plants*, Academic Press, Cambridge, 1995.

Mateucci, S., The vegetation of Falcon state, Venezuela, *Vegetatio*, 70, 67, 1987.

Matson, P.A. and Vitousek, P.M., Nitrogen trace gas emissions in a tropical dry forest ecosystem, in *Seasonally Dry Tropical Forests*, Bullock, S.H., Mooney, H.A., and Medina, E., Eds, Cambridge University Press, Cambridge, 1995, 384.

McDonald, M.A. and McLaren, K.P., Coppice regrowth in a disturbed tropical dry limestone forest in Jamaica, *Forest Ecol. Manag.*, 180, 99, 2003.

Medina, E., 1995. Diversity of life forms of higher plants in neotropical dry forests, in *Seasonally Dry Tropical Forests*, Bullock, S.H., Mooney, H.A., and Medina, E., Eds, Cambridge University Press, Cambridge, 1995, 221.

Medina, E. and Cuevas, E., Propiedades fotosintéticas y eficiencia de uso de agua de plantas leñosas del bosque decíduo de Guánica: consideraciones generales y resultados preliminares, *Acta Científica*, 4, 25, 1990.

Miller, P.M. and Kaufman, J.B., Effects of slash and burn agriculture on species abundance and composition of a tropical deciduous forest, *Forest Ecol. Manag.*, 103, 191, 1998.

Molina Colón, S., *Long-term recovery of a Caribbean dry forest after abandonment of different land uses in Guánica, Puerto Rico*, PhD Dissertation, University of Puerto Rico, Puerto Rico, 1998.

Murphy, P.G. and Lugo, A.E., Ecology of tropical dry forest, *Annu. Rev. Ecol. Syst.*, 17, 67, 1986a.

Murphy, P.G. and Lugo, A.E., Structure and biomass of a subtropical dry forest in Puerto Rico, *Biotropica*, 18, 89, 1986b.

Murphy, P.G. and Lugo, A.E., Dry forests of Central America and the Caribbean, in *Seasonally Dry Tropical Forests*, Bullock, S.H., Mooney, H.A., and Medina, E., Eds, Cambridge University Press, Cambridge, 1995, 9.

Myers, N. et al., Biodiversity hotspots for conservation priorities, *Nature*, 403, 853, 2000.

Neumann, C.J. et al., *Tropical cyclones of the North Atlantic Ocean, 1871–1977*, National Climatic Center, US Department of Commerce, National Oceanic and Atmospheric Administration, Ashville, North Carolina, 1978.

Oliveira-Filho, A.T. and Ratter, J.A., Vegetation physiognomies and woody flora of the Cerrado Biome, in *The Cerrados of Brazil*, Oliveira, P.S. and Marquis. R.J., Eds, Columbia University Press, New York, USA, 2002, 91.

Olson, D. et al., Terrestrial ecoregions of the world: a new map of life on Earth, *Bioscience*, 51, 933, 2001.

Ramjohn, I.A., *The role of disturbed Carribean dry forest fragments in the survival of native plant diversity*, PhD Dissertation, Michigan State University, Department of Botany and Plant Pathology, East Lansing, 2004, 288.

Ratter, J. A. et al., Observations on the vegetation of northeastern Mato Grosso. I. The woody vegetation types of the Xavantina-Cachimbo expedition area, *Phil. Trans. R. Soc. Lond.* B, 266, 449, 1973.

Reich, P.B. and Borchert, R., Water stress and tree phenology in a tropical dry forest in the lowlands of Costa Rica, *J. Ecol.*, 72: 61, 1984.

Rosen, B.R., Empiricism and the biogeographical black box: concepts and methods in marine palaeobiogeography, *Palaeogeography Palaeoclimatology Palaeoecology*, 92, 171, 1992.

Rosen, B.R. and Smith, A.B., Tectonics from fossils? Analysis of reef-coral and sea-urchin distributions from late Cretaceous to Recent, using a new method, in *Gondwana and Tethys*, Audeley-Charles, M.G. and Hallman, A., Eds, Geological Society, Special Publication (London) 37, 1988, 275.

Rouse, I., *The Tainos: Rise and Decline of the People who Greeted Columbus,* Yale University Press, New Haven, 1992.

Samek, V., Fitorregionalización del Caribe, *Revista del Jardín Botánico Nacional (Cuba)*, 9, 25, 1988.

Sánchez-Azofeifa, G.A. et al., Need for integrated research for a sustainable future in tropical dry forests, *Conserv. Biol.*, 19, 285, 2005.

Sarmiento, G., A conceptual model relating environmental factors and vegetation formations in the lowlands of tropical South America, in *Nature and Dynamics of Forest-Savanna Boundaries*, Furley, P.A., Proctor, J., and Ratter, J.A., Eds, Chapman and Hall, London, 1992, 583.

Sauleda, R.P., A revision of the genus *Psychilis* Rafinesque (Orchidaceae), *Phytologia*, 65, 1, 1988.

Serrano, A.E., Ortiz, C., and Aponte, M., Vegetación y factores edáficos en el bosque de Cambalache, in *Los Bosques de Puerto Rico*, Lugo, A.E., Ed, Servicio Forestal de los Estados Unidos de América y Departamento de Recursos Naturales de Puerto Rico, San Juan, 1983, 131.

Stoffers, A.L., Dry coastal ecosystems of the West Indies, in *Dry Coastal Ecosystems: Africa, America, Asia and Oceania*, Van der Maarel, E., Ed, Elsevier, Amsterdam, The Netherlands, 1993, 407.

Tanner, E.V.J., Four montane forests of Jamaica: a quantitative characterization of the floristics, soils, and foliar mineral levels and interactions, *J. Ecol.*, 65, 883, 1977.

Trejo-Torres, J.C. and Ackerman, J.D., Biogeography of the Antilles based on a parsimony analysis of orchid distributions, *J. Biogeogr.*, 28, 775, 2001.

Trejo-Torres, J.C. and Ackerman, J.D., Composition patterns of Caribbean limestone forests: Are parsimony, classification, and ordination analyses congruent? *Biotropica*, 34, 502, 2002.

Van Bloem, S.J., Murphy, P.G., and Lugo, A.E., Subtropical dry forest trees with no apparent damage sprout following a hurricane, *Trop. Ecol.*, 44, 1, 2003.

Van der Maarel, E., *Dry Coastal Ecosystems: Africa, America, Asia and Oceania,* Elsevier, Amsterdam, The Netherlands, 1993.

Wadsworth, F.H., Plantaciones forestales en el Bosque Estatal de Guánica, *Acta Científica*, 4, 61, 1990.

Weaver, P.L., *Tree growth in several tropical forests of Puerto Rico*, USDA Forest Service Research Paper SO-152, Southern Forest Experiment Station, New Orleans, 1979, 1.

Wright, S.J., Seasonal drought, soil fertility and species density of tropical forest plant communities, *Trends Ecol. Evol.*, 7, 260, 1992.

Wunderle, J.M., Lodge, D.J., and Waide, R.B., Short-term effects of Hurricane Gilbert on terrestrial bird populations on Jamaica, *Auk*, 109,148, 1992.

Zanoni, T.A. et al., La flora y vegetación de Los Haitises, República Dominicana, *Moscosoa*, 6, 46, 1990.

16 Diversity, Biogeography and Conservation of Woody Plants in Tropical Dry Forest of South Florida

Thomas W. Gillespie

CONTENTS

ABSTRACT

Tropical dry forests in south Florida occur as fragments in the Florida Keys, Miami-Dade county and the Florida Everglades. Native woody plants were examined in 23 plots (500 m²) in mature stands of tropical dry forest and 48 reserves in south Florida. In total, 64 native woody species were recorded whose primary habitat is dry forest. Tropical dry forests in south Florida contain low diversity and endemism, but the floristic composition and forest structure are similar to those of dry forests in the Bahamas and coastal Cuba. Dry forests in south Florida and the Caribbean have significantly different natural history characteristics (flower size, dispersal types and canopy phenology) compared with most neotropical dry forests on the mainland. There are 23 species listed as endangered and nine species listed as threatened. However, the conservation status of remaining fragments in south Florida can be viewed as excellent based on the number of reserves and reserve management.

16.1 TROPICAL DRY FORESTS OF SOUTH FLORIDA

Tropical dry forests, called tropical hardwood hammocks or hardwood hammocks in Florida, historically occurred on limestone outcrops in extreme southern Florida. Today, tropical dry forests occur on islands in the Keys, on elevated limestone along the Atlantic coast and as fragments within the more extensive pine savannas of the Florida Everglades (Snyder et al., 1990; Horvitz et al., 1998). Native tropical trees from Caribbean dry forests dominate these forests, although the mainland and Upper Keys are periodically exposed to short-term frosts that do not fit the technical climatic classifications of tropical dry forests as forests in frost-free regions (Holdridge et al., 1971; Ross et al., 1992). However, the forest structure is identical to that of the tropical dry forests of the Bahamas and coastal Cuba, and the flora is a subset of species that can withstand extremely rare frost events (Patterson and Stevenson, 1977; Tomlinson, 1980; Bisse, 1988). Like tropical dry forests of the Bahamas, there is little topographical relief in the region and limestone is the dominant substrate. Tropical dry forests in south Florida occur on Miami and Key Largo limestone, which has skeletal organic soils with minor mineral components and rarely exceeds 20 cm in depth (Snyder et al., 1990; Ross et al., 2001). Mean annual temperatures in the tropical dry forest region range from 23°C (74°F) in the north to 26°C (77°F) in the Lower Keys. Precipitation primarily occurs from June to October and ranges from 1650 mm along the Atlantic coast, decreasing southwards, to less than 1000 mm in the Lower Keys (Snyder et al., 1990; Ross et al., 1992).

16.2   HISTORICAL BIOGEOGRAPHY

The flora of tropical dry forests in south Florida is most likely derived from recent colonization since the Wisconsin glacial maximum 18,000 years ago. Although the extent of south Florida was significantly greater during the Wisconsin glaciation as a result of lowered sea levels, modelled climatic data and pollen evidence suggest lower temperatures and precipitation in the region, so that it may have been too cold and dry to support most tropical dry forest plants (Webb, 1990). During the last glacial maximum, south Florida was dominated by Florida scrub, swamps and mangroves on the southern tip of the peninsula (Webb, 1990). The presence of mangroves during this time does suggest that some tropical dry forest plants could persist in the region during the Pleistocene. During the mid Holocene, temperature and precipitation regimes were probably adequate for the current flora to colonize the region. Most species probably colonized south Florida by drifting from Cuba and the Bahamas. A majority of tropical dry forest trees in south Florida have drupes with relatively durable exocarps that protect the seed for a short period of time in the salt water. These trees may have dispersed to south Florida in the jet stream or during the hurricanes that are frequent in the region.

When Europeans first arrived in south Florida, they reported large expanses of pine savanna (*Pinus elliottii* Engelm. var. *densa* Little & Dorman) on limestone outcrops in south Florida. Records from the 1700s and 1800s suggest a high incidence of fire in south Florida from activities of Native Americans, and a high frequency of lighting strike fires in the region (Craighead, 1974). During the first half of the twentieth century, anthropogenic disturbance in the form of agriculture, logging, cattle grazing and fire had a significant impact on the extent and quality of tropical dry forests (Ross et al., 2001). Historical agricultural activity was an important anthropogenic impact on dry forests in south Florida. Many mature patches of forest in the Keys today are located in areas where intensive agriculture occurred as recently as 1935; however, most agriculture in the Keys was abandoned after a 1935 hurricane that destroyed the railroad that ran from Miami to Key West (Ross et al., 2001). Logging and cattle grazing were common in south Florida, although both have rarely occurred over the past 50 years.

Fire is required for the maintenance of pine savannas that are in close proximity to many tropical dry forests in south Florida. Fires in pine savannas are generally surface fires that consume litter and some understorey vegetation. These surface fires usually go out within a matter of minutes when they reach the margin of tropical dry forests (Robertson, 1953). However, soil fires that can proceed for weeks have occurred in tropical dry forests in south Florida during extreme drought periods (Robertson, 1953; Craighead, 1974). Historically, fire has been the single most important variable in determining the boundary between pine savannas and tropical dry forest boundaries in south Florida. I have observed a similar pattern for subtropical dry forests in Thailand, Cuba and Mexico in the northern hemisphere, and in the dry forests of New Caledonia and Australia in the southern hemisphere, where the savannas are dominated by fire-resistant Myrtaceae species. Currently, there are few fires in the tropical dry forests of south Florida outside the Florida Everglades and Big Pine Key because of active fire suppression programmes in Miami and the Florida Keys. This has resulted in a decrease in the extent of pine savanna, as woody dry forest species colonize unburned areas.

16.3 OBJECTIVES AND METHODS

Tropical dry forests of south Florida are within the Caribbean biodiversity hotspot because of extensive deforestation in the region (Myers et al., 2000). However, there have been few regional-scale studies that examine species richness, floristic composition and forest structure of remaining fragments of dry forests in south Florida (Ross et al., 2001). This research examines patterns of species richness, floristic composition, natural history characteristics and forest structure of the tropical dry forests of south Florida, and examines the conservation status of native woody plants in stands and reserves in south Florida. Results are compared with those from tropical dry forests in the Caribbean and Neotropics to identify general biogeographical patterns that occur in the region.

Research was undertaken in 23 stands of tropical dry forest in 20 protected areas (Figure 16.1). Eleven stands were located on islands in the Florida Keys, nine stands were located within an urban-agricultural

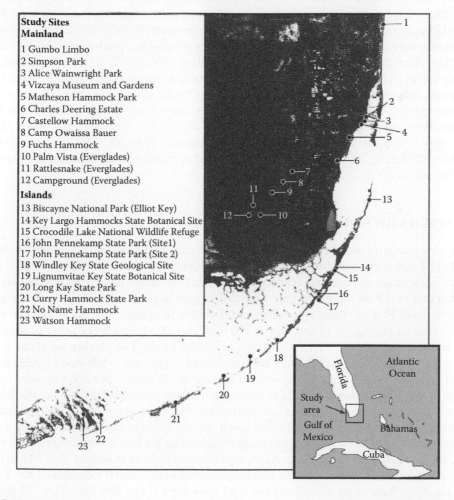

Study Sites
Mainland

1 Gumbo Limbo
2 Simpson Park
3 Alice Wainwright Park
4 Vizcaya Museum and Gardens
5 Matheson Hammock Park
6 Charles Deering Estate
7 Castellow Hammock
8 Camp Owaissa Bauer
9 Fuchs Hammock
10 Palm Vista (Everglades)
11 Rattlesnake (Everglades)
12 Campground (Everglades)

Islands

13 Biscayne National Park (Elliot Key)
14 Key Largo Hammocks State Botanical Site
15 Crocodile Lake National Wildlife Refuge
16 John Pennekamp State Park (Site1)
17 John Pennekamp State Park (Site 2)
18 Windley Key State Geological Site
19 Lignumvitae Key State Botanical Site
20 Long Kay State Park
21 Curry Hammock State Park
22 No Name Hammock
23 Watson Hammock

FIGURE 16.1 The location of 23 mature stands of tropical dry forest in south Florida. Shaded areas are land and white areas are water.

matrix in Miami-Dade and Broward counties, and three were located in fragments in the more extensive pine savannas of the Everglades National Park. All sites are protected at the federal, state or local levels (such as city parks) or managed by non-profit organizations. All sites contain forest greater than 2 ha that is dominated by at least 95% native tropical tree species, and they have a similar geological substrate with no evidence of surface water or flooding during the wet season. These sites represent the largest and best-protected tropical dry forest sites in south Florida.

Woody plant species richness, floristic composition and structure were quantified at the stand and patch level in tropical dry forests of south Florida. At the stand level, permanent belt transects totalling 500 m^2 were established to quantify woody plant species richness at each site in the summers of 2000, 2001 and 2002. Each sample consisted of five belt transects (2 m × 50 m), 10 m apart, in which all plants ≥ 2.5 cm in diameter at breast height (dbh) rooted in the sample area were recorded (Gentry, 1988). Gentry's original transect method included 10 transects, but only 5 were used in south Florida because of relatively low diversity, the small size of a number of fragments and extremely high mosquito densities in the summer. Tree height was measured into 2-m cohorts (i.e. 0–2 m, 2–4 m and 4–6 m), with a retractable pole for all individuals within transects. Transects were located in mature stands or stands judged to be in an advanced state of regeneration following Gentry (1988), and were established at least 20 m from the edge.

At the patch level, presence or absence data for native tropical woody plants were collected from 48 protected areas that contain tropical dry forest. Only native tropical woody plants able to reach a dbh of 2.5 cm or greater that occur in mature stands of tropical dry forests in south Florida were included in this analysis. Woody plants that occur primarily in pine savannas, mangroves and coastal dunes were excluded from this analysis. Species richness per patch was based on extensive species lists from the Institute for Regional Conservation (IRC, 2003). The IRC has undertaken extensive surveys of all plant species in a number of habitats and protected areas in south Florida over the past 10 years. I undertook systematic searches for all species with dbh 2.5 cm or greater at six sites (Simpson, Matheson, Vizcaya, John Pennekcamp, Windley Key and the single largest site, Elliot Key) in order to assess the accuracy of the species lists from the IRC. Plant lists from the IRC appear to be very accurate, with an overall accuracy of 99.5% when compared with my field survey (Gillespie, 2005). All species I encountered in transects at all sites had been identified by the IRC as occurring at each study site.

16.4 SPECIES RICHNESS

Tropical dry forests in south Florida contain low diversity and endemism compared with other tropical dry forest regions in the Neotropics, which is contrary to findings that neotropical forests at higher latitudes are more diverse (Gentry, 1988). There are approximately 64 woody plants able to reach a dbh of 2.5 cm whose primary habitat is mature tropical dry forest, none of which are endemic to south Florida. There is low standard deviation of species richness within 100-m^2 belt transects, with an average of 11 species per 100 m^2 (Table 16.1). Species richness is surprisingly similar in mature stands of tropical dry forest in south Florida. The Florida mainland averages about 18 species per 500 m^2, with a standard deviation of 2.7 species, while stand species richness is slightly higher in the Florida Keys, with an average of 20 species per 500 m^2 and a standard deviation of 4.3 species. However, there is greater variation in patch species richness. Patch species richness is greatest in the Upper Keys and declines in the north along the Florida mainland and south towards the Lower Keys. Variation in patch species richness can be best explained by the size of the tropical dry forest fragments in south Florida, with the Upper Keys containing a number of large fragments (Elliott Key, 371 ha; Key Largo Hammocks State Botanical Site, 544 ha), while the forest fragments in the urban areas (Gumbo Limbo Environmental Complex, 6 ha; Simpson Park, 4 ha; Alice Wainwright Park, 7 ha) and the Lower Keys (Long Key State Park, 19 ha; Curry Hammock State Park, 33 ha) are significantly smaller (Gillespie, 2005).

TABLE 16.1
Woody Plant Species Richness in Stands (500 m²) and Patches (Presence or Absence Data for 64 Native Tropical Woody Plants) of Tropical Dry Forest in South Florida

Study Site	Mean no. of Spp. per 100-m² Transect (SD)	Stand Species Richness	Patch Species Richness
Gumbo Limbo Environmental Complex	7.4 (1.1)	15	21
Simpson Park	9.6 (1.8)	19	43
Alice Wainwright Park	9.0 (1.9)	18	31
Vizcaya Museum and Gardens	7.4 (1.5)	14	31
Matheson Hammock Park	9.4 (2.3)	19	39
Charles Deering Estate	10.4 (1.1)	19	41
Castellow Hammock	11.0 (1.4)	17	34
Camp Owaissa Bauer	10.2 (0.4)	18	34
Fuchs Hammock	8.4 (1.8)	15	27
Everglades National Park (Palm Vista)	11.2 (1.3)	19	—
Everglades National Park (Rattlesnake)	15.4 (1.8)	24	—
Everglades National Park (campground)	11.8 (1.3)	20	—
Biscayne National Park (Elliott Key)	13.8 (1.8)	27	50
Key Largo Hammocks State Botanical Site	10.8 (1.5)	21	51
Crocodile Lake National Wildlife Refuge	9.4 (1.7)	20	49
John Pennekamp State Park (site 1)	12.0 (0.7)	19	49
John Pennekamp State Park (site 2)	12.8 (1.5)	19	49
Windley Key State Geological Site	14.8 (1.6)	24	43
Lignumvitae Key State Botanical Site	13.4 (1.3)	23	40
Long Key State Park	13.8 (1.6)	25	37
Curry Hammock State Park	10.0 (0.7)	14	29
No Name Hammock	9.6 (1.7)	21	—
Watson Hammock	6.8 (1.8)	13	—

Remote sensing imagery from Landsat ETM+ and ASTER can be used to calculate the extent of tropical dry forests, and they can be used to predict stand and patch level species richness based on forest fragment area and spectral indices such as the normalized difference vegetation index (NDVI). NDVI is the ratio between the red and infrared bands, and is an estimate of vegetation greenness. The higher greenness values are associated with higher diversity. Combining forest patch area with NDVI significantly improved the prediction of stand and patch species richness in south Florida. NDVI has also been used to predict patch species richness in tropical dry forest fragments in Lake Guri, Venezuela (Feeley et al., 2005). This suggests that a first-order approximation of woody plant species richness in stands and patches of tropical dry forest is possible in biodiversity hotspots using multiple sensors (Gillespie, 2005).

16.5 FLORISTIC COMPOSITION

The most species-rich families in the tropical dry forests of south Florida are Myrtaceae, Sapotaceae, Rubiaceae, Rhamnaceae, Leguminosae (Fabaceae), Euphorbiaceae and Sapindaceae, which is similar to the case in tropical dry forests in the Bahamas and Cuba (Patterson and Stevenson, 1977; Tomlinson, 1980; Bisse, 1988). However, compared with tropical dry forests in the mainland Neotropics, there is a notable lack of Bignoniaceae, Apocynaceae and Flacourtiaceae (Gentry, 1995). The most speciose genera in south Florida are *Eugenia* (four spp.), *Bourrreia* (three spp.) and *Colubrina* (three spp.). Six species account for 63% of the native woody plants identified in

TABLE 16.2
**The Floristic Composition of the 20 Most Common Native Woody Plants from
23 Mature Stands of Tropical Dry Forest in South Florida, Incidence in 48 Reserves
and Incidence by Counties in Florida**

Family	Scientific Name	Individuals from Transects	Incidence in Reserves	Incidence in Counties
Polygonaceae	*Coccoloba diversifolia* Jacq.	830	42	9
Myrtaceae	*Eugenia axillaris* (Sw.) Willd.	555	45	20
Lauraceae	*Nectandra coriacea* (Sw.) Griseb.	539	33	11
Euphorbiaceae	*Ateramnus lucida* (Sw.) Rothm.	508	23	2
Myrtaceae	*Eugenia foetida* Pers.	488	37	12
Burseraceae	*Bursera simaruba* (L.) Sarg.	337	46	14
Myrtaceae	*Eugenia confusa* DC.	194	15	3
Anacardiaceae	*Metopium toxiferum* (L.) Krug. & Urb.	186	40	5
Rhamnaceae	*Krugiodendron ferreum* (Vahl) Urb.	153	35	7
Sapotaceae	*Dipholis salicifolia* (L.) A. DC.	123	32	6
Sapotaceae	*Sideroxylon foetidissimum* (Jacq.) H.J. Lam	112	36	13
Myrsinaceae	*Ardisia escallonioides* Schltdl. & Cham.	107	43	20
Nyctaginaceae	*Guapira discolor* (Spreng.) Little	94	36	7
Rosaceae	*Prunus myrtifolia* (L.) Urb.	87	13	1
Simaroubaceae	*Simarouba glauca* DC.	85	38	8
Leguminosae	*Lysiloma latisiliquum* (L.) Benth.	83	25	3
Rutaceae	*Amyris elemifera* L.	68	26	9
Palmae	*Thrinax radiata* Lodd. ex Schult. & Schult. f.	67	24	3
Meliaceae	*Swietenia mahagoni* (L.) Jacq.	57	20	3
Leguminosae	*Piscidia piscipula* (L.) Sarg.	50	27	6

23 stands of dry forest in south Florida (Table 16.2). These six species — *Coccoloba diversifolia* Jacq. (16%), *Eugenia axillaris* (Sw.) Willd. (11%), *Nectandra coriacea* (Sw.) Griseb. (10%), *Ateramnus lucida* (Sw.) Rothm. (10%), *Eugenia foetida* Pers. (9%) and *Bursera simaruba* (L.) Sarg. (7%) — are extremely common in tropical dry forests of south Florida and are common in the Bahamas, Cuba and Puerto Rico (Patterson and Stevenson, 1977; Bisse, 1988). Based on species incidence in 48 reserves that contain tropical dry forest, *Ficus aurea* Nutt. (48), *Bursera simaruba* (46), *Eugenia axillaris* (45), *Ardisia escallonioides* Schltdl. & Cham. (43), *Coccoloba diversifolia* (42) and *Chiococca alba* (L.) Hitchc. (42) are the most widely distributed native woody plants in tropical dry forests of south Florida.

There are a number of tropical dry forest plants that have large ranges within Florida based on their incidence in Florida's 67 counties. The most widely distributed species from tropical dry forests, based on incidence by counties, are *Psychotria nervosa* Sw. (33), *Zanthoxylum fagara* (L.) Sarg. (25), *Chiococca alba* (24), *Ardisia escallonioides* (20), *Eugenia axillaris* (20), *Ficus aurea* (19) and *Bursera simaruba* (14). These trees are generally confined to coastal thickets along the Gulf and Atlantic coast of Florida in a U-shaped pattern that corresponds nicely with the average temperature for the month of January (Tomlinson, 1980).

16.6 NATURAL HISTORY CHARACTERISTICS

Tropical dry forests in south Florida and the Caribbean have significantly different natural history characteristics of life forms, flower sizes, dispersal types and canopy phenology compared with tropical dry forests in the mainland Neotropics (see also Lugo et al., Chapter 15). There is a notable lack of lianas in south Florida and the Caribbean (Gentry, 1995). Most dry forest lianas on the mainland are wind dispersed and unable to disperse long distances over water, and many liana seeds may not be able to germinate after long periods in salt water. The most commonly encountered native lianas in tropical dry forests in Florida are *Pisonia aculeata* L. in the Florida Keys and *Vitis* spp. in the mainland. Two-thirds to three-quarters of the trees, shrubs and lianas of tropical dry forest in the mainland Neotropics have conspicuous flowers (>2 cm) (Gentry, 1982, 1995); however, only 14% of tropical dry forests trees and shrubs in south Florida have large showy flowers greater than 2 cm. The small flowers that occur in south Florida and the Caribbean may be a result of the general lack of specialized pollinators in isolated regions.

Tropical dry forest trees of Central America are composed of approximately 32% anemochory (wind-dispersed seeds), and tropical dry forests in South America are composed of approximately 26% anemochory (Gentry, 1995; Gillespie et al., 2000). However, tropical dry forest trees of south Florida contain only 5% anemochoric species, while the vast majority of species are zoochoric (90%). It appears that tropical dry forests in Florida, the Caribbean and other island regions are primarily composed of zoochoric trees. Even trees in tropical dry forests on isolated oceanic islands such as Hawaii and New Caledonia are dominated by zoochory (97% and 96%, respectively) (Wagner et al., 1990; Jaffre et al., 1993). Wind dispersal may not be as advantageous as zoochory in tropical dry forests in the Caribbean, because anemochory has been identified as a selective disadvantage on islands in the tropics (Carlquist, 1974).

The phenology of tropical dry forests is also significantly different from that of tropical dry forests in the mainland Neotropics. In particular, there are very few truly deciduous trees in south Florida. All tropical dry forests have a pronounced dry season, when a number of canopy species lose their leaves. The percentage of the canopy species around the world that lose their leaves varies between 10% and 90%, but in south Florida and the Caribbean early successional forests are generally dominated by semideciduous species and late successional forest are dominated by evergreen species (Murphy and Lugo, 1986; Ross et al., 2001; Lugo et al., Chapter 15). This has been well documented in tropical dry forest of the Florida Keys, where all dominant species in stands of forest <50 years old were semideciduous, 16% of the dominant species in stands 50–75 years old were semideciduous, and no semideciduous species were dominant in stands >75 years old (Ross et al., 2001). Ross et al. (2001) suggest that the replacement of deciduous by evergreen species may reflect soil development through an increase in buffering from periodic moisture stress during the dry season. This is in sharp contrast to the case in tropical dry forests on the Pacific coast of Mexico and Central America, where the vast majority of trees in mature stands of tropical dry forest lose their leaves during the dry season (Gentry, 1995).

16.7 FOREST STRUCTURE

A total of 5379 stems $\geq$2.5 cm dbh were recorded from 23 stands in south Florida (Table 16.3). There was an average of 47 stems per 100-m^2 transect, with a standard deviation of 15 stems. Stand density ranged from 72 stems per 500 m^2 at Gumbo Limbo in the northern region of the study area to 408 stems per 500 m^2 at Long Key State Park in the southern region of the study area, with an average number of 234 stems per 500 m^2 (standard deviation 73) or approximately 4680 stems $\geq$2.5 cm dbh per hectare. Stand basal area ranged from 17.2 m^2/ha at Camp Owaissa Bauer to 45.9 m^2/ha at Key Largo Hammock, and averaged 33.5 m^2/ha (standard deviation 7.6) for all 23 stands of tropical dry forest in south Florida. Mean tree height in the tropical dry forests of south Florida was 7.4 m and ranged between 5.8 m and 9.2 m. Maximum tree height ranged from a low of 10 m in the lower Florida Keys to a high of 24 m in the upper Florida Keys. Study sites with higher temperatures and lower precipitation resulted in forests with greater stand density, while study sites with higher precipitation and lower temperatures result in greater canopy heights (Gillespie et al., 2005).

TABLE 16.3
Summary of Forest Structure Characteristics for Native Woody Plants ≥2.5 cm in Diameter at Breast Height from 500-m² Samples in 23 Tropical Dry Forest Stands in South Florida

Study Site	No. of Stems	Stems per 100 m² (SD)	Basal Area (m²/ha)	Mean Height (m)	Maximum Height (m)
Gumbo Limbo Environmental Complex	72	14.4 (3.3)	38.6	7.93	12
Simpson Park	161	32.2 (4.6)	45.1	8.65	18
Alice Wainwright Park	175	35.0 (7.3)	38.0	7.93	18
Vizcaya Museum and Gardens	146	29.2 (3.5)	37.2	8.62	18
Matheson Hammock Park	167	33.4 (7.2)	33.1	6.73	16
Charles Deering Estate	192	38.4 (4.0)	31.8	6.62	18
Castellow Hammock	239	47.8 (7.7)	23.7	7.86	18
Camp Owaissa Bauer	223	44.6 (10.1)	17.3	7.26	20
Fuchs Hammock	184	36.8 (7.6)	37.5	8.70	26
Everglades National Park (Palm Vista)	205	41.0 (3.5)	26.6	7.40	18
Everglades National Park (Rattlesnake)	274	54.8 (7.8)	24.3	7.85	16
Everglades National Park (campground)	188	37.6 (3.8)	27.6	8.67	18
Biscayne National Park (Elliott Key)	303	60.6 (15.0)	26.4	6.44	14
Key Largo Hammocks State Botanical Site	321	64.2 (11.3)	45.9	7.80	16
Crocodile Lake National Wildlife Refuge	212	42.4 (4.5)	34.7	6.31	14
John Pennekamp State Park (site 1)	324	64.8 (10.5)	35.5	8.25	22
John Pennekamp State Park (site 2)	277	55.4 (6.1)	41.9	9.22	24
Windley Key State Geological Site	315	63.0 (12.2)	36.1	7.21	16
Lignumvitae Key State Botanical Site	263	52.6 (6.2)	45.2	6.36	18
Long Key State Park	408	81.6 (15.7)	31.8	6.07	10
Curry Hammock State Park	264	52.8 (8.6)	27.0	5.85	10
No Name Hammock	244	52.8 (2.5)	28.3	6.04	16
Watson Hammock	222	44.4 (10.5)	37.8	6.69	14

Tropical dry forests in Florida and the Caribbean had a greater density of individual stems and shorter canopy heights than tropical dry forests in the mainland Neotropics (Gillespie et al., 2006). Hurricanes, called cyclones and typhoons in other tropical regions, can have a significant long-term impact on tropical dry forest structure, and hurricane occurrence in tropical regions has been hypothesized to reduce canopy height and increase density in tropical dry forests (Quigley and Platt, 2003).

16.8 CONSERVATION

16.8.1 CONSERVATION STATUS OF SPECIES

Of the 64 native woody plants identified from tropical dry forests of south Florida, there are 23 species that have been listed as endangered and nine species listed as threatened by the Florida Fish and Wildlife Service (Table 16.4). However, when compared with their distribution and abundance in south Florida, it is clear that some species are more endangered than others. There are five endangered tropical dry forest trees restricted to only one reserve each in south Florida. *Bourreria radula* (Poir.) G. Don is restricted to Little Hammaca City Park in Key West, *Cupania glabra* Sw. is restricted to Watson Hammock on Big Pine Key, *Licaria triandra* (Sw.) Kosterm. is restricted

TABLE 16.4
Endangered and Threatened Woody Plants from Tropical Dry Forest of South Florida

Family	Scientific Name	Status*	Incidence in 48 Reserves	Incidence from Transects	Restricted to Caribbean Hotspot?
Boraginaceae	*Bourreria radula* (Poir.) G. Don	E	1	0	Yes
Sapindaceae	*Cupania glabra* Sw.	E	1	1	—
Lauraceae	*Licaria triandra* (Sw.) Kosterm.	E	1	0	—
Nyctaginaceae	*Pisonia rotundata* Griseb.	E	1	0	Yes
Palmae	*Pseudophoenix sargentii* H. Wendl. ex Sarg.	E	1	0	—
Simaroubaceae	*Alvaradoa amorphoides* Liebm.	E	5	0	—
Myrtaceae	*Calyptranthes zuzygium* (L.) Sw.	E	5	2	—
Rhamnaceae	*Colubrina cubensis* (Jacq.) Brongn.	E	5	0	Yes
Rhamnaceae	*Colubrina elliptica* (Sw.) Brizicky & Stern	E	5	7	—
Myrtaceae	*Eugenia rhombea* Krug & Urb. ex Urb.	E	5	0	—
Sapindaceae	*Hypelate trifoliata* Sw.	E	5	3	Yes
Celastraceae	*Gyminda latifolia* (Sw.) Urb.	E	6	6	—
Rubiaceae	*Exostema caribaeum* (Jacq.) Roem. & Schult.	E	7	0	—
Palmae	*Roystonea regia* (Kunth) O.F. Cook	E	7	2	Yes
Zygophyllaceae	*Guaiacum sanctum* L.	E	9	2	—
Simaroubaceae	*Picramnia pentandra* Sw.	E	9	14	—
Canellaceae	*Canella winterana* (L.) Gaertn.	E	11	0	Yes
Rhamnaceae	*Colubrina arborescens* (Mill.) Sarg.	E	11	0	—
Celastraceae	*Schaefferia frutescens* Jacq.	E	12	4	—
Euphorbiaceae	*Drypetes diversifolia* Krug. & Urb.	E	14	25	Yes
Boraginaceae	*Bourreria obvata* Miers	E	15	23	—
Myrtaceae	*Eugenia confusa* DC.	E	15	194	Yes
Palmae	*Thrinax radiata* H. Wendl.	E	24	67	—
Aquifoliaceae	*Ilex krugiana* Loes.	T	12	22	Yes
Rosaceae	*Prunus myrtifolia* (L.) Urb.	T	13	87	—
Sapotaceae	*Manilkara jaimiqui* (L.) Cronquist	T	16	29	Yes
Rhamnaceae	*Reynosia septentrionalis* Urb.	T	20	43	Yes
Meliaceae	*Swietenia mahagoni* (L.) Jacq.	T	20	57	Yes
Euphorbiaceae	*Drypetes lateriflora* (Sw.) Krug & Urb.	T	24	13	—
Myrtaceae	*Calyptranthes pallens* (Poir.) Griseb.	T	25	16	—
Sapotaceae	*Chrysophyllum oliviforme* L.	T	27	22	—
Leguminosae	*Pithecellobium keyense* Britton	T	30	31	—

*E, endangered; T, threatened.

to Simpson Hammock in the centre of downtown Miami, *Pisonia rotundata* Griseb. is restricted to the National Key Deer Refuge and *Pseudophoenix sargentii* H. Wendl. ex Sarg. is restricted to Elliot Key in Biscayne National Park. Based on transect data from stands, only one individual of *Cupania glabra* was encountered, suggesting that these species are indeed rare in south Florida. After these extremely rare species, there are six species restricted to five reserves and the remaining 12 species are restricted to between six and 24 reserves. Of the nine species listed as threatened, all occur in 12 or more reserves and were represented in transects within many of the study sites. Overall, it appears that the Florida Fish and Wildlife Service has done a good job of identifying endangered and threatened species in the tropical dry forests of south Florida.

TABLE 16.5
Species Richness and Endangerment for Tropical Dry Forests Based on Gentry's Transect Method

Biodiversity Hotspot (Region)	No. of Sites	Species Richness	No. of Threatened Spp.	IUCN Classes*
Caribbean (Florida)	2	34, 24	0	—
Mesoamerica (Mexico)	2	89, 83	3	2 E, 1 V
Mesoamerica (Costa Rica)	2	75, 65	0	—
Chocó–Darién–Western Ecuador (Ecuador)	2	61, 72	3	1 E, 2 R
Tropical Andes (Bolivia)	2	86, 79	4	4 R
Madagascar	2	46, 42	0	—
Coastal forest of Tanzania and Kenya (Tanzania)	1	89	3	2 V, 1 R
Western Ghats (India)	1	15	0	—
New Caledonia	2	55, 34	10	1 E, 9 V
Hawaii	2	7, 5	2	1 E, 1 R

* E, endangered; V, vulnerable; R, rare.

Source: Modified from Gillespie and Jaffre, 2003

Tropical dry forests of south Florida are part of the Caribbean biodiversity hotspot (Myers et al., 2000). The Caribbean biodiversity hotspot contains a number of isolated tropical dry forest regions, with large expanses of tropical dry forest in Cuba and the Bahamas; isolated fragments in protected areas on the Virgin Islands, Puerto Rico and Jamaica; and severely fragmented and degraded forest in the Lesser Antilles. The contribution of south Florida to the diversity and endemism in the hotspot is minimal compared with that of islands such as Cuba, which contains the highest diversity of tropical dry forest trees in the region and the highest endemism (Bisse, 1988). There are no endemic species in south Florida, which somewhat lessens the regional and global conservation status of the region. Compared with other tropical dry forests in biodiversity hotspots around the world, the tropical dry forests of Florida have no species listed by the IUCN and have some of the lowest species richness levels of any region (Table 16.5). It is interesting to note that the five very rare and endangered species in south Florida are common in other regions in the Caribbean (Patterson and Stevenson, 1977; Bisse, 1988).

16.8.2 CONSERVATION STATUS OF FRAGMENTS

The conservation status of remaining fragments in Florida at a global spatial scale can be viewed as excellent based on the number of reserves in the region, the lack of anthropogenic disturbance and reserve management. There is no question that tropical dry forest in Miami-Dade county and Key West has been affected by urban and agricultural development. Indeed, there are over two million people in the tropical dry forest region of south Florida. However, there are a number of federal (National Park Service, National Wildlife Refuge), state (state parks, state botanical sites), local (city parks), non-profit organization (educational, museums, the Nature Conservancy, botanical gardens) and private groups (estates) that have done an excellent job of protecting and managing remaining tropical dry forest fragments.

The tropical dry forests that occur as fragments within pine savannas in the Florida Everglades are extremely well protected by the National Park Service. The world's largest restoration project is underway in the Everglades National Park, at the cost of over one billion dollars. Part of this restoration project has been to replant tropical dry forests and pine savannas in areas that were historically converted to agriculture, and on the abandonment of agriculture invaded by pure stands of *Schinus terebinthifolius* Raddi. Restoration of these areas has been undertaken by mechanically removing all exotic vegetation and soils down to the limestone substrate. New soils are then brought in and native pine savanna and tropical dry forest species replanted.

Most tropical dry forest in Miami-Dade county has been converted to urban housing developments. The remaining fragments are all currently protected in a number of reserves and are also extremely well managed. In particular, the greatest threat to these remaining fragments is the invasion of exotic trees and lianas. This was best seen after hurricane Andrew in 1992, when a number of exotic trees and lianas were dispersed into conservation areas and began to out-compete native forest species (Horvitz et al., 1998). However, exotic species removal programmes were initiated, and currently there are few exotics within reserves. There were only six exotic species (23 individual plants) recorded in transects from 23 stands of tropical dry forest because of an active exotic species removal programme in the region.

The Florida Keys are currently the region where the extent and quality of tropical dry forest are the most threatened, especially surrounding urbanized areas. There are a number of tropical dry forest patches that occur in private parcels in the Upper and Middle Keys. These parcels are being colonized by exotic plants and converted to exotic forests because of lack of management. However, the Nature Conservancy, along with a number of federal and state agencies, is in the process of purchasing some of the largest and best remaining fragments of tropical dry forest in the area.

Tropical dry forests in south Florida are some of the most extensively surveyed and managed tropical dry forests in the world. Although a significant shift in government policy is always possible, it does appear that the people and the federal, state and local governments have done an outstanding job of protecting the only tropical forest that occurs in the continental USA. In contrast, tropical dry forests of Hawaii are the most under-protected, degraded and possibly endangered dry forest in the world. Furthermore, there are currently no comparative data on dry forest reserves or species composition in reserves in Hawaii. Therefore the conservation of dry forests in south Florida is not something that can be attributed to being part of the USA. Results from this research suggest that strict nature reserves that permit people to visit them, and the resources to protect and manage such reserves, are needed in order to protect mature stands of tropical dry forest at a global spatial scale. Strict reserves and the resulting benefits from local, national and international tourism may outweigh the negative impacts and cost of reserve establishment, infrastructure development and management in the long term. This certainly appears to be the case in south Florida.

REFERENCES

Bisse, J., *Arboles de Cuba*, Editorial Cientifico-Técnica, Havana, 1988, 384.

Carlquist, S., *Island Life*, Natural History Press, New York, 1974, 660.

Craighead, F.C., Hammocks of south Florida, in *Environments of South Florida: Present and Past*, Gleason, P.J., Ed, Miami Geological Society, Miami, 1974, 53.

Feeley, K.J., Gillespie, T.W., and Terborgh, J.W., The utility of spectral indices from Landsat ETM+ for measuring the structure and composition of tropical dry forests, *Biotropica*, 37, 508, 2005.

Gentry, A.H., Patterns of neotropical plant species diversity, *Evol. Biol.*, 15, 1, 1982.

Gentry, A.H., Changes in plant community diversity and floristic composition on environmental and geographical gradients, *Ann. Missouri Bot. Gard.*, 75, 1, 1988.

Gentry, A.H., Diversity and floristic composition of neotropical dry forests, in *Seasonally Dry Tropical Forests*, Bullock, S.H., Mooney, H.A., and Medina, E., Eds, Cambridge University Press, Cambridge, 1995, 146.

Gillespie, T.W., Predicting plant species richness in tropical dry forests: a case study from south Florida, *Ecol. Appl.*, 15, 27, 2005.

Gillespie, T.W. and Jaffre, T., Tropical dry forests of New Caledonia, *Biodivers. Conservation*, 12, 1687, 2003.

Gillespie, T.W., Grijalva, A., and Farris, C., Diversity, composition and structure of tropical dry forest in Central America, *Pl. Ecol.*, 147, 37, 2000.

Gillespie, T.W. et al., Predicting and quantifying the structure of tropical dry forests in South Florida and the Neotropics, *Glob. Ecol. Biogeogr.*, in press, 2006.

Holdridge, L.R. et al., *Forest Environments in Tropical Life Zones: A Pilot Study*, Pergamon Press, Oxford, 1971, 747.

Horvitz, C. et al., Functional role of invasive non-indigenous plants in hurricane-affected subtropical hardwood forest, *Ecology*, 8, 947, 1998.

Institute for Regional Conservation, *Floristic Inventory of South Florida*, http://www.regionalconservation.org/ircs/DatabaseChoice.cfm, 2003

Jaffre, T.P., Morat, P., and Veillon, J.M., Etude floristique et phytogéographique de la forêt sclérophylle de Nouvelle-Calédonie, *Bull. Mus. Natl. Hist. Nat., Ser. 3, Bot.*, 15, 107, 1993.

Murphy, P.G. and Lugo, A.E., Ecology of tropical dry forest, *Annual Rev. Ecol. Syst.*, 17, 67, 1986.

Myers, N. et al., Biodiversity hotspots for conservation priorities, *Nature*, 403, 853, 2000.

Patterson, J. and Stevenson, G., *Native Trees of the Bahamas*, J. Patterson, Hope Town, 1977, 128.

Quigley, M.F. and Platt, W.J., Composition and structure of seasonally deciduous forests in the Americas, *Ecol. Monogr.*, 73, 87, 2003.

Robertson, W.B., Jr, *A Survey of the Effects of Fire in the Everglades National Park*, National Park Service, Homestead, 1953, 68.

Ross, M.S., O'Brien, J., and Flynn, L., Ecological site classification of Florida Keys terrestrial habitats, *Biotropica*, 24, 488, 1992.

Ross, M.S. et al., Forest succession in tropical hardwood hammocks of the Florida Keys: effects of direct mortality from Hurricane Andrew, *Biotropica*, 33, 23, 2001.

Snyder, J.R., Herndon, A., and Robertson, W.B., South Florida rockland, in *Ecosystems of Florida*, Myers, R.L. and Ewel, J.J., Eds, University of Central Florida Press, Orlando, 1990, 230.

Tomlinson, P.B., *The Biology of Trees Native to Tropical Florida*, Harvard University Printing Office, Cambridge, 1980, 210.

Wagner, W.L., Herbst, D.R., and Sohmer, S.H., *Manual of the Flowering Plants of Hawai'i*, University of Hawaii Press and Bishop Museum Press, Honolulu, 1990, 1853.

Webb, T., Historical biogeography, in *Ecosystems of Florida*, Myers, R.L. and Ewel, J.J., Eds, University of Central Florida Press, Orlando, 1990, 70.

17 The Late Quaternary Biogeographical History of South American Seasonally Dry Tropical Forests: Insights from Palaeo-Ecological Data

Francis E. Mayle

CONTENTS

ABSTRACT

A review of previously published palaeo-ecological data reveals a complex biogeographical history of seasonally dry tropical forests (SDTF) in South America over the Late Quaternary. Some dry forest communities (e.g. Colombian Andes) have undergone considerable re-assortment of species through the Holocene, whilst others (e.g. south-eastern Brazil) appear to have been more stable. In contrast to prediction from the dry forest refugia hypothesis, the key dry forest species, *Anadenanthera colubrina*, has only been an important component of the Bolivian Chiquitano dry forest since the early Holocene, and appears to have been absent during the last glacial period. Thorn-scrub caatinga vegetation probably dominated north-eastern Brazil over most of the Late Quaternary period, although it diminished during wetter intervals, e.g. towards the end of the last glacial period (*c.*15,500–11, 800 carbon-14 yr BP (18,500–13,800 cal yr BP)), when semi-deciduous dry forest expanded. Although the distribution of SDTF across South America during the last glacial period remains unclear, there is consistent evidence from most dry forest regions for a mid–late Holocene peak in SDTF cover, centred around 6000–3000 carbon-14 yr BP (6800–3200 cal yr BP), due to widespread reduction in precipitation across tropical South America. Rather than assuming that the current disjunct distribution of SDTF taxa across South America is attributable to vicariance (i.e. fragmentation) of once continuous populations spanning these dry forest areas, I propose population migration/dispersal as an alternative explanation.

17.1 INTRODUCTION

The responses of neotropical ecosystems to Late Quaternary environmental changes have long been the subject of considerable debate and controversy. Scientific interest, however, has focused almost exclusively on the biogeographical histories of rain forest versus savanna ecosystems, especially with respect to Amazonia (e.g. Haffer, 1969; Colinvaux et al., 1996, 2000; Haffer and Prance, 2001), with little attention paid to seasonally dry tropical forests (SDTF), perhaps because so little of this forest type remains intact today (Pennington et al., 2000). Consequently, the biogeographical history of SDTF is poorly understood. It should be noted that the term SDTF is used here in a general sense for the Neotropics, following Murphy and Lugo (1995), encompassing tall, closed-canopy semi-deciduous forest on moister sites (e.g. the Bolivian Chiquitano forest) as well as cactus thorn scrub in more arid areas (e.g. the Brazilian caatinga).

The aim of this chapter, which expands upon Mayle (2004), is to review previously published palaeo-ecological data (predominantly fossil pollen records) from the South American tropics to determine how SDTF have responded to Late Quaternary environmental changes, and thereby better understand the development of the current disjunct pattern of dry forest distributions, and the floristic links between them. In particular, the validity of the dry forest refugia hypothesis, proposed by Prado and Gibbs (1993) and Pennington et al. (2000) will be examined.

17.2 DRY FOREST REFUGIA HYPOTHESIS

Seasonally dry tropical forests exhibit highly fragmentary distributions across South America (Figure 17.1). Small, isolated patches of dry forest occupy dry valleys in the tropical Andes between Colombia and Bolivia. However, the largest areas of SDTF occur south of the Amazon rain forests, forming a discontinuous, disjunct distribution, connected by cerrado savannas and chaco woodlands, which together form a 'dry diagonal' (Prado and Gibbs, 1993) of woody vegetation between the caatinga in the north-east and the Andean piedmont dry forests in the south-west (Figure 17.1).

Prado and Gibbs (1993) and Pennington et al. (2000) compared the current distributions of dry forest species across the South American tropics and showed that over 100 phylogenetically unrelated species have similar geographical patterns, forming four disjunct (i.e. separated) dry forest blocks or nuclei (caatinga, misiones, chiquitano, and Andean piedmont nuclei) arranged diagonally south of Amazonia. A key taxon is *Anadenanthera colubrina* (Vell.) Brenan, which was found to be present, and in some areas dominant, in all the major dry forest nuclei of South America (Figure 17.2), with the exception of the Caribbean coasts of Colombia and Venezuela.

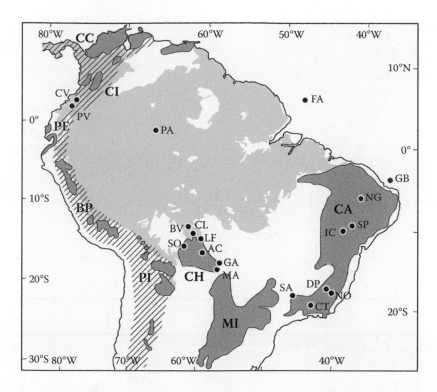

FIGURE 17.1 Map showing the distribution of seasonally dry tropical forest (SDTF) in South America and pollen records discussed in the text. CA: caatinga nucleus; MI: misiones nucleus; CH: Chiquitano dry forest nucleus; PI: piedmont nucleus; BP: Bolivian and Peruvian inter-Andean valleys; PE: Pacific coastal Ecuador; CL: Laguna Chaplin (14°28S, 61°04W) (Burbridge et al., 2004); BV: Laguna Bella Vista (13°37S, 61°33W) (Burbridge et al., 2004); LF: Los Fierros vegetation plot (14°34 50S, 66°49 48W); AC: Acuario vegetation plot (15°14 58S, 61°14 42W); GA: Laguna La Gaiba (17°47S, 57°43W); MA: Laguna Mandiore (18°05 31S, 57°33 46W); SO: Laguna Socorros (16°08 30S, 63°07 00W); GB: Marine core GeoB 3104-1 (3°40S, 37°43W) (Behling et al., 2000); IC: Icatu River Valley (10°24S, 43°13W) (De Oliveira et al., 1999); SP: Speleothems (10°10S, 40°50W) (Wang et al., 2004); NG: noble gases (7°S, 41.5°W) (Stute et al., 1995); SA: Salitre (19°S, 46°46W) (Ledru et al., 1996); CT: Catas Altas (20°05S, 43°22W) (Behling and Lichte, 1997); DP: Lago do Pires (17°57S, 42°13W) (Behling, 1995); NO: Lagoa Nova (17°58S, 42°12W) (Behling, 2003); CV: Cauca Valley (Quilichao-1 [3°6'N, 76°31W] and La Teta-2 [3°5'N, 76°32'W]) (Berrio et al., 2002); PV: Patía Valley (2°02N, 77°W) (Vélez et al., 2005); PA: Lake Pata (0°16N, 66°41W) (Colinvaux et al., 1996; Bush et al., 2002); FA: Amazon Fan (5°12.7N, 47°1.8W) (Haberle and Maslin, 1999). (Modified from Pennington et al., 2000, and Mayle, 2004).

The authors interpreted these repeated patterns of fragmented populations as evidence that the dry forest nuclei constitute remnants (i.e. refugia) of a much larger, single formation (the Pleistocene dry forest arc) that spanned these disjunct dry forest regions during drier periods of the Pleistocene (e.g. the Last Glacial Maximum (LGM), *c.*18,000 carbon-14 yr BP (radiocarbon years before present), or 21,000 cal yr BP (calendar years before present) (Stuiver et al., 1998), extending into regions now covered by Amazon rain forest and cerrado savannas.

Prado and Gibbs (1993) hypothesized that the increasing temperatures and precipitation over the last glacial-Holocene transition caused contraction and fragmentation of this Pleistocene dry forest arc into the present-day disjunct distribution of dry forest nuclei. They argued that increased precipitation would have favoured replacement of dry forests by rain forests in southern Amazonia.

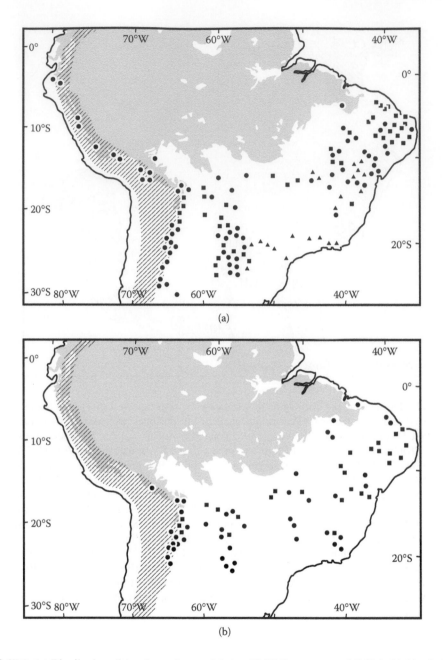

FIGURE 17.2 (a) Distribution of *Anadenanthera colubrina* (Vell.) Brenan *var. cebil* (Griseb) Altschul (circles: herbarium specimens or citations in recent monographs; squares: citations in floristic lists) and *Anadenanthera colubrina var. colubrina* (triangles). (b) Distribution of *Astronium urundeuva* (circles: herbarium specimens or citations in recent monographs; squares: citations in floristic lists). Shaded area: Amazonian moist evergreen rain forest; hatched area: Andes mountains. (Modified from Prado and Gibbs, 1993; after Mayle, 2004).

Further south, Ratter et al. (1988) postulated that progressive leaching of nutrients under the more humid climate would have led to acidification of the poorer soils associated with much of the pre-Cambrian shield, favouring expansion of cerrado (upland) savannas at the expense of dry forests. According to this hypothesis, dry forests would be expected to persist continuously since the LGM only in those areas (i.e. the contemporary SDTF refugia) where base-rich soils could be continuously replenished by weathering of parent calcareous rocks (Ratter et al., 1988).

Pennington et al. (2000) further showed that some dry forest species (e.g. *Commiphora leptophloeos* (Mart.) Gillet and *Aspidosperma pyrifolium* Mart.) are not entirely confined to the disjunct dry forest areas, but also occur scattered throughout the Amazon Basin, albeit very sparsely and at very low densities amongst the evergreen rain forest. The authors considered these sporadic Amazonian distributions to be refugia or relicts of larger populations during the last glacial period, arguing that a drier climate during the LGM, in combination with a lower water table associated with lowered sea levels, would have reduced seasonal flooding and nutrient leaching in riverine areas. This would have favoured replacement of *varzea* (riverine) rain forest by SDTF, perhaps resulting in an essentially contiguous, dendritic distribution of populations of SDTF species along the major river courses of the Amazon basin. Therefore, Pennington et al. (2000) proposed that, during the LGM, SDTF did not merely form a 'Pleistocene arc' around the periphery of the Amazon basin, but may also have occurred throughout much of the Amazon interior, albeit sporadically at low densities, confined to areas of fertile soil.

Irrespective of what the precise spatial distribution of SDTF during the LGM may have been, two key tenets of the hypothesis that this paper addresses are (1) the present-day disjunct distributions constitute geographical refugia resulting from contraction and fragmentation (i.e. vicariance) of a previously more expansive single distribution, and (2) consequently, although not explicitly stated by the authors, the floristic composition of the current dry forest areas has remained relatively unchanged since the LGM. Prado and Gibbs (1993) and Pennington et al. (2000) argued that this hypothesis provides a more parsimonious explanation for the current biogeographical patterns than population migration by long-distance dispersal.

Although there are insufficient palaeodata to provide a definitive test of the dry forest refugia hypothesis, it is instructive to examine whether or not the available palaeodata support its predictions or not.

17.3 LATE QUATERNARY VEGETATION HISTORY OF SOUTH AMERICAN SDTF REGIONS

17.3.1 SDTF POLLEN SIGNAL

Distinguishing between SDTF and rain forest pollen spectra is problematic, given the floristic similarity between these forest types at the genus and family levels (Pennington et al., 2000). However, analysis of modern pollen rain data collected from lake surface sediment samples and artificial pollen traps from permanent vegetation study plots (Gosling et al., 2003; Gosling, 2004) shows that *Anadenanthera* pollen (Figure 17.3) serves as a reliable indicator taxon of the Bolivian Chiquitano semi-deciduous dry forest (Figure 17.4; Mayle et al., 2004). Although pollen of this

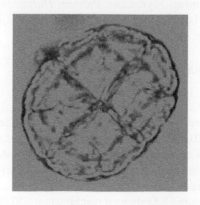

FIGURE 17.3 Pollen grain of *Anadenanthera colubrina* (photo courtesy of W.G. Gosling).

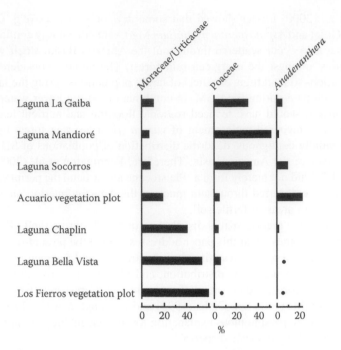

FIGURE 17.4 Modern pollen rain percentages of Moraceae/Urticaceae, Poaceae and *Anadenanthera* pollen from semideciduous dry forest and humid evergreen forest sites in lowland Bolivia (after Mayle et al., 2004). The Acuario (15°14 58S, 61°14 42W) and Los Fierros (14°34 50S, 66°49 48W) sites are 20 × 500-m permanent vegetation study plots. Pollen data from artificial pollen traps from these two plots (Gosling et al., 2003; Gosling, 2004) constitute mean values of counts from 10 evenly spaced traps from each plot sampled over three consecutive years. Pollen samples from the five lakes were collected from the uppermost centimetre of sediment at the sediment-water interface. See Burbridge et al. (2004) for more detailed pollen data of the surface samples from Laguna Chaplin and Laguna Bella Vista.

genus cannot be identified to species level, it most likely belongs to the key dry forest species, *A. colubrina*, because this is much more common than the other species of this genus (*A. peregrina* (L.) Speg.) in the Chiquitano forest (Killeen et al., 1998).

Corroboratory palynological evidence for SDTF, although not necessarily in all SDTF regions, comes from pollen of *Astronium urundeuva* (Allemão) Engl. (Figures 17.2 and 17.5), *Gallesia* and *Erythroxylum* (Figure 17.5), *Tabebuia* and *Malouetia* (Figure 17.9). Furthermore, pollen assemblages dominated by Leguminosae, Myrtaceae, Palmae, *Didymopanax* and Moraceae/Urticaceae (<25%) are typical of SDTF, but should be interpreted with caution, as they are also consistent with other tropical forest types, such as evergreen rain forest.

17.3.2 The Bolivian Chiquitano Dry Forest

17.3.2.1 Laguna Chaplin

The Chiquitano dry forest in lowland eastern Bolivia constitutes the largest remaining intact block of SDTF in the Neotropics (120,000 km²) (Parker et al., 1993) and is located at the central core of the current disjunct pattern of SDTF distribution across South America (Figure 17.1). Laguna Chaplin (14°28S, 61°04W) is located ~30 km beyond the northern ecotone of this dry forest and is surrounded by humid evergreen rain forest (Figure 17.1).

Burbridge et al. (2004) (Figure 17.5) showed that between ~50,000 and 10,000 carbon-14 yr BP (11,400 cal yr BP) (i.e. the entire Pleistocene portion of the sequence, including the LGM) the local

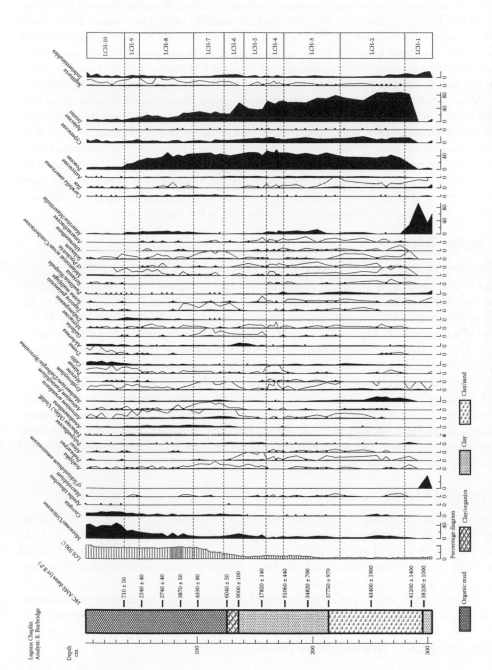

FIGURE 17.5 Laguna Chaplin summary pollen percentage diagram showing the most common taxa from the full complement of 290 pollen types. Dots on the curves denote <0.5%. Curves showing ×10 exaggeration are depicted for selected taxa which have low percentages. Total land pollen (TLP) sums are >300 grains per sample, except for the four pollen spectra in zone LCH-1 which had lower TLP sums of 100–160, due to low pollen concentrations. (After Burbridge et al., 2004. With permission.)

vegetation surrounding Laguna Chaplin consisted of a mosaic of semi-deciduous dry forest and seasonally flooded savannas. High percentages of Poaceae (>40%) and Cyperaceae pollen (~15%), together with the consistent presence of *Mauritia/Mauritiella* pollen, are indicative of seasonally flooded termite savannas and savanna marsh in low-lying areas, whilst continuous presence of typical dry forest taxa, such as *Astronium urundeuva, Gallesia, Paullinia/Roupala, Serjania*, Myrtaceae and *Erythroxylum* (albeit at relatively low percentages), together with low values of Moraceae pollen (<10%), are consistent with SDTF occupying higher ground. (N.B. This Moraceae pollen is unlikely to have originated from rain forest, as modern pollen-vegetation studies (Gosling et al., 2005) demonstrate that the latter is characterized by at least 40% Moraceae pollen). However, these Pleistocene dry forests differed floristically from those of the Holocene and present-day Chiquitano dry forest. Most significantly, *Anadenanthera* pollen is absent from the Pleistocene sediments (except for two pollen grains). Semi-deciduous dry forests, floristically similar to those of today (i.e. containing *Anadenanthera*), only appeared ~8000 carbon-14 yr BP (9000 cal yr BP), after which they were continuously present throughout most of the Holocene, in combination with seasonally flooded savannas, until they were replaced by rain forest within the last millennium. Furthermore, *Astronium urundeuva, A. fraxinifolium* Schott, Leguminosae (Mimosoideae) and *Didymopanax* were most abundant in the Holocene dry forests, whilst *Paullinia/Roupala, Serjania* and *Machaerium/Dalbergia/Byrsonima* were most abundant in the Pleistocene.

17.3.2.2 Laguna Bella Vista

Laguna Bella Vista (Burbridge et al., 2004) (13°37S, 61°33W) is located 100 km north of Laguna Chaplin and is also surrounded by humid evergreen rain forest (Figure 17.1). There is a 28,000-year sedimentary hiatus at this site, spanning the LGM, between 39,000 and 11,000 carbon-14 yr BP (13,000 cal yr BP), most likely caused by desiccation of the lake under a drier climate. Pollen assemblages in the underlying Pleistocene sediments older than 39,000 carbon-14 yr BP are dominated by cf. *Talisia, Machaerium/Dalbergia/Byrsonima*, Leguminosae (Papilionoideae), *Sapium, Alchornea* and Myrtaceae, which may represent SDTF, but could also reflect a combination of evergreen gallery forests and seasonally flooded savannas (Burbridge et al., 2004).

As with Laguna Chaplin, SDTF communities, floristically similar to those of the Chiquitano dry forest today (e.g. *Anadenanthera, Astronium*), together with seasonally flooded savannas, grew around Laguna Bella Vista throughout most of the Holocene. However, these dry forests were replaced by evergreen moist forest *c*.1000 years earlier at this site (*c*.1650 carbon-14 yr BP [1550 cal yr BP]) than at Laguna Chaplin (*c*.700 carbon-14 yr BP [650 cal yr BP]).

17.3.2.3 Laguna La Gaiba

Although there are no published fossil pollen records from within the Chiquitano dry forest, preliminary fossil pollen data from Laguna La Gaiba (F.Mayle, unpublished data) (Figure 17.1) suggest that the eastern limit of this dry forest nucleus was treeless during the last glacial period and that arboreal taxa (including Moraceae/Urticaceae, *Anadenanthera* and *Astronium*) have only surrounded this site since the early Holocene.

These data suggest that at least part of the area currently covered by Chiquitano semi-deciduous dry forest was instead covered by open herbaceous vegetation during the last glacial period.

17.3.3 THE CAATINGA OF NORTH-EASTERN BRAZIL

Since *Anadenanthera colubrina* and *Astronium urundeuva* are absent from other South American pollen diagrams, the origin of the disjunct distributions of these key dry forest taxa (Figure 17.2) is unclear. However, difficulty in distinguishing the pollen of these taxa from that of other genera in the Mimosoideae and Anacardiaceae, respectively, means that the pollen of these dry forest species may have been overlooked by other palynologists, especially if modern pollen reference material of these species has not been examined.

Notwithstanding these limitations, pollen data, together with other palaeo-environmental proxies, have provided important insights into the Late Quaternary vegetation history of other SDTF regions, such as the caatinga region of north-eastern Brazil (Figure 17.1). The latter is characterized by a very long dry season (6–11 months) and mean annual precipitation of only 250–750 mm pa. Consequently, xerophytic vegetation dominates the region, comprising a mix of semi-deciduous/deciduous gallery forest, open thorn scrub and patches of cerrado.

17.3.3.1 Fossil Pollen Data

The marked precipitation deficit means that suitable environments for the preservation of long, continuous fossil pollen records (i.e. permanent lakes) are extremely rare. The only Pleistocene pollen record for the region that spans the LGM comes from marine core GeoB 3104-1 (Figures 17.1 and 17.6) (3°40S, 37°43W) (Behling et al., 2000), located on the upper continental slope 90 km offshore from the coastal city of Fortaleza. Most of the pollen reaching these marine sediments originates from the Rio Jaguaribe, which flows into the Atlantic Ocean c.80 km south of the core site (Behling et al., 2000). This river is 700 km long, and therefore constitutes an extensive pollen catchment for the caatinga vegetation. Although pollen taxa exclusively found in SDTF (e.g. *Anadenanthera colubrina*) are absent from this pollen record, the proportions of key pollen types (e.g. Poaceae, Asteraceae, *Borreria*, Melastomataceae/Combretaceae, Palmae, Myrtaceae) are relatively constant through most of the sequence and are generally consistent with semi-arid vegetation, broadly similar to that of today, dominating the Rio Jaguaribe catchment of north-eastern Brazil over most of the period between 8500 carbon-14 yr BP (9550 cal yr BP) and at least 42,000 carbon-14 yr BP. However, high concentrations of pollen, particularly fern spores, indicate short-term increases in fluvial input (i.e. higher precipitation) c.40,000, 33,000, and 24,000 carbon-14 yr BP, with maximum precipitation during the late-glacial between 15,500 and 11,800 carbon-14 yr BP (18,500 and 13,800 cal yr BP). Coincident peaks in *Alchornea* pollen and fern spores suggest there was expansion of forest at these times, although it is unclear whether the latter was semi-deciduous dry forest or moist evergreen forest. Not only were there marked fluctuations in precipitation regime in the caatinga region during the Late Quaternary, but also significant changes in temperature. A palaeotemperature record derived from noble gases dissolved in ground water in the heart of the caatinga region (Stute et al., 1995) (Figure 17.1) reveals that temperatures during the last glacial period (between c.35,000 and 10,000 carbon-14 yr BP (11,400 cal yr BP)) were c.5°C colder than during the Holocene. Cooler Pleistocene temperatures would be expected to have reduced water deficits associated with a semi-arid climate via lower evapotranspiration.

The only complete Holocene pollen record for the caatinga region comes from a *Mauritia* palm swamp forest (10°24S, 43°13W) bordering the Icatu river valley (Figure 17.1), a tributary of the middle São Francisco river (De Oliveira et al., 1999). This valley swamp is bordered by a narrow strip of semi-deciduous gallery forest; its river catchment extends to the western limit of 'The Little Sahara' fossil sand dunes (located in Bahia, 50 km east of the study site) which are presently covered by a mix of different types of caatinga vegetation. Peaks in tree taxa such as Palmae, *Alchornea*, *Cecropia* and Myrtaceae, are indicative of greater forest cover, and therefore wetter conditions, at the end of the late-glacial (c.11,000–10,500 carbon-14 yr BP [13,000–12, 600 cal yr BP]) compared with today. Furthermore, pollen of *Pouteria*, *Protium*, *Simarouba*, *Symphonia* and *Trichilia* (albeit <5%) suggest that the climate may have been wet enough at this time to support evergreen rain forest communities, perhaps as gallery forest lining the rivers. Although typical caatinga and cerrado taxa (e.g. *Byrsonima*, *Cuphea*, *Mimosa*) occur throughout the sequence, there is a progressive increase in their abundance through the Holocene at the expense of tropical forest species, with present-day climatic and vegetation conditions established by c.4,240 carbon-14 yr BP (4,800 cal yr BP). This vegetation history is supported by thermoluminescence (TL) dating of the nearby sand dune sediments (Barreto, 1996), which shows a minimum in dune activity between 10,500 and 9000 yr BP (coincident with maximum precipitation and forest growth) and maximum dune growth between 4500 and 1700 yr BP (coincident with maximum aridity and expansion of caatinga/cerrado vegetation).

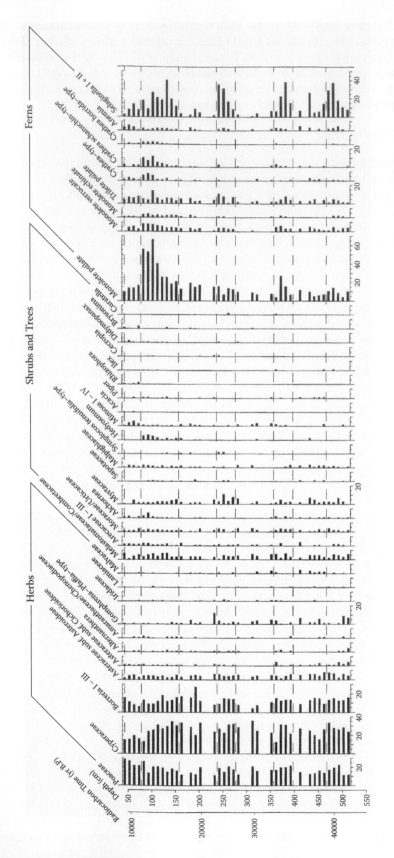

FIGURE 17.6 GeoB 3104-1 marine core pollen percentage diagram, showing the most frequent pollen and spore taxa. The pollen sum, upon which the percentage calculations are based, comprises herbs, shrubs and trees. (After Behling et al., 2000. With permission.)

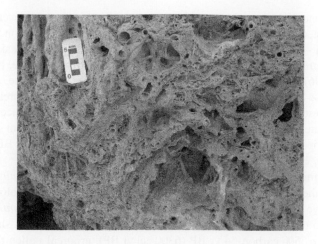

FIGURE 17.7 Travertine deposit from Salitre river valley, north-eastern Brazil, containing leaf, trunk and root casts characteristic of dense semideciduous forest. Modern caatinga vegetation comprises cactus thorn scrub. (After Wang et al., 2004. With permission.)

17.3.3.2 Speleothems and Travertines

The strongest evidence that north-eastern Brazil experienced past climates significantly wetter than today comes from speleothem and travertine deposits (Figures 17.1 and 17.7) in northern Bahia state (centred around 10°S, 40°W in the heart of the semi-arid caatinga region). Current precipitation in the region is insufficient to allow these deposits to form today. Their presence is therefore unequivocal evidence of wetter conditions in the past. Using the U/Th method to date these deposits, Auler et al. (2004) and Wang et al. (2004) showed that these past pluvial episodes occurred over the last 210,000 years, but were highly episodic and short-term in nature, lasting for only a few hundred to a few thousand years, alternating with much longer periods of semi-arid climate. During the last glacial period, pluvial phases, determined from the speleothem record, occurred at 15,000, 39,000, and 48,000 cal yr BP (in broad agreement with the Pleistocene pollen record [Behling et al., 2000]), whilst six further wet phases occurred between 60,000 and 70,000 cal yr BP. The travertine record reveals additional, less intense, wet episodes during the Younger Dryas chronozone (11,700 to 12,100 cal yr BP) and spanning the LGM (16,700 to 21,700 cal yr BP). Wang et al. (2004) argued that these pluvial periods were most likely caused by increased southerly penetration of the intertropical convergence zone (ITCZ) across north-eastern Brazil during times of high austral autumn insolation at 10°S.

Over 45 morphological types of leaf cast have been identified from the travertine deposits (Auler et al., 2004) (Figure 17.7), which provide further corroborative evidence that during these past pluvial phases semi-deciduous forest grew in localities presently dominated by cactus thorn scrub.

17.3.4 THE SDTF OF SOUTH-EASTERN BRAZIL

South-eastern Brazil experiences a much wetter climate than that of the caatinga region of north-eastern Brazil, with mean annual precipitation typically between 1000 and 1500 mm and a 3–5 month dry season. Consequently, the dominant forest type is tall (20–30 m), dense semideciduous forest. However, only scattered relics remain on hilltops, most of the forest having been replaced with agriculture. The Salitre site (Figure 17.1) (19°S, 46°46W) (Ledru et al., 1996) constitutes the most complete Late Quaternary pollen record from the SDTF region of south-eastern Brazil, spanning most of the last 50,000 years, although the LGM is not represented due to a sedimentary hiatus between 29,000 and 17,000 carbon-14 yr BP (20,400 cal yr BP). This site is an infilled lake, located at the dry forest/savanna ecotone of Minas Gerais with the surrounding catchment covered by a mosaic of semideciduous forest, cerradão (thickly wooded savanna), and swamp vegetation. The arboreal pollen assemblage between ~40,000 and 33,000 carbon-14 yr BP, characterized by peak pollen percentages of *Alchornea, Casearia, Celtis, Gallesia*, Melastomataceae and Mimosoideae, has been interpreted by Ledru et al. (1996) as evidence for maximum extent of SDTF at this time, although, as argued earlier, most of these taxa are also represented in other types of tropical forest. Furthermore, the regional extent of these Pleistocene SDTF in south-eastern Brazil is unclear, since pollen assemblages from the dry forest site of Catas Altas (Figure 17.1) (20°05S, 43°22W), ~250 km south-east of Salitre, are consistent with open grassland between >48,000 and 18,000 carbon-14 yr BP (21,400 cal yr BP) (Behling and Lichte, 1997) rather than forest.

Other plant communities (e.g. seasonally flooded Myrtaceae forests, *Araucaria* forests) subsequently dominated the Salitre site (i.e. after 33,000 yr BP) until the early Holocene (8500 carbon-14 yr BP [9550 cal yr BP]) when pollen assemblages containing *Tabebuia, Zanthoxylum, Alchornea* and Palmae point to renewed development of SDTF (Ledru, 1993). After a short-term expansion of herbaceous taxa *c*.5500 carbon-14 yr BP (6300 cal yr BP), arboreal pollen taxa consistent with SDTF once again increase *c*.4350 carbon-14 yr BP (4850 cal yr BP).

Holocene pollen records from two other dry forest sites in south-eastern Brazil, Lago do Pires (Behling, 1995) and Lagoa Nova (Behling, 2003) (Figures 17.1 and 17.8), corroborate the Salitre data, showing that SDTF communities, floristically similar to those surrounding the sites today (e.g. *Alchornea, Myrsine*, Palmae, *Anacardium*-type, Myrtaceae, *Celtis, Gallesia*), have existed in the region since *c*.8500 carbon-14 yr BP (9500 cal yr BP). Furthermore, increases in pollen of Myrtaceae, *Gallesia*, Palmae and *Psychotria alba* type in these two records indicate that these dry forests expanded their distribution over the last millennium, most likely due to increasing precipitation.

17.3.5 The SDTF of the Colombian Andes

17.3.5.1 The Cauca Valley

Pollen data from two Andean dry forest sites in the Cauca valley of Colombia (Quilichao-1, 3°6N, 76°31W, and La Teta-2, 3°5N, 76°32W; Figure 17.1) reveal that SDTF communities have occupied the valley since the late-glacial period, *c*.11,500 carbon-14 yr BP (13,450 cal yr BP) (Berrio et al., 2002), although there have been changes in their floristic composition and cover over this time.

Between *c*.11,500 and 10,500 carbon-14 yr BP (13,450 and 12,600 cal yr BP, towards the end of the late-glacial period) pollen assemblages dominated by *Crotalaria, Celtis, Alchornea* and Moraceae/Urticaceae are consistent with dry forest communities growing in the valley. This floristic association most closely resembles that of the dry forests in the Cesar Valley of the Colombian Caribbean (Rangel et al., 1995). Dry forest cover diminished in the early Holocene (prior to 7700 carbon-14 yr BP [8400 cal yr BP]) at Quilichao-1, most likely due to reduced precipitation, and underwent floristic changes, *Erythrina* becoming more important. Comparison with the La Teta-2 record (Figure 17.9), which begins *c*.8850 carbon-14 yr BP (10,100 cal yr BP), demonstrates early Holocene floristic variations in SDTF communities across the valley. The La Teta-2 pollen record is characterized by taxa such as *Malouetia*, Sapindaceae, *Tabebuia* and Papilionoideae in the early Holocene (*c*.8850–7560 carbon-14 yr BP [10,100–8400 cal yr BP]), representing a dry forest floristically similar to that of the present-day vegetation within the Cauca valley.

A 5000-year period of mid-Holocene climatic aridity, between *c*.7700 and 2700 carbon-14 yr BP (8400 and 2800 cal yr BP), caused the Quilichao site to dry up completely and influx of sand into the La Teta-2 basin, most likely due to destabilization of the catchment slopes caused by increased bare ground. SDTF communities reached their maximum extent around the La Teta-2 site during this arid interval, and were dominated by Apocynaceae, *Crotalaria*-type, Anacardiaceae, *Tabebuia*-type, Annonaceae, *Croton*, Sapindaceae, *Alchornea*, Moraceae/Urticaceae, Myrtaceae and

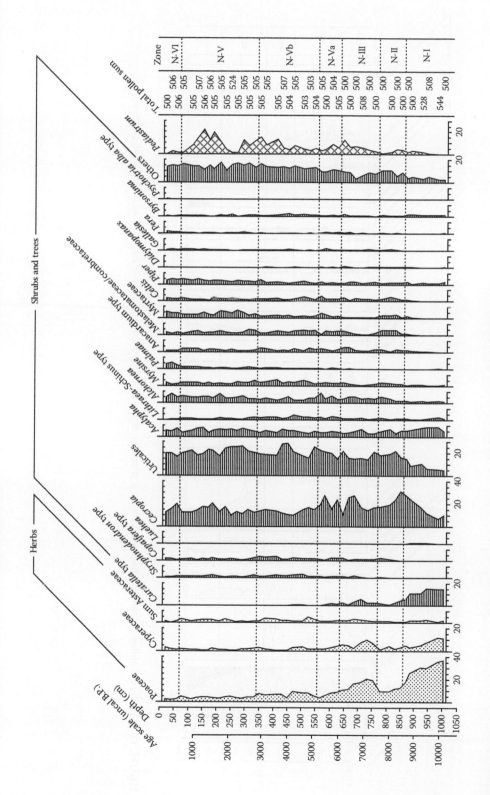

FIGURE 17.8 Lagoa Nova pollen percentage diagram of the most frequent taxa. (After Behling, 2003. With permission.)

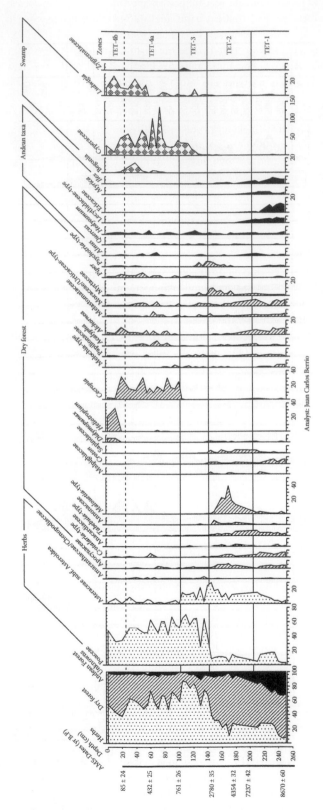

FIGURE 17.9 La Teta-2 pollen percentage diagram showing a selection of the most important taxa. (After Berrío et al., 2002. With permission.)

Psychotria-type. Interestingly, these forests appear to have been dominated almost exclusively by *Malouetia* (Apocynaceae) *c.*4300 carbon-14 yr BP (4850 cal yr BP), with pollen percentages reaching 40%.

Although the last three millennia were marked by a return to more humid conditions, SDTF were very limited in extent. This can be attributed to deforestation by palaeo-Indians, revealed by archaeological evidence (Gnecco and Salgado, 1989) and *Zea mays* pollen, which resulted in a late Holocene landscape alternating between open grassy vegetation and disturbed/secondary forest dominated by *Cecropia*.

17.3.5.2 The Patía Valley

Velez et al. (2005) examined the pollen records of two swamps, Patía-1 and Patía-2 (2°02N, 77°0W; Figure 17.1), 2 km apart from one another, within the neighbouring Patía Valley, *c.*160 km south of the Quilichao and La Teta sites. These authors demonstrate that SDTF has grown in this valley since the early Holocene (*c.*8000 carbon-14 yr BP [9000 cal yr BP]), although there have been marked changes in species composition, both spatially (i.e. between the two sites) and temporally. For example, *Mimosa* is a dominant taxon at Patía-1 (Figure 17.10), whilst *Piper* is dominant at Patía-2. At Patía-1, *Cecropia* and *Mimosa* are most abundant in the early Holocene (prior to 6000–7000 carbon-14 yr BP (6800–7800 cal yr BP]), whilst *Celtis, Chamaesyce*-type and Apocynaceae are most abundant in the mid/late Holocene. At Patía-2, *Piper* is restricted to the early/mid Holocene (*c.*7500–3800 carbon-14 yr BP [8350–4200 cal yr BP]), whilst *Portulaca, Acalypha, Alchornea, Celtis* and *Malouetia* are most abundant in the late Holocene (3800–2000 carbon-14 yr BP [4200–2000 cal yr BP]), and *Chamaesyce*-type and *Dalbergia*-type are most abundant in the last millennium.

As in the Cauca Valley, SDTF was most extensive in the early/mid Holocene, between *c.*7500 and 2500– 3000 carbon-14 yr BP (8350 and 2700–3200 cal yr BP), coinciding with sedimentological evidence for mid-Holocene aridity. Dry forests became less extensive in the late Holocene due to anthropogenic activity, i.e. clearance for agriculture (e.g. maize cultivation).

17.4 LATE QUATERNARY VEGETATION HISTORY OF THE AMAZON RAIN FOREST

The responses of Amazonian rain forest communities to climatic and atmospheric changes since the LGM are discussed in detail by Mayle et al. (2004). Pollen records from sites near the margins of Amazonia (e.g. north-eastern Bolivia (Burbridge et al., 2004) and eastern Colombia [Behling and Hooghiemstra, 1999, 2000]) show that ecotonal areas currently occupied by rain forest were instead covered by savanna and/or SDTF between the LGM and the end of the Pleistocene. However, pollen data from the Amazon Fan (5°12.7N, 47°1.8W, Haberle and Maslin, 1999) and Lake Pata in central Amazonia (0°16N, 66°41W, Colinvaux et al., 1996; Bush et al., 2002) (Figure 17.1) suggest that most of the Amazon Basin remained forested during the last glacial period, although the kind of forest that existed at this time remains controversial. Colinvaux et al. (1996) and Bush et al. (2002) inferred from the Lake Pata pollen data that moist evergreen forest has dominated Amazonia since at least 170,000 cal yr BP, albeit undergoing marked changes in floristic composition. For example, Andean taxa (e.g. *Podocarpus* and *Alnus*) formed mixed communities with lowland rain forest taxa under colder glacial climates.

Pennington et al. (2000) pointed out that inventories of SDTF (e.g. Ratter et al., 1988; Oliveira-Filho and Ratter, 1995) include species from 32 of the 40 genera found in the Lake Pata pollen record and that, furthermore, the four taxa identified by Colinvaux et al. (1996) as 'strongly suggestive of a tropical rain forest' (i.e. *Cedrela, Clusia, Didymopanax* and *Bombacaceae*) contain species which grow in SDTF. Therefore, interpreting the palaeo-ecological significance of these fossil pollen assemblages, in terms of semi-deciduous dry forest versus evergreen rain forest, is problematic. However, newly available modern pollen rain data (Gosling, 2004; Gosling et al.,

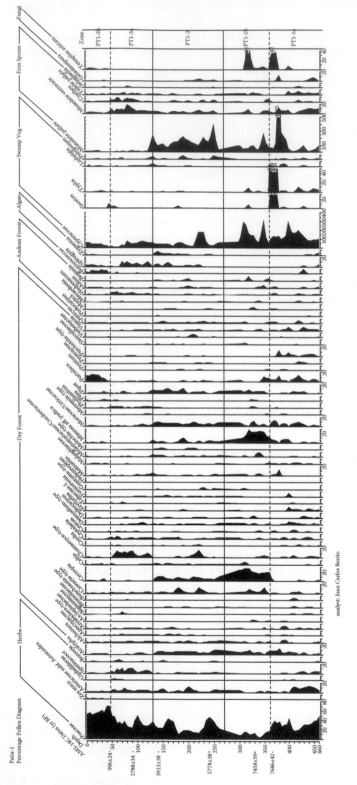

FIGURE 17.10 Patia-1 summary pollen percentage diagram. (After Vélez et al., 2005, with permission.)

2005) provide the potential for reliably differentiating between the fossil pollen spectra of these two forest types in the future, at least with respect to south-western Amazonia.

17.5 SYNTHESIS OF PALAEO-ECOLOGICAL DATA AND IMPLICATIONS FOR THE PLEISTOCENE DRY FOREST REFUGIA HYPOTHESIS AND PATTERNS OF PLANT SPECIES DIVERSITY AND ENDEMISM

17.5.1 THE BOLIVIAN CHIQUITANO DRY FOREST

There are insufficient palaeodata to determine the distribution of SDTF in eastern Bolivia during the last glacial period. However, the pollen evidence is consistent with increased water stress (due to a combination of a longer dry season and reduced atmospheric CO_2 levels; Burbridge et al., 2004) causing northerly expansion of semideciduous tree species into areas now dominated by rain forest, and expansion of savannas into areas of Bolivia now dominated by semideciduous trees. Such a scenario is supported by a dynamic vegetation model simulation for the LGM (Mayle, 2004; Mayle and Beerling, 2004; Mayle et al., 2004). Irrespective of the distribution of these Pleistocene SDTF in eastern Bolivia, present-day populations of *Anadenanthera colubrina* in the Chiquitano dry forest cannot be considered as refugial populations or remnants of a formerly larger Pleistocene distribution because this key dry forest taxon is largely absent from Pleistocene pollen spectra in Bolivia (Figure 17.5). In contrast to prediction from the dry forest refugia hypothesis, the distribution and abundance of this taxon has increased, rather than decreased, since the early Holocene. The pollen spectra of *Anadenanthera colubrina* and *Astronium* spp. demonstrate that dry forest species were most widespread in the mid/late Holocene (*c.*8000–1650 carbon-14 yr BP [9000–1530 cal yr BP]), extending at least 130 km beyond the current northern ecotone of the Chiquitano dry forest. This mid/late Holocene dry forest maximum in lowland Bolivia can be attributed to drier conditions at this time, especially between 6000 and 3000 carbon-14 yr BP (6800 and 3200 cal yr BP), for which there is extensive independent palaeoclimatic evidence (reviewed by Mayle et al., 2004), e.g. charcoal peaks in the Laguna Chaplin and Bella Vista records (Burbridge et al., 2004) and Lake Titicaca water levels 90 m below present (D'Agostino et al., 2002). Subsequent increasing precipitation over the past two millennia caused expansion of Amazonian rain forest at the expense of dry forest and savanna (Mayle et al., 2000; Burbridge et al., 2004), at least at the northern Chiquitano forest ecotone.

This hypothesis of a 'young' Chiquitano dry forest, that only became established in the Holocene, is supported by Killeen et al. (Chapter 9), whose floristic comparisons of 118 permanent plots show that this dry forest is (a) only moderately diverse, in terms of α-diversity, (b) low in spatial heterogeneity (β-diversity) compared to adjacent regions and (c) most significantly, has surprisingly low levels of endemism, with only three endemic species in the entire woody flora of the Chiquitano forest.

17.5.2 THE CAATINGA OF NORTH-EASTERN BRAZIL

The limited pollen data available (De Oliveira et al., 1999; Behling et al., 2000; Figure 17.6) suggest that the thorn-scrub caatinga vegetation, which currently dominates semi-arid north-eastern Brazil, has persisted through the last glacial-interglacial cycle, supporting the Pleistocene dry forest refugia hypothesis (Prado and Gibbs, 1993). This Quaternary caatinga vegetation history is consistent with the high number of endemic plant taxa that have been recorded (over 40% of caatinga flowering plant species [Prado, 1991; De Queiroz, Chapter 6]), which is indicative of a long geological history for the caatinga flora. Further evidence that caatinga SDTF has existed over at least the past several million years comes from molecular phylogenetic studies of *Coursetia* (Leguminosae; Lavin, Chapter 19).

Despite this multidisciplinary evidence for continued presence of caatinga taxa in north-eastern Brazil through the Quaternary period, the pollen and macrofossil data suggest that the population sizes of these caatinga species varied considerably in response to past climate changes. For example, the geographic cover of caatinga thorn scrub expanded progressively since the early Holocene, at least in the vicinity of the Icatu river valley (De Oliveira et al., 1999), and forest cover (probably semideciduous dry forest) in north-eastern Brazil was likely to have been much more extensive during the wetter conditions of the late-glacial period (15,500–10,500 carbon-14 yr BP [18,500–12,600 cal yr BP]) and short-term wet periods earlier in the Pleistocene, including the LGM (Behling et al., 2000; Auler et al., 2004; Wang et al., 2004). The peak in dune activity between 4500 and 1700 yr BP (Barreto, 1996), coincident with a marked expansion of thorn scrub caatinga vegetation c.4240 carbon-14 yr BP (4800 cal yr BP), indicates that north-eastern Brazil also experienced a mid/late Holocene peak in aridity, roughly coincident with that of eastern Bolivia, suggestive of common climatic forcing. The cessation of dune activity c.1700 yr BP is coincident with the expansion of rain forest at the expense of SDTF in north-eastern Bolivia and suggests that increasing precipitation in the last two millennia allowed caatinga vegetation to become sufficiently dense to stabilize the land surface. It should be noted, however, that some areas of present-day caatinga thorn scrub may not necessarily constitute natural vegetation, but may instead be a function of anthropogenic degradation of semideciduous/deciduous dry forest by farmers over the past several centuries (J. Ratter, Royal Botanic Garden Edinburgh, pers. comm.).

17.5.3 THE SDTF OF SOUTH-EASTERN BRAZIL

The extent of SDTF during the Pleistocene in south-eastern Brazil is unclear, due to the paucity of sites and incomplete and conflicting pollen data from those available. The Salitre site was surrounded by SDTF c.40,000–33,000 carbon-14 yr BP, whilst the Catas Altas record 250 km away was surrounded by open grassland. However, the Holocene sites indicate that SDTF communities, floristically similar to those of today, have existed since the early Holocene, c.8500 carbon-14 yr BP (9500 cal yr BP), consistent with the evidence from the Chiquitano dry forest of eastern Bolivia.

17.5.4 THE SDTF OF THE COLOMBIAN ANDES

Pollen records from the southern Colombian Andes do not extend beyond 11,500 carbon-14 yr BP (13,450 cal yr BP), so the distribution and abundance of Andean SDTF during most of the last glacial period is unknown. However, macrofossil evidence from Ecuador suggests that SDTF communities have grown in inter-Andean valleys since at least 10 my BP (Ma) (e.g. Hughes, 2005), corroborated by phylogenetic studies of the dry forest legume *Coursetia* by Lavin (Chapter 19) and Lavin et al. (2004). Lavin (Chapter 19) argues that the narrowly confined crown clades of sister species of *Coursetia* are evidence of floristic stability of SDTF communities since at least 5–7 Ma, in line with the dry forest refugia hypothesis (Prado and Gibbs, 1993). However, the pollen data from the Colombian Andes (e.g. Berrío et al., 2002; Figure 17.9) instead show that there were significant changes in the floristic composition of SDTF communities over the course of the Holocene, demonstrating that the present-day dry forest species associations in the Cauca and Patia valleys are unlikely to be refugial dry forest communities dating from the last glacial period. Unfortunately the studies of Lavin (Chapter 19) and Lavin et al. (2004) do not include species endemic to these Colombian valleys. However, this fossil pollen evidence for successive immigration and re-assortment of taxa in SDTF communities through the Holocene, instead of long-term floristic stability, suggests that the phylogenetic evidence for floristic stability of *Coursetia* species (Lavin, Chapter 19) may not be typical of other dry forest genera or all dry forest areas.

With respect to the changing geographic cover of SDTF in the Colombian Andes over the Holocene, this forest type was most extensive during the mid Holocene between c.8000 and 3000 carbon-14 yr BP (9000 and 3200 cal yr BP), in common with SDTF of eastern Bolivia, signifying a common dry forest response to increased aridity across tropical South America.

17.6 VICARIANCE VERSUS MIGRATION/LONG-DISTANCE DISPERSAL HYPOTHESES

17.6.1 VICARIANCE?

Prado and Gibbs (1993) and Pennington et al. (2000) argued that the most parsimonious explanation for the current disjunct distributions of neotropical dry forest species (e.g. *Anadenanthera colubrina*; Figure 17.2) is vicariance (i.e. fragmentation) of a previously much larger, single dry forest distribution that spanned these widely separated geographical regions during the LGM, whereby increasing precipitation, temperatures and atmospheric CO_2 concentrations since the LGM resulted in population contraction and fragmentation. Pennington et al. (2000) pointed out that, whilst there are numerous widespread dry forest species common to several dry forest nuclei, there are also a number of taxonomically unrelated genera (e.g. *Loxopterygium, Pereskia, Ruprechtia, Coursetia* and *Chaetocalyx-Nissolia*) which exhibit high levels of endemicity in these different dry forest regions, strongly suggestive of allopatric speciation according to the vicariance hypothesis.

Pennington et al. (2004) used evolutionary rates analysis to determine the ages of the endemic species within these genera and cladistic vicariance analysis to test the hypothesis that they arose by vicariance of a common, widespread Pleistocene distribution. Significantly, these authors found that (a) not only did these South American genera diversify into endemic species prior to the Quaternary, i.e. during the late Miocene and Pliocene, but also (b) the cladistic analyses failed to reveal any common patterns in area relationships among these unrelated genera, contrary to expectation from allopatric speciation by vicariance. This lack of evidence for Quaternary allopatric speciation by vicariance associated with glacial-interglacial climatic changes, at least with respect to South American dry forest species, raises the possibility that vicariance, as a process, may also not have been the underlying cause for the disjunct distributions of widespread non-endemic dry forest species discussed earlier (e.g. *Anadenanthera colubrina*).

17.6.2 MIGRATION/LONG-DISTANCE DISPERSAL?

Due to the paucity of palaeo-ecological data, it remains unclear whether or not the current disjunct populations of SDTF taxa were formerly connected during drier intervals of the last glacial period, forming much larger, uninterrupted distributions, as postulated by the dry forest refugia hypothesis. The data from eastern Bolivia, however, suggest this hypothesis could be false, at least with respect to *Anadenanthera*, which is virtually absent from Pleistocene pollen records. The climatic and atmospheric conditions of tropical South America during the last glacial period were markedly different from those of today. Temperatures were 5°C cooler and atmospheric CO_2 concentrations were c.35% below pre-industrial values during the LGM (Bush and Silman, 2004; Mayle et al., 2004). Levels of precipitation and its seasonal, spatial and temporal distribution through the Quaternary are far more complex and less well understood (Bush and Silman, 2004). For example, during the LGM precipitation was higher than present in north-eastern Brazil (Wang et al., 2004) and the Bolivian/Peruvian altiplano (Baker et al., 2001) and lower than present in lowland north-eastern Bolivia (Burbridge et al., 2004) and eastern Colombia (Behling and Hooghiemstra, 1999). Consequently, even though the geographical distributions of SDTF species during the LGM are poorly understood, one would expect them to have been quite different from those of today.

Irrespective of their past distributions, population migration and long-distance transport may be sufficient to explain the current disjunct dry forest distributions, rather than vicariance. It has long been recognized from fossil pollen data from Europe (e.g. Huntley and Birks, 1983) and eastern North America (e.g. Davis, 1983) that maximum rates of population expansion of temperate tree species during the Holocene (300–1000 m pa) were at least an order of magnitude greater than those observed today, and that, contrary to expectation, differences in Holocene migration rates were unrelated to the life-history characteristics of the species (e.g. wind-adapted seeds vs. animal/bird adapted seeds) (Wilkinson, 1997). Such high Holocene migration rates point to rare long-distance dispersal events by animals, storms, and especially migratory birds, as the most likely dispersal agent over Quaternary timescales (e.g. Webb, 1986), even for species with wind-adapted seeds (Wilkinson, 1997). These unusually high rates of spread are typical, not only of species expanding into open, recently deglaciated landscapes (e.g. *Betula, Picea, Pinus*), but also species expanding their ranges in forested landscapes south of formerly glaciated areas and/or later in the Holocene. If tropical tree species displayed similarly high population migration rates over glacial–interglacial timescales, they could have spread throughout much of tropical South America over the course of the late-glacial and Holocene, irrespective of their geographical location at the LGM. The current disjunct dry forest distribution patterns could simply be explained by common climatic and/or edaphic requirements of the constituent species, rather than evidence of fragmentation of a once-continuous dry forest community into isolated refugia.

It is important to point out, however, that long-distance population expansion may not necessarily need to be invoked to account for SDTF floristic changes in all dry forest regions. Certain dry forest nuclei with sufficient topographic and/or edaphic complexity (e.g. inter-Andean valleys and the caatinga of north-eastern Brazil) may have contained a wide enough spectrum of habitats with favourable microclimates to maintain viable populations of most dry forest taxa (albeit at population sizes too low to be detectable palynologically), even when subject to glacial/interglacial climatic changes. This idea is supported by the high numbers of endemic plant species in SDTF nuclei such as the caatinga of north-eastern Brazil (De Queiroz, Chapter 6).

Some dry forest nuclei (e.g. Colombian Andes) exhibit considerable variation in the timing and sequence of arrival of different species over the course of the Holocene, demonstrating individualistic species responses to past environmental changes. The dynamic nature of these Andean dry forest communities, which have been undergoing continual species reassortment, may reflect the ecotonal position of the sites, close to the climatic thresholds of most dry forest taxa. In contrast, other dry forest communities (e.g. Lagoa Nova, south-eastern Brazil) appear to have been much more stable and uniform over most of the Holocene, signifying perhaps that Holocene climatic changes have been well within the range of tolerance of most taxa in these areas.

Therefore, I argue that it may not be necessary to invoke vicariance to explain the current disjunct geographic pattern of SDTF species across South America, and that dispersal alone could be a sufficient explanation.

17.7   CONCLUSIONS

Despite difficulties in distinguishing between SDTF and other kinds of tropical forest palynologically, and the paucity of palaeo-ecological records from dry forest areas, a review of previously published data (predominantly fossil pollen assemblages from lake sediments) reveals a complex biogeographical history of SDTF in South America over the Quaternary. For example, some dry forest communities (e.g. the inter-Andean Colombian dry forests) have undergone considerable floristic changes through the Holocene, whilst others (e.g. south-eastern Brazil) appear to have been more stable floristically.

Anadenanthera colubrina is an important component of the Chiquitano dry forest, but only became established in the early Holocene, and, in contrast to prediction from the dry forest refugia hypothesis (Prado and Gibbs, 1993), is absent from Pleistocene pollen records.

Although thorn scrub caatinga vegetation has grown in north-eastern Brazil during both glacial and interglacial periods, it has fluctuated in geographical extent over this time. For example, increased precipitation during the late-glacial period caused expansion of tropical forest at the expense of caatinga thorn scrub.

Whilst the distribution of SDTF during the last glacial period remains poorly known, it is clear that dry forests of Andean Colombia and lowland eastern Bolivia (considered over the Holocene) were most extensive during the mid-Holocene, especially between *c.*6000 and 3000 carbon-14 yr BP (6800 and 3200 cal yr BP), which also correlates broadly with expansion of thorn-scrub caatinga in north-eastern Brazil. These SDTF and caatinga expansions are indicative of a widespread reduction in precipitation across tropical South America at this time.

Rather than attributing the current disjunct distribution of SDTF taxa to vicariance, i.e. fragmentation of once continuous populations spanning these dry forest areas, I suggest population migration/long-distance dispersal as an alternative explanation.

ACKNOWLEDGEMENTS

This chapter was written during the tenure of a Leverhulme Trust Research Fellowship. I am grateful to Toby Pennington for inviting me to write this chapter and present some of these data and ideas at the Tropical Savannas and Seasonally Dry Forests conference at the Royal Botanic Garden Edinburgh, 2003. I also thank Toby Pennington, Jim Ratter, Rob Marchant, and Sarah Metcalfe, whose comments and suggestions improved the manuscript.

REFERENCES

Auler, A.S. et al., Quaternary ecological and geomorphic changes associated with rainfall events in presently semiarid northeastern Brazil, *J. Quat. Sci.*, 19, 693, 2004.

Baker, P.A. et al., The history of South American tropical precipitation for the past 25,000 years, *Science*, 291, 640, 2001.

Barreto, A.M.F., *Interpretação paleoambiental do sistema de dunas fixas do médio Rio São Francisco, Bahia*, PhD thesis, University of São Paulo, São Paulo, 1996.

Behling, H., A high resolution Holocene pollen record from Lago do Pires, SE Brazil: vegetation, climate and fire history, *J. Palaeolimnol.*, 14, 253, 1995.

Behling, H., Late glacial and Holocene vegetation, climate and fire history inferred from Lagoa Nova in the southeastern Brazilian lowland, *Veget. Hist. Archaeobot.*, 12, 263, 2003.

Behling, H. and Hooghiemstra, H., Environmental history of the Colombian savannas of the Llanos Orientales since the Last Glacial Maximum from lake records El Pinal and Carimagua, *J. Paleolim.*, 21, 461, 1999.

Behling, H. and Hooghiemstra, H., Holocene Amazon rain forest-savanna dynamics and climatic implications: high-resolution pollen record from Laguna Loma Linda in eastern Colombia, *J. Quat. Sci.*, 15, 687, 2000.

Behling, H. and Lichte, M., Evidence of dry and cold climatic conditions at glacial times in tropical southeastern Brazil, *Quat. Res.*, 48, 348, 1997.

Behling, H. et al., Late Quaternary vegetational and climatic dynamics in northeastern Brazil, inferences from marine core GeoB 3104-1, *Quat. Sci. Rev.*, 19, 981, 2000.

Berrio, J.C. et al., Late-glacial and Holocene history of the dry forest area in the south Colombian Cauca Valley, *J. Quat. Sci.*, 17, 667, 2002.

Burbridge, R.E., Mayle, F.E., and Killeen, T.J., Fifty-thousand-year vegetation and climate history of Noel Kempff Mercado National Park, Bolivian Amazon, *Quat. Res.*, 61, 215, 2004.

Bush, M.B. et al., Orbital-forcing signal in sediments of two Amazonian lakes, *J. Paleolim.*, 27, 341, 2002.

Bush, M.B. and Silman, M.R., Observations on Late Pleistocene cooling and precipitation in the lowland Neotropics, *J. Quat. Sci.*, 19, 677, 2004.

Colinvaux, P.A., De Oliveira, P.E., and Bush, M.B., A long pollen record from lowland Amazonia: forest and cooling in glacial times, *Science*, 274, 85, 1996.

Colinvaux, P.A, De Oliveira, P.E, and Bush, M.B., Amazonian and neotropical plant communities on glacial time-scales: the failure of the aridity and refuge hypotheses, *Quat. Sci. Rev.*, 19, 141, 2000.

D'Agostino, K. et al., Late-Quaternary lowstands of Lake Titicaca: evidence from high-resolution seismic data, *Palaeogeogr. Palaeoclimatol. Palaeoecol.*, 179, 97, 2002.

Davis, M.B., Quaternary history of deciduous forests of eastern North America and Europe, *Ann. Missouri Bot. Garden*, 70, 550, 1983.

De Oliveira, P.E., Barreto, A.M.F., and Suguio, K., Late Pleistocene/Holocene climatic and vegetational history of the Brazilian caatinga: the fossil dunes of the middle São Francisco River, *Palaeogeogr. Palaeoclimatol. Palaeoecol.*, 152, 319, 1999.

Gnecco, C. and Salgado, H., Adaptaciones pre-ceramicas en el suroccidente de Colombia, *Boletín del Museo del Oro*, 24, 35, 1989.

Gosling, W.D., *Characterization of neotropical forest and savannah ecosystems by their modern pollen spectra*, PhD thesis, University of Leicester, 2004.

Gosling, W.D. et al., A simple and effective methodology for sampling modern pollen rain in tropical environments, *The Holocene*, 13(4), 613, 2003.

Gosling, W.D. et al., Modern pollen rain characteristics of tall *terra firme* moist evergreen forest, southern Amazonia, *Quat. Res.*, 64, 284, 2005.

Haberle, S.G. and Maslin, M.A., Late Quaternary vegetation and climate change in the Amazon basin based on a 50,000 year pollen record from the Amazon fan, PDP site 932, *Quat. Res.*, 51, 27, 1999.

Haffer, J., Speciation in Amazonian forest birds, *Science*, 165, 131, 1969.

Haffer, J. and Prance, G.T., Climatic forcing of evolution in Amazonia during the Cenozoic: on the refuge theory of biotic differentiation, *Amazoniana*, 16, 579, 2001.

Hughes, C.E., Four new legumes in forty-eight hours, *Oxford Plant Systematics*, 12, 6, 2005.

Huntley, B. and Birks, H.J.B., *An Atlas of Past and Present Pollen Maps for Europe: 0-13000 Years Ago*, Cambridge University Press, Cambridge, 1983.

Killeen, T.J. et al., Diversity, composition and structure of tropical semi deciduous forest in the Chiquitanía region of Santa Cruz, Bolivia, *J. Trop. Ecol.*, 14, 803, 1998.

Lavin, M. et al., Metacommunity process rather than continental tectonic history better explains geographically structured phylogenies in legumes, *Phil. Trans. R. Soc. Lond. B*, 359, 1509, 2004.

Ledru, M-P., Late Quaternary environmental and climatic changes in central Brazil, *Quat. Res.*, 39, 90, 1993.

Ledru, M-P. et al., The last 50,000 years in the Neotropics (Southern Brazil): evolution of vegetation and climate, *Palaeogeogr. Palaeoclimatol. Palaeoecol.*, 123, 239, 1996.

Mayle, F.E., Assessment of the Neotropical dry forest refugia hypothesis in the light of palaeoecological data and vegetation model simulations, *J. Quat. Sci.*, 19, 713, 2004.

Mayle, F.E. and Beerling, D.J., Late Quaternary changes in Amazonian ecosystems and their implications for global carbon cycling, *Palaeogeogr. Palaeoclimatol. Palaeoecol.*, 214, 11, 2004.

Mayle, F.E., Burbridge, R., and Killeen, T.J., Millennial-scale dynamics of southern Amazonian rain forests, *Science*, 290, 2291, 2000.

Mayle, F.E. et al., Responses of Amazonian ecosystems to climatic and atmospheric CO_2 changes since the Last Glacial Maximum, *Phil. Trans. R. Soc., Lond. B*, 359, 499, 2004.

Murphy, P. and Lugo, A.E., Dry forests of Central America and the Caribbean, in *Seasonally Dry Tropical Forests*, Bullock, S.H., Mooney, H.A., Medina, E., Eds, Cambridge University Press, Cambridge, 1995, 9.

Oliveira-Filho, A.T. and Ratter, J.A., A study of the origin of central Brazilian forests by the analysis of the plant species distributions, *Edinburgh J. Botany*, 52, 141, 1995.

Parker, T.A. III et al., *The lowland dry forests of Santa Cruz, Bolivia: a global conservation priority*, RAP Working Papers 4, Conservation International, Washington DC, 1993.

Pennington, R.T. et al., Neotropical seasonally dry forests and Quaternary vegetation changes, *J. Biogeogr.*, 27, 261, 2000.

Pennington, R.T. et al., Historical climate change and speciation: neotropical seasonally dry forest plants show patterns of both Tertiary and Quaternary diversification, *Phil. Trans. R. Soc., Lond. B*, 359, 515, 2004.

Prado, D.E., *A critical evaluation of the floristic links between chaco and caatinga vegetation in South America*, PhD thesis, University of St Andrews, 1991.

Prado, D.E. and Gibbs, P.E., Patterns of species distributions in the dry seasonal forests of South America, *Ann. Missouri Bot. Garden*, 80, 902, 1993.

Rangel, J.O. et al., Región Andina, in *Colombia Diversidad Biótica, 1*, Rangel, J.O., Ed, Instituto de Ciencias Naturales, Universidad Nacional de Colombia: Bogota, 1995, 121.

Ratter, J.A. et al., Observations on woody vegetation types in the Pantanal and at Corumbá, Brazil, *Notes Roy. Bot. Garden Edinb.*, 45, 503, 1988.

Stuiver, M. et al., INTCAL98 radiocarbon age calibration, 24,000 – 0 cal BP, *Radiocarbon*, 40, 1041, 1998.

Stute, M. et al., Cooling of tropical Brazil (5°C) during the Last Glacial Maximum, *Science*, 269, 379, 1995.

Velez, M.I. et al., Palaeoenvironmental changes during the last ca. 8590 calibrated yr (7800 radiocarbon yr) in the dry forest ecosystem of the Patía Valley, Southern Colombian Andes: a multiproxy approach, *Palaeogeogr. Palaeoclimatol. Palaeoecol.*, 216, 279, 2005.

Wang, X. et al., Wet periods in northeastern Brazil over the past 210 kyr linked to distant climate anomalies, *Nature*, 432, 740, 2004.

Webb, S.L., Potential role of passenger pigeons and other vertebrates in the rapid Holocene migrations of nut trees, *Quat. Res.*, 26, 367, 1986.

Wilkinson, D.M., Plant colonization: are wind dispersed seeds really dispersed by birds at larger spatial and temporal scales? *J. Biogeogr.*, 24, 61, 1997.

18 Population Genetics and Inference of Ecosystem History: An Example Using Two Neotropical Seasonally Dry Forest Species

Y. Naciri, S. Caetano, R.T. Pennington, D. Prado
and R. Spichiger

CONTENTS

ABSTRACT

In South America, many seasonally dry tropical forest (SDTF) species are found in different areas all around the Amazon basin and separated by thousands of kilometres. This pattern suggests that there may have been a wider expansion of SDTF in the cool dry climates of the Quaternary. Alternatively, these current distribution patterns of SDTF may have been caused by long-distance dispersal events. In this chapter, we explain how population genetics and coalescence theory can help in inferring SDTF history by means of the study of two representative forest trees: *Astronium urundeuva* (Fr. Allemão) Engl. (Anacardiaceae) and *Geoffroea spinosa* Jacq. (Leguminosae).

The results reveal a higher differentiation level for *G. spinosa* than for *A. urundeuva* using microsatellite markers ($\theta = 0.151$ versus 0.064; $P < 0.01$), which may reflect different dispersal abilities and historical events. The preliminary analysis of three chloroplast markers on *A. urundeuva* does not allow discrimination between the two hypotheses of vicariance and dispersal.

18.1 INTRODUCTION

Fluctuations in climate during the Quaternary epoch caused major ice ages, which resulted in great changes in the distribution of living organisms. These changes had different genetic consequences in boreal, temperate and tropical zones (Hewitt, 2000). Because of the availability of pollen deposits and the ease of sampling organisms for genetic studies, Europe and North America are the best-studied areas and show some similar trends as well as specific features resulting from their geographical peculiarities. In these areas, the past decade has provided numerous examples of genetic signatures of postglacial dispersion and colonization in plant species, as exemplified by Petit et al. (1993, 1997, 2003), Démesure et al. (1996), Hewitt (1999), Grivet and Petit (2002) and Heuertz et al. (2004), among others.

In Europe, numerous pollen cores indicate that species now inhabiting the boreal and temperate regions had their ice age refugia south of the permafrost in Spain, Italy and the Balkans (Ferris et al., 1998; Petit, 1999). These records show that postglacial recolonization was extraordinarily rapid for a significant number of species. Populations at the northern limit of the ice age refugia are suspected to have expanded to large, open zones cleared by ice and permafrost thaw. This rapid expansion was probably mediated by long-distance dispersers, which established colonies and filled out the areas before others arrived. At the genetic level, such rapid expansions are expected to have reduced the level of heterozygosity (but see Comps et al., 2001), with a concomitant reduction in allelic diversity. This pattern of dispersal should lead to large homogenous patches, as suggested by theoretical and simulation studies (Ibrahim et al., 1996; Le Corre et al., 1997). This is confirmed by numerous empirical studies that show lower genetic diversity in the northern part of the distribution areas for species that colonized new northern territories after the last glacial maximum (Démesure et al., 1996; Sewell et al., 1996; Hewitt, 1996, 1999; but see Walter and Epperson, 2001 for contrasting results).

The situation is, by far, more poorly understood in the tropics when compared with in the northern hemisphere, and this is especially the case in South America (Coltrinari, 1993; Servant et al., 1993). In the Neotropics, conditions during the last glacial age have been suggested to be both drier (e.g. Van der Hammen and Absy, 1994; Hooghiemstra, 1997) and cooler (e.g. Colinvaux et al., 1996), and there is currently disagreement over which factor affected species distributions most significantly (Pennington et al., 2000). South America differs significantly from both Europe and North America, because the ice sheets and the permafrost were restricted to high altitudes and never covered the lowlands. The climatic changes during the Pleistocene in South America therefore must have driven changes between vegetation types (e.g. rain forest, montane forest and drier vegetation) rather than recolonization of new empty territories. This process might have influenced the species demography, because postglacial recolonization might have occurred at a slower rate in South America than in Europe, without population flushes or from rare and long-distance dispersal of seeds.

Discussions of historical vegetation changes in the Neotropics over the past three decades have focused on rain forests and, in particular, on explanations for the enormous species diversity in the Amazon basin. Refuge theory explained the observed diversity by fragmentation of the ranges of rain forest species followed by speciation within areas of rain forest that were isolated by drier vegetation during Pleistocene glaciations (Haffer, 1969, 1974, 1982; Prance, 1982). Most workers, for example Bush (1994) and Hoorn (1997), have assumed that the dry-adapted vegetation was a form of grass-rich savanna.

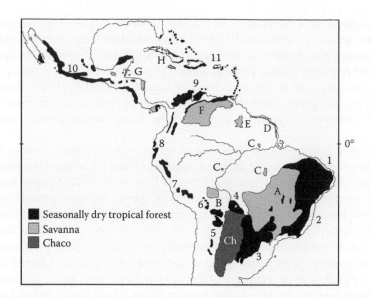

FIGURE 18.1 Schematic distribution of seasonally dry forests and savannas in the Neotropics, highlighting areas mentioned in the text. Seasonally dry forest: 1. caatingas; 2. south-east Brazilian seasonal forests; 3. Misiones nucleus; 4. Chiquitano; 5. Piedmont nucleus; 6. Bolivian inter-Andean valleys; 7. Peruvian and Ecuadorean inter-Andean valleys; 8. Pacific coastal Peru and Ecuador; 9. Caribbean coast of Colombia and Venezuela; 10. Mexico and Central America; 11. Caribbean Islands (small islands coloured black are not necessarily entirely covered by seasonally dry forests); 12. Florida. Savannas: (a). cerrado; (b) Bolivian; (c) Amazonian (smaller areas not represented); (d) coastal (Amapá, Brazil to Guyana); (e). Rio Branco-Rupununi; (f) Llanos; (g). Mexico and Central America; (h) Cuba. Ch: Chaco. Modified after Pennington et al. (2000) and Huber et al. (1987).

The lack of grass pollen in the few pollen cores from the Amazon basin (Colinvaux et al., 1996; Haberle, 1997) has led authors to conclude that the rain forest cannot have been fragmented by savanna (e.g. Connor, 1986; Colinvaux et al., 1996, 2001; Colinvaux, 1997a,b; Hoorn, 1997; see also Lynch, 1988), but no other vegetation type has been considered instead. For instance, Bush (1994) and Colinvaux et al. (1996) typically neglected seasonally dry tropical forest (SDTF) as a possible vegetation type that replaced the rain forests. Yet it happens that close floristic links exist between widely distinct areas of SDTF, suggesting that they are fragments of a previously much more extensive area of SDTF (Prado, 1991; Prado and Gibbs 1993). These links are shown by individual species that are distributed in two or more of the separate seasonally dry forest areas, as shown in Figure 18.1.

This observation has led some authors (Prado and Gibbs, 1993; Pennington et al., 2000; Prado, 2000) to suggest that another form of dry-adapted vegetation, SDTF, may have spread widely during the last ice age. This idea is supported by the pollen core from north-west Amazon, which was claimed to represent rain forest vegetation (Colinvaux et al., 1996) but may be consistent with a SDTF (Pennington et al., 2000). However, it is possible that the split of a widespread seasonal forest could have occurred much earlier than the Pleistocene, because fossils of contemporary seasonally dry forest genera dating from 10–20 Ma have been discovered (reviewed in Burnham and Graham, 1999). Other authors (e.g. Gentry, 1982; Mayle, Chapter 17) clearly state that the current disjunct distributions of SDTF species have been caused by recent events of long-distance dispersal.

The hypothesis of a drier and cooler climate during the last glacial maximum was suggested by Markgraf and Bradbury (1982), Markgraf (1991) and Behling (1997a,b, 2002), as well as Soubies et al. (1991). Behling (2001) quantified the dryness to have been –30% and the coldness

–5°C when compared with the present. According to the ecological requirements that can be deduced from the present distribution of SDTF, a simulated distribution was computed for several characteristic species (Spichiger et al., Chapter 8). For a number of simulations, species distributions for a cooler and drier climate correspond to a large and continuous territory centred in the Amazon basin, as suggested by the hypothesis of Prado and Gibbs (1993). These simulations, with their inherent limitations, suggest that for a number of species the existence of a large and ancestral SDTF during the last ice ages, from which the actual ones are derived, can receive some support.

The purpose here is to present the tools used to test whether the current areas of SDTF are truly refugial fragments of a previously larger and more continuous forest (vicariance hypothesis, Prado, 1991; Prado and Gibbs, 1993), which may have spread across large areas of the Amazon basin (Pennington et al., 2000), or whether the wide distribution of SDTF species has been attained by long-distance dispersal (dispersal hypothesis). Our approach contrasts with studies conducted in Europe and North America, which largely focused on species whose distributions were restricted to southern latitudes during the last glacial maximum and recolonized empty territories that were previously under the ice sheet. We will, in contrast, study a case where species may have become restricted to smaller areas placed variously within or at the periphery of the original larger area, with concomitant population demographic changes such as bottlenecks and founder effects. Using chloroplast and nuclear markers in two chosen representative tree species, this study aims to address the following questions.

- Is there any evidence of a former larger and more continuous population during the Pleistocene? Or have some disjunct populations resulted from rare long-distance dispersal events?
- If the current distributions were attained by vicariance rather than dispersal, did this fragmentation occur in the Quaternary ice ages, or is it more ancient?
- What does the current genetic structure within and between areas tell us about gene flow between the putative refugia?
- Is there any evidence of differential structuring using nuclear or chloroplast markers?
- What can be inferred about the dispersal of pollen and seeds?

It should be noted that this population genetic study is conducted at a scale that has never been used for a South American plant species. It is the first time, to our knowledge, that such a continental study has been conducted in the Neotropics on two different species in parallel. All other published studies usually focused on smaller geographical ranges (Dutech et al., 2000, 2003, 2004; Trapnell and Hamrick, 2004) or used only one type of marker (Degen et al., 2001; Lira et al., 2003).

18.2 MATERIAL AND METHODS

18.2.1 SPECIES CHOICE

Ideal study species would have a wide distribution throughout South American SDTF; be taxonomically well defined and easy to recognize in the field in the absence of flowers and fruits; relatively abundant, with a good habitat specificity; and finally, unlikely to be dispersed by humans. Candidate species that we considered included *Anadenanthera colubrina* (Vell.) Brenan, *Geoffroea spinosa* Jacq., *Astronium urundeuva* (Fr. Allemão) Engl, *Cedrela fissilis* Vell., *Tabebuia impetiginosa* (Mart. ex DC.) Toledo and *Aspidosperma polyneuron* Müll. Arg. Among these, we selected *G. spinosa* (Leguminosae) and *A. urundeuva* (Anacardiaceae) because they best match the required characteristics.

Geoffroea spinosa is the only species with vertebrate-dispersed fruits, and it can thus provide a contrast with wind-dispersed species. It is also distributed in all putative present-day SDTF refuges, and has been the subject of a recent taxonomic revision that demonstrated it to be clearly distinct from the only other species in the genus, *G. decorticans* (Gillies ex Hook. & Arn.) Burkart (Ireland and Pennington, 1999). It is also abundant in most areas where it grows.

Astronium urundeuva is a good model for wind-dispersed species that are characteristic of SDTF, easy to recognize and relatively abundant. This species is especially well represented in the Misiones nucleus (Spichiger et al., 1995, 2004, Chapter 8). Two congeneric species, *A. fraxinifolium* Schott and *A. balansae* Engl., are also predominant in other types of Paraguayan vegetation, but they are distinct, on the basis of the last taxonomic revison (Muñoz, 1990).

18.2.2 MOLECULAR TOOLS

Two types of markers were selected to analyse the structure of the above two species: microsatellite nuclear markers and chloroplast markers.

18.2.2.1 Microsatellite Markers

Microsatellite markers are tandemly repeated sequences whose unit of repetition is between one and five base pairs. They are now widely used in population genetic studies. They have proved to be highly polymorphic (Jeffreys et al., 1988), scattered throughout the genome (Wong et al., 1990; Wintero et al., 1992) and neutral in most cases (see Jarne and Lagoda, 1996, for a review). The dinucleotide repeats are especially useful because they are generally more variable than tri- or tetranucleotide repeats. Microsatellite analyses can provide insights into population genetic structure (Estoup et al., 1995; Slatkin, 1995; Estoup and Cornuet, 1999; Goldstein and Schlötterer, 1999, for a review), kinship investigations (Queller et al., 1993) or demographic inference (Kimmel et al., 1998; Luikart and Cornuet, 1998; Beaumont, 1999; King et al., 2000).

Dinucleotide microsatellites were identified in the two model species. For *A. urundeuva*, nine microsatellites were characterized, among which seven were polymorphic (Caetano et al., 2005, Table 18.1). These microsatellites were analysed in two populations from north and east Paraguay (Cerro León and Mbaracayú) and one population from north-east Argentina (El Rey). For *G. spinosa* (Naciri-Graven et al., 2005), 11 microsatellites were characterized, of which 10 were polymorphic (Table 18.2). Six populations were analysed, three from Peru, Amazonas (east), Cajamarca (east) and Andes (western slope of the Andes); two from Paraguay, Cerro León (north) and Fortin Garrapatal (north-east); and one from north Argentina, Antequera.

TABLE 18.1
Characteristics of the Nine Microsatellites in *Astronium urundeuva* Tested on Two Populations from Paraguay and One Population from Argentina

Name	Motif	GenBank (N)	Size Range (bp)	Populations Screened (N)	Total Sample Size	N_a[a]	R_t[a]
Auru.B209	$(TA)_{12}(GA)_{11}$	AY509817	184–208	3	39	10	7.2
Auru.A392	$(CT)_{17}TT(CT)_3$	AY640260	179–201	3	46	11	9.3
Auru.C072	$(CT)_{11}ATGT(AT)_9$	AY640264	150–186	3	46	14	7.4
Auru.D167	$(GA)_{10}$	AY640268	134–140	3	46	4	2.8
Auru.D282	$(CT)_{11}(AT)_{11}$	AY640270	185–248	3	47	20	11.1
Auru.D094.	$(GA)_{13}$	AY640267	108–120	3	46	8	4.3
Auru.E062	$(CT)_8$	AY640273	80–90	3	46	4	3.2
Auru.J185	$(TA)_3T(TA)_2$	AY509819	181	3	23	1	1
Auru.D200	$(TC)_8GT(CT)_2$	AY640269	195	3	23	1	1

[a] Number of alleles over all populations.
[b] A measure of allelic richness over all populations.

Source: Caetano, S. et al., *Molec. Ecol. Notes*, 5, 21, 2005. With permission.

TABLE 18.2
Characteristics of the 11 Optimized Microsatellites in *Geoffroea spinosa* Tested on Three Populations from Peru, Two from Paraguay and One from North Argentina

Name	Motif(s)	GenBank (N)	Size Range (bp)	Populations Screened (N)	Total Sample Size	N_a[a]	R_t[b]
Gsp.A104	$(GT)_{11}$ and $(GA)_8$	AY644736	197–213	6	79	9	4.6
Gsp.B458	$(CA)_{10}(TA)_6$	AY644742	121–147	6	75	13	8.4
Gsp.B331	$(CA)_{14}TG(CA)_4$	AY644734	312–350	6	87	14	7.8
Gsp.G226	$(TC)_{22}$	AY644746	174–122	6	68	19	10.6
Gsp.F119	$(CA)_{10}$	AY644745	155–185	6	88	13	6.0
Gsp.A149	$(GA)_{15}$	AY644737	199–223	6	83	12	8.0
Gsp.B284	$(TC)_8(TATG)_2$	AY644740	188–200	6	82	7	2.8
Gsp.A021	$(TA)_4(TG)_{14}$	AY644738	216–238	3	32	8	3.5
Gsp.I168	$(CT)_{34}$	AY644748	143–193	3	37	18	7.6
Gsp.B264	$(CT)_{14}(AT)_2$	AY644733	175–193	3	27	9	5.3
Gsp.B291	$(CT)_{37}(CA)_{10}$ CTCA$(CT)_3$	AY644741	313	3	21	1	1.0

[a]Number of alleles over all populations.
[b]A measure of allelic richness over all populations.

Source: Naciri-Graven, Y. et al., *Molec. Ecol. Notes*, 5, 45, 2005. With permission.

18.2.2.2 Chloroplast Markers

The chloroplast organelle is generally maternally inherited in most angioperms (Corriveau and Coleman, 1988; but see Reboud and Zeyl, 1994; Testolin and Cipriani, 1997), whereas it seems to be paternally inherited in gymnosperms (Wagner, 1992; Cato and Richardson, 1996). This uniparental inheritance allows for gene genealogies to be reconstructed for phylogeographical purposes (Avise, 1994) or for demographic inference using coalescence theory (Hudson, 1990). Indeed, it has been demonstrated that chloroplast DNA (cpDNA) is variable enough to be suitable for infraspecific analyses using sequences, polymerase chain reaction (PCR) restriction fragment length polymorphisms (Taberlet et al., 1991; Démesure et al., 1995; Sewell et al., 1996; Dumoulin-Lapègue et al., 1997; Hamilton, 1999a; Rendell and Ennos, 2003) or chloroplast microsatellites (Powell et al., 1995; Vendramin et al., 1996; Provan et al., 1998, for the pioneers). The usefulness of cpDNA markers in analyses of population structuring, either alone to study population structuring through seeds or coupled with nuclear markers, has been well emphasized (McCauley, 1994, 1995; Hamilton, 1999b; Dutech et al., 2000).

In the present study, results have been obtained only for *A. urundeuva*. For this species, three chloroplast spacers, namely *trn*H–*psb*A, *trn*S–*trn*G (Hamilton, 1999a) and *trn*L–*trn*F (Taberlet et al., 1991) were fully sequenced, using an ABI 377 automated sequencer and standard protocols, on four populations in Argentina and 17 populations in Paraguay, for a total of 92 individuals (Table 18.3).

18.2.3 STATISTICAL ANALYSES

18.2.3.1 Population Genetics

Using population genetic tools, it is possible to get information about the forces that have actually had an impact on the present genetic variation, i.e. mutation, recombination, migration, genetic drift and selection, and the way these forces have shaped the genetic variation (see Hartl and Clark, 1997, for a complete view, and Wakeley, 2005, for a critical commentary). Among the many tools developed for population genetics, only the F-statistics developed by Wright (1921, 1965) will be used in this chapter.

F-statistics, by means of the F_{ST} index, allow the measure of genetic differences among populations; that is, the extent to which the studied populations, taken as a whole, move away from panmixia (also known as Hardy-Weinberg equilibrium, HWE) and differ from each other. F_{ST} is therefore a measure of genetic distance between populations. This index is calculated as $(H_T - H_S)/H_T$, where H_T is the expected heterozygosity under HWE when all individuals from all populations are considered together, and H_S is the average expected heterozygosity under HWE within the different populations. If all the populations behave as a unique panmictic population, F_{ST} will be 0 ($H_T = H_S$). As soon as allele frequencies among populations become different because of random processes, selection or migration, F_{ST} will be positive.

At the population level, a second linked index, known as F_{IS}, gives a measure of inbreeding level. It is defined as $(H_S - H_O)/H_S$, where H_S and H_O are, respectively, the expected heterozygosity under HWE and the observed heterozygosity in the considered population. When the population is at HWE, $F_{IS} = 0$, whereas F_{IS} is found positive ($H_S > H_O$) when an excess of homozygotes exists, for instance as a result of selfing.

Another index, θ, initially defined by Weir and Cockerham (1984), is an equivalent estimate of F_{ST}. It is computed using allele frequencies at each locus in an analysis of variance framework that allows the computation of the genetic variance between populations, σ^2_b. θ (or F_{ST}) is therefore defined as σ^2_b/σ^2_T, where σ^2_T is the total genetic variance over all individuals. This mode of calculation, also known as analysis of molecular variance (AMOVA) (Excoffier et al., 1992), is particularly interesting when dealing with haplotype markers or to get an index over loci. This approach is also useful when populations can be assigned to different groups. The F_{CT} index then gives the proportion of the total variance that results from genetic differences among groups (σ^2_a). F_{CT} is then given by σ^2_a/σ^2_T, and a slightly different formula is then used for F_{ST} that incorporates σ^2_a (Michalakis and Excoffier, 1996).

For both species, F_{ST} and F_{IS} were computed for microsatellite data at each locus using FSTAT software (Goudet, 2001). The same program was used to get θ over loci. For *A. urundeuva* sequence data, an AMOVA was computed using ARLEQUIN software (Schneider et al., 2000), a program that is more appropriate for dealing with haplotype markers, as well as with hierarchical structuring. Two groups of populations were defined, the first one with all the Paraguayan populations (Misiones nucleus) and the second one with the Argentinean populations (Piedmont nucleus). F_{ST} and F_{CT} were calculated accordingly.

18.2.3.2   Coalescence Theory

Other tools, derived from the coalescence theory, can also be used for answering the questions identified. Coalescence theory has been largely used on humans and other mammals (Harding et al., 1997; Excoffier and Schneider, 1999; Harris and Hey, 1999; among others) but seldom applied to plants (but see Schaal and Olsen, 2000, for a review), particularly because of the persistent but erroneous idea that cpDNA is not sufficiently variable at the specific level.

The coalescence precepts were developed more recently than population genetic ones (see Kingman, 2000, for a historical commentary), mainly during the 1970s by J.F.C. Kingman, followed by R.R. Hudson, using the works of mathematicians interested in stochastic mechanisms, and the studies of geneticists such as W. Ewens (1972), P. Moran (1975), and T. Ohta and M. Kimura (1973), who were interested in neutral evolution in finite populations (Kingman, 2000). Coalescence theory aims to reconstruct the genealogical history of genes, be they alleles (for nuclear markers) or haplotypes (for mitochondrial, chloroplast or Y chromosome markers), until reaching the most recent common ancestor (MRCA) (Kingman, 1982; Hudson, 1990). Because of this genealogical approach, the coalescence is a sister theory to molecular phylogenetics, but with the fundamental difference that it takes into account population genetic features such as the demographic history (Tajima, 1989a; Rogers and Harpending, 1992; Harpending et al., 1998). Other characteristics that make coalescence different from molecular phylogenetics are that some assumptions of the latter theory are relaxed: coalescence theory takes into account the existence of ancestor alleles that can be shared by different taxa, does not assimilate gene history to population history and breaks the obligate hierarchical relationship between taxa that is usually assumed by phylogenetics (Posada and Crandall, 2001).

One major insight of coalescence theory was to show that the date of the MRCA is fundamentally different from the date of speciation. Edwards and Beerli (2000) have shown, using a rather simple model, that the date inferred from the phylogeny can be, on average, 50% older than the real date of speciation (which is defined here as the absence of gene flow between the two new species). Coalescence theory also provides several statistical tools that allow us to test for marker neutrality, because the whole theory is based on the assumption that the analysed genes are not subject to selection (Tajima, 1989b; Fu, 1997; Li et al., 2003). Finally, another major insight of this theory concerns the relationships between the different sampled genes, which are depicted in a network framework rather than in a hierarchical tree framework. Networks can take into account extant ancestral alleles, or the fact that each gene can lead to several other genes through different independent mutations.

A median-joining network analysis was carried out on chloroplast markers for *A. unrundeuva* using NETWORK software (Bandelt et al., 1999). For this purpose, and assuming that the chloroplast does not recombine in this species, or at least that heteroplasmy is absent, the haplotypes found for *trn*H–*psb*A, *trn*S–*trn*G and *trn*L–*trn*F were combined for each individual to form superhaplotypes.

18.3 RESULTS

18.3.1 NUCLEAR MARKERS: MICROSATELLITES FOR THE TWO SPECIES

For the seven polymorphic microsatellites identified in *A. urundeuva*, the total number of alleles ranged between four and 20, depending on the locus, for a minimum sample size of 23 individuals per locus (Table 18.1; Caetano et al., 2005). A significant but small amount of genetic structure was found over all loci for the three populations studied from Argentina and Paraguay, with $\theta = 0.064$ ($P < 0.01$). Among the 11 microsatellites characterized in *G. spinosa*, four loci amplified only on the Peruvian population, one of them being monomorphic (Table 18.2; Naciri-Graven et al., 2005). The total number of alleles ranged between seven and 19, depending on the locus, for a minimum sample size of 21 individuals per marker. The genetic differentiation over all loci was found to be as high as $\theta = 0.212$ ($P < 0.01$) over the six populations studied from Peru, Argentina and Paraguay.

The results for the two species can be compared using the populations from Paraguay and Argentina, because the geographical distances between them are approximately equivalent for the two sampling schemes. The average differentiation among the three Paraguayan-Argentinean populations of *G. spinosa* dropped to $\theta = 0.151$ ($P < 0.01$) versus $\theta = 0.212$, but remained significantly higher than what was found for *A. urundeuva* ($\theta = 0.064$, $P < 0.01$). Allelic richness per locus allows the comparison of marker polymorphisms in a way that they are independent from population size. This index was slightly higher, although not significantly, for *G. spinosa* ($2.8 < R_t < 10.6$, $\mu = 6.89 \pm 2.42$) than for *A. urundeuva* ($2.8 < R_t < 11.1$, $\mu = 6.47 \pm 2.92$) when computed over the seven loci that respectively amplified on the Paraguayan-Argentinean populations of the two species.

For both species, significant positive F_{IS} values were found. The presence of a significant inbreeding level in a population is expected to give positive F_{IS} values for all loci in that particular population, which is not the case for both *A. urundeuva* and *G. spinosa*. When dealing with microsatellite markers that exhibit high mutation rates, a common explanation of such positive F_{IS} values found in some populations, at some loci, is the presence of null alleles (i.e. alleles that do not amplify because of a mutation in the primer regions used for PCR). The number of significant and positive F_{IS} values suggests that this phenomenon exists in both species, although it is more pronounced in *G. spinosa* than in *A. urundeuva*.

TABLE 18.3

Combined Haplotypes for *Astronium urundeuva* Using trnH–psbA, trnS–trnG and trnL–trnF Sequences in 21 Populations from Paraguay and Argentina

Country	Population Name	Individuals (N)	AAA	ABA	ABC	BAA	BAB	CAA
Paraguay	St Louis 1	3	3	—	—	—	—	—
Paraguay	St Louis 2	6	5	—	—	—	—	1
Paraguay	St Louis 3	6	—	—	—	6	—	—
Paraguay	St Louis 4	5	1	—	—	4	—	—
Paraguay	St Louis 5	2	—	—	—	2	—	—
Paraguay	St Louis 6	6	—	—	—	6	—	—
Paraguay	St Louis 7	6	6	—	—	—	—	—
Paraguay	Cordillera	5	—	—	—	2	3	—
Paraguay	Presidente Hayes, Estancia Riochito	6	2	—	—	4	—	—
Paraguay	Altos	3	—	—	—	3	—	—
Paraguay	Mbaracayu 1	2	—	—	—	2	—	—
Paraguay	Mbaracayu 2	3	2	—	—	1	—	—
Paraguay	Mbaracayu 3	9	2	—	—	7	—	—
Paraguay	Mbaracayu 4	5	5	—	—	—	—	—
Paraguay	Jardin Botanique Asunción	4	—	—	—	4	—	—
Paraguay	Eusebio-Aguaity	3	—	—	—	3	—	—
Paraguay	Cerro Leon 2	4	4	—	—	—	—	—
Argentina	Las Lajitas 2	2	—	—	2	—	—	—
Argentina	Ledesma	4	—	4	—	—	—	—
Argentina	Oran	4	—	4	—	—	—	—
Argentina	Embarcacion	4	—	4	—	—	—	—
Total	21	92	30	12	2	44	3	1

18.3.2 CHLOROPLAST MARKERS FOR *ASTRONIUM URUNDEUVA*

Three different haplotypes were found for *trn*H–*psb*A (A, B and C), two for *trn*S–*trn*G (A and B) and three for *trn*L–*trn*F (A, B and C). Combining the different haplotypes led to six superhaplotypes (Table 18.3). Six populations out of 21 were polymorphic, with two haplotypes each. These six polymorphic populations were all located in Paraguay, mainly in the north and east parts of the country. The overall genetic differentiation was very high, with $F_{ST} = 0.83$ ($P < 0.001$). The differentiation between the Paraguayan populations (Misiones nucleus) on the one hand and the Argentinean ones (Piedmont nucleus) on the other was also very high, and accounted for 58% of the total genetic differentiation ($F_{CT} = 0.48$, $P < 0.001$). It is difficult to compare this figure with the equivalent one obtained for microsatellites, because the genetic differentiation was computed on many more populations. Reducing the sample size to the one analysed with microsatellites, F_{ST} dropped to 0.56 ($P < 0.001$) but remained eight to nine times higher than the genetic differentiation found for microsatellites.

The median-joining network of haplotypes (Figure 18.2) shows that the most common haplotype is BAA (47.8%), a haplotype that also shares the highest number of connections with other ones (three links). All other haplotypes derive from BAA by one or more mutations. Four haplotypes were found to be exclusively Paraguayan (BAA, AAA, 32.6%; BAB, 3.3%; and CAA, 1.1%), whereas two were exclusively located in Argentina (ABA, 13.0%, and ABC, 2.2%).

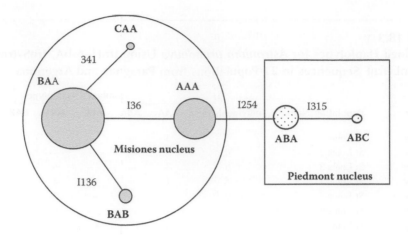

FIGURE 18.2 Median-joining network for the six haplotypes (circles) obtained for *Astronium urundeuva*, with three chloroplast spacers, *trn*H–*psb*A, *trn*S–*trn*G and *trn*L–*trn*F on 92 individuals from 17 populations from Paraguay and four populations from Argentina. The size of the circles is proportional to the haplotypes frequencies. Haplotypes that are specific to Argentina (Piedmont nucleus) and Paraguay (Misiones nucleus) are indicated in dotted grey and in grey, respectively. Mutations (in grey) are positioned between haplotypes and referred to by their position. A preceding 'I' indicates that the mutation is an insertion-deletion.

18.4 DISCUSSION

18.4.1 WHAT HISTORY FOR THE TWO SPECIES?

The preliminary results presented here show that, even with a low number of populations, it is already possible to infer some hypotheses about the two species using population genetics and coalescence theory. The two types of markers give us different and complementary information on *A. urundeuva* and *G. spinosa*, mainly a behavioural signal (e.g. reproductive strategy and inbreeding level, as well as demography) when using microsatellite markers, and a phylogeographical and demographic signal when using chloroplast markers.

The first insight concerns the different levels of differentiation obtained for the two species. The higher differentiation observed for *G. spinosa* than for *A. urundeuva* ($\theta = 0.151$ versus 0.064) can be a result of an artefact of a higher average number of alleles linked to different mutational processes, to specific historical events or to contrasted dispersal capacities. It is a general feature of microsatellites to mutate rapidly, which tends to create homoplasious alleles, i.e. alleles that are identical but originating independently. Homoplasy can lead to the underestimation of differentiation levels, because it erases information. The homoplasy hypothesis would suggest that the lower differentiation found in *A. urundeuva* is a result of a higher homoplasy in this species. However, the equivalent allelic richness recorded for the two species suggests similar overall mutation rates; more data are needed to confirm this (see Estoup and Cornuet, 1999). Consequently, the higher differentiation level for *G. spinosa* is suspected to be rather the consequence of dispersal capacities or the result of specific historical events. *Geoffroea spinosa* is a species dispersed by small vertebrates, whereas *A. urundeuva* is wind dispersed. More genetic exchanges between populations are thus expected for *A. urundeuva* than for *G. spinosa*, in agreement with the observed differentiation rates.

The second, non-exclusive, explanation suggests that the separated populations of *G. spinosa* reflect more ancient vicariance events. This makes sense, because *G. spinosa* is much more widely distributed than *A. urundeuva* around the Amazon basin. This last hypothesis is supported by the non-amplification of 4 microsatellites out of 11 in the non-Peruvian populations, and by the lack of amplification of some individuals scattered in these populations (the existence of so-called null alleles).

Such a result means that the flanking regions used to amplify the microsatellites are subject to mutations, a signature of populations of more ancient divergence.

The second insight is related to the difference observed for *A. urundeuva* differentiation when different types of markers are used. For a hermaphrodite species, the effective population size of the nuclear genome (diploid) is expected to be twice the effective size of the chloroplast genome (haploid). If equal dispersion of seeds and pollen are postulated, this means that the chloroplast differentiation should be twice that of the nuclear one. *A. urundeuva* is, however, suspected to be either polygamodioecious, such as other Anacardiaceae species (D. Prado, Rosario University, Argentina, pers. obs.), with female and male flowers in different trees as well as hermaphroditic flowers present in both trees (Muñoz, 1990), or dioecious (Santin and de Freitas Leitão, 1991). This singularly complicates the theoretical expectation, because the ratio of chloroplast to nuclear differentiation should therefore vary between four (strict dioecious species with equal sex ratio, equal dispersion of seeds and pollen) to two (complete hermaphrodite plants), instead of the eight to nine times we detected. This suggests a higher dispersion of pollen compared with seed, with pollen migrating at least about twice as far as seeds, or unequal sex ratio, with fewer female than male trees, or both. Nothing is known about pollinators, so it is difficult, for the moment, to correlate this result with biological features. It can, however, be noticed that such discrepancies between chloroplast and nuclear F_{ST} are quite common for tree species (Petit, 1999).

The third insight is deduced from the *A. urundeuva* chloroplast network. Coalescence theory predicts that network topology depends greatly on population demography (Tajima, 1989a; Harpending et al., 1998). For instance, demographic expansions lead to networks with a starlike pattern. The theoretical results also show that there is a direct link between gene age and gene frequency; a gene widely distributed geographically and exhibiting a high frequency has a high probability of being ancient. In the same way, ancient genes share more mutational connections with other genes than more recent ones, which will be seen at the network periphery (Posada and Crandall, 2001). Using these predictions, the somewhat starlike pattern of the network, which still needs to be confirmed by more data, could suggest a demographic expansion for the species.

Moreover, the Paraguayan misiones BAA haplotype can be assumed to be an ancestor because of both a high frequency and the number of mutational connections it shares. Haplotypes found in the Piedmont (the ABA and ABC haplotypes) are located at the network periphery and should therefore be more recent than BAA. The network topology thus suggests that north-west Argentina (which belongs to the Piedmont nucleus) was colonized from the east, i.e. from Paraguay in the Misiones nucleus (BAA, AAA, CAA and BAB haplotypes). This conclusion is also supported by the fact that nearly all polymorphic populations (five of six) are found in the northern and eastern parts of Paraguay.

18.4.2 INFERENCES RELATING TO ECOSYSTEM HISTORY

Two main issues can be deduced from the preceding discussion. First, *A. urundeuva* and *G. spinosa* must have experienced different historical events, with the latter being suspected to have a more ancient history of differentiation, whatever the influence of their respective dispersal abilities. The second issue concerns the two hypotheses tested in this study: vicariance versus dispersal as the main explanation for SDTF distribution. From the *A. urundeuva* network and the localization of the polymorphic populations, it is tempting to conclude that the results best match the dispersal hypothesis. Such an inference would be too hasty, because we still do not know where, in other parts of South America, the so-called more ancestral haplotype BAA is actually found. More samples are needed, especially from Brazil, Argentina and Bolivia, to have a better understanding of the history of this species. Analysing the chloroplast network for *G. spinosa* in addition to *A. urundeuva* will also shed a probable different light on the past history of SDTF.

Despite the preliminary nature of some of the analyses presented here, it is clear that the joint use of population genetics and coalescence theory on nuclear and chloroplast markers for two or more species is an informative method to infer species histories and, consequently, to better understand the ecosystem to which they belong.

ACKNOWLEDGEMENTS

This research is currently funded by the Swiss National Science Foundation (grant no. 3100A0-100806/1 to YNG). We wish to thank Dr Ary Oliveira Filho (University of Lavras, Brazil) for his commitment to the project, and Professor F. Mereles (University of Asunción, Paraguay), Dr S. Beck (Botanical Garden of La Paz, Bolivia) and Dr L. Rodríguez (National Herbarium Caracas, Venezuela) for their help in getting the collection permits. We are also indebted to Dr Andrès Amarilla (University of Asunción, Paraguay); Mónica Soloaga, Nathalie Rasolofo and Rachele Martini (Conservatoire et Jardin botaniques); Luis Oakley (Rosario University, Argentina); Sam Bridgewater (Royal Botanic Garden Edinburgh, UK); Aniceto Daza and Carlos Reynel (La Molina University, Peru); and Karem Elizeche for sampling support, and to Philippe Busso, Lara Turin, Bérivan Polat and Hélène Geser for technical assistance.

REFERENCES

Avise, J.C., *Molecular Markers, Natural History and Evolution*, Chapman and Hall, New York, 1994.

Bandelt, H.-J., Forster, P., and Röhl, A., Median-joining networks for inferring intraspecific phylogenies, *Molec. Biol. Evol.*, 16, 37, 1999.

Beaumont, M.A., Detecting population expansion and decline using microsatellites, *Genetics*, 153, 2013, 1999.

Behling, H., Late Quaternary vegetation, climate and fire history from the tropical mountain region of Morro de Itapeva, SE Brazil, *Palaeogeogr. Palaeoclimatol. Palaeoecol.*, 129, 407, 1997a.

Behling, H., Late Quaternary vegetation, climate and fire history in the Araucaria forest and campos region from Serra Campos Gerais (Paraná), S Brazil, *Rev. Paleobot. Palynol.*, 97, 109, 1997b.

Behling, H., South and southeast Brazilian grassland during late Quaternary times: a synthesis, *Palaeogeogr. Palaeoclimatol. Palaeoecol.*, 177, 19, 2001.

Behling, H., Carbon storage increases by major forest ecosystems in tropical South America since the Last Glacial Maximum and the early Holocene, *Glob. Planet. Change*, 33, 107, 2002.

Burnham, R.J. and Graham, A., The history of Neotropical vegetation: new developments and status, *Ann. Missouri Bot. Gard.*, 86, 546, 1999.

Bush, M.B., Amazonian speciation: a necessarily complex model, *J. Biogeogr.*, 21, 5, 1994.

Caetano, S. et al., Identification of microsatellite markers in a neotropical seasonally dry forest tree, *Astronium urundeuva* (Anacardiaceae), *Molec. Ecol. Notes*, 5, 21, 2005.

Cato, A. and Richardson, T.E., Inter- and intraspecific polymorphism at chloroplast SSR loci and the inheritance of plastids in *Pinus radiata* D. Don., *Theor. Appl. Genet.*, 93, 587, 1996.

Colinvaux, P.A., An arid Amazon? *Trends Ecol. Evol.*, 12, 318, 1997a.

Colinvaux, P.A., *The Ice-Age Amazon and the Problem of Diversity*, NWO—Huygenslezing, The Hague, 1997b, 7.

Colinvaux, P.A. et al., A long pollen record from lowland Amazonia: forest and cooling in glacial times, *Science*, 274, 85, 1996.

Colinvaux, P.A. et al., A paradigm to be discarded: geological and paleoecological data falsify the Haffer and Prance refuge hypothesis of Amazonian speciation, *Amazoniana*, 16, 609, 2001.

Coltrinari, L., Global Quaternary changes in South America, *Glob. Planet. Change*, 7, 11, 1993.

Comps, B. et al., Diverging trends between heterozygosity and allelic richness during postglacial colonization in the European beech, *Genetics*, 157, 389, 2001.

Connor, E.F., The role of Pleistocene forest refugia in the evolution and biogeography of tropical biotas, *Trends Ecol. Evol.*, 1, 165, 1986.

Corriveau, J.L. and Coleman, A.W., Rapid screening method to detect potential biparental inheritance of plastid DNA and results over 200 angiosperm species, *Amer. J. Bot.*, 75, 1443, 1988.

Degen, B. et al., Fine-scale spatial genetic structure of eight tropical tree species as analysed by RAPDs, *Heredity*, 87, 497, 2001.

Démesure, B., Sodzi, N., and Petit, R.J., A set of universal primers for amplification of polymorphic noncoding regions of mitochondrial and chloroplast DNA in plants, *Molec. Ecol.*, 4, 129, 1995.

Démesure, B., Comps, B., and Petit, R.J., Chloroplast DNA phylogeography of the common Beech (*Fagus sylvatica* L.) in Europe, *Evolution*, 50, 2515, 1996.

Dumoulin-Lapègue, S., Pemonge, M.-H., and Petit, R.J., An enlarged set of consensus primers for the study of organelle DNA in plants, *Molec. Ecol.*, 6, 393, 1997.

Dutech, C., Maggia, L., and Joly, H.I., Chloroplast diversity in *Vouacapoua americana* (Caesalpiniaceae), a neotropical forest tree, *Molec. Ecol.*, 9, 1427, 2000.

Dutech, C. et al., Tracking a genetic signal of extinction–recolonization events in a neotropical tree species: *Vouacapoua americana* Aublet in French Guiana, *Evolution*, 57, 2753, 2003.

Dutech, C., Joly, H.I., and Jarne, P., Gene flow, historical population dynamics and genetic diversity within French Guianan populations of a rain forest tree species, *Vouacapoua americana*, *Heredity*, 92, 69, 2004.

Edwards, S.V. and Beerli, P., Perspective: gene divergence, population divergence, and the variance in coalescence time in phylogeographic studies, *Evolution*, 54, 1839, 2000.

Estoup, A. and Cornuet, J.-M., Microsatellite evolution: inferences from population data, in *Microsatellites: Evolution and Applications*, Goldstein, D.B. and Schlötterer, C., Eds, Oxford University Press, Oxford, 1999, 49.

Estoup, E. et al., Microsatellite variation in honey bee populations: hierarchical genetic structure and test of the infinite allele and stepwise mutation models, *Genetics*, 140, 679, 1995.

Ewens, W.J., The sampling theory of selectively neutral alleles, *Theor. Popul. Biol.*, 6, 143, 1972.

Excoffier, L. and Schneider, S., Why hunter-gatherer populations do not show signs of Pleistocene demographic expansions, *Proc. Natl. Acad. Sci. USA*, 96, 10597, 1999.

Excoffier, L., Smouse, P., and Qauttro, J., Analysis of molecular variance inferred from metric distances among DNA haplotypes: application to human mitochondrial DNA restriction data, *Genetics*, 131, 479, 1992.

Ferris, C. et al., Chloroplast DNA recognizes three refugial sources of European oaks and suggests independent eastern and western immigrations to Finland, *Heredity*, 80, 584, 1998.

Fu, Y.-X., Statistical tests of neutrality of mutations against population growth, hitchhiking and background selection, *Genetics*, 147, 915, 1997.

Gentry, A.H., Neotropical floristic diversity: phytogeographical connections between Central and South America, Pleistocene climatic fluctuations, or an accident of the Andean orogeny, *Ann. Missouri Bot. Gard.*, 69, 557, 1982.

Goldstein, D.B. and Schlötterer, C., *Microsatellites: Evolution and Applications*, Oxford University Press, Oxford, 1999.

Goudet, J., *FSTAT, a program to estimate and test gene diversities and fixation indices (version 2.9.3)*, http://www.unil.ch/izea/softwares/fstat.html, 2001.

Grivet, D. and Petit, R.J., Phylogeography of the common ivy (*Hedera* sp.) in Europe: genetic differentiation through space and time, *Molec. Ecol.*, 11, 1351, 2002.

Haberle, S., Upper Quaternary vegetation and climate history of the Amazon basin: correlating marine and terrestrial pollen records, *Proc. ODP, Sci. Results*, 155, 381, 1997.

Haffer, J., Speciation in Amazonian forest birds, *Science*, 165, 131, 1969.

Haffer, J., Avian speciation in tropical South America, *Publ. Nuttall Ornithol. Club*, 14, 1974.

Haffer, J., General aspects of the Refuge Theory, in *Biological Diversification in the Tropics*, Prance, G.T., Ed, Columbia University Press, New York, 1982, 6.

Hamilton, M.B., Four primer pairs for the amplification of chloroplastic intergenic regions with intraspecific variation, *Molec. Ecol.*, 8, 513, 1999a.

Hamilton, M.B., Tropical tree gene flow and seed dispersal, *Nature*, 401, 129, 1999b.

Harding, R.M. et al., Archaic African and Asian lineages in the genetic ancestry of modern humans, *Amer. J. Hum. Genet.*, 60, 772, 1997.

Harpending, H.C. et al., Genetic traces of ancient demography, *Proc. Natl. Acad. Sci. USA*, 95, 1961, 1998.

Harris, E.E. and Hey, J., X chromosome evidence for ancient human histories. *Proc. Natl. Acad. Sci. USA*, 96, 3320, 1999.

Hartl, D.L. and Clark, A.G., *Principles of Population Genetics*, 3rd ed., Sinauer Associates, Sunderland, 1997.

Heuertz, M. et al., Chloroplast DNA variation and postglacial recolonization of common ash (*Fraxinus excelsior* L.) in Europe, *Molec. Ecol.*, 13, 3437, 2004.

Hewitt, G., Some genetic consequences of ice ages, and their role in divergence and speciation, *Biol. J. Linn. Soc.*, 58, 247, 1996.

Hewitt, G., Post-glacial re-colonization of European biota, *Biol. J. Linn. Soc.*, 68, 87, 1999.

Hewitt, G., The genetic legacy of the Quaternary ice ages, *Nature*, 405, 907, 2000.

Hooghiemstra, H., *Tropical Rain Forest Versus Savanna: Two Sides of a Precious Medal*, NWO–Huygenslezing, The Hague, 1997, 32.

Hoorn, C., Palynology of the Pleistocene glacial/interglacial cycles of the Amazon fan (holes 940A, 944A, and 946A), *Proc. ODP, Sci. Results*, 155, 397, 1997.

Hudson, R.R., Gene genealogies and the coalescent process, in *Gene Genealogies and the Coalescent Process, vol. D*, Futuyma, J. and Antonovics, J.D., Eds, Oxford University Press, New York, 1990, 1.

Ibrahim, K.M., Nichols, R.A., and Hewitt, G., Spatial patterns of genetic variation generated by different forms of dispersal during range expansion, *Heredity*, 77, 282, 1996.

Ireland, H. and Pennington, R.T., A revision of *Geoffroea* Leguminosae–Dalbergieae, *Edinburgh J. Bot.*, 56, 329, 1999.

Jarne, P. and Lagoda, P.J.L., Microsatellites, from molecules to populations and back, *Trends Ecol. Evol.*, 11, 424, 1996.

Jeffreys, A.J. et al., Spontaneous mutation rates to new alleles at tandem repetitive hypervariable loci in human DNA, *Nature*, 322, 278, 1988.

Kimmel, M. et al., Signature of population expansion in microsatellite repeat data, *Genetics*, 148, 1921, 1998.

King, J.P., Kimmel, M., and Chakraborty, R., A power analysis of microsatellite-based statistics for inferring past population growth, *Molec. Biol. Evol.*, 17, 1859, 2000.

Kingman, J.F.C., The coalescent, *Stochastic Process. Appl.*, 13, 235, 1982.

Kingman, J.F.C., Origin of the coalescent: 1974–1982, *Genetics*, 156, 1461, 2000.

Le Corre, V. et al., Colonization with long-distance seed dispersal and genetic structure of maternally inherited genes in forest trees: a simulation study, *Genet. Res. Camb.*, 69, 117, 1997.

Li, H. et al., Neutrality tests using DNA polymorphism from multiple samples, *Genetics*, 163, 1147, 2003.

Lira, C.F. et al., Long-term population isolation in the endangered tropical tree species *Caesalpinia echinata* Lam. revealed by chloroplast microsatellites, *Molec. Ecol.*, 12, 3219, 2003.

Luikart, G. and Cornuet, J.M., Empirical evaluation of a test for identifying recently bottlenecked populations from allele frequency data, *Conservation Biol.*, 12, 228, 1998.

Lynch, J.D., Refugia, in *Analytical Biogeography*, Myers, A.A. and Giller, P.S., Eds, Chapman and Hall, New York, 1988, 311.

Markgraf, V., Younger *Dryas* in southern South America? *Boreas (Oslo)*, 20, 63, 1991.

Markgraf, V. and Bradbury, J.P., Holocene climatic history of South America, *Striae*, 16, 40, 1982.

McCauley, D.E., Contrasting the distribution of chloroplast DNA and allozyme polymorphisms among local populations of *Silene alba*: implications for studies of gene flow in plants, *Proc. Natl. Acad. Sci. USA*, 91, 8127, 1994.

McCauley, D.E., The use of chloroplast DNA polymorphism in studies of gene flow in plants, *Trends Ecol. Evol.*, 10, 198, 1995.

Michalakis, Y. and Excoffier, L., A generic estimation of population subdivision using distances between alleles with special reference to microsatellite loci, *Genetics*, 142, 1061, 1996.

Moran, P.A.P., Wandering distributions and the electrophoretic profile, *Theor. Popul. Biol.*, 8, 318, 1975.

Muñoz, J.D., Anacardiaceae, in *Flora del Paraguay*, Spichiger, R. and Ramella, L., Eds, Conservatoire et Jardin Botaniques de la Ville de Genève and Missouri Botanical Garden, Geneva, 1990.

Naciri-Graven, Y. et al., Development and characterization of microsatellite markers in a widespread Neotropical seasonally dry forest tree species, *Geoffroea spinosa* Jacq. (Leguminosae), *Molec. Ecol. Notes*, 5, 45, 2005.

Ohta, T. and Kimura, M., A model of mutation appropriate to estimate the number of electrophoretically detectable alleles in a finite population, *Genet. Res.*, 22, 201, 1973.

Pennington, R.T., Prado, D.E., and Pendry, C.A., Neotropical seasonally dry forests and Quaternary vegetation changes, *J. Biogeogr.*, 27, 261, 2000.

Petit, R.J., *Diversité génétique et histoire des populations d'arbres forestiers, dossier d'habilitation à diriger des recherches*, Université Paris-Sud, Orsay, 1999.

Petit, R.J., Kremer, A., and Wagner, D.B., Geographic structure of chloroplast DNA polymorphisms in European oaks, *Theor. Appl. Genet.*, 87, 122, 1993.

Petit, R.J. et al., Chloroplast DNA footprints of postglacial recolonization by oaks, *Proc. Natl. Acad. Sci. USA*, 94, 9996, 1997.

Petit, R.J. et al., Glacial refugia: hotspots but not melting pots of genetic diversity, *Science*, 300, 1563, 2003.

Posada, D. and Crandall, K.A., Intraspecific gene genealogies: trees grafting into networks, *Trends Ecol. Evol.*, 16, 37, 2001.

Powell, W. et al., Hypervariable microsatellites provide a general source of polymorphic DNA markers for the chloroplast genome, *Curr. Biol.*, 5, 1023, 1995.

Prado, D.E., *A critical evaluation of the floristic links between chaco and caatingas vegetation in South America*, PhD thesis, University of St Andrews, 1991.

Prado, D.E., Seasonally dry forests of tropical South America: from forgotten ecosystems to a new phytogeographic unit, *Edinburgh J. Bot.*, 57, 437, 2000.

Prado, D.E. and Gibbs P.E., Patterns of species distribution in the dry seasonal forests of South America, *Ann. Missouri Bot. Gard.*, 80, 902, 1993.

Prance, G.T., Forest refugies: evidences from woody angiosperms, in *Biological Diversification in the Tropics*, Prance, G.T., Ed, Columbia University Press, New York, 1982, 137.

Provan, J. et al., Gene-pool variation in Caledonian and European Scots pines (*Pinus sylvestris* L.) revealed by chloroplast simple-sequence repeats, *Proc. R. Soc. Lond. B Biol. Sci.*, 265, 1697, 1998.

Queller, D.C., Strassmann, J.E., and Hughes, C.R., Microsatellites and kinship, *Trends Ecol. Evol.*, 8, 285, 1993.

Reboud, X. and Zeyl, C., Organelle inheritance in plants, *Heredity*, 72,132, 1994.

Rendell, S. and Ennos, R.A., Chloroplast DNA diversity of the dioecious European tree *Ilex aquifolium* L., *Molec. Ecol.*, 12, 2681, 2003.

Rogers, A.R. and Harpending, H., Population growth makes waves in the distribution of pairwise genetic differences, *Molec. Biol. Evol.*, 9, 552, 1992.

Santin, D.A. and de Freitas Leitão, F., Restabelecimento e revisão taxonômica do gênero *Myracrodruon* Freire Allemão (Anacardiaceae), *Rev. Brasil. Bot.*, 14, 133, 1991.

Schaal, B.A. and Olsen, K.M., Gene genealogies and population variation in plants, *Proc. Natl. Acad. Sci. USA*, 97, 7024, 2000.

Schneider, S., Roessli, D., and Excoffier, L., arlequin version 2000: a Software for Population Genetic Data Analysis, University of Geneva, Geneva, 2000.

Servant, M. et al., Tropical forest changes during the late Quaternary in African and South American lowlands, *Glob. Planet. Change*, 7, 25, 1993.

Sewell, M.M., Parks, C.R., and Chase M.W., Intraspecific chloroplast DNA variation and biogeography of North American *Liriodendron* L. (Magnoliaceae), *Evolution*, 50, 1147, 1996.

Slatkin, M., A measure of population subdivision based on microsatellite allele frequencies, *Genetics*, 139, 457, 1995.

Soubies, F. et al., The Quaternary deposits of the Serra dos Carajas (State of Pará, Brazil) – ages and other preliminary results, *Bol. Inst. Geosci. Univ. São Paulo*, 8, 223, 1991.

Spichiger, R. et al., Origin, affinities and diversity hot-spots of the Paraguayan dendrofloras, *Candollea*, 50, 515, 1995.

Spichiger, R., Calenge, C., and Bise B., The geographical zonation in the Neotropics of tree-species characteristic of the Paraguay–Paraná Basin, *J. Biogeogr.*, 31, 1489, 2004.

Taberlet, P. et al., Universal primers for amplification of three non-coding regions of chloroplast DNA, *Pl. Molec. Ecol.*, 17, 1105, 1991.

Tajima, F., The effect of change in population size on DNA polymorphism, *Genetics*, 123, 597, 1989a.

Tajima, F., Statistical method for testing the neutral mutation hypothesis by DNA polymorphism, *Genetics*, 123, 585, 1989b.

Testolin, R. and Cipriani, G., Paternal inheritance of chloroplast DNA and maternal inheritance of mitochondrial DNA in the genus *Actinidia*, *Theor. Appl. Genet.*, 94, 897, 1997.

Trapnell, D.W. and Hamrick, J.L., Partitioning nuclear and chloroplast variation at multiple spatial scales in the neotropical epiphytic orchid, *Laelia rubescens*, *Molec. Ecol.*, 13, 2655, 2004.

Van der Hammen, T. and Absy, M.L., Amazonia during the last glacial, *Palaeogeogr. Palaeoclimatol. Palaeoecol.*, 109, 247, 1994.

Vendramin, G.G. et al., A set of primers for the amplification of 20 chloroplast microsatellites in Pinaceae, *Molec. Ecol. Notes*, 5, 595, 1996.

Wagner, D.B., Nuclear, chloroplast and mitochondrial DNA polymorphisms as biochemical markers in population genetic analyses of forest trees, *New For.*, 6, 373, 1992.

Wakeley, J., The limits of theoretical population genetics, *Genetics*, 169, 1, 2005.

Walter, R. and Epperson, K., Geographic pattern of genetic variation in *Pinus resinosa*: area of greatest diversity is not the origin of postglacial populations, *Molec. Ecol.*, 10, 103, 2001.

Weir, B.S. and Cockerham, C.C., Estimating F-statistics for the analysis of population structure, *Evolution*, 38, 1358, 1984.

Wintero, A.K., Fredholm, M., and Thomsen, P.D., Variable $(dG–dT)_n \cdot (dC–dA)_n$ sequences in the porcine genome, *Genomics*, 12, 281, 1992.

Wong, A.K. et al., Distribution of CT-rich tracts inversed in vertebrate chromosomes, *Chromosoma*, 99, 344, 1990.

Wright S., Systems of mating, *Genetics*, 6, 111, 1921.

Wright S., The interpretation of population structure by F-statistics with special regard to systems of mating, *Evolution*, 19, 395, 1965.

19 Floristic and Geographical Stability of Discontinuous Seasonally Dry Tropical Forests Explains Patterns of Plant Phylogeny and Endemism

Matt Lavin

CONTENTS

ABSTRACT

The controversies over explanations of endemism and the patchiness of South American seasonally dry tropical forests are potentially resolved with molecular phylogenetics. The genera *Coursetia* and *Inga*, for example, each show an ecological predilection to a different forest type. Sister species of *Coursetia* are often confined to the same narrow geographical region of dry forest, in contrast to *Inga*, in which closely related species often are widely distributed throughout neotropical rain forests. If these patterns are general, then the geographical phylogenetic structure of dry forest clades can be explained by low dispersal rates among isolated forest patches, a rate that does not replace the accumulation of endemic species by resident speciation. Recent critiques of vicariance and refugia hypotheses, the usual historical explanations for endemism and patchiness, are compelling. However, clades of sister species narrowly confined to dry forest patches commonly trace their ancestry back in time to at least several millions of years. Such minimum ages suggest that dry forest patches may be more persistent floristically and geographically than the critics of refugia hypotheses contend. If the phylogeny of *Inga* is representative of rain forest clades, then it is the rain forests rather than the dry forests that are more dynamic floristically and geographically.

19.1 INTRODUCTION

South American seasonally dry tropical forests (SDTF) are characterized by Prado and Gibbs (1993) and Pennington et al. (2000, 2004) as having fertile soils, being dominated by plant families such as Leguminosae and Bignoniaceae, having a sparse grass cover often accompanied by a conspicuous representation of succulents (e.g. Cactaceae), lacking a frost period and being not prone to fire. These dry forests have been identified by overlaying maps of the many species with an affinity for this habitat (Prado and Gibbs, 1993; Pennington et al., 2000). The result has been a clearer identification of specific regions in South America that harbour the floristically and structurally distinctive SDTF. These regions include the caatinga, the Misiones, the Piedmont, the Chiquitano, scattered inter-Andean valleys from Peru to Venezuela, coastal Ecuador, and the Caribbean coastal regions of Venezuela and Colombia. Schrire et al. (2005) identified taxonomically and structurally similar vegetation in other parts of the world that were also dominated by legumes and succulents (e.g. Euphorbiaceae) but not grasses. Such patches of SDTF, thorn scrub or bush thicket include the Pacific coastal and Tamaulipan regions of Mexico and adjacent USA, the large Caribbean islands, the Somalia-Masai region in the Horn of Africa to north-west India, and the Karoo-Namib region of southern Africa.

Although Prado and Gibbs (1993) and Pennington et al. (2000) suggested that the South American patches of SDTF could be refugia or vicariant remnants of a once more widespread Quaternary dry forest, Pennington et al. (2004) did not find support for this, possibly because the age of forest fragmentation was too recent to effect allopatric speciation. The ages of species diversifications (crown clades) restricted to dry forests were discovered to be often much older than the Quaternary. Furthermore, no vicariant pattern could be distilled from the phylogenies of several groups showing high endemism in these forests. Thus Quaternary events do not readily explain at least the ages of endemic species associated with these dry forest patches.

Mayle (2004, Chapter 17) argues soundly against refugia and vicariance explanations of the endemism and patchy distributions of South American dry forests. Pollen data combined with vegetation model simulations strongly suggest that dry forests are floristically, structurally and geographically dynamic over time periods extending back to just the Holocene. For example, key present-day South American dry forest species such as *Anadenanthera colubrina* (Vell.) Brenan may represent Holocene immigrations into at least the Chiquitano dry forests, according to pollen data. Also, dry forests were located north of the Chiquitano–Piedmont–Misiones regions during the Quaternary, according to the vegetation models.

In this study, two phylogenies of legume genera that have diversified in South America are presented to illustrate how an analysis of phylogenetic tree shape and molecular age estimation can illuminate vegetation dynamics at evolutionary timescales. The most important point to be demonstrated here is that narrowly confined patches of SDTF often harbour sister species whose most recent common ancestry traces back in time to at least several million years. This finding can be explained only by a floristic and geographical stability of dry forest patches during at least this timeframe.

19.2 METHODS

The genera *Coursetia* and *Inga* were chosen because they may well represent a general difference between clades confined largely to dry forests versus those primarily in wet forests. The conclusions drawn are tentative, but at least the phylogenetic patterns distinctive to each of these genera are representative of a larger sample analysed by Lavin et al. (2004). Although species sampling in *Coursetia* has been much more exhaustive, additional sampling in *Inga* will not change the conclusions drawn in this chapter, according to a larger taxon sampling of *Inga* (Richardson et al., 2001) that included not only the nuclear ribosomal internal transcribed spacer-5.8S (ITS) but also chloroplast *trn*L-F sequences. The comparison of *Coursetia* and *Inga* was made with only sequences from the nuclear ribosomal ITS-5.8S region because of the nucleotide substitution rates at this locus, which are highly similar within and among these two genera (see results).

19.2.1 RATE AND AGE ESTIMATION

The analysis of rates of substitution and ages of lineages has been facilitated by theoretical developments of 'relaxed' molecular clock methods (e.g. Thorne et al., 1998; Sanderson, 2002). The assumption that rates of substitution of DNA sequences have to be strictly clocklike is now replaced by the assumption that rates of evolution are constrained to various degrees from ancestors to descendants. With this assumption, various rate-smoothing approaches can be used to simultaneously find the best fit of all ancestor-descendant rate changes over an entire phylogenetic tree. If one or more branching points of a phylogenetic tree can be constrained with an age derived from the fossil record (e.g. the fossil possesses shared derived traits unique to that particular branching point), then relative rates of substitution can be converted to absolute rates, which then allow age estimates to be made for all the branching points in a phylogeny.

For each of *Coursetia* and *Inga*, the ITS region evolves in a non-clocklike rate. A likelihood ratio test for clocklike rates of substitution in the *Coursetia* ITS phylogeny yielded a likelihood ratio of 542 ($P = 0.000000$, d.f. = 80), whereas for *Inga* a likelihood ratio of 151 ($P = 0.000000$, maximum d.f. = 41). Thus the program r8s (Sanderson, 2002, 2003) was used to accommodate such rate variation (cf. Lavin et al., 2004; 2005). Bayesian trees produced with MRBAYES (Huelsenbeck and Ronquist, 2001) were rate-smoothed using the penalized likelihood, Langley-Fitch (rate constant), and non-parametric options in r8s to produce chronograms (phylogenies scaled to time). To convert relative rates of substitution to absolute rates, and thus estimate ages of most recent common ancestors (MRCAs), root nodes within the *Coursetia* and *Inga* phylogenies were constrained by the ages estimated for these very nodes in Lavin et al. (2003, 2004, 2005). This included a minimum age for the *Coursetia* crown clade of 17 Ma (from Lavin et al., 2004) and a fixed age of the MRCA of *Coursetia* and outgroups of 33.7 Ma (fossil-constrained node M in Lavin et al., 2005). For the mimosoid crown clade including *Inga* and closely related Ingeae (node 12 in Lavin et al., 2005), the age estimate of 23.9 ± 3.1 Ma was fixed at 24 Ma, but varying this root age from 21 to 27 Ma did not qualitatively affect the results presented here.

19.2.2 MEASURE OF GEOGRAPHICAL PHYLOGENETIC STRUCTURE

An evolving lineage splitting into multiple sublineages will inevitably be shaped by its geographical and ecological setting. For example, African lineages tend to remain in Africa, and tropical lineages tend to remain in the tropics (e.g. Wiens and Donoghue, 2004). Over time, such geographical and niche conservatism of lineages results in phylogenetic relationships in which sister species are highly similar geographically or ecologically (Figure 19.1). The degree to which geographical (or ecological) setting predicts phylogenetic relatedness defines geographical (or ecological) phylogenetic structure (e.g. Irwin, 2002). The degree to which community membership predicts relatedness defines community phylogenetic structure (Webb, 2000).

In this study, geographical phylogenetic structure was quantified by assigning each terminal to its geographical area of distribution, and then analysing with MACCLADE (Maddison and Maddison, 2005), PAUP (Swofford, 2001) and PHYLOCOM (Webb, 2000; Webb et al., 2002, 2005). With MACCLADE and PAUP, geographical areas were treated as character states optimized on phylogenetic trees. The number of area transformations (steps) was then compared with the same area character mapped on to 1000 random trees produced in MACCLADE. Observed area transformations that involved significantly fewer steps than expected on random trees were taken as evidence of geographical phylogenetic structure. Webb's approach to quantifying community phylogenetic structure was co-opted for purposes of quantifying geographical phylogenetic structure. The net relatedness index (NRI) and nearest taxa index (NTI) were each estimated from the Bayesian consensus, rate-smoothed Bayesian consensus,

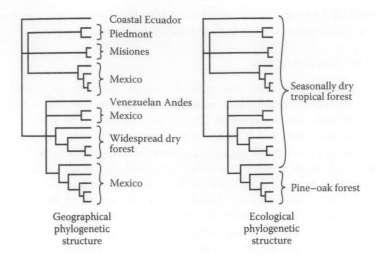

Geographical
phylogenetic
structure

Ecological
phylogenetic
structure

FIGURE 19.1 Different forms of phylogenetic structure. Geographical phylogenetic structure emerges when dispersal is limited over evolutionary time; geography becomes a good predictor of phylogenetic relationships. Ecological phylogenetic structure arises when dispersal is confined ecologically over evolutionary time; ecological setting becomes a good predictor of phylogenetic relationships. Community phylogenetic structure (not shown) arises when dispersal is confined ecologically, but that ecological filtering occurs over evolutionary time such that closely related species come to reside within the same community; community membership becomes a good predictor of phylogenetic relationships (Webb, 2000).

and parsimony phylogenies of *Coursetia* and *Inga*. Higher values of NRI and NTI indicate sister taxa with a high probability of coming from the same geographical area (aggregated), and lower values indicate sister taxa most likely occupying any geographical areas (random). Sister taxa preferentially occupying different geographical areas (over-dispersed) show negative values of NRI and NTI. Significance of actual NRI and NTI values is determined by generating a frequency distribution of these values calculated for 10,000 randomly assigned geographical areas on the given phylogeny (using the 'clust' option in PHYLOCOM).

The area assigned to each of the terminal taxa in these biogeographical analyses involves some uncertainty, such as how narrowly to define the area of endemism occupied by a given taxon. This uncertainty can be taken advantage of by introducing bias against the favoured hypothesis. In this study, one hypothesis is that the distribution of *Coursetia* among the highly fragmented patches of seasonally dry forests has occurred as a result of low dispersal rates that allow phylogenetic structure to emerge. In contrast, the second hypothesis is that the distribution of *Inga* among the more continuous neotropical rain forests has occurred as a result of higher rates of dispersal that ultimately reduce the level of phylogenetic structure (i.e. immigrant lineages ultimately outnumber and then replace resident lineages). Thus areas should be assigned to terminal taxa in a manner that biases against geographical structure in *Coursetia* and biases for structure in *Inga*. Results supporting the hypotheses described here can then be considered robust.

Biasing against structure was introduced into the *Coursetia* analysis in four ways.

1. By omitting duplicate accessions of the same species from analysis (i.e. they are likely to come from the same geographical area).
2. By designating as narrowly as possible the geographical position of each species (e.g. designating sister species as one coming from the Peruvian Andes and the other from the Piedmont did not contribute to geographical structure compared with the case in which designating them both as from the Andes would contribute to geographical structure).

3. Biogeographical changes were mapped on to a Bayesian consensus tree in which poly-tomies are likely to be more frequent, and thus phylogenetic structure less readily detected.
4. Widespread species (e.g. *Coursetia caribaea* (Jacq.) Lavin var. *caribaea*) were coded for all areas in which they occur, which adds real ambiguity to the analysis because a clade containing widespread species is not so distinctly geographically distributed.

Biasing for geographical phylogenetic structure in the *Inga* analysis was achieved in the following ways.

1. By including duplicate accessions of the same species from the same geographical area (i.e. they probably come from the same geographical area).
2. By designating the geographical position of each species using broad political boundaries (e.g. designating species as coming from south-eastern Brazil or Mexico probably con-tributes to geographical structure compared with the case in which species are designated as coming from more narrowly restricted areas, such as Atlantic rain forest).
3. Biogeographical changes were mapped on to a single most parsimonious tree in which polytomies are likely to be less frequent and phylogenetic structure thus more easily detected.
4. Widespread species (e.g. *Inga edulis* Mart.) were coded for only the area from which they were sampled, which subtracts real ambiguity from the analysis because a clade may then be distinctly geographically distributed, but only as an artefact of not including all possible areas of occurrence for the constituent species.

The biases described were employed in the analyses involving MacClade and paup but not in phylocom, which relies on multiple samples per unit area (i.e. more than one species has to come from a single area of endemism). The results from phylocom, therefore, should be considered provisional and exploratory as to their application at higher levels in biogeographical analysis (i.e. above the plant community).

19.3 RESULTS

The *Coursetia* chronogram (Figure 19.2) shows many hierarchically nested and well-supported species clades that are mostly confined to adjacent dry forest patches. For example, *Coursetia brachyrhachis* Harms, endemic to the Piedmont, is sister to a clade of species comprising *C. fruticosa* J.F. Macbr., *C. cajamarcana* Lavin and *C. maraniona* Lavin, which are endemic to the inter-Andean dry forest of Peru. This pattern is reiterated by the outgroup *Poissonia hypoleuca* (Speg.) Lillo from the piedmont and *P. orbicularis* Hauman (along with unsampled *P. eriantha* Hauman) from Peru. From estimated rates of $5.0–8.2 \times 10^{-9}$ substitutions per site per year (s/s/y) (Table 19.1), the ages of crown clades comprising more than one species confined to a narrow region of dry forest range from 5.3 ± 1.2 Ma in the Coursetia rostrata clade (ROST; Table 19.2, Figure 19.2) to 7.4 ± 1.2 Ma in the Coursetia fruticosa clade (FRUT; Table 19.2, Figure 19.2).

The *Inga* chronogram (Figure 19.3) shows little hierarchy of nested clades and few well-supported lineages. Furthermore, a species sampled from a particular region within Brazil is as likely to be as closely related to one from Costa Rica as it is to other Brazilian species. From estimated rates of $7.1–7.9 \times 10^{-9}$ s/s/y (Table 19.1), the age of the *Inga* crown clade is probably less than 2 Ma, according to the penalized likelihood rate-smoothing estimates (Table 19.2). Only two well-supported subclades are resolved by Bayesian analysis, and the estimated ages of these are essentially the same as that of the *Inga* crown clade (results not shown).

The estimated rates of substitution for the *Coursetia* and *Inga* ITS data-sets are very similar, with a maximum in each of about 8×10^{-9} s/s/y (Table 19.1). Even after adjusting for ITS rate

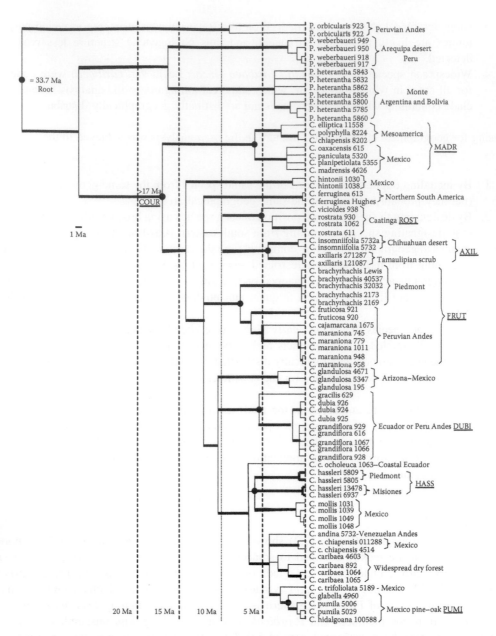

FIGURE 19.2 Chronogram of *Coursetia* (a penalized likelihood rate-smoothed Bayesian consensus tree). The root was fixed at 33.7 Ma, as derived from Lavin et al. (2003, 2005), whereas the Coursetia crown clade was constrained to a minimum of 17 Ma, as derived from Lavin et al. (2003, 2004). The thicker horizontal lines represent branches supported by posterior probabilities of 0.90–1.00, whereas the thinner horizontal lines represent branches supported by 0.60–0.89 (no branches below 0.60 were resolved in the Bayesian consensus). The labels refer to the fragments of seasonally dry tropical forest or other vegetation types to which the corresponding clade is endemic. These areas of endemism include the pine–oak woodlands throughout Mexico; the inter-Andean dry forests of Colombia and adjacent Venezuela (Mérida); the dry forests of coastal Ecuador; the inter-Andean dry forests of Ecuador and adjacent northern Peru; the inter-Andean dry forests of central and southern Peru; the Arequipa desert of southern coastal Peru; the monte temperate desert of northern Argentina and adjacent Bolivia; the dry forests of Mesoamerica (Chiapas, Mexico, to Panama); the dry forests primarily of Pacific coastal Mexico extending north into the Sonoran desert; the Caribbean coastal dry forest of Colombia and Venezuela; the caatinga of Brazil; the Tamaulipan thorn scrub and adjacent Chihuahuan desert in north-eastern Mexico; the Piedmont dry forest of Argentina and Bolivia; and the Misiones dry forest of Argentina, Paraguay and Brazil. Crown clades: COUR, *Coursetia*; MADR, *C. madrensis* Micheli group; AXIL, *C. axillaris* J.M. Coult. & Rose group; ROST, *C. rostrata* Benth. group; FRUT, *C. fruticosa* J.F. Macbr. group; DUBI, *C. dubia* DC. group; HASS, *C. hassleri* Chodat group; PUMI, *C. pumila* (Rose) Lavin group.

TABLE 19.1

Rate Estimates Derived from An r8s Analysis Using Penalized Likelihood, Globally Constant and Non-Parametric Forms of Rate-Smoothing

| Crown Clade[a] | Group | Rate ± SD (substitutions per site per year × 10⁻⁹)[b] | | |
		Penalized Likelihood	Globally Constant	Non-Parametric
COUR	*Coursetia*	5.58 ± 0.9	5.43 ± 0.8 (rate constant)	16.61 ± 3.2
MADR	*C. madrensis* Micheli	5.00 ± 0.9		7.57 ± 1.4
AXIL	*C. axillaris* J.M. Coult. & Rose	6.90 ± 1.3		7.04 ± 1.6
ROST	*C. rostrata* Benth.	6.90 ± 1.2		5.56 ± 1.3
FRUT	*C. fruticosa* J.F. Macbr.	6.53 ± 1.1		7.15 ± 1.5
DUBI	*C. dubia* DC.	7.78 ± 1.2		8.76 ± 1.8
HASS	*C. hassleri* Chodat	8.18 ± 1.2		7.70 ± 1.6
PUMI	*C. pumila* (Rose) Lavin	7.93 ± 1.2		5.11 ± 1.5
OUTG	*Inga* outgroups	7.06 ± 2.6	7.94 ± 0.8 (rate constant)	9.26 ± 3.5
INGA	*Inga*	7.88 ± 0.8		3.44 ± 1.0

[a] *Coursetia* and *Inga* crown clades correspond to the labels in Figures 19.2 and 19.3, respectively.

[b] Means and standard deviations derived from 100 Bayesian trees sampled at likelihood stationarity for each of the *Coursetia* and *Inga* data-sets, as in Lavin et al. (2005).

variation within each of the two data-sets, the MRCA of the entire *Inga* diversification is markedly younger than that for each of the narrowly confined subclades with *Coursetia* (Table 19.2, Figures 19.2 and 19.3).

The degree of geographical phylogenetic structure analysed with MACCLADE and PAUP revealed that the bias to reduce structure in the *Coursetia* phylogeny had no effect. The 14 geographical areas assigned to the *Coursetia* phylogeny showed significant geographical phylogenetic structure

TABLE 19.2

Age Estimates Derived from An r8s Analysis Using Penalized Likelihood, Globally Constant and Non-Parametric Forms of Rate-Smoothing

| Crown Clade[a] | Group | Age ± SD (Ma)[b] | | |
		Penalized Likelihood	Globally Constant	Non-Parametric[c]
COUR	*Coursetia*	17.1 ± 0.2	17.9 ± 1.1	21.7 ± 1.3
MADR	*C. madrensis* Micheli	6.7 ± 1.6	6.1 ± 1.1	17.6 ± 2.7
AXIL	*C. axillaris* J.M. Coult. & Rose	5.3 ± 1.5	6.2 ± 1.4	9.0 ± 2.9
ROST	*C. rostrata* Benth.	5.3 ± 1.2	6.5 ± 1.3	9.8 ± 2.1
FRUT	*C. fruticosa* J.F. Macbr.	7.4 ± 1.2	8.8 ± 1.3	14.0 ± 1.6
DUBI	*C. dubia* DC.	5.4 ± 1.0	7.0 ± 1.2	9.3 ± 1.4
HASS	*C. hassleri* Chodat	5.9 ± 0.9	8.1 ± 1.1	9.9 ± 1.3
PUMI	*C. pumila* (Rose) Lavin	1.9 ± 0.5	2.7 ± 0.6	4.2 ± 1.2
OUTG	*Inga* outgroups	21.5 ± 1.5	21.5 ± 1.5	20.7 ± 1.9
INGA	*Inga*	1.6 ± 0.4	1.5 ± 0.4	9.1 ± 2.6

[a] *Coursetia* and *Inga* crown clades correspond to the labels in Figures 19.2 and 19.3, respectively.

[b] Means and standard deviations derived from 100 Bayesian trees sampled at likelihood stationarity for each of the *Coursetia* and *Inga* data-sets, as in Lavin et al. (2005).

[c] Typically biased towards older ages.

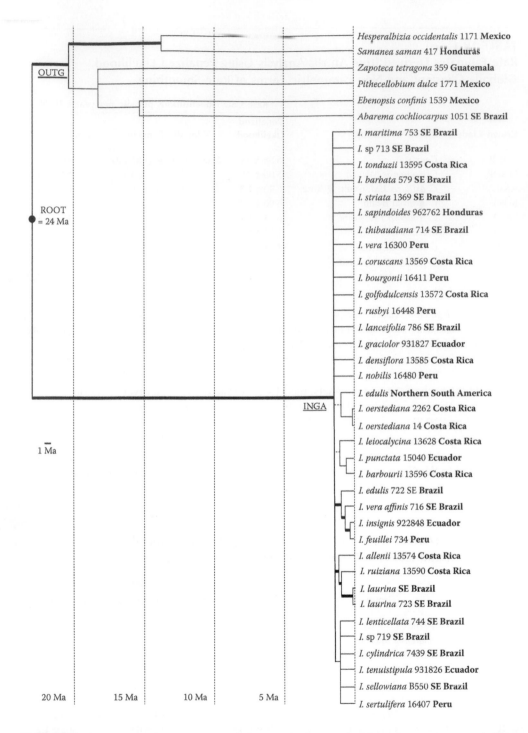

FIGURE 19.3 Chronogram of *Inga* (a penalized likelihood rate-smoothed Bayesian consensus tree). The root was fixed at 24 Ma, as derived from Lavin et al. (2005). The thicker solid horizontal lines represent branches supported by posterior probabilities of 0.90–1.00, the thinner solid horizontal lines represent branches supported by 0.60–0.89 and the dashed horizontal lines represent branches with posterior probabilities of below 0.60. The labels refer to the area from which the sample was taken (not the geographical distribution of the species). Crown clades: OUTG, *Inga* outgroups; INGA, *Inga*.

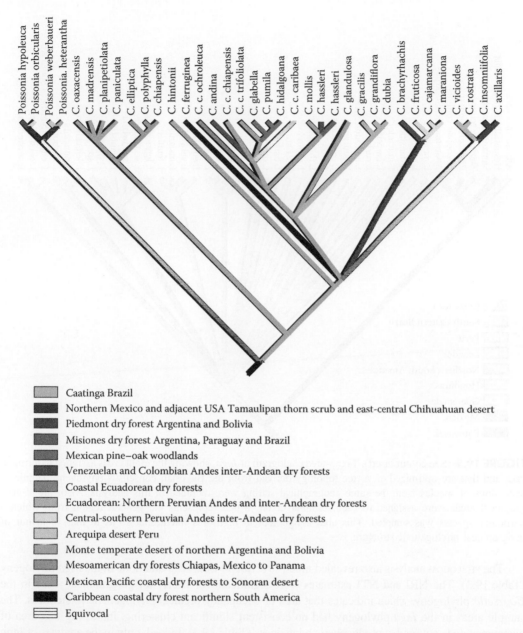

Caatinga Brazil

Northern Mexico and adjacent USA Tamaulipan thorn scrub and east-central Chihuahuan desert

Piedmont dry forest Argentina and Bolivia

Misiones dry forest Argentina, Paraguay and Brazil

Mexican pine–oak woodlands

Venezuelan and Colombian Andes inter-Andean dry forests

Coastal Ecuadorean dry forests

Ecuadorean: Northern Peruvian Andes and inter-Andean dry forests

Central-southern Peruvian Andes inter-Andean dry forests

Arequipa desert Peru

Monte temperate desert of northern Argentina and Bolivia

Mesoamerican dry forests Chiapas, Mexico to Panama

Mexican Pacific coastal dry forests to Sonoran desert

Caribbean coastal dry forest northern South America

Equivocal

FIGURE 19.4 (See colour insert following page 208). Taxon-area cladogram of *Coursetia*, showing the 14 areas assigned to terminal taxa and that are optimized on a tree topology derived from the Bayesian consensus phylogeny. Duplicate accessions of species from the same geographical setting were omitted from this analysis. Narrow areas of endemism were assigned to the terminal taxa, and much geographical structure was thus lost. For example, the genus *Poissonia* (the outgroup) is confined to the southern Andes, even though each of the four species of this genus is coded for a separate area of endemism. Similarly, the two lineages of *Coursetia hassleri* Chodat occur in adjacent dry forest patches, the Piedmont and Misiones, in South America.

(Figure 19.4), whereas the eight geographical areas assigned to the *Inga* phylogeny did not (Figure 19.5); see also Figure 19.6a and Figure 19.6b, respectively. Sister species in *Coursetia* have a high likelihood of being confined to the same narrow geographical region, whereas sister species in *Inga* are much less likely to be narrowly confined geographically.

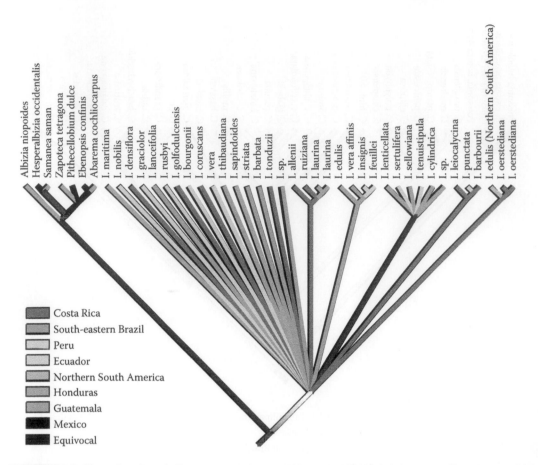

FIGURE 19.5 (See colour insert). Taxon-area cladogram of *Inga*, showing the eight areas assigned to terminal taxa and that are optimized on a tree topology derived from the Bayesian consensus phylogeny. Duplicate accessions of species from the same geographical setting were included in this analysis. Also, broad geographical areas were assigned to each terminal taxon, which represents the general region from which a particular species was sampled. This area assignment in *Inga* was intentional so as to bias in favour of geographical phylogenetic structure.

The PHYLOCOM analysis also revealed much structure in *Coursetia* and none in the *Inga* phylogeny (Table 19.3). The NRI and NTI estimates were significant for nearly all geographical areas in the *Coursetia* phylogeny, which indicates that area of occupancy predicts phylogenetic relatedness. The sample areas in the *Inga* phylogeny had no consistent significant clustering, indicating that area of occupancy does not predict phylogenetic relatedness (Table 19.3). Indeed, only in the analysis of *Inga* were negative values of NRI and NTI detected, further suggesting little geographical phylogenetic structure (i.e. closely related species tend not to occur in the same narrow geographical region).

19.4 DISCUSSION

Why are the phylogenies of *Coursetia* and *Inga* so different in almost every respect, especially in regard to the different ages of extant species diversifications and the degree of geographical phylogenetic structure? In the case of *Coursetia*, the geographical position of many narrowly confined species is a good predictor of phylogenetic relatedness. Perhaps correlated to this is the observation that only one taxon within the genus *Coursetia* (*C. caribaea* var. *caribaea*), out of a total of 46 (2%), has a widespread distribution (Lavin, 1988; Lavin and Sousa, 1995). All other species are confined to a narrow region of SDTF or desert (or pine-oak woodlands in a few Mexican species).

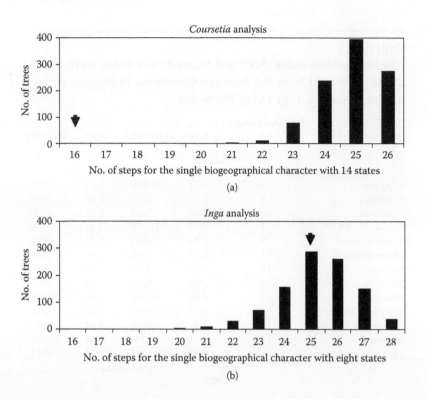

FIGURE 19.6 Histograms showing the number of steps for the biogeographical character states optimized via parsimony on 1000 random equiprobable phylogenetic trees of *Coursetia* or *Inga*. (a) The *Coursetia* analysis optimized the 14 biogeographical areas with a minimum of 16 transformations on the Bayesian consensus tree, whereas optimizations on each of 1000 random trees yielded a range of 21–26 steps for the biogeographical character. The 16 observed transformations (arrow) lie outside this distribution and are thus considered non-random and indicative of significant geographical structure. (b) The *Inga* analysis optimized the eight biogeographical areas with a minimum of 25 transformations on each of the most parsimonious trees, whereas optimizations on each of 1000 random trees ranged 20–28 steps. The 25 observed transformations (arrow) lie well within the frequency distribution of random values on the *Inga* phylogeny and are thus not indicative of geographical structure.

The commonly restricted species distributions in *Coursetia*, combined with a greater age of the MRCA of the narrowly confined crown clades, suggest that species of *Coursetia* disperse at a rate much less than those of *Inga*. The community structure of *Coursetia* may also be related to this and very different from that of *Inga*, in that most species are uncommon components of dry forests (Lavin, 1988 and pers. obs.). In contrast, 74 of the approximately 260 taxa of *Inga* (28%) are widely distributed across the Andes, through Amazonia into other basins, or across the Panamanian isthmus (derived from the distribution maps in Pennington, 1997). The entire genus traces back to an MRCA less than 2 Ma in age, and groups of two to four most closely related taxa in *Inga* (i.e. putative crown clades) commonly span a broad area of South America, if not both American continents (inferred from the distribution maps in Pennington, 1997). In addition, the community structure of *Inga* includes many abundant and often sympatric species (Pennington, 1997; Reynel and Pennington, 1997; R.T. Pennington, Royal Botanic Garden Edinburgh, pers. comm.). By all indications, species of *Inga* have experienced a much greater dispersal rate within the Neotropics than have those of *Coursetia*.

According to Hubbell (2001), rates of dispersal influence the shape of a phylogeny. In this regard, lineages that comprise the SDTF community must undergo relatively slow rates of dispersal among

TABLE 19.3
Net Relatedness Index (NRI) and Nearest Taxa Index (NTI; Webb, 2000) Estimated from the Bayesian Consensus Phylogenies of Coursetia and Inga Using Phylocom

Area Sampled	Taxa (Accessions) (n)	NRI[a]	Quantile[b]	NTI[c]	Quantile[b]
Coursetia					
Arequipa	4	4.02	10,000	3.15	9994
Caatinga	4	3.16	9996	2.59	9937
Caribbean	2	2.20	9944	2.20	9944
Ecuador Andes	9	6.30	10,000	3.36	9999
Ecuador Coast	3	2.53	9963	2.57	9974
Mesoamerica	3	3.07	9993	2.87	9993
Mexico	18	3.95	9999	3.40	9999
Misiones	2	2.19	9917	2.19	9917
Monte	7	5.66	10,000	3.46	10,000
North-east Mexico	4	3.23	9995	3.10	9990
Peru Andes	11	1.91	9506	3.10	9994
Piedmont	11	1.21	8758	4.05	10,000
Pine–oak	4	3.77	9999	2.90	9971
Inga					
Costa Rica	10	1.49	9087	1.47	9121
Ecuador	4	0.88	7172	0.84	7436
Guatemala	2	−2.17	458	−2.17	458
Honduras	3	−0.96	1965	−0.68	1830
Mexico	3	−2.59	144	−3.19	180
Peru	6	1.19	9016	1.09	8145
South-east Brazil	15	1.17	8663	1.36	9133

[a] Quantifies the degree of aggregation of the areas sampled on the phylogeny; higher values indicate phylogenetic structure.

[b] The number of random trees in which the NRI and NTI measures were lower than the actual estimate (9750 and above is significant).

[c] Quantifies the degree to which a given taxon has its sister from the same sample area; higher values indicate phylogenetic structure.

Estimates from the rate-smoothed Bayesian consensus and maximum parsimony trees yielded essentially the same results. All but two areas in the *Coursetia* phylogeny (Peruvian Andes and Piedmont) had a significant NRI, whereas none did in the *Inga* phylogeny. All areas in the *Coursetia* phylogeny had a significant NTI, whereas none did in the *Inga* phylogeny. This analysis is biased in favour of structure for *Coursetia*, because multiple accessions of the same species were sampled from the same area (PHYLOCOM requires multiple samples). However, bias in favour of structure in the *Inga* phylogeny is also introduced by sampling fewer and much more broadly defined geographical areas. See Figures 19.2 and 19.3 for areas designated for *Coursetia* and *Inga*, respectively. *Source:* Webb et al., 2005

forest patches. Resident lineages (e.g. endemic clades of sister species of *Coursetia*) are able to evolve undisturbed by immigration. Because immigration is potentially open to all lineages predisposed to inhabiting a dry forest patch, localized resident lineages would probably be replaced by constant immigration from all the adjoining patches. That patches of SDTF harbour resident clades of species is a prediction of low immigration among patches and persistence of individual patches (Lavin et al., 2004).

In contrast, lineages that comprise the tropical rain forest community must undergo relatively faster rates of dispersal among forest patches, perhaps because such patches are not so small or fragmented as dry forest patches (Schrire et al., 2005). Highly localized resident lineages (e.g. narrowly confined sister species of *Inga*) are not likely to evolve because immigration is potentially open to many lineages predisposed to inhabiting rain forest environments, and a new resident species is likely to be replaced by one of the many immigrants. In the case of tropical rain forests, the prediction derived from the *Inga* example is that clades of narrowly confined sister species will not be common.

Differences in life history traits among *Coursetia* and *Inga* species do not explain the findings reported here. That is, the young age estimates for *Inga* cannot be ascribed to a shorter generation time, which putatively results in the evolution of a faster substitution rate — a questionable relationship, as pointed out by Whittle and Johnston (2003). Coursetias range from small shrubs to trees, whereas Ingas are all trees. Species of *Inga*, like those of *Coursetia*, flower when only a few years old (Richardson et al., 2001; M. Lavin, pers. obs.). Regardless, both genera have very similar rates of nucleotide substitution in the ITS region (Table 19.1). In the original analysis of *Inga* (Richardson et al., 2001), a single non-parametric rate-smoothing estimate of 2.34×10^{-9} s/s/y for the ITS region translated to a 9.8 Ma age estimate for the *Inga* diversification. Their estimate of 1.30×10^{-9} s/s/y for the chloroplast DNA *trn*L-F region translated to a 1.6 Ma age estimate. While faster rates of substitution at the ITS locus are estimated here, a similar non-parametric rate-smoothing age estimate is made for the Inga crown clade (Table 19.2). Although age estimates derived from non-parametric rate-smoothing are biased towards older ages (Sanderson, 2003; Lavin et al., 2005), the relative age comparisons reveal that, no matter what method is used, the age of the narrowly confined species diversifications within *Coursetia* are as old to much older than the entire *Inga* diversification (Table 19.2).

The phylogenies of *Coursetia* and *Inga* may represent extremes in the degree of phylogenetic structure. Nevertheless, the patterns detailed for each of these phylogenies may well be general. Most of the phylogenies analysed by Lavin et al. (2004) and Pennington et al. (2004) show a high degree of geographical structure and represent groups that are confined to SDTF or bush thicket (e.g. *Chaetocalyx*, *Nissolia* and *Ruprechtia*). In contrast, plant groups confined largely to wet forests (or savannas) lack the high-level geographical structuring exemplified by *Coursetia*, for example *Andira* (Pennington, 1996; C. Skema and R.T. Pennington, Royal Botanic Garden Edinburgh, unpublished ITS data) and *Clusia* (Gustafsson and Bittrich, 2003). Although *Andira* includes a clade of species generally confined to south-east Brazilian rain and restinga forests (clade III in Pennington, 1996), a rates analysis suggests that this is of very recent origin (C. Skema and R.T. Pennington, unpublished ITS data). Future phylogenetic work on clades with an ecological predilection to a particular forest type will bear on this putative general difference.

Mayle's (2004) conclusion of discounting vicariance or refugia explanations for the patchiness of SDTF finds support in this analysis. That is, age estimates of the narrowly confined crown clades never coincide with age estimates of the Quaternary historical events that could have shaped them. Similar conclusions were reached by Lavin et al. (2004) and Pennington et al. (2004). Furthermore, singular historical events, such as the putative fragmentation of the South American seasonally dry forests, the Andean orogeny, the location of refugia or the formation of the Panamanian isthmus (e.g. Pennington et al., 2000) do not make good explanations for general phylogenetic patterns. If phylogenies of clades largely confined to dry forests differ qualitatively and quantitatively from those mainly of wet forests, then general ecological explanations are needed.

19.5 CONCLUSION

The antiquity of narrowly confined crown clades of more than one species is the essence that distinguishes the *Coursetia* from the *Inga* phylogeny. The question is not whether clades disbursed widely throughout SDTF are older than those of the rain forest, but rather how narrowly confined are sister species and how far back in time do such sister groups trace to their MRCA? For the dry forest *Coursetia*, the answer is that narrowly confined crown clades are common and in the order of 5–7 Ma in age.

Such geographical phylogenetic structure may be determined less by singular historical events than by long-term ecosystem dynamics, including the rates of extinction and speciation, as well as immigration (cf. Hubbell 2001). Why dry forests may be less prone to immigration among the local communities can at present only be open to speculation. Dry forests are highly fragmented and may have smaller average patch sizes compared with those of tropical wet forests (e.g. Schrire et al., 2005). Additionally, given a high dependency on regular moisture availability, tropical wet forests may be prone to higher rates of resident plant mortality (e.g. Stokstad, 2005), which according to Hubbell's (2001) theory would elevate the rates of immigration into and among local wet forest communities. Perhaps seasonally dry forests are more stable in community composition because they are less subject to disturbance; for example, they are not prone to fire and are resistant to desiccation (Pennington et al., 2000). This is indicated by the prevalence of succulent families such as Cactaceae in dry forests (Pennington et al., 2000; Schrire et al., 2005).

If the estimated ages of the *Coursetia* crown clades are general for dry forest clades, as suggested by Lavin et al. (2004), then dry forests and bush thicket communities probably have had a long history of discontinuous distributions and a floristic stability of at least 5–7 Ma. The fossil record of dry forest plants in Ecuador suggests a floristic stability of at least 10 Ma (e.g. Burnham, 1995; Burnham and Carranco, 2004; summarized in Hughes, 2005), which contrasts with the fossil record of rain forest plants in Amazonia, where dynamic ecosystem change is revealed within a 40,000-year timeframe by immigration from montane elements, for example (Colinvaux et al., 1996). Thus, Mayle's (2004) assertion that dry forests are ephemeral even within time periods encompassing the Holocene does not find support with these new interpretations of geographical phylogenetic structure.

ACKNOWLEDGEMENTS

Gwilym Lewis, Toby Pennington and Brian Schrire provided critical comments that greatly improved the manuscript. This study was supported in part by a grant from the National Science Foundation (DEB-0075202).

REFERENCES

Burnham, R.J., A new species of winged fruit from the Miocene of Ecuador: *Tipuana ecuatoriana*, *Amer. J. Bot.*, 82, 1599, 1995.

Burnham, R.J. and barranco, N.L., Miocene winged fruits of *Loxopterygium* (Anacardiaceae) from the Ecuadorian Andes, *Amer. J. Bot.*, 91, 1767, 2004.

Colinvaux, P.A. et al., A long pollen record from lowland Amazonia: forest and cooling in glacial times, *Science*, 274, 85, 1996.

Gustafsson, M.H.G. and Bittrich, V., Evolution of morphological diversity and resin secretion in flowers of *Clusia* (Clusiaceac): insights from ITS sequence variation, *Nord. J. Bot.*, 2, 183, 2003.

Hubbell, S., *The Unified Neutral Theory of Biodiversity and Biogeography*, Princeton University Press, Princeton, 2001, 448.

Huelsenbeck, J.P. and Ronquist, F.R., MrBayes: Bayesian inference of phylogeny, *Bioinformatics*, 17, 754, 2001.

Hughes, C.E., Four new legumes in forty-eight hours, *Oxford Pl. Syst.*, 12, 6, 2005.

Irwin, D.E., Phylogeographic breaks without geographic barriers to gene flow, *Evolution*, 56, 2383, 2002.

Lavin, M., Systematics of *Coursetia* (Leguminosae–Papilionoideae), *Syst. Bot. Monogr.*, 21, 1, 1988.

Lavin, M. and Sousa, M., Phylogenetic systematics and biogeography of the tribe Robinieae (Leguminosae), *Syst. Bot. Monogr.*, 45, 1, 1995.

Lavin, M. et al., Phylogeny of robinioid legumes (Fabaceae) revisited: *Coursetia* and *Gliricidia* recircum-scribed, and a biogeographical appraisal of the Caribbean endemics, *Syst. Bot.*, 28, 387, 2003.

Lavin, M. et al., Metacommunity process rather than continental tectonic history better explains geographically structured phylogenies in legumes, *Philos. Trans., Ser. B*, 359, 1509, 2004.

Lavin, M., Herendeen, P., and Wojciechowski, M.F., Evolutionary rates analysis of Leguminosae implicates a rapid diversification of lineages during the Tertiary, *Syst. Biol.*, 54, 530, 2005.

Maddison, D.R. and Maddison, W.P, MacClade, *version 4.07*, Sinauer Associates, Sunderland, 2005.

Mayle, F.E., Assessment of the Neotropical dry forest refugia hypothesis in the light of palaeoecological data and vegetation model simulations, *J. Quat. Sci.*, 19, 713, 2004.

Pennington, R.T., Molecular and morphological data provide phylogenetic resolution at different hierarchical levels in *Andira*, *Syst. Biol.*, 45, 496, 1996.

Pennington, R.T., Prado, D.E., and Pendry, C.A., Neotropical seasonally dry forests and Quaternary vegetation changes, *J. Biogeogr.*, 27, 261, 2000.

Pennington, R.T. et al., Historical climate change and speciation: neotropical seasonally dry forest plants show patterns of both Tertiary and Quaternary diversification, *Philos. Trans., Ser. B*, 359, 515, 2004.

Pennington, T.D., *The Genus* Inga: *Botany*, Royal Botanic Gardens, Kew, Richmond, 1997, 844.

Prado D.E. and Gibbs, P.E., Patterns of species distributions in the dry seasonal forests of South America, *Ann. Missouri Bot. Gard.*, 80, 902, 1993.

Reynel, C. and Pennington, T.D., *El Genero* Inga *en el Peru. Morfologia, Distribucion y Usos*, Royal Botanic Gardens, Kew, Richmond, 1997, 236.

Richardson, J.A. et al., Recent and rapid diversification of a species-rich genus of Neotropical trees, *Science*, 293, 2242, 2001.

Sanderson, M.J., Estimating absolute rates of molecular evolution and divergence times: a penalized likelihood approach, *Molec. Biol. Evol.*, 19, 101, 2002.

Sanderson, M.J., *r8s, version 1.6, User's Manual (April 2003)*, http://ginger.ucdavis.edu/r8s/, University of California, Davis, 2003.

Schrire, B.D., Lavin, M., and Lewis, G.P., Global distribution patterns of the Leguminosae: insights from recent phylogenies, *Biologiske Skrifter*, 55, 375, 2005.

Stokstad, E., Experimental drought predicts grim future for rainforest, *Science*, 308, 346, 2005.

Swofford, D., PAUP*: *Phylogenetic Analysis Using Parsimony (*and Other Methods), version 4.0β10*, Sinauer Associates, Sunderland, 2001.

Thorne, J., Kishino, H., and Painter, I.S., Estimating the rate of evolution of the rate of molecular evolution, *Molec. Biol. Evol.*, 15, 1647, 1998.

Webb, C.O., Exploring the phylogenetic structure of ecological communities: an example for rain forest trees, *Amer. Naturalist*, 156, 145, 2000.

Webb, C.O. et al., Phylogenies and community ecology, *Annual Rev. Ecol. Syst.*, 33, 475, 2002.

Webb, C.O., Ackerly, D.D., and Kembel. S., PHYLOCOM *User's Manual, version 3.22*, http://www.phylodiversity.net/phylocom/, 2005.

Weins, J.J. and Donoghue, M.J., Historical biogeography, ecology, and species richness. *Trends Ecol. Evol.*, 19, 639, 2004.

Whittle C.A. and Johnston M.O., Broad-scale analysis contradicts the theory that generation time affects molecular evolutionary rates in plants, *J. Molec. Evol.*, 56, 223, 2003.

20 The Seasonally Dry Vegetation of Africa: Parallels and Comparisons with the Neotropics

J. Michael Lock

CONTENTS

ABSTRACT

The definitions of seasonally dry vegetation, savanna and seasonally dry forest are discussed. The basic geography, patterns of climate and human history, as well as the effects of fire, wild mammals and domestic animals, in Africa and South America are described. Key differences include the longer history of human occupation and animal domestication, and the related greater frequency of fire in Africa and the much more diverse large mammal fauna that did not suffer major extinctions during the Pleistocene. The major physiognomic vegetation types (seasonally dry forests, woodlands, bushland and thicket and grasslands) are outlined for each continent, and the terms used for equivalent vegetation in the two continents are compared. Some comparisons are also made of the composition of the major vegetation types, by plant family, in the two continents.

20.1 INTRODUCTION

Africa and South America both span the Equator. Both have large blocks of tropical rain forest in their Equatorial regions. Both have very extensive areas of tropical seasonal vegetation. There are, however, substantial differences in the distribution of major vegetation types in the two continents. These arise from the different positions of the continents relative to the Equator, from differences in the position and distribution of high ground and mountain ranges and also to the position of neighbouring continents. Superimposed on these basic geographical factors are others, such as the disparate mammalian faunas, the very large differences in human settlement history and, probably related to the last, differences in the historical fire regimes.

This chapter attempts to list some of the factors that make the vegetation of the two continents different, to compare apparently similar vegetation types in the two, and to provide a guide to the main names used for physiognomically similar vegetation. It is, at best, a preliminary account and should be regarded as a beginner's guide. However, if it encourages workers on the seasonally dry vegetation of the two continents to look more at each other's regions, it will have achieved its objective.

Walter (1979; Figure 1) illustrated the vegetation of an 'average continent'. At the Equator, and for a few degrees to the north and south, there is tropical rain forest. Between about 25 and 35°N and S, there is a belt of very dry country supporting desert or near-desert vegetation. Between these two lie two belts, north and south of the Equator, of tropical seasonal vegetation. This chapter is concerned with the vegetation types of these. It considers only the tropical parts of Africa and concentrates on South America, with little mention of the tropical parts of Central America.

20.2 SOME DEFINITIONS

20.2.1 SEASONALLY DRY VEGETATION

Most of Africa shows some kind of seasonality. Virtually all the climadiagrams reproduced by White (1983) show a dry season of some kind. Only a few stations close to the Equator in the Congo basin have almost evenly distributed rainfall. In South America, Gentry (1995) and Graham and Dilcher (1995) have defined seasonally dry tropical forests as occurring where the annual rainfall is less than 1600 mm/year, with at least 5–6 months receiving less than 100 mm. This definition serves perfectly well in Africa. In West Africa, trees can grow over the entire countryside, instead of only along watercourses, when annual rainfall exceeds 300–350 mm (Gillet, 1986), but the present chapter makes little mention of areas with annual rainfall below 600 mm, nor does it consider soil features in the two continents.

20.2.2 Savanna

Pennington et al. (2000) draw a distinction between tropical dry forests, with a more or less continuous canopy and grasses as a minor component, and savannas that possess a xeromorphic, fire-tolerant grass layer. Here we immediately encounter semantic problems: *savanna* has been widely rejected, particularly in East Africa, as a useful term for describing vegetation, because it had come to be used indiscriminately for almost any vegetation from closed forest to open grassland. White (1983) thought it best used only in a general sense for vegetation in which both grasses and trees are prominent. However, in West Africa the classification scheme of Chevalier (1933) into Guinea, Sudan and Sahel savanna is very well established and well understood, and the terms have generally been used in a fairly precise way (e.g., see Lawson, 1966; Sanford and Isichei, 1986). South of the Equator, the term savanna has been less used and, perhaps because woody plants are more prominent over large areas, the term *woodland* is well established in English-speaking countries. Francophone countries, on the other hand, have used the term *forêt claire* (open forest). This is unfortunate, because it implies that woodland is a kind of tropical forest (in French *forêt dense*, dense or closed forest), although the two are completely different in physiognomy, structure and species composition (e.g., see Hopkins, 1965).

The term savanna has been used in a more restricted sense by Scholes and Walker (1993). They exclude treeless grasslands and denser woodlands (crown cover > 80%) from their definition, which otherwise encompasses vegetation types (sensu White, 1983) ranging from more open types of woodland through wooded grassland to grassland with 5–10% woody cover. Scholes and Walker (1993) do not mention fire in their 'set of rules for deciding whether a given vegetation is a savanna or not'. The main criteria that they use are as follows.

1. Monthly mean temperature > 10°C throughout the year.
2. At least 60 consecutive days per year with insufficient water for plant growth.
3. At least 60 days per year when there is sufficient water for plant growth.
4. The wet period warmer than the dry period.
5. Aerial cover of woody plants > 5% but < 80%.
6. Woody canopy above 2 m.
7. Grass aerial cover at least 5% in a year with average rainfall and grazing.
8. Grasses and trees not spatially separated.

If the occurrence of a fire-tolerant grass layer, usually with an aerial cover much more than 5%, is added to the Scholes and Walker definition, we arrive at approximate equivalence between their savanna and that of Pennington et al. (2000).

In South America, and particularly in Brazil, the term *cerrado* (short for *campo cerrado*, closed field) was often used in a similar broad sense to include everything from dry forest (strictly cerradão) to open grassland (*campo limpo*, clear, or clean, field). The divisions are defined by, among others, Eiten (1972), Furley (1999) and Oliveira-Filho and Ratter (2002), and the subdivisions are now generally used more precisely. They are discussed in more detail later.

20.2.3 Seasonally Dry Forests

Pennington et al. (2000) regarded seasonally dry tropical forests as occurring where annual rainfall is less than 1600 mm, with at least 5–6 months receiving less than 100 mm. The vegetation is dominated by trees, with a complete or almost complete crown cover. A fire-tolerant grass layer is absent. Tropical dry forests sensu Pennington et al. (2000) are similarly referred to as dry forests in Africa. Most are simpler in structure than tropical rain forests, lower in stature, and differ in species composition. However, such dry forests cover only a small area of Africa; some examples are given later.

Pennington et al. (2000) also map the caatinga regions of north-eastern Brazil as seasonally dry tropical forest. Descriptions of this vegetation suggest that the most similar vegetation in Africa is the deciduous bushland and thicket (sensu White, 1983) that occupies large areas of the Horn of Africa. The environment of the two regions is similar; in particular, rainfall tends to be bimodal and both rainfall peaks are extremely variable. Fires are very rare or absent. The vegetation is largely woody and composed mainly of bushy species with several stems from the base, or of trees that branch very low down. Many of the woody species are spiny; most are facultatively deciduous, producing leaves only when soil moisture conditions are favourable; there is little grass; and succulents with either rosettes of leaves or succulent stems are frequent.

Fixing the dry limit of savannas and dry forests in Africa is more difficult than in South America because of the gradual transition to desert. In South America, the transition between the tropical and temperate zones is occupied by chaco vegetation, which is a dry woodland (Bucher, 1982). If the Scholes and Walker criteria are followed, then much of the Sahel savanna zone (see later) falls outside their definition of savanna, because the growing season is too short.

Using these criteria, a map (Figure 20.1) has been produced that shows vegetation types that are approximately comparable with those shown in the map of Pennington et al. (2000), here reproduced as Figure 20.2. Figure 20.1 differs from the map of Scholes and Walker (1993) in that it distinguishes areas of savanna from areas of dry forest. It also excludes those areas that they mapped as transitional between savannas and arid shrublands.

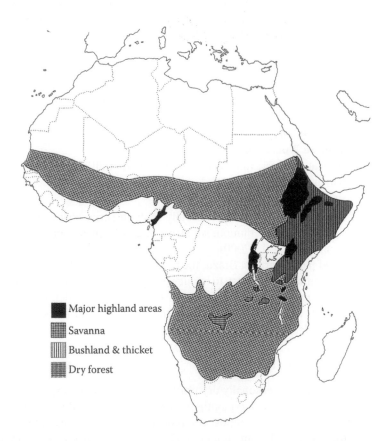

FIGURE 20.1 The seasonally dry vegetation of tropical Africa. The map shows dry forests, savannas (mainly woodlands) and bushland and thicket. Dry forests are shown only approximately; most areas are too small to map at this scale. Grasslands are also not sufficiently extensive to map at this scale; most occur as part of a mosaic or catena within woodlands. Data from White (1983).

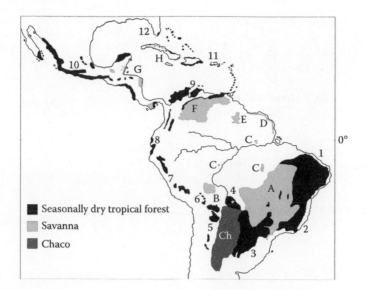

FIGURE 20.2 Schematic distribution of seasonally dry forests and savannas in the Neotropics. Seasonally dry forest: 1. caatingas; 2. south-east Brazilian seasonal forests; 3. Misiones nucleus; 4. Chiquitano; 5. Piedmont nucleus; 6. Bolivian inter-Andean valleys; 7. Peruvian and Ecuadorean inter-Andean valleys; 8. Pacific coastal Peru and Ecuador; 9. Caribbean coast of Colombia and Venezuela; 10. Mexico and Central America; 11. Caribbean Islands (small islands coloured black are not necessarily entirely covered by seasonally dry forests); 12. Florida. Savannas: (A) cerrado; (B) Bolivian; (C) Amazonian (smaller areas represented); (D) coastal (Amapá, Brazil to Guyana); (E) Rio Branco-Rupununi; (F) Llanos; (G) Mexico and Central America; (H) Cuba. Ch: Chaco. Modified after Pennington et al. (2000) and Huber et al. (1987).

Rough estimates of the area of seasonally dry tropical vegetation falling into the woodland and the bushland and thicket categories (see later) can be made using the maps prepared by White (1983). They are: West Africa, 7.4×10^6 km² (mainly woodland); east and north-east Africa, 1.9×10^6 km² (mainly bushland and thicket); and southern Africa, 4.5×10^6 km² (mainly woodland). By contrast, the total area of the Brazilian cerrados is around 2×10^6 km² (Ratter et al., 1997; Chapter 2) and that of the Venezuelan Llanos around 0.24×10^6 km² (Huber et al., Chapter 5). No estimate of the area of seasonally dry forest in Africa and South America is attempted here because these forests tend to occur in narrow bands and small patches that do not lend themselves to easy mapping.

20.3 SOME INTERCONTINENTAL COMPARISONS

20.3.1 BASIC GEOGRAPHY

The positions of the two continents in relation to the Equator are shown in Figure 20.3. In Africa, the Equator passes to the south of the 'centre of gravity' of the continent, through Gabon and Lake Victoria. The tropic of Cancer bisects the Sahara, and the tropic of Capricorn passes through southern Mozambique, northern South Africa, Botswana and Namibia. The southern tip of the continent lies close to 34°S, the latitude of Los Angeles and Beirut in the northern hemisphere. The Equator passes through the northern half of South America; the tropic of Cancer (23°28′N) passes through southern Mexico, far to the north of the continent of South America; and the tropic of Capricorn (23°28′S) passes through southern Brazil, Paraguay, northern Argentina and northern Chile. The southern tip of the continent lies at 55°S, the same equivalent latitude as, in the northern hemisphere, Moscow or extreme south-eastern Alaska. The northernmost mainland of South America, on the other hand, lies at 12°N.

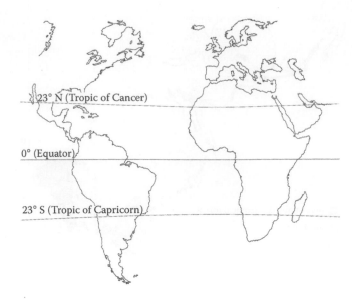

FIGURE 20.3 The relative positions of the continents of Africa and South America.

The two continents therefore differ greatly in their relation to the Equator. Africa is more or less symmetrical, while South America has its 'centre of gravity' well to the south. The southern portion of South America shows a complete transition from tropical to cool temperate climates and vegetation, while Africa has, in essence, no temperate vegetation other than on mountains.

20.3.2 PATTERNS OF CLIMATE

The basic pattern of Walter's (1979) 'ideal continent' — single dry seasons north and south of the Equator, and either a double wet season or continuously wet conditions close to it — is subject to some variation, because continents are neither symmetrical nor level. Adjacent continents, cold currents close to the shore, and mountain ranges can all modify the basic pattern. In East Africa, the presence of the dry Arabian landmass to the north-east means that winds from that direction are dry. The Horn of Africa is therefore much drier than would be expected from its proximity to the sea. (This is clearly not the whole story, because the corresponding area in South America, north-eastern Brazil, also has a drier climate than might otherwise be expected.) In south-western Africa, the cold waters of the Benguela current immediately offshore mean that the onshore winds are not saturated with water, so rainfall is very low and, again, somewhat unpredictable. This is paralleled in South America in northern coastal Chile and Peru, where the cold Humboldt current offshore leads to extreme aridity onshore and thus the Atacama desert, one of the driest places in the world. In both these areas, much of the available moisture comes from mists. The Senegal current produces a similar but less marked effect in north-western Africa.

20.3.3 HUMAN SETTLEMENT AND HISTORY

Humans evolved in Africa and have been in the continent for several million years. In South America, on the other hand, humans have been present for at most 40,000 years and probably much less. There is also the question of when humans moved from being part of the natural ecosystem to modifying entire ecosystems to suit themselves. In Africa, civilizations such as those of the Nile Valley, the western Sahel and Zimbabwe must have had substantial effects on their environments, and the earliest date to at least 7000 BP. In South America, there is, in general, only local evidence of extensive ecosystem modification before European colonization (in contrast to in Central America, where the downfall of some civilizations may have been caused by environmental degradation). Some inter-Andean valleys close to major indigenous centres such as Cusco show clear pre-conquest terracing, and there were also high population densities and extensive sophisticated agriculture on the floodplains of major Amazon basin rivers (R.T. Pennington, Royal Botanic Garden Edinburgh, pers. comm.).

20.3.4 FIRE

Fire occurs naturally in tropical seasonal vegetation types, started by lightning strikes, by volcanic activity and by such unlikely events as sparks struck by falling boulders. However, the frequency of such natural fires is low. Human activity must have increased fire frequency in three phases: when humans learned to use fire; when they learned to make fire; and, finally, with the advent of the match (Lock, 1998). There is evidence from Africa (Brain and Sillen, 1988) that humans have been using fire for at least one million years. Fire has long been used as a vegetation management tool by cultivators for clearance before cultivation; by pastoralists for removing coarse unpalatable vegetation and stimulating nutritious young growth; and by hunters for improving visibility, for attracting game, and as a hunting weapon (driving and ring firing) in itself (see review for Africa by West, 1965).

In South America, the shorter period of human occupation is likely to mean that, in the long term, fire frequency has been lower than in Africa. However, many of the trees in both African savannas and South American cerrado show similar features, such as thick corky bark, that can be interpreted as adaptations that increase fire resistance.

Lock (1998) has shown that the proportion of fire-adapted grass species is lower in South America than in Africa, and has suggested that this reflects the differing fire histories of the two regions. In Africa, many of the common savanna grasses belong to the tribe Andropogoneae and have seeds with long hygroscopic awns. The movements of these awns in response to alternating low and high humidity can help to bury the seeds in the soil, where they are protected from fires (Lock and Milburn, 1971).

Jackson (1968, 1974) found that the seedlings of some trees of regularly burned West African savanna woodlands showed features that act to protect the growing point beneath the soil, where it is protected from the heat of fires. In species such as *Vitellaria paradoxa* Gaertn. f. (Sapotaceae), the two petioles of the cotyledons are fused for some distance above the terminal bud, so that the latter lies at the bottom of a tube. The fused petioles elongate and push the growing point below soil level. The growing point then breaks out laterally through the wall of the tube formed by the fused petioles of the cotyledons, and grows upwards to the surface. The young shoot bears reduced leaves on its subterranean portion, and the buds in the axils of these can produce new shoots if the terminal bud is destroyed by fire. Jackson (1968, 1974) found similar but less marked behaviour in species of *Combretum*, *Guiera* and *Quisqualis* (Combretaceae), *Lophira* (Ochnaceae), *Gardenia* (Rubiaceae), and *Pterocarpus* and *Piliostigma* (Leguminosae). I have not seen descriptions of similar behaviour in woody cerrado species.

The possible long-term effects of fire on species composition are described later.

20.3.5 DOMESTIC ANIMALS

In Africa, cattle, goats, sheep, camels and donkeys were all domesticated several thousand years ago, and the increase in their numbers, and land management for their benefit, are likely to have had significant impacts on vegetation. However, the tsetse fly and the disease-causing trypanosomes for which it is the vector have limited the expansion of both humans and domestic animals over large areas. In South America, by contrast, domestication of local species has been minimal and is largely confined to the highlands. Only with the introduction of cattle, horses, sheep and goats from Europe have effects on vegetation become significant.

20.3.6 Wild Mammals

Africa has a very rich mammalian fauna, of which the members coexist by exploiting different niches in the savanna ecosystem (e.g., see Lamprey, 1963; Jarman and Sinclair, 1979). Often, also, seasonal variations in forage availability are circumvented by migration, so that different resources are exploited at different times of the year (e.g., see Maddock, 1979; Cobb et al., 1988). Total numbers of wild mammals have decreased in the past 200–300 years, and increasing human populations and settlement have disrupted migration routes. However, large mammals still remain a potent force for vegetation change in Africa, in particular the 'bulldozer herbivores': elephant, hippopotamus and the two rhinoceros species (Kortlandt, 1984). In South America, the present-day large mammal fauna is impoverished, but we know from fossil evidence that it was formerly much more numerous and diverse, and that it included species such as giant sloths (*Megatherium*) that would undoubtedly fall into the bulldozer herbivore category. The precise timing and causes of these extinctions are still a matter of dispute but seem to have occurred in two phases: from about 3 Ma, when the land bridge to Central and North America became established, and about 10,000–20,000 years ago, when humans arrived (Ceballos, 1995).

20.4 MAJOR SEASONALLY DRY VEGETATION TYPES

Table 20.1 lists the main vegetation types of seasonally dry regions in Africa and the Neotropics. The first column gives the recommended general term, usually based on that used by White (1983). The second column attempts a brief definition of the climatic conditions under which the main vegetation type is found. The third column gives terms that have been widely used in African literature, and the final column the same for South America. White (1983: 269–270) has a table of vernacular terms that have appeared in African ecological literature.

Some notes are now given on each of the main categories.

20.4.1 Seasonally Dry Forests

Seasonally dry forests have been defined earlier. True dry forests, in which there is a long dry season, no grass layer, and therefore no fires, are rare in Africa north of the Equator. Some are found in specialized habitats, perhaps as relics. The forests of *Gillettiodendron glandulosum* (Portères) J. Léonard (Leguminosae: Caesalpinioideae) in Mali occupy rocky slopes where there is protection from fire and probably water seepage from underground (Jaeger, 1956; White, 1983: 103). Hall and Swaine (1981) surveyed the whole range of forests occurring within Ghana. Their southern marginal type occurs mainly on hills with shallow soil under annual rainfall of 1000–1250 mm, generally weakly bimodal in distribution. Many of the common tree species are gregarious; they include *Hildegardia barteri* (Mast.) Kosterm. (Sterculiaceae), *Manilkara obovata* (Sabine and G. Don) J.H. Hemsl. (Sapotaceae), *Hymenostegia afzelii* (Oliv.) Harms and *Talbotiella gentii* Hutch. & Greenway (both Leguminosae: Caesalpinioideae). Their driest forest category, south-eastern outliers, are found on isolated rocky hills standing in the coastal grasslands of Ghana, with annual rainfall of 750–1000 mm, tending to be distributed bimodally. They are often dominated by *Millettia thonningii* (Schum. & Thonn.) Baker (Leguminosae: Papilionoideae), with *Diospyros abyssinica* (Hiern) F. White and *D. mespiliformis* Hochst. ex A. DC. (Ebenaceae); *Talbotiella gentii* also occurs. Hall and Swaine (1981) also describe a fire zone subtype of their dry semideciduous forest that may represent the closest approach to true dry forest north of the Equator in Africa. The structure of these fire-zone forests is affected by occasional fires that burn through the litter layer at intervals of 10–20 years. They kill many young saplings and also favour species such as oil palm (*Elaeis guineensis* Jacq., Palmae) that are relatively fire resistant. Stand curves from these forests are therefore deficient in the smaller stem girth classes (Hall and Swaine, 1981). Aubréville (1949) believed that dry forests similar to those described by Hall and Swaine formerly covered much of the savanna zones, and that the present savannas are the

TABLE 20.1

Terminology of Tropical Seasonal Vegetation in Africa and South America

Recommended Term*	Environment and Structure	African Term(s)	South American Term(s): Approximate Equivalents
Dry forest	Single dry season, 3–6 months. Fire absent or rare. Tree-dominated vegetation, usually with more than one layer; woody climbers and epiphytes usually present. Grasses not xeromorphic, scarce.	Dry forest	Cerradão or seasonally dry tropical forest s.l.
Woodland	Single dry season > 4 months. Fires regular, often annual. Tree-dominated vegetation; crown cover at least 40%. Usually only one main tree layer. Woody climbers and epiphytes absent or very scarce. Grasses narrow-leaved, tussock-forming, often xeromorphic.	Guinea savanna, savanna woodland, *Brachystegia* woodland, Sudan savanna, miombo woodland, mopane woodland, tall grass–low tree savanna, forêt claire, miombo savanna	Cerrado sensu stricto
Wooded grassland	Single dry season > 4 months. Trees with crown cover < 40%, > 10%. One tree layer. Grasses narrow-leaved, tussock-forming, often xeromorphic.	Scattered tree grassland, wooded grassland	Campo cerrado, sabana arbolada
Bushed grassland	Single dry season > 4 months. Bushes (multistemmed, short stature) < 40%, > 10%. One shrub layer. Grasses narrow-leaved, usually tussock-forming and xeromorphic.	Open bushland, bushed grassland, savanna bushland, bush savanna	Campo sujo, sabana arbustiva
Grassland	Single dry season > 4 months. Woody plants with canopy cover > 10%. Grasses usually tussock-forming and xeromorphic, at least in Africa. Fires regular. Natural grasslands often in sites with seasonal waterlogging, shallow soil or high metallic ion concentrations.	Grass savanna, savanna grassland	Campo limpo (no large woody plants), campo sujo, sabana abierta, sabana lisa
Bushland and thicket	More than one dry season. Rainfall low and variable both between wet seasons and between years. Fires rare except in very wet years (when productivity and standing crop of the grass layer is high). Trees and shrubs deciduous, often spiny, with open canopy (but dense and difficult to penetrate in 'thicket'). Grasses annual and sparse.	*Acacia* bushland, *Acacia-Commiphora* bushland, itigi thicket, Sahel savanna (some types), bushland thicket, succulent bushland	Caatinga or seasonally dry tropical forest s.l.

* These terms largely reflect those used by White (1983). For definitions of South American terms, see Pennington et al. (Chapter 1) and Ratter et al. (Chapter 2).

product of burning and clearance for agriculture. Hall and Swaine (1981) disagree, and believe that fire is a factor that is an integral part of the environment of these forests, which can persist as long as fires occur only at long intervals. Hall and Swaine (1981) found that species diversity declined from their wettest to their driest forest types, with up to 200 species per 0.0625-ha plot in the wettest forest but as few as 20 in the driest.

South of the Equator, in White's (1983) Zambezian regional centre of endemism, dry forests are widespread, although often much reduced and relict. Evergreen dry forest, in which *Berlinia giorgii* De Wild., *Cryptosepalum pseudotaxus* Baker f. and *Daniellia alsteeniana* Duvign. (all Leguminosae: Caesalpinioideae) are common components, are widespread. On the deep, sandy, nutrient-poor soils developed on Kalahari sands, evergreen dry forests dominated by *Cryptosepalum pseudotaxus* (Leguminosae: Caesalpinioideae) grow in wetter areas, while deciduous dry forests of *Baikiaea plurijuga* Harms (Leguminosae: Caesalpinioideae) are likewise extensive under lower annual rainfalls (600–1000 mm; Huckaby, 1986). The Kalahari sands were laid down under desert conditions during the Pleistocene or earlier; they cover a huge area of the upper Zambezi basin and extend north to the Congo river in the Kinshasa area.

Near the Equator, in regions of East Africa where there tend to be two wet and two dry seasons each year, semi-evergreen dry forests of species of *Olea* (Oleaceae), *Diospyros* (Ebenaceae), *Teclea* (Rutaceae), *Cordia* (Boraginaceae), *Turraea* (Meliaceae) and *Euphorbia* (Euphorbiaceae) (among other genera) occur in various scattered localities. These include northern Kenya (Herlocker, 1979), northern Tanzania (Herlocker, 1975) and western Uganda (Lock, 1977).

As some of the examples described here suggest, members of Leguminosae are often prominent in African dry forests. It is clear from the table of woody species given by Gentry (1995, Table 7.3) that the same is true in South America. Of the 25 sites for which he gives the composition by family, Leguminosae is the most speciose family in 19 and comes second in a further four. Bignoniaceae is the next highest ranking family, most speciose in five sites and in second place in 13. This is certainly not the case in tropical Africa, where there are very few species of Bignoniaceae.

In Brazil, Ratter (1992) described several forms of the transition between cerrado and Amazonian forest. In most places, there is a distinct transition zone occupied by cerradão dry forest. In Africa, by contrast, a true transition zone between forest and savanna is often almost absent. Forest and savanna, of almost completely different species composition (e.g., see Hopkins, 1965, and papers cited therein), are often separated by only a few tens of metres, or forest may be bounded by secondary tall grassland of the type described later in this chapter. In Uganda, the spiny shrub *Acanthus pubescens* Engl. often dominates this narrow interface. It seems possible that the much greater incidence of fire in Africa may have sharpened the boundary between forest and savanna, to the detriment of dry forests.

20.4.2 WOODLAND

Woodland, as defined in Africa by White (1983), consists of a more or less continuous tree layer over a herbaceous understorey in which grasses are dominant. The trees are mostly deciduous, and many of them flower towards the end of the dry season before the first rains. Well-developed woodlands have a virtually continuous tree canopy but still have a grass layer beneath, which burns in most years (see profile diagrams in Langdale-Brown et al., 1964, Figure 12; Pratt et al., 1966, Figure 3a; Menaut et al., 1995).

The *miombo* woodlands that cover large areas in southern Tanzania, Zambia and southern Congo (Kinshasa) are of this type. Species of *Brachystegia*, *Isoberlinia* and *Julbernardia* (all Leguminosae: Caesalpinioideae) are the most abundant tree genera. Miombo woodland would appear to be the equivalent of the taller and less frequently burned forms of cerrado, sometimes verging on cerradão. North of the Equator, such woodlands are widespread but much more heavily cleared and cultivated. Species of *Combretum* and *Terminalia* (both Combretaceae) are more abundant than they are south of the Equator. *Isoberlinia doka* Craib & Stapf (Leguminosae: Caesalpinioideae) is widespread and often conspicuous, but species of *Brachystegia* and *Julbernardia*, so characteristic of the southern miombo woodlands, are absent.

Experiments have shown that at least the wetter types of woodland change in structure and species composition if fire frequency is reduced or if the timing of burning is altered from late in the dry season (hot fires) to early (cooler, less complete fires) (e.g., Trapnell, 1959; Charter and Keay, 1960; Chidumayo, 1988; see Lock, 1998, for a summary of experimental evidence). Some of these experiments were set up in what was regularly burned woodland, and, with time, fire-protected and early-burned plots have changed in the direction of dry forest. The main changes include the decline of fire-resistant grasses and the appearance of lianas and of thin-barked, fire-sensitive tree species. Similar changes in response to protection from fire have been recorded in cerrado vegetation in Brazil (Ratter, 1992; Durigan, Chapter 3; J. Ratter, Royal Botanic Garden Edinburgh, pers. comm.), where open cerrado sensu stricto (see later) changes towards the more closed cerradão within 10 years of fire protection. It is possible that dry forests are less extensive in Africa because the long history of human settlement and fire use, and consequent greater fire frequency, have led to a wider loss of dry forests than appears to be the case in South America.

Particularly in the wetter types, where grass growth is greater and fires therefore hotter, many of the trees are low in stature (<15 m) and have many contorted branches. Many have thick bark that probably helps to protect the delicate cambial tissues from the heat of fires (for profile diagrams of such woodlands see Langdale-Brown et al., 1964, Figures 13–16). Similar features are found in many cerrado trees (Oliveira-Filho and Ratter, 2002).

Woody climbers (lianas) and vascular epiphytes are usually absent or very uncommon. Shrubs and young individuals of the canopy trees are found in the understorey, and they may form dense thickets in sites such as termite mounds, where fire pressure is reduced (see Section 20.3.4 for notes on seed biology in relation to fire). The grasses are mostly perennial, often form loose tussocks and, in Africa, many belong to the tribe Andropogoneae (see Section 20.3.4 for discussion of germination biology).

Both in Africa and in South America, only a few of the associated broad-leaved herbs are annuals. Most are perennials, of which a proportion produce annual shoots from an underground woody rootstock, which is often extensive and much branched to form a virtual subterranean tree. Such geoxylic suffrutices are a feature of both African woodlands and wooded grasslands (White, 1977), and of cerrado environments, where the terms *recurrent shrubs*, *semishrubs*, *hemixyles* and *geoxyles* have been used (Filgueiras, 2002). In Africa, White (1977) regarded geoxylic suffrutices as characteristic of seasonally waterlogged sites, but this has been challenged by Vollesen (1981), who observed that they are mostly confined to well-drained sandy soils. Vollesen (1981) also pointed out that, in at least some species, the geoxylic habit is facultative and is developed much more where fires are regular and severe.

In Africa, the composition of the tree layer by family is similar north and south of the Equator. Table 20.2 shows the percentages of species in the major families in West Africa (Geerling, 1982), in Zambia (White, 1962) and in Brazil (Mendonça et al., 1998). The samples are large, and the compositions of woodlands north and south of the Equator in Africa are remarkably similar. Leguminosae provide most species in both Africa and South America, but there are striking differences in the less abundant families, with Rubiaceae, Euphorbiaceae, Combretaceae and Capparidaceae commoner in Africa, and Myrtaceae, Palmae, Vochysiaceae and Annonaceae commoner in South America. Within Leguminosae, there are also differences (Table 20.3), with Mimosoideae and Caesalpinioideae commonest in Africa, and Papilionoideae and Caesalpinioideae commonest in South America.

The composition of vegetation by family provides one comparison between the woodlands of Africa and South America, but takes no account of the relative abundance of individuals. In southern Africa, members of Leguminosae subfamily Caesalpinioideae, particularly of the genera *Brachystegia*, *Isoberlinia* and *Julbernardia*, very often dominate the vegetation in terms of individuals. In Brazilian

TABLE 20.2
Family Composition of African Woodland and South American Cerrado (Trees and Shrubs Only)

	West Africa[a]		Zambia[b]		Brazilian Cerrado[c]	
	No. of Spp.	Percentage	No. of Spp.	Percentage	No. of Spp.	Percentage
Leguminosae	78	21	102	21	79	17
Rubiaceae	41	11	48	10	27	6
Euphorbiaceae	27	7	35	7	10	2
Combretaceae	22	6	29	6	13	3
Moraceae	19	5	14	3	1	0.2
Capparidaceae	14	4	20	4	—	—
Anacardiaceae	13	4	16	3	10	2
Myrtaceae	3	1	3	0.5	21	4
Palmae	6	2	4	0.5	13	3
Vochysiaceae	—	—	—	—	22	5
Annonaceae	3	1	11	2	8	2
Melastomataceae	—	—	—	—	16	3
Bombacaceae	3	1	1	0.2	8	2
Malpighiaceae	—	—	—	—	13	3
Bignoniaceae	3	1	5	1	15	3
Apocynaceae	9	2	5	1	19	—

[a] Figures taken from Geerling (1982) (sample size 371 spp.).
[b] Figures taken from White (1962) (sample size 479 spp.).
[c] Figures taken from Mendonça et al. (1998) (sample size 477 spp.).

cerrado, on the other hand, members of Vochysiaceae, particularly the genus *Qualea*, are by far the commonest trees (Oliveira-Filho and Ratter, 2002).

20.4.3 BUSHLAND AND THICKET

In much of the Horn of Africa, and in some parts of south-western Africa, rainfall is both low and irregular. While climadiagrams often show two clear rainfall peaks each year, the variability of the rainfall is very high, and in any one year one or both of these wet seasons may fail. The amount falling in any wet season is also extremely variable. Fires are rare or absent from these regions.

TABLE 20.3
Subfamilies of Woody Leguminosae in African and South American Savanna

	West Africa		Zambia		Brazilian Cerrado	
	No. of Spp.	Percentage	No. of Spp.	Percentage	No. of Spp.	Percentage
Caesalpinioideae	20	26	44	43	24	30
Mimosoideae	34	44	35	34	21	26
Papilionoideae	24	30	23	23	34	43

[a] Figures taken from Geerling (1982) (total 78 spp.).
[b] Figures taken from White (1962) (total 102 spp.).
[c] Figures taken from Mendonça et al. (1998) (total 79 spp.).

Vegetation development of woody plants is essentially facultative: leaves, flowers and fruit are produced not to any regular seasonal pattern, but in response to heavy rains. Grasses are not prominent in the vegetation, which is dominated by woody plants of low stature, often spiny and often with grey trunks and branches. The flora of the Horn of Africa is rich in species, including many endemics, and is very different in species and familial composition to neighbouring regions with a unimodal rainfall pattern. White (1983) used the term *bushland and thicket* for this physiognomic vegetation type.

20.4.3.1 Somalia-Masai Deciduous Bushland and Thicket

Somalia-Masai deciduous bushland and thicket (sensu White, 1983) is dominated by spiny shrubs and small trees, often with grey bark. Species of *Acacia* (Leguminosae: Mimosoideae) and *Commiphora* (Burseraceae) are the commonest larger woody plants. The shrubby layer includes, among others, species of *Capparis*, *Maerua*, *Boscia* and *Cadaba* (all Capparidaceae), shrubby Acanthaceae, *Sericocomopsis* (Amaranthaceae) and *Grewia* (Tiliaceae). Candelabriform species of *Euphorbia* and the swollen-trunked baobab tree (*Adansonia digitata* L., Bombacaceae) provide parallels with growth forms also found in the caatinga.

Succulent-leaved species of *Sansevieria* (Agavaceae) form low thickets, and the succulent rosettes of *Aloe* (Aloeaceae) grow either singly or in dense patches.

20.4.3.2 Itigi Thicket

A second type of African thicket (*Itigi thicket* sensu White, 1983) is made up of densely interlaced deciduous shrubs and small trees forming a canopy 3–5 m in height. The woody plants are thornless and cast a dense shade that excludes almost all grasses. Woody climbers are virtually absent. Fires do not penetrate such thickets. These thickets appear to occupy areas of sandy soil overlying granite (see White, 1983, and references therein).

20.4.3.3 Caatinga

In South America, similar climates with often bi- or even trimodal rainfall patterns are found in north-eastern Brazil and coastal Venezuela. Here again, the main feature of the rainfall is its extreme variability and unpredictability. Rainfall at a single station can vary from 1000 mm in good years to zero in drought ones (Hueck, cited by Bucher, 1982). Although there is much variation from place to place, the most typical vegetation is made up of grey-stemmed thorny shrubs and small trees, and is referred to in Brazil as caatinga. Again as in Africa, the flora is rich in endemics and different in composition to surrounding regions (Oliveira-Filho et al., Chapter 7; Queiroz, Chapter 6). Localized patches of rather similar vegetation are found in eastern Bolivia (J.R.I. Wood, University of Oxford, pers. comm.), extending into the Corumbá region of western Mato Grosso do Sul, Brazil; however, the rainfall here is higher, and its occurrence more predictable (J. Ratter, pers. comm.).

The caatingas of north-eastern Brazil include vegetation of various physiognomies (see Chapter 7), but the most characteristic is a dense spiny thicket (Sampaio, 1995). According to Rodal and colleagues (quoted in Sampaio, 1995), the most abundant woody family is Leguminosae, which includes 118 (35%) of the 339 woody species recorded. The subfamilies Caesalpinioideae and Mimosoideae are the most frequent. Euphorbiaceae and Cactaceae are the next most abundant woody families. Trees such as *Cavanillesia arborea* K. Schum. and *Ceiba glaziovii* (Kuntze) K. Schum. (both Bombacaceae) have swollen water-storing trunks. Tall genera of Cactaceae such as *Pilosocereus* are frequent. African shrubby and arborescent succulent species of *Euphorbia* (Euphorbiaceae) are replaced in South America by cacti of the *Cereus* group (Cactaceae). Small cacti and bromeliads form low thickets. Shrubby species of *Capparis* (Capparidaceae) are often characteristic (J. Ratter, pers. comm.).

The parallels between the caatingas of South America and some forms of bushland and thicket in Africa are striking. The physiognomy of the vegetation of both is very similar, although there are no species and very few genera in common.

20.4.4 GRASSLANDS

White (1983) uses the term grassland not only for vegetation completely lacking woody plants, but also for grassland in which there is up to 10% woody plant cover. The South American term *campo sujo* can thus include vegetation that would, in Africa, just be called grassland; pure grassland is *campo limpo*. Both in Africa and in South America, pure or almost pure grasslands tend to be found in areas where there is great variation in water availability through the year. The term grassland is used here in a physiognomic sense; plants from families of similar appearance — such as Cyperaceae and, particularly in nutrient-poor sites, Xyridaceae — may also form a significant proportion of the standing crop.

20.4.4.1 Valley Grasslands

In Africa, natural grasslands often occupy seasonally waterlogged valleys within woodlands (*dambo* grasslands; see White, 1983). The soils here are clay-rich and dark, and shrink when dry and swell when wet. The vegetation is a tussocky grassland, and during the wet season a range of small ephemeral plants in families such as Cyperaceae, Droseraceae, Lentibulariaceae, Xyridaceae and Scrophulariaceae often appear in the spaces between the tussocks. A few dwarf tree species (<3 m), such as *Acacia drepanolobium* Harms ex Sjöstedt, may grow within the grassland. The descriptions of Oliveira-Filho and Ratter (2002) do not seem to provide a precise South American equivalent for the southern African dambo, a broad shallow valley often without a clear central watercourse or riverine forest. The *veredas* that they describe may be similar but seem to occur in sites with higher rainfall. The vereda pattern of forest on hilltops, grassland on slopes, and riverine forest in the valley bottom (see also diagram in Furley, 1999) occurs on fine-textured soils in Ghana (West Africa). On soils over sandstones, the forests are confined to the valleys, with the slopes and hilltops bearing woodland or wooded grassland (Swaine et al., 1976).

20.4.4.2 Seasonally Inundated Grasslands

The extensive seasonally flooded grasslands in the Sudd region along the White Nile in southern Sudan may be compared with those of the Llanos in Venezuela. The grasslands of the Sudd region have been described by Lock (1988). They occur on almost-level alluvial plains with dark clay soils. During the wet season, these soils become sealed and rainwater accumulates on the surface. At the same time, the river rises and spills out over the surrounding land. The river discharge, and therefore the amount of spillage and flooding, vary enormously from year to year and also show longer period fluctuations (Sutcliffe and Parks, 1988). A distinction can be drawn between seasonally rain-flooded grasslands in which *Hyparrhenia rufa* (Nees) Stapf is the dominant grass, and seasonally river-flooded grasslands dominated by species of *Echinochloa*, *Vossia cuspidata* Griff. and *Oryza longistaminata* Chev. & Roehr. The two former grow upwards with the rising water to form floating mats; the latter is rhizomatous and tends to grow in areas that are less deeply flooded (Lock, 1988). Within the Sudd region, there are extensive areas of semipermanent swamp (19,000 km^2 in 1980) dominated by *Cyperus papyrus* L. and *Typha domingensis* Pers., the former occurring close to the river courses and the latter further away from them, probably reflecting differences in nutrient supply. Similar but less extensive seasonally flooded grasslands occur elsewhere in Africa (see White, 1983).

The Llanos, by contrast, lack extensive areas of permanent swamp. Huber and colleagues (Chapter 5) distinguish between rain-flooded and river-flooded savannas (which are almost pure grasslands in their usage). The river-flooded areas are dominated by large floating grasses such as

Hymenachne amplexicaulis (Rudge) Nees and *Paspalum fasciculatum* Flügge. Wetter but not river-flooded sites bear a mixture of grass genera including *Panicum*, *Paspalum*, *Mesosetum* and *Schizachyrium*. *Trachypogon spicatus* (L. f.) Kuntze occupies the driest sites. According to Huber and coworkers (Chapter 5), forests of various kinds are widespread within the Llanos, mainly as gallery forests along rivers and as clumps occupying raised ground. The Sudd region proper is devoid of true forest, but open woodlands, usually virtually monospecific, of *Acacia seyal* Delile, *Acacia polyacantha* Willd. and *Balanites aegyptiaca* Delile are widespread (Lock, 1988).

Similar grasslands occur in other parts of northern South America, such as the Rupununi region of Guyana.

20.4.4.3 Grasslands on Very Shallow Soils

Grasslands are also found on sites with very shallow soils over rock. Such soils are waterlogged during the wet season yet extremely dry at other times. Eiten (1983, Figure 99) shows open grassland in Brazil on shallow soil over a hard laterite platform (*campo litossólico*). Such sites are widespread in West Africa, where they are known as lateritic hardpan or lateritic ironstone (in French, *cuirasse*, or carapace); they are also sometimes given the local name *bowal*. The very shallow soils and extreme dryness for part of each year produce a sparse short annual grassland, brief details of which are given by Lawson et al. (1968), Rose Innes (1977) and White (1983).

20.4.4.4 Grasslands on Metal-Rich Soils

In southern Africa, grasslands are found in sites where the soil has a high content of metallic elements such as nickel, copper or cobalt (Wild, 1978). Similar grasslands probably occur around Niquelândia, Goiás, Brazil (J. Ratter, pers. comm.).

20.4.4.5 Secondary Grasslands

In Africa, close to the boundary between forest and savanna, under annual rainfalls of 1500–1800 mm and a dry season of 3–4 months, grass production during the wet season is high. This biomass dries out during the dry season and provides abundant fuel for fires. Under these conditions, any regeneration of forest is prevented by regular intense fires, and secondary grasslands are formed (White, 1983). These may be dominated by giant, almost bamboo-like grasses such as *Pennisetum purpureum* Schumach., which regularly attains a height of 5 m, or shorter species such as *Imperata cylindrica* (L.) P. Beauv., whose underground rhizomes make it very fire resistant and invasive. While there can be little doubt that grasslands of this kind are of recent origin, the status of grasslands dominated by species of *Hyparrhenia* and *Andropogon* is less clear. These are often wooded grasslands or even open woodlands in which the trees are typical woodland species. There is, however, little doubt that secondary grasslands are extensive in Africa.

20.4.5 PATTERNS AND MOSAICS IN SAVANNA VEGETATION

20.4.5.1 Campo de Murundus

An illustration in Eiten (1983, Figure 106) shows a vegetation type (*campo de murundus*) found in the Pantanal region; such vegetation is widespread in central Brazil. The tops of regularly arranged low mounds are occupied by thicket vegetation, while the lower parts of the mounds and the areas between them support grassland. The origin of these mounds has been studied by Oliveira-Filho (1992). He considered two alternative hypotheses, differential erosion and termite activity, and concluded that the latter was the most likely explanation in the area that he studied. The former hypothesis may, however, be correct elsewhere (Furley, 1986). Vegetation of this type is frequent in Uganda, where it is referred to by Langdale-Brown et al. (1964) as 'seasonal swamp grassland with semi-deciduous thicket on anthills'. White (1983) refers to it as 'bush-group' grassland. Variations in relief in such types of vegetation may be caused partly by termite activity, and perhaps by the activities of fossorial mammals (Cox and Gakahu, 1985). However, work in southern Sudan by, for example, ILACO (1981) has shown that shrinking and swelling of soil clays in response to seasonal variations in moisture content may produce the original microrelief, which is then accentuated by differential termite activity. On level ground, the soil movements produce a regular pattern of low mounds, but on slopes, similar movements produce parallel ridges running up and down the slope. Soils of this type (*gilgai* soils) were originally described in Australia by Hallsworth et al. (1955); see also the summary and further references in Young (1976).

20.4.5.2 Campo Rupestre

Campo rupestre is a complex mosaic of grassy and often marshy valleys between mountain ridges that bear a saxicolous flora. In the Pico das Almas area of Bahia, Brazil, at least, the underlying rocks are metamorphosed sandstones (quartzites) and conglomerates, producing highly acidic, nutrient-poor soils. The climate is highly seasonal, with a single peak of rainfall. The vegetation is species-rich, with numerous endemics (Stannard, 1995). Many species, particularly in Ericaceae and Melastomataceae, are small shrubs with very small thick leaves, a heather-like growth form (Stannard, 1995).

Secondary upland grasslands, derived from forest by clearance and burning, are widespread on tropical African mountains; upland rocky grasslands like campos rupestres are, however, relatively uncommon, perhaps because most of the mountains are either recent volcanics or of gentle relief. Nevertheless, rather similar vegetation is found in the Chimanimani mountains on the Zimbabwe–Mozambique border. Very hard quartzitic rocks produce much open vegetation with grassland, thicket patches and seasonally waterlogged areas. There are a number of endemic species. Heather-like species in Ericaceae and Proteaceae are widespread. The vegetation has been described by Phipps and Goodier (1962) and the endemics discussed by Wild (1964).

20.5 CONCLUSIONS

It is not surprising that similar climates and similar soils in the seasonally dry regions of Africa and South America bear vegetation that is similar in physiognomy, and also in the ways in which vegetation types of different physiognomy are arranged in relation to topography. What is striking is that, while the vegetations of the two continents are physiognomically similar, they are made up of different assemblages of families and genera. Few workers in either continent have paid much attention to the physiognomically similar vegetation in the other. In South America, huge areas of seasonal tropical vegetation are being converted to industrial agriculture, a process that has hardly started in Africa.

ACKNOWLEDGEMENTS

I am extremely grateful to the editors and to those who read this chapter and provided many useful suggestions. In particular, I would like to thank Jimmy Ratter, Royal Botanic Garden Edinburgh, for his many very useful suggestions and for calculating and providing the cerrado composition figures in Table 20.2. I am also most grateful to the organizers of the *Tropical Savannas and Seasonally Dry Forests* international conference (Edinburgh, September 2003) for inviting me to contribute on this topic.

REFERENCES

Aubréville, A., *Climats, Forêts et Désertification de l'Afrique Tropicale*, Société d'Éditions Géographiques Maritimes et Coloniales, Paris, 1949.

Brain, C.K. and Sillen, A., Evidence from the Swaartkrans cave for the earliest use of fire, *Nature*, 336, 464, 1988.

Bucher, E.H., Chaco and caatinga — South American arid savannas, woodlands and thickets, in *Ecology of Tropical Savannas (Ecological Studies 42)*, Huntley, B.J. and Walker, B.H., Eds, Springer-Verlag, Berlin, 1982, 42.

Ceballos, G., Vertebrate diversity, ecology and conservation in neotropical dry forests, in *Seasonally Dry Tropical Forests*, Bullock, S.H., Mooney, H.A., and Medina, E., Eds, Cambridge University Press, Cambridge, 1995, 195.

Charter, J.R. and Keay, R.W.J., Assessment of the Olokemeji fire-control experiment (investigation 254) 28 years after establishment, *Nigerian Forest Inform. Bull.*, 3, 1, 1960.

Chevalier, A., Le territoire géobotanique de l'Afrique tropicale nord-occidentale et ses subdivisions, *Bull. Soc. Bot. Fr.*, 80, 4, 1933.

Chidumayo, E.N., A re-assessment of effects of fire on miombo vegetation in the Zambian copperbelt, *J. Trop. Ecol.*, 4, 361, 1988.

Cobb, S., Lock, J.M., and Fison, T., Wildlife, in *The Jonglei Canal: Impact and Opportunity*, Howell, P.P., Lock, J.M., and Cobb, S., Eds, Cambridge University Press, Cambridge, 1988, 350.

Cox, G.W. and Gakahu, C.G., Mima mound microtopography and vegetation pattern in Kenyan savannas, *J. Trop. Ecol.*, 1, 23, 1985.

Eiten, G., The cerrado vegetation of Brazil, *Bot. Rev.*, 38, 201, 1972.

Eiten, G., *Classificação da Vegetação do Brasil*, CNPq–Coordenação Editorial, Brasília, 1983.

Filgueiras, T.S., Herbaceous plant communities, in *The Cerrados of Brazil: Ecology and Natural History of a Neotropical Savanna*, Oliveira, P.O. and Marquis, R.J., Eds, Columbia University Press, New York, 2002, 121.

Furley, P.A., Classification and distribution of murundus in the cerrado of central Brazil, *J. Biogeogr.*, 13, 265, 1986.

Furley, P.A., The nature and diversity of neotropical savanna vegetation with particular reference to the Brazilian cerrados, *Glob. Ecol. Biogeogr.*, 8, 223, 1999.

Geerling, C., Guide de terrain des ligneux Sahéliens et Soudano–Guinéens, *Meded. Landbauwhogesch. Wageningen*, 82, 1982.

Gentry, A.H., Diversity and floristic composition of neotropical dry forests, in *Seasonally Dry Tropical Forests*, Bullock, S.H., Mooney, H.A., and Medina, E., Eds, Cambridge University Press, Cambridge, 1995, 146.

Gillet, H., Desert and Sahel, in *Plant Ecology in West Africa*, Lawson, G.W., Ed, John Wiley, Chichester, 1986, 151.

Graham, A. and Dilcher, D., The Cenozoic record of tropical dry forest in northern Latin America and the southern United States, in *Seasonally Dry Tropical Forests*, Bullock, S.H., Mooney, H.A., and Medina, E., Eds, Cambridge University Press, Cambridge, 1995, 124.

Hall, J.B. and Swaine, M.D., *Distribution and Ecology of Vascular Plants in a Tropical Rain Forest: Forest Vegetation in Ghana*, W. Junk, The Hague, 1981.

Hallsworth, E.G., Robertson, F.R., and Gibbons, F.R., Studies in pedogenesis in New South Wales. VII. The 'gilgai' soils, *J. Soil. Sci.*, 6, 1, 1955.

Herlocker, D., *Woody vegetation of the Serengeti National Park*, Texas A&M University, College Station, 1975.

Herlocker, D., *Vegetation of southwestern Marsabit District, Kenya*, IPAL technical report no. 1, UNEP, Nairobi, 1979.

Hopkins, B., *Forest and Savanna*, Heinemann, London, 1965.

Huckaby, J.D., The geography of Zambezi teak, in *The Zambezi Teak Forests*, Piearce, G.D., Ed, Forest Department, Ndola, 1986, 5.

ILACO, *Pengko Plain Development Study. Vol. II: Technical Annexes*, Arnhem, the Netherlands, 1981.

Jackson, G., Notes on West African vegetation — III. The seedling morphology of *Butyrospermum paradoxum* (Gaertn.) Hepper, *J. W. African Sci. Assoc.*, 13, 215, 1968.

Jackson, G., Cryptogeal germination and other seedling adaptations to the burning of vegetation in savanna regions: the origin of the pyrophytic habit, *New Phytol.*, 73, 771, 1974.

Jaeger, P., Contribution à l'étude des forêts reliques du Soudan occidental, *Bull. IFAN, Sér. A*, 18, 993, 1956.

Jarman, P.J. and Sinclair, A.R.E., Feeding strategy and the pattern of resource partitioning in ungulates, in *Serengeti — Dynamics of an Ecosystem*, Sinclair, A.R.E. and Norton-Griffiths, M., Eds, University of Chicago Press, Chicago, 1979, 130.

Kortlandt, A., Vegetation research and the 'bulldozer' herbivores of tropical Africa, in *Tropical Rain-Forests. The Leeds Symposium*, Chadwick, A.C. and Sutton, S.L., Eds, Leeds Philosophical and Literary Society, Leeds, 1984, 205.

Lamprey, H.F., Ecological separation of the large mammal species in the Tarangire Game Reserve, Tanganyika, *E. African Wildlife J.*, 1, 63, 1963.

Langdale-Brown, I., Osmaston, H.A., and Wilson, J.G., *The Vegetation of Uganda and its Bearing on Land Use*, Government Printer, Entebbe, 1964.

Lawson, G.W., *Plant Life in West Africa*, Oxford University Press, Oxford, 1966.

Lawson, G.W., Jeník, J., and Armstrong-Mensah, K.O., A study of a vegetation catena in Guinea savanna at Mole Game Reserve (Ghana), *J. Ecol.*, 56, 505, 1968.

Lock, J.M., The vegetation of Rwenzori National Park, Uganda, *Bot. Jahrb. Syst.*, 98, 372, 1977.

Lock, J.M., Ecology of plants in the swamps and floodplains, in *The Jonglei Canal: Impact and Opportunity*, Howell, P.P., Lock, J.M., and Cobb, S., Eds, Cambridge University Press, Cambridge, 1988, 146.

Lock, J.M., Aspects of fire in tropical African vegetation, in *Chorology, Taxonomy and Ecology of the Floras of Africa and Madagascar*, Huxley, C.R., Lock, J.M., and Cutler, D.F., Eds, Royal Botanic Gardens, Kew, Richmond, 1998, 181.

Lock, J.M. and Milburn, T.R., The seed biology of *Themeda triandra* in relation to fire, in *The Scientific Management of Animal and Plant Communities for Conservation*, Duffey, E. and Watt, A.S., Eds, Blackwell Scientific Publications, Oxford, 1971, 337.

Maddock, L., The 'migration' and grazing succession, in *Serengeti — Dynamics of an Ecosystem*, Sinclair, A.R.E. and Norton-Griffiths, M., Eds, University of Chicago Press, Chicago, 1979, 104.

Menaut, J.C., Lepage, M., and Abbadie, L., Savannas, woodlands and dry forests in Africa, in *Seasonally Dry Tropical Forests*, Bullock, S.H., Mooney, H.A., and Medina, E., Eds, Cambridge University Press, Cambridge, 1995, 64.

de Mendonça, R.C. et al., Flora vascular do cerrado, in *Cerrado: Ambiente e Flora*, Sano, S.M. and de Almeida, S.P., Eds, EMBRAPA, Planaltina, 1998, 289.

Oliveira-Filho, A.T., Floodplain 'murundus' of central Brazil: evidence for the termite-origin hypothesis, *J. Trop. Ecol.*, 8, 1, 1992.

Oliveira-Filho, A.T. and Ratter, J.A., Vegetation physiognomy, in *History of a Neotropical Savanna*, Oliveira, P.S. and Marquis, R.J., Eds, Columbia University Press, New York, 2002, 91.

Pennington, R.T., Prado, D.E., and Pendry, C.A., Neotropical seasonally dry forests and Quaternary vegetation changes, *J. Biogeogr.*, 27, 261, 2000.

Phipps, J.B. and Goodier, R., A preliminary account of the plant ecology of the Chimanimani Mountains, *J. Ecol.*, 50, 291, 1962.

Pratt, D.J., Greenway, P.J., and Gwynne, M.D., A classification of East African rangeland, with an appendix on terminology, *J. Appl. Ecol.*, 3, 369, 1966.

Ratter, J.A., Transitions between cerrado and forest vegetation in Brazil, in *Nature and Dynamics of Forest–Savanna Boundaries*, Furley, P.A., Proctor, J., and Ratter, J.A., Eds, Chapman and Hall, London, 1992, 417.

Ratter, J.A., Ribeiro, J.F., and Bridgewater, S., The Brazilian cerrado vegetation and the threats to its biodiversity, *Ann. Bot.*, 80, 223, 1997.

Rose Innes, R., *A Manual of Ghana Grasses*, Ministry of Overseas Development, Surbiton, 1977.

Sampaio, E.V.S.B., Overview of the Brazilian caatinga, in *Seasonally Dry Tropical Forests*, Bullock, S.H., Mooney, H.A., and Medina, E., Eds, Cambridge University Press, Cambridge, 1995, 35.

Sanford, W.W. and Isichei, A.O., Savanna, in *Plant Ecology in West Africa*, Lawson G.W., Ed, John Wiley, Chichester, 1986, 95.

Scholes, R.J. and Walker, B.H., *An African Savanna: Synthesis of the Nylsvley Study*, Cambridge University Press, Cambridge, 1993.

Stannard, B.L., Ed, *Flora of the Pico das Almas, Chapada Diamantina — Bahia, Brazil*, Royal Botanic Gardens, Kew, Richmond, 1995.

Sutcliffe, J. and Parks, Y., Hydrology of the Bahr el Jebel swamps, in *The Jonglei Canal: Impact and Opportunity*, Howell, P.P., Lock, J.M., and Cobb, S., Eds, Cambridge University Press, Cambridge, 1988, 100.

Swaine, M.D., Hall, J.B., and Lock, J.M., The forest–savanna boundary in west-central Ghana, *Ghana J. Sci.*, 16, 35, 1976.

Trapnell, C.G., Ecological results of woodland burning experiments in Northern Rhodesia, *J. Ecol.*, 47, 129, 1959.

Vollesen, K., *Catunaregam pygmaeum* (Rubiaceae) — a new geoxylic suffrutex from the woodlands of SE Tanzania, *Nord. J. Bot.*, 1, 735, 1981.

Walter, H., *Vegetation of the Earth and Ecological Systems of the Geo-biosphere*, 2nd English ed., Springer-Verlag, New York, 1979.

West, O., *Fire in Vegetation and its Use in Pasture Management with Special Reference to Tropical and Subtropical Africa*, Mimeo Publication no. 1/1965, Commonwealth Bureau of Pastures and Field Crops, Farnham Royal, 1965.

White, F., *Forest Flora of Northern Rhodesia*, Oxford University Press, Oxford, 1962.

White, F., The underground forests of Africa: a preliminary review, *Gard. Bull. Singapore*, 29, 55, 1977.

White, F., The Vegetation of Africa. A Descriptive Memoir to Accompany the UNESCO/AETFAT/UNSC Vegetation Map of Africa, UNESCO, Paris, 1983.

Wild, H., The endemic species of the Chimanimanian Mountains and their significance, *Kirkia*, 4, 125, 1964.

Wild, H., The vegetation of heavy metal and other toxic soils, in *Biogeography and Ecology of Southern Africa*, Werger, M.J.A., Ed, W. Junk, The Hague, 1978, 1301.

Young, A., *Tropical Soils and Soil Survey*, Cambridge University Press, Cambridge, 1976.

Index

Systematics Association Publications

1. Bibliography of Key Works for the fIdentification of the British Fauna and Flora, 3rd edition (1967)[†]
   Edited by G.J. Kerrich, R.D. Meikie and N. Tebble

2. Function and Taxonomic Importance (1959)[†]
   Edited by A.J. Cain

3. The Species Concept in Palaeontology (1956)[†]
   Edited by P.C. Sylvester-Bradley

4. Taxonomy and Geography (1962)[†]
   Edited by D. Nichols

5. Speciation in the Sea (1963)[†]
   Edited by J.P. Harding and N. Tebble

6. Phenetic and Phylogenetic Classification (1964)[†]
   Edited by V.H. Heywood and J. McNeill

7. Aspects of Tcthyan biogeography (1967)[†]
   Edited by C.G. Adams and D.V. Ager

8. The Soil Ecosystem (1969)[†]
   Edited by H. Sheals

9. Organisms and Continents through Time (1973)[†]
   Edited by N.F. Hughes

10. Cladistics: A Practical Course in Systematics (1992)[*]
    P.L. Forey, C.J. Humphries, I.J. Kitching, R.W. Scotland, D.J. Siebert and D.M. Williams

11. Cladistics: The Theory and Practice of Parsimony Analysis (2nd edition)(1998)[*]
    I.J. Kitching, P.L. Forey, C.J. Humphries and D.M. Williams

[*] Published by Oxford University Press for the Systematics Association
[†] Published by the Association (out of print)

SYSTEMATICS ASSOCIATION SPECIAL VOLUMES

1. The New Systematics (1940)
   Edited by J.S. Huxley (reprinted 1971)

2. Chemotaxonomy and Serotaxonomy (1968)[*]
   Edited by J.C. Hawkes

3. Data Processing in Biology and Geology (1971)[*]
   Edited by J.L. Cutbill

4. Scanning Electron Microscopy (1971)[*]
   Edited by V.H. Heywood

5. Taxonomy and Ecology (1973)*
   Edited by V.H. Heywood

6. The Changing Flora and Fauna of Britain (1974)*
   Edited by D.L. Hawksworth

7. Biological Identification with Computers (1975)*
   Edited by R.J. Pankhurst

8. Lichenology: Progress and Problems (1976)*
   Edited by D.H. Brown, D.L. Hawksworth and R.H. Bailey

9. Key Works to the Fauna and Flora of the British Isles and Northwestern Europe, 4th
   edition (1978)*
   Edited by G.J. Kerrich, D.L. Hawksworth and R.W. Sims

10. Modern Approaches to the Taxonomy of Red and Brown Algae (1978)
    Edited by D.E.G. Irvine and J.H. Price

11. Biology and Systematics of Colonial Organisms (1979)*
    Edited by C. Larwood and B.R. Rosen

12. The Origin of Major Invertebrate Groups (1979)*
    Edited by M.R. House

13. Advances in Bryozoology (1979)*
    Edited by G.P. Larwood and M.B. Abbott

14. Bryophyte Systematics (1979)*
    Edited by G.C.S. Clarke and J.G. Duckett

15. The Terrestrial Environment and the Origin of Land Vertebrates (1980)
    Edited by A.L. Pachen

16 Chemosystematics: Principles and Practice (1980)*
    Edited by F.A. Bisby, J.G. Vaughan and C.A. Wright

17. The Shore Environment: Methods and Ecosystems (2 volumes)(1980)*
    Edited by J.H. Price, D.E.C. Irvine and W.F. Farnham

18. The Ammonoidea (1981)*
    Edited by M.R. House and J.R. Senior

19. Biosystematics of Social Insects (1981)*
    Edited by P.E. House and J.-L. Clement

20. Genome Evolution (1982)*
    Edited by G.A. Dover and R.B. Flavell

21. Problems of Phylogenetic Reconstruction (1982)
    Edited by K.A. Joysey and A.E. Friday

22. Concepts in Nematode Systematics (1983)*
    Edited by A.R. Stone, H.M. Platt and L.F. Khalil

23. Evolution, Time And Space: The Emergence of the Biosphere (1983)*
    Edited by R.W. Sims, J.H. Price and P.E.S. Whalley

24. Protein Polymorphism: Adaptive and Taxonomic Significance (1983)*
    Edited by G.S. Oxford and D. Rollinson

25. Current Concepts in Plant Taxonomy (1983)*
    Edited by V.H. Heywood and D.M. Moore

26. Databases in Systematics (1984)*
    Edited by R. Allkin and F.A. Bisby

49. Plant Galls: Organisms, Interactions, Populations (1994)[‡]
    Edited by M.A.J. Williams

50. Systematics and Conservation Evaluation (1994)[‡]
    Edited by P.L. Forey, C.J. Humphries and R.I. Vane-Wright

51. The Haptophyte Algae (1994)[‡]
    Edited by J.C. Green and B.S.C. Leadbeater

52. Models in Phylogeny Reconstruction (1994)[‡]
    Edited by R. Scotland, D.I. Siebert and D.M. Williams

53. The Ecology of Agricultural Pests: Biochemical Approaches (1996)[**]
    Edited by W.O.C. Symondson and J.E. Liddell

54. Species: the Units of Diversity (1997)[**]
    Edited by M.F. Claridge, H.A. Dawah and M.R. Wilson

55. Arthropod Relationships (1998)[**]
    Edited by R.A. Fortey and R.H. Thomas

56. Evolutionary Relationships among Protozoa (1998)[**]
    Edited by G.H. Coombs, K. Vickerman, M.A. Sleigh and A. Warren

57. Molecular Systematics and Plant Evolution (1999)
    Edited by P.M. Hollingsworth, R.M. Bateman and R.J. Gornall

58. Homology and Systematics (2000)
    Edited by R. Scotland and R.T. Pennington

59. The flagellates: Unity, Diversity and Evolution (2000)
    Edited by B.S.C. Leadbeater and J.C. Green

60. Interrelationships of the Platyhelminthes (2001)
    Edited by D.T.J. Littlewood and R.A. Bray

61. Major Events in Early Vertebrate Evolution (2001)
    Edited by P.E. Ahlberg

62. The Changing Wildlife of Great Britain and Ireland (2001)
    Edited by D.L. Hawksworth

63. Brachiopods Past and Present (2001)
    Edited by H. Brunton, L.R.M. Cocks and S.L. Long

64. Morphology, Shape and Phylogeny (2002)
    Edited by N. MacLeod and P.L. Forey

65. Developmental Genetics and Plant Evolution (2002)
    Edited by Q.C.B. Cronk, R.M. Bateman and J.A. Hawkins

66. Telling the Evolutionary Time: Molecular Clocks and the Fossil Record (2003)
    Edited by P.C.J. Donoghue and M.P. Smith

67. Milestones in Systematics (2004)[§]
    Edited by D.M. Williams and P.L. Forey

68. Organelles, Genomes and Eukaryote Phylogeny
    Edited by R.P. Hirt and D.S. Horner

*Published by Academic Press for the Systematics Association
[†]Published by the Palaeontological Association in conjunction with Systematics Association
[‡]Published by the Oxford University Press for the Systematics Association
[**]Published by Chapman & Hall for the Systematics Association
[§]Published by CRC Press for the Systematics Association

Printed and bound by CPI Group (UK) Ltd, Croydon, CR0 4YY

23/10/2024

01778251-0009